AF362053

Antiviral Agents

Antiviral Agents

Special Issue Editor
Catherine Adamson

MDPI • Basel • Beijing • Wuhan • Barcelona • Belgrade • Manchester • Tokyo • Cluj • Tianjin

Special Issue Editor
Catherine Adamson
University of St Andrews
UK

Editorial Office
MDPI
St. Alban-Anlage 66
4052 Basel, Switzerland

This is a reprint of articles from the Special Issue published online in the open access journal *Viruses* (ISSN 1999-4915) (available at: https://www.mdpi.com/journal/viruses/special_issues/antiviral_drugs).

For citation purposes, cite each article independently as indicated on the article page online and as indicated below:

LastName, A.A.; LastName, B.B.; LastName, C.C. Article Title. *Journal Name* **Year**, *Article Number, Page Range*.

ISBN 978-3-03928-953-0 (Hbk)
ISBN 978-3-03928-954-7 (PDF)

Contents

About the Special Issue Editor

Catherine Adamson is a Principal Investigator and Lecturer in the School of Biology at the University of St. Andrews in Scotland, U.K. Before her appointment at St. Andrews, Dr. Adamson spent six years as international research fellow at the NIH in the United States. Catherine's laboratory focuses on virology with a specific interest in antiviral drugs. Her internationally recognized work encompasses multiple important aspects of the drug discovery process including innovative cell-based assay design, development, and optimization; execution of industry-standard high-throughput screens to discover antiviral compounds against novel viral targets; hit antiviral compound validation and characterization; determination of lead compound mechanism-of-action, antiviral activity, evolution of drug resistance, and understanding performance in clinical trials; and using antiviral compounds as chemical tools to provide new fundamental knowledge in virology to drive further drug discovery. More specifically, Dr. Adamson's work has focused on HIV-1 maturation inhibitors, a new class of antiretroviral compounds with a novel mechanism of action. A second major research interest is targeting viral interferon antagonists; her laboratory has developed an industry-standard modular cell-based screening platform to target a viral interferon antagonist of choice. The Adamson laboratory has successfully used the screening platform to discover novel compounds that specifically inhibit the interferon antagonist activity of the human respiratory syncytial virus NS2 protein. Dr. Adamson is also involved in a collaborative project that has identified compounds with potent antiviral activity against human cytomegalovirus, which block major immediate early gene transcription.

Editorial

Antiviral Agents: Discovery to Resistance

Catherine S. Adamson

School of Biology, Biomedical Sciences Research Complex, University of St Andrews, St Andrews KY16 9ST, Scotland, UK; csa21@st-andrews.ac.uk

Received: 3 April 2020; Accepted: 4 April 2020; Published: 7 April 2020

In the midst of the SARS-CoV-2/Covid-19 outbreak the need for research into, and development of, antiviral agents is brought into sharp focus worldwide for scientists, governments and the public alike. This special issue includes primary research into new antiviral agents with activity against a range of clinically and economically important viruses. New antiviral strategies are discussed in six review articles.

SARS-CoV-2 is a newly emerged β-Coronavirus with sustained human-to-human transmission via the respiratory tract. Influenza viruses are another group of respiratory viruses, which account for 300–600 K deaths annually and represent an ever-present pandemic threat. Emergence of drug-resistant influenza strains is driving research into new anti-influenza drug candidates with different modes of action. In this issue, two papers report compounds with anti-influenza activity against a range of influenza A viruses that inhibit early steps in influenza replication not targeted by clinically approved anti-influenza drugs [1,2]. A further study reports that respiratory syncytial virus is inhibited in vitro and in vivo by natural product plant extracts in which phenolic glycoside acetoside was the active chemical component [3].

Drug resistance is an ongoing issue in the global rollout of combination antiretroviral therapy for human immunodeficiency virus (HIV) prevention and treatment. Tadesse et al. report the prevalence of pretreatment drug resistance among HIV-infected children in the resource-limited setting of Ethiopia [4]. Lectins are experimental HIV-1 entry inhibitors. Shahid et al. engineered the lectin microvirin to contain two carbohydrate-binding domains, this reduced its anti-HIV potency but anti-hepatitis C virus (HCV) activity was demonstrated [5]. Sherpa and Le Grice review recent research into the structural fluidity of the HIV Rev response element (RRE) and its potential as a target for therapeutic intervention [6].

Antiviral agents are not available against many emerging and neglected viruses. This issue features four research papers reporting new antiviral compound candidates against zika virus [7,8], hantaan virus [9] and herpes B virus [10]. Measles virus is considered a re-emerging virus due to recent decline in vaccine coverage. Ferren et al. provide a comprehensive review on measles virus with a focus on central nervous system complications and the most advanced therapeutic approaches [11].

In addition to study into the zoonotic herpes B virus, this issue also includes two primary research papers investigating antiviral approaches targeting reactivation of latent herpesvirus infection. Anderson et al. demonstrated that the antipsychotic drug cloazapine inhibits epstein-barr virus (EBV) lytic reactivation [12]. In contrast, Chen et al. explore the potential application of genome-editing transcription activator-like effector nucleases (TALENs) against latent murine cytomegalovirus (CMV) infections [13]. The special issue also includes a review of the current antiviral strategies against human CMV [14] and a more focused review by Adamson and Nevels, which argues the case for the development of novel anti-CMV compounds by targeting major immediate-early (IE) gene expression or protein function [15]. RNA viruses can also establish persistent infections and this issue includes a study investigating the acquisition of drug resistance to fluoxetine during treatment of pancreatic cell-lines persistently infected with coxsackievirus B4 [16].

Research into antiviral agents against economically important animal viruses is also included in this issue, with three primary papers reporting antiviral agents with inhibitory activity against porcine circovirus type 2 (PCV2) [17] and porcine reproductive and respiratory syndrome virus (PRRSV) [18,19]. The study by Yu et al. demonstrates that antiviral effects of Ginsenoside Rg1 against PRRSV can be attributed to reductions in levels of host cell pro-inflammatory cytokines and suppression of NF-κB signalling [18].

As mentioned, this special issue includes reviews into antiviral strategies against HIV, measles and CMV [6,11,14,15]. In addition, the issue contains reviews that discuss sphingolipids as potential therapeutic targets against enveloped human RNA viruses [20] and human antimicrobial peptides as therapeutics for viral infections [21].

Finally, I would like to thank all the authors, reviewers and editors, who made this issue possible, both at Viruses and in the wider academic community.

References

1. Ye, M.; Liao, Y.; Wu, L.; Qi, W.; Choudhry, N.; Liu, Y.; Chen, W.; Song, G.; Chen, J. An oleanolic acid derivative inhibits hemagglutinin-mediated entry of influenza a virus. *Viruses* **2020**, *12*, 225. [CrossRef] [PubMed]
2. Zhang, X.; Xia, Y.; Yang, L.; He, J.; Li, Y.; Xia, C. Brevilin a, a sesquiterpene lactone, inhibits the replication of influenza a virus in vitro and in vivo. *Viruses* **2019**, *11*, 835. [CrossRef] [PubMed]
3. Chathuranga, K.; Kim, M.S.; Lee, H.C.; Kim, T.H.; Kim, J.H.; Gayan Chathuranga, W.A.; Ekanayaka, P.; Wijerathne, H.; Cho, W.K.; Kim, H.I.; et al. Anti-respiratory syncytial virus activity of plantago asiatica and clerodendrum trichotomum extracts in vitro and in vivo. *Viruses* **2019**, *11*, 604. [CrossRef] [PubMed]
4. Tadesse, B.T.; Tsai, O.; Chala, A.; Chaka, T.E.; Eromo, T.; Lapointe, H.R.; Baraki, B.; Shahid, A.; Tadesse, S.; Makonnen, E.; et al. Prevalence and correlates of pre-treatment hiv drug resistance among hiv-infected children in ethiopia. *Viruses* **2019**, *11*, 877. [CrossRef]
5. Shahid, M.; Qadir, A.; Yang, J.; Ahmad, I.; Zahid, H.; Mirza, S.; Windisch, M.P.; Shahzad-Ul-Hussan, S. An engineered microvirin variant with identical structural domains potently inhibits human immunodeficiency virus and hepatitis c virus cellular entry. *Viruses* **2020**, *12*, 199. [CrossRef]
6. Sherpa, C.; Grice, S. Structural fluidity of the human immunodeficiency virus rev response element. *Viruses* **2020**, *12*, 86. [CrossRef]
7. Bernatchez, J.A.; Coste, M.; Beck, S.; Wells, G.A.; Luna, L.A.; Clark, A.E.; Zhu, Z.; Hecht, D.; Rich, J.N.; Sohl, C.D.; et al. Activity of selected nucleoside analogue protides against zika virus in human neural stem cells. *Viruses* **2019**, *11*, 365. [CrossRef]
8. Imran, M.; Saleemi, M.K.; Chen, Z.; Wang, X.; Zhou, D.; Li, Y.; Zhao, Z.; Zheng, B.; Li, Q.; Cao, S.; et al. Decanoyl-arg-val-lys-arg-chloromethylketone: An antiviral compound that acts against flaviviruses through the inhibition of furin-mediated prm cleavage. *Viruses* **2019**, *11*, 1011. [CrossRef]
9. Sanna, G.; Piras, S.; Madeddu, S.; Busonera, B.; Klempa, B.; Corona, P.; Ibba, R.; Murineddu, G.; Carta, A.; Loddo, R. 5,6-dichloro-2-phenyl-benzotriazoles: New potent inhibitors of orthohantavirus. *Viruses* **2020**, *12*, 122. [CrossRef]
10. LeCher, J.C.; Diep, N.; Krug, P.W.; Hilliard, J.K. Genistein has antiviral activity against herpes b virus and acts synergistically with antiviral treatments to reduce effective dose. *Viruses* **2019**, *11*, 499. [CrossRef]
11. Ferren, M.; Horvat, B.; Mathieu, C. Measles encephalitis: Towards new therapeutics. *Viruses* **2019**, *11*. [CrossRef] [PubMed]
12. Anderson, A.G.; Gaffy, C.B.; Weseli, J.R.; Gorres, K.L. Inhibition of epstein-barr virus lytic reactivation by the atypical antipsychotic drug clozapine. *Viruses* **2019**, *11*, 450. [CrossRef] [PubMed]
13. Chen, S.J.; Chen, Y.C. Potential application of talens against murine cytomegalovirus latent infections. *Viruses* **2019**, *11*, 414. [CrossRef] [PubMed]
14. Chen, S.J.; Wang, S.C.; Chen, Y.C. Antiviral agents as therapeutic strategies against cytomegalovirus infections. *Viruses* **2019**, *12*, 21. [CrossRef]
15. Adamson, C.S.; Nevels, M.M. Bright and early: Inhibiting human cytomegalovirus by targeting major immediate-early gene expression or protein function. *Viruses* **2020**, *12*, 110. [CrossRef]
16. Alidjinou, E.K.; Bertin, A.; Sane, F.; Caloone, D.; Engelmann, I.; Hober, D. Emergence of fluoxetine-resistant variants during treatment of human pancreatic cell cultures persistently infected with coxsackievirus b4. *Viruses* **2019**, *11*, 486. [CrossRef]

17. Li, J.; Song, D.; Wang, S.; Dai, Y.; Zhou, J.; Gu, J. Antiviral effect of epigallocatechin gallate via impairing porcine circovirus type 2 attachment to host cell receptor. *Viruses* **2020**, *12*, 176. [CrossRef]

18. Yu, Z.Q.; Yi, H.Y.; Ma, J.; Wei, Y.F.; Cai, M.K.; Li, Q.; Qin, C.X.; Chen, Y.J.; Han, X.L.; Zhong, R.T.; et al. Ginsenoside rg1 suppresses type 2 prrsv infection via nf-kappab signaling pathway in vitro, and provides partial protection against hp-prrsv in piglet. *Viruses* **2019**, *11*, 1045. [CrossRef]

19. Wang, J.; Li, J.; Wang, N.; Ji, Q.; Li, M.; Nan, Y.; Zhou, E.M.; Zhang, Y.; Wu, C. The 40 kda linear polyethylenimine inhibits porcine reproductive and respiratory syndrome virus infection by blocking its attachment to permissive cells. *Viruses* **2019**, *11*, 876. [CrossRef]

20. Yager, E.J.; Konan, K.V. Sphingolipids as potential therapeutic targets against enveloped human rna viruses. *Viruses* **2019**, *11*, 912. [CrossRef]

21. Ahmed, A.; Siman-Tov, G.; Hall, G.; Bhalla, N.; Narayanan, A. Human antimicrobial peptides as therapeutics for viral infections. *Viruses* **2019**, *11*, 704. [CrossRef] [PubMed]

Review

Structural Fluidity of the Human Immunodeficiency Virus Rev Response Element

Chringma Sherpa [†] and Stuart F. J. Le Grice *

Basic Research Laboratory, National Cancer Institute, Frederick, MD 21701, USA; chringma@gmail.com
* Correspondence: legrices@mail.nih.gov; Tel.: +1-(301)-846-5256; Fax: +1-(301)-846-6013
† Current address: Division of Cellular and Gene Therapies, Center for Biologics Evaluation and Research, U.S. Food and Drug Administration, Silver Spring, MD 20933, USA.

Received: 3 December 2019; Accepted: 9 January 2020; Published: 11 January 2020

Abstract: Nucleocytoplasmic transport of unspliced and partially spliced human immunodeficiency virus (HIV) RNA is mediated in part by the Rev response element (RRE), a ~350 nt cis-acting element located in the envelope coding region of the viral genome. Understanding the interaction of the RRE with the viral Rev protein, cellular co-factors, and its therapeutic potential has been the subject of almost three decades of structural studies, throughout which a recurring discussion theme has been RRE topology, i.e., whether it comprises 4 or 5 stem-loops (SLs) and whether this has biological significance. Moreover, while in vitro mutagenesis allows the construction of 4 SL and 5 SL RRE conformers and testing of their roles in cell culture, it has not been immediately clear if such findings can be translated to a clinical setting. Herein, we review several articles demonstrating remarkable flexibility of the HIV-1 and HIV-2 RREs following initial observations that HIV-1 resistance to trans-dominant Rev therapy was founded in structural rearrangement of its RRE. These observations can be extended not only to cell culture studies demonstrating a growth advantage for the 5 SL RRE conformer but also to evolution in RRE topology in patient isolates. Finally, RRE conformational flexibility provides a target for therapeutic intervention, and we describe high throughput screening approaches to exploit this property.

Keywords: HIV; Rev response element; chemical footprinting; SHAPE; drug discovery; branched peptides

---

1. Introduction

Conformational "fluidity" of RNA allows it to mediate a variety of biological functions, examples of which include (a) catalyzing cleavage by the hammerhead ribozyme of satellite RNAs and viroids [1]; (b) bacterial riboswitches [2]; (c) RNA thermometers [3]; and (d) tRNA-dependent control of specific aminoacylation and translation regulation [4]. In the case of human immunodeficiency virus (HIV), cis-acting sequences encoded in its (+)RNA genome are central to transcription of the integrated provirus, nucleocytoplasmic transport of unspliced and partially spliced RNAs, initiation of reverse transcription, genome dimerization/packaging, and ribosomal frameshifting [5]. A comprehensive understanding of the structural dynamics of these regulatory elements would be predicted to accelerate development of small molecules [6], oligonucleotides [7], peptide nucleic acids [8] evolved RNA recognition motifs [9], and nucleic acid aptamers [10] as novel therapeutic modalities to complement existing anti-HIV agents. With these objectives in mind, the goal of this review was to highlight conformational flexibility of the HIV-1 RRE (and its HIV-2 counterpart) through a series of experiments that extend in vitro structural analysis to their in vivo outcome in cell culture systems and, finally, in sequential viral isolates in a clinical setting.

Figure 1A provides a summary of the functional requirement for the HIV-1 RRE. Productive HIV infection produces three types of viral transcripts, i.e., unspliced, partially spliced, and fully spliced

RNAs (http://hivinsite.ucsf.edu). Early in the viral life cycle, fully spliced viral RNAs encoding the regulatory proteins Rev, Tat, and Nef are exported and translated in the cytoplasm. However, the presence of introns in unspliced and partially spliced viral RNAs results in their nuclear retention by host RNA surveillance mechanisms that normally restrict nucleo-cytoplasmic export of intron-retaining mRNAs [11,12]. Later in the life cycle, Rev, through its nuclear localization signal (NLS), is imported into the nucleus [13,14]. The NLS domain is a basic, arginine-rich motif (ARM) that also serves as an RNA-binding domain (RBD) that binds specifically to the RRE (Figure 1B) [15]. Inside the nucleus, Rev binds cooperatively to the RRE present in all intron-retaining viral RNAs through a process involving both protein–protein and protein–RNA interactions [15–24]. The Rev–RRE complex is recognized by CRM1 and RAN-GTP forming an export competent ribonucleoprotein (RNP) complex [25] allowing unspliced and partially spliced viral RNAs to circumvent host cellular restriction and transit to the cytoplasm, where they are either translated or packaged into assembling virions [26–28].

(A)

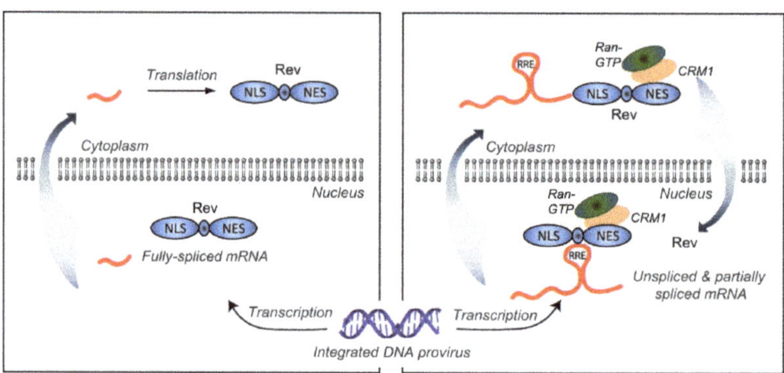

(B)

Figure 1. (**A**) Functional requirement of the HIV RRE: Early in the viral lifecycle, fully spliced viral RNAs are exported from the nucleus in a Rev/RRE-independent manner. Among these, Rev mRNA are translated and Rev is imported into the nucleus. In the late phase of the lifecycle, nuclear RRE-containing RNAs recruit Rev and cellular nuclear-export machinery, allowing them to circumvent splicing and transit to the cytoplasm, where they are either translated or packaged into assembling virions (**B**) Organization of the 116 aa HIV-1 Rev and amino acid changes in the trans-dominant M10 variant. NLS; nuclear localization signal, NES, nuclear export signal. Pink areas flanking the NLS represent Rev oligomerization domains.

A logical first step in understanding molecular details of the Rev/RRE complex was defining the topology of the RRE, an ~350 nt, RNA comprising multiple stem-loops and bulges. Combining computational modeling with chemical and enzymatic footprinting led to the proposal of a 5 stem-loop (SL) structure with a central SL-1 branching into SL-II (the primary Rev binding site), SL-III, SL IV, and SL-V [29–31]. The SL-II consists of stem IIA branching out of the central loop and opening into a three-way junction. The junction opens into two stem-loops IIB and IIC [19,24,32–34]. In contrast, a 4 SL structure, differing in rearrangement of SL-III and -IV, has been reported [23,35], wherein SL-III and -IV of the 5 SL RRE combine to form a single SL-III/IV off of the central loop. Despite these differences, the 4 SL and 5 SL RRE conformers preserve SL-II topology. Since the majority of these structures were derived from different in vitro probing methodologies, it cannot be ruled out that such differences reflect subtle alterations in buffer probing conditions and are not truly reflective of the biological system. This review summarizes several papers, including analysis of patient isolates that collectively suggest both that HIV-1 and HIV-2 RRE possess sufficient flexibility to adopt alternative conformations, and that for HIV-1, at least, stabilizing these by in vitro mutagenesis confers a growth advantage for the 5 SL conformer. Lastly, we present data suggesting that RRE conformational flexibility might be exploited therapeutically.

2. Resistance to Trans-Dominant RevM10 Therapy Induces a Conformational Change in the HIV-1 RRE

The prototype Rev (NL4-3) from HIV clade B is a ~18 kD 116 amino acid phospho-protein [36]. Thee Rev protein comprises several well-characterized functional domains central to Rev-RRE-mediated nucleocytoplasmic transport of unspliced and partially spliced mRNAs [13]. One of these, aa 34–50 constitute the arginine-rich nuclear localization signal (NLS), the primary contact point to the RRE. The NLS is flanked by oligomerization domains that are required for Rev multimerization. Oligomerization is mediated mainly by Leu 12, Val16, Leu18, Leu55, and Leu60 [37]. Towards the carboxy-terminal is a leucine-rich nuclear export signal (NES) (aa 75–84) that mediates interactions with host nuclear export factors (Figure 1B, [38]). During a functional delineation study of Rev by site-directed mutagenesis, Malim et al. [39] constructed a defective variant of Rev, designated RevM10, by replacing two critical NES residues (Figure 1B). When expressed in transfected cells, this mutant Rev protein successfully inhibited the function of its wild-type counterpart. That RevM10 serves as an effective trans-dominant inhibitor of Rev function was further validated by others [40–44] thereby catapulting RevM10-based gene therapy into Phase I and II clinical trials [45]. However, a subsequent study by Hamm et al. [46], involving passage of HIV-1 in a T-cell line constitutively expressing RevM10, reported rapid emergence of a RevM10-resistant virus [46]. Surprisingly, escape mutations were not associated with Rev, but rather the RRE. The same study demonstrated that only two silent mutations (with respect to viral envelop protein) outside the primary Rev-binding region of the RRE, namely, G164 > A164 (at the base of SL-III/IV) and G245 > A245 (in the central loop) sufficed to confer RevM10 resistance, suggesting a conformational change in the RRE was responsible. With this in mind, RRE structures of wild-type and RevM10-resistant HIV variants containing these two RRE mutations (RRE-61) were examined by selective 2'-OH acylation monitored by primer extension (SHAPE) [47]. In this study, the chemical reactivity profile of the wild-type RRE (Figure 2A) predicted the 4 SL conformer (Figure 2B). In contrast, the two silent RRE-61 mutations, inducing RevM10 resistance, introduced significant alterations in chemical reactivity over ~60 nt (Figure 2A), deconvolution of which predicted the 5 SL conformer (Figure 2C). Constructing individually-mutated RRE variants, G164 > A164 (at the base of SL-III/IV) and G245 > A245 (in the central loop), allowed their contribution to be evaluated, where a combination of in vitro replication kinetics, non-denaturing polyacrylamide gel electrophoresis, and chemical footprinting indicated the G245 > A245 mutation sufficed to mimic the structure and activity of RRE-61. Interestingly, in this study, RevM10 resistance could not be attributed to differential/reduced Rev binding to RRE-61. Therefore, it is likely that the stable SL-IV structure in the 5 SL RRE61 provides binding sites for cellular or viral factors that mediate RevM10 resistance (see later). Nevertheless, the study by Legiewicz et al. [47] demonstrated a considerable degree of conformation flexibility inasmuch as a relatively modest nucleotide change in the RRE central loop had a major impact on its topology.

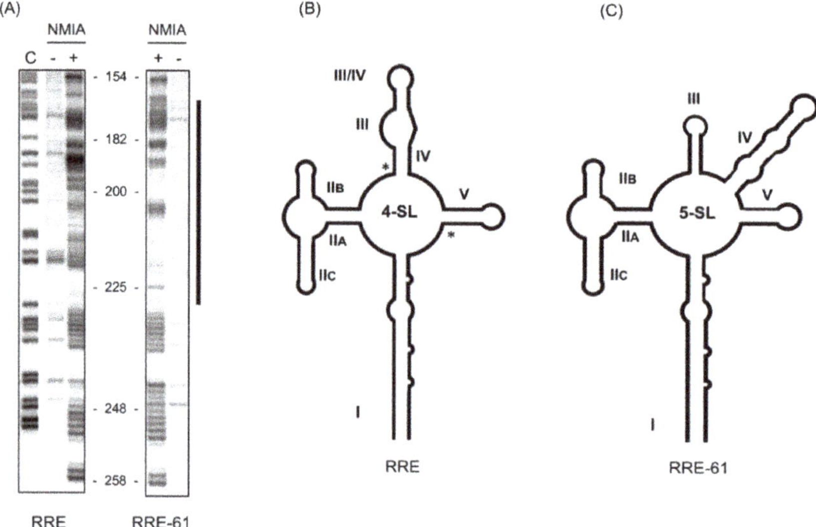

Figure 2. Mutations conferring resistance to trans-dominant RevM10 therapy induce a conformational change in the HIV-1 RRE. (**A**) Structural probing of the wild-type RRE (left) and RevM10-resistant RRE-61 (right). (**C**) Control DNA sequencing lane from which nucleotide numbering was derived. Designations – and + refer to untreated and NMIA-treated RNA, respectively. Major alterations in chemical reactivity are indicated by the bar. (**B,C**) Cartoons depicting SHAPE-derived conformations of wild-type RRE and RRE-61. SL, stem-loop. Positions of RRE point mutations, inducing RevM10 resistance, are indicated by asterisks in (**B**). Modified from Reference [47].

3. Structural Conformers of the Wild-Type HIV-1 RRE

Since the discovery of Rev [48,49] and the RRE [26–28,50] about 30 years ago, a wealth of studies on the HIV-1 Rev-RRE system have advanced our understanding of the structural details of the system. A consensus secondary structure of the RRE, however, has been lacking. Although the minimal size of a functional RRE was initially reported to be 234 nt [28], later studies [23,31,32] indicated that a fully functional RRE is ~350 nt. Early studies on RRE secondary structure were performed with the subtype B HXB2 234 nt RRE [29,30] which was reported to assume a 5 SL structure. As NL4-3 increasingly become the standard HIV molecular clone studied in laboratories, subsequent structural studies of the RRE were performed mostly on this isolate. Interestingly, studies on the 351 nt NL4-3 RRE also supported a 4 SL conformer [23,35]. It was then speculated that the presence of a longer stem-loop I used might have facilitated the alternative structure. However, contrary to this speculation, Watt et al. [31] subsequently showed that the full-length RRE of RNA purified from HIV-1 NL4-3 virions adopts a 5 SL conformation. This study, therefore, showed that at least, in NL4-3 RRE, the longer stem-loop I was not the determinant of the alternative structure. This finding was further supported by another study that used SHAPE to show that an in vitro transcribed 232 nt NL4-3 RRE formed the alternative 4 SL structure [47]. Besides HXB2 and NL4-3, the secondary structure of another HIV-I molecular clone (ARV-2/SF2) RRE (354 nt), differing from NL4-3 sequence at 13 nucleotides, has also been reported to adopt a 5 SL like conformation. This variant differs from the canonical 5 SL structure in that the nucleotides to the left of the top stem region of SL I/I′ base pair with nucleotides from the central loop, bridging SL-IV and SL-V and forming a stem region which opens into individual SL-IV and SL-V.

Since chemical probing techniques used in these studies provide ensemble-average structural information, it could not be excluded that RRE structural heterogeneity might give rise to such discrepancies. In-gel probing, in contrast, offers an alternative strategy to examine conformationally heterogeneous RNAs, providing that they can be separated by non-denaturing strategies. As an example, Kenyon et al. [51] applied in-gel SHAPE to define the structure of monomeric and dimeric species of the HIV-1 packaging signal RNA, supporting a structural switch model of RNA genomic dimerization and packaging [51]. In light of (i), HIV-2 RRE conformational heterogeneity (see later) and (ii), the observation that a single nucleotide alteration sufficed to stabilize the 5 SL HIV-1 RRE conformer [47], we showed that by extended non-denaturing polyacrylamide gel electrophoresis that the HIV-1 RRE could be resolved into two closely migrating species (Figure 3A) [52]. Subsequent in-gel SHAPE verified the slower migrating RNA as the 4 SL conformer and the faster migrating RNA as the 5 SL conformer (Figure 3B,C, respectively), suggesting that in vivo, the wild-type HIV-1 RRE could exist in a conformational equilibrium.

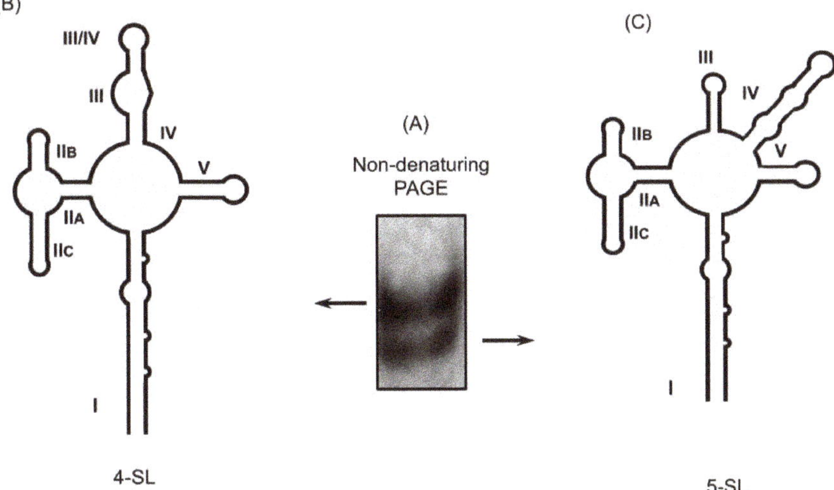

Figure 3. The HIV-1 RRE exists in a conformational equilibrium. (**A**) Following extended non-denaturing PAGE, slow and fast migrating RRE conformers were observed. Subjecting these RNAs to in-gel SHAPE defines these as 4 SL (**B**) and 5 SL conformers (**C**). Note that, despite their conformational heterogeneity, the topology of SL-II, the primary Rev binding suite, is preserved. Modified from Sherpa et al. [52]. The 232 nt HIV-1 RRE RNAs appended with a 3′ structure cassette were prepared for analysis by in vitro transcription.

To investigate the function of these alternate HIV-1 RRE conformers, Sherpa et al. [52] created stable 4 SL and 5 SL RRE variants by in vitro mutagenesis to determine their Rev-RRE activity. The RRE Mutant M1 was predicted to disrupt base pairing at the base of the combined SL-III/IV structure in the 4 SL conformer but maintain base pairing in SL-IV in the 5 SL variant and, thus, likely to adopt only the latter structure. Conversely, the mutant RRE M3 was expected to disrupt the base pairing in both SL-III and SL-IV of the 5 SL structure but keep the combined SL-III/IV intact and, thus, likely adopt only a 4 SL conformation. As illustrated in Figure 4A,B, these predictions were borne out experimentally since the initial RRE population was resolved into two stable conformers. The SHAPE analysis indicated that the mutant RREs preserved the structure predicted by mutagenesis. More importantly, electrophoretic mobility shift experiments indicated that Rev binding to the mutant RREs was largely unaffected, demonstrating that their global topology had not been affected by mutagenesis. Growth competition assays were next performed in a T-cell line (SupT1) to assess whether these RRE conformers conferred a selective growth advantage. In this heteroduplex tracking analysis strategy [53], viruses with a different RRE were added to the same culture of SupT1 cells at an equal multiplicity of infection (MOI) and allowed to replicate and spread throughout the cultures for several days, after which cellular DNA was recovered. Relative amounts of integrated pro-viral DNA produced by each virus was measured using a PCR-based heteroduplex tracking. Since sequence mismatches within the wt/M1 and wt/M3 heteroduplexes caused them to migrate differently from each other and the perfectly wt/wt homoduplex, each heteroduplex could be resolved by native gel electrophoresis. Figure 4C indicates that the stable 5 SL RRE conformer displayed a selective growth advantage over its stable 4 SL counterpart as well virus encoding the wt RRE. Supporting this observation, HIV gag/pol expression assays demonstrated that the stabilized 5 SL conformer was functionally superior to wt and the 4 SL RRE. Thus, structural plasticity of the RRE promoting similar [54] or different levels of Rev-RRE function demonstrates how the Rev-RRE regulatory axis might function as a "replication rheostat" rather than a simple on/off switch.

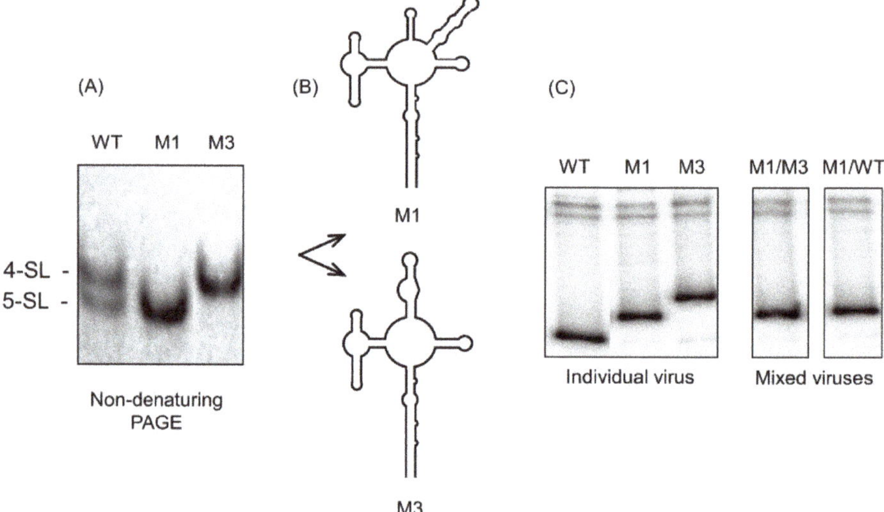

Figure 4. Alternative HIV-1 RRE conformers promote different rates of virus replication. (**A**) Conformer construction (see text). (**B**) Chemical acylation (SHAPE) confirms a 5 SL conformation of mutant M1 and a 4 SL conformation of mutant M3. (**C**) Heteroduplex tracking analysis. In both M1/wt and M1/M3 mutant co-infections, the stabilized 5 SL M1 conformer displays a replicative growth advantage. The full experimental background is provided in Reference [52].

4. Conformational Flexibility of HIV-1 RRE SL-I and Rev Sequestration

Although detailed structural data are available for the HIV-1 RRE in either the absence or presence of Rev [18,20,55], understanding how the Rev-mediated nuclear export complex assembles has proven a challenge. In the model proposed by Frankel and co-workers, initial Rev binding to SL-IIB promotes its recruitment to an additional secondary site on SL 1, with at least six copies of Rev ultimately driving formation of the functional ribonucleoprotein complex [19].

In this model, an additional role for SL-I beyond providing the accessory Rev-binding site had not been considered. Using time-resolved SHAPE and SAXS, Bai et al. [32] demonstrated that SL-I flexibility makes an important contribution to the formation of the export complex, illustrating that Rev binding affects chemical acylation levels/structure at three distinct regions (I, II, II) in the RRE. Region I maps to the high affinity Rev binding site and the stem II three-way junction. Region II covers the central loop and the top region of SL-I that includes the previously identified secondary

Rev binding site [19], while Region III lies at the center of SL-I of the 351 nt RRE. Figure 5A depicts the SAXS-derived A-like structure of the RRE reported by Feng et al. [55] with an RNA construct whose SL-I was truncated. In this model, the maximum diameter of the RRE was calculated ~195 A°. Surprisingly, this value did not change when an RRE containing the full-length SL-1 was re-examined by SAXS [32], suggesting that rather than adopting an extended "tail", SL-I folds into and interacts with the RRE core to adopt a compact structure. Proof of this notion was provided by hybridizing an antisense oligonucleotide to either of the interacting partners. In both cases, an increase in the maximum diameter ~300 A° was observed, suggesting disruption of the long-range SL-1-mediated interaction. Based on their findings, Bai et al. [32] propose a model where a Rev dimer binds to Region I of a pre-organized RRE which is immediately followed by binding of another Rev dimer in Region II. This induces tertiary long-range interaction between Region III and the central loop region, exposing a cryptic Rev binding site in SL-I, to which additional Rev molecules can bind [23] (Figure 5B,C). While once more demonstrating remarkable flexibility of the RRE, data from Bai et al. [32] raises the intriguing notion of developing multi-dentate ligands to target long-range interaction critical to Rev/RRE assembly as a therapeutic strategy. Data from a later section address this possibility.

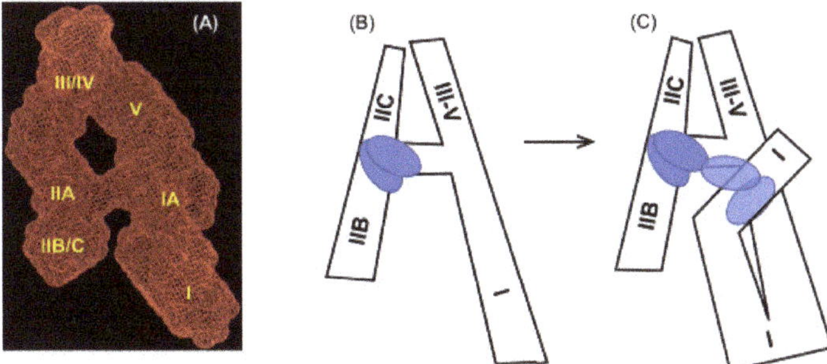

Figure 5. Conformational flexibility of HIV-1 RRE SL-I contributes towards assembly of the Rev-mediated export complex. (**A**) Molecular envelope of the RRE RNA, drawn in mesh and derived by SAXS [32]. The spatial resolution of the envelope is 21 A°. (**B**) Cartoon representation of the RRE, depicting assembly initiating via a single nucleation point in SL-II for two Rev molecules (blue). (**C**) Through an SL-I conformational change, "coupling" of SL-I and SL-II Rev-binding sites promote a tetrameric intermediate complex proposed to serve as a specificity checkpoint. Rev and the RRE could thereafter simultaneously sample a number of interaction conformations until an optimal binding state for Crm1 binding and nuclear export is attained.

5. Interchanging HIV-1 RRE Conformers in Patient Isolates

Several lines of in vitro evidence discussed thus far suggest the RRE as a dynamic structure capable of existing as a mixture of conformers or assuming alternate conformations in response to minor nucleotide changes (e.g., RRE-61). Such observations raise the question whether this might also occur in a clinical setting in the course of HIV infection. Indeed, nucleotide changes over the course of virus infection have been linked with enhanced RRE activity and a more rapid CD4 decline [56], while decreased Rev activity has been linked to slower disease progression and reduced susceptibility to T-cell killing [57]. Efforts to develop new therapeutic interventions directed at the Rev/RRE axis would likely benefit from studies of its structural and functional evolution in the course of natural infection. To address this, Sloan et al. [58] examined evolution the Rev/RRE axis in the blood plasma of a single patient using samples collected from initial detection of p24 antibodies within 6 months of infection (designated visit 10 or V10-2) through the subsequent 6 years of infection in the absence of antiretroviral therapy (designated V20-1). Functionally, this study demonstrated that V20-1 RRE promoted Rev multimerization at a lower Rev concentration than its V10-2 counterpart. In a follow-up study, a structural analysis of the V10-2 and V20-1 RREs was undertaken by Sherpa et al. [59]. As shown in Figure 6, non-denaturing gel electrophoresis differentiated among these two RREs based on their migration properties, suggesting alternate conformers. Chemical probing analysis indicated that a portion of the V10-2 central loop between SL-III and SL-IV paired with nucleotides from the upper stem of SL-I, forming a stem that bridges the central loop and SL-IV and -V. In contrast, V20-1 RRE formed the canonical 5 SL structure, with SL-I to SL-V radiating directly from the single-stranded central loop. This is the first detailed long-term study to examine longitudinal RRE evolution in a patient following infection, highlighting that selection pressures impart an influence on the RRE sequence, with a tendency toward increased functional activity. Increased activity has been be explained by large-scale conformational changes within the RRE and a decrease in base-pairing stability at the initial Rev binding site of SL-II. Although selective pressure on the HIV *Env* gene during disease progression in terms of immune evasion and replication efficiency may well be contributing factors, functional differences in Rev-RRE activity likely also contribute to viral fitness.

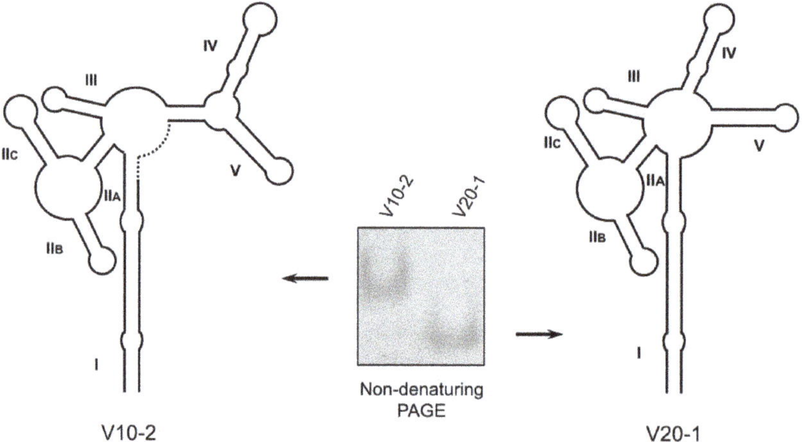

Figure 6. Patient-derived HIV-1 RREs from early and late time-points post-infection exhibit different secondary structures. Secondary structures of V10-2 RRE (an early isolate, (**left**) and V20-1 RRE (a late isolate, (**right**) determined by SHAPE-MaP. (**Center**) differential migration rate of V10-2 and V20-1 RRE, following non-denaturing PAGE and UV shadowing, is suggestive of alternate conformers/ Adapted from Sherpa et al. [59].

This study, also for the first time, showed experimentally that structural fluidity exists in the SL-II region of primary HIV isolates which can modulate Rev-RRE activity. A more recent paper [60] further explored the structural flexibility of RRE SL-II region using NMR to highlight that in vitro synthesized wt (NL4-3) SL-II exists in dynamic equilibrium of three different conformers which includes two non-native excited states (ES1 and ES2) that remodel key structural elements required for Rev binding and one ground state (GS). These ES populations constitute around 20% of the SL-II structural ensemble and bound Rev peptides with 15 to 80 fold weaker affinity. Such studies highlight the need to consider structural flexibility of SL-II regions in developing anti-HIV therapeutics targeting the RRE as traditional approaches that rely on high throughput screening and/or rational design of small molecules/peptides/agents that bind to the GS RRE II. Agents that lock the RRE in the less active ES forms should therefore be explored as new avenues for anti-HIV drug design.

It is also important that structural flexibility of regions of the RRE outside the primary Rev binding site be considered during anti-HIV drug design. A good example of this notion is reflected in development of drug resistance against ENF (enfutivirtide or T20), the first fusion inhibitor used for HIV treatment. T20 acts by binding to a region of gp41 subunit of HIV Env and has been reported to select for secondary mutations in Rev and the RRE [61]. The primary mutations associated with

ENF resistance were located within the ENF target region and map to gp41 aa 36–45 which lies within the RRE. Secondary mutations were found to restore the RRE structure predicted to be disrupted by the primary mutations. Such "structure conservation mutations" were observed in SL-IIC [61] and SL-III [62], underscoring the importance of conformational fluidity beyond the primary Rev binding site. A thorough molecular understanding of the various alternative RRE conformers in primary isolates will therefore be pivotal in designing more effective anti-HIV drugs that delay/prevent the onset of RRE structural flexibility-mediated drug resistance.

6. Conformational Changes Underlying "Maturation" of the HIV-2 RRE

An intriguing issue is whether observations and models suggesting structural fluidity are unique to the HIV-1 RRE or whether its' HIV-2 counterpart is likewise conformationally heterogeneous. Early mutational studies of the HIV-2 RRE [63] indicated that (i) the interaction with its cognate Rev was more dependent on maintenance of secondary structure than primary nucleotide sequence and (ii) HIV-2 RRE structures permitting interaction with HIV-1 Rev, while coinciding with those required for HIV-2 Rev binding, were dissimilar in structure and nucleotide sequence. Prior to performing HIV-2 RRE characterization by SHAPE, data from Figure 7A raised a formidable challenge, since in the absence of any binding partner this too displayed unexpected conformational flexibility. Although denaturing polyacrylamide gel electrophoresis indicated a single RNA species following in vitro transcription, subsequent non-denaturing electrophoresis identified three conformers that gradually "coalesced" into a single species upon prolonged renaturation [64]. Since SHAPE requires that the target RNA adopt a uniform structure in solution [65], understanding this unexpected stepwise HIV-2 RRE folding required a mathematical model to be developed that extracted the contributions of individual conformers from ensemble chemical reactivity values. The model makes two assumptions. Firstly, it assumes that each ensemble SHAPE reactivity value obtained from a population of RNA conformers equals the sum of reactivity values of the contributing conformers, weighted according to their fractional contribution to the total RNA population. Secondly, if ensemble reactivity values and fractional contributions of individual conformations changed with differing conditions (e.g., folding time), and these values could be determined for a number of conditions equal to or greater than the total conformer content of the mixture, specific chemical reactivity values for each conformer could be mathematically derived. By adopting this strategy, the HIV-2 RRE folding program depicted in Figure 7B was proposed.

The most significant differences among these structures involved the central junction, changes within which affect positioning of the substructures relative to each other. Transition from the open to intermediate conformation requires several concerted substructure translations/rotations, most notably rotation of SL-IIB/IIC, caused by base pair formation in S-L IIA. Pairing of the bridge helix defines the subsequent transition from the intermediate to closed conformer, which also involves inward rotation of the SL-IV/V substructure, folding of the SL-IIB apical loop toward the central junction and formation of mutually stabilizing contacts between SL-IIB and SL-V. The SL-IIB is arranged orthogonally to SL-IIC and coaxially with SL-I, consistent with assembly models, whereby Rev initially binds to high affinity sites on SL-IIB and SL-I, then multimerizes linearly along an SL-IIB/SL-I axis. Given the close contact between SL-IIB and SL-V in the closed conformer, rotational/translational flexibility of the SL-IV/V substructure would be required to create space for Rev binding to SL-IIB.

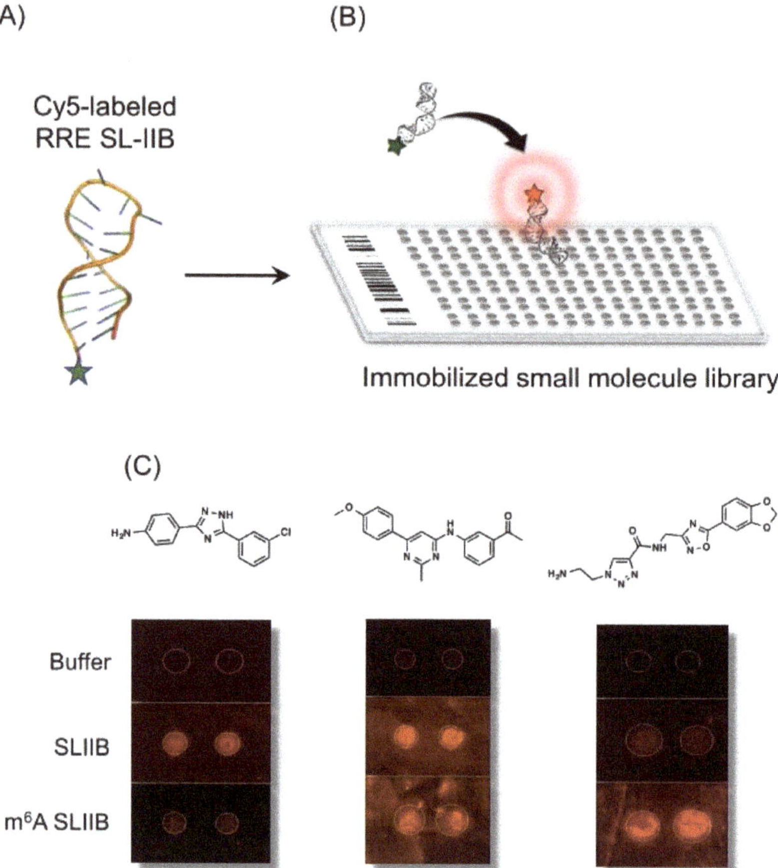

Figure 7. Time-dependent conformational rearrangement of the HIV-2 RRE. Analyses were performed on a 216 nt RRE derived from HIV-2$_{ROD}$ by in vitro transcription. (**A**) Native gel electrophoresis as a function of incubation time indicates the HIV-2 RRE comprises a mixture of "open", "intermediate", and "closed" conformers (**A–C**, respectively) at short incubation times and whose ratio varies with time, with the closed conformer ultimately predominating. (**B**) SHAPE-derived conformations of the open, intermediate, and closed HIV-2 RRE forms, respectively. Secondary structural motifs are indicated and color-coded as follows: SL-I, red; S-IIA, dark green, SL-IIB, -IIC and adjacent connecting loops, magenta; SL-III, yellow; SL-IV, blue; SL-V, orange. Modified from Reference [64].

7. Targeting RRE Conformations: the HIV Epitranscriptome.

Over several decades, post-transcriptional chemical modifications that impact RNA metabolism, function, and localization have been recorded, the most common associated with tRNA and ribosomal RNA. Of the >100 post-transcriptional modifications identified to date, N^6-methyladenosine (m⁶A) has been the most extensively studied with respect to its importance in health and disease [66]. Although the functional role remains to be fully elucidated, a role for post-transcriptional modification of the retroviral RNA genome was first proposed from studies by Kane et al. [67]. Three recent studies have established m⁶A modification of the HIV-1 genome [68–70], of which Lichinchi et al. [69] have suggested that methylation of conserved adenosines in RRE SL-II (A7877 and A7883) enhanced Rev binding and influenced nuclear export of viral RNA.

Of possible roles ascribed to post-transcriptional modification, alterations to RNA structure to up- or downregulate interactions with either host or viral proteins would seem likely [71]. An extension of this notion would be the potential to develop small molecules that specifically recognized m^{6}A-induced structural alterations within the RNA genome. As a step in this direction, a small molecule microarray (SMM) strategy was developed to allow facile and rapid screening for RNA-binding chemotypes [72–74]. The SMM strategy is outlined in Figure 8A, wherein a short, structured RNA (40–60 nt), appended with a fluorophore, is flowed over a library of covalently immobilized small molecules. Ligands interacting with the RNA are recorded as a fluorescent signal (Figure 8B). To determine whether differential recognition of modified RNA could be achieved, unmethylated and m^{6}A-modified RREs were screened. As indicated in Figure 8C, chemotypes specific for SL-IIB and m^{6}A SL-IIB were identified, in addition to a third class that recognized both RNAs. Although recent study of Chu et al. [75] employing several biochemical and biophysical strategies concluded that the stability, structure, and dynamics of RRE SL-IIB are only marginally affected by m^{6}A modification, SMM screening implies that these might be sufficient for selective ligand recognition. While high throughput screening strategies for RNA are in their infancy, data reported here suggests the notion of targeting viral epitranscriptomes should not be overlooked.

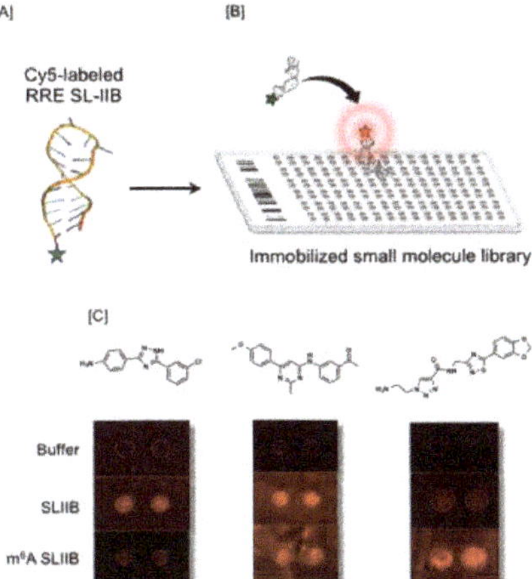

Figure 8. Screening for small molecules that recognize m^{6}A-modifed HIV-1 RRE SL-IIB. (**A**) A synthetic RNA fragment harboring the m^{6}A modifications reported in SL-IIB is labeled with Cy5 on its 3′ terminus. (**B**) Labeled SL-II RNA was flowed over microtiter plates containing covalently immobilized small molecules. Binding of RRE SL-II RNA to candidate ligands was recorded via a fluorescence signal. (**C**) HTS screening suggests specificity of small molecules for unmethylated (left) and methylated SL-IIB (right). The central panel highlights a ligand that recognizes both SL-II forms.

8. Exploiting RRE Conformational Flexibility with Branched Peptides

The unique 3D architecture of RNA, often achieved by assuming a combination of local structures including bulges, stem-loops, pseudoknots, and turns, presents an attractive therapeutic opportunity. While exemplified by antimicrobial agents, such as aminoglycosides, macrolide, oxazolidinone, and tetracycline, that interfere with ribosomal RNA function, the field of small molecule RNA therapeutics, despite advances in high throughput screening technologies, has not fully matured and would benefit from innovative strategies.

As an alternative strategy that considers molecules of intermediate size, branched peptides offer the possibility of supporting multivalent interactions with different regions of a highly flexible RNA such as the RRE (Figure 9A) and would be predicted to enhance their selectivity and affinity. Support for branched peptides was provided by Bryson et al. [76], whose on-bead screening of a 4000 molecule library identified ligands that spanned the bulge and apical loop of HIV-1 TAR RNA with binding affinities in the low micromolar range. Subsequently, to enhance selectivity/affinity Dai et al. [77] synthesized a ~46,000 compound on-bead library of branched peptides composed of unnatural amino acids (Figure 9B). These included L-guanidinoproline and D-aminoproline as electrostatic mimics of arginine and lysine, respectively, 1-naphthalene to promote π—π stacking with nucleobases and pyrazine which acts as a hydrogen bond donor/acceptor. High throughput screening identified the branched peptide 4A5 (Figure 8B) as a high affinity ligand for a synthetic RRE SL-IIB mimic (K_d = 0.88 $\pm$ 0.02 μM). The SHAPE analysis indicated that at a 1:1 RNA/4A5 ratio, nucleotides of a loop constituting the secondary SL-1 Rev binding site were protected from modification, while at a higher branched peptide/RNA ratio, nucleotides within SL-II became refractory to acylation. From this study, roles for 4A5 in occupying the active Rev binding sites and/or acting allosterically to prevent the SL-I-driven conformational change in the RRE were proposed. Although the exact mechanism remains to be elucidated, the SHAPE data lend credence to the notion of multivalent binding of branched peptides to the RRE. Finally, in the same study, 4A5 was demonstrated to inhibit Rev-RRE function in cell culture using HEK 293T cells transiently transfected with a Rev-expressing plasmid and a CMV promoter-driven GagPol-RRE plasmid (Figure 9C).

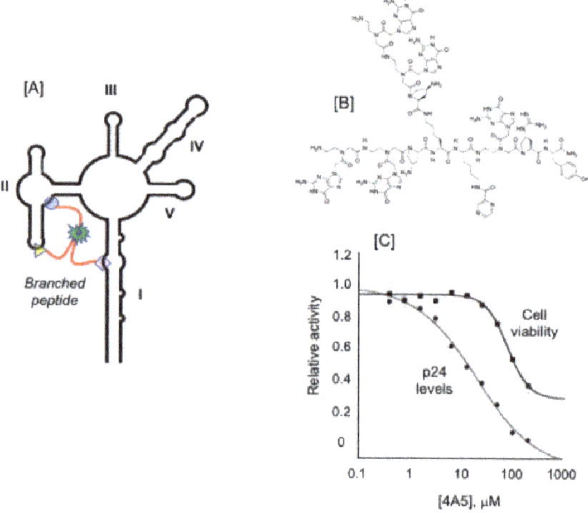

Figure 9. Restricting RRE conformational flexibility with multivalent branched peptides. (**A**) Cartoon depicting the branched peptide strategy, i.e., binding in a multivalent fashion to enhance affinity and selectivity toward the RNA target. (**B**) Structure of branched peptide 4A5. (**C**) Inhibition of Rev-RRE function in vivo using HEK 293T cells transiently transfected with a Rev-expressing plasmid and a CMV promoter-driven GagPol-RRE plasmid. Modified from Dai et al. [77].

9. Cellular Factors Interacting with the Rev and the RRE

While this review has concentrated on the dynamic nature of the HIV-1 and HIV-2 RREs, it is important to emphasize that concerted binding of various cellular factors is required for Rev-RRE function, most of which promote shuttling of Rev and RRE across the nuclear membrane. Two key factors are Crm1 [25,78–80] and importin β [81,82]. Crm1, or Exportin-1, is a member of the β-importin family of transport receptors and has been demonstrated to specifically interact with many proteins containing NES domains. Normally employed for the nucleocytoplasmic transport of proteins, snRNAs and rRNAs, the Crm1 pathway is hijacked by HIV to facilitate export of intron-retaining RNAs. Crm1 serves as an adaptor for binding of Ran-GTP [83–85] and nucleoporins such as Rip/Rab [86] to Rev, forming a functional export complex that ferries the Rev/RRE RNP across the nucleopore complex into the cytoplasm. In the cytoplasm, RanGAP1 and RanBP1 hydrolyze Ran-GTP to Ran-GDP, releasing

Crm1 from the Rev/RRE complex [86]. Importin-β then binds to the Rev NLS region, facilitating displacement of the RRE for translation of viral transcripts. The resulting Rev/importin-β complex interacts with Ran-GDP, allowing translocation of the Rev complex into the nucleus. Inside the nucleus, RCC1 catalyzes conversion of Ran-GDP to Ran-GTP, leading to dissociation of Rev from importin-β [87]. This dissociation unmasks the Rev NLS domain, promoting interaction of its NLS/RBD region with the RRE. Several members of the DEAD box RNA helicase family, such as RHA [88], DDX3 [89], DDX1 [90,91], DDX5 [92] and DDX24 [93], have also been reported as cellular co-factors of Rev, although their precise role in the Rev/RRE pathway remains to be elucidated. Most of these are nucleocytoplasmic shuttling proteins some of which have been implicated in promoting HIV genome packaging, restriction of Rev function in astrocytes, premature release of intron-retaining HIV transcripts from the splicing machinery. Similarly, cellular factors such as eIF-5A, SF2/ASF, B23, p32, Sam68, and ATM kinase have been proposed to modulate Rev/RRE function, though their roles are either unclear or controversial [38,94].

10. Conclusions and Outlook

Although the role of the RRE in directing nucleocytoplasmic transport of unspliced and partially spliced viral RNAs has been unchallenged for over three decades, the structural flexibility that mediates these steps in HIV replication is only now coming to light, in many cases, following the development of new RNA probing strategies. For example, the generally accepted notion that RRE SL-I played a structurally "passive" role has now been challenged by chemical and biophysical techniques suggesting it folds into the central RRE core, possibly as a check for Rev loading. By extending early data on the structure of RRE-61, we now have evidence that the wild-type RRE may, in fact, exist as an equilibrium of conformers, each of which confers distinct growth properties in cell culture that may extend as far as their evolution in HIV-infected individuals. One might speculate that RRE structural fluidity would confer growth/fitness advantage to the virus by acting as "replication rheostat", regulating the level of HIV gene expression by allowing RRE to adopt various functional states. A less active RRE would shield HIV from a functional host immune system during early stages of infection, allowing the virus to establish itself within the host. Conversely, highly active RREs would promote a higher rate of virus replication without CTL killing in immunocompromised individuals. Unlike mechanisms that affect overall transcription and translation, down-modulation of gene expression by less active RRE conformers primarily affects viral structural protein expression and is not expected to affect HIV Nef expression. Thus, this mechanism would allow the immune evasion activities of Nef to persist, while the lower structural protein levels would make the infected cells less susceptible to CTL killing, As we uncover additional aspects of RRE conformational dynamics, exploiting this vis-à-vis developing novel high throughput screening strategies, exemplified here with branched peptides and small molecules that recognize the viral epitranscriptome, promises to open new and exciting avenues for targeting viral RNA genomes.

Acknowledgments: SFJLG and CS were supported by the Intramural Research Program of the National Cancer Institute, National Institutes of Health, Department of Health and Human Services.

Conflicts of Interest: The authors declare no conflict of interest.

References

1. Forster, A.C.; Davies, C.; Sheldon, C.C.; Jeffries, A.C.; Symons, R.H. Self-cleaving viroid and newt RNAs may only be active as dimers. *Nature* **1988**, *334*, 265–267. [CrossRef] [PubMed]
2. Barrick, J.E.; Breaker, R.R. The distributions, mechanisms, and structures of metabolite-binding riboswitches. *Genome Biol.* **2007**, *8*, R239. [CrossRef] [PubMed]
3. Loh, E.; Righetti, F.; Eichner, H.; Twittenhoff, C.; Narberhaus, F. RNA Thermometers in Bacterial Pathogens. *Microbiol. Spectr.* **2018**, *6*. [CrossRef]

4. Meng, F.; He, Z.; Tang, X.; Zheng, J.; Jin, X.; Zhu, Y.; Ren, X.; Zhou, M.; Wang, M.; Gong, S.; et al. Contribution of the tRNA(Ile) 4317A–>G mutation to the phenotypic manifestation of the deafness-associated mitochondrial 12S rRNA 1555A–>G mutation. *J. Biol. Chem.* **2018**, *293*, 3321–3334. [CrossRef]

5. Le Grice, S.F. Targeting the HIV RNA genome: High-hanging fruit only needs a longer ladder. *Curr. Top Microbiol. Immunol.* **2015**, *389*, 147–169. [CrossRef]

6. Connelly, C.M.; Moon, M.H.; Schneekloth, J.S., Jr. The Emerging Role of RNA as a Therapeutic Target for Small Molecules. *Cell Chem. Biol.* **2016**, *23*, 1077–1090. [CrossRef]

7. Bennett, C.F. Therapeutic Antisense Oligonucleotides Are Coming of Age. *Annu. Rev. Med.* **2019**, *70*, 307–321. [CrossRef]

8. Nielsen, P.E. Gene targeting and expression modulation by peptide nucleic acids (PNA). *Curr. Pharm. Des.* **2010**, *16*, 3118–3123. [CrossRef]

9. Crawford, D.W.; Blakeley, B.D.; Chen, P.H.; Sherpa, C.; Le Grice, S.F.; Laird-Offringa, I.A.; McNaughton, B.R. An Evolved RNA Recognition Motif That Suppresses HIV-1 Tat/TAR-Dependent Transcription. *ACS Chem. Biol.* **2016**, *11*, 2206–2215. [CrossRef]

10. Held, D.M.; Kissel, J.D.; Patterson, J.T.; Nickens, D.G.; Burke, D.H. HIV-1 inactivation by nucleic acid aptamers. *Front. Biosci.* **2006**, *11*, 89–112. [CrossRef]

11. Brogna, S.; Wen, J. Nonsense-mediated mRNA decay (NMD) mechanisms. *Nat. Struct. Mol. Biol.* **2009**, *16*, 107–113. [CrossRef] [PubMed]

12. Buckley, P.T.; Khaladkar, M.; Kim, J.; Eberwine, J. Cytoplasmic intron retention, function, splicing, and the sentinel RNA hypothesis. *Wiley Interdiscip. Rev. RNA* **2014**, *5*, 223–230. [CrossRef] [PubMed]

13. Berger, J.; Aepinus, C.; Dobrovnik, M.; Fleckenstein, B.; Hauber, J.; Bohnlein, E. Mutational analysis of functional domains in the HIV-1 Rev trans-regulatory protein. *Virology* **1991**, *183*, 630–635. [CrossRef]

14. Cochrane, A.W.; Perkins, A.; Rosen, C.A. Identification of sequences important in the nucleolar localization of human immunodeficiency virus Rev: Relevance of nucleolar localization to function. *J. Virol.* **1990**, *64*, 881–885. [CrossRef]

15. Zapp, M.L.; Hope, T.J.; Parslow, T.G.; Green, M.R. Oligomerization and RNA binding domains of the type 1 human immunodeficiency virus Rev protein: A dual function for an arginine-rich binding motif. *Proc. Natl. Acad. Sci. USA* **1991**, *88*, 7734–7738. [CrossRef] [PubMed]

16. Cook, K.S.; Fisk, G.J.; Hauber, J.; Usman, N.; Daly, T.J.; Rusche, J.R. Characterization of HIV-1 REV protein: Binding stoichiometry and minimal RNA substrate. *Nucleic Acids Res.* **1991**, *19*, 1577–1583. [CrossRef] [PubMed]

17. Daly, T.J.; Doten, R.C.; Rennert, P.; Auer, M.; Jaksche, H.; Donner, A.; Fisk, G.; Rusche, J.R. Biochemical characterization of binding of multiple HIV-1 Rev monomeric proteins to the Rev responsive element. *Biochemistry* **1993**, *32*, 10497–10505. [CrossRef] [PubMed]

18. Daugherty, M.D.; Booth, D.S.; Jayaraman, B.; Cheng, Y.; Frankel, A.D. HIV Rev response element (RRE) directs assembly of the Rev homooligomer into discrete asymmetric complexes. *Proc. Natl. Acad. Sci. USA* **2010**, *107*, 12481–12486. [CrossRef]

19. Daugherty, M.D.; D'Orso, I.; Frankel, A.D. A solution to limited genomic capacity: Using adaptable binding surfaces to assemble the functional HIV Rev oligomer on RNA. *Mol. Cell* **2008**, *31*, 824–834. [CrossRef] [PubMed]

20. Daugherty, M.D.; Liu, B.; Frankel, A.D. Structural basis for cooperative RNA binding and export complex assembly by HIV Rev. *Nat. Struct. Mol. Biol.* **2010**, *17*, 1337–1342. [CrossRef]

21. DiMattia, M.A.; Watts, N.R.; Stahl, S.J.; Rader, C.; Wingfield, P.T.; Stuart, D.I.; Steven, A.C.; Grimes, J.M. Implications of the HIV-1 Rev dimer structure at 3.2 A resolution for multimeric binding to the Rev response element. *Proc. Natl. Acad. Sci. USA* **2010**, *107*, 5810–5814. [CrossRef] [PubMed]

22. Malim, M.H.; Cullen, B.R. HIV-1 structural gene expression requires the binding of multiple Rev monomers to the viral RRE: Implications for HIV-1 latency. *Cell* **1991**, *65*, 241–248. [CrossRef]

23. Mann, D.A.; Mikaelian, I.; Zemmel, R.W.; Green, S.M.; Lowe, A.D.; Kimura, T.; Singh, M.; Butler, P.J.; Gait, M.J.; Karn, J. A molecular rheostat. Co-operative rev binding to stem I of the rev-response element modulates human immunodeficiency virus type-1 late gene expression. *J. Mol. Biol.* **1994**, *241*, 193–207. [CrossRef] [PubMed]

24. Zemmel, R.W.; Kelley, A.C.; Karn, J.; Butler, P.J. Flexible regions of RNA structure facilitate co-operative Rev assembly on the Rev-response element. *J. Mol. Biol.* **1996**, *258*, 763–777. [CrossRef]

25. Fornerod, M.; Ohno, M.; Yoshida, M.; Mattaj, I.W. CRM1 is an export receptor for leucine-rich nuclear export signals. *Cell* **1997**, *90*, 1051–1060. [CrossRef]
26. Hadzopoulou-Cladaras, M.; Felber, B.K.; Cladaras, C.; Athanassopoulos, A.; Tse, A.; Pavlakis, G.N. The rev (trs/art) protein of human immunodeficiency virus type 1 affects viral mRNA and protein expression via a cis-acting sequence in the env region. *J. Virol.* **1989**, *63*, 1265–1274. [CrossRef]
27. Hammarskjold, M.L.; Heimer, J.; Hammarskjold, B.; Sangwan, I.; Albert, L.; Rekosh, D. Regulation of human immunodeficiency virus env expression by the rev gene product. *J. Virol.* **1989**, *63*, 1959–1966. [CrossRef]
28. Malim, M.H.; Hauber, J.; Le, S.Y.; Maizel, J.V.; Cullen, B.R. The HIV-1 rev trans-activator acts through a structured target sequence to activate nuclear export of unspliced viral mRNA. *Nature* **1989**, *338*, 254–257. [CrossRef]
29. Dayton, E.T.; Powell, D.M.; Dayton, A.I. Functional analysis of CAR, the target sequence for the Rev protein of HIV-1. *Science* **1989**, *246*, 1625–1629. [CrossRef]
30. Kjems, J.; Brown, M.; Chang, D.D.; Sharp, P.A. Structural analysis of the interaction between the human immunodeficiency virus Rev protein and the Rev response element. *Proc. Natl. Acad. Sci. USA* **1991**, *88*, 683–687. [CrossRef]
31. Watts, J.M.; Dang, K.K.; Gorelick, R.J.; Leonard, C.W.; Bess, J.W., Jr.; Swanstrom, R.; Burch, C.L.; Weeks, K.M. Architecture and secondary structure of an entire HIV-1 RNA genome. *Nature* **2009**, *460*, 711–716. [CrossRef] [PubMed]
32. Bai, Y.; Tambe, A.; Zhou, K.; Doudna, J.A. RNA-guided assembly of Rev-RRE nuclear export complexes. *Elife* **2014**, *3*, e03656. [CrossRef]
33. Jayaraman, B.; Crosby, D.C.; Homer, C.; Ribeiro, I.; Mavor, D.; Frankel, A.D. RNA-directed remodeling of the HIV-1 protein Rev orchestrates assembly of the Rev-Rev response element complex. *Elife* **2014**, *3*, e04120. [CrossRef] [PubMed]
34. Van Ryk, D.I.; Venkatesan, S. Real-time kinetics of HIV-1 Rev-Rev response element interactions. Definition of minimal binding sites on RNA and protein and stoichiometric analysis. *J. Biol. Chem.* **1999**, *274*, 17452–17463. [CrossRef]
35. Charpentier, B.; Stutz, F.; Rosbash, M. A dynamic in vivo view of the HIV-I Rev-RRE interaction. *J. Mol. Biol.* **1997**, *266*, 950–962. [CrossRef] [PubMed]
36. Bohnlein, E.; Berger, J.; Hauber, J. Functional mapping of the human immunodeficiency virus type 1 Rev RNA binding domain: New insights into the domain structure of Rev and Rex. *J. Virol.* **1991**, *65*, 7051–7055. [CrossRef] [PubMed]
37. Jain, C.; Belasco, J.G. Structural model for the cooperative assembly of HIV-1 Rev multimers on the RRE as deduced from analysis of assembly-defective mutants. *Mol. Cell* **2001**, *7*, 603–614. [CrossRef]
38. Pollard, V.W.; Malim, M.H. The HIV-1 Rev protein. *Annu. Rev. Microbiol.* **1998**, *52*, 491–532. [CrossRef] [PubMed]
39. Malim, M.H.; Bohnlein, S.; Hauber, J.; Cullen, B.R. Functional dissection of the HIV-1 Rev trans-activator—derivation of a trans-dominant repressor of Rev function. *Cell* **1989**, *58*, 205–214. [CrossRef]
40. Bevec, D.; Dobrovnik, M.; Hauber, J.; Bohnlein, E. Inhibition of human immunodeficiency virus type 1 replication in human T cells by retroviral-mediated gene transfer of a dominant-negative Rev trans-activator. *Proc. Natl. Acad. Sci. USA* **1992**, *89*, 9870–9874. [CrossRef]
41. Bonyhadi, M.L.; Moss, K.; Voytovich, A.; Auten, J.; Kalfoglou, C.; Plavec, I.; Forestell, S.; Su, L.; Bohnlein, E.; Kaneshima, H. RevM10-expressing T cells derived in vivo from transduced human hematopoietic stem-progenitor cells inhibit human immunodeficiency virus replication. *J. Virol.* **1997**, *71*, 4707–4716. [CrossRef] [PubMed]
42. Escaich, S.; Kalfoglou, C.; Plavec, I.; Kaushal, S.; Mosca, J.D.; Bohnlein, E. RevM10-mediated inhibition of HIV-1 replication in chronically infected T cells. *Hum. Gene Ther.* **1995**, *6*, 625–634. [CrossRef]
43. Liu, J.; Woffendin, C.; Yang, Z.Y.; Nabel, G.J. Regulated expression of a dominant negative form of Rev improves resistance to HIV replication in T cells. *Gene Ther.* **1994**, *1*, 32–37. [PubMed]
44. Plavec, I.; Agarwal, M.; Ho, K.E.; Pineda, M.; Auten, J.; Baker, J.; Matsuzaki, H.; Escaich, S.; Bonyhadi, M.; Bohnlein, E. High transdominant RevM10 protein levels are required to inhibit HIV-1 replication in cell lines and primary T cells: Implication for gene therapy of AIDS. *Gene Ther.* **1997**, *4*, 128–139. [CrossRef] [PubMed]

45. Woffendin, C.; Ranga, U.; Yang, Z.; Xu, L.; Nabel, G.J. Expression of a protective gene-prolongs survival of T cells in human immunodeficiency virus-infected patients. *Proc. Natl. Acad. Sci. USA* **1996**, *93*, 2889–2894. [CrossRef]

46. Hamm, T.E.; Rekosh, D.; Hammarskjold, M.L. Selection and characterization of human immunodeficiency virus type 1 mutants that are resistant to inhibition by the transdominant negative RevM10 protein. *J. Virol.* **1999**, *73*, 5741–5747. [CrossRef]

47. Legiewicz, M.; Badorrek, C.S.; Turner, K.B.; Fabris, D.; Hamm, T.E.; Rekosh, D.; Hammarskjold, M.L.; Le Grice, S.F. Resistance to RevM10 inhibition reflects a conformational switch in the HIV-1 Rev response element. *Proc. Natl. Acad. Sci. USA* **2008**, *105*, 14365–14370. [CrossRef]

48. Feinberg, M.B.; Jarrett, R.F.; Aldovini, A.; Gallo, R.C.; Wong-Staal, F. HTLV-III expression and production involve complex regulation at the levels of splicing and translation of viral RNA. *Cell* **1986**, *46*, 807–817. [CrossRef]

49. Sodroski, J.; Goh, W.C.; Rosen, C.; Dayton, A.; Terwilliger, E.; Haseltine, W. A second post-transcriptional trans-activator gene required for HTLV-III replication. *Nature* **1986**, *321*, 412–417. [CrossRef]

50. Rosen, C.A.; Terwilliger, E.; Dayton, A.; Sodroski, J.G.; Haseltine, W.A. Intragenic cis-acting art gene-responsive sequences of the human immunodeficiency virus. *Proc. Natl. Acad. Sci. USA* **1988**, *85*, 2071–2075. [CrossRef]

51. Kenyon, J.C.; Prestwood, L.J.; Le Grice, S.F.; Lever, A.M. In-gel probing of individual RNA conformers within a mixed population reveals a dimerization structural switch in the HIV-1 leader. *Nucleic Acids Res.* **2013**, *41*, e174. [CrossRef]

52. Sherpa, C.; Rausch, J.W.; Le Grice, S.F.; Hammarskjold, M.L.; Rekosh, D. The HIV-1 Rev response element (RRE) adopts alternative conformations that promote different rates of virus replication. *Nucleic Acids Res.* **2015**, *43*, 4676–4686. [CrossRef]

53. Nelson, J.A.; Fiscus, S.A.; Swanstrom, R. Evolutionary variants of the human immunodeficiency virus type 1 V3 region characterized by using a heteroduplex tracking assay. *J. Virol.* **1997**, *71*, 8750–8758. [CrossRef]

54. Cunyat, F.; Beerens, N.; Garcia, E.; Clotet, B.; Kjems, J.; Cabrera, C. Functional analyses reveal extensive RRE plasticity in primary HIV-1 sequences selected under selective pressure. *PLoS ONE* **2014**, *9*, e106299. [CrossRef] [PubMed]

55. Fang, X.; Wang, J.; O'Carroll, I.P.; Mitchell, M.; Zuo, X.; Wang, Y.; Yu, P.; Liu, Y.; Rausch, J.W.; Dyba, M.A.; et al. An unusual topological structure of the HIV-1 Rev response element. *Cell* **2013**, *155*, 594–605. [CrossRef] [PubMed]

56. Phuphuakrat, A.; Auewarakul, P. Functional variability of Rev response element in HIV-1 primary isolates. *Virus Genes* **2005**, *30*, 23–29. [CrossRef] [PubMed]

57. Bobbitt, K.R.; Addo, M.M.; Altfeld, M.; Filzen, T.; Onafuwa, A.A.; Walker, B.D.; Collins, K.L. Rev activity determines sensitivity of HIV-1-infected primary T cells to CTL killing. *Immunity* **2003**, *18*, 289–299. [CrossRef]

58. Sloan, E.A.; Kearney, M.F.; Gray, L.R.; Anastos, K.; Daar, E.S.; Margolick, J.; Maldarelli, F.; Hammarskjold, M.L.; Rekosh, D. Limited nucleotide changes in the Rev response element (RRE) during HIV-1 infection alter overall Rev-RRE activity and Rev multimerization. *J. Virol.* **2013**, *87*, 11173–11186. [CrossRef]

59. Sherpa, C.; Jackson, P.E.H.; Gray, L.R.; Anastos, K.; Le Grice, S.F.J.; Hammarskjold, M.L.; Rekosh, D. Evolution of the HIV-1 Rev Response Element during Natural Infection Reveals Nucleotide Changes That Correlate with Altered Structure and Increased Activity over Time. *J. Virol.* **2019**, *93*. [CrossRef]

60. Chu, C.C.; Plangger, R.; Kreutz, C.; Al-Hashimi, H.M. Dynamic ensemble of HIV-1 RRE stem IIB reveals non-native conformations that disrupt the Rev-binding site. *Nucleic Acids Res.* **2019**, *47*, 7105–7117. [CrossRef]

61. Nameki, D.; Kodama, E.; Ikeuchi, M.; Mabuchi, N.; Otaka, A.; Tamamura, H.; Ohno, M.; Fujii, N.; Matsuoka, M. Mutations conferring resistance to human immunodeficiency virus type 1 fusion inhibitors are restricted by gp41 and Rev-responsive element functions. *J. Virol.* **2005**, *79*, 764–770. [CrossRef] [PubMed]

62. Svicher, V.; Alteri, C.; D'Arrigo, R.; Lagana, A.; Trignetti, M.; Lo Caputo, S.; Callegaro, A.P.; Maggiolo, F.; Mazzotta, F.; Ferro, A.; et al. Treatment with the fusion inhibitor enfuvirtide influences the appearance of mutations in the human immunodeficiency virus type 1 regulatory protein rev. *Antimicrob. Agents Chemother.* **2009**, *53*, 2816–2823. [CrossRef] [PubMed]

63. Dillon, P.J.; Nelbock, P.; Perkins, A.; Rosen, C.A. Function of the human immunodeficiency virus types 1 and 2 Rev proteins is dependent on their ability to interact with a structured region present in env gene mRNA. *J. Virol.* **1990**, *64*, 4428–4437. [CrossRef] [PubMed]

64. Lusvarghi, S.; Sztuba-Solinska, J.; Purzycka, K.J.; Pauly, G.T.; Rausch, J.W.; Grice, S.F. The HIV-2 Rev-response element: Determining secondary structure and defining folding intermediates. *Nucleic Acids Res.* **2013**, *41*, 6637–6649. [CrossRef] [PubMed]

65. Weeks, K.M.; Mauger, D.M. Exploring RNA structural codes with SHAPE chemistry. *Acc. Chem. Res.* **2011**, *44*, 1280–1291. [CrossRef]

66. Chi, K.R. The RNA code comes into focus. *Nature* **2017**, *542*, 503–506. [CrossRef]

67. Kane, S.E.; Beemon, K. Precise localization of m6A in Rous sarcoma virus RNA reveals clustering of methylation sites: Implications for RNA processing. *Mol. Cell. Biol.* **1985**, *5*, 2298–2306. [CrossRef]

68. Kennedy, E.M.; Bogerd, H.P.; Kornepati, A.V.R.; Kang, D.; Ghoshal, D.; Marshall, J.B.; Poling, B.C.; Tsai, K.; Gokhale, N.S.; Horner, S.M.; et al. Posttranscriptional m(6)A Editing of HIV-1 mRNAs Enhances Viral Gene Expression. *Cell Host Microbe* **2017**, *22*, 830. [CrossRef]

69. Lichinchi, G.; Gao, S.; Saletore, Y.; Gonzalez, G.M.; Bansal, V.; Wang, Y.; Mason, C.E.; Rana, T.M. Dynamics of the human and viral m(6)A RNA methylomes during HIV-1 infection of T cells. *Nat. Microbiol.* **2016**, *1*, 16011. [CrossRef]

70. Tirumuru, N.; Zhao, B.S.; Lu, W.; Lu, Z.; He, C.; Wu, L. N(6)-methyladenosine of HIV-1 RNA regulates viral infection and HIV-1 Gag protein expression. *Elife* **2016**, *5*. [CrossRef]

71. Liu, N.; Zhou, K.I.; Parisien, M.; Dai, Q.; Diatchenko, L.; Pan, T. N6-methyladenosine alters RNA structure to regulate binding of a low-complexity protein. *Nucleic Acids Res.* **2017**, *45*, 6051–6063. [CrossRef] [PubMed]

72. Abulwerdi, F.A.; Le Grice, S.F.J. Recent Advances in Targeting the HIV-1 Tat/TAR Complex. *Curr. Pharm. Des.* **2017**, *23*, 4112–4121. [CrossRef] [PubMed]

73. Abulwerdi, F.A.; Shortridge, M.D.; Sztuba-Solinska, J.; Wilson, R.; Le Grice, S.F.; Varani, G.; Schneekloth, J.S., Jr. Development of Small Molecules with a Noncanonical Binding Mode to HIV-1 Trans Activation Response (TAR) RNA. *J. Med. Chem.* **2016**, *59*, 11148–11160. [CrossRef] [PubMed]

74. Abulwerdi, F.A.; Xu, W.; Ageeli, A.A.; Yonkunas, M.J.; Arun, G.; Nam, H.; Schneekloth, J.S., Jr.; Dayie, T.K.; Spector, D.; Baird, N.; et al. Selective Small-Molecule Targeting of a Triple Helix Encoded by the Long Noncoding RNA, MALAT1. *ACS Chem. Biol.* **2019**, *14*, 223–235. [CrossRef] [PubMed]

75. Chu, C.C.; Liu, B.; Plangger, R.; Kreutz, C.; Al-Hashimi, H.M. m6A minimally impacts the structure, dynamics, and Rev ARM binding properties of HIV-1 RRE stem IIB. *PLoS ONE* **2019**, *14*, e0224850. [CrossRef]

76. Bryson, D.I.; Zhang, W.; McLendon, P.M.; Reineke, T.M.; Santos, W.L. Toward targeting RNA structure: Branched peptides as cell-permeable ligands to TAR RNA. *ACS Chem. Biol.* **2012**, *7*, 210–217. [CrossRef]

77. Dai, Y.; Wynn, J.E.; Peralta, A.N.; Sherpa, C.; Jayaraman, B.; Li, H.; Verma, A.; Frankel, A.D.; Le Grice, S.F.; Santos, W.L. Discovery of a Branched Peptide That Recognizes the Rev Response Element (RRE) RNA and Blocks HIV-1 Replication. *J. Med. Chem.* **2018**, *61*, 9611–9620. [CrossRef]

78. Fukuda, M.; Asano, S.; Nakamura, T.; Adachi, M.; Yoshida, M.; Yanagida, M.; Nishida, E. CRM1 is responsible for intracellular transport mediated by the nuclear export signal. *Nature* **1997**, *390*, 308–311. [CrossRef]

79. Mattaj, I.W.; Englmeier, L. Nucleocytoplasmic transport: The soluble phase. *Annu. Rev. Biochem.* **1998**, *67*, 265–306. [CrossRef]

80. Stade, K.; Ford, C.S.; Guthrie, C.; Weis, K. Exportin 1 (Crm1p) is an essential nuclear export factor. *Cell* **1997**, *90*, 1041–1050. [CrossRef]

81. Henderson, B.R.; Percipalle, P. Interactions between HIV Rev and nuclear import and export factors: The Rev nuclear localisation signal mediates specific binding to human importin-beta. *J. Mol. Biol.* **1997**, *274*, 693–707. [CrossRef] [PubMed]

82. Truant, R.; Cullen, B.R. The arginine-rich domains present in human immunodeficiency virus type 1 Tat and Rev function as direct importin beta-dependent nuclear localization signals. *Mol. Cell. Biol.* **1999**, *19*, 1210–1217. [CrossRef] [PubMed]

83. Askjaer, P.; Jensen, T.H.; Nilsson, J.; Englmeier, L.; Kjems, J. The specificity of the CRM1-Rev nuclear export signal interaction is mediated by RanGTP. *J. Biol. Chem.* **1998**, *273*, 33414–33422. [CrossRef] [PubMed]

84. Daelemans, D.; Costes, S.V.; Lockett, S.; Pavlakis, G.N. Kinetic and molecular analysis of nuclear export factor CRM1 association with its cargo in vivo. *Mol. Cell. Biol.* **2005**, *25*, 728–739. [CrossRef]

85. Kuersten, S.; Ohno, M.; Mattaj, I.W. Nucleocytoplasmic transport: Ran, beta and beyond. *Trends Cell Biol.* **2001**, *11*, 497–503. [CrossRef]

86. Neville, M.; Stutz, F.; Lee, L.; Davis, L.I.; Rosbash, M. The importin-beta family member Crm1p bridges the interaction between Rev and the nuclear pore complex during nuclear export. *Curr. Biol.* **1997**, *7*, 767–775. [CrossRef]

87. Bischoff, F.R.; Ponstingl, H. Catalysis of guanine nucleotide exchange on Ran by the mitotic regulator RCC1. *Nature* **1991**, *354*, 80–82. [CrossRef]

88. Li, J.; Tang, H.; Mullen, T.M.; Westberg, C.; Reddy, T.R.; Rose, D.W.; Wong-Staal, F. A role for RNA helicase A in post-transcriptional regulation of HIV type 1. *Proc. Natl. Acad. Sci. USA* **1999**, *96*, 709–714. [CrossRef]

89. Yedavalli, V.S.; Neuveut, C.; Chi, Y.H.; Kleiman, L.; Jeang, K.T. Requirement of DDX3 DEAD box RNA helicase for HIV-1 Rev-RRE export function. *Cell* **2004**, *119*, 381–392. [CrossRef]

90. Fang, J.; Acheampong, E.; Dave, R.; Wang, F.; Mukhtar, M.; Pomerantz, R.J. The RNA helicase DDX1 is involved in restricted HIV-1 Rev function in human astrocytes. *Virology* **2005**, *336*, 299–307. [CrossRef]

91. Fang, J.; Kubota, S.; Yang, B.; Zhou, N.; Zhang, H.; Godbout, R.; Pomerantz, R.J. A DEAD box protein facilitates HIV-1 replication as a cellular co-factor of Rev. *Virology* **2004**, *330*, 471–480. [CrossRef] [PubMed]

92. Zhou, X.; Luo, J.; Mills, L.; Wu, S.; Pan, T.; Geng, G.; Zhang, J.; Luo, H.; Liu, C.; Zhang, H. DDX5 facilitates HIV-1 replication as a cellular co-factor of Rev. *PLoS ONE* **2013**, *8*, e65040. [CrossRef] [PubMed]

93. Ma, J.; Rong, L.; Zhou, Y.; Roy, B.B.; Lu, J.; Abrahamyan, L.; Mouland, A.J.; Pan, Q.; Liang, C. The requirement of the DEAD-box protein DDX24 for the packaging of human immunodeficiency virus type 1 RNA. *Virology* **2008**, *375*, 253–264. [CrossRef] [PubMed]

94. Suhasini, M.; Reddy, T.R. Cellular proteins and HIV-1 Rev function. *Curr. HIV Res.* **2009**, *7*, 91–100. [CrossRef] [PubMed]

 *viruses*

Review

Measles Encephalitis: Towards New Therapeutics

Marion Ferren *, Branka Horvat and Cyrille Mathieu *

CIRI, International Center for Infectiology Research, INSERM U1111, University of Lyon, University Claude Bernard Lyon 1, CNRS, UMR5308, Ecole Normale Supérieure de Lyon, France; branka.horvat@inserm.fr
* Correspondence: marion.ferren@inserm.fr (M.F.); cyrille.mathieu@inserm.fr (C.M.)

Received: 30 September 2019; Accepted: 31 October 2019; Published: 2 November 2019

Abstract: Measles remains a major cause of morbidity and mortality worldwide among vaccine preventable diseases. Recent decline in vaccination coverage resulted in re-emergence of measles outbreaks. Measles virus (MeV) infection causes an acute systemic disease, associated in certain cases with central nervous system (CNS) infection leading to lethal neurological disease. Early following MeV infection some patients develop acute post-infectious measles encephalitis (APME), which is not associated with direct infection of the brain. MeV can also infect the CNS and cause sub-acute sclerosing panencephalitis (SSPE) in immunocompetent people or measles inclusion-body encephalitis (MIBE) in immunocompromised patients. To date, cellular and molecular mechanisms governing CNS invasion are still poorly understood. Moreover, the known MeV entry receptors are not expressed in the CNS and how MeV enters and spreads in the brain is not fully understood. Different antiviral treatments have been tested and validated in vitro, ex vivo and in vivo, mainly in small animal models. Most treatments have high efficacy at preventing infection but their effectiveness after CNS manifestations remains to be evaluated. This review describes MeV neural infection and current most advanced therapeutic approaches potentially applicable to treat MeV CNS infection.

Keywords: measles virus; central nervous system; tropism; treatments

---

1. Measles Virus Epidemiology

Measles virus (MeV) is the etiologic agent responsible for measles disease. Humans are the only known reservoir for MeV. Despite the availability of a very efficient vaccine [1], measles remains one of the most contagious diseases with a R0 ranking from 12 to 18 [2] meaning that (in a fully susceptible population) an infected patient will on average transmit the infection to 12 to 18 individuals. This propagation rate may even increase among people with low or compromised immunity [3]. Viral transmission generally occurs from person to person through aerosols [3] and precedes onset of skin rash, making the disease even more difficult to contain. After decades of emergences mainly restricted to the poorest countries, measles has made a strong comeback and re-emerged in industrialized countries [4] where access to the vaccine was supposed to be easier. Measles killed more than 100,000 people every year [5] since 2010. In 2017, 110,000 people died from measles, mostly children under five years old [3]. Indeed, in the absence of vaccination, children are the main targets of MeV [6], although adults can be infected as well [3]. Last year, WHO documented 268,038 confirmed cases. Nevertheless, according to other estimations, there are 7 to 20 million people getting infected by measles each year [7,8].

In most developed countries measles was considered eliminated, in recent years. However the rate of vaccination decreased due to a vaccination hesitancy, and as consequence the decreased herd immunity led to large outbreaks and today measles is considered re-emerged [4,9]. This year, in many developed countries including USA and France, there is a 300% increase in reported MeV cases compared to last year [10]. Notably, 1250 cases have been reported in the USA in 2019 (from January to

October) [11]. Those outbreaks confirm the re-emergence of measles, already announced by the NIAID following MeV epidemics in 2014 (CDC).

2. Virus

MeV belongs to the Morbillivirus genus within the *Paramyxoviridae* family and *Mononegavirales* order. This enveloped virus produces pleiomorphic viral particles with an average size ranging from 150 to 300 nm and up to 900 nm [12]. Its genome is a negative-sense, single stranded RNA of 15,894 nucleotides that encodes six structural proteins: The nucleocapsid (N) protein, the phosphoprotein (P), the matrix (M) protein, the fusion (F) protein, the haemagglutinin (H) protein, and the polymerase (large, L) protein. Two non-structural proteins, V and C are produced from the P gene [13] and mainly alter the innate immune sensing and response [14–17].

Wild type MeV strains use signaling lymphocytic activation molecule 1 (SLAMF1, also called SLAM or CD150) and nectin-4 receptors to infect target cells [18–20]. MeV vaccine strains use the ubiquitously expressed CD46 molecule as an additional entry receptor in vitro [21,22]. MeV entry is pH-independent and occurs directly at the cell surface [23]. However, MeV entry may also occur by endocytosis mediated by SLAM in B-lymphoblastoid cells or A549-SLAM cells [24], and through a nectin-4-mediated macropinocytosis pathway, in breast and colon cancer cell lines (MCF7, HTB-20, and DLD-1) [25]. It was also suggested that MeV Edmonston or Hallé strains could use a macropinocytosis-like pathway in non-lymphoid and lymphoid cells when SLAM and CD46 are engaged but this remains poorly documented [26,27].

To initiate the infection of the main target cells, the MeV H protein binds to entry receptor on the surface. This attachment triggers the F protein and leads to exposure of its hydrophobic fusion peptide that then inserts into the host cell membrane. The F protein undergoes serial conformational changes allowing the merge of the host and viral membranes creating a fusion pore allowing the ribonucleocapsid (RNP) delivery in the cytoplasm (Figure 1A,B) [28,29]. Infection also spreads efficiently via cell-to-cell contact [30,31].

Transcription by the RNA-dependent RNA polymerase (RdRp) starts from a single promoter resulting in a transcriptional gradient from the most abundant mRNA for N to the least abundant mRNA for L in order to allow efficient viral cycles. These mRNAs are then translated into viral proteins. The accumulation of N and P leads to viral genome replication into positive stranded RNA anti-genome that will allow further synthesis of negative sense RNA strands that will be encapsidated by newly synthesized N, P, and L proteins [32]. Viral RNA synthesis and assembly are regulated through the interaction between M and N [33]. Viral proteins assemble to the plasma membrane and the budding of new virions can occur (Figure 1A). Alternatively, the surface glycoproteins are transported to the plasma membrane and allow cell-to-cell dissemination.

The viral RNA is encapsidated by the protein N and forms the helical nucleocapsid [34]. Each N protein covers six nucleotides, hence the genome length has to follow the "rule of 6" for being fully protected [35,36]. Together, the proteins L and P form the viral RdRp. That polymerase interacts with the nucleocapsid to progress on the viral RNA: Altogether they form the RNP.

The M protein generally ensures the viral particle integrity. The M protein also orchestrates the viral assembly at the plasma membrane and the budding of the new infectious viral particles [23].

The H and F proteins constitute the viral fusion complex that is responsible for the viral entry into the host cell. The H protein is a tetramer organized as a dimer of dimers responsible for the binding to the entry receptor. The F protein mediates the fusion between the virus and the host plasma membranes. The F is a trimer first produced as a precursor F0 that is cleaved in the trans-Golgi by a furin protease in F1 plus F2 subunits linked by a disulfide bond. The extracellular domain is constituted by the F1 and F2 subunits containing the fusion peptide at the N terminus followed by two complementary heptad repeat domains, respectively at the N terminus (HRN) and the C terminus (HRC). While the crystal structure of the prefusion form of the F protein has been described [37], the exact delimitations of the F sub-domains are still not completely defined [38–40].

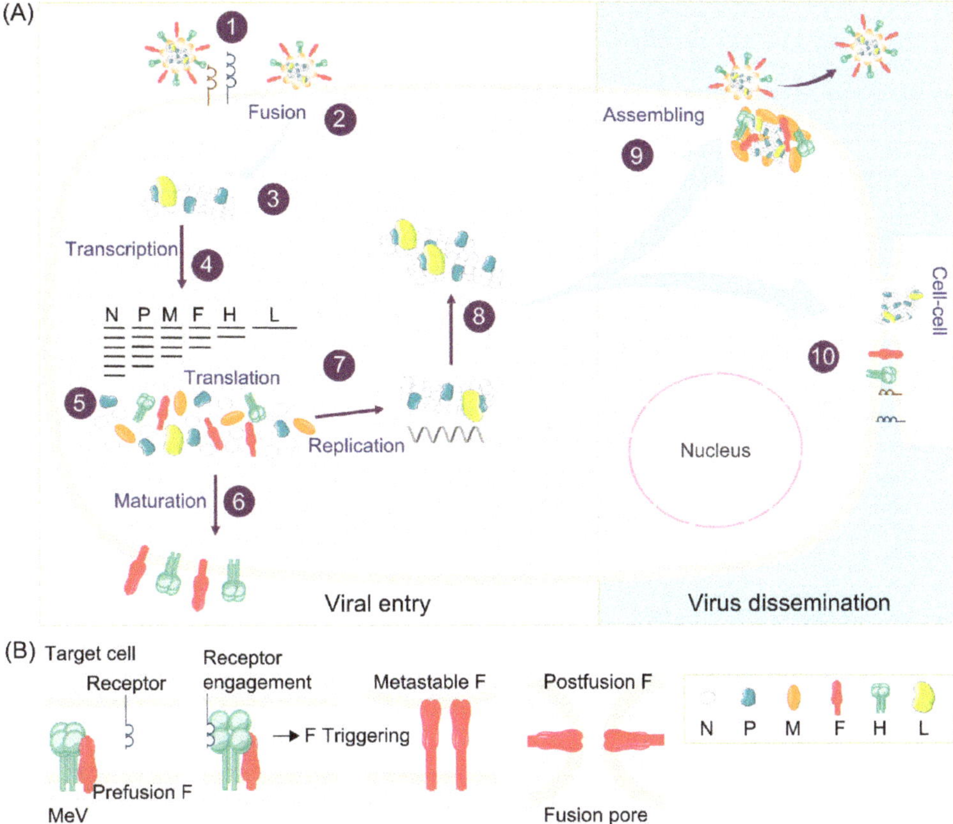

Figure 1. Measles Virus (MeV) replication cycle. (**A**) In order to infect a susceptible and permissive cell, MeV binds to its entry receptors on the cell surface (1) and initiates the virus-cell membrane fusion (2), as described in detail in (**B**). Virus and cell membranes fusion leads to genome delivery into the cytoplasm (3). Viral RNA is transcribed in mRNA (4) that is further translated into viral proteins (5). Viral glycoproteins mature during their transport to the cell surface (6). The replication of positive stranded anti-genomic RNA starts in the cytoplasm (7) and serves as a template for synthesis of new negative stranded genomic RNA (8). Viral proteins assemble at the cell surface, leading either to budding of new virions (9) or cell-to-cell fusion (10). (**B**) The haemagglutinin (H) protein binds to the MeV receptor at the cell surface, allowing the triggering of fusion (F) which reaches a metastable conformation. Then, F protein anchors its fusion peptide in the target cell membrane, F undergoes serial conformational changes bringing the two membranes close enough to merge and form a pore throughout which the viral ribonucleocapsid (RNP) is delivered to the cytoplasm.

Based on bioinformatic tools the HRN domain encompasses residues 116/138 to residue 190 and the HRC domain is included between residues 438 and 488/489. The current crystal structure however shows the region between 438 and 458 as disorganized while a canonical heptad repeat is shown after residue 458 [41,42].

3. Vaccines

A highly efficient live-attenuated virus vaccine is available to prevent measles outbreaks. MeV transmissibility is very high and 95% of the population needs to possess anti-measles immunity for disease eradication [43]. In 1997, during a meeting co-sponsored by the World Health Organization

(WHO), the Pan American Health Organization (PAHO), and the Centers for Disease Control (CDC), the experts agreed that measles eradication was technically feasible by 2005–2010. Nevertheless, vaccination coverage decreased and led to a re-emergence of measles infection. Nowadays, measles global eradication is one of the top priorities of the expanded program on immunization (EPI) supported by the WHO. The Global Vaccine Action Plan aims to eliminate measles in five WHO Regions by 2020. Based on confirmed cases reported by the WHO, the countries with the most measles cases in 2018 were India, Ukraine, Philippines, Brazil, and Yemen. Recently, measles strongly re-emerged in industrialized countries due to the significant decrease in vaccination coverage [4,44].

Different MeV strains have been used for vaccine purpose starting with the Edmonston strain isolated in 1954 that was very reactogenic. Five vaccines were derived from Edmonston: Edmonston-Zagreb, AIK-C, Moraten, Schwarz, and Edmonston-B [45]. Some of them such as Edmonston-B remained too reactogenic. The Edmonston vaccine was replaced by the more attenuated Schwarz vaccine strain in early 60s and Moraten vaccine strains in 1968. Years later, studies have shown that Schwarz and Moraten were in fact the same virus [45]. Other vaccines derived from other strains have also been developed. Leningrad strain (isolated in 1957) attenuation successively led to Leningrad 4 and more recently to the Chinese vaccine Changchung-47. Shanghai isolate (1960) attenuation allowed production of shanghai-191 vaccine while Cam-70 which was currently produced and used in Indonesia and Japan, derived from the Tanabe (Japan, 1968) strain. All vaccines strains belong to the measles virus genotype A [45]. Measles vaccine is usually combined with mumps and rubella vaccines, known as MMR (Measles, Mumps, and Rubella) vaccine, or with mumps, rubella, and varicella (chickenpox) vaccines, called MMRV (Measles, Mumps, Rubella, and Varicella) vaccine. MMR is a live-attenuated measles virus [46]. MMR vaccination is given in a two-dose schedule, with a first dose generally administered to 12–15 months old children, and a second one three to five years later [4]. While MMR vaccine cannot be used in immunocompromised patients (with low CD4+ cell count, or severely immunedepressed), the WHO strongly recommends the vaccination of human immunodeficiency virus (HIV) positive patients without severe immunosuppression [47].

Generally, vaccinated people develop a strong humoral and cellular immunity. Only 2–10% of people who received the two vaccine doses do not produce protective measles antibodies. However, most of them remain protected by their T cell immunity [48,49].

Taken together, the too low vaccination coverage combined with the increasing proportion of immunocompromised and other non-vaccinable people call for the development of an efficient, preventive, and/or curative treatment.

4. Disease/Generalities

4.1. Symptoms and Complications

During the acute phase of MeV infection, the patients develop several symptoms, including fever, cough, nasal congestion, characteristic erythematous maculopapular rash, conjunctivitis, and pathognomonic Koplik spots on oral mucosa. Diarrhea and vomiting are often observed in infected children during the disease [50,51] or appear as a complication following the disease [5,52]. Additionally, MeV infection leads to a strong immunosuppression that can last for several months and lead to severe secondary infections [53,54]. Moreover, MeV seems to impact FoxP3 T regulatory cells homeostasis by increasing their frequency and attenuating the hypersensitivity cellular response [55]. A more recent study suggests a MeV-induced immune amnesia relying on the depletion of pre-existing memory lymphocytes [50].

MeV infection can lead to several complications such as pneumonia, which is the main cause of measles mortality [56] or to central nervous system (CNS) complications, and to a lower extent to thrombocytopenia, blindness, or hearing loss [57]. Briefly, interstitial pneumonitis associated with mucosal inflammation due to large syncytia formations in the lungs are mainly observed in immunocompromised patients (Hecht's pneumonia) [56,58]. This cytopathic effect leads to bronchio-epithelial destruction generally resolved within few days of hospitalization (Figure 2A).

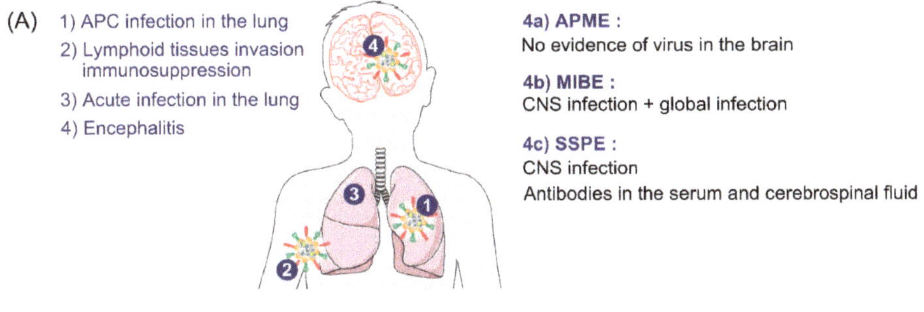

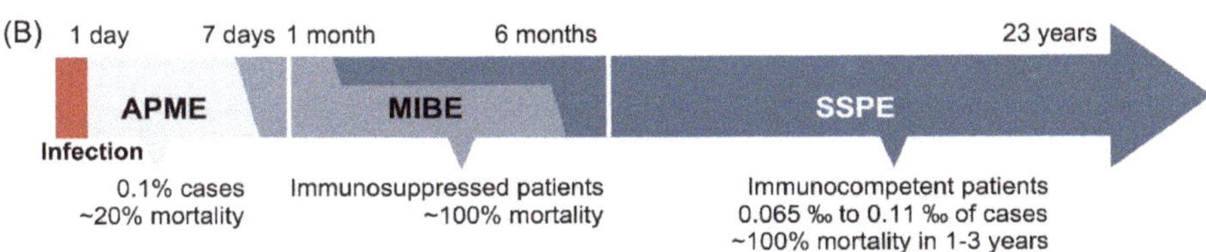

Figure 2. Course of MeV infection leading to measles encephalitis. (**A**) Initially, MeV infects myeloid cells in the respiratory tract. Then, MeV-infected lymphocytes disseminate the infection via the lymphatic and vascular systems. As a consequence of transient immunosuppression or autoimmunity, patients can develop acute post-infectious measles encephalitis (APME) shortly after exposure without systematic central nervous system (CNS) infection. However, measles inclusion-body encephalitis (MIBE) and subacute sclerosing panencephalitis (SSPE) are associated with MeV infection of the CNS. (**B**) The occurrence of MeV encephalitis may range from one day to 15 years following initial infection.

4.2. Associated Factors (Age/Nutrition)

Multiple factors such as malnutrition and vitamin A deficiency seem to increase measles associated morbidity and mortality. Indeed, regardless of vaccination coverage, MeV-infected people in poorest countries are more likely to develop complications leading to severe disease [3,59–61].

4.3. Pathogenesis

Pathogenesis starts with MeV infection of myeloid cells in the respiratory tract. As mentioned in Section 2, the two known entry receptors for MeV wild-type strains are SLAM/CD150 and nectin-4 [18,19]. Wild-type (wt) viruses generally target lung resident macrophages and/or dendritic cells, expressing SLAM [62–64]. These antigen presenting cells (APCs) migrate to the lymph nodes and transmit the viral infection to SLAM expressing lymphocytes with subsequent spread of the virus in the lymphatic and vascular systems (viremia). During the late stages of the infection, circulating infected immune cells that reach the respiratory tract and the skin can transmit the infection in cis to epithelial cells expressing nectin-4 on their basolateral side [20,65–67]. Then, the virions produced at the apical membrane can be shed into the respiratory mucus or aerosolized in the respiratory tract through coughing [68].

5. Disabilities and Nervous System

MeV can also cause damages to the nervous system.

5.1. Hearing Loss

MeV can induce hearing loss [69]. Before the introduction of mass vaccination, hearing loss was observed in 5% to 10% of measles cases in the USA. This remains highly frequent in under-developed countries where vaccination coverage is low [70]. One possible explanation is that otitis associated with measles in up to 25% of infected patients could cause hearing loss [71]. This pathology seems related to a super infection due to MeV-related transient immune-suppression.

Alternatively, hearing loss can occur immediately after the acute phase of the infection or later following measles acute encephalitis (described in paragraph 6.1) with typical bilateral and moderate to profound sensorineural hearing loss [69]. The mechanism associated with MeV-induced hearing loss remains unclear since neither viral antigen nor RNA have been detected in samples from the inner ear [57].

5.2. Blindness

Eye related symptoms such as conjunctivitis or corneal inflammation (keratitis) are commonly associated with measles [57]. Corneal complications are often more serious when superinfection (bacterial or viral) occurs during MeV-induced immune-suppression. However, there is a correlation between vitamin A deficiency and measles-induced blindness. Indeed, vitamin A deficiency is associated with severe keratitis and considerably increases the risk of xerophthalmia, corneal ulceration, and blindness [72]. This may explain why measles related blindness is more common in areas where children are already suffering of malnutrition.

Viral RNA can be detected in tear secretions [73]. In addition, human ex vivo cornea rim tissue is susceptible to MeV infection on its basolateral pole but neither syncytium formation nor released infectious particle have been found [74].

The relationship between MeV infection of ocular epithelial cells and the potential relationship with neural cell infection with cases of blindness is still unclear.

6. Central Nervous System (CNS) Infection

How the virus enters the CNS remains unclear since the known MeV receptors are not expressed. While its expression in the CNS seems to be only transient, nectin-4 has also been suggested to play a crucial role in MeV neuroinvasion based on observations made on closely related canine distemper virus whose neurotropism directly depends on nectin-4 specific patterns of expression [65,75–77]. Nectin-1 positive cells have recently been shown to be able to capture membranes and their cytoplasm from the surface of adjacent cells expressing nectin-4 at their surface via a trans-endocytosis mechanism [78]. In this context, viral RNP could transit from nectin-4 positive cells in nasal turbinate or meninges to neural cells expressing nectin-1 in olfactory bulb or brain parenchyma, respectively. The following key elements involved in the neural cell-to-cell dissemination and successful CNS invasion remain to be investigated (Figure 2A).

Three main CNS complications are associated with measles: The acute and the chronic forms, the latter being subdivided in two sub-types, the first as a measles inclusion-body encephalitis (MIBE) in immunocompromised patients and the second as a subacute sclerosing panencephalitis (SSPE) occurring in immunocompetent patients [79,80] (Figure 2B).

6.1. Acute Encephalitis

The acute post-infectious measles encephalitis (APME) occurs in 0.1% of measles cases, about a week following the appearance of first clinical signs. The APME is also called post-infection encephalitis (PIE), acute demyelinating encephalomyelitis, or acute disseminated encephalomyelitis.

APME is associated with 20% mortality and severe neurological sequelae, mainly in adults. Symptoms include fever, headaches, seizures, and consciousness alterations. APME is a complication associated with MeV infection that seems to be related to an auto-immune reaction against the myelin basic protein mainly expressed by oligodendrocytes [81–83]. APME causes CNS lesions in both white and grey matters and is characterized by brain inflammation and perivenous demyelination [68,84–86]. Moreover, APME is often associated with more immunological abnormalities such as high levels of IgE antibodies in the serum [87]. The binding of infected leukocytes to brain microvascular endothelial cells, or a direct infection of endothelial cells themselves in the brain may also partially contribute to this inflammatory immune reaction [88]. Overall, MeV acute encephalitis is poorly described in the recent literature. Note that there is a total lack of evidence of the virus presence in the brain parenchyma compared to that in the blood circulation. Based on the absence of virus detection in certain cases, multiple groups have suggested that the encephalitis could be caused by an autoimmune-like response [89]. While the presence of myelin basic protein (MBP) in the cerebrospinal fluid (CSF) suggests autoimmune-mediated encephalitis, oligodendrocytes viability and neurons myelination have not been explored yet [89].

6.2. MIBE

MIBE occurs in immunosuppressed patients ranging from three weeks to six months following wild-type MeV infection or in some rare cases after inappropriate vaccination with former vaccine strains [90–92]. MIBE is characterized by the presence of intracytoplasmic or intranuclear inclusion bodies composed of nucleocapsids, mainly in neurons, oligodendrocytes, and astrocytes [93,94]. Patients develop febrile focal seizures and behavior disorders before lapsing into coma. At a molecular level, mutations have sometimes been observed in the intracytoplasmic domain of MeV F protein and lead to the expression of hyperfusogenic viral phenotypes. Some mutations similar to those observed in SSPE have also been detected in the N gene and it has been hypothesized that MIBE and SSPE might be very similar, apart from the more rapid development of MIBE in immunocompromised subjects [95]. Recently, other MIBE- associated mutant viruses have also been described and present an hyperfusogenic phenotype [42,92]. Notably, the mutation L454W in the HRC domain of the F protein emerged in two patients that contracted MIBE in South Africa. This mutation confers the ability of entry without the presence of known receptor even at 25 °C. The mutation L454W leads to a highly unstable F protein potentially due to a lower interaction with the H protein which loses its protection role from random triggering of the fusion protein. This finding suggests that hyperfusogenicity of these neurotropic variants allows better viral dissemination, without the need of H binding to a high affinity receptor [37,96,97].

6.3. SSPE

Subacute sclerosing panencephalitis (SSPE) cases occur in 6.5 to 11 cases per 100,000 [97,98] in immunocompetent patients that contracted measles in their childhood, with a mortality rate close to 100%. Within children infected by MeV before the age of 12 months, the incidence for SSPE rises to 1/609, while reaching 1/1367 for children under five years old [99]. There is a latency period ranging from one to 15 years following primary infection and before appearance of symptoms [100,101]. In addition, because of the non-specificity of the first symptoms, SSPE diagnosis is generally delayed [102]. In most of the cases, patients do not survive more than 1–3 years following appearance of the symptoms associated with important neurological signs and dementia. Patients are developing severe physical and mental impairments but also a loss of motor control that tends to evolve in myoclonic jerks and spasms, seizures, and coma. Patients that underwent primary infection below the age of two are more at risk of developing SSPE. It was suggested that an immature immune system before two years old could contribute to persistent brain infection [83]. A dual viral hit was suggested to play a role in SSPE development. In this model, authors proposed that during classical first exposure to MeV immunocompetent patients do not develop encephalitis. However, a first exposure to a virus different from MeV, but capable of inducing an immunosuppression, which is followed with MeV infection later in the life, may favor development of CNS disease such as SSPE, as shown in the model of transgenic mice susceptible to MeV infection [103]. Most epidemiological studies are pointing that young boys are more often affected by SSPE than girls [17,104]. In one study in Germany from 2003 to 2009, the authors counted 21 males within 31 SSPE cases. SSPE is characterized by an excessive intrathecal synthesis of MeV specific antibodies. Most of the time, in the brain of SSPE patients, the genes that are encoding for MeV matrix protein (M), fusion protein (F), and attachment protein (H) are mutated [28,105,106] (Figure 3A).

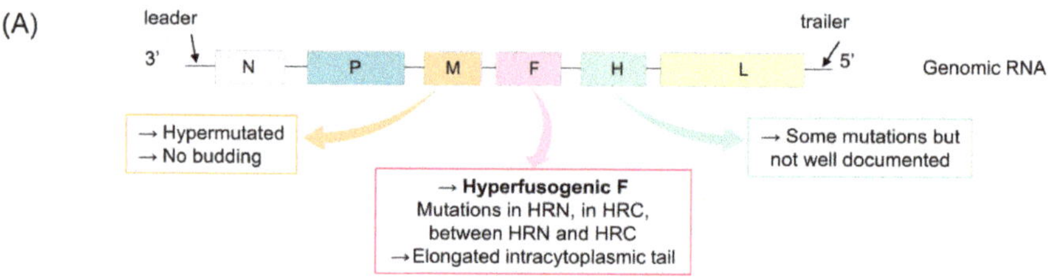

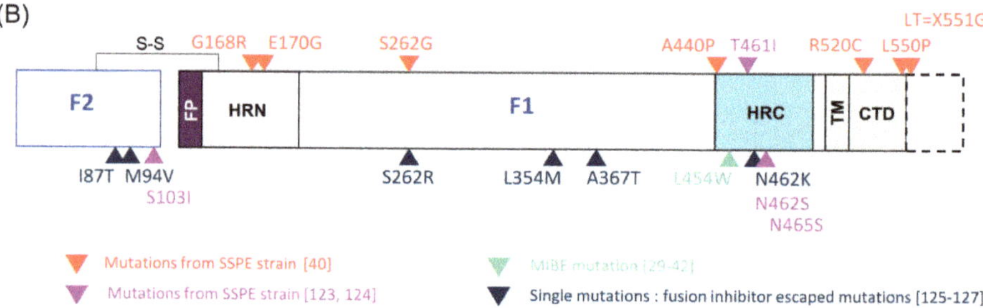

Figure 3. MeV F gene mutations related to CNS infection. (**A**) Schematic of MeV genome showing the most common mutations found in SSPE cases. (**B**) Details of MeV mutations in F protein leading to a hyperfusogenic phenotype and/or CNS infection.

7. Mutations Associated with MeV CNS Infection

7.1. M Protein

In SSPE, uridine-to-cytidine biased hypermutations of M protein are characteristic [107]. Studies have shown that MeV can evade the innate immunity control by taking advantage of the adenosine deaminase acting on RNA 1 (ADAR1), an IFN-stimulated gene that binds double-stranded RNA and converts adenosine to inosine by deamination [108]. The biased hypermutations in M (and other) gene in SSPE (or MIBE) cases might also be related to ADAR1 activity. Hypermutation of M protein leads to an unstable and defective M protein in viral particles assembly [109]. As a result, the virus is defective in budding from the plasma membranes and cannot produce viral particles. Among the large number of mutations in mRNA, the lack of the AUG initiation codon is leading to a low expression of M protein [110]. Nevertheless, in the context of brain invasion, the hypermutated M gene still allows MeV to replicate, spread, and cause disease [111,112]. Indeed, M protein negatively regulates the viral polymerase activity and thus to impact mRNA transcription and genome replication [113]. One of the roles of M protein is the distribution of both F and H glycoproteins at the apical cell surface [114]. Thereby, mutations in M protein could impact the virus fusion (and F stabilization), through association with surface glycoproteins tails, and thus influence the virus dissemination through the brain. Although in transgenic mice the infection with a M hypermutated MeV induces a more fusogenic phenotype despite attenuated budding, resulting in a more suitable virus for brain infection [111]. Other mutations impact interactions with the viral nucleocapsid and surface glycoproteins [115,116]. This provides another explanation for the absence of viral particle productions in SSPE-patient brains. This lack of budding is a key property highlighting that patients are non-contagious [93]. While numerous studies report the isolation of SSPE infectious viral particles from patient brains, none of them have physically shown whether classic infectious viral particles or virus RNP-containing apoptotic bodies expressing surface glycoproteins were effectively isolated [107,117–120].

7.2. F protein

The F proteins observed in SSPE cases present several mutations conferring a hyperfusogenic phenotype. F is produced as metastable protein in its pre-fusion state. This pre-fusion state is generally less stable in the CNS isolates. The F can also fuse without H engagement to any known receptor. Thus, it is suggested that these mutations facilitate CNS spread [40].

Mutations can occur in the HRC domain (T461I, A440P, N462S, N465S, and L454W), in the HRN domain (G168R/E170G), in between HRC and HRN domains (S262G), in the cytoplasmic tail domain (CTD) (R520C, L550P), and in the F2 subunit of F protein. Among the mutations found in the F-SSPE sequence from South African patient (G168R/E170G/S262G/A440P/R520C/L550P and X551G), only the mutation S262G (position already associated to hyperfusogenicity with a mutation S262R) located at the interface of three protomers, involved in fusion activation, may independently confer an hyperfusogenic phenotype to F without needing any other mutation. The functional analysis of MeV_IC323 virus carrying this F-SSPE with all seven mutations confirmed the finding that an SSPE strain can disseminate via cell-to-cell spreading in Vero cells, in the absence of known receptors [40] (Figure 3B).

The mutation of stop codon (X551G) in F-SSPE strains has been frequently observed previously [107] and leads to an elongated cytoplasmic tail (called LT for Long Tail) that can enhance the incorporation of F and nonspecific cellular protein in the virion [121,122].

Other mutations found in F extracellular domain from SSPE sequences isolated from patients brain (T461I and S103I/N462S/N465S) also confer hyperfusogenicity and can spread in human neuroblastoma cell lines and suckling hamster brains in the absence of known MeV receptors [123,124].

Fusion inhibitors such as 3G or FIP are tested on MeV and it has been documented that several mutations emerged in F protein in order to escape the treatment. The impact of these mutations (I87T, M94V, S262R, L354M, A367T, N462K) on the fusion machinery is of great interest [125–127]. One of the most interesting mutations that emerged is located at the residue 262. The escape mutation S262R confers hyperfusogenicity, as well as the mutant S262G that has been described in a real case of SSPE [40,123]. These data highlight the fact that emergence of mutations under a selective pressure can lead to viral adaptation to CNS. This can also allow a better design of inhibitors that could counteract these adaptation mutations.

As discussed in Section 6.2. the hyperfusogenicity correlates with a lower thermal stability of the pre-fusion state of F [28,125]. As an example, the L454W F is highly unstable and this characteristic could be sufficient to trigger F in a postfusion state by itself, allowing the fusion to occur without any receptor engagement. In the context of a circulating viral particle outside the brain, that property might not be an advantage for the virus, which could explain why any hyperfusogenic form of MeV has never been found in circulating viruses.

7.3. H Protein

The H protein of SSPE strains is often mutated as well and contributes to neurovirulence [128]. In a recent study, three mutations were found in the H gene of South African SSPE strain, in the cytoplasmic tail, the stalk domain, and β5 blade of the head domain, associated with substitutions R7Q, R62Q, and D530E, respectively [40]. The residue D530 is necessary for cell entry through SLAM, so the mutation D530E could compromise the use of infection through SLAM [129,130].

In a modified Edmonston strain expressing a murine-adapted H protein from a neurovirulent strain CAM/RB, the substitutions G195R and S200N lead to complete loss of neurovirulence in mice

C57BL/B6 [83,131,132]. Due to questionable strains and animal model used in this study, these data have to be considered carefully and these findings might be difficult to transpose to human SSPE cases. Nevertheless, it highlights the potential existence of a specific site in H involved in neurovirulence or a site of an unknown neuron-specific receptor.

C-terminus elongation of the H protein due to single-point mutation at the stop codon have also been reported multiple times in SSPE cases [75,133]. Contrarily to deletions of the cytoplasmic tail of H which were shown to enhance fusion activity [121], elongation of the extracellular domain of H seemed to impact binding, targeting, and may explain at least partially the high level of antibodies in SSPE cases [75,133].

Unlike SSPE, mutations in H gene of MIBE virus sequences seem to be less frequent and further investigations for their potential impact in CNS infection is required [92].

7.4. Mutations in Other Genes

In SSPE cases, some mutations have also been found in N, P, and L proteins but most of the recent studies focused on F and M proteins. Some P genes from SSPE cases exhibit an impaired editing system that lead to less V protein production. Most of the time, the viral cycle does not seem compromised but the lower expression level of V could contribute to the viral persistence by reduced inhibition of interferon (IFN) response [134]. Is has also been shown that the P gene of the multi-mutated rodent brain-adapted strain CAMR40 is largely involved in neurovirulence, suggesting that MeV P gene could also play a role in CNS infection [135].

8. Animal Models for Neuro-Invasion Studies

Humans are the only natural reservoir for MeV. Thus, the choice of the best animal model remains a challenge and depends on the type of scientific questions asked, to be faithfully representative of the CNS infection in humans. A summary of the most used small animal models and their related application is presented in Table 1. Several genetically modified mice have been used, mainly to study tropism, dissemination, and to develop new treatments. Historically, the Lewis rat was commonly used to study viral tropism and dissemination through the CNS [136]. More recently, the Golden Hamsters are preferred to study MeV neurovirulence because of the similarities in the brain lesions observed by MeV in this model compared to human cases of SSPE. Moreover, unlike mice, suckling hamsters are naturally susceptible to MeV infection, especially in the brain, despite the lack of expression of any known receptors as reported in human [123,137] (Table 1). Numerous studies have been done using neurotropic strains obtained following multiple passages in Hamster brains. Nowadays, these strains, supposed to mimic persistent infection in the brain, are not used anymore. Indeed, the strains CAM/RB or HNT were highly virulent in mice, rat and hamster but the induced infection was not representative of a persistent MeV infection in the brain. These hypermutated neuro-adapted strains led to an acute infection in the brain that was not representative of the slow and progressive infection seen in SSPE [136,138,139]. Such type of infection cannot be representative of an APME or MIBE since there is no CNS infection in the first case, and there are very distinctive inclusion bodies in the brains in the second one. Nevertheless, it may allow a better understanding of the behavior of MeV once these mutations have emerged in the CNS.

Multiple murine models have been developed to address specific question about MeV pathogenesis, CNS invasion, antiviral treatments, and persistence (Table 1). Notably, MeV persistence has been demonstrated in mice infected with the Edmonston strain or a recombinant MeV expressing H from CAM/RB strain up to two months [140,141] and in nude mice with Edmonston strain highlighting the emergence of mutations [142]. SLAM transgenic (tg) ant CD46 tg mice models and derivates expressing stably and ubiquitously or not the human receptors for wt or vaccine MeV strains were also extensively used [143,144]. When these receptors are ubiquitously expressed (notably in the CNS) these very artificial models highly facilitate MeV entry. In SLAM transgenic suckling mice infected intranasally, MIBE-related mutants such as MeV F L454W were able to propagate in lungs, meninges, and neural cells in brain parenchyma confirming the maintenance of its ability to infect a host from the respiratory tract [42]. Additionally, such animal models allow not only the study of the key factors of the cells permissiveness independently of the entry step, but also to validate the efficacy of antiviral drugs in the most stringent context, since the virus spread is the most difficult to block [29,96,145,146] (Table 1).

Non-human primates represent faithful models of measles since they are fully susceptible to wild-type MeV infection [147]. Thus, rhesus and cynomolgus macaques or squirrel monkey are often used mainly for studies focusing on the acute pathogenesis [20,65,66,68,148–150]. These studies highlighted numerous similarities between measles pathogenesis in humans and primates. Particularly, they allowed confirming the essential role of nectin-4 for the shedding and inter-human transmission of MeV, but symptoms related to CNS infection have not been reported so far. Accidental transmission of the circulating MeV strain from human to primate have occurred notably causing five deaths out of 21 cases in rhesus monkey [151]. In this study CNS infection was not investigated and all deaths were due to secondary infection related to MeV-induced immunosuppression. In 1999, another natural outbreak led to the death of 12 Japanese macaques out of 53 cases. In the brain, demyelination was observed in one monkey and two monkeys showed neuronal inclusions with measles antigens [152] but no infectious viral particle has been isolated. In order to better characterize the CNS infection, rhesus monkeys were infected intracerebrally with a SSPE derived virus but animals did not develop any visible symptom and the virus was not detectable after three weeks, suggesting the resolution of the infection [153]. Another study reported that rhesus monkeys infected intracerebrally with hamster-brain-adapted strain developed encephalitis with morphological characteristics similar to those observed in the brain of human SSPE cases. However, as already observed in rodent, these brain infections induced by the hamster-brain-adapted strain evolved during the acute phase of infection and do not reflect the slow progression observed in the SSPE [154]. MeV CNS infection still has to be characterized in this model.

More recently, comparative analysis of MeV infection, tropism, and spread in human to canine distemper virus (CDV) in natural host species such as dog and ferrets suggested that studies of this closely related morbillivirus infection could shed light on key elements of MeV pathogenesis [155].

The tamarin (*Saguinus mystax*), often called marmoset in the literature, has been shown to be susceptible to MeV infection with Edmonston and JM strains [156,157]. The JM strain was highly pathogenic in this model, especially following cerebral inoculation [158].

Table 1. Small animal models used to study MeV infection. IFNAR stands for interferon alpha/beta receptor, Rag for recombination activating gene, and TLR for Toll-like receptor.

Animal	Purpose	Route of Infection
MICE		
NSE-CD46: Expression of BC1 isoform of human CD46 under the control of the neuron-specific enolase (NSE) [159,160] YAC-CD46: Similar expression level and location than in human [159] CD46 [143,161,162] CD46-IFNAR$^{-/-}$ [163]	Vaccinal MeV behavior in the brain. Ability to disseminate in the brain. Pathogenesis of MeV infection in the CNS. Permissiveness. Immune response.	Intranasally (i.n.)Intracranially (i.c.)
SLAM: Ubiquitous expression [144] SLAM: Dendritic cell expression only [164,165] CD46/TLR induced CD150 [166] CD46/TLR induced CD150-IFNAR$^{-/-}$	Innate immune response Spreading and pathogenicity of Edmonston and wild-type Ichinose (IC) strains	i.n. i.c.
SLAM IFNAR$^{-/-}$ [167] (Figure in Section 9.3) IFNAR$^{-/-}$ [42,168] (Figure in Section 9.3)	Tropism Dissemination within CNS	Intraperitoneal (i.p.) i.n.
SLAM$^{+/+}$/Stat 1$^{-/-}$: Same expression level than human [169]	Innate immune response	i.p. i.n.
CD46 IFN-α/βKO [163]	Induction of MeV encephalitis with Edmonston	i.c.
CD46 RagKO [96,103]	Study of the establishment of SSPE Role of the immunosuppression in the MeV persistence Drug testing	Multiple
CD46 Neurokinin-1 KO [96]	Trans-synaptic viral dissemination	i.c.
C57BL/6 [170] (Figure in Section 9.3)	Viral persistence Tropism	i.c.
RAT		
Lewis rat [136,139,171]	Tropism	i.c.
Cotton rat (Sigmodon hispidus) [12,172,173]	Treatment development. Respiratory infection MeV induced immunosuppression CNS infection Immune suppression	i.n.
CD46 Sprague-Dawley rat [174]	permissiveness	Multiple
Brown Norway rat [175]	Immune response in MeV associated neurologic disease	i.c.
HAMSTER		
Syrian Golden Hamster [123] (Figure in Section 9.3)	Tropism Dissemination and invasion by mutated viruses	i.c.
FERRET		
Ferret [97,155,176,177]	To mimic SSPE Transmission Pathogenesis of CDV infection to model MeV infection	i.c. i.n.

9. MeV Tropism

Although MeV is primary a lymphotropic virus it could also infect the CNS. One of the ways the virus might enter into CNS could be through the hematogenous way by crossing the blood-brain barrier (BBB) [17]. Since endothelial cells are susceptible to infection in vitro, in vivo and in SSPE cases, their infection at the BBB could also give an opportunity for MeV to reach the CNS [88,178]. In addition, lymphocytes are also able to pass the BBB meaning that MeV-infected lymphocytes could carry the virus across the BBB [179,180]. However, the specific mechanisms allowing the virus to enter the CNS remain unclear [88,105,181]. The hyperfusogenic phenotype seems to be necessary to allow viral dissemination through neurons even in the absence of known receptor. To date, the early tropism and dissemination of *Paramyxoviridae* within the CNS during early stages of infection remain poorly documented. There are also very few available data on cellular and molecular mechanisms governing CNS invasion. To date, investigations are mainly limited to clinical symptoms, serology, RNA sequencing, and tissue immunostaining. Moreover, most of the studies have been performed with MeV vaccinal strains such as Edmonston B strain or neuro-adapted strains in Hamster, using several wild type or transgenic rodents, or other in vitro models such as primary or immortalized neural cultures. Nowadays, whether these viruses and models perfectly reflect what occurs in human remains questionable, they allow addressing specific questions in obtaining important information regarding the tropism of MeV infection in the brain.

9.1. Post-Mortem Studies

Post-mortem analyses of MeV-infected human brains show lesions in almost all areas (Figure 4A). In the same late context, studies of brain infection in human and animal models described the cell types harboring viral antigens in the CNS, nevertheless the early targeted and permissive cells need to be clarified.

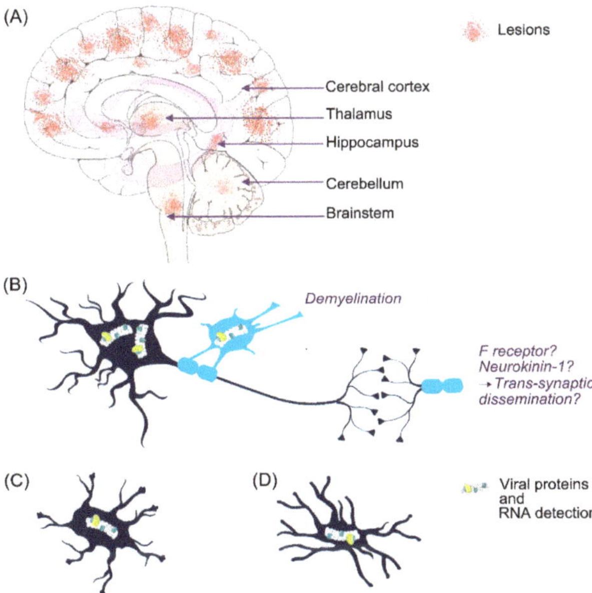

Figure 4. MeV central nervous system infection. Lesion areas are found in the brain of SSPE and MIBE patients but the specific areas associated with RNA detection are still poorly documented (**A**). Generally, MeV infects neurons and oligodendrocytes in humans (**B**). Occasionally, MeV RNA is also found in astrocytes (**C**) and microglia (**D**).

In the CNS, MeV infection occurs mainly in neurons but also in oligodendrocytes, astrocytes, and microglia [17,182,183] (Figure 4B–D). In MIBE and in SSPE cases, viral antigens and RNA have been found in neurons and oligodendrocytes [181]. In human SSPE cases, neurons are the main target with evidence of transneuronal viral spread [97]. Infected oligodendrocytes are often located near infected neurons, suggesting oligodendrocytes infection as a secondary infection from axons. The infection of oligodendrocytes is highly related to their demyelination. The authors suggest that MeV induces demyelination that could be a hallmark of SSPE (Figure 4A,B).

Viral genome and antigen have also been found in the perinuclear cytoplasm of astrocytes, albeit with lower frequency [181].

In a study using Edmonston B strain, infection of organotypic cultures of rat hippocampus ex vivo showed that the virus can infect neurons in the absence of CD46 receptor [139].

Meninges infection has been observed following intracranial MeV inoculation in ferrets [184] and hamsters [185], as well as following intranasal infection of SLAM transgenic mice [42]. Interestingly, MeV strains and mutants used in these studies were all known as hyperfusogenic. However, meninges infection has not been reported in humans yet.

9.2. Early Events in MeV Infection?

It is strongly suggested that MeV may use a third receptor or co-receptor yet unknown to enter the CNS. A parallel could be done with studies of CNS invasion with the closely related CDV conducted in dog and showing that astrocytes are neither expressing SLAM nor nectin-4, but remains permissive to the infection [76].

For MeV, the hypothesis that single or combination of mutations would be sufficient to confer adaptation in brain tissues for invasion without the engagement of any receptor is also relevant. Indeed, highly unstable F mutants such as L454W, observed in MIBE cases, do not need any communication with the H for triggering and fusion and thus cell-to-cell dissemination [28]. Alternatively, there is no proof that such a virus would be able to attach to any cell in absence of H and thus go through the first event allowing the entry in the CNS. Additionally, other hyperfusogenic mutants more stable and also observed in encephalitis cases were shown to conserve there dependence on H for F triggering [40], reinforcing the idea that at least a low affinity neural receptor should allow the initial entry in a CNS cell [186].

To date, the very first cell target of MeV infection in CNS, is unknown. A recent study focused on cell susceptibility during MeV infection in the CNS using hippocampus organotypic brain cultures (OBC) from IFNAR deficient genetically modified C57BL/6 mice expressing human SLAM receptor [168]. While all cell types were susceptible to infection in the absence of IFN-I response, the permissiveness of astrocytes and microglia strongly decreased when astrogliosis was observed in immunocompetent OBC. Astrogliosis and microgliosis have been observed in MeV encephalitis [144,187,188]. These data could explain why infection of astrocytes and microglia in post-mortem analysis are barely detectable.

9.3. Models to Study Tropism?

The main obstacle to study early tropism of MeV and other *Paramyxoviridae* is the lack of adequate models that could faithfully represent the infection in human brains. To date, ex vivo models seem to be a good compromise [189]. Organotypic brain cultures from mice, hamsters, and rats can be generated with several brain substructures such as cerebellum, cerebral cortex, or hippocampus [139,189,190]. The advantages of this model are the presence of all four cell types in the CNS (neurons, astrocytes, oligodendrocytes, and microglia), the possibility to produce OBC from any transgenic animal, and the unique opportunity to have a direct visibility of the CNS as an open window. Moreover, several slices can be made from each substructure. Therefore, a large number of conditions can be tested with a very limited number of animals, making this model ethically preferable, compared to in vivo experiments. The main weaknesses of OBC are the lack of a vascular system with circulating leukocytes and the decreasing susceptibility to infection through time concomitant to the development of astrogliosis [168].

Murine OBC offer many possibilities but mice are not susceptible to infection so their OBC would not be suitable to study early tropism. On the other hand, golden hamsters are susceptible to MeV infection. Thus, hamster OBC might be a more relevant *ex vivo* models but the lack of tools and available antibodies for this species still strongly slows down the study of the early tropism in this model.

Organotypic cerebellar cultures (OCC) from suckling SLAM-IFNARKO mice (Figure 5A), IFNARKO mice (Figure 5B), wild-type C57BL/6 mice (Figure 5C), and Syrian Hamster (Figure 5D) allows highlighting the hyperfusogenic phenotype of MeV-IC323 bearing a L454W or T461I mutated F protein compared to the wild-type in a CNS context. The fluorescence signal is used for tracking the infection and shows the massive dissemination of the viruses MeV-IC323-eGFP-F-L454W and MeV-IC323-eGFP-F-T461I in OCC even in the absence of known entry receptor (Figure 5B–D) while the MeV-IC323-eGFP-F-wt needs the expression of SLAM in order to disseminate efficiently in the OCC (Figure 5A).

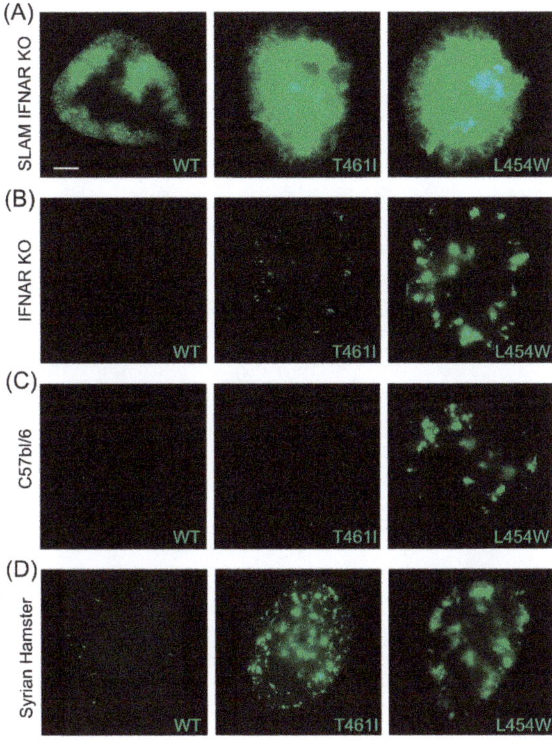

Figure 5. Wild-type and hyperfusogenic MeV growth in organotypic cerebellar cultures (OCC). OCC from suckling SLAM-IFNARKO mice (**A**), IFNARKO mice (**B**), wild-type C57BL/6 mice (**C**), and Syrian Hamster (**D**) were prepared as described elsewhere [189] and infected on the day of slicing with 10^3 PFU per slice with MeV-IC323-eGFP-F-wild-type (left side images), MeV-IC323-eGFP-F-L454W (right side images) and MeV-IC323-eGFP-F-T461I (middle images). Pictures were taken at day three post infection (dpi) and reconstituted using the Stitching plug-in with ImageJ software [191]. Scale bars, 1 mm.

10. MeV Dissemination in the CNS

In SSPE brain tissue, extracellular MeV has not been detected, suggesting that neuron-to-neuron viral dissemination can occur without released infectious viral particle [182]. MeV spread in mice and rat neurons is based on cell-to-cell contact [139,192,193]. The functional analysis of hyperfusogenic

MeV bearing a mutated F protein T461I confirmed this theory by being able to disseminate exclusively from cell-to-cell in human primary neurons [124,128]. The combination of mutations found in SSPE strains seems to enable viral fitness in the brain and neurovirulence [128]. Viruses with these mutations can spread in the brain of genetically modified mice [111].

It is suggested that MeV dissemination can be mediated by the microfusion at synaptic membranes [97,128]. In this theory, the F protein may interact with Neurokinin-1, the receptor of the P substance [96,139] (Figure 5B). This interaction would lead to the formation of a fusion micropore, allowing viral RNP to pass disseminate through neurons without the need of neither budding nor other receptor engagement. This could also explain the lack of syncytia formation in human primary neurons following infection with hyperfusogenic MeV forms. It has also been hypothesized that some supporting cells of myelinated nerves could block cell-to-cell contact between neurons and trans-synaptic spread in the brain could be the only way to allow viral dissemination [38].

It is strongly suggested that neurovirulent MeV strains are using a third receptor or co-receptor yet unknown. Nevertheless, the hypothesis that single or combination of mutations would be sufficient to confer adaptation in brain tissues for infection and dissemination without the engagement of any receptor is also relevant.

Models to Study the Dissemination?

Neuronal cell lines such as human cells NT2, human astrocytoma cells, or mouse neuroblastoma cells were also used, but their relevance remains difficult to appreciate when considering the important variation of behavior of cells out of their tissue context [31,97,192,194–196]. Primary neurons or neural polycultures were also often used [97] but are poorly representative of the neural population in human brain. In many studies, these cultures have been useful to investigate both intra and inter-neuronal spread of MeV [96], especially because they can be made from the brain of any transgenic mice.

The recently developed three-dimensional (3D) human brain organoid model has a high potential in order to investigate viral dissemination and evolution in the brain. The 3D brain organoids are generated from human pluripotent stem cells or human embryonic stem cells. This more ethical model offers a unique opportunity in generating relevant data that could be transposed faithfully to brain infection in humans [197]. Human brain organoids still require further development in order to overcome the lack of microglia and vascularization, but also their high cost and variability of the system [198]. However, to date, this model can be very useful in combination with ex vivo models, especially to test the efficacy of inhibitors in the context of brain infection, to follow viral dissemination and highlight the emergence of mutations.

11. Treatments

11.1. Symptomatic Treatment

Very few treatments are available against MeV infection and there is no therapeutic treatment for MeV-related encephalitis. The very first therapy administered after initial signs of infection are mainly supportive and focus on symptoms such as fever, dehydration, and diarrhea. Then, most of the treatments are generally dedicated to prevent or to cure from super infection such as pneumonia, often observed in infected patients. Antibiotics are commonly used to treat the complications related to bacterial superinfection [199].

11.2. Treatment Based on the Enhancement of Immune Response

In order to enhance the immune response, ribavirin, interferon alpha (IFN-α), and immune serum globulin can also be used clinically to treat MeV infection.

11.2.1. Immune Serum Globulin

From the 1940s, intramuscular injection of immune serum globulin was reported to confer up to 79% protection to unvaccinated patients having close contact with measles infected patients [200]. More recently, effectiveness of immune serum globulin as post-exposure prophylaxis was estimated from 50% to 69% during the 2014 measles outbreak in British Columbia in Canada. However, this estimation is highly controversial because many other factors could have contributed to prevent the appearance of the disease. Indeed, the potential pre-exposure immune status as well as the unknown exposure intensity and timing make the effectiveness of the immune serum globulin very difficult to quantify [201]. Moreover, the level of measles-specific antibodies has been shown to be lower when induced by the vaccine compared to the acquisition from a wild type measles infection [202]. This led to the necessity to increase the doses of immune serum globulin in order to maintain a protective level of measles antibodies [201]. However, as mentioned in paragraph 7.3, SSPE seems to develop mainly when the exposure to MeV occurs during the first years of age before the immune system is completely mature and when maternal antibodies are still lasting [17]. Additionally, administration of immunoglobulin may have led to SSPE cases [203] and the use of MeV-specific antibodies to treat rodents after infection via intracerebral route led to persistency of MeV infection and encephalitis [204–206]. Thus, the use of immunoglobulins to treat measles infection should be very carefully thought before introduction in therapies and would greatly benefit from the combination with other antivirals acting at different levels of the viral replication cycle in order to cure the infection instead of inducing persistency.

11.2.2. Ribavirin, IFN-α, Isoprinosine

Ribavirin is an antiviral drug with a broad antiviral activity, initially used for treatment of HCV [207]. It is a nucleic acid analog derived from guanosine and its main antiviral activity shown in vivo is its incorporation as a mutagenic nucleoside by the viral RNA polymerase [208]. The use of ribavirin and immune serum globulin seems to decrease respiratory symptoms in MeV-infected patients [209] but to date there is no standard protocol and doses recommended to treat patients.

IFN-α, ribavirin, and inosine pranobex are also used for SSPE treatment, with relative long-term effectiveness [210]. Many clinical reports show that Ribavirin can decrease measles antibody titers in cerebrospinal fluid (CSF) of SSPE patients and improve neurologic symptoms without side effects [211,212], especially when combined with IFN-α. In rare cases, long term IFN-α treatment stabilizes clinical symptoms of SSPE patients for years [213]. A recent study suggests also that continuous intraventricular administration of ribavirin and interferon-α in CSF by using a subcutaneous infusion pump, combined with oral administration of inosine pranobex, could limit the progression of SSPE [214]. Intrathecal IFN-α treatment combined with oral isoprinosine could also be effective to treat SSPE patients and is the most common treatment used nowadays [215,216]. Isoprinosine is a derivative of inosine and aims at blocking viral replication, probably through an immunoregulatory activity. Again, these treatments have rarely been shown to recover loss of function but they can stabilize the disease for several years [98,213,217]. Despite the benefits of IFN-α treatments, its use can be associated with side effects and could lead to interferonopathies [218]. Alternatively, there is induction of IFNα/β in vivo with MeV infection. This induction is associated at least partially to the presence of defective interfering (DI) particles which are also reducing the viral replication by occupying the proteins from the replication machinery and may thus constitute helpful complementary tool for treatments [219,220].

11.2.3. Vitamin A

Vitamin A deficiency is highly related to measles complications and the supplementation of vitamin A has been shown to decrease the morbidity and mortality related to MeV infection in children [59–61]. Vitamin A is also mainly used to prevent blindness due to MeV infection in children [72,221]. Thus, WHO recommends immediate vitamin A administration to MeV-infected children with two repeated doses of 200,000 IU especially as vitamin A deficiency is a public health problem [3,222–225]. Nevertheless, vitamin A is also encouraged to be given in all severe cases, regardless of the country or patient age [225]. In severe cases of measles, the combination of vitamin A with ribavirin treatment can also decrease the morbidity [226].

At the beginning of the infection, the innate immune response relies on the detection of PAMPs (pathogen-associated molecular pattern) by pathogens recognition receptors (PRR) such as the RIG-I like Receptors (RLRs) in the cytoplasm [227]. This pathway allows the synthesis and the secretion of type-I interferon. Among the RLRs recognizing the double stranded RNA patterns for activation of the type I interferon response, RIG-I (Retinoic acid-inducible gene I) is activated by several RNA viruses including MeV [228–230].

Retinoic acid is a metabolic product of vitamin A (retinol) that inhibits MeV replication in vitro via a retinoid nuclear receptor-dependent pathway [231] and a type I interferon (IFN)-dependent mechanism [232].

The mechanism of action of vitamin A as an antiviral still needs to be better understood. Nevertheless, RIG-I is required for MeV inhibition by retinoids [233], suggesting an implication of RIG-I in the efficacy of vitamin A treatment.

11.2.4. Interferon-Stimulating Genes (ISGs) and Other Treatments

The antiviral response is mediated by the interferon-stimulated genes (ISGs) that lead to the cell-intrinsic immunity. Recently, the overexpression of the bone marrow stromal antigen 2 proteins, also called BST2, Tetherin, or CD317, have been shown to inhibit Morbilliviruses cell to cell fusion in vitro by targeting the H protein [234]. In addition, the interferon-inducible transmembrane protein 1 (IFTIM1) has been shown to inhibit infection by several RNA viruses *in vitro*. While MeV enters via the plasma membrane, the effect of IFTIM1 on MeV replication is low compared to other *paramyxoviruses* such as the respiratory syncytial virus (RSV) but might be of interest in combination with other treatments [235].

Numerous other treatments, such as immunomodulators, carbamazepineamantadine, steroids, cimetidine, and plasmapheresis have been tested to treat SSPE but their efficacy seems to be case-dependent and need to be confirmed [216,217]. In addition, several alternative inhibitors such as antisense molecules, adenosine, and guanosine nucleosides, including ring-expanded "fat" nucleoside analogues, brassinosteroids, coumarins, modulators of cholesterol synthesis, and a variety of natural products have been investigated on MeV-infected patients. All these inhibitors showed relative efficacy or toxicity in vitro and in vivo and remain to be improved [236]. Among patients who received two doses of vaccine after initial infection some developed SSPE suggesting that the vaccine may not act as a therapeutic cure and prevent from encephalitis in this particular case [237].

11.3. Transcription/Replication Inhibitors

In order to inhibit the MeV growth, a strategy is to silence mRNAs encoding one of the key polymerase complex, namely N, P or L using small interfering RNAs (siRNAs) or shRNAs, as synthetic oligonucleotides, encoded by plasmids, or transduced using lentiviral vectors. siRNAs targeting the mRNA of either L [238] or N [239] or P [240], or the three in combination [241] have shown their efficiency in preventing virus growth over few days without cytotoxic effect. However, MeV finally escapes the silencing even in cells that constitutively express the siRNAs without acquiring any mutation even those that could disrupt the siRNA target sequence [240]. This likely reflects the remarkable long half-life of the polymerase brought by the incoming virus particles that last at least over 24 h [242,243] and the saturation of the siRNA linked to the continuous viral mRNA synthesis by the incoming polymerases.

As mentioned in Section 2, MeV P interacts with L protein. Although this interaction is independent of heat shock proteins such as the heat shock protein 90 (HSP90), both MeV P and HSP90 are necessary to fold and stabilize functional MeV L proteins able to enter the polymerase complex [243]. This transient requirement of HSP90 constitutes a potential target for transcription inhibition. Indeed, Geldanamycin and derivates such as 17-DMAG blocking HSP90 chaperon activity by entering its ATP pocket. These compounds showed the ability to block the viral transcription in preventive and post infection treatment in vitro and ex vivo in organotypic brain cultures [243]. Moreover, it is unlikely that a HSP90 inhibitor leads to the emergence of escape mutant virus [244]. While already used in cancer treatment, antivirals directly targeting the chaperon activity of HSP90 might be too toxic for human application. Nevertheless, molecules interfering between HSP90 and L and thus its functional folding are of interest for antiviral cure development.

Nucleoside analogs such as Remdesivir (GS-5734) and R1479 exhibit a broad spectrum activity against *paramyxoviruses* infections, including MeV [245]. Briefly, in cells, Remdesivir is metabolically converted to active nucleoside triphosphate. Obtained metabolite specifically inhibits several polymerases from different *Mononegavirales* such as Filoviruses and Henipaviruses, but not host polymerases. Recently, Remdesivir has been shown to inhibit Nipah virus polymerase activity by delaying the chain termination synthesis, notably in vivo in the African green monkey model [245,246]. Based on the huge conservation of the polymerases among *Mononegavirales*, there is a high probability that Remdesivir may also inhibit measles virus polymerase activity. Interestingly, pharmakokinetic studies performed in non-human primates showed high and persistent levels of the active metabolite in peripheral blood mononuclear cells (PBMCs) mainly targeted by wt MeV during the early stages of the pathogenesis [247]. Additionally, Remdesivir and subsequent active nucleoside seem to be able to reach the brain and may thus also inhibit CNS adapted variants of MeV observed in MIBE and SSPE cases.

Finally, the compound 16677 (1-methyl-3-trifluoromethyl-5-pyrazolecarboxylic acid) has been described as a non-nucleoside inhibitor of the RNA-dependent RNA polymerase complex activity [248]. The way this compound interacts with the replication machinery as well as the emergence of resistant variants remain poorly documented. Nevertheless, when tested in combination with an entry inhibitor increasing the stability of the fusion protein, the use of such replication inhibitor offered a high potential as a specific treatment against MeV. More recently, the same group has shown that compound AS-136A, analog to 16677, was able to block viral RNA synthesis by targeting L protein. This compound has also been associated to three candidates' hotspots of mutation increasing the knowledge of L sequence adaptation [249]. In order to face its poor solubility in water, known to influence the antiviral activity, structure-activity relationship investigations were driven to discover analogs which could be used in vivo and resulted in the generation of orally bioavailable compound 2O (ERDRP-00519) more potent and aqueous soluble than former generation [250]. As the former candidates, this antiviral remains quite cytotoxic but could be particularly efficient in combination with fusion inhibitors or antiviral immune response activators.

11.4. Inhibitors of MeV Fusion and Entry

As mentioned in Section 2, the first step of the infection relies on entry of the virus into its target cell. Briefly, H protein engages entry receptor and triggers F protein. F exposes its highly hydrophobic fusion peptide which inserts into host cell plasma membrane. This transient intermediate stage is highly unstable. Consequently, F undergoes serial conformational changes leading to the interaction between the two heptad repeat domains that brings the two membranes close enough to merge and form the fusion pore. The viral RNP can thus enter in the cell host cytoplasm. In order to prevent viral entry, the main target is to block fusion of the virus. Blocking the interaction with the receptor or F serial conformational changes are the two mainly considered possibilities.

The receptor binding site of MeV H is considered as a potential neutralizing target. Indeed, the insertion of any compound in the H pocket responsible for the binding to the receptor could either prevent from the virus attachment to the host cell or pre-trigger the F protein leading to fusion dead viral particles. Several neutralizing antibodies targeting the H protein have been proposed mainly resulting in the emergence of resistant mutants not anymore able to bind either SLAM or nectin-4 [251]. While this loss of function should not exist in the wild, the question of the ability of such variants to invade the CNS which does not express SLAM or nectin-4 receptor under this selective pressure still needs to be investigated. More recently, neutralizing antibody-derived molecules such as single chain variable fragments targeting the H protein represent a major advance in the field of therapeutics design [252]. As for the corresponding neutralizing antibodies, the ability of such molecules to penetrate the brain parenchyma and to block hyperfusogenic variants depending less on the receptor engagement as those commonly observed in CNS infection has never been tested.

As described in Section 10, Neurokinin-1 has been shown to be a potential receptor for MeV F. As an antagonist of Neurokinin-1, Aprepitant has been shown to drastically limit the viral dissemination of vaccine strain in the brain of CD46+/RAG-2ko mice [96].

The fusion inhibitor peptide (FIP), Z-D-Phe-L-Phe Gly that is a small hydrophobic peptide and other small molecules such as AS-48 or 3G (an analogue of AS-48) can block the membrane fusion in vitro [125,126,253]. These inhibitors are known to stabilize the prefusion state of the F protein. Nevertheless the use of these inhibitors leads to the emergence of mutations in the HRC of the F that can evade their efficacy leading to the selection of MeV hyperfusogenic variants [254].

In contrast, HRC-derived peptides, aim at blocking the fusion by capturing the F protein in the post-triggering state and freezing the fusion process at an early stage (Figure 6A–D). The so-called HRC4 peptide is a MeV F HRC-derived dimeric peptide that interacts with the HRN domain during the structural transition of F (Figure 6C,D). Briefly, HRC4 peptide is a dimer constituted of the HRC derived peptide linked with two chains of PEG that acts as a spacer each conjugated with a molecule of cholesterol. The cholesterol allows the fusion peptide to anchor into the host membrane and thus increases the antiviral potency of the HRC-derived peptide by two logs [255]. The HRC-derived peptides conjugated to cholesterol as well as tocopherol have shown high efficacy in vitro, ex vivo and in vivo, even in the context of the CNS infection by crossing the blood-brain barrier [29,42,146]. Notably, dissemination of viruses bearing the L454W mutation in F can be efficiently blocked in vitro and in vivo by F HRC-derived fusion inhibitors, regardless of the presence of SLAM [42]. To date, these fusion inhibitors are the only system already tested against both the wt and hyperfusogenic variants observed in CNS infection, and figure thus among the priority candidates for preclinical studies, to test alone and in combination with treatments targeting other viral functions.

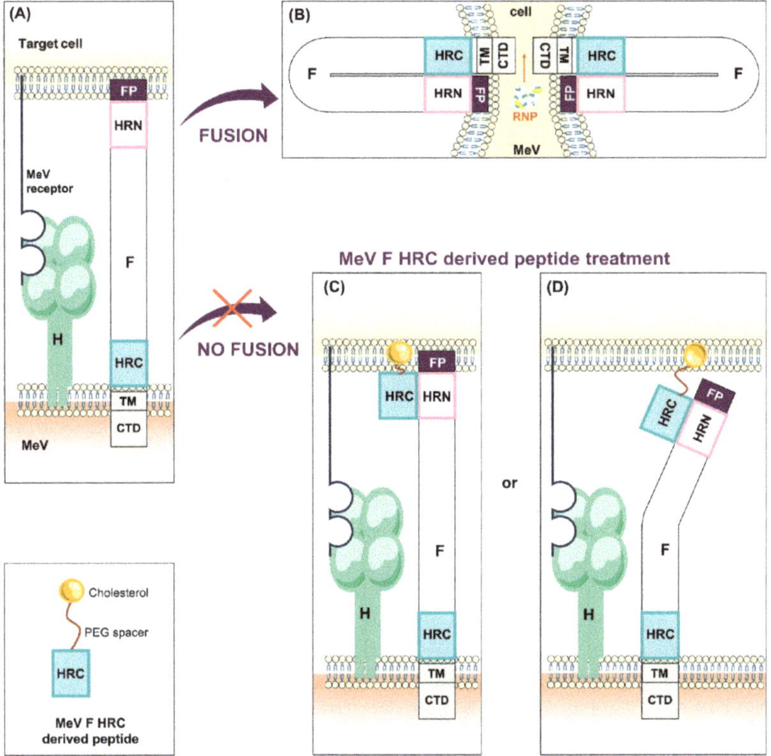

Figure 6. MeV F heptad repeats at the C terminal domain (HRC)-derived peptide. Following its engagement with any MeV receptor, H triggers F which inserts its fusion peptide in the host membrane (**A**). Then, F undergoes serial conformational changes to reach its post fusion state, bringing the two membranes close enough to form a fusion pore (**B**). MeV F HRC-derived peptides interact with MeV F HRN and catch the intermediate states of MeV F to block the fusion, regardless of the insertion of the fusion peptide in the host membrane (**C,D**).

12. Conclusions

A better understanding of MeV CNS invasion remains a priority in the field of MeV studies, especially because of the recent re-emergences of measles and the increasing number of associated fatal encephalitis [44,256]. While the vaccine remains the most efficient prevention against MeV infection, the decreasing coverage combined to the increasing number of immunocompromised people difficult to vaccinate confirm the necessity to develop efficient antiviral strategies.

To date, the emergence of the mutations observed in the brain of SSPE or MIBE patients is still poorly understood. These mutations could have emerged through an adaptation to the brain, leading to SSPE or MIBE, or through a selection of pre-existing mutations as a polymorphism among the circulating strains. Since MIBE only concerns immunocompromised patients and occurs usually very shortly after the primary MeV infection, one could speculate that it is more likely that a minor population of MeV, bearing the mutations that allow the virus grow in a neural context, gets selected and take the advantage of this immunological status for further propagation in the brain. Regardless of the type of encephalitis or MeV variant invading the brain, the high mortality rate associated to measles virus CNS complication highlight the requirement to validate antiviral molecules against these variants.

While the number of tested potential antiviral therapeutics keeps growing, a single molecule or treatment capable to block the major viral cycle steps is still not available. Ultimately, a combination of the treatments that could block the viral entry, the dissemination, the replication, and stimulate the immune system seems to be the most promising solution to prevent and cure MeV systemic infection and will be even more critical for the treatment of CNS infection.

Author Contributions: Conceptualization, M.F. and C.M.; methodology, M.F.; validation, C.M. and B.H.; investigation, M.F.; resources, C.M. and B.H..; writing—original draft preparation, M.F.; writing—review and editing, M.F., C.M. and B.H.; visualization, M.F.; supervision, C.M; project administration, B.H. and C.M.; funding acquisition, B.H. and C.M.

Funding: The work was supported by grant from NIH RO1-NS091263, from the French National Research Agency (ANR) NITRODEP (ANR-13-PDOC-0010-01) to C.M., from Region Auvergne Rhone Alpes (Pack Ambition Recherche) and LABEX ECOFECT (ANR-11-LABX-0048) of Lyon University, within the program "Investissements d'Avenir" (ANR-11-IDEX-0007) operated by the French National Research Agency (ANR) to BH.

Acknowledgments: The authors wish to thank M. Porotto and D. Gerlier for precious scientific advising and M. Iampietro for English proof-reading of the manuscript. We are grateful to SERVIER Medical Art, for their image bank which helped to create Figures 1, 2A, 4 and 6. SERVIER Medical Art is licensed by Creative Commons 3.0 -https://creativecommons.org/licenses/by/3.0/.

References

1. Holzmann, H.; Hengel, H.; Tenbusch, M.; Doerr, H.W. Eradication of Measles: Remaining Challenges. *Med. Microbiol. Immunol.* **2016**, *205*, 201–208. [CrossRef] [PubMed]
2. Guerra, F.M.; Bolotin, S.; Lim, G.; Heffernan, J.; Deeks, S.L.; Li, Y.; Crowcroft, N.S. The Basic Reproduction Number (R0) of Measles: A Systematic Review. *Lancet. Infect. Dis.* **2017**, *17*, e420–e428. [CrossRef]
3. Moss, W.J. Measles. *Lancet* **2017**, *390*, 2490–2502. [CrossRef]
4. Brechot, C.; Bryant, J.; Endtz, H.; Garry, R.F.; Griffin, D.E.; Lewin, S.R.; Mercer, N.; Osterhaus, A.; Picot, V.; Vahlne, A.; et al. 2018 International Meeting of the Global Virus Network. *Antivir. Res.* **2019**, *163*, 140–148. [CrossRef]
5. Strebel, P.M.; Orenstein, W.A. Measles. *N. Engl. J. Med.* **2019**, *381*, 349–357. [CrossRef]
6. Waaijenborg, S.; Hahné, S.J.M.; Mollema, L.; Smits, G.P.; Berbers, G.A.M.; van der Klis, F.R.M.; de Melker, H.E.; Wallinga, J. Waning of Maternal Antibodies against Measles, Mumps, Rubella, and Varicella in Communities with Contrasting Vaccination Coverage. *J. Infect. Dis.* **2013**, *208*, 10–16. [CrossRef]
7. LaVito, A.; John, S.W. Measles Infected up to 4 million a year in US before Vaccine. Available online: https://www.cnbc.com/2019/05/24/measles-infected-up-to-4-million-a-year-in-us-before-vaccine.html (accessed on 24 October 2019).
8. Tesini, B.L. Measles-Pediatrics—Merck Manuals Professional Edition. Available online: https://www.merckmanuals.com/professional/pediatrics/miscellaneous-viral-infections-in-infants-and-children/measles (accessed on 24 October 2019).
9. Paules, C.I.; Marston, H.D.; Fauci, A.S. Measles in 2019—Going Backward. *N. Engl. J. Med.* **2019**, *380*, 2185–2187. [CrossRef]
10. WHO. New Measles Surveillance Data for 2019. Available online: https://www.who.int/immunization/newsroom/measles-data-2019/en/ (accessed on 23 October 2019).
11. Measles Cases and Outbreaks|CDC. Available online: https://www.cdc.gov/measles/cases-outbreaks.html (accessed on 15 September 2019).
12. Griffin, D.E. Measles Virus. In *Fields Virology*; Knipe, D.M., Ed.; Wolters Kluwer: Philadelphia, PA, USA, 2013; Volume 6.
13. Bellini, W.J.; Englund, G.; Rozenblatt, S.; Arnheiter, H.; Richardson, C.D. Measles Virus P Gene Codes for Two Proteins. *J. Virol.* **1985**, *53*, 908–919.
14. Shaffer, J.A.; Bellini, W.J.; Rota, P.A. The C Protein of Measles Virus Inhibits the Type I Interferon Response. *Virology* **2003**, *315*, 389–397. [CrossRef]
15. Schuhmann, K.M.; Pfaller, C.K.; Conzelmann, K.-K. The Measles Virus V Protein Binds to P65 (RelA) to Suppress NF-KappaB Activity. *J. Virol.* **2011**, *85*, 3162–3171. [CrossRef]

16. Gotoh, B.; Komatsu, T.; Takeuchi, K.; Yokoo, J. Paramyxovirus Accessory Proteins as Interferon Antagonists. *Microbiol. Immunol.* **2001**, *45*, 787–800. [CrossRef] [PubMed]
17. Griffin, D.E.; Lin, W.-H.; Pan, C.-H. Measles Virus, Immune Control, and Persistence. *FEMS Microbiol. Rev.* **2012**, *36*, 649–662. [CrossRef] [PubMed]
18. Tatsuo, H.; Ono, N.; Tanaka, K.; Yanagi, Y. SLAM (CDw150) Is a Cellular Receptor for Measles Virus. *Nature* **2000**, *406*, 893–897. [CrossRef] [PubMed]
19. Noyce, R.S.; Bondre, D.G.; Ha, M.N.; Lin, L.-T.; Sisson, G.; Tsao, M.-S.; Richardson, C.D. Tumor Cell Marker PVRL4 (Nectin 4) Is an Epithelial Cell Receptor for Measles Virus. *PLoS Pathog.* **2011**, *7*, e1002240. [CrossRef] [PubMed]
20. Mühlebach, M.D.; Mateo, M.; Sinn, P.L.; Prüfer, S.; Uhlig, K.M.; Leonard, V.H.J.; Navaratnarajah, C.K.; Frenzke, M.; Wong, X.X.; Sawatsky, B.; et al. Adherens Junction Protein Nectin-4 Is the Epithelial Receptor for Measles Virus. *Nature* **2011**, *480*, 530–533. [CrossRef] [PubMed]
21. Naniche, D.; Varior-Krishnan, G.; Cervoni, F.; Wild, T.F.; Rossi, B.; Rabourdin-Combe, C.; Gerlier, D. Human Membrane Cofactor Protein (CD46) Acts as a Cellular Receptor for Measles Virus. *J. Virol.* **1993**, *67*, 6025–6032.
22. Dörig, R.E.; Marcil, A.; Chopra, A.; Richardson, C.D. The Human CD46 Molecule Is a Receptor for Measles Virus (Edmonston Strain). *Cell* **1993**, *75*, 295–305. [CrossRef]
23. Lamb, R.A.; Parks, G.D. *Paramyxoviridae: The Viruses and Their Replication*; Lippincott, Williams, and Wilkins: Philadelphia, PA, USA, 2007; pp. 1449–1496.
24. Gonçalves-Carneiro, D.; McKeating, J.A.; Bailey, D. The Measles Virus Receptor SLAMF1 Can Mediate Particle Endocytosis. *J. Virol.* **2017**, *91*. [CrossRef]
25. Delpeut, S.; Sisson, G.; Black, K.M.; Richardson, C.D. Measles Virus Enters Breast and Colon Cancer Cell Lines through a PVRL4-Mediated Macropinocytosis Pathway. *J. Virol.* **2017**, *91*. [CrossRef]
26. Crimeen-Irwin, B.; Ellis, S.; Christiansen, D.; Ludford-Menting, M.J.; Milland, J.; Lanteri, M.; Loveland, B.E.; Gerlier, D.; Russell, S.M. Ligand Binding Determines Whether CD46 Is Internalized by Clathrin-Coated Pits or Macropinocytosis. *J. Biol. Chem.* **2003**, *278*, 46927–46937. [CrossRef]
27. Frecha, C.; Lévy, C.; Costa, C.; Nègre, D.; Amirache, F.; Buckland, R.; Russell, S.J.; Cosset, F.-L.; Verhoeyen, E. Measles Virus Glycoprotein-Pseudotyped Lentiviral Vector-Mediated Gene Transfer into Quiescent Lymphocytes Requires Binding to Both SLAM and CD46 Entry Receptors. *J. Virol.* **2011**, *85*, 5975–5985. [CrossRef] [PubMed]
28. Jurgens, E.M.; Mathieu, C.; Palermo, L.M.; Hardie, D.; Horvat, B.; Moscona, A.; Porotto, M. Measles Fusion Machinery Is Dysregulated in Neuropathogenic Variants. *MBio* **2015**, *6*. [CrossRef] [PubMed]
29. Mathieu, C.; Huey, D.; Jurgens, E.; Welsch, J.C.; DeVito, I.; Talekar, A.; Horvat, B.; Niewiesk, S.; Moscona, A.; Porotto, M. Prevention of Measles Virus Infection by Intranasal Delivery of Fusion Inhibitor Peptides. *J. Virol.* **2015**, *89*, 1143–1155. [CrossRef] [PubMed]
30. Singh, B.K.; Li, N.; Mark, A.C.; Mateo, M.; Cattaneo, R.; Sinn, P.L. Cell-to-Cell Contact and Nectin-4 Govern Spread of Measles Virus from Primary Human Myeloid Cells to Primary Human Airway Epithelial Cells. *J. Virol.* **2016**, *90*, 6808–6817. [CrossRef] [PubMed]
31. Duprex, W.P.; McQuaid, S.; Hangartner, L.; Billeter, M.A.; Rima, B.K. Observation of Measles Virus Cell-to-Cell Spread in Astrocytoma Cells by Using a Green Fluorescent Protein-Expressing Recombinant Virus. *J. Virol.* **1999**, *73*, 9568–9575.
32. Rima, B.K.; Duprex, W.P. The Measles Virus Replication Cycle. *Curr. Top. Microbiol. Immunol.* **2009**, *329*, 77–102.
33. Iwasaki, M.; Takeda, M.; Shirogane, Y.; Nakatsu, Y.; Nakamura, T.; Yanagi, Y. The Matrix Protein of Measles Virus Regulates Viral RNA Synthesis and Assembly by Interacting with the Nucleocapsid Protein. *J. Virol.* **2009**, *83*, 10374–10383. [CrossRef]
34. Bhella, D.; Ralph, A.; Murphy, L.B.; Yeo, R.P. Significant Differences in Nucleocapsid Morphology within the Paramyxoviridae. *J. Gen. Virol.* **2002**, *83*, 1831–1839. [CrossRef]
35. Radecke, F.; Spielhofer, P.; Schneider, H.; Kaelin, K.; Huber, M.; Dötsch, C.; Christiansen, G.; Billeter, M.A. Rescue of Measles Viruses from Cloned DNA. *EMBO J.* **1995**, *14*, 5773–5784. [CrossRef]
36. Calain, P.; Roux, L. The Rule of Six, a Basic Feature for Efficient Replication of Sendai Virus Defective Interfering RNA. *J. Virol.* **1993**, *67*, 4822–4830.

37. Hashiguchi, T.; Fukuda, Y.; Matsuoka, R.; Kuroda, D.; Kubota, M.; Shirogane, Y.; Watanabe, S.; Tsumoto, K.; Kohda, D.; Plemper, R.K.; et al. Structures of the Prefusion Form of Measles Virus Fusion Protein in Complex with Inhibitors. *Proc. Natl. Acad. Sci. USA* **2018**, *115*, 2496–2501. [CrossRef] [PubMed]

38. Watanabe, S.; Shirogane, Y.; Sato, Y.; Hashiguchi, T.; Yanagi, Y. New Insights into Measles Virus Brain Infections. *Trends Microbiol.* **2018**. [CrossRef] [PubMed]

39. Plattet, P.; Alves, L.; Herren, M.; Aguilar, H.C. Measles Virus Fusion Protein: Structure, Function and Inhibition. *Viruses* **2016**, *8*, 112. [CrossRef] [PubMed]

40. Angius, F.; Smuts, H.; Rybkina, K.; Stelitano, D.; Eley, B.; Wilmshurst, J.; Ferren, M.; Lalande, A.; Mathieu, C.; Moscona, A.; et al. Analysis of a Subacute Sclerosing Panencephalitis (SSPE) Genotype B3 Virus from the 2009/10 South African Measles Epidemic Shows Hyperfusogenic F Proteins Contribute to Measles Virus Infection in the Brain. *J. Virol.* **2018**. [CrossRef] [PubMed]

41. Lambert, D.M.; Barney, S.; Lambert, A.L.; Guthrie, K.; Medinas, R.; Davis, D.E.; Bucy, T.; Erickson, J.; Merutka, G.; Petteway, S.R. Peptides from Conserved Regions of Paramyxovirus Fusion (F) Proteins Are Potent Inhibitors of Viral Fusion. *Proc. Natl. Acad. Sci. USA* **1996**, *93*, 2186–2191. [CrossRef]

42. Mathieu, C.; Ferren, M.; Jurgens, E.; Dumont, C.; Rybkina, K.; Harder, O.; Stelitano, D.; Madeddu, S.; Sanna, G.; Schwartz, D.; et al. Measles Virus Bearing Measles Inclusion Body Encephalitis-Derived Fusion Protein Is Pathogenic after Infection via the Respiratory Route. *J. Virol.* **2019**, *93*. [CrossRef]

43. Coughlin, M.M.; Beck, A.S.; Bankamp, B.; Rota, P.A. Perspective on Global Measles Epidemiology and Control and the Role of Novel Vaccination Strategies. *Viruses* **2017**, *9*, 11. [CrossRef]

44. Melenotte, C.; Zandotti, C.; Gautret, P.; Parola, P.; Raoult, D. Measles: Is a New Vaccine Approach Needed? *Lancet. Infect. Dis.* **2018**, *18*, 1060–1061. [CrossRef]

45. Bankamp, B.; Takeda, M.; Zhang, Y.; Xu, W.; Rota, P.A. Genetic Characterization of Measles Vaccine Strains. *J. Infect. Dis.* **2011**, *204*, S533–S548. [CrossRef]

46. Vaccine for Measles (MMR Shot)|CDC. Available online: https://www.cdc.gov/measles/vaccination.html (accessed on 15 September 2019).

47. McLean, H.Q.; Fiebelkorn, A.P.; Temte, J.L.; Wallace, G.S. Centers for Disease Control and Prevention. Prevention of Measles, Rubella, Congenital Rubella Syndrome, and Mumps, 2013: Summary Recommendations of the Advisory Committee on Immunization Practices (ACIP). *Morb. Mortal. Wkly.* **2013**, *62*, 1–34.

48. Haralambieva, I.H.; Ovsyannikova, I.G.; O'Byrne, M.; Pankratz, V.S.; Jacobson, R.M.; Poland, G.A. A Large Observational Study to Concurrently Assess Persistence of Measles Specific B-Cell and T-Cell Immunity in Individuals Following Two Doses of MMR Vaccine. *Vaccine* **2011**, *29*, 4485–4491. [CrossRef] [PubMed]

49. Haralambieva, I.H.; Kennedy, R.B.; Ovsyannikova, I.G.; Whitaker, J.A.; Poland, G.A. Variability in Humoral Immunity to Measles Vaccine: New Developments. *Trends Mol. Med.* **2015**, *21*, 789–801. [CrossRef] [PubMed]

50. Laksono, B.M.; de Vries, R.D.; Verburgh, R.J.; Visser, E.G.; de Jong, A.; Fraaij, P.L.A.; Ruijs, W.L.M.; Nieuwenhuijse, D.F.; van den Ham, H.-J.; Koopmans, M.P.G.; et al. Studies into the Mechanism of Measles-Associated Immune Suppression during a Measles Outbreak in the Netherlands. *Nat. Commun.* **2018**, *9*, 4944. [CrossRef] [PubMed]

51. Sindhu, T.G.; Geeta, M.G.; Krishnakumar, P.; Sabitha, S.; Ajina, K.K. Clinical Profile of Measles in Children with Special Reference to Infants. *Trop. Doct.* **2019**, *49*, 20–23. [CrossRef] [PubMed]

52. Ben-Chetrit, E.; Oster, Y.; Jarjou'i, A.; Megged, O.; Lachish, T.; Cohen, M.J.; Stein-Zamir, C.; Ivgi, H.; Rivkin, M.; Milgrom, Y.; et al. Measles-Related Hospitalizations and Associated Complications in Jerusalem, 2018-2019. *Clin. Microbiol. Infect.* **2019**. [CrossRef] [PubMed]

53. Marie, J.C.; Saltel, F.; Escola, J.-M.; Jurdic, P.; Wild, T.F.; Horvat, B. Cell Surface Delivery of the Measles Virus Nucleoprotein: A Viral Strategy To Induce Immunosuppression. *J. Virol.* **2004**, *78*, 11952–11961. [CrossRef] [PubMed]

54. Mina, M.J.; Metcalf, C.J.E.; de Swart, R.L.; Osterhaus, A.D.M.E.; Grenfell, B.T. Long-Term Measles-Induced Immunomodulation Increases Overall Childhood Infectious Disease Mortality. *Science* **2015**, *348*, 694–699. [CrossRef] [PubMed]

55. Sellin, C.I.; Jégou, J.-F.; Renneson, J.; Druelle, J.; Wild, T.F.; Marie, J.C.; Horvat, B. Interplay between Virus-Specific Effector Response and Foxp3+ Regulatory T Cells in Measles Virus Immunopathogenesis. *PLoS ONE* **2009**, *4*, e4948. [CrossRef]

56. Moss, W.J.; Griffin, D.E. Measles. *Lancet* **2012**, *379*, 153–164. [CrossRef]

57. Rima, B.K.; Duprex, W.P. Morbilliviruses and Human Disease. *J. Pathol.* **2006**, *208*, 199–214. [CrossRef]
58. Enders, J.F.; McCarthy, K.; Mitus, A.; Cheatham, W.J. Isolation of Measles Virus at Autopsy in Cases of Giant-Cell Pneumonia without Rash. *N. Engl. J. Med.* **1959**, *261*, 875–881. [CrossRef] [PubMed]
59. D'Souza, R.M. Vitamin A for the Treatment of Children with Measles—A Systematic Review. *J. Trop. Pediatr.* **2002**, *48*, 323–327. [CrossRef] [PubMed]
60. Ellison, J.B. Intensive Vitamin Therapy in Measles. *Br. Med. J.* **1932**, *2*, 708–711. [CrossRef] [PubMed]
61. Imdad, A.; Mayo-Wilson, E.; Herzer, K.; Bhutta, Z.A. Vitamin A Supplementation for Preventing Morbidity and Mortality in Children from Six Months to Five Years of Age. *Cochrane Database Syst. Rev.* **2017**. [CrossRef]
62. Avota, E.; Koethe, S.; Schneider-Schaulies, S. Membrane Dynamics and Interactions in Measles Virus Dendritic Cell Infections. *Cell. Microbiol.* **2013**, *15*, 161–169. [CrossRef]
63. Ferreira, C.S.A.; Frenzke, M.; Leonard, V.H.J.; Welstead, G.G.; Richardson, C.D.; Cattaneo, R. Measles Virus Infection of Alveolar Macrophages and Dendritic Cells Precedes Spread to Lymphatic Organs in Transgenic Mice Expressing Human Signaling Lymphocytic Activation Molecule (SLAM, CD150). *J. Virol.* **2010**, *84*, 3033–3042. [CrossRef]
64. De Vries, R.D.; Mesman, A.W.; Geijtenbeek, T.B.; Duprex, W.P.; de Swart, R.L. The Pathogenesis of Measles. *Curr. Opin. Virol.* **2012**, *2*, 248–255. [CrossRef]
65. Delpeut, S.; Sawatsky, B.; Wong, X.-X.; Frenzke, M.; Cattaneo, R.; von Messling, V. Nectin-4 Interactions Govern Measles Virus Virulence in a New Model of Pathogenesis, the Squirrel Monkey (Saimiri Sciureus). *J. Virol.* **2017**, *91*. [CrossRef]
66. Frenzke, M.; Sawatsky, B.; Wong, X.X.; Delpeut, S.; Mateo, M.; Cattaneo, R.; von Messling, V. Nectin-4-Dependent Measles Virus Spread to the Cynomolgus Monkey Tracheal Epithelium: Role of Infected Immune Cells Infiltrating the Lamina Propria. *J. Virol.* **2013**, *87*, 2526–2534. [CrossRef]
67. Gourru-Lesimple, G.; Mathieu, C.; Thevenet, T.; Guillaume-Vasselin, V.; Jégou, J.-F.; Boer, C.G.; Tomczak, K.; Bloyet, L.-M.; Giraud, C.; Grande, S.; et al. Measles Virus Infection of Human Keratinocytes: Possible Link between Measles and Atopic Dermatitis. *J. Dermatol. Sci.* **2017**, *86*, 97–105. [CrossRef]
68. Lemon, K.; de Vries, R.D.; Mesman, A.W.; McQuaid, S.; van Amerongen, G.; Yüksel, S.; Ludlow, M.; Rennick, L.J.; Kuiken, T.; Rima, B.K.; et al. Early Target Cells of Measles Virus after Aerosol Infection of Non-Human Primates. *PLoS Pathog.* **2011**, *7*, e1001263. [CrossRef] [PubMed]
69. Cohen, B.E.; Durstenfeld, A.; Roehm, P.C. Viral Causes of Hearing Loss: A Review for Hearing Health Professionals. *Trends Hear.* **2014**, *18*, 233121651454136. [CrossRef] [PubMed]
70. Dunmade, A.; Segun-Busari, S.; Olajide, T.; Ologe, F. Profound Bilateral Sensorineural Hearing Loss in Nigerian Children: Any Shift in Etiology? *J. Deaf Stud. Deaf Educ.* **2006**, *12*, 112–118. [CrossRef] [PubMed]
71. Stephenson, J. Will the Current Measles Vaccines Ever Eradicate Measles? *Expert Rev. Vaccines* **2002**, *1*, 355–362. [CrossRef] [PubMed]
72. Semba, R.D.; Bloem, M.W. Measles Blindness. *Surv. Ophthalmol.* **2004**, *49*, 243–255. [CrossRef] [PubMed]
73. Shinoda, K.; Kobayashi, A.; Higashide, T.; Shirao, Y.; Sakurai, M.; Shirota, Y.; Kagaya, M. Detection of Measles Virus Genomic RNA in Tear Samples from a Patient with Measles Keratitis. *Cornea* **2002**, *21*, 610–612. [CrossRef] [PubMed]
74. Ludlow, M.; Rennick, L.J.; Sarlang, S.; Skibinski, G.; McQuaid, S.; Moore, T.; de Swart, R.L.; Duprex, W.P. Wild-Type Measles Virus Infection of Primary Epithelial Cells Occurs via the Basolateral Surface without Syncytium Formation or Release of Infectious Virus. *J. Gen. Virol.* **2010**, *91*, 971–979. [CrossRef]
75. Muñoz-Alía, M.Á.; Muller, C.P.; Russell, S.J. Hemagglutinin-Specific Neutralization of Subacute Sclerosing Panencephalitis Viruses. *PLoS ONE* **2018**, *13*, e0192245. [CrossRef]
76. Pratakpiriya, W.; Ping Teh, A.P.; Radtanakatikanon, A.; Pirarat, N.; Thi Lan, N.; Takeda, M.; Techangamsuwan, S.; Yamaguchi, R. Expression of Canine Distemper Virus Receptor Nectin-4 in the Central Nervous System of Dogs. *Sci. Rep.* **2017**, *7*, 349. [CrossRef]
77. Pratakpiriya, W.; Seki, F.; Otsuki, N.; Sakai, K.; Fukuhara, H.; Katamoto, H.; Hirai, T.; Maenaka, K.; Techangamsuwan, S.; Lan, N.T.; et al. Nectin4 Is an Epithelial Cell Receptor for Canine Distemper Virus and Involved in Neurovirulence. *J. Virol.* **2012**, *86*, 10207–10210. [CrossRef]
78. Generous, A.R.; Harrison, O.J.; Troyanovsky, R.B.; Mateo, M.; Navaratnarajah, C.K.; Donohue, R.C.; Pfaller, C.K.; Alekhina, O.; Sergeeva, A.P.; Indra, I.; et al. Trans-Endocytosis Elicited by Nectins Transfers Cytoplasmic Cargo, Including Infectious Material, between Cells. *J. Cell Sci.* **2019**, *132*, jcs235507. [CrossRef] [PubMed]

79. Griffin, D.E. Measles Virus and the Nervous System. In *Handbook of Clinical Neurology*; Elsevier: Amsterdam, The Netherlands, 2014; Volume 123, pp. 577–590. [CrossRef]

80. Zachariah, P.; Stockwell, M.S. Measles Vaccine: Past, Present, and Future. *J. Clin. Pharmacol.* **2016**, *56*, 133–140. [CrossRef] [PubMed]

81. Johnson, R.T.; Griffin, D.E.; Hirsch, R.L.; Wolinsky, J.S.; Roedenbeck, S.; Lindo de Soriano, I.; Vaisberg, A. Measles Encephalomyelitis—Clinical and Immunologic Studies. *N. Engl. J. Med.* **1984**, *310*, 137–141. [CrossRef] [PubMed]

82. Moench, T.R.; Griffin, D.E.; Obriecht, C.R.; Vaisberg, A.J.; Johnson, R.T. Acute Measles in Patients with and without Neurological Involvement: Distribution of Measles Virus Antigen and RNA. *J. Infect. Dis.* **1988**, *158*, 433–442. [CrossRef] [PubMed]

83. Reuter, D.; Schneider-Schaulies, J. Measles Virus Infection of the CNS: Human Disease, Animal Models, and Approaches to Therapy. *Med. Microbiol. Immunol.* **2010**, *199*, 261–271. [CrossRef]

84. Garg, R.K. Acute Disseminated Encephalomyelitis. *Postgrad. Med. J.* **2003**, *79*, 11–17. [CrossRef]

85. Miller, H.G.; Stanton, J.B.; Gibbons, J.L. Acute Disseminated Encephalomyelitis and Related Syndromes. *Br. Med. J.* **1957**, *1*, 668–672. [CrossRef]

86. Ludlow, M.; Kortekaas, J.; Herden, C.; Hoffmann, B.; Tappe, D.; Trebst, C.; Griffin, D.E.; Brindle, H.E.; Solomon, T.; Brown, A.S.; et al. Neurotropic Virus Infections as the Cause of Immediate and Delayed Neuropathology. *Acta Neuropathol.* **2016**, *131*, 159–184. [CrossRef]

87. Griffin, D.E.; Cooper, S.J.; Hirsch, R.L.; Johnson, R.T.; Lindo de Soriano, I.; Roedenbeck, S.; Vaisberg, A. Changes in Plasma IgE Levels during Complicated and Uncomplicated Measles Virus Infections. *J. Allergy Clin. Immunol.* **1985**, *76*, 206–213. [CrossRef]

88. Esolen, L.M.; Takahashi, K.; Johnson, R.T.; Vaisberg, A.; Moench, T.R.; Wesselingh, S.L.; Griffin, D.E. Brain Endothelial Cell Infection in Children with Acute Fatal Measles. *J. Clin. Investig.* **1995**, *96*, 2478–2481. [CrossRef]

89. Hosoya, M. Measles Encephalitis: Direct Viral Invasion or Autoimmune-Mediated Inflammation? *Intern. Med.* **2006**, *45*, 841–842. [CrossRef] [PubMed]

90. Baldolli, A.; Dargère, S.; Cardineau, E.; Vabret, A.; Dina, J.; de La Blanchardière, A.; Verdon, R. Measles Inclusion-Body Encephalitis (MIBE) in a Immunocompromised Patient. *J. Clin. Virol.* **2016**, *81*, 43–46. [CrossRef] [PubMed]

91. Hughes, I.; Jenney, M.E.; Newton, R.W.; Morris, D.J.; Klapper, P.E. Measles Encephalitis during Immunosuppressive Treatment for Acute Lymphoblastic Leukaemia. *Arch. Dis. Child.* **1993**, *68*, 775–778. [CrossRef] [PubMed]

92. Hardie, D.R.; Albertyn, C.; Heckmann, J.M.; Smuts, H.E.M. Molecular Characterisation of Virus in the Brains of Patients with Measles Inclusion Body Encephalitis (MIBE). *Virol. J.* **2013**, *10*, 283. [CrossRef] [PubMed]

93. Roos, R.P.; Graves, M.C.; Wollmann, R.L.; Chilcote, R.R.; Nixon, J. Immunologic and Virologic Studies of Measles Inclusion Body Encephalitis in an Immunosuppressed Host: The Relationship to Subacute Sclerosing Panencephalitis. *Neurology* **1981**, *31*, 1263–1270. [CrossRef] [PubMed]

94. Bitnun, A.; Shannon, P.; Durward, A.; Rota, P.A.; Bellini, W.J.; Graham, C.; Wang, E.; Ford-Jones, E.L.; Cox, P.; Becker, L.; et al. Measles Inclusion-Body Encephalitis Caused by the Vaccine Strain of Measles Virus. *Clin. Infect. Dis.* **1999**, *29*, 855–861. [CrossRef]

95. Ohuchi, M.; Ohuchi, R.; Mifune, K.; Ishihara, T.; Ogawa, T. Characterization of the Measles Virus Isolated from the Brain of a Patient with Immunosuppressive Measles Encephalitis. *J. Infect. Dis.* **1987**, *156*, 436–441. [CrossRef]

96. Makhortova, N.R.; Askovich, P.; Patterson, C.E.; Gechman, L.A.; Gerard, N.P.; Rall, G.F. Neurokinin-1 Enables Measles Virus Trans-Synaptic Spread in Neurons. *Virology* **2007**, *362*, 235–244. [CrossRef]

97. Young, V.A.; Rall, G.F. Making It to the Synapse: Measles Virus Spread in and among Neurons. *Curr. Top. Microbiol. Immunol.* **2009**, *330*, 3–30.

98. Mekki, M.; Eley, B.; Hardie, D.; Wilmshurst, J.M. Subacute Sclerosing Panencephalitis: Clinical Phenotype, Epidemiology, and Preventive Interventions. *Dev. Med. Child Neurol.* **2019**, dmcn.14166. [CrossRef]

99. Wendorf, K.A.; Winter, K.; Zipprich, J.; Schechter, R.; Hacker, J.K.; Preas, C.; Cherry, J.D.; Glaser, C.; Harriman, K. Subacute Sclerosing Panencephalitis: The Devastating Measles Complication That Might Be More Common Than Previously Estimated. *Clin. Infect. Dis.* **2017**, *65*, 226–232. [CrossRef] [PubMed]

100. Nakamura, Y.; Iinuma, K.; Oka, E.; Nihei, K. Epidemiologic Features of Subacute Sclerosing Panencephalitis from Clinical Data of Patients Receiving a Public Aid for Treatment. *No to hattatsu = Brain Dev.* **2003**, *35*, 316–320.

101. Miller, C.; Andrews, N.; Rush, M.; Munro, H.; Jin, L.; Miller, E. The Epidemiology of Subacute Sclerosing Panencephalitis in England and Wales 1990-2002. *Arch. Dis. Child.* **2004**, *89*, 1145–1148. [CrossRef]

102. Prashanth, L.K.; Taly, A.B.; Ravi, V.; Sinha, S.; Rao, S. Long Term Survival in Subacute Sclerosing Panencephalitis: An Enigma. *Brain Dev.* **2006**, *28*, 447–452. [CrossRef] [PubMed]

103. Oldstone, M.B.A.; Dales, S.; Tishon, A.; Lewicki, H.; Martin, L. A Role for Dual Viral Hits in Causation of Subacute Sclerosing Panencephalitis. *J. Exp. Med.* **2005**, *202*, 1185–1190. [CrossRef]

104. Bellini, W.J.; Rota, J.S.; Lowe, L.E.; Katz, R.S.; Dyken, P.R.; Zaki, S.R.; Shieh, W.-J.; Rota, P.A. Subacute Sclerosing Panencephalitis: More Cases of This Fatal Disease Are Prevented by Measles Immunization than Was Previously Recognized. *J. Infect. Dis.* **2005**, *192*, 1686–1693. [CrossRef]

105. Forcić, D.; Baricević, M.; Zgorelec, R.; Kruzić, V.; Kaić, B.; Della Marina, B.M.; Sojat, L.C.; Tesović, G.; Mazuran, R. Detection and Characterization of Measles Virus Strains in Cases of Subacute Sclerosing Panencephalitis in Croatia. *Virus Res.* **2004**, *99*, 51–56. [CrossRef]

106. Moulin, E.; Beal, V.; Jeantet, D.; Horvat, B.; Wild, T.F.; Waku-Kouomou, D. Molecular Characterization of Measles Virus Strains Causing Subacute Sclerosing Panencephalitis in France in 1977 and 2007. *J. Med. Virol.* **2011**, *83*, 1614–1623. [CrossRef]

107. Cattaneo, R.; Schmid, A.; Eschle, D.; Baczko, K.; ter Meulen, V.; Billeter, M.A. Biased Hypermutation and Other Genetic Changes in Defective Measles Viruses in Human Brain Infections. *Cell* **1988**, *55*, 255–265. [CrossRef]

108. Pfaller, C.K.; Donohue, R.C.; Nersisyan, S.; Brodsky, L.; Cattaneo, R. Extensive Editing of Cellular and Viral Double-Stranded RNA Structures Accounts for Innate Immunity Suppression and the Proviral Activity of ADAR1p150. *PLoS Biol.* **2018**, *16*, e2006577. [CrossRef] [PubMed]

109. Rima, B.K.; Duprex, W.P. Molecular Mechanisms of Measles Virus Persistence. *Virus Res.* **2005**, *111*, 132–147. [CrossRef] [PubMed]

110. Liebert, U.G.; Baczko, K.; Budka, H.; ter Meulen, V. Restricted Expression of Measles Virus Proteins in Brains from Cases of Subacute Sclerosing Panencephalitis. *J. Gen. Virol.* **1986**, *67*, 2435–2444. [CrossRef]

111. Cathomen, T.; Mrkic, B.; le Spehner, D.; Drillien, R.; Naef, R.; Pavlovic, J.; Aguzzi, A.; ABilleter, M.; Cattaneo, R. A Matrix-Less Measles Virus Is Infectious and Elicits Extensive Cell Fusion: Consequences for Propagation in the Brain. *EMBO J.* **1998**, *17*, 3899–3908. [CrossRef] [PubMed]

112. Patterson, J.B.; Cornu, T.I.; Redwine, J.; Dales, S.; Lewicki, H.; Holz, A.; Thomas, D.; Billeter, M.A.; Oldstone, M.B. Evidence That the Hypermutated M Protein of a Subacute Sclerosing Panencephalitis Measles Virus Actively Contributes to the Chronic Progressive CNS Disease. *Virology* **2001**, *291*, 215–225. [CrossRef]

113. Reuter, T.; Weissbrich, B.; Schneider-Schaulies, S.; Schneider-Schaulies, J. RNA Interference with Measles Virus N, P, and L MRNAs Efficiently Prevents and with Matrix Protein MRNA Enhances Viral Transcription. *J. Virol.* **2006**, *80*, 5951–5957. [CrossRef]

114. Maisner, A.; Klenk, H.; Herrler, G. Polarized Budding of Measles Virus Is Not Determined by Viral Surface Glycoproteins. *J. Virol.* **1998**, *72*, 5276–5278. [PubMed]

115. Kühne, M.; Brown, D.W.G.; Jin, L. Genetic Variability of Measles Virus in Acute and Persistent Infections. *Infect. Genet. Evol.* **2006**, *6*, 269–276. [CrossRef]

116. Kweder, H.; Ainouze, M.; Cosby, S.L.; Muller, C.P.; Lévy, C.; Verhoeyen, E.; Cosset, F.-L.; Manet, E.; Buckland, R. Mutations in the H, F, or M Proteins Can Facilitate Resistance of Measles Virus to Neutralizing Human Anti-MV Sera. *Adv. Virol.* **2014**, *2014*, 205617. [CrossRef]

117. Homma, M.; Tashiro, M.; Konno, H.; Ohara, Y.; Hino, M.; Takase, S. Isolation and Characterization of Subacute Sclerosing Panencephalitis Virus (Yamagata-1 Strain) from a Brain Autopsy. *Microbiol. Immunol.* **1982**, *26*, 1195–1202. [CrossRef]

118. Ito, N.; Ayata, M.; Shingai, M.; Furukawa, K.; Seto, T.; Matsunaga, I.; Muraoka, M.; Ogura, H. Comparison of the Neuropathogenicity of Two SSPE Sibling Viruses of the Osaka-2 Strain Isolated with Vero and B95a Cells. *J. Neurovirol.* **2002**, *8*, 6–13. [CrossRef]

119. Makino, S.; Sasaki, K.; Nakagawa, M.; Saito, M.; Shinohara, Y. Isolation and Biological Characterization of a Measles Virus-like Agent from the Brain of an Autopsied Case of Subacute Sclerosing Panencephalitis (SSPE). *Microbiol. Immunol.* **1977**, *21*, 193–205. [CrossRef]

120. Ogura, H.; Ayata, M.; Hayashi, K.; Seto, T.; Matsuoka, O.; Hattori, H.; Tanaka, K.; Tanaka, K.; Takano, Y.; Murata, R. Efficient Isolation of Subacute Sclerosing Panencephalitis Virus from Patient Brains by Reference to Magnetic Resonance and Computed Tomographic Images. *J. Neurovirol.* **1997**, *3*, 304–309. [CrossRef]

121. Cathomen, T.; Naim, H.Y.; Cattaneo, R. Measles Viruses with Altered Envelope Protein Cytoplasmic Tails Gain Cell Fusion Competence. *J. Virol.* **1998**, *72*, 1224–1234. [PubMed]

122. Schmid, A.; Spielhofer, P.; Cattaneo, R.; Baczko, K.; ter Meulen, V.; Billeter, M.A. Subacute Sclerosing Panencephalitis Is Typically Characterized by Alterations in the Fusion Protein Cytoplasmic Domain of the Persisting Measles Virus. *Virology* **1992**, *188*, 910–915. [CrossRef]

123. Watanabe, S.; Shirogane, Y.; Suzuki, S.O.; Ikegame, S.; Koga, R.; Yanagi, Y. Mutant Fusion Proteins with Enhanced Fusion Activity Promote Measles Virus Spread in Human Neuronal Cells and Brains of Suckling Hamsters. *J. Virol.* **2013**, *87*, 2648–2659. [CrossRef] [PubMed]

124. Watanabe, S.; Ohno, S.; Shirogane, Y.; Suzuki, S.O.; Koga, R.; Yanagi, Y. Measles Virus Mutants Possessing the Fusion Protein with Enhanced Fusion Activity Spread Effectively in Neuronal Cells, but Not in Other Cells, without Causing Strong Cytopathology. *J. Virol.* **2015**, *89*, 2710–2717. [CrossRef]

125. Ader, N.; Brindley, M.; Avila, M.; Örvell, C.; Horvat, B.; Hiltensperger, G.; Schneider-Schaulies, J.; Vandevelde, M.; Zurbriggen, A.; Plemper, R.K.; et al. Mechanism for Active Membrane Fusion Triggering by Morbillivirus Attachment Protein. *J. Virol.* **2013**, *87*, 314–326. [CrossRef] [PubMed]

126. Avila, M.; Alves, L.; Khosravi, M.; Ader-Ebert, N.; Origgi, F.; Schneider-Schaulies, J.; Zurbriggen, A.; Plemper, R.K.; Plattet, P. Molecular Determinants Defining the Triggering Range of Prefusion F Complexes of Canine Distemper Virus. *J. Virol.* **2014**, *88*, 2951–2966. [CrossRef] [PubMed]

127. Doyle, J.; Prussia, A.; White, L.K.; Sun, A.; Liotta, D.C.; Snyder, J.P.; Compans, R.W.; Plemper, R.K. Two Domains That Control Prefusion Stability and Transport Competence of the Measles Virus Fusion Protein. *J. Virol.* **2006**, *80*, 1524–1536. [CrossRef]

128. Sato, Y.; Watanabe, S.; Fukuda, Y.; Hashiguchi, T.; Yanagi, Y.; Ohno, S. Cell-to-Cell Measles Virus Spread between Human Neurons Is Dependent on Hemagglutinin and Hyperfusogenic Fusion Protein. *J. Virol.* **2018**, *92*. [CrossRef] [PubMed]

129. Mateo, M.; Navaratnarajah, C.K.; Cattaneo, R. Structural Basis of Efficient Contagion: Measles Variations on a Theme by Parainfluenza Viruses. *Curr. Opin. Virol.* **2014**, *5*, 16–23. [CrossRef]

130. Vongpunsawad, S.; Oezgun, N.; Braun, W.; Cattaneo, R. Selectively Receptor-Blind Measles Viruses: Identification of Residues Necessary for SLAM- or CD46-Induced Fusion and Their Localization on a New Hemagglutinin Structural Model. *J. Virol.* **2004**, *78*, 302–313. [CrossRef]

131. Moeller-Ehrlich, K.; Ludlow, M.; Beschorner, R.; Meyermann, R.; Rima, B.K.; Duprex, W.P.; Niewiesk, S.; Schneider-Schaulies, J. Two Functionally Linked Amino Acids in the Stem 2 Region of Measles Virus Haemagglutinin Determine Infectivity and Virulence in the Rodent Central Nervous System. *J. Gen. Virol.* **2007**, *88*, 3112–3120. [CrossRef]

132. Moeller, K.; Duffy, I.; Duprex, P.; Rima, B.; Beschorner, R.; Fauser, S.; Meyermann, R.; Niewiesk, S.; ter Meulen, V.; Schneider-Schaulies, J. Recombinant Measles Viruses Expressing Altered Hemagglutinin (H) Genes: Functional Separation of Mutations Determining H Antibody Escape from Neurovirulence. *J. Virol.* **2001**, *75*, 7612–7620. [CrossRef]

133. Hotta, H.; Nihei, K.; Abe, Y.; Kato, S.; Jiang, D.-P.; Nagano-Fujii, M.; Sada, K. Full-Length Sequence Analysis of Subacute Sclerosing Panencephalitis (SSPE) Virus, a Mutant of Measles Virus, Isolated from Brain Tissues of a Patient Shortly after Onset of SSPE. *Microbiol. Immunol.* **2006**, *50*, 525–534. [CrossRef]

134. Millar, E.L.; Rennick, L.J.; Weissbrich, B.; Schneider-Schaulies, J.; Duprex, W.P.; Rima, B.K. The Phosphoprotein Genes of Measles Viruses from Subacute Sclerosing Panencephalitis Cases Encode Functional as Well as Non-Functional Proteins and Display Reduced Editing. *Virus Res.* **2016**, *211*, 29–37. [CrossRef]

135. Arai, T.; Terao-Muto, Y.; Uchida, S.; Lin, C.; Honda, T.; Takenaka, A.; Ikeda, F.; Sato, H.; Yoneda, M.; Kai, C. The P Gene of Rodent Brain-Adapted Measles Virus Plays a Critical Role in Neurovirulence. *J. Gen. Virol.* **2017**, *98*, 1620–1629. [CrossRef]

136. Jehmlich, U.; Ritzer, J.; Grosche, J.; Härtig, W.; Liebert, U.G. Experimental Measles Encephalitis in Lewis Rats: Dissemination of Infected Neuronal Cell Subtypes. *J. Neurovirol.* **2013**, *19*, 461–470. [CrossRef]

137. Parhad, I.M.; Johnson, K.P.; Wolinsky, J.S.; Swoveland, P. Measles Retinopathy. A Hamster Model of Acute and Chronic Lesions. *Lab Investig.* **1980**, *43*, 52–56.

138. Burnstein, T.; Jensen, J.H.; Waksman, B.H. The Development of a Neurotropic Strain of Measles Virus in Hamsters and Mice. *J. Infect. Dis.* **1964**, *114*, 265–272. [CrossRef]

139. Ehrengruber, M.U.; Ehler, E.; Billeter, M.A.; Naim, H.Y. Measles Virus Spreads in Rat Hippocampal Neurons by Cell-to-Cell Contact and in a Polarized Fashion. *J. Virol.* **2002**, *76*, 5720–5728. [CrossRef]

140. Neighbour, P.A.; Rager-Zisman, B.; Bloom, B.R. Susceptibility of Mice to Acute and Persistent Measles Infection. *Infect. Immun.* **1978**, *21*, 764–770.

141. Schubert, S.; Möller-Ehrlich, K.; Singethan, K.; Wiese, S.; Duprex, W.P.; Rima, B.K.; Niewiesk, S.; Schneider-Schaulies, J. A Mouse Model of Persistent Brain Infection with Recombinant Measles Virus. *J. Gen. Virol.* **2006**, *87*, 2011–2019. [CrossRef]

142. Abe, Y.; Hashimoto, K.; Watanabe, M.; Ohara, S.; Sato, M.; Kawasaki, Y.; Hashimoto, Y.; Hosoya, M. Characteristics of Viruses Derived from Nude Mice with Persistent Measles Virus Infection. *J. Virol.* **2013**, *87*, 4170–4175. [CrossRef]

143. Horvat, B.; Rivailler, P.; Varior-Krishnan, G.; Cardoso, A.; Gerlier, D.; Rabourdin-Combe, C. Transgenic Mice Expressing Human Measles Virus (MV) Receptor CD46 Provide Cells Exhibiting Different Permissivities to MV Infections. *J. Virol.* **1996**, *70*, 6673–6681.

144. Sellin, C.I.; Davoust, N.; Guillaume, V.; Baas, D.; Belin, M.-F.; Buckland, R.; Wild, T.F.; Horvat, B. High Pathogenicity of Wild-Type Measles Virus Infection in CD150 (SLAM) Transgenic Mice. *J. Virol.* **2006**, *80*, 6420–6429. [CrossRef]

145. Figueira, T.N.; Palermo, L.M.; Veiga, A.S.; Huey, D.; Alabi, C.A.; Santos, N.C.; Welsch, J.C.; Mathieu, C.; Horvat, B.; Niewiesk, S.; et al. In Vivo Efficacy of Measles Virus Fusion Protein-Derived Peptides Is Modulated by the Properties of Self-Assembly and Membrane Residence. *J. Virol.* **2017**, *91*. [CrossRef]

146. Welsch, J.C.; Talekar, A.; Mathieu, C.; Pessi, A.; Moscona, A.; Horvat, B.; Porotto, M. Fatal Measles Virus Infection Prevented by Brain-Penetrant Fusion Inhibitors. *J. Virol.* **2013**, *87*, 13785–13794. [CrossRef]

147. Van Binnendijk, R.S.; van der Heijden, R.W.J.; van Amerongen, G.; UytdeHaag, F.G.C.M.; Osterhaus, A.D.M.E. Viral Replication and Development of Specific Immunity in Macaques after Infection with Different Measles Virus Strains. *J. Infect. Dis.* **1994**, *170*, 443–448. [CrossRef]

148. Devaux, P.; Hudacek, A.W.; Hodge, G.; Reyes-del Valle, J.; McChesney, M.B.; Cattaneo, R. A Recombinant Measles Virus Unable To Antagonize STAT1 Function Cannot Control Inflammation and Is Attenuated in Rhesus Monkeys. *J. Virol.* **2011**, *85*, 348–356. [CrossRef]

149. Leonard, V.H.J.; Sinn, P.L.; Hodge, G.; Miest, T.; Devaux, P.; Oezguen, N.; Braun, W.; McCray, P.B.; McChesney, M.B.; Cattaneo, R.; et al. Measles Virus Blind to Its Epithelial Cell Receptor Remains Virulent in Rhesus Monkeys but Cannot Cross the Airway Epithelium and Is Not Shed. *J. Clin. Investig.* **2008**, *118*, 2448–2458. [CrossRef]

150. El Mubarak, H.S.; Yuksel, S.; van Amerongen, G.; Mulder, P.G.H.; Mukhtar, M.M.; Osterhaus, A.D.M.E.; de Swart, R.L. Infection of Cynomolgus Macaques (Macaca Fascicularis) and Rhesus Macaques (Macaca Mulatta) with Different Wild-Type Measles Viruses. *J. Gen. Virol.* **2007**, *88*, 2028–2034. [CrossRef]

151. Remfry, J. A Measles Epizootic with 5 Deaths in Newly-Imported Rhesus Monkeys (*Macaca Mulatta*). *Lab. Anim.* **1976**, *10*, 49–57. [CrossRef]

152. Choi, Y.K.; Simon, M.A.; Kim, D.Y.; Yoon, B.I.; Kwon, S.W.; Lee, K.W.; Seo, I.B.; Kim, D.Y. Fatal Measles Virus Infection in Japanese Macaques (Macaca Fuscata). *Vet. Pathol.* **1999**, *36*, 594–600. [CrossRef]

153. Schumacher, H.P.; Albrecht, P.; Clark, R.G.; Kirschstein, R.L.; Tauraso, N.M. Intracerebral Inoculation of Rhesus Monkeys with a Strain of Measles Virus Isolated from a Case of Subacute Sclerosing Panencephalitis. *Infect. Immun.* **1971**, *4*, 419–424.

154. Albrecht, P.; Shabo, A.L.; Burns, G.R.; Tauraso, N.M. Experimental Measles Encephalitis in Normal and Cyclophosphamide-Treated Rhesus Monkeys. *J. Infect. Dis.* **1972**, *126*, 154–161. [CrossRef]

155. da Fontoura Budaszewski, R.; von Messling, V. Morbillivirus Experimental Animal Models: Measles Virus Pathogenesis Insights from Canine Distemper Virus. *Viruses* **2016**, *8*, 274. [CrossRef]

156. Lorenz, D.; Albrecht, P. Susceptibility of Tamarins (Saguinus) to Measles Virus. *Lab. Anim. Sci.* **1980**, *30*, 661–665.

157. Albrecht, P.; Lorenz, D.; Klutch, M.J.; Vickers, J.H.; Ennis, F.A. Fatal Measles Infection in Marmosets Pathogenesis and Prophylaxis. *Infect. Immun.* **1980**, *27*, 969–978.

158. Albrecht, P.; Lorenz, D.; Klutch, M.J. Encephalitogenicity of Measles Virus in Marmosets. *Infect. Immun.* **1981**, *34*, 581–587.

159. Oldstone, M.B.; Lewicki, H.; Thomas, D.; Tishon, A.; Dales, S.; Patterson, J.; Manchester, M.; Homann, D.; Naniche, D.; Holz, A. Measles Virus Infection in a Transgenic Model: Virus-Induced Immunosuppression and Central Nervous System Disease. *Cell* **1999**, *98*, 629–640. [CrossRef]
160. Rall, G.F.; Manchester, M.; Daniels, L.R.; Callahan, E.M.; Belman, A.R.; Oldstone, M.B. A Transgenic Mouse Model for Measles Virus Infection of the Brain. *Proc. Natl. Acad. Sci. USA* **1997**, *94*, 4659–4663. [CrossRef] [PubMed]
161. Blixenkrone-Møller, M.; Bernard, A.; Bencsik, A.; Sixt, N.; Diamond, L.E.; Logan, J.S.; Wild, T.F. Role of CD46 in Measles Virus Infection in CD46 Transgenic Mice. *Virology* **1998**, *249*, 238–248. [CrossRef] [PubMed]
162. Evlashev, A.; Moyse, E.; Valentin, H.; Azocar, O.; Trescol-Biemont, M.-C.; Marie, J.C.; Rabourdin-Combe, C.; Horvat, B. Productive Measles Virus Brain Infection and Apoptosis in CD46 Transgenic Mice. *J. Virol.* **2000**, *74*, 1373–1382. [CrossRef] [PubMed]
163. Mrkic, B.; Pavlovic, J.; Rülicke, T.; Volpe, P.; Buchholz, C.J.; Hourcade, D.; Atkinson, J.P.; Aguzzi, A.; Cattaneo, R. Measles Virus Spread and Pathogenesis in Genetically Modified Mice. *J. Virol.* **1998**, *72*, 7420–7427.
164. Hahm, B.; Arbour, N.; Oldstone, M.B. Measles Virus Interacts with Human SLAM Receptor on Dendritic Cells to Cause Immunosuppression. *Virology* **2004**, *323*, 292–302. [CrossRef]
165. Hahm, B.; Cho, J.-H.; Oldstone, M.B.A. Measles Virus-Dendritic Cell Interaction via SLAM Inhibits Innate Immunity: Selective Signaling through TLR4 but Not Other TLRs Mediates Suppression of IL-12 Synthesis. *Virology* **2007**, *358*, 251–257. [CrossRef]
166. Shingai, M.; Inoue, N.; Okuno, T.; Okabe, M.; Akazawa, T.; Miyamoto, Y.; Ayata, M.; Honda, K.; Kurita-Taniguchi, M.; Matsumoto, M.; et al. Wild-Type Measles Virus Infection in Human CD46/CD150-Transgenic Mice: CD11c-Positive Dendritic Cells Establish Systemic Viral Infection. *J. Immunol.* **2005**, *175*, 3252–3261. [CrossRef]
167. Ohno, S.; Ono, N.; Seki, F.; Takeda, M.; Kura, S.; Tsuzuki, T.; Yanagi, Y. Measles Virus Infection of SLAM (CD150) Knockin Mice Reproduces Tropism and Immunosuppression in Human Infection. *J. Virol.* **2007**, *81*, 1650–1659. [CrossRef]
168. Welsch, J.C.; Charvet, B.; Dussurgey, S.; Allatif, O.; Aurine, N.; Horvat, B.; Gerlier, D.; Mathieu, C. Type I Interferon Receptor Signaling Drives Selective Permissiveness of Astrocytes and Microglia to Measles Virus during Brain Infection. *J. Virol.* **2019**, *93*. [CrossRef]
169. Welstead, G.G.; Iorio, C.; Draker, R.; Bayani, J.; Squire, J.; Vongpunsawad, S.; Cattaneo, R.; Richardson, C.D. Measles Virus Replication in Lymphatic Cells and Organs of CD150 (SLAM) Transgenic Mice. *Proc. Natl. Acad. Sci. USA* **2005**, *102*, 16415–16420. [CrossRef]
170. Chan, S.P.K. Induction of Chronic Measles Encephalitis in C57BL/6 Mice. *J. Gen. Virol.* **1985**, *66*, 2071–2076. [CrossRef]
171. Liebert, U.G. Measles Virus Infections of the Central Nervous System. *Intervirology* **1997**, *40*, 176–184. [CrossRef] [PubMed]
172. Bankamp, B.; Brinckmann, U.G.; Reich, A.; Niewiesk, S.; ter Meulen, V.; Liebert, U.G. Measles Virus Nucleocapsid Protein Protects Rats from Encephalitis. *J. Virol.* **1991**, *65*, 1695–1700. [PubMed]
173. Niewiesk, S. Cotton Rats (Sigmodon Hispidus): An Animal Model to Study the Pathogenesis of Measles Virus Infection. *Immunol. Lett.* **1999**, *65*, 47–50. [CrossRef]
174. Niewiesk, S.; Schneider-Schaulies, J.; Ohnimus, H.; Jassoy, C.; Schneider-Schaulies, S.; Diamond, L.; Logan, J.S.; ter Meulen, V. CD46 Expression Does Not Overcome the Intracellular Block of Measles Virus Replication in Transgenic Rats. *J. Virol.* **1997**, *71*, 7969–7973.
175. Liebert, U.G.; Meulen, V.T. Virological Aspects of Measles Virus-Induced Encephalomyelitis in Lewis and BN Rats. *J. Gen. Virol.* **1987**, *68*, 1715–1722. [CrossRef]
176. Katz, M.; Rorke, L.B.; Masland, W.S.; Brodano, G.B.; Koprowski, H. Subacute Sclerosing Panencephalitis: Isolation of a Virus Encephalitogenic for Ferrets. *J. Infect. Dis.* **1970**, *121*, 188–195. [CrossRef]
177. Thormar, H.; Mehta, P.D.; Lin, F.H.; Brown, H.R.; Wisniewski, H.M. Presence of Oligoclonal Immunoglobulin G Bands and Lack of Matrix Protein Antibodies in Cerebrospinal Fluids and Sera of Ferrets with Measles Virus Encephalitis. *Infect. Immun.* **1983**, *41*, 1205–1211.
178. Cosby, S.L.; Brankin, B. Measles Virus Infection of Cerebral Endothelial Cells and Effect on Their Adhesive Properties. *Vet. Microbiol.* **1995**, *44*, 135–139. [CrossRef]

179. Laksono, B.M.; de Vries, R.D.; McQuaid, S.; Duprex, W.P.; de Swart, R.L. Measles Virus Host Invasion and Pathogenesis. *Viruses* **2016**, *8*, 210. [CrossRef]
180. Bechmann, I.; Galea, I.; Perry, V.H. What Is the Blood-Brain Barrier (Not)? *Trends Immunol.* **2007**, *28*, 5–11. [CrossRef]
181. Allen, I.V.; McQuaid, S.; McMahon, J.; Kirk, J.; McConnell, R. The Significance of Measles Virus Antigen and Genome Distribution in the CNS in SSPE for Mechanisms of Viral Spread and Demyelination. *J. Neuropathol. Exp. Neurol.* **1996**, *55*, 471–480. [CrossRef] [PubMed]
182. Delpeut, S.; Noyce, R.S.; Siu, R.W.C.; Richardson, C.D. Host Factors and Measles Virus Replication. *Curr. Opin. Virol.* **2012**, *2*, 773–783. [CrossRef] [PubMed]
183. McQuaid, S.; Cosby, S.L. An Immunohistochemical Study of the Distribution of the Measles Virus Receptors, CD46 and SLAM, in Normal Human Tissues and Subacute Sclerosing Panencephalitis. *Lab. Investig.* **2002**, *82*, 403–409. [CrossRef] [PubMed]
184. Thormar, H.; Brown, H.R.; Goller, N.L.; Barshatzky, M.R.; Wisniewski, H.M. Transmission of Measles Virus Encephalitis to Ferrets by Intracardiac Inoculation of a Cell-Associated SSPE Virus Strain. *APMIS* **1988**, *96*, 1125–1128. [CrossRef]
185. Woyciechowska, J.; Breschkin, A.M.; Rapp, F. Measles Virus Meningoencephalitis. Immunofluorescence Study of Brains Infected with Virus Mutants. *Lab. Investig.* **1977**, *36*, 233–236.
186. Shirogane, Y.; Hashiguchi, T.; Yanagi, Y. Weak Cis and Trans Interactions of the Hemagglutinin with Receptors Trigger Fusion Proteins of Neuropathogenic Measles Virus Isolates. *J. Virol.* **2019**. [CrossRef]
187. Hofman, F.M.; Hinton, D.R.; Baemayr, J.; Weil, M.; Merrill, J.E. Lymphokines and Immunoregulatory Molecules in Subacute Sclerosing Panencephalitis. *Clin. Immunol. Immunopathol.* **1991**, *58*, 331–342. [CrossRef]
188. Manchester, M.; Eto, D.S.; Oldstone, M.B. Characterization of the Inflammatory Response during Acute Measles Encephalitis in NSE-CD46 Transgenic Mice. *J. Neuroimmunol.* **1999**, *96*, 207–217. [CrossRef]
189. Welsch, J.; Lionnet, C.; Terzian, C.; Horvat, B.; Gerlier, D.; Mathieu, C. Organotypic Brain Cultures: A Framework for Studying CNS Infection by Neurotropic Viruses and Screening Antiviral Drugs. *Bio-Protocol* **2017**, *7*. [CrossRef]
190. Sheppard, R.D.; Raine, C.S.; Burnstein, T.; Bornstein, M.B.; Feldman, L.A. Cell-Associated Subacute Sclerosing Panencephalitis Agent Studied in Organotypic Central Nervous System Cultures: Viral Rescue Attempts and Morphology. *Infect. Immun.* **1975**, *12*, 891–900. [PubMed]
191. Preibisch, S.; Saalfeld, S.; Tomancak, P. Globally Optimal Stitching of Tiled 3D Microscopic Image Acquisitions. *Bioinformatics* **2009**, *25*, 1463–1465. [CrossRef] [PubMed]
192. Lawrence, D.M.; Patterson, C.E.; Gales, T.L.; D'Orazio, J.L.; Vaughn, M.M.; Rall, G.F. Measles Virus Spread between Neurons Requires Cell Contact but Not CD46 Expression, Syncytium Formation, or Extracellular Virus Production. *J. Virol.* **2000**, *74*, 1908–1918. [CrossRef]
193. O'Donnell, L.A.; Conway, S.; Rose, R.W.; Nicolas, E.; Slifker, M.; Balachandran, S.; Rall, G.F. STAT1-Independent Control of a Neurotropic Measles Virus Challenge in Primary Neurons and Infected Mice. *J. Immunol.* **2012**, *188*, 1915–1923. [CrossRef]
194. Duprex, W.P.; McQuaid, S.; Roscic-Mrkic, B.; Cattaneo, R.; McCallister, C.; Rima, B.K. In Vitro and in Vivo Infection of Neural Cells by a Recombinant Measles Virus Expressing Enhanced Green Fluorescent Protein. *J. Virol.* **2000**, *74*, 7972–7979. [CrossRef]
195. Ludlow, M.; McQuaid, S.; Cosby, S.L.; Cattaneo, R.; Rima, B.K.; Duprex, W.P. Measles Virus Superinfection Immunity and Receptor Redistribution in Persistently Infected NT2 Cells. *J. Gen. Virol.* **2005**, *86*, 2291–2303. [CrossRef]
196. McQuaid, S.; Campbell, S.; Wallace, I.J.; Kirk, J.; Cosby, S.L. Measles Virus Infection and Replication in Undifferentiated and Differentiated Human Neuronal Cells in Culture. *J. Virol.* **1998**, *72*, 5245–5250.
197. Brawner, A.T.; Xu, R.; Liu, D.; Jiang, P. Generating CNS Organoids from Human Induced Pluripotent Stem Cells for Modeling Neurological Disorders. *Int. J. Physiol. Pathophysiol. Pharmacol.* **2017**, *9*, 101–111.
198. Qian, X.; Nguyen, H.N.; Song, M.M.; Hadiono, C.; Ogden, S.C.; Hammack, C.; Yao, B.; Hamersky, G.R.; Jacob, F.; Zhong, C.; et al. Brain-Region-Specific Organoids Using Mini-Bioreactors for Modeling ZIKV Exposure. *Cell* **2016**, *165*, 1238–1254. [CrossRef]
199. Sabella, C. Measles: Not Just a Childhood Rash. *Cleve. Clin. J. Med.* **2010**, *77*, 207–213. [CrossRef] [PubMed]

200. Ordman, C.W.; Jennings, C.G.; Janeway, C.A. Chemical, Clinical, and Immunological Studies on the Products of Human Plasma Fractionation. XII. The Use of Concentrated Normal Human Serum Gamma Globulin (Human Immune Serum Globulin) in the Prevention and Attenuation of Measles. *J. Clin. Investig.* **1944**, *23*, 541–549. [CrossRef] [PubMed]
201. Bigham, M.; Murti, M.; Fung, C.; Hemming, F.; Loadman, S.; Stam, R.; Van Buynder, P.; Lem, M. Estimated Protective Effectiveness of Intramuscular Immune Serum Globulin Post-Exposure Prophylaxis during a Measles Outbreak in British Columbia, Canada, 2014. *Vaccine* **2017**, *35*, 2723–2727. [CrossRef] [PubMed]
202. Maldonado, Y.A.; Lawrence, E.C.; DeHovitz, R.; Hartzell, H.; Albrecht, P. Early Loss of Passive Measles Antibody in Infants of Mothers with Vaccine-Induced Immunity. *Pediatrics* **1995**, *96*, 447–450. [PubMed]
203. Rammohan, K.W.; McFarland, H.F.; McFarlin, D.E. Subacute Sclerosing Panencephalitis after Passive Immunization and Natural Measles Infection: Role of Antibody in Persistence of Measles Virus. *Neurology* **1982**, *32*, 390–394. [CrossRef] [PubMed]
204. Liebert, U.G.; Schneider-Schaulies, S.; Baczko, K.; ter Meulen, V. Antibody-Induced Restriction of Viral Gene Expression in Measles Encephalitis in Rats. *J. Virol.* **1990**, *64*, 706–713. [PubMed]
205. Rammohan, K.W.; McFarland, H.F.; McFarlin, D.E. Induction of Subacute Murine Measles Encephalitis by Monoclonal Antibody to Virus Haemagglutinin. *Nature* **1981**, *290*, 588–589. [CrossRef]
206. Wear, D.J.; Rapp, F. Latent Measles Virus Infection of the Hamster Central Nervous System. *J. Immunol.* **1971**, *107*, 1593–1598.
207. Mori, K.; Hiraoka, O.; Ikeda, M.; Ariumi, Y.; Hiramoto, A.; Wataya, Y.; Kato, N. Adenosine Kinase Is a Key Determinant for the Anti-HCV Activity of Ribavirin. *Hepatology* **2013**, *58*, 1236–1244. [CrossRef]
208. Crotty, S.; Cameron, C.; Andino, R. Ribavirin's Antiviral Mechanism of Action: Lethal Mutagenesis? *J. Mol. Med.* **2002**, *80*, 86–95. [CrossRef]
209. Stogner, S.W.; King, J.W.; Black-Payne, C.; Bocchini, J. Ribavirin and Intravenous Immune Globulin Therapy for Measles Pneumonia in HIV Infection. *South. Med. J.* **1993**, *86*, 1415–1418. [CrossRef]
210. Solomon, T.; Hart, C.A.; Vinjamuri, S.; Beeching, N.J.; Malucci, C.; Humphrey, P. Treatment of Subacute Sclerosing Panencephalitis with Interferon-Alpha, Ribavirin, and Inosiplex. *J. Child Neurol.* **2002**, *17*, 703–705. [CrossRef]
211. Hosoya, M.; Mori, S.; Tomoda, A.; Mori, K.; Sawaishi, Y.; Kimura, H.; Shigeta, S.; Suzuki, H. Pharmacokinetics and Effects of Ribavirin Following Intraventricular Administration for Treatment of Subacute Sclerosing Panencephalitis. *Antimicrob. Agents Chemother.* **2004**, *48*, 4631–4635. [CrossRef]
212. Tomoda, A.; Nomura, K.; Shiraishi, S.; Hamada, A.; Ohmura, T.; Hosoya, M.; Miike, T.; Sawaishi, Y.; Kimura, H.; Takashima, H.; et al. Trial of Intraventricular Ribavirin Therapy for Subacute Sclerosing Panencephalitis in Japan. *Brain Dev.* **2003**, *25*, 514–517. [CrossRef]
213. Kwak, M.; Yeh, H.-R.; Yum, M.-S.; Kim, H.-J.; You, S.J.; Ko, T.-S. A Long-Term Subacute Sclerosing Panencephalitis Survivor Treated with Intraventricular Interferon-Alpha for 13 Years. *Korean J. Pediatr.* **2019**, *62*, 108–112. [CrossRef]
214. Miyazaki, K.; Hashimoto, K.; Suyama, K.; Sato, M.; Abe, Y.; Watanabe, M.; Kanno, S.; Maeda, H.; Kawasaki, Y.; Hosoya, M. Maintaining Concentration of Ribavirin in Cerebrospinal Fluid by a New Dosage Method; 3 Cases of Subacute Sclerosing Panencephalitis Treated Using a Subcutaneous Continuous Infusion Pump. *Pediatr. Infect. Dis. J.* **2019**, *38*, 496–499. [CrossRef]
215. Gokcil, Z.; Odabasi, Z.; Demirkaya, S.; Eroglu, E.; Vural, O. Alpha-Interferon and Isoprinosine in Adult-Onset Subacute Sclerosing Panencephalitis. *J. Neurol. Sci.* **1999**, *162*, 62–64. [CrossRef]
216. Gutierrez, J.; Issacson, R.S.; Koppel, B.S. Subacute Sclerosing Panencephalitis: An Update. *Dev. Med. Child Neurol.* **2010**, *52*, 901–907. [CrossRef]
217. Ravikumar, S.; Crawford, J.R. Role of Carbamazepine in the Symptomatic Treatment of Subacute Sclerosing Panencephalitis: A Case Report and Review of the Literature. *Case Rep. Neurol. Med.* **2013**, *2013*, 327647. [CrossRef]
218. Blank, T.; Prinz, M. Type I Interferon Pathway in CNS Homeostasis and Neurological Disorders. *Glia* **2017**, *65*, 1397–1406. [CrossRef]
219. Rima, B.K.; Davidson, W.B.; Martin, S.J. The Role of Defective Interfering Particles in Persistent Infection of Vero Cells by Measles Virus. *J. Gen. Virol.* **1977**, *35*, 89–97. [CrossRef] [PubMed]

220. Yount, J.S.; Gitlin, L.; Moran, T.M.; López, C.B. MDA5 Participates in the Detection of Paramyxovirus Infection and Is Essential for the Early Activation of Dendritic Cells in Response to Sendai Virus Defective Interfering Particles. *J. Immunol.* **2008**, *180*, 4910–4918. [CrossRef] [PubMed]
221. Bello, S.; Meremikwu, M.M.; Ejemot-Nwadiaro, R.I.; Oduwole, O. Routine Vitamin A Supplementation for the Prevention of Blindness Due to Measles Infection in Children. *Cochrane Database Syst. Rev.* **2016**. [CrossRef]
222. Barclay, A.J.; Foster, A.; Sommer, A. Vitamin A Supplements and Mortality Related to Measles: A Randomised Clinical Trial. *Br. Med. J. (Clin. Res. Ed).* **1987**, *294*, 294–296. [CrossRef]
223. Coutsoudis, A.; Broughton, M.; Coovadia, H.M. Vitamin A Supplementation Reduces Measles Morbidity in Young African Children: A Randomized, Placebo-Controlled, Double-Blind Trial. *Am. J. Clin. Nutr.* **1991**, *54*, 890–895. [CrossRef]
224. Hussey, G.D.; Klein, M. Routine High-Dose Vitamin A Therapy for Children Hospitalized with Measles. *J. Trop. Pediatr.* **1993**, *39*, 342–345. [CrossRef]
225. WHO. Measles Vaccines: WHO Position Paper—April 2017. *Wkly Epidemiol Rec.* **2017**, *92*, 205–227.
226. Bichon, A.; Aubry, C.; Benarous, L.; Drouet, H.; Zandotti, C.; Parola, P.; Lagier, J.-C. Case Report. *Medicine (Baltimore)* **2017**, *96*, e9154. [CrossRef]
227. Loo, Y.-M.; Gale, M. Immune Signaling by RIG-I-like Receptors. *Immunity* **2011**, *34*, 680–692. [CrossRef]
228. Gerlier, D.; Lyles, D.S. Interplay between Innate Immunity and Negative-Strand RNA Viruses: Towards a Rational Model. *Microbiol. Mol. Biol. Rev.* **2011**, *75*, 468–490. [CrossRef]
229. Plumet, S.; Herschke, F.; Bourhis, J.-M.; Valentin, H.; Longhi, S.; Gerlier, D. Cytosolic 5′-Triphosphate Ended Viral Leader Transcript of Measles Virus as Activator of the RIG I-Mediated Interferon Response. *PLoS ONE* **2007**, *2*, e279. [CrossRef]
230. Yoneyama, M.; Onomoto, K.; Jogi, M.; Akaboshi, T.; Fujita, T. Viral RNA Detection by RIG-I-like Receptors. *Curr. Opin. Immunol.* **2015**, *32*, 48–53. [CrossRef]
231. Trottier, C.; Chabot, S.; Mann, K.K.; Colombo, M.; Chatterjee, A.; Miller, W.H.; Ward, B.J. Retinoids Inhibit Measles Virus in Vitro via Nuclear Retinoid Receptor Signaling Pathways. *Antivir. Res.* **2008**, *80*, 45–53. [CrossRef]
232. Trottier, C.; Colombo, M.; Mann, K.K.; Miller, W.H.; Ward, B.J. Retinoids Inhibit Measles Virus through a Type I IFN-Dependent Bystander Effect. *FASEB J.* **2009**, *23*, 3203–3212. [CrossRef]
233. Soye, K.J.; Trottier, C.; Richardson, C.D.; Ward, B.J.; Miller, W.H. RIG-I Is Required for the Inhibition of Measles Virus by Retinoids. *PLoS ONE* **2011**, *6*, e22323. [CrossRef]
234. Kelly, J.T.; Human, S.; Alderman, J.; Jobe, F.; Logan, L.; Rix, T.; Gonçalves-Carneiro, D.; Leung, C.; Thakur, N.; Birch, J.; et al. BST2/Tetherin Overexpression Modulates Morbillivirus Glycoprotein Production to Inhibit Cell–Cell Fusion. *Viruses* **2019**, *11*, 692. [CrossRef]
235. Smith, S.E.; Busse, D.C.; Binter, S.; Weston, S.; Diaz Soria, C.; Laksono, B.M.; Clare, S.; Van Nieuwkoop, S.; Van den Hoogen, B.G.; Clement, M.; et al. Interferon-Induced Transmembrane Protein 1 Restricts Replication of Viruses That Enter Cells via the Plasma Membrane. *J. Virol.* **2018**, *93*. [CrossRef]
236. Barnard, D.L. Inhibitors of Measles Virus. *Antivir. Chem. Chemother.* **2004**, *15*, 111–119. [CrossRef]
237. Schönberger, K.; Ludwig, M.-S.; Wildner, M.; Weissbrich, B. Epidemiology of Subacute Sclerosing Panencephalitis (SSPE) in Germany from 2003 to 2009: A Risk Estimation. *PLoS ONE* **2013**, *8*, e68909. [CrossRef]
238. Otaki, M.; Sada, K.; Kadoya, H.; Nagano-Fujii, M.; Hotta, H. Inhibition of Measles Virus and Subacute Sclerosing Panencephalitis Virus by RNA Interference. *Antivir. Res.* **2006**, *70*, 105–111. [CrossRef]
239. Keita, D.; Servan de Almeida, R.; Libeau, G.; Albina, E. Identification and Mapping of a Region on the MRNA of Morbillivirus Nucleoprotein Susceptible to RNA Interference. *Antivir. Res.* **2008**, *80*, 158–167. [CrossRef]
240. Brunel, J.; Chopy, D.; Dosnon, M.; Bloyet, L.-M.; Devaux, P.; Urzua, E.; Cattaneo, R.; Longhi, S.; Gerlier, D. Sequence of Events in Measles Virus Replication: Role of Phosphoprotein-Nucleocapsid Interactions. *J. Virol.* **2014**, *88*, 10851–10863. [CrossRef]
241. Zinke, M.; Kendl, S.; Singethan, K.; Fehrholz, M.; Reuter, D.; Rennick, L.; Herold, M.J.; Schneider-Schaulies, J. Clearance of Measles Virus from Persistently Infected Cells by Short Hairpin RNA. *J. Virol.* **2009**, *83*, 9423–9431. [CrossRef]
242. Plumet, S.; Duprex, W.P.; Gerlier, D. Dynamics of Viral RNA Synthesis during Measles Virus Infection. *J. Virol.* **2005**, *79*, 6900–6908. [CrossRef]

243. Bloyet, L.-M.; Welsch, J.; Enchery, F.; Mathieu, C.; de Breyne, S.; Horvat, B.; Grigorov, B.; Gerlier, D. HSP90 Chaperoning in Addition to Phosphoprotein Required for Folding but Not for Supporting Enzymatic Activities of Measles and Nipah Virus L Polymerases. *J. Virol.* **2016**, *90*, 6642–6656. [CrossRef]

244. Geller, R.; Andino, R.; Frydman, J. Hsp90 Inhibitors Exhibit Resistance-Free Antiviral Activity against Respiratory Syncytial Virus. *PLoS ONE* **2013**, *8*, e56762. [CrossRef]

245. Lo, M.K.; Jordan, R.; Arvey, A.; Sudhamsu, J.; Shrivastava-Ranjan, P.; Hotard, A.L.; Flint, M.; McMullan, L.K.; Siegel, D.; Clarke, M.O.; et al. GS-5734 and Its Parent Nucleoside Analog Inhibit Filo-, Pneumo-, and Paramyxoviruses. *Sci. Rep.* **2017**, *7*, 43395. [CrossRef]

246. Jordan, P.C.; Liu, C.; Raynaud, P.; Lo, M.K.; Spiropoulou, C.F.; Symons, J.A.; Beigelman, L.; Deval, J. Initiation, Extension, and Termination of RNA Synthesis by a Paramyxovirus Polymerase. *PLoS Pathog.* **2018**, *14*, e1006889. [CrossRef]

247. Warren, T.K.; Jordan, R.; Lo, M.K.; Ray, A.S.; Mackman, R.L.; Soloveva, V.; Siegel, D.; Perron, M.; Bannister, R.; Hui, H.C.; et al. Therapeutic Efficacy of the Small Molecule GS-5734 against Ebola Virus in Rhesus Monkeys. *Nature* **2016**, *531*, 381–385. [CrossRef]

248. White, L.K.; Yoon, J.-J.; Lee, J.K.; Sun, A.; Du, Y.; Fu, H.; Snyder, J.P.; Plemper, R.K. Nonnucleoside Inhibitor of Measles Virus RNA-Dependent RNA Polymerase Complex Activity. *Antimicrob. Agents Chemother.* **2007**, *51*, 2293–2303. [CrossRef]

249. Yoon, J.-J.; Krumm, S.A.; Ndungu, J.M.; Hoffman, V.; Bankamp, B.; Rota, P.A.; Sun, A.; Snyder, J.P.; Plemper, R.K. Target Analysis of the Experimental Measles Therapeutic AS-136A. *Antimicrob. Agents Chemother.* **2009**, *53*, 3860–3870. [CrossRef]

250. Ndungu, J.M.; Krumm, S.A.; Yan, D.; Arrendale, R.F.; Reddy, G.P.; Evers, T.; Howard, R.; Natchus, M.G.; Saindane, M.T.; Liotta, D.C.; et al. Non-Nucleoside Inhibitors of the Measles Virus RNA-Dependent RNA Polymerase: Synthesis, Structure-Activity Relationships, and Pharmacokinetics. *J. Med. Chem.* **2012**, *55*, 4220–4230. [CrossRef]

251. Tahara, M.; Ohno, S.; Sakai, K.; Ito, Y.; Fukuhara, H.; Komase, K.; Brindley, M.A.; Rota, P.A.; Plemper, R.K.; Maenaka, K.; et al. The Receptor-Binding Site of the Measles Virus Hemagglutinin Protein Itself Constitutes a Conserved Neutralizing Epitope. *J. Virol.* **2013**, *87*, 3583–3586. [CrossRef]

252. Tadokoro, T.; Jahan, M.L.; Ito, Y.; Tahara, M.; Chen, S.; Imai, A.; Sugimura, N.; Yoshida, K.; Saito, M.; Ose, T.; et al. Biophysical Characterization and Single-chain Fv Construction of a Neutralizing Antibody to Measles Virus. *FEBS J.* **2019**, febs.14991. [CrossRef]

253. Plemper, R.K.; Erlandson, K.J.; Lakdawala, A.S.; Sun, A.; Prussia, A.; Boonsombat, J.; Aki-Sener, E.; Yalcin, I.; Yildiz, I.; Temiz-Arpaci, O.; et al. A Target Site for Template-Based Design of Measles Virus Entry Inhibitors. *Proc. Natl. Acad. Sci. USA* **2004**, *101*, 5628–5633. [CrossRef]

254. Ha, M.N.; Delpeut, S.; Noyce, R.S.; Sisson, G.; Black, K.M.; Lin, L.-T.; Bilimoria, D.; Plemper, R.K.; Privé, G.G.; Richardson, C.D. Mutations in the Fusion Protein of Measles Virus That Confer Resistance to the Membrane Fusion Inhibitors Carbobenzoxy-d-Phe-l-Phe-Gly and 4-Nitro-2-Phenylacetyl Amino-Benzamide. *J. Virol.* **2017**, *91*. [CrossRef]

255. Porotto, M.; Yokoyama, C.C.; Palermo, L.M.; Mungall, B.; Aljofan, M.; Cortese, R.; Pessi, A.; Moscona, A. Viral Entry Inhibitors Targeted to the Membrane Site of Action. *J. Virol.* **2010**, *84*, 6760–6768. [CrossRef]

256. Griffin, D.E. Emergence and Re-Emergence of Viral Diseases of the Central Nervous System. *Prog. Neurobiol.* **2010**, *91*, 95–101. [CrossRef]

Review

Antiviral Agents as Therapeutic Strategies Against Cytomegalovirus Infections

Shiu-Jau Chen [1,2,†], Shao-Cheng Wang [3,4,†] and Yuan-Chuan Chen [5,*]

[1]   Department of Neurosurgery, Mackay Memorial Hospital, Taipei 10491, Taiwan; chenshiujau@gmail.com
[2]   Department of Medicine, Mackay Medicine College, Taipei 25245, Taiwan
[3]   Jianan Psychiatric Center, Ministry of Health and Welfare, Tainan 71742, Taiwan; WShaocheng@gmail.com
[4]   Department of Mental Health, Johns Hopkins Bloomberg School of Public Health, Baltimore, MD 21205, USA
[5]   Program in Comparative Biochemistry, University of California, Berkeley, CA 94720, USA
*   Correspondence: yuchuan1022@gmail.com
†   These authors contributed equally to this work.

Received: 28 November 2019; Accepted: 20 December 2019; Published: 23 December 2019

Abstract: Cytomegalovirus (CMV) is a threat to human health in the world, particularly for immunologically weak patients. CMV may cause opportunistic infections, congenital infections and central nervous system infections. CMV infections are difficult to treat due to their specific life cycles, mutation, and latency characteristic. Despite recent advances, current drugs used for treating active CMV infections are limited in their efficacy, and the eradication of latent infections is impossible. Current antiviral agents which target the UL54 DNA polymerase are restricted because of nephrotoxicity and viral resistance. CMV also cannot be prevented or eliminated with a vaccine. Fortunately, letermovir which targets the human CMV (HCMV) terminase complex has been recently approved to treat CMV infections in humans. The growing point is developing antiviral agents against both lytically and latently infected cells. The nucleic acid-based therapeutic approaches including the external guide sequences (EGSs)-RNase, the clustered regularly interspaced short palindromic repeats (CRISPR)/CRISPR-associated protein 9 (Cas9) system and transcription activator-like effector nucleases (TALENs) are being explored to remove acute and/or latent CMV infections. HCMV vaccine is being developed for prophylaxis. Additionally, adoptive T cell therapy (ACT) has been experimentally used to combate drug-resistant and recurrent CMV in patients after cell and/or organ transplantation. Developing antiviral agents is promising in this area to obtain fruitful outcomes and to have a great impact on humans for the therapy of CMV infections.

Keywords: cytomegalovirus; acute/latent infection; congenital infection; antiviral agent; therapeutic strategies; nucleic acid-based therapeutic approach; HCMV vaccine; adoptive cell therapy

---

1. Introduction

1.1. Cytomegalovirus Overview

Cytomegalovirus (CMV) is a genus of *Herpesvirus* in the order Herpesvirales, in the family Herpesviridae, and in the subfamily Betaherpesvirinae. There are nine distinct human herpesvirus (HHV) species known to cause human diseases such as HHV-1, HHV-2, HHV-3, HHV-4, HHV-5, HHV-6A, HHV-6B, HHV-7, HHV-8 [1,2]. Human cytomegalovirus (HCMV, HHV-5), with a double-stranded DNA genome of about 230 kb, is the most studied one among all CMV. HCMV usually causes moderate or subclinical diseases in immunocompetent adults; however, it may lead to opportunistic infections to affect individuals whose immune functions are compromised or immature [3,4]. The primary target cells of HCMV are monocytes, lymphocytes, and epithelial cells, and its major sites of latency are peripheral monocytes and CD34+ progenitor cells. HCMV infection causes a broad range of diseases such as pneumonia, retinitis, gastrointestinal diseases, mental retardation and vascular disorders, and is a major cause of morbidity and mortality for humans [5–7]. After infection, HCMV is recurrent and competent to remain latent within the body over long periods [5,6]. In all patients, the reactivation of latent HCMV can damage tissues and lead to organ disease, and reactivated CMV may trigger indirect immunomodulatory effects to cause detrimental outcomes, including increased mortality and graft rejection of organ transplantation in recipients [8]. Furthermore, congenital infection is a major problem with HCMV in that it can result in a severe cytomegalic inclusion disease of the neonate, mucoepidermoid carcinoma, and other malignancies eventually [7,9].

1.2. CMV Molecular Biology

CMV structure mainly consist of DNA core, capsid, tegument and envelope from inside to outside. The genome is complexed helically to form a DNA core, which is enclosed in a capsid composed of a total of 162 capsomere protein subunits. The capsid with a diameter of 100 nm is surrounded by the tegument. The tegument is enclosed by a lipid bilayer envelope containing viral glycoproteins to give a final diameter of about 180 nm for mature infectious viral particles (virions) [10]. The tegument compartment contains most of the viral proteins, with the most abundant one being the lower matrix phosphoprotein 65 (pp65) which is also referred as unique long 83 (UL83). The function of the tegument proteins can be classified as follows: (1) proteins that play an important role for the assembly of virions during proliferation and the disassembly of the virions during entry (structural use) and (2) proteins that modulate the host cell responses for viral infection (non-structural use) [11]. The viral envelope surrounding the tegument contains more than 20 glycoproteins that are involved in the attachment and penetration of host cells. These structural proteins include glycoprotein B, H, L, M, N, and O. CMV productive infection results in the coordinated synthesis of proteins in three overlapping phases according to the time of synthesis after infection, that is, immediate-early (0 to 2 h), early (<24 h), and late (>24 h) viral proteins which are expressed by immediate-early, early and late genes, respectively [11]. Immediately after CMV infections, the immediate early genes transcribe and ensure the transcription of early genes, which encode proteins required for the viral replication. The late genes mainly code for structural proteins.

1.3. CMV Life Cycle

CMV infection will start once a virion attaches a host cell with specific receptors on the cellular surface. For a lytic infection pathway, following linking of viral envelope glycoproteins to host cell membrane receptors, the virions enter the hosts by receptor-mediated endocytosis and membrane fusion. The viral capsid decomposes to release viral DNA genome to manipulates host enzyme systems to make new virions. During symptomatic infection, infected cells express lytic genes to demonstrate a lytic pathway [12]. Nevertheless, instead of this, some viral genes may transcribe latency associated transcripts to accumulate in host cells. In this pattern, viruses can persist in host cells indefinitely to have a latent infection pathway. The primary infection may be accompanied by limited illness and long-term latency is often asymptomatic. For the lysogenic pathway, the viruses are persistent in the host, not causing any adverse reactions, but can be transmitted to other hosts by direct contact. When CMV are stimulated by explanation or their host immune system is suppressed, the dormant viruses can reactivate to begin generating large number of viral progenies to cause symptoms and diseases, described as the lytic life cycle [12,13].

Viral latency can be divided into two models, namely proviral latency and episomal latency. CMV is defined as the episomal latency model which is essentially quiescent in myeloid progenitor cells. It can be reactivated by differentiation, inflammation, immunosuppression or critical diseases [14]. Latency is a specific phase in CMV life cycles in which virions stop producing posterior to infection, but the viral genome has not been entirely removed from host cells, that is, CMV latency is referred to as the absence of virions, despite the detection of viral DNA in hosts. In some clinical cases, the reactivation of latent infections is likely to lead to health risk. The molecular mechanisms by which latency is established and maintained have been explored. However, our understanding of the biology of CMV latency and reactivation at the molecular level would be significantly strengthened through analyses of both experimental and natural latency using systematic approaches [14].

2. CMV Infection

2.1. Signs, Symptoms and Complications

Most CMV infections are silent and CMV rarely causes signs or symptoms in healthy people. Though CMV infection is usually ignored in healthy people, the diseases can be life-threatening for the immunocompromised, immunosuppressed and immunonaive patients, such as newborn infants, the elderly, the sick, acquired immunodeficiency syndromes (AIDS) patients, and organ transplant recipients. A mother who acquires an acute CMV infection during pregnancy can transmit viruses to her baby, and thereby the baby might experience signs and symptoms. People at higher risk of CMV infections encompass newborns infected through their mothers before birth, babies infected through breast milk and people with weakened immune systems such as organ transplantation recipients or immunodeficient patients. The major signs, symptoms and complications which CMV influence all individuals including healthy adults, people with weakened immunity and babies are shown in Table 1.

Table 1. CMV influence on individuals.

Individual	Major Signs and Symptoms	Complications
Healthy adult	Fatigue, fever, sore throat, muscle aches	problems with the digestive system, liver, brain and nervous system
People with weakened immunity	Problems affecting eyes, lungs, liver, esophagus, stomach, intestines, brain	Vision loss due to the retinitis inflammation, digestive system problems including inflammation of the colon, esophagus and liver, nervous system problems including encephalitis and myelitis, pneumonia
Baby	Premature birth, low birth weight, jaundice (yellow skin and eyes), enlarged and poor liver function, purple skin splotches and/or rashes, microencephaly (abnormally small head), enlarged spleen, pneumonia, seizures	Hearing loss, intellectual disability, vision problems, seizures, lack of coordination, muscle weakness

2.2. Congenital Infection and Sequelae

CMV is transmitted by close interpersonal contact such as saliva, semen, urine, breast milk, or vertically transmission which viruses pass the placenta and directly infect the fetus [15,16]. CMV is the leading cause of congenital viral infection [17–20]. CMV infection is mostly or mildly asymptomatic among the general population (85%–90%). However, around 10%–15% of infants with the congenital infection may be at risk of sequelae such as mental retardation, jaundice, hepatosplenomegaly, microcephaly, hearing impairment and thrombocytopenia [21–24]. Among the above sequelae, the most devastating one is the central nervous system (CNS) sequelae related to neurodevelopment in that CNS injury is irreversible and persists for life, including mental retardation, seizures, hearing loss, ocular abnormalities and cognitive impairment [25–27]. That means the asymptomatic newborns with CMV infection still have an increased risk for long-term sequelaes, especially, mental retardation and sensorineural hearing loss (SNHL) [28–31], making CMV the leading nonhereditary cause of SNHL [24,32]. CMV can undermine both adaptive and innate immunity, silencing natural killer (NK) cells and inhibiting T cells to present viral antigens [33–35].

3. CMV Anti-Viral Drugs

At present, some antiviral drugs have been approved for the treatment of CMV infections clinically. Current available drugs for antiviral therapy of CMV infections include the inhibitors of viral DNA polymerase, such as the nucleoside analog ganciclovir, the nucleotide analog cidofovir, and the pyrophosphate analogue foscarnet [36]. All these drugs have low oral bioavailability and dose-related toxicities, and therefore new antiviral agents with improved efficacy and fewer side effects need to be developed. Several drugs with anti-HCMV activity are preclinically or clinically evaluated, including a series of benzimidazole riboside compounds showing efficient inhibition in the process of HCMV replication such as genomic DNA maturation. Another attractive inhibitor candidate was the phosphorothioate oligonucleotide fomivirsen, which specifically binds to sequences complementary to CMV major immediate-early transcription sites so that it inhibits the viral gene expression. However, these inhibitor compounds are currently waiting for further examination before they can be used in clinics [17].

Currently, ganciclovir is still the first treatment of choice for CMV infections. Letermovir has been approved for the prophylaxis of CMV infections in patients. Several new drugs were developed but still failed in the phase III and more clinical trials would be needed, including maribavir and brincidofovir [37]. Valnoctamide, a neuroactive mood stabilizer which inhibits CMV infection in the developing brain and attenuates neurobehavioral dysfunctions, was shown to have anti-CMV potential [38].

3.1. Letermovir

Letermovir is a novel antiviral drug which has been approved by the USA Food and Drug Administration (FDA) through a fast track procedure and granted as an orphan drug by the European Medicines Agency (EMA). The drug was tested in CMV infected patients and likely be useful for other patients who had organ transplantation or human immunodeficient virus (HIV) infections [39]. It has been clinically applied for CMV prophylaxis or treatment in hematopoietic stem cell recipients, thoracic organ recipients and lung transplantation recipients [40–42]. Letermovir has several advantages over conventional CMV antiviral agents as follows. Firstly, it can be given orally, so hospitalization and intravenous injection are not needed. Secondly, it is mild in toxicity, not related to myelotoxicity and nephrotoxicity [43,44]. Thirdly, it targets the CMV terminase complex instead of CMV DNA polymerase, so there is no risk to induce cross-resistance with existing anti-CMV drugs [44]. However, the CMV antiviral therapy will finally fail and acquired antiviral drug resistance is not avoidable if there is no immune control [45]. It should be noted that more data are required to provide insights of the mutations detected in vivo, interpretation of genotyping results, and outcomes of the clinical correlation. To provide useful information, it would be recommended to establish databases for the surveillance and interpretation of resistance for CMV [36].

3.2. Maribavir

Maribavir is a promising anti-HCMV compound which is administered orally; however, it is still under advanced clinical trials. The drug targets the viral kinase UL97 which is crucial for the formation of viral teguments and assembly complexes for virion releasing [36]. However, it is not recommended to co-administer maribavir and ganciclovir both in that maribavir is an inhibitor of the UL97 enzyme which is required for the assimilation of ganciclovir. Maribavir potentially substitutes for other traditional anti-HCMV drugs because of its reduced haematotoxicity and nephrotoxicity compared with ganciclovir and valganciclovir [36].

Maertens et al. used maribavir (dose-blinded) to treat cytomegalovirus reactivation preemptively for recipients of hematopoietic cell or solid organ transplants (SOT) (≥18 years old) with CMV reactivation in a phase II and open-label clinical trial [46]. The results showed that maribavir at a dose of at least 400 mg twice daily had efficacy like that of valganciclovir for removing CMV viremia. Though a higher incidence of gastrointestinal adverse events were found in the maribavir -treated group, the neutropenia incidence was lower [46].

Papanicolaou et al. used dose-blinded maribavir 400, 800, or 1200 mg twice-daily for up to 24 weeks to treat hematopoietic-cell or SOT recipients (≥12 years old) with refractory or resistant CMV infections in a phase II and double-blind clinical trial [47]. The result revealed that it was active to against refractory or resistant CMV infections using maribavir more than 400 mg twice daily in transplant recipients and no new safety signals were identified in this trial [47].

4. CMV Inhibition by Nucleic Acid-Based Therapeutic Approaches

The treatment of diseases caused by CMV is quite challenging because of high mutation rates and latency. Thus, infection is still a serious threat to humans. Fortunately, external guide sequences (EGSs), transcription activator-like effectors nucleases (TALENs) and the clustered regularly interspaced short palindromic repeats (CRISPRs)/CRISPR-associated 9 (Cas9) nuclease system might provide effective therapeutic strategies to treat diseases caused by CMV through designing a specific DNA or RNA sequence that target essential genes for viral growth. However, the effective modification of the viral genome avoiding off-target effects and the option of escape variants ignoring the editing of these approaches are required for successful clinical application.

4.1. EGS-RNase

Ribonuclease P (RNase P) is a unique RNases in that it is a ribozyme – an RNA that acts as a catalyst which is somewhat like a protein enzyme. Its function is to cut an extra or precursor sequence of RNA on transfer RNA (tRNA) molecules; that is, to catalyze the cleavage of precursor tRNA into active tRNA without any protein component. RNase P has the activity in cleaving the 5′ leader sequence of precursor tRNA. EGSs signify the short RNAs that induce RNase P to specifically cleave a target mRNA by forming a precursor tRNA-like complex. Therefore, EGS technology probably acts as an effective strategy for gene-targeting therapy.

Deng et al. reported that engineered EGS variants induced RNase P to efficiently hydrolyze target mRNAs which code for HCMV major capsid protein [48]. In vitro, the engineered EGS variant was more efficient in inducing human RNase P-mediated cleavage of the target mRNA than a natural tRNA-derived EGS by about 80-fold. In cells infected with HCMV, the EGS variant and natural EGSs resulted in HCMV gene expression reduction rate by about 98% and 73%, and the viral growth was inhibited by about 10,000 and 200-fold, respectively. The results showed that the EGS variant has higher efficiency in blocking the expression of HCMV genes and viral growth, compared with the natural EGS [48].

Li et al. explored the antiviral effects of an engineered EGS variant in targeting the shared mRNA sequence which codes for capsid scaffolding proteins (mCSP) and assemblins of murine CMV (MCMV) in the animals [49]. In vitro, the EGS variant was more active in directing RNase P cleavage of the target mRNA than a natural tRNA by 60-fold. In MCMV-infected cells, the EGS variant decreased mCSP expression by about 92% and inhibited viral growth by about 8000-fold. In MCMV-infected mice, the EGS variants were more effective in reducing mCSP expression, decreasing viral production, and increasing animal survival, compared with the natural EGS. The results demonstrated that the EGS variant with higher targeting activity in vitro are also more effective in inhibiting MCMV gene expression in mice [49].

4.2. CRISPR/Cas9

In CRISPR/Cas9 system, CRISPR is used to build RNA-guided genes drives to target a specific DNA sequence. By the Cas proteins and a specifically designed single-guiding RNA (sgRNA), the genome can be cut at most locations with only the limitation of a protospacer adjacent motif (PAM) sequence (NGG) existing in the target site [12]. CRISPR/Cas9 has been extensively used as an effective technique of gene editing for engineering or modifying specific genes. It was shown to successfully work as an efficient genome editing tool in a wide range of organisms including HCMV [50]. Consequently, it hints that the CRISPR/Cas9 can be a potential antiviral agent for the treatment of CMV infections.

Gergen et al. designed two CRISPR/Cas9 systems which contain three sgRNAs to target the HCMV UL122/123 gene crucial for the regulation of lytic replication and reactivation from latency [51]. Both systems caused mutations in the target gene and an accompanying reduction of immediate early gene expression in primary fibroblasts. The singleplex strategy caused 50% of insertions and/or deletions (indels) in the viral genome to appear in further detailed analyses in U-251 MG cells, resulting in a reduction in immediately early protein production. The multiplex strategy cleaved the immediate early gene in 90% of viral genomes and thereby inhibited immediate early gene expression. Therefore, viral genome replication and late protein expression were reduced by 90%. The multiplex CRISPR/Cas9 system can target the HCMV UL122/123 gene efficiently and prevent viral replication significantly [51].

van Diemen et al. observed that the clear depletion emerged in the anti-HCMV sgRNA expressing cells targeting essential genes UL57 and UL70 (1.3% and 4.6% mutants with frameshifts, respectively), compared with the sgRNAs targeting the nonessential genes US7 and US11 (83.5% and 85.8% mutants with frameshifts, respectively) [52]. The sequence complexity of the mutants selected upon UL57 and UL70 targeting was low, this suggested the selection of few suitable variants and subsequent expansion of infectious mutants need a lot of time. The results showed that the CRISPR/Cas9 is a promising strategy to restrict HCMV replication [52].

4.3. TALENs

Transcription activator-like effectors (TALEs) are crucial virulence factors that function as transcriptional activators in the cell nucleus of plants, where they directly bind to DNA via a central domain of tandem repeats [53]. Currently, TALENs were shown to be an effective tool for precise genome engineering with low toxicity and could be engineered to adapt for an antiviral strategy [54]. Hence, TALENs are likely to become part of a new approach for the treatment of CMV infections.

Chen et al. utilized three pairs of TALEN plasmids (MCMV1–2, 3–4, and 5–6) to target the MCMV M80 and M80.5 overlapping (M80/80.5) sequence to test their efficacy in blocking MCMV lytic replication in NIH3T3 cells [54]. Using lipofectamine or a specific lipoid NKS11 as transfection reagents, TALEN plasmids could specifically target the M80/80.5 sequence and effectively inhibit MCMV growth in cell culture when the plasmid transfection is prior to the MCMV infection. Using NKS11 which was previously proved to be nontoxic to mice as a transfection reagent, the most specific pairs of TALEN plasmids (MCMV3–4) showed that its competency to inhibit the replication and gene expression of latent MCMV in immunocompetent Balb/c mice. The administration of MCMV3–4 plasmids resulted in significant reduction in the copy number level of immediately early gene-1 DNA which is key to viral latency in mice, compared with the controls. Additionally, the innate immune DNA-sensing pathways of host might be involved in the induction of cytokine secretion such as type I interferon (IFN) (mainly IFN α and β) to fight against invading viruses. The result hinted that TALENs were able to provide an effective strategy to clear latent MCMV in animals [54].

5. HCMV Vaccines

The vaccines against HCMV are still being developed, no licensed vaccine is available so far. It is necessary to have a specific and strong antibody and cell-mediated immunity to confer protection against HCMV primary infection through the analysis of the immune response to HCMV. Many efforts have been made to produce an HCMV vaccine for years, but a successful vaccine candidate has not yet to be developed, probably due to what immune responses needed for protecting against HCMV infections are still poorly understood [36]. To develop an effective HCMV vaccine, immune responses required to fight against HCMV and how to enhance these specific immune responses requires further study.

Choi et al. used the guinea pig which is a small-animal model to develop a CMV vaccine. A glycoprotein pentamer complex encoded by guinea pig cytomegalovirus (GPCMV) is essential for viral entry into non-fibroblast cells to enable congenital CMV [55]. Like HCMV, GPCMV needs a guinea pig specific cell receptor (platelet-derived growth factor receptor α) for fibroblast entry, but other receptors are required for non-fibroblast cells. A disabled infectious GPCMV vaccine strain induced humoral immune responses against viral pentamers to promote neutralization on non-fibroblast cells; thus, the vaccinated guinea pigs were protected from congenital CMV infections. The design including the pentamer complex as a part of vaccines may significantly enhance efficacy. This new finding lays stress on the importance of the immune response to the pentamer complex in contributing to the protection against congenital CMV and has opened a new era for the development of CMV vaccines [55].

Liu et al. hypothesized that a vaccine candidate able to elicit immune responses analogous to those of HCMV-seropositive subjects may confer protection against congenital HCMV [56]. The V160 vaccine has been shown to be safe and immunogenic in HCMV-seronegative humans, inducing both humoral and cell-mediated immune responses. In this study, they further demonstrated that sera from V160-immunized HCMV-seronegative subjects had similar quality attributes to those from seropositive subjects, including high avidity antibodies to viral antigens. This vaccine is a promising candidate against HCMV, but further evaluation in clinics for the prevention of congenital HCMV is required to warrant its safety, efficiency, and effectiveness [56].

6. Adoptive Cell Therapy for CMV Infections

It is known that the infection-related morbidity and mortality will be increased, if the T cell-mediated immune responses are impaired in transplantation recipients. Virus-specific T cells capable of targeting a variety of pathogens in patients after hematopoietic stem cell transplantation (HSCT) have demonstrated potential efficacy for multiple viruses such as CMV, Epstein-Barr Virus (EBV) and adenovirus [57]. Adoptive T cell therapy (ACT), a type of immunotherapy in which T cells are given to a patient to treat diseases, has been developed to fight against drug resistant and recurrent CMV in SOT recipients. Therefore, ACT has become one of the therapeutic strategies for CMV reactivations in patients undergoing allogeneic HSCT and SOT.

Faist et al. used the peptide specific proliferation assay (PSPA) to study CMV specific central memory T cells (TCM) repertoires and determined their functional and reproductive abilities in vitro [58]. In the animal model, the pathogen-specific TCM has demonstrated to have protective ability even at low numbers and could survive for long-term, proliferate extensively and show high plasticity after adoptive transfer. Though the clinical data showed that minimal doses of purified human CMV epitope-specific T cells are competent to remove viremia, it is still necessary to evaluate whether the human virus-specific TCM shows the same characteristic for ACT as mice. The results concluded that TCM had potential for prophylactic low-dose ACT. These good manufacturing practice (GMP)-compatible TCM could be used as a broad-spectrum antiviral T cell prophylaxis in allogeneic HSCT patients. In addition, PSPA would be a necessary tool for TCM characterization further during simultaneous immune monitoring [58].

Smith et al. applied high-throughput T cell receptor $V\beta$ sequencing and T cell functional profiling to demonstrate the influence of ACT on T cell repertoire remodeling in the pretherapy immunity and ACT products [59]. The clinical response was consistent with significant changes in the T cell receptor $V\beta$ landscape after therapy. This reconstitution was related to the emergence of effector memory T cells in responding patients, while nonresponding patients showed dramatic pretherapy T cell expansions with minimal change following ACT. The results revealed that immunological modulation following ACT required significant repertoire remodeling which might be damaged in nonresponding patients on account of the preexisting immune environment. Immunological interventions which controlled this environment were likely to improve clinical outcomes. ACT appears to be an advantageous strategy to restore immunological control against CMV affecting immunosuppressed patients such as SOT recipients [59].

7. Conclusions

CMV is the most frequent etiological factor for congenital infections and its infections are still global health problems of humans. Though CMV infections are often opportunistic, they sometimes cause serious diseases in healthy adults with weakened immunity and babies. Currently, no drugs are available for asymptomatic infants and for infants with CMV congenital infections to reduce related morbidity during the neonatal period. Some traditional antiviral drugs (e.g., ganciclovir, valganciclovir, cidofovir, foscarnet) are applied for the treatment of CMV acute infections, however, their efficacy is limited by side effects, cross-resistance and others. There is no effective cure for CMV infections, especially for latent infections. Promisingly, many novel antiviral drugs/agents and preventive/therapeutic strategies have been approved for clinical application (e.g., letermovir) or are being developed (e.g., maribavir, EGS-RNase, CRISPR/Cas9, TALENs, HCMV vaccine, ACT). We should investigate the interactions between CMV and hosts thoroughly to understand how antiviral agents or therapeutic strategies affect CMV infection outcomes. Moreover, the development and implications of novel antiviral agents and preventive/therapeutic strategies should be explored as extensively as possible. The future research tendency and application of these new insights should also be a highlight and could potentially become a promising milestone in the development of therapeutic strategies for CMV infections.

Author Contributions: S.-J.C., S.-C.W. and Y.-C.C. wrote this article. Y.-C.C. designed, organized and reviewed this article. All authors have read and agreed to the published version of the manuscript.

Funding: This research received no external funding.

Acknowledgments: We are grateful for the grant support from Mackay Memorial Hospital in publishing this article.

Conflicts of Interest: The authors declare no conflict of interests.

Abbreviations

CMV	cytomegalovirus
EGSs	external guide sequences
CRISPRs	clustered regularly interspaced short palindromic repeats
TALENs	transcription activator-like effector nucleases
ACT	adoptive T cell therapy
HHV	herpesvirus
AIDS	acquired immunodeficiency syndromes
HIV	human immunodeficiency virus
SOT	solid organ transplant
CNS	central nervous system
SNHL	sensorineural hearing loss
mCSP	mRNA sequence which codes for capsid scaffolding proteins
sgRNA	single-guiding RNA
PAM	protospacer adjacent motif
GPCMV	guinea pig cytomegalovirus
HSCT	hematopoietic stem cell transplantation
PSPA	peptide specific proliferation assay
TCM	specific central memory T cells
GMP	good manufacturing practice

References

1. Adams, M.J.; Carstens, E.B. Ratification vote on taxonomic proposals to the International Committee on Taxonomy of Viruses. *Arch. Virol.* **2012**, *157*, 1411–1422. [CrossRef] [PubMed]
2. Murray, P.R.; Rosenthal, K.S.; Pfaller, M.A. *Medical Microbiology*, 5th ed.; Elsevier Mosby: Maryland Heights, MO, USA, 2005; ISBN 978-0-323-03303-9.
3. Zhan, X.; Lee, M.; Xiao, J.; Liu, F. Construction and characterization of murine cytomegaloviruses that contain transposon insertions at open reading frames m09 and M83. *J. Virol.* **2000**, *74*, 7411–7421. [CrossRef] [PubMed]
4. Lee, M.; Abenes, G.; Zhan, X.; Dunn, W.; Haghjoo, E.; Tong, T.; Tam, A.; Chan, K.; Liu, F. Genetic analyses of gene function and pathogenesis of murine cytomegalovirus by transposon-mediated mutagenesis. *J. Clin Virol.* **2002**, *25* (Suppl. 2), S111–S122. [CrossRef]

5. Dunn, W.; Chou, C.; Li, H.; Hai, R.; Patterson, D.; Stolc, V.; Zhu, H.; Liu, F. Functional profiling of a human cytomegalovirus genome. *Proc. Natl. Acad. Sci. USA* **2003**, *100*, 14223–14228. [CrossRef]

6. Scholz, M.; Doerr, H.W.; Cinatl, J. Inhibition of cytomegalovirus immediate early gene expression: A therapeutic option? *Antivir. Res.* **2001**, *49*, 129–145. [CrossRef]

7. Marschall, M.; Freitag, M.; Weiler, S.; Sorg, G.; Stamminger, T. Recombinant green fluorescent protein-expressing human cytomegalovirus as a tool for screening antiviral agents. *Antimicrob. Agents Chemother.* **2000**, *44*, 1588–1597. [CrossRef]

8. La, Y.; Kwon, D.E.; Yoo, S.G.; Lee, K.H.; Han, S.H.; Song, Y.G. Human cytomegalovirus seroprevalence and titers in solid organ transplant recipients and transplant donors in Seoul, South Korea. *BMC Infect. Dis.* **2019**, *19*, 948. [CrossRef]

9. Melnick, M.; Sedghizadeh, P.P.; Allen, C.M.; Jaskoll, T. Human cytomegalovirus and mucoepidermoid carcinoma of salivary glands: Cell-specific localization of active viral and oncogenic signaling proteins is confirmatory of a causal relationship. *Exp. Mol. Pathol.* **2012**, *92*, 118–125. [CrossRef]

10. Griffiths, P.D.; Grundy, J.E. Molecular biology and immunology of cytomegalovirus. *Biochem. J.* **1987**, *241*, 313–324. [CrossRef] [PubMed]

11. Crough, T.; Khanna, R. Immunobiology of human cytomegalovirus: From bench to bedside. *Clin. Microbiol. Rev.* **2009**, 76–98. [CrossRef] [PubMed]

12. Chen, Y.-C.; Sheng, J.; Trang, P.; Liu, F. Potential application of the CRISPR/Cas9 system against herpesvirus infections. *Viruses* **2018**, *10*, 291. [CrossRef] [PubMed]

13. Porter, K.R.; Starnes, D.M.; Hamilton, J.D. Reactivation of latent murine cytomegalovirus from kidney. *Kidney Int.* **1985**, *28*, 922–925. [CrossRef] [PubMed]

14. Dupont, L.; Reeves, M.B. Cytomegalovirus latency and reactivation: Recent insights into an old age problem. *Rev. Med. Virol.* **2016**, *26*, 75–89. [CrossRef] [PubMed]

15. Fowler, K.B.; Stagno, S.; Pass, R.F. Maternal immunity and prevention of congenital cytomegalovirus infection. *JAMA* **2003**, *289*, 1008–1011. [CrossRef]

16. Yamamoto, A.Y.; Mussi-Pinhata, M.M.; Boppana, S.B.; Novak, Z.; Wagatsuma, V.M.; de Frizzo Oliveira, P.; Duarte, G.; Britt, W.J. Human cytomegalovirus reinfection is associated with intrauterine transmission in a highly cytomegalovirus-immune maternal population. *Am. J. Obstet. Gynecol.* **2010**, *202*, 297-e1. [CrossRef]

17. Dollard, S.C.; Grosse, S.D.; Ross, D.S. New estimates of the prevalence of neurological and sensory sequelae and mortality associated with congenital cytomegalovirus infection. *Rev. Med. Virol.* **2007**, *17*, 355–363. [CrossRef]

18. Kenneson, A.; Cannon, M.J. Review and meta-analysis of the epidemiology of congenital cytomegalovirus (CMV) infection. *Rev. Med. Virol.* **2007**, *17*, 253–276. [CrossRef]

19. Nyholm, J.L.; Schleiss, M.R. Prevention of maternal cytomegalovirus infection: Current status and future prospects. *Int. J. Women's Health* **2010**, *2*, 23. [CrossRef]

20. Leung, A.K.; Sauve, R.S.; Davies, H.D. Congenital cytomegalovirus infection. *J. Natl. Med. Assoc.* **2003**, *95*, 213.

21. Dreher, A.M.; Arora, N.; Fowler, K.B.; Novak, Z.; Britt, W.J.; Boppana, S.B.; Ross, S.A. Spectrum of disease and outcome in children with symptomatic congenital cytomegalovirus infection. *J. Pediatrics* **2014**, *164*, 855–859. [CrossRef]

22. Swanson, E.C.; Schleiss, M.R. Congenital cytomegalovirus infection: New prospects for prevention and therapy. *Pediatric Clin.* **2013**, *60*, 335–349. [CrossRef]

23. Yamamoto, A.Y.; Mussi-Pinhata, M.M.; Isaac, M.d.L.; Amaral, F.R.; CARVALHEIRO, C.G.; Aragon, D.C.; MANFREDI, A.K.D.S.; Boppana, S.B.; Britt, W.J. Congenital cytomegalovirus infection as a cause of sensorineural hearing loss in a highly immune population. *Pediatric Infect. Dis. J.* **2011**, *30*, 1043. [CrossRef] [PubMed]

24. Fowler, K.B.; Boppana, S.B. Congenital cytomegalovirus (CMV) infection and hearing deficit. *J. Clin. Virol.* **2006**, *35*, 226–231. [CrossRef] [PubMed]

25. Boppana, S.B.; Ross, S.A.; Fowler, K.B. Congenital cytomegalovirus infection: Clinical outcome. *Clin. Infect. Dis.* **2013**, *57*, S178–S181. [CrossRef] [PubMed]

26. Mussi-Pinhata, M.M.; Yamamoto, A.Y.; Brito, R.M.M.; Isaac, M.d.L.; de Carvalhoe Oliveira, P.F.; Boppana, S.; Britt, W.J. Birth prevalence and natural history of congenital cytomegalovirus infection in a highly seroimmune population. *Clin. Infect. Dis.* **2009**, *49*, 522–528. [CrossRef] [PubMed]

27. Pass, R.F.; Fowler, K.B.; Boppana, S.B.; Britt, W.J.; Stagno, S. Congenital cytomegalovirus infection following first trimester maternal infection: Symptoms at birth and outcome. *J. Clin. Virol.* **2006**, *35*, 216–220. [CrossRef]

28. Nassetta, L.; Kimberlin, D.; Whitley, R. Treatment of congenital cytomegalovirus infection: Implications for future therapeutic strategies. *J. Antimicrob. Chemother.* **2009**, *63*, 862–867. [CrossRef]

29. Barbi, M.; Binda, S.; Caroppo, S.; Calvario, A.; Germinario, C.; Bozzi, A.; Tanzi, M.L.; Veronesi, L.; Mura, I.; Piana, A. Multicity Italian study of congenital cytomegalovirus infection. *Pediatric Infect. Dis. J.* **2006**, *25*, 156–159. [CrossRef]

30. Coll, O.; Benoist, G.; Ville, Y.; Weisman, L.E.; Botet, F.; Greenough, A.; Gibbs, R.S.; Carbonell-Estrany, X. Guidelines on CMV congenital infection. *J. Perinat. Med.* **2009**, *37*, 433–445. [CrossRef]

31. Schleiss, M.R. Congenital cytomegalovirus infection: Update on management strategies. *Curr. Treat. Options Neurol.* **2008**, *10*, 186–192. [CrossRef]

32. Nance, W.E.; Lim, B.G.; Dodson, K.M. Importance of congenital cytomegalovirus infections as a cause for pre-lingual hearing loss. *J. Clin. Virol.* **2006**, *35*, 221–225. [CrossRef] [PubMed]

33. Babić, M.; Krmpotić, A.; Jonjić, S. All is fair in virus–host interactions: NK cells and cytomegalovirus. *Trends Mol. Med.* **2011**, *17*, 677–685. [CrossRef] [PubMed]

34. Wilkinson, G.; Aicheler, R.J.; Wang, E.C. Natural killer cells and human cytomegalovirus. In *Cytomegaloviruses: From Molecular Biology to Intervention*; Caister Academic Press: Wymondham, UK, 2013; Volume 2, pp. 173–191.

35. Vidal, S.; Krmpotic, A.; Pyzik, M.; Jonjic, S. Innate immunity to cytomegalovirus in the murine model. In *Cytomegaloviruses from Molecular Pathogenesis to Intervention*; Reddehase, M.J., Ed.; Caister Academic Press: Norfolk, UK, 2013; pp. 192–214. [CrossRef]

36. Krishna, B.A.; Wills, M.R.; Sinclair, J.H. Advances in the treatment of cytomegalovirus. *Br. Med. Bull.* **2019**, ldz031. [CrossRef] [PubMed]

37. Griffiths, P. New vaccines and antiviral drugs for cytomegalovirus. *J. Clin Virol.* **2019**, *116*, 58–61. [CrossRef]

38. Ornaghi, S.; Hsieh, L.S.; Bordey, A.; Vergani, P.; Paidas, M.J.; van den Pol, A.N. Valnoctamide Inhibits Cytomegalovirus Infection in Developing Brain and Attenuates Neurobehavioral Dysfunctions and Brain Abnormalities. *J. Neurosci.* **2017**, *37*, 6877–6893. [CrossRef]

39. Popping, S.; Dalm, V.A.S.H.; Lübke, N.; Cristanziano, V.D.; Kaiser, R.; Boucher, C.A.B.; Van Kampen, J.J.A. Emergence and persistence of letermovir-resistant cytomegalovirus in a patient with primary immunodeficiency. In *Open Forum Infectious Diseases*; Oxford University Press: Oxford, UK, 2019; Volume 6, no. 9; p. ofz375. [CrossRef]

40. Lin, A.; Maloy, M.; Su, Y.; Bhatt, V.; DeRespiris, L.; Griffin, M.; Lau, C.; Proli, A.; Barker, J.; Shaffer, B.; et al. Letermovir for primary and secondary cytomegalovirus prevention in allogeneic hematopoietic cell transplant recipients: Real-world experience. *Transpl. Infect. Dis* **2019**, e13187. [CrossRef]

41. Aryal, S.; Katugaha, S.B.; Cochrane, A.; Brown, A.W.; Nathan, S.D.; Shlobin, O.A.; Ahmad, K.; Marinak, L.; Chun, J.; Fregoso, M.; et al. Single-center experience with use of letermovir for CMV prophylaxis or treatment in thoracic organ transplant recipients. *Transpl Infect. Dis.* **2019**, e13166. [CrossRef]

42. Veit, T.; Munker, D.; Kauke, T.; Zoller, M.; Michel, S.; Ceelen, F.; Schiopu, S.; Barton, J.; Arnold, P.; Milger, K.; et al. Letermovir for difficult to treat cytomegalovirus infection in lung transplant recipients. *Transplantation* 2019. [CrossRef]

43. Chemaly, R.F.; Ullmann, A.J.; Stoelben, S.; Richard, M.P.; Bornhäuser, M.; Groth, C.; Einsele, H.; Silverman, M.; Mullane, K.M.; Brown, J.; et al. Letermovir for cytomegalovirus prophylaxis in hematopoietic-cell transplantation. *N. Engl. J. Med.* **2014**, *370*, 1781–1789. [CrossRef]

44. Marty, F.M.; Ljungman, P.; Chemaly, R.F.; Maertens, J.; Dadwal, S.S.; Duarte, R.F.; Haider, S.; Ullmann, A.J.; Katayama, Y.; Brown, J.; et al. Letermovir prophylaxis for cytomegalovirus in hematopoietic-cell transplantation. *N. Engl. J. Med.* **2017**, *377*, 2433–2444. [CrossRef]

45. Herling, M.; Schröder, L.; Awerkiew, S.; Chakupurakal, G.; Holtick, U.; Kaiser, R.; Pfister, H.; Scheid, C.; Di Cristanziano, V. Persistent CMV infection after allogeneic hematopoietic stem cell transplantation in a CMV-seronegative donorto-positive recipient constellation: Development of multidrug resistance in the absence of anti-viral cellular immunity. *J. Clin Virol.* **2016**, *74*, 57–60. [CrossRef]

46. Maertens, J.; Cordonnier, C.; Jaksch, P.; Poiré, X.; Uknis, M.; Wu, J.; Wijatyk, A.; Saliba, F.; Witzke, O.; Villano, S. Maribavir for preemptive treatment of cytomegalovirus reactivation. *N. Engl. J. Med.* **2019**, *381*, 1136–1147. [CrossRef]

47. Papanicolaou, G.A.; Silveira, F.P.; Langston, A.A.; Pereira, M.R.; Avery, R.K.; Uknis, M.; Wijatyk, A.; Wu, J.; Boeckh, M.; Marty, F.M.; et al. Maribavir for refractory or resistant cytomegalovirus infections in hematopoietic-cell or solid-organ transplant recipients: A randomized, dose-ranging, double-blind, phase 2 Study. *Clin. Infect. Dis.* **2019**, *68*, 1255–1264. [CrossRef]

48. Deng, Q.; Liu, Y.; Li, X.; Yan, B.; Sun, X.; Tang, W.; Trang, P.; Yang, Z.; Gong, H.; Wang, Y.; et al. Inhibition of human cytomegalovirus major capsid protein expression and replication by ribonuclease P-associated external guide sequences. *RNA* **2019**, *25*, 645–655. [CrossRef]

49. Li, W.; Sheng, J.; Xu, M.; Vu, G.P.; Yang, Z.; Liu, Y.; Sun, X.; Trang, P.; Lu, S.; Liu, F. Inhibition of murine cytomegalovirus infection in animals by RNase P-associated external guide sequences. *Mol. Nucleic Acids* **2017**, *9*, 322–332. [CrossRef]

50. King, M.W.; Munger, J. Editing the human cytomegalovirus genome with the CRISPR/Cas9 system. *Virology* **2019**, *529*, 186–194. [CrossRef]

51. Gergen, J.; Coulon, F.; Creneguy, A.; Elain-Duret, N.; Gutierrez, A.; Pinkenburg, O.; Verhoeyen, E.; Anegon, I.; Nguyen, T.H.; Halary, F.A.; et al. Multiplex CRISPR/Cas9 system impairs HCMV replication by excising an essential viral gene. *PLoS ONE* **2018**, *13*, e0192602. [CrossRef]

52. van Diemen, F.R.; Kruse, E.M.; Hooykaas, M.J.; Bruggeling, C.E.; Schürch, A.C.; van Ham, P.M.; Imhof, S.M.; Nijhuis, M.; Wiertz, E.J.; Lebbink, R.J. CRISPR/Cas9-mediated genome editing of herpesviruses limits productive and latent Infections. *PloS Pathog.* **2016**, *12*, e1005701. [CrossRef]

53. Boch, J.; Scholze, H.; Schornack, S.; Landgraf, A.; Hahn, S.; Kay, S.; Lahaye, T.; Nickstadt, A.; Bonas, U. Breaking the code of DNA binding specificity of TAL-type III effectors. *Science* **2009**, *326*, 1509–1512. [CrossRef]

54. Chen, S.J.; Chen, Y.C. Potential application of TALENs against murine cytomegalovirus latent infections. *Viruses* **2019**, *11*, 414. [CrossRef]

55. Choi, K.Y.; El-Hamdi, N.S.; McGregor, A. Inclusion of the viral pentamer complex in a vaccine design greatly improves protection against congenital cytomegalovirus in the guinea pig model. *J. Virol.* **2019**, *93*, e01442–e014519. [CrossRef]

56. Liu, Y.; Freed, D.C.; Li, L.; Tang, A.; Li, F.; Murray, E.M.; Adler, S.P.; McVoy, M.A.; Rupp, R.E.; Barrett, D.; et al. A replication defective human cytomegalovirus vaccine elicits humoral immune responses analogous to those with natural infection. *J. Virol.* **2019**, JVI.00747–JVI.00819. [CrossRef]

57. Fatic, A.; Zhang, N.; Keller, M.D.; Hanley, P.J. The pipeline of antiviral T-cell therapy: What's in the clinic and undergoing development. *Transfusion* **2019**. [CrossRef]

58. Faist, B.; Schlott, F.; Stemberger, C.; Dennehy, K.M.; Krackhardt, A.; Verbeek, M.; Grigoleit, G.U.; Schiemann, M.; Hoffmann, D.; Dick, A.; et al. Targeted in-vitro-stimulation reveals highly proliferative multi-virus-specific human central memory T cells as candidates for prophylactic T cell therapy. *PLoS ONE* **2019**, *14*, e0223258. [CrossRef]

59. Smith, C.; Corvino, D.; Beagley, L.; Rehan, S.; Neller, M.A.; Crooks, P.; Matthews, K.K.; Solomon, M.; Le Texier, L.; Campbell, S.; et al. T cell repertoire remodeling following post-transplant T cell therapy coincides with clinical response. *J. Clin Invest.* **2019**, *129*, 5020–5032. [CrossRef]

Review

Bright and Early: Inhibiting Human Cytomegalovirus by Targeting Major Immediate-Early Gene Expression or Protein Function

Catherine S. Adamson * and Michael M. Nevels *

School of Biology, Biomedical Sciences Research Complex, University of St Andrews,
St Andrews KY16 9ST, Scotland, UK
* Correspondence: csa21@st-andrews.ac.uk (C.S.A.); mmn3@st-andrews.ac.uk (M.M.N.)

Received: 19 December 2019; Accepted: 8 January 2020; Published: 16 January 2020

Abstract: The human cytomegalovirus (HCMV), one of eight human herpesviruses, establishes lifelong latent infections in most people worldwide. Primary or reactivated HCMV infections cause severe disease in immunosuppressed patients and congenital defects in children. There is no vaccine for HCMV, and the currently approved antivirals come with major limitations. Most approved HCMV antivirals target late molecular processes in the viral replication cycle including DNA replication and packaging. "Bright and early" events in HCMV infection have not been exploited for systemic prevention or treatment of disease. Initiation of HCMV replication depends on transcription from the viral major immediate-early (IE) gene. Alternative transcripts produced from this gene give rise to the IE1 and IE2 families of viral proteins, which localize to the host cell nucleus. The IE1 and IE2 proteins are believed to control all subsequent early and late events in HCMV replication, including reactivation from latency, in part by antagonizing intrinsic and innate immune responses. Here we provide an update on the regulation of major IE gene expression and the functions of IE1 and IE2 proteins. We will relate this insight to experimental approaches that target IE gene expression or protein function via molecular gene silencing and editing or small chemical inhibitors.

Keywords: herpesvirus; cytomegalovirus; immediate-early; IE1; IE2; antiviral; ribozyme; RNA interference; CRISPR/Cas; small molecule

---

1. Introduction

Human cytomegalovirus (HCMV), also known as human herpesvirus 5, is a member of the *Betaherpesvirinae*, a subfamily of the *Herpesviridae*. Infectious HCMV particles are composed of a polymorphic lipid envelope containing viral glycoproteins, a tegument layer consisting mainly of viral phosphoproteins and an icosahedral protein capsid encasing the viral genome [1,2]. The HCMV genome comprises roughly 235,000 base pairs of double-stranded DNA in a single chromosome. By harnessing cellular RNA polymerase II, the viral genome gives rise to a highly complex transcriptome encompassing both mRNAs with more than 700 translated open reading frames as well as non-coding RNAs [3–9]. Upon infection of permissive cells, the HCMV genome is expressed and replicated in three sequential steps referred to as immediate-early (IE), early and late. The viral major IE gene, expressed within hours of infection, and the corresponding IE proteins will be at the center of this review. Major IE proteins inhibit intrinsic and innate host cell responses and initiate transcription from viral early genes [10–15]. Early gene products regulate host cell functions to facilitate virus replication and contribute to late events including viral DNA replication and packaging. Typical early viral proteins include the DNA polymerase (pUL54), phosphotransferase (pUL97) and components of the terminase (pUL51, pUL52, pUL56, pUL77, pUL89, pUL93, pUL104), which are all targets of approved anti-HCMV drugs [16–18]. Finally, late genes are expressed after viral DNA replication has commenced and encode mostly structural proteins of the capsid, tegument or envelope required for the assembly and egress of progeny virions [19–21]. HCMV replicates in a wide variety of differentiated cell types, and targets select types of poorly differentiated cells including myeloid progenitors for latent infection with limited viral gene expression [22–26]. Viral reactivation from latency is brought about by cellular differentiation and/or stimulation and contributes greatly to pathogenesis in vulnerable hosts [27–29].

HCMV is the cause of an ongoing "silent pandemic" affecting 40% to 100% of people in populations around the world. Co-evolution over millions of years has resulted in latent or low-level productive HCMV infection that persists for the life of the host in the absence of major disease symptoms. This type of persistence is due to a fine-tuned balance between our intrinsic, innate and adaptive immune responses and manifold viral countermeasures. Developmental or acquired immune system defects disrupt the delicate balance between virus and host and can result in severe disease outcomes. HCMV infection is the most common congenital (present at birth) infection worldwide, with an estimated incidence in developed countries between 0.6% and 0.7% of all live births. This incidence results in approximately 60,000 neonates born every year with congenital HCMV infection in the United States and the European Union combined [30–33]. Since congenital HCMV infection parallels maternal seroprevalence, the estimated incidence in developing countries is even higher, between 1% and 5% of all live births [34,35]. More than 10% of congenitally infected children will suffer neurodevelopmental damage and other disorders present at birth or long-term sequelae including hearing loss. Consequently, HCMV has been recognized as a leading cause of birth defects. HCMV reactivation from latency or primary infection also remain a major source of morbidity and mortality in immunosuppressed individuals including recipients of solid organ and haematopoietic stem cell allografts, people with acquired immunodeficiency syndrome (AIDS) and other critically ill patients. For example, HCMV infections are diagnosed in roughly 50% of all allograft recipients [36–38]. Cytomegaloviruses are highly species-specific, but certain aspects of HCMV infection and pathogenesis are replicated in animal models including mice infected with murine cytomegalovirus (MCMV) [39,40].

HCMV is spread through various routes including sexual contact, organ and stem cell transplantation, breast milk and from mother to baby (transplacental) during pregnancy. Women can reduce HCMV transmission through practicing appropriate hygiene behaviors [41–44]. In seropositive pregnant women HCMV hyperimmunoglobulin is applied as passive immunization to improve the adaptive immune response and reduce the risk of congenital infection. However, the value of this treatment is controversial with limited data supporting improved clinical outcomes [45–49]. The development of active immunization for HCMV is a major public health priority, and a number of candidate vaccines have been evaluated in clinical trials as well as preclinical models. However, no effective vaccine for HCMV is currently available [50–56].

Since cell-mediated adaptive immunity is believed to be key in counteracting HCMV infection, adoptive transfer of virus-specific T cells holds promise for antiviral therapy [57–59]. In addition, a multitude of antiviral agents from a wide diversity of chemical classes are known to be active against HCMV. The exact mechanism of action is unknown for most of these antivirals, and only a small subset has been tested in clinical trials. Seven anti-HCMV drugs have received approval for various indications: Ganciclovir (GCV), Valganciclovir, Acyclovir, Foscarnet, Cidofovir, Letermovir and Fomivirsen [60–63]. Fomivirsen is an antisense oligonucleotide targeting expression of a major IE protein and will be discussed in Section 5.1. GCV, an acyclic analogue of deoxyguanosine, was the first drug approved for the prevention and treatment of HCMV disease. The prodrug Valganciclovir is an orally applicable valine ester that is rapidly metabolized to GCV. Inside cells, GCV is converted to the active triphosphate via initial phosphorylation by the HCMV phosphotransferase (pUL97) and subsequent phosphorylation steps by cellular kinases. GCV inhibits the HCMV DNA polymerase (pUL54) by competing with deoxyguanosine triphosphate for the enzyme's active site, thus preventing nucleotide incorporation into the elongating viral DNA [60,64]. A closely related nucleoside analogue, Acyclovir, is potent against members of the *Alphaherpesvirinae* but exhibits very modest antiviral activity for HCMV [60,65]. Thus, GCV and Valganciclovir have been the first line choice for prevention and treatment of HCMV disease. However, GCV is associated with serious toxicity including neutropenia, thrombocytopenia or anaemia [66,67]. In addition, the development of GCV-resistant HCMV strains associated with prolonged exposure, severe immunosuppression, suboptimal GCV doses and high viral loads poses a serious challenge. In 90% of all cases, resistance to GCV arises from mutations in conserved regions of either pUL97 or pUL54 [67,68]. In such cases, Foscarnet or Cidofovir are the usual alternative treatments. The two drugs do not require phosphorylation by pUL97 for activation and exhibit broad spectrum antiviral activity against DNA viruses. Foscarnet, a pyrophosphate analogue, directly inhibits pUL54 by blocking the enzyme's pyrophosphate binding site via a non-competitive mechanism. By this mechanism, Foscarnet interferes with cleavage of the pyrophosphate moiety from the nucleotide triphosphate substrate during incorporation into the nascent DNA chain. Cidofovir is an acyclic nucleoside phosphonate and an analogue of deoxycytidine monophosphate. After phosphorylation to the active diphosphate (a deoxycytidine triphosphate analogue) by cellular kinases, competitive incorporation into the elongating DNA chain by pUL54 inhibits viral genome replication. Resistance to Foscarnet and Cidofovir occurs with rates similar to GCV, and the two drugs can select for mutations conferring cross-resistance to GCV. Moreover, lack of oral bioavailability as well as serious side effects including nephrotoxicity have limited their clinical use [66,67]. Brincidofovir is a lipid ester prodrug of cidofovir with improved oral bioavailability and reduced toxicity that demonstrated promising results in clinical trials with allogeneic stem cell transplant recipients seropositive for HCMV [67,69]. Likewise, the benzimidazole L-riboside Maribavir, a highly specific inhibitor of pUL97, has been successfully tested in clinical trials with a similar group of patients [67,69]. Recently, a number of molecules have been discovered that inhibit the packaging of viral DNA into preformed capsids by the HCMV terminase complex. Letermovir is the first of this class to be approved and reduced the levels of HCMV DNA in stem cell transplant patients in the absence of myelotoxic effects [18,70]. Although successful in immunosuppressed patients, none of the anti-HCMV drugs described above have been approved for use during pregnancy because of their teratogenic or embryotoxic effects in animal studies [60,71].

Due to the medical importance of HCMV, the absence of effective ways to prevent infection and the shortcomings of existing therapeutic drugs, it is imperative to develop novel antiviral strategies involving new molecular targets and mechanisms of action. All approved drugs currently available to prevent or treat HCMV disease target viral enzymes expressed in the early phase and critical for late processes in the infection cycle. In contrast, molecular events before the onset of HCMV DNA replication have been largely neglected with respect to antiviral approaches. This review will outline our current understanding of the regulation at the HCMV major IE gene and the functional characteristics of IE proteins derived from this gene. We will further highlight past, present and future antiviral strategies aimed at IE gene expression and protein function for improved intervention with HCMV infection and disease.

2. Major IE Gene Expression

2.1. Transcriptional Control of the Major IE Gene

The outcome of HCMV infection is believed to depend largely on the level and timing of expression from the major IE gene [23,72,73]. This is the first viral gene to be transcribed following initial infection, and likely during reactivation from latency, in a process that does not require de novo viral protein synthesis [13,23,74]. Expression of the major IE gene is highly dynamic with transcription levels ranging from extremely high to negligibly low depending on the type and differentiation or activation state of the infected cell. While productive HCMV infection is linked to activated transcription, viral latency is characterized by transcriptional repression at this gene. The major IE gene is located in the unique long (UL) segment of the viral genome, close to the internal repeat elements. The organization of this gene is unusually complex, not just by viral standards, as multiple promoters and numerous transcripts including both sense and antisense RNAs have been identified in this region [3,75–83].

Some of these promoters appear to have a specific role during latent infection or reactivation from latency [75,79,84]. However, the combined major IE enhancer and promoter (MIEP) is considered the principal driver of IE transcription during productive HCMV infection. It contains an extremely strong enhancer which has been widely utilized in heterologous expression systems. The MIEP is bidirectional and has been roughly divided into four functional entities: a core promoter (+1 to −40 nucleotides from the transcription start site), an enhancer (−40 to −550 nucleotides), a unique region (−550 to −750 nucleotides) and a modulator (−750 to −1140 nucleotides) [85,86] (Figure 1). The modulator's role is largely unknown, although a cell-type specific regulatory function has been suggested [87–89]. "Rightward" transcription from the MIEP is suppressed by the unique region which binds cellular homeobox proteins and appears to function as an insulator between the enhancer and UL127 [90–95] (Figure 1). The core promoter is sufficient, yet not required, for low-level transcription to the "leftward" direction of the major IE gene [76,96]. It contains a TATA-box as well as the *cis*-repressive sequence (crs) that serves as a binding site for IE2 dimers (see Section 3.1). The enhancer hugely augments transcription from the major IE gene, in part via a number of small *cis*-acting repeat sequences (18-bp, 19-bp and 21-bp), and is required for viral replication. It may be further divided into proximal and distal enhancer halves (−40 to −300 nucleotides and −300 to −550 nucleotides, respectively) that differ in structural makeup, yet function jointly by contributing multiple *cis*-acting elements to provide efficient MIEP activation and viral replication. Accordingly, a long list of activating cellular transcription factors have been shown or proposed to bind to the *cis*-acting elements in the enhancer, unique region and modulator. In addition, binding of several repressive cellular transcription factors to the enhancer and modulator has been reported (Figure 1). The vast number of transcription factors that may activate or repress the MIEP is thought to account for much of the highly dynamic expression observed at the major IE gene [13,23,74].

The complexity of MIEP regulation further amplifies when considering transcription in the chromatin or "epigenetic" context. The MIEP may undergo limited DNA methylation, especially in systems for transgene expression, and the major IE gene exhibits CpG dinucleotide suppression [97–101]. Beyond these observations, there is little evidence that MIEP activity or IE transcription are regulated by DNA methylation following HCMV or MCMV infection [102–104]. Nuclear HCMV genomes form nucleosomes, octamers of core histones H2A, H2B, H3 and H4 wrapped with just under 150 bp of DNA, resembling host chromatin structure [9,105]. Consequently, the chromatin of HCMV and other DNA viruses that replicate in the nucleus is subject to regulation by nucleosome occupancy, histone composition and post-translational histone modification [106–108]. Nucleosome occupancy on the MIEP is believed to be low during productive infection, but likely increases during establishment of latency based on findings from the mouse model and by analogy to other herpesviruses [9,105,109]. Numerous studies have shown correlations between activating or repressive histone modifications associated with the major IE gene and the levels of viral gene expression. For example, association of the MIEP with H3K4me2, H3K4me3, H3K9/14ac, H3S10ph or H4Kac has been linked to high levels of IE (or transgene) transcription and productive infection or reactivation from latency [72,110–120]. By contrast, the presence of H3K9me2, H3K9me3 or H3K27me3 at the MIEP generally correlates with low levels of IE transcription and either latent or the onset (pre-IE phase) of productive infection [110–112,115,116] (Figure 1). In agreement with these observations, histone modifying enzymes and enzyme complexes including histone acetyltransferases (e.g., KAT6A/MOZ), histone deacetylases (e.g., HDAC1, HDAC3), histone methyltransferases (e.g., EHMT2/G9A, EZH2, SETDB1, SUV39H1), histone demethylases (e.g., KDM1A/LSD1, KDM4A/JMJD2, KDM6B/JMJD3) and histone kinases (e.g., MSK family) have all been implicated in regulating transcription from the MIEP [72,119–129] (Figure 1). The histone modifying proteins are typically recruited by transcription factors bound to the MIEP including cAMP responsive element binding protein 1 (CREB1), ETS2 repressor factor (ERF) and Ying Yang 1 transcription factor (YY1). In turn, chromatin modifications lead to the recruitment of further activators or repressors that may affect IE expression making for a complex hierarchy of transcriptional regulation [107,130,131].

Histone deacetylases, histone demethylases and other proteins conferring repressive histone modifications to HCMV chromatin may be considered components of the intrinsic cellular immune system also known as restriction factors [132]. Many of the best known restriction factors for HCMV reside in nuclear organelles referred to as nuclear domain 10 or promyelocytic leukaemia (PML) bodies [133–135]. While PML bodies may confer transcriptional repression as a whole, constituents of these organelles including alpha thalassemia/mental retardation syndrome X-linked protein (ATRX), death domain-associated protein (DAXX), PML protein and SP100 nuclear antigen have been shown or proposed to act as repressors of major IE gene expression in part via chromatin-based mechanisms [136–138]. More recently, cellular proteins that mediate foreign or damaged DNA sensing and signalling, including cyclic guanosine monophosphate-adenosine monophosphate (cGAMP) synthase, interferon (IFN) gamma-inducible protein 16 (IFI16) and stimulator of IFN genes (STING), have been identified as restriction factors of HCMV and other DNA viruses [139–142]. These proteins are known or predicted to restrict IE transcription, at least indirectly, although IFI16 may activate rather than repress the MIEP [143,144].

Expression of the major IE gene also varies with the activity of cellular signalling pathways that connect the extra- and intracellular environment to the nucleosomes and transcription factors associated with the HCMV genome including the MIEP. The virus has been shown to activate, rewire or inhibit numerous of these signalling pathways. HCMV infection triggers both pathways considered to be proviral as well as pathways linked to innate immune responses resulting in the production of proinflammatory and antiviral cytokines. In fact, many signalling pathways appear to exhibit both pro- and antiviral potential, and the net effect on the virus depends on various factors including cell type and stage of infection. Binding of HCMV to receptor proteins on the cell surface initiates the first wave of signalling. The virus engages various cellular entry receptors, several of which activate similar pathways relevant to the IE phase of infection [145–149]. In particular, epidermal growth factor receptor (EGFR), platelet-derived growth factor receptor alpha and integrins independently trigger the phosphatidylinositol 3-phosphate and protein kinase B (PI3K/AKT) pathway [150–152]. The PI3K/AKT pathway is central to many cellular properties including motility, proliferation and survival [153–155]. Transient induction of this pathway triggered by receptor signalling appears to be followed by more sustained activation involving the viral major IE proteins [156–159]. Initial PI3K/AKT activation is required for efficient viral entry as well as optimal replication in fibroblasts and establishment of latency in monocytes [156,158,160–163]. However, at later times during infection inhibition of EGFR or PI3K seems to favour viral replication and reactivation from latency suggesting a negative regulatory role at this point [164–167]. Besides PI3K/AKT signalling, various other kinase pathways are known to be activated very early during HCMV infection. These pathways include mitogen-activated kinase (MAPK) signalling both via extracellular signal-regulated kinase (ERK) 1 and 2 including RAF1 (MAPKKK upstream of ERK) as well as via p38 MAPK [168–172]. Other kinases thought to be involved in the IE phase of HCMV infection include adenosine monophosphate-activated protein kinase (AMPK) [173], hematopoietic cell kinase (a src family kinase) [174], cyclin-dependent kinases (CDKs) [175], protein kinase A [176] and mitogen and stress activated kinase (MSK) [128]. The activation of kinase signalling pathways in the initial infection phase comes with multiple, mostly beneficial consequences for the virus including major IE gene activation. For example, ERK mediates induction of major IE gene expression via binding of CREB to the MIEP and recruitment of MSK. In turn, MSK-mediated histone H3 phosphorylation promotes histone demethylation and the subsequent exit of HCMV from latency [128]. One of the most crucial transcription factors linked to the PI3K/AKT, MAPK and other signalling pathways relevant to HCMV infection is nuclear factor kappa B (NF-κB). Canonical NF-κB activation requires degradation of inhibitor of NF-κB (IκB), which depends on phosphorylation by a three-subunit IκB kinase (IKK). IKK-mediated phosphorylation of IκB is triggered as early as five minutes after exposure of cells to HCMV particles resulting in activation of preformed NF-κB [177,178]. This first phase of the NF-κB response to HCMV infection may facilitate IE expression via binding sites in the proximal enhancer of the MIEP (Figure 1). However, the requirement of

NF-κB for efficient IE expression varies widely with cell type, virus strain and other conditions of infection [179–182]. A second phase of NF-κB activation due to initiation of NF-κB transcription allows for continued expression throughout infection. While NF-κB activation benefits HCMV replication, at least under certain conditions, it also comes with adverse effects for the virus. NF-κB, along with IFN regulator factor 3 (IRF3), binds to promoters and stimulates transcription of numerous cytokine and chemokine genes. Some of these genes encode antiviral proteins including type I IFNs. HCMV gene products including tegument proteins (e.g., pUL35, pUL82/pp71, pUL83/pp65) and IE proteins as well as non-coding RNAs target the IFN response and other signalling pathways, adding an additional layer of complexity. Targeting of host cell signalling by HCMV will be discussed below in the context of IE proteins (see Section 3.4), but is otherwise beyond the scope of this review. For a comprehensive and detailed account of this topic, the reader is referred to several other recent reviews [183–185].

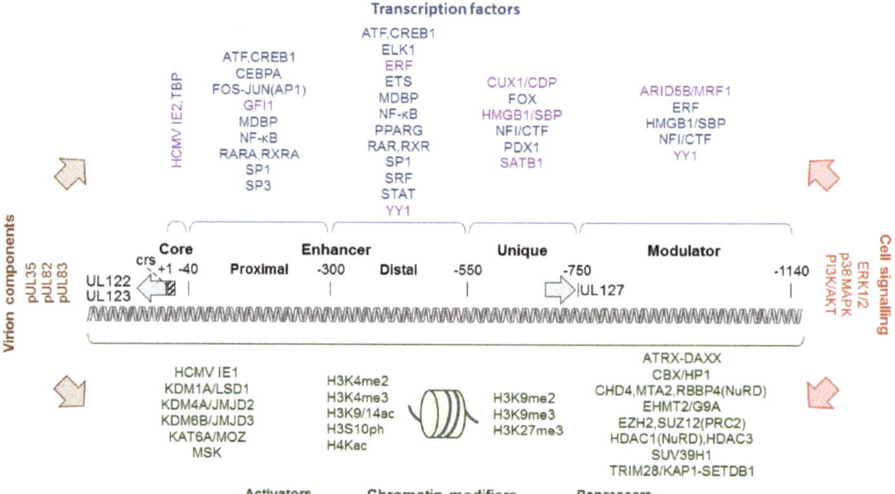

Figure 1. Organisation of the human cytomegalovirus (HCMV) major IE enhancer and promoter (MIEP) and select protein factors involved in its regulation. The MIEP is composed of a core promoter containing a TATA-box and the crs that mediates repression by IE2, an enhancer with proximal and distal parts, a unique element and a modulator. Nucleotide positions relative to the transcription start sites and the direction of transcription (grey arrows) are indicated. "Leftward" transcription results in mRNAs encoding the IE1 and IE2 proteins ("rightward" transcription results in uncharacterized mRNAs containing the UL127 open reading frame). Transcription factors known or predicted to bind to the individual parts of the MIEP are shown above (repressors are shown in purple). Chromatin modifiers and histone tail modifications reported to activate or repress the MIEP are shown below. A few examples of virion components and cell signalling pathways known to activate the MIEP are shown at the left and right side, respectively, of the diagram. ARID5B/MRF1, AT-rich interaction domain 5B protein; ATF, activating transcription factor family; CBX/HP1, heterochromatin protein 1; CEBPA, CCAAT enhancer binding protein alpha; CHD4, chromodomain helicase DNA binding protein 4, nucleosome remodeling and deacetylase (NuRD) subunit; CUX1/CDP, cut-like homeobox 1 protein; ELK1, ETS transcription factor Elk1; ETS, Ets proto-oncogene transcription factor; EZH2, enhancer of zeste 2 polycomb repressive complex 2 (PRC2) subunit; FOS, Fos proto-oncogene, activator protein 1 (AP-1) transcription factor subunit; FOX, forkhead transcription factor family; GFI1, growth factor-independent 1 transcriptional repressor; HMGB1/SBP, high mobility group box 1 protein; JUN, Jun proto-oncogene, AP-1 transcription factor subunit; KAT6A/MOZ, lysine acetyltransferase 6A; KDM1A/LSD1, lysine demethylase 1A; KDM4A/JMJD2, lysine demethylase 4A; KDM6B/JMJD3, lysine

demethylase 6B; MDBP, methylated DNA binding protein family; MTA2, metastasis-associated 1 family member 2, NuRD subunit; NFI/CTF, nuclear factor 1 family; PDX1, pancreatic and duodenal homeobox 1 protein; PPARG, peroxisome proliferator-activated receptor gamma; RARA, retinoic acid receptor alpha; RBBP4, Rb binding protein 4 chromatin remodelling factor, NuRD subunit; RXRA, retinoic X receptor alpha; SATB1, special AT-rich sequence binding homeobox 1 protein; SETDB1, SET domain bifurcated histone lysine methyltransferase 1; SP1, Sp1 transcription factor; SP3, Sp3 transcription factor; SRF, serum response factor; SUZ12, suppressor of zeste 12 PRC2 subunit; TBP, TATA-box binding protein; TRIM28/KAP1, tripartite motif containing 28 protein. See main text for other abbreviations.

2.2. Post-Transcriptional and Translational Control of the Major IE Gene

The primary transcript derived from the MIEP is subject to extensive regulation at the post-transcriptional and translational level. It undergoes alternative splicing and polyadenylation to generate multiple mRNA species assigned to either the IE1 or IE2 family [12,13,74]. This differential post-transcriptional regulation is believed to involve the cellular 65-kDa U2-associated factor and ubiquitin-dependent segregase valosin containing protein p97 [186,187]. RNA sequencing showed increased IE1 and decreased IE2 splicing following p97 knockdown [187]. The processed IE1 and IE2 mRNAs accumulate with different kinetics and share the first three exons [186–188]. However, IE1 mRNAs contain exon 4 while IE2 mRNAs contain exon 5 sequences.

Translation of the IE1 and IE2 mRNAs is subject to control by viral non-coding RNAs [189–191]. For example, the HCMV long non-coding RNA 4.9 has been reported to bind to the MIEP and recruit repressor complex PRC2 to this region [120]. In addition, HCMV miRNA miR-UL112-1 was shown to target the IE1 mRNA and to reduce the corresponding protein levels by translational inhibition [192–194]. Likewise, HCMV miR-UL25-1 and miR-UL25-2 appear to be linked to reduced IE1 protein levels, although most likely indirectly via cellular targets [195].

The major IE mRNAs ultimately give rise to the IE1 (UL123) and IE2 (UL122) families of proteins with several members each. The largest, most abundant and by far best studied family members are the 72-kDa (491 amino acids) IE1 protein, also known as IE72, and the 86-kDa (579 amino acids) IE2 protein, also known as IE86. The two proteins share 85 amino acids encoded by exon 3 at their amino termini but are otherwise unrelated. For simplicity, they are referred to as IE1 and IE2 in this review.

2.3. Post-Translational Control of the Major IE Proteins

IE1 and IE2 are both believed to exist as dimers, while IE2 may also form higher order oligomers [196–200]. Both IE proteins can undergo at least two types of post-translational modification, phosphorylation at serine or threonine residues [201,202] and conjugation to small ubiquitin-like modifiers (SUMOylation) at lysine residues [203–206]. Various positive or negative regulatory effects on IE protein function and HCMV replication have been ascribed to these modifications [205,207–214]. While IE1 is a metabolically highly stable protein with an estimated intracellular half-life between 21 and >30 h [215,216], IE2 exhibits a much shorter half-life of approximately 2.5 h in cells [197,215]. Alongside post-transcriptional mechanisms (see Section 2.2), the differences in metabolic stability contribute to the much higher steady-state levels of IE1 compared to IE2 observed during productive HCMV infection. Nuclear localization signals in IE1 and IE2 target the proteins to the cell nucleus, where they are found in various compartments including PML bodies, chromatin and the nucleoplasm [11,13,14].

2.4. Summary

Highly complex interactions between a multitude of cellular and viral components at the level of DNA, chromatin and upstream signalling pathways determine the initiation and magnitude of transcription from the HCMV MIEP. The highly dynamic transcription from the major IE gene is complemented by post-transcriptional processing and translational regulation, ultimately controlling the synthesis of the IE1 and IE2 families of predominantly nuclear proteins. The major IE proteins are subject to post-translational regulation and are thought to activate the viral replicative cycle during both initial infection and reactivating from latency. It is therefore believed that the eventual outcome of HCMV infection depends on the level and timing of IE1 and IE2 expression.

3. Major IE Protein Function

3.1. Role in Activation and Repression of Transcription

IE1 and IE2 were initially identified as activators of transcription in reporter assays using transiently transfected plasmids [11,13,14]. In these assays, the IE proteins were shown to activate the HCMV MIEP (positive auto-regulation) and various viral early gene promoters. In addition, IE2 turned out to be a repressor of the MIEP (negative auto-regulation). In fact, IE2 sequence-specifically binds to the crs in the core promoter (Figure 1) to block RNA polymerase II occupancy at the transcription start site. Furthermore, several heterologous viral promoters as well as cellular promoters proved to be responsive to activation by the IE proteins. The impact of IE1 and IE2 on transcription from a broad variety of promoters in transient transfection assays earned them the title "promiscuous transactivators". IE2 usually appeared as the stronger activator compared to IE1 and, depending on the reporter construct, the two proteins often acted in an additive or synergistic manner. Activation by IE1 and IE2 was mapped to both upstream elements as well as core promoter regions including the TATA-box. Accordingly, numerous specific and basic transcription factors or transcription factor complexes were reported to interact with IE1 (e.g., CEBP, E2F1-5, SP1) and IE2 (e.g., AP1, CREB1, EGR1, SP1, TAF4, TBP, TFIIB, TFIID).

Many key findings from transient transfection assays about the impact of IE1 and IE2 on HCMV transcription were later corroborated by studies involving mutant viruses and global transcriptome analyses. These findings confirmed positive autoregulation at the MIEP by IE1 [217], crs-dependent repression of the MIEP by IE2 and activation of viral early genes by IE1 and IE2 [9,218–221]. In contrast, "promiscuous transactivation" by IE1 and IE2 was not replicated in transcriptome analyses of endogenous human genes. Instead of showing broad activation of gene expression from the human genome, the differential transcript profiles from cells individually expressing IE1 or IE2 were rather distinct with little or no overlap to the genes activated by the IE proteins in assays with transfected reporter plasmids. Following expression under conditions closely mimicking the situation during productive infection, IE1 turned out to be as significant a repressor as it is an activator of host gene expression in growth-arrested human fibroblasts [222,223]. Cells induced to express IE1 exhibited global repression of interleukin 6 (IL6)- and oncostatin M-responsive signal transducer and activator of transcription (STAT) 3 target genes. This repression was followed by STAT1-dependent activation of type II IFN-stimulated genes (ISGs), normally induced by IFN-γ, many of which encode immune-stimulatory proteins including proinflammatory chemokines [222–224]. Moreover, in the presence of IFN-α or IFN-β, IE1 was found to inhibit the induction of type I ISGs by the trimeric complex of STAT1, STAT2 and IRF9 known as ISG factor 3 (ISGF3) [225,226]. The effects IE1 exerts on the human transcriptome are thought to result largely from direct physical interactions with STAT2 and STAT3 (see Section 3.4). While transcriptional regulation by IE1 appears to be dominated by pathways depending on proteins of the STAT family, IE2 has been shown to inhibit the induction of IFN and other antiviral cytokine genes via a mechanism involving NF-kB and STING (see Section 3.4). However, the transcriptome profile for IE2 in cycling human fibroblasts was dominated by genes regulating the cell cycle and DNA replication many of which are E2F-responsive [227]. This finding likely reflects the impact of IE2 on the cell cycle. IE2 promotes cell cycle progression from G0/G1 to G1/S and arrests cells at the G1/S interface, inhibiting cellular DNA synthesis, or at the G2/M interface [13,14,228]. Many human genes activated or repressed by isolated expression of IE1 or IE2 were also shown to be differentially regulated during productive HCMV infection.

3.2. Role in Chromatin-Based Epigenetic Regulation

IE2 is known to bind sequence-specifically to DNA, but there is no convincing evidence for direct DNA binding by IE1 [11,108,131]. However, IE1 associates with chromatin via core histones. IE1 exhibits two physically separable histone interacting regions with differential binding specificities for H2A-H2B dimers and H3-H4 dimers or tetramers. The H2A-H2B binding region was mapped to an evolutionarily conserved nucleosome binding motif (amino acids 479–488) within the chromatin tethering domain (CTD) at the C-terminus [229,230]. This motif docks with the acidic patch formed by H2A-H2B on the nucleosome surface [229,230]. The consequences of the IE1-nucleosome interaction have not been fully elucidated, but they appear to include alterations to higher order chromatin structure [230]. Histone binding by IE1 might also be linked to the overall low nucleosome levels and temporal reorganization of nucleosomes across viral genomes observed during productive HCMV infection [9,105]. To our knowledge, transcriptional regulation via the IE1 CTD has not been experimentally addressed. Instead, it has been reported that IE1x4, a small variant form of IE1 expressed from a promoter internal to major IE exon 4, facilitates viral genome maintenance and replication during HCMV latency via a CTD-dependent mechanism. Despite a lack of experimental evidence, the mechanism is predicted to involve HCMV episome tethering to host mitotic chromosomes via nucleosome binding by IE1x4 resulting in nuclear retention and partitioning of viral genomes across latently infected dividing cells [75]. Although IE2 appears to have relatively little affinity for histones, both IE proteins have been shown to be present in complexes with nucleosome modifying host cell proteins. For example, IE1 binds to histone deacetylase (HDAC) 3 [124], while IE2 binds to HDAC1-3 [124,126,231], lysine acetyltransferases CREB binding protein (CBP, KAT3A), p300 (KAT3B) and p300/CBP-associated factor (KAT2B) [232,233], and lysine methyltransferases G9A (euchromatic histone lysine methyltransferase 2, EHMT2) and suppressor of variegation 3-9 homolog 1 (SUV39H1) [126]. Accordingly, transcriptional activation of viral IE and early genes by IE1 correlates with histone acetylation, while transcriptional repression of the MIEP by IE2 involves histone deacetylation and methylation [130,131,234].

3.3. Role in Inhibition of Intrinsic Immunity

Intrinsic or cell-autonomous immunity is considered the first intracellular line of defence against viral attack. Intrinsic immunity confers (partial) resistance to viruses via constitutively produced cellular inhibitors of viral replication known as restriction factors [137,235,236].

Consistent with its presence during the very early stages of HCMV infection, IE1, along with several viral tegument proteins, has been recognized as a viral antagonist of cellular intrinsic immunity [10,12,137]. Specifically, IE1 has been shown to target three restriction factors based in nuclear organelles known as PML bodies (see Section 2.1): PML (tripartite motif 19) proteins, SP100A and DAXX. Although various activities have been linked to these restriction factors, they all seem to mediate transcriptional repression of HCMV gene expression in part via chromatin-based mechanisms [134,137,237]. Both IE1 orthologues of animal cytomegaloviruses as well as HCMV IE1 were shown to associate with DAXX [238–240]. The sites of interaction in the two proteins have not been mapped, and it remains unclear whether binding is direct. However, transcriptional activation of the HCMV latent undefined nuclear antigen (LUNA) gene depends on relief from DAXX-mediated repression conferred by IE1 [238]. Most of the proteins IE1 targets remain metabolically stable, but a subset appears to be subject to proteolytic degradation [241–243]. IE1 was shown to interact physically with the N-terminal domain of SP100A and to target the restriction factor for degradation via the proteasome. This finding explains the loss of SP100 observed in the late phase of productive HCMV infection [241,242,244]. The relevance of IE1-mediated SP100 degradation for HCMV replication remains to be determined. Finally, it has been established that IE1 binds to PML proteins via its central core domain (amino acids 25–378) predicted to exhibit an all α-helical, femur-shaped fold [196,245]. This interaction appears to interfere with PML oligomerization and de novo poly-SUMOylation [203,246–248]. SUMOylated PML isoforms are the central organizers of PML bodies, and inhibition of SUMOylation by IE1 correlates with organelle disruption resulting in diffuse nuclear distribution of the associated restriction factors [203,249–251]. The loss of PML body integrity adds an additional layer to inhibition of intrinsic immunity by IE1 that extends beyond the mere targeting of individual restriction factors. Despite limited experimental evidence, PML targeting and disruption of PML bodies are considered to be key to IE1 function and HCMV replication, especially at low multiplicity of infection [216,252].

Preceding disruption by IE1, IE2 co-localizes with PML bodies, most likely as a consequence of binding to the viral genome which also associates with these organelles [251,253,254]. However, IE2 is not currently considered an antagonist of PML bodies. Having said that, both IE1 and IE2 target histone modifying enzymes (see Section 3.2), some of which are bona fide restriction factors, and more cellular mediators of intrinsic antiviral immunity targeted by the IE proteins will likely emerge in the future.

Finally, both IE1 and IE2 inhibit apoptotic cell death, which may be considered part of the intrinsic antiviral defence system [157,233,255–258]. It appears that each IE protein can block extrinsic apoptosis pathways via activation of PI3K/AKT signalling, although additional mechanisms likely contribute including complex formation between IE2 and p53 [259–261]. Despite the fact that the antiapoptotic potential of the two IE proteins has been clearly established in several overexpression settings, its true relevance to HCMV infection remains to be determined.

3.4. Role in Inhibition of Innate Immunity

Post-attachment events associated with HCMV entry and the recognition of virion components by pattern recognition receptors including foreign DNA sensors trigger the induction of cytokine and chemokine genes [262–264]. Many of these cytokines and chemokines are important components of our innate immune system, especially type I, II and III IFNs. The synthesis and secretion of these IFNs activates signalling pathways that involve the phosphorylation of STAT family members including STAT1 and STAT2. Activated STAT proteins form homodimers or heteromeric complexes that subsequently bind to and stimulate transcription from promoters of ISGs many of which encode proteins that interfere with viral replication at various points.

IE1 confers increased type I IFN resistance to HCMV [225]. This phenotype was largely attributed to nuclear complex formation between IE1 and STAT2 depending on amino acids 410–420 in the presumably disordered "acidic domain" of the viral protein downstream from the central core domain and upstream of the CTD [209,222,225,226]. The IE1-STAT2 interaction causes reduced sequence-specific DNA binding by ISGF3 and diminished activation of type I ISGs (e.g., CXCL10, IFIT2, ISG15, MX1) [209,225,226,265,266]. The C-terminal part of IE1 has also been reported to disrupt type II ISG activation by STAT1 homodimers, although IE1 is not believed to bind to STAT1 directly (only indirectly via STAT heterodimers) [222,225,226,267]. The ability of IE1 to inhibit type I ISG induction via STAT2 interaction facilitates HCMV replication and appears to be conserved among IE1 homologs of other betaherpesviruses [209,226,268]. Besides STAT2 interaction, complex formation between PML and IE1 (see Section 3.3) may also contribute to the inhibition of ISG induction during HCMV infection [216,269].

IE2 is not known to interact with STAT family members, but this protein limits HCMV-induced expression of antiviral cytokine and proinflammatory chemokine genes (e.g., IFNB1, CCL3, CCL5, CCL8, CXCL8, CXCL9) [270,271]. A very recent report has also shown that IE2 targets interleukin 1 beta (IL1B) at both the transcript and protein level [272]. The underlying mechanism appears to involve inhibition of virus- or tumor necrosis factor alpha-induced binding of NF-κB to the IFN-β promoter, resulting in attenuated target gene expression [273]. Another recent report has demonstrated that IE2 inhibits IFN-β promoter activation induced by STING, a critical sensor of intracellular DNA and adaptor for type I IFN signalling. IE2 facilitated the proteasome-dependent degradation of STING and inhibited cGAMP-mediated induction of IFNB1 and CXCL10 [274]. Taken together, these studies suggest that IE2 targets STING (and likely other proteins) post-translationally resulting in inhibition of

NF-κB-dependent induction of cytokine and chemokine genes relevant to the innate immune response to HCMV infection.

3.5. Role in Inflammation and Adaptive Immunity

HCMV reactivation and replication are typically linked to a strong inflammatory host response that involves numerous cytokines and chemokines, which often contributes to pathogenesis [27,28,275]. Despite their roles as intrinsic and innate immune antagonists (see Sections 3.3 and 3.4), the major IE proteins may also facilitate inflammation, most obviously by driving HCMV replication. However, IE1 and IE2 may promote inflammation even in the absence of viral replication [223,276–281]. Consistent with this idea, the IE1-specific host cell transcriptome is largely characterized by downregulation of genes responsive to IL6-type cytokines and upregulation of ISGs normally induced by IFN-γ (see Section 3.1) [222,223]. IE1-dependent gene activation proved to be independent of IFN-γ and other IFNs, yet required phosphorylated STAT1. Accordingly, IE1 induced phosphorylation, nuclear accumulation and binding of STAT1 to type II ISG promoters. Moreover, the repression of STAT3- and the activation of STAT1-responsive genes by IE1 turned out to be coupled. By targeting STAT3, IE1 rewires upstream STAT3 to downstream STAT1 signalling. Consequently, genes normally induced by IL6 are repressed while genes normally induced by IFN-γ become responsive to IL6 in the presence of IE1. Thus, IE1 merges two central cellular signalling pathways diverting cytokine responses relevant to inflammation and (neuro)pathogenesis [222,282].

Adaptive antibody as well T cell responses are thought to be important for long-term control of HCMV. Studies on T cell immunity in HCMV have traditionally focused on pUL83/pp65 and IE1. However, it has become clear that both IE1 and IE2 are highly immunogenic CD4+ and CD8+ T cell antigens adding to their complex roles in the immune response to HCMV infection [283–285]. Based on the stimulatory effect IE1 exerts on cellular adaptive immunity, the protein has been utilized in the development of both diagnostic assays as well as vaccine candidates [286–288].

3.6. Role in Viral Replication, Latency and Reactivation

Various highly differentiated cell types including primary human fibroblasts are susceptible to HCMV infection and permissive for viral replication. The importance of IE1 in the viral productive cycle was first highlighted by studying laboratory-adapted high passage HCMV strains (Towne/Toledo and Towne) from which major IE exon 4 had been specifically deleted. Mutant virus replication in fibroblasts was (almost) normal at high but profoundly impaired at low multiplicity of infection [217,219]. The IE1-specific phenotype was eventually attributed to a broad reduction in viral early gene expression and a failure to form replication compartments [218,219]. Nonetheless, more recent studies of IE1-deficient viruses in the background of both high (AD169) and particularly low passage HCMV strains (TB40E) demonstrated substantial attenuation even at high input multiplicity [9,216]. Thus, IE1 is important for efficient HCMV replication in cellulo, albeit not essential. In contrast, IE2 is considered to be indispensable for viral replication at any multiplicity of infection in cultured fibroblasts [220,221,289,290].

While robust expression of the major IE gene is crucial for productive HCMV infection, the absence or low levels of IE proteins are linked to the establishment of latent infection. HCMV establishes latency in a subset of typically poorly differentiated susceptible cells including cells of the myeloid lineage. The MIEP is largely repressed in these cell types, although low levels of IE1 (and even IE2) may still be produced. A study led by the late Greg Pari proposed that a variant form of IE1 referred to as IE1x4 rather than the full-length protein is expressed in latently HCMV-infected haematopoietic progenitor cells [75]. Their results suggest that IE1x4 is required for latent viral genome replication and maintenance involving interactions with the cellular transcription factor SP1 and topoisomerase IIβ. This report is in line with the idea that the IE1 CTD binds to mitotic chromosomes via the acidic patch formed by histones H2A-H2B on the nucleosome core particle (see Section 3.2). The presence and function of IE1x4 during HCMV latency await independent confirmation.

Although it is generally assumed that IE1 and IE2 are required for HCMV reactivation from latency, there is little experimental evidence to confirm this notion. In a promonocytic cell-line, ectopic expression of IE1 and IE2 was sufficient for induction of viral early gene expression but not for production of infectious virus [129]. Studies in the mouse and rat models concluded that the IE1 orthologs are not even required for viral reactivation from latency [291–293]. Thus, while IE2 is almost certainly necessary for HCMV reactivation (being essential for viral replication) the importance of IE1 in this process remains ambiguous.

3.7. Summary

IE1 and IE2 are nuclear localized HCMV proteins expressed at the beginning of infection. They autoregulate the MIEP, activate viral early genes and modulate expression of cellular genes many of which are involved in the cytokine and chemokine response to infection. Regulation of viral gene expression by the IE proteins appears to result in part from chromatin-based mechanisms including histone modification, and at least IE2 shares properties with factors that actively control transcription. In addition, IE1 and IE2 regulate transcription more passively by targeting signalling effectors upstream of the genome such as STAT2/3 and STING, respectively. Both IE proteins are powerful antagonists of intrinsic and innate immunity predicted to be individually essential for HCMV replication in vivo. That said, IE1 and IE2 may contribute to HCMV pathogenesis even in the absence of viral replication.

4. Case for Antiviral Targeting of Major IE Gene Expression or Protein Function

Antiviral strategies for HCMV have long relied on a single molecular target, the viral DNA polymerase. Even more recently approved antivirals and candidate drugs under development are directed at viral targets involved in late molecular processes of HCMV replication including DNA packaging. At this late stage, infection is fully established and adverse immune-related effects including inflammation have been triggered. In fact, immunopathogenic rather than cytopathogenic origins have been proposed for some HCMV disease including pneumonitis in allogeneic transplant recipients [275–277]. Similarly, in mouse models of pneumonitis MCMV replication was not sufficient to cause disease [276–279]. Conversely, MCMV caused pneumonitis in the absence of viral replication [280]. Likewise, HCMV retinitis in AIDS patients was proposed to be partly due to immunopathogenesis triggered by IE gene expression, as disease progressed in the absence of replicating virus [281,294]. Along these lines, the IE1 protein was shown to induce pro-inflammatory gene expression and chemokine secretion [222,223]. The chemokines produced upon IE1 expression included C-X-C motif chemokine receptor 3 (CXCR3) ligands CXCL9, CXCL10 and CXCL11, which have been implicated in a large variety of inflammatory and other immune-related disorders including transplant dysfunction or rejection [278,279]. This evidence links IE gene expression to HCMV pathogenesis.

We consider the major IE gene and proteins promising alternative or complementing targets for anti-HCMV strategies for various reasons. Targeting the expression or function of IE1/2 would interfere with infection at a "bright and early" stage before all other currently approved systemic drugs. MIEP- or IE1/2-targeted drugs are predicted to prevent or dampen inflammation even before viral replication commences. Since IE1/2 are also powerful antagonists of intrinsic immunity and the IFN response, compounds targeting their expression or function are expected to confer susceptibility to these host responses providing a novel mechanism of action. In addition, the IE1/2-targeted drugs exhibit potential for "epigenetic therapy" as both viral proteins exert their functions in part via histone modifications, again providing a novel mechanism of action. These drugs are expected to interfere not only with an ongoing productive infection but also with early stages of reactivation, since both the MIEP and IE1/2 function are likely required for this process. Conceivably, even viral persistence may be inhibited based on the observation that IE1x4 mediates viral genome replication and maintenance during latency. Finally, IE1/2-directed drugs are not expected to confer cross-resistance to or interfere with the activity of existing compounds approved for HCMV monotherapy. They may therefore be combined with these drugs for combination therapies with improved efficacy.

5. Inhibition of Major IE Gene Expression by Gene Silencing or Editing

5.1. IE Gene Silencing

Silencing IE gene expression is expected to exert significant pleiotropic antiviral effects due to the multi-functional roles played by IE gene products in HCMV replication, latency and pathogenicity (see Sections 2 and 3). Molecular approaches can efficiently target IE gene expression (Figure 2), and initial feasibility of this approach has been demonstrated via the antisense oligonucleotide Fomivirsen (also known as ISIS 2922 or Vitravene). Fomivirsen is a 21-base synthetic oligonucleotide complementary to IE2 mRNA sequence with phosphorothioate linkages to enhance nuclease resistance. It exhibits potent HCMV antiviral activity with EC_{50} values in the sub-micromolar range [295,296]. Fomivirsen's mechanism of action is primarily thought to block IE2 gene expression by sequence-dependent hybridization to its target mRNA that results in reduced IE2 protein levels due to mRNA degradation via RNaseH recognition of the DNA:RNA hybrid complex [295,297]. This is not, however, the sole mechanism of action, as other sequence-dependent and sequence-independent effects have been reported to contribute to its antiviral activity [295,297,298]. Fomivirsen, developed by Isis Pharmaceuticals in collaboration with Novartis Opthalmics, was in 1998 the first oligonucleotide-based therapy to be approved for clinical use by the FDA [299]. It was approved for treatment of HCMV-induced retinitis in HIV/AIDS patients via local intravitreous injection, and its clinical effectiveness was demonstrated in small-scale clinical trials [300–302]. Fomivirsen is no longer marketed, due to a significant decline in HCMV-induced retinitis cases in HIV/AIDS patients following the successful implementation of antiretroviral therapy and the availability of alternative treatments [299]. Despite its discontinuation for commercial reasons, Fomivirsen's development has provided convincing proof-of-concept evidence that inhibition of IE gene expression can be an effective HCMV antiviral therapeutic strategy.

An alternative approach to targeting IE mRNA and hence IE gene silencing, is to utilize gene-targeting ribozymes, which are catalytically active RNA molecules that specifically cleave target mRNA sequences. M1GS ribozyme technology has been used to target both IE1 and IE2 by utilizing the shared mRNA region of these genes [303–306]. Target IE1/2 mRNA sequences have been selected by determining accessibility for M1GS binding via dimethyl sulfate mapping [303–306]. M1GS is partially derived from the M1 RNA catalytic subunit of the *E.coli* RNase P ribozyme, which mediates tRNA maturation [307,308]. M1 RNA can be converted into an M1SG sequence-specific ribozyme by covalently linking it to an external guide sequence (EGS) that contains nucleotides complementary to the target mRNA sequence [307,308]. The tertiary structure generated upon hybridization between the mRNA substrate and the EGS is required for recognition and cleavage by the ribozyme active site [307,308]. The initial IE1/2-targeted study used a wild-type M1 sequence and IE1/2 exon 3 as the cleavage site. This IE1/2-targeted ribozyme reduced IE1/2 gene expression by 75–80% and inhibited HCMV replication 150-fold in cell culture [303]. A protein engineering and selection strategy has subsequently been employed to identify various highly active M1SG variants containing mutations in M1 that enhance substrate binding and cleavage rates [304–306,309]. The most potent variant reported to-date, F-R228-IE, reduced IE1/2 expression by 98%–99% and inhibited HCMV replication 50,000-fold in cell culture [306]. F-R228-IE uses nucleotide position 43 downstream from the IE1/2 initiation codon as the designated cleavage site and contains three M1 RNA point mutations (G59A, C123U, C326U). However, the mechanism by which these mutations enhance cleavage is currently unknown [306]. Whilst M1SG technology has potential for HCMV therapeutic applications, to the best of our knowledge it has not yet been clinically tested.

RNA interference (RNAi) offers an alternative approach to targeting IE gene expression. RNAi is a cellular gene-silencing pathway that results in sequence-specific degradation of the target mRNA via complementary short-interfering (siRNA) molecules. Various siRNA or short-hairpin RNA (shRNA, processed into siRNA) molecules targeting IE1/2 mRNA have been shown to cause significant inhibitory effects on HCMV replication in cell culture. These antiviral effects correlated with reductions in IE1/2 mRNA and protein levels [310–313]. In addition, IE1/2 siRNA treatment offset some consequences of HCMV infection for the host cell, by retaining PML body integrity and preventing DNA damage response signalling [310]. Treatment of cells with IE-targeted siRNA after HCMV infection resulted in a modest antiviral effect; this is a valuable observation as therapeutic treatment of patients after establishment of HCMV infection would be an important clinical application [310]. Although RNAi technology has potential for anti-HCMV applications, this technology along with antisense oligonucleotides and gene-targeting ribozymes, may be superseded by the recent development of genome-editing techniques.

5.2. IE Gene Editing

Instead of targeting IE gene expression at the mRNA level, genome-editing technology could be used to directly target the HCMV DNA genome to disrupt the UL122/123 gene responsible for major IE transcription. At the time of writing, one study has reported UL122/123 gene-editing, using the Clustered Regularly Interspaced Short Palindromic Repeats (CRISPR)/CRISPR-associated protein 9 (Cas9) system [314]. CRISPR/Cas9 is a new powerful technology that targets specific DNA sequences in eukaryotic cells for cleavage, via double-stranded DNA breaks, using a Cas9 endonuclease and a guide RNA (gRNA) that determines target specificity. Double-stranded DNA breaks are repaired by host mechanisms, such as non-homologous end joining, which are error prone and can introduce small insertion/deletion mutations at the targeted location, or larger deletions if multiple breaks are introduced. A multiplex strategy using three gRNAs targeting UL122/123 successfully excised the UL122/123 gene in 90% of all viral genomes in an HCMV-infected cell population and resulted in a significant decrease in IE protein production and 90% reduction in HCMV replication [314]. Multiplex approaches have been developed to overcome acquisition of resistance mutations in the target sequences. This study provides proof-of-concept that a multiplex anti-UL122/123 CRISPR/Cas9 system can efficiently target the HCMV genome. This system may be useful for targeting HCMV in latently infected cells, where viral gene expression is low or absent and thus mRNA is not available for other molecular approaches discussed in this section and viral proteins (e.g., DNA polymerase) are not available as a target for conventional antiviral drugs.

5.3. Summary

Molecular techniques offer a promising future for development of new anti-HCMV approaches. However, a number of drawbacks must be addressed including reduction of toxicity as well as off-target and immunostimulatory effects combined with improvements in the mode, stability and efficiency of delivery vehicles and methods [315,316].

6. Inhibition of Major IE Gene Expression by Small Molecule Chemical Inhibitors

6.1. Introduction

Inhibition of major IE gene expression can be achieved by identification of small molecules that directly or indirectly inhibit activation of the MIEP and thus prevent IE gene transcription and translation (Figure 2). The complexity of MIEP regulation and its dependence on host cell signalling pathways and transcription machinery (see Section 2.1) means that numerous host factors are potential drug targets. Activation of key host cell signalling pathways is dependent on post-translational phosphorylation events mediated by kinases, and thus kinase inhibitors have been largely implicated in IE gene expression inhibition. Host "epigenetic" factors are also potential targets to lock down IE transcription and hence inhibit viral replication and reactivation from latency. In addition, viral proteins are known to be directly involved in IE gene expression or regulation of host cell signalling pathways exploited by HCMV to activate IE gene expression. However, compounds targeting these viral proteins are considered beyond the scope of this review, as they would contribute to their own specific drug classes.

Small molecules that inhibit IE gene expression have been identified using a variety of strategic approaches; (i) exploitation of existing compounds that inhibit cell signalling pathways modulated by HCMV infection to facilitate MIEP activation and IE gene expression, (ii) targeted-screening of compound libraries composed of molecules that inhibit key cellular signalling components, e.g., kinase inhibitors, (iii) cell-based assays designed to discover novel molecules that target the early steps of HCMV replication and (iv) testing of compounds that have anecdotal evidence suggesting that they may have antiviral activity. Groups of related compounds have been identified using a combination of these approaches. Key groups of molecules are discussed below, although it should be noted that, in general, the small molecules that have thus far been identified have not had their mechanism of action fully elucidated.

6.2. Artemisinin and Derivatives

Testing compounds that have anecdotal evidence suggesting that they may have antiviral activity often revolves around natural products. Various natural products have been reported to have anti-HCMV activity linked to inhibition of IE expression or function, but most remain largely unsubstantiated beyond initial observations. However, a considerable body of evidence has been generated with respect to the anti-HCMV activity of natural product artemisinin, its semi-synthetic derivative artesunate and various related compounds. Artemisinin is a natural product derived from the plant *Artemisia annua* (Sweet Wormwood), a herb used in traditional Chinese medicine [317,318]. Artemisinin and its derivatives are best known for effective antimalarial activity and treatment [317,318], which provided the premise for testing artesunate for anti-HCMV activity [319]. Artesunate, along with various related compounds, exhibit in vitro inhibitory activity against laboratory, clinical and drug-resistant strains of HCMV in a range of cell types with EC_{50} values generally in the low micromolar to sub-micromolar range [319–329]. Chemically linking artemisinin-related molecules into dimers and trimers significantly improves antiviral potency [320,324–326,330,331]. Examples include artemisinin-derived dimer diphenyl phosphate (838), a potent, selective HCMV inhibitor with irreversible activity [332,333] and trimeric artesunate derivative TF27 [326], which exhibits potent in vitro and in vivo activity in the MCMV model [329]. Hybridization of artemisinin-derivatives with bioactive molecules, such as quinazoline, has produced novel compounds with potent anti-HCMV activity significantly better than parental compounds and GCV [334–337].

The mechanism of action by which artesunate and the various derivatives generate their anti-HCMV activity has not been fully elucidated, but the general consensus is that artesunate primarily interferes with the NF-κB pathway [319,326,331]. The NF-κB pathway is stimulated upon HCMV infection and activates the MIEP, driving expression of IE proteins and hence subsequent steps in HCMV lytic replication and pathogenesis [179,338,339]. Indeed, artesunate, along with many derivatives, has been shown to block the IE phase of HCMV replication via a reduction in expression levels of IE2, and to a lesser extent IE1 [319,320,326,328,331,332]. Artesunate, and dimer/trimer derivatives such as TF27, have been shown to interfere with the NF-κB pathway, which is proposed to occur via a direct interaction of the compound with NF-κB subunit RelA/p65 [319,326,331]. Interaction of artesunate with a host cell factor leads to the expectation that acquisition of drug resistance would be less likely; indeed, attempts to generate artesunate drug-resistant isolates in vitro have thus far been unsuccessful [326,340]. Alternative modes of action, implicating other cell signalling pathways and modulation of cell cycle progression, have also been proposed for artesunate compounds [319,340].

Clinical use of artesunate for management of drug-resistant HCMV infections in stem cell or solid organ transplant recipients is considered feasible due the documented anti-HCMV activity discussed above, favourable results in a rodent animal model study [341] and the long and safe clinical history of artesunate treatment in malaria patients [317]. The first use of artesunate in a clinical setting was a success with artesunate being reported as an effective inhibitor of HCMV replication in the treated patient [342]. However, subsequent studies reported either mixed success or that artesunate was ineffective in controlling HCMV infection [343–345]. Further studies are required to fully determine the differences in clinical outcomes for artesunate-treated patients, and studies with more potent artesunate derivatives may hold future promise. For example, the trimeric derivative TF27 has recently been demonstrated to display antiviral efficacy in the mouse model. MCMV replication was significantly reduced and restricted to the site of infection, preventing organ dissemination without adverse effects [329].

6.3. NF-κB Inhibitors

HCMV infection modulates several cell signalling pathways, including the NF-κB and PI3K/AKT pathways, in order to facilitate MIEP activation and IE gene expression (see Section 2.1) [183]. Artemsinin and derivatives, which interfere with the NF-κB pathway, were discussed in Section 6.2. The mode of action of these compounds was identified after testing for anti-HCMV activity based on their anti-malarial properties. An alternative strategy for identification of anti-HCMV compounds that inhibit major IE gene expression, is to exploit existing compounds already known to inhibit cell signalling pathways modulated by HCMV. A rich source of compounds that could be repurposed as anti-HCMV compounds are the numerous NF-κB pathway inhibitors that have been identified for reasons unrelated to HCMV [346,347]. For example, IKK2 inhibitor AS602868 targets a crucial step in NF-κB pathway activation: the phosphorylation and subsequent degradation of IκB by the IKK complex [348,349]. Testing of AS602868 showed that this compound prevents HCMV mediated NF-κB pathway activation, resulting in significant inhibition of IE gene expression, HCMV replication and HCMV-induced host cell inflammatory response without cytotoxicity [350]. HCMV infection also up-regulates the PI3K/AKT pathway leading to activation of NF-κB in a PI3K-dependent manner. LY294002, a PI3K inhibitor, significantly reduces HCMV IE1/2 expression, viral DNA replication and viral titers [156,351]. Disruption of the PI3K pathway and subsequent AKT and NF-κB activation has been suggested as a possible mechanism of action for heat shock protein 90 (hsp90) inhibitors geldanamycin and 17AAG, which significantly inhibit HCMV replication by affecting IE protein production and hence subsequent steps in HCMV productive replication [351,352]. The examples discussed above demonstrate the value of repurposing existing cell signalling pathway inhibitors for targeting HCMV by inhibition of major IE gene expression.

6.4. Kinase Inhibitors

A general theme in host cell signalling pathway inhibitors is to target kinases, which mediate regulatory post-translational phosphorylation modifications of pathway components. There is an abundance of kinase inhibitors, which have been identified and developed for a wide range of applications particularly cancer treatment [353], which can potentially be repurposed as anti-HCMV compounds. Indeed, examples of kinase inhibitors (AS602868, LY294002) with activity against HCMV have already been discussed in Section 6.3. A further example of a kinase inhibitor repurposed for anti-HCMV testing is the multi-targeted anti-cancer tyrosine kinase inhibitor sorafenib (Nexavar), which has been shown to inhibit MIEP activity, IE expression and also later stages of HCMV replication [172]. The mechanism by which sorafenib inhibits HCMV was not fully elucidated due to its multitude of known kinase targets. However, inhibition of RAF1 activation was implicated but via a mechanism independent of MAPK/ERK signalling [172]. Inhibitors of CDKs also have potential as antiviral drug candidates; for example, the CDK7 inhibitor LDC4297 blocked HCMV replication with EC_{50} values in the nanomolar range [175]. The compound's mode of action was concluded to be multifaceted but occurs at the level of IE gene expression and interferes with HCMV-mediated inactivation of the retinoblastoma (Rb) protein, which controls progression through the G1 phase of the cell cycle via its phosphorylation state and ability to bind transcription factor complexes [175]. Promisingly, LDC4297 has been shown to possess in vivo antiviral activity in the mouse model. MCMV replication was significantly reduced and restricted to the site of infection, preventing organ dissemination without adverse effects [329].

In addition to directly repurposing known kinase inhibitors, a number of cell-based screens have been performed against various targeted kinase inhibitor compound libraries [354–357]. Compound-treated HCMV-infected cells were monitored for antiviral effects via expression of a green fluorescent protein (GFP) reporter [354] or late viral protein pp28 from the HCMV genome [355–357]. These screens have identified a number of interesting kinase inhibitors with anti-HCMV activity against laboratory and clinical strains that target a variety of cellular kinases without causing significant cytotoxicity. The lack of kinase inhibitor target specificity has made full elucidation of mechanism of action challenging, although in all cases discussed here antiviral activity has been linked to interference with IE expression or protein production without affecting viral entry. Several c-Jun N-terminal kinase (JNK) inhibitors were identified following a 600 compound kinase inhibitor library screen and the SP600125 inhibitor was shown to inhibit JNK activation and suppress IE gene transcription [354]. XMD7 5-aminopyrazine compounds were identified upon screening of the Gray kinase inhibitor library. These compounds target a range of cellular protein kinases to inhibit HCMV replication via a reduction in genome-wide transcription and a defect in the production of certain HCMV proteins including IE2 (86kDa, 60kDa and 40kDa species) [355]. The proposed mechanism of action for XMD7 compounds is consistent with IE2's role as a viral transcriptional activator, but it is not clear why only a subset of HCMV proteins is affected. CMGC kinase inhibitor RO0504985, an oxindole compound with anti-HCMV activity identified by screening a Roche kinase inhibitor library, also inhibited IE2 and pp28 protein levels [356]. Screening of the GlaxoSmithKline kinase inhibitor set identified SB-734117, a furazan benzimidazole compound, which inhibits several proteins from the AGC and CMCG kinase groups [357]. SB-734117 inhibited IE protein production and reduced phosphorylation of host cell transcription factor CREB and histone H3. However, disappointingly these effects did not lead to any defects in transcription from the MIEP and thus the compound's mechanism of action remains undetermined [357]. Overall, the wealth of preexisting kinase inhibitors and accompanying knowledge offers good potential for the identification and development of novel anti-HCMV compounds.

6.5. Histone Modifying Enzyme Inhibitors

Major IE gene expression is also regulated by "epigenetic" modifications, including histone post-translational methylation, which can result in repressed gene expression [358]. Histone demethylases are required to remove repressive "epigenetic" marks to promote IE gene expression and hence HCMV productive infection or reactivation from latency [358]. Histone demethylase inhibitors (e.g., ML324, a JMJD2 demethylase family inhibitor) have been shown to potently inhibit HCMV IE gene expression [123,359,360]. These demethylase inhibitors also repress IE gene expression from the related Herpes simplex virus type 1 (HSV-1), and importantly they have been shown to potently inhibit HSV-1 infection and reactivation from latency [123,359,360]. These results suggest that compounds that target histone demethylases and possibly other histone modifying or chromatin remodeling enzymes may have potential as HCMV inhibitors [358].

6.6. Cardiac Glycosides

Discovery of novel small molecules that inhibit major IE gene expression can be accomplished using cell-based assays designed to target the early steps of HCMV replication including IE expression but also virus attachment, entry and capsid transport steps [361,362]. One such assay utilizes an engineered variant of the HCMV laboratory strain AD169 that expresses IE2 with a C-terminal yellow fluorescent protein tag (AD169$_{\text{IE2-YFP}}$) [362]. IE2-YFP levels in the nucleus of infected cells are quantified using high-content confocal microscopy and hit inhibitory compounds identified via a decrease in nuclear fluorescent signal and therefore a decrease in IE2-YFP protein levels. This IE2-YFP cell-based reporter assay was used to screen a 2080 bioactive compound library and identified one lead compound, the cardiac glycoside convallatoxin. This compound exhibited potent anti-HCMV activity (EC$_{50}$ values in the low nanomolar range) without significant cellular cytotoxicity [362,363]. However, it should be noted that convallatoxin has been discounted as a hit from a different screen due to toxicity [364].

Interestingly, other cardiac glycosides (ouabain, β-antiarin, digoxin, digitoxin) have also been reported to exhibit anti-HCMV activity [363,365–367]. Inhibition of HCMV by cardiac glycosides is effective against clinical and GCV-resistant strains and exhibits additive activity when administered to cells in combination with GCV [362,363,365,366]. Members of this compound family have been used clinically for the treatment of heart conditions such as congestive heart failure, although toxicity and dosage issues mean that they are increasingly replaced with synthetic drugs such as ACE inhibitors and beta-blockers [368]. Clinical development as antiviral drugs has not yet been undertaken, but medicinal chemistry approaches have demonstrated the ability to improve antiviral activity and selectivity [369].

A common feature of cardiac glycoside treatment is reduction in IE1/2 protein levels [362,363,365]. Mechanism of action studies demonstrated that convallatoxin does not inhibit IE2 mRNA levels but instead inhibits global translation of viral and host cell proteins [363]. Cellular translation machinery is not directly inhibited by convallatoxin; instead the compound reduces methionine transport into the cell, limiting the intracellular pool of this essential amino acid for translation. Convallatoxin has been proposed to mediate this indirect mechanism of translation inhibition by its ability to bind to and inhibit the cellular sodium-potassium ATP pump (NA^+,K^+-ATPase) [363]. In this model, inhibition of the pump causes a reduction in the sodium gradient across the cell membrane, leading to a decrease in sodium-dependent methionine transport [363]. Despite inhibition of global translation, minimal cellular cytotoxicity was observed at the nanomolar concentrations of convallatoxin required for antiviral effect. This observation suggests that, whilst the cell can tolerate a reduction in protein synthesis, HCMV is unable to compensate for reductions in viral protein levels, particularly in IE proteins which are required for early and late protein production and are thus essential for virus replication [363]. Convallatoxin-induced inhibition of viral protein translation by methionine transport reduction is not the only mechanism attributed to the antiviral activity of cardiac glycosides. Alternate mechanisms of action are based on the ability of these compounds to modulate cell signalling pathways [370]. For example, cardiac glycoside digitoxin has been reported to inhibit HCMV through induction of cellular autophagy following activation of the regulatory kinase AMPK via a novel NA^+,K^+-ATPase subunit α1-AMPK-ULK1 pathway [173]. In addition to inhibiting HCMV, cardiac glycosides act as antivirals against a range of clinically important DNA and RNA viruses. This broad-spectrum activity has been attributed to a range of host-directed mechanisms [371].

6.7. Novel Miscellaneous Compounds

A cell-based screen designed to monitor IE2 nuclear translocation was used to identify the cardiac glycoside convallatoxin discussed in Section 6.6. A second screening approach targeting IE2 gene expression utilized a reporter cell-line in which the IE2-activated HCMV TRL4 promoter drives luciferase expression [361]. The reporter cell-line assay was used to screen a 9600 compound library for inhibitors of early phase HCMV infection [361]. Two hit compounds arising from the screen, 1-(3,5-dichloro-4-pyridyl)piperidine-4-carboxamide and 2,4-diamino-6-(4-methoxyphenyl)pyrimidine termed DPCC and 35C10, respectively, have been demonstrated to potently inhibit HCMV replication as effectively as GCV [361,372]. DPCC was also independently identified as having potent anti-HCMV activity in an unrelated high throughput screen designed to target IE1 IFN antagonist function (assay concept described in Section 7.2) [373]. In this screen, DPPC was alternatively termed StA-IE1-3, and a further novel hit compound with similar anti-HCMV activity termed StA-IE1-2 (1-(3-nitrophenyl)-2-(pyrido[3,2-d][1,3]thiazol-2-ylthio)ethan-1-one) was also identified [373]. All three structurally diverse compounds act after viral entry but before IE expression, with significantly decreased IE1 and IE2 expression at both the mRNA and protein levels [361,372,373]. Like the majority of compounds that have been shown to inhibit HCMV IE gene expression, the precise mechanism of action of these three compounds has not been elucidated. However, it has been postulated that they may target a cellular transcription factor or upstream signalling protein required for activation of the HCMV MIEP [373].

6.8. Summary

Overall, a variety of approaches utilizing existing knowledge to repurpose known compounds or screening campaigns to discover novel compounds has successfully identified a wide variety of IE gene expression inhibitors. These inhibitors exploit host cell factors and signalling pathways utilized by HCMV to activate the MIEP and thus offer the hypothetical advantage of a reduced risk of drug resistance. A major challenge associated with the development of these compounds is the complexity in elucidating their relevant host cell targets and mechanism of action. Characterization of these compounds has been predominantly conducted in vitro. However, a few compounds have been progressed to in vivo testing using the MCMV model, and clinical testing of artesunate produced mixed clinical outcomes that warrant further investigation. Overall, compounds that inhibit HCMV IE gene expression merit future investigation and development as potential antivirals.

7. Inhibition of Major IE Protein Functions

7.1. IE2 Inhibitors

Compounds that inhibit IE2 function have been identified and offer promise as a potential new class of HCMV inhibitors (Figure 2). IE2 has been targeted, as it is an essential multifunctional protein that regulates critical events in HCMV replication including transactivation of early and late genes and auto-regulation of the MIEP. IE2 has also been linked to broad dysregulation of host gene expression affecting cell cycle progression, immunomodulation and pathogenesis (see Section 3.1). The first compound demonstrated to directly inhibit IE2 function was WC5, a 6-aminoquinolone derivative [374]. WC5 was tested based on evidence that compounds within this chemical group exhibit antiviral activity against HIV-1 by inhibiting Tat transactivation [375,376]. WC5 specifically inhibits HCMV but not a selection of other herpesviruses [376,377]. In addition to IE, early and late gene expression profiles, which suggest inhibition of IE2 function, WC5 has been shown to directly inhibit IE2's transactivating activity via a cell-based assay in which an EGFP reporter gene was placed under the control of IE2-dependent early gene promoters [374]. In these assays, WC5 significantly inhibited IE2-mediated transcriptional activation of early gene promoters UL54 and UL112/113. A minimal region of the UL54 promoter composed of a 150-bp segment upstream of the transcriptional start site has been demonstrated to be sufficient to mediate the inhibitory activity of WC5 [378]. Within this 150-bp segment is the IR-1 signal (8-bp inverted repeat element 1), a *cis*-acting sequence with an established role in IE2-dependent transactivation, yet the IR-1 signal has been shown not to be required for WC5's inhibitory activity [378]. In addition, two key protein interactions, IE2 dimerization and its interaction with TBP, known to be involved in IE2-dependent transactivation of viral promoters have also been discounted as WC5's target [378]. Intriguingly, WC5's activity appears to be specifically confined to the regulation of HCMV promoters, as the compound exhibits no effect on a variety of cellular promoters regulated either by IE2 protein interactions or a direct IE2 interaction with promoter DNA [378]. In addition to inhibiting IE2 transactivation of viral promoters, a second mechanism of action by which WC5 inhibits a different IE2 function has been identified [378]. WC5 specifically disrupts IE2's direct interaction with the crs within the MIEP (Figure 1). Disruption of the IE2-crs interaction abolishes IE2's auto-repression of its own promoter, a function essential for viral replication. Although WC5 has been demonstrated to inhibit two IE2 functions, transactivation of viral early and late genes and MIEP auto-regulation, the exact molecular mechanisms require further elucidation.

WC5's unique activity offers the possibility to develop a new mechanistic class of anti-HCMV compounds. Towards this goal, WC5 potently and selectively inhibits HCMV replication in the sub-micromolar range irrespective of testing against laboratory or clinical isolates, and its activity is comparable to GCV [376]. Unsurprisingly, given WC5's novel mechanisms of action, the compound similarly inhibits isolates resistant to clinically approved anti-herpesvirus DNA polymerase inhibitors [376]. Further, when WC5 is combined with GCV, synergistic activity against HCMV replication was observed without significant increases in cellular cytotoxicity [374]. WC5 also inhibits

MCMV replication, albeit with ~10-fold lower activity compared to HCMV [376]. Importantly, WC5's mechanism of action against HCMV and MCMV appears to be conserved, as it has been shown to block MCMV early gene transactivation mediated by the MCMV IE2 homolog ie3 [378]. Thus, it has been suggested that the murine model may be used to test WC5 activity in vivo as a prerequisite to clinical development [378]. Structure–activity relationship studies have been conducted with the aim of improving WC5's potency [374,379]. These studies gained insight into chemical groups required for WC5 activity, and identified an analogue with an improved selectivity index compared to WC5 without compromising antiviral activity. However, analogues with significantly improved potency were not identified [379].

WC5's discovery together with its novel mechanism of action provided proof-of-principle that IE2 is a valid target for drug discovery and encouraged a screening campaign to identify new compounds targeting IE2 [364,380]. A screen has been performed employing essentially the same cell-based assay used to determine WC5's mechanism of action as an inhibitor of IE2-mediated transactivation of early gene expression [374]. Assay optimization identified conditions using the stable cell-line that expresses EGFP under the control of the IE2-dependent UL54 early promoter as suitable for screening purposes [364]. A 2320 bioactive compound library including all FDA-approved drugs was screened and six hit compounds have so far been selected for further study [364,381,382]. These hit compounds are deguelin (DGN), nitazoxanide (NTZ), thioguanosine (TGN), alexidine dihydrochloride (AXN), manidipine dihydrochloride (MND) and berberine (BBR). All hits inhibited HCMV replication with EC_{50} values in the low micromolar range and lacked significant toxicity. This antiviral activity was observed for laboratory, clinical and drug-resistant HCMV isolates. Further, MND was shown to be inactive against a selection of other DNA and RNA viruses and is thus likely to be a specific anti-HCMV compound [381]. The antiviral mechanism of these compounds was confirmed to be inhibition of IE2-mediated viral early gene transactivation and, like with WC5, a minimal 150-bp segment of the UL54 promoter is sufficient for inhibitory activity. However, the precise mechanism of action has not been elucidated, although prior knowledge of these bioactive compounds has led to the proposal that they are likely to interfere with pathways in HCMV-infected cells that are required for the switch from the IE to early phase of viral replication [364]. Despite the lack of a precise mechanism of action, repurposing of bioactive compounds for anti-HCMV activities may allow compound development to be fast-tracked, especially in the case of MND, as it is already an FDA-approved drug used in the treatment of hypertension [381].

7.2. IE1 Inhibitors

Identified inhibitors of IE2-dependent transactivation do not inhibit IE1-dependent transactivation and are thus specific to IE2 not IE1 function [364,382,383]. IE1's function as an IFN antagonist has been targeted for drug discovery via a modular cell-based screening platform designed to identify compounds that inhibit a viral IFN antagonist choice [373]. The platform utilizes two reporter cell-lines that provide a simple method to detect activation of IFN induction or signalling via an EGFP gene placed under the control of the IFN-β or an ISRE-containing promoter, respectively. IE1 counteracts type I IFN signalling by binding directly to STAT2 thereby preventing the ISGF3 transcription factor from binding ISRE elements in the promoters of ISGs (see Section 3.4) [225]. Therefore, a derivative of the IFN signalling reporter cell-line expressing IE1 was generated that blocks EGFP expression upon IFN signalling pathway activation. This reporter cell-line was used to screen a 16,000 compound library to identify compounds that release the IE1-imposed IFN response block and hence restore EGFP expression [373]. Two hit compounds, StA-IE1-2 and StA-IE1-3, were identified and demonstrated to be potent inhibitors of HCMV replication [373]. Compound characterization revealed that, instead of identifying anticipated IE1 IFN antagonist function inhibitors that target at the protein level, the compounds act at the mRNA level and interfere with IE1/2 transcription. The likely explanation is that IE1 expression was driven by MIEP sequences in the lentiviral vector used to create the IE1 reporter cell-line derivative. StA-IE1-2 and -3 are therefore also described in Section 6.7 concerning inhibition of

IE gene expression. Despite the unexpected results, the assay platform had previously been used to identify compounds that specifically target the IFN antagonist function of Respiratory Syncytial Virus non-structural protein 2 [373]. To the best of our knowledge, no other IE1-specific drug discovery screens have been undertaken.

7.3. Summary

Overall, strategies to target IE2 function have identified a number of interesting compounds, although their exact mechanism of action has not been fully elucidated. To date, compounds that target IE1 function have not been identified, and IE1 remains an important but underexploited potential drug target.

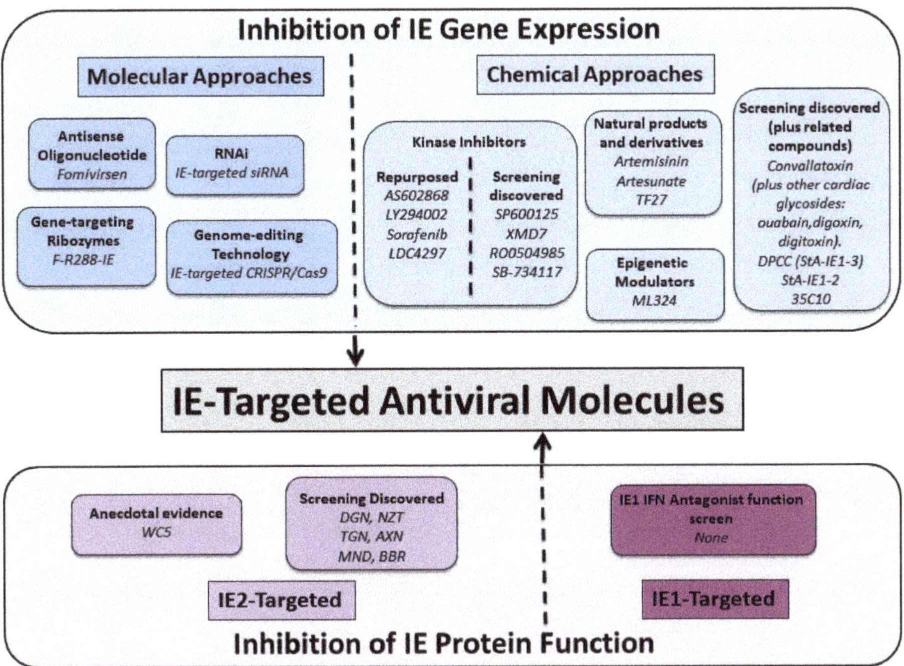

Figure 2. Schematic of molecular and chemical approaches used to target major IE gene expression and IE protein function. Key groups of molecules are listed by category, and examples of molecules within each category given in italics. DGN, deguelin; NZT, nitazoxanide; TGN, thioguanosine; AXN, alexidine dihydrochloride; MND, manidipine dihydrochloride.

8. Conclusions and Future Perspectives

This review provides an update on the regulation of HCMV major IE gene expression and IE1 and IE2 protein functions. We discussed existing clinically approved therapies and why major IE gene expression and IE1/2 protein functions are considered potential alternative targets for anti-HCMV strategies. We outlined the various molecular and chemical approaches that are being used to target major IE gene expression and protein function. Advances in molecular approaches, particularly genome-editing technology, are opening up new promising strategies for targeting HCMV. However, further research and development are required before this novel technology can be translated clinically. Key groups of small chemical inhibitors targeting major IE gene expression or IE1/2 protein function are highlighted in the review. A major challenge associated with the vast majority of these compounds is the complexity in elucidating their relevant targets, many of which are host cell proteins, and mechanism of action. Basic research to determine this knowledge will highlight the value of these compounds as chemical tools to further understand regulation of major IE gene expression and/or IE1/2 protein functions. It will also promote further in vivo and clinical testing of these molecules, which is currently limited to only a few studies and compounds. A key advantage of targeting major IE gene expression and IE1/2 function is the potential to inhibit reactivation from latency, a property that existing therapies, which target viral replication, do not achieve. However, testing key compounds for this attribute is mostly lacking due to the specialized methodology and limitations of in vitro cell-based latency models. Yet, the identification of compounds that repress latency is becoming more pressing as organ and haematopoietic stem cell transplantation has become more common. Existing anti-HCMV drugs are also not approved for use during pregnancy because of their teratogenic and embryotoxic effects in animal studies, yet the need for antiviral therapy in congenitally infected neonates for improved long-term outcomes is increasingly appreciated. Finally, IE1/2-directed drugs are not expected to confer cross-resistance to or interfere with the activity of existing drugs currently approved for HCMV monotherapy. They may therefore be combined with these drugs for combination therapies with improved efficacy. Overall, "bright and early" events in HCMV infection deserve more attention as a promising antiviral strategy against HCMV.

Author Contributions: C.S.A. and M.M.N. equally contributed to the conceptualization and writing of this review. All authors have read and agreed to the published version of the manuscript.

Funding: This research was funded by the Medical Research Council, grant number MR/P022146/1, to M.M.N.

Acknowledgments: We thank Christina Paulus (University of St Andrews) for helpful discussions and apologize to all those colleagues whose relevant work was not cited.

Conflicts of Interest: The authors declare no conflict of interest.

References

1. Yu, X.; Jih, J.; Jiang, J.; Zhou, Z.H. Atomic structure of the human cytomegalovirus capsid with its securing tegument layer of pp150. *Science* **2017**, *356*, 6892. [CrossRef] [PubMed]
2. Liu, F.; Zhou, Z.H. Comparative virion structures of human herpesviruses. In *Human Herpesviruses: Biology, Therapy, and Immunoprophylaxis*; Arvin, A., Campadelli-Fiume, G., Mocarski, E., Moore, P.S., Roizman, B., Whitley, R., Yamanishi, K., Eds.; Cambridge University Press: Cambridge, UK, 2007.
3. Stern-Ginossar, N.; Weisburd, B.; Michalski, A.; Le, V.T.; Hein, M.Y.; Huang, S.X.; Ma, M.; Shen, B.; Qian, S.B.; Hengel, H.; et al. Decoding human cytomegalovirus. *Science* **2012**, *338*, 1088–1093. [CrossRef]
4. Balazs, Z.; Tombacz, D.; Szucs, A.; Csabai, Z.; Megyeri, K.; Petrov, A.N.; Snyder, M.; Boldogkoi, Z. Long-read sequencing of human cytomegalovirus transcriptome reveals RNA isoforms carrying distinct coding potentials. *Sci. Rep.* **2017**, *7*, 15989. [CrossRef]
5. Gatherer, D.; Seirafian, S.; Cunningham, C.; Holton, M.; Dargan, D.J.; Baluchova, K.; Hector, R.D.; Galbraith, J.; Herzyk, P.; Wilkinson, G.W.; et al. High-resolution human cytomegalovirus transcriptome. *Proc. Natl. Acad. Sci. USA* **2011**, *108*, 19755–19760. [CrossRef] [PubMed]
6. Shnayder, M.; Nachshon, A.; Krishna, B.; Poole, E.; Boshkov, A.; Binyamin, A.; Maza, I.; Sinclair, J.; Schwartz, M.; Stern-Ginossar, N. Defining the transcriptional landscape during cytomegalovirus latency with single-cell RNA sequencing. *MBio* **2018**, *9*, e00013-18. [CrossRef] [PubMed]
7. Cheng, S.; Caviness, K.; Buehler, J.; Smithey, M.; Nikolich-Zugich, J.; Goodrum, F. Transcriptome-wide characterization of human cytomegalovirus in natural infection and experimental latency. *Proc. Natl. Acad. Sci. USA* **2017**, *114*, E10586–E10595. [CrossRef] [PubMed]
8. Balazs, Z.; Tombacz, D.; Szucs, A.; Snyder, M.; Boldogkoi, Z. Dual platform long-read RNA-sequencing dataset of the human cytomegalovirus lytic transcriptome. *Front. Genet.* **2018**, *9*, 432. [CrossRef]
9. Zalckvar, E.; Paulus, C.; Tillo, D.; Asbach-Nitzsche, A.; Lubling, Y.; Winterling, C.; Strieder, N.; Mücke, K.; Goodrum, F.; Segal, E.; et al. Nucleosome maps of the human cytomegalovirus genome reveal a temporal switch in chromatin organization linked to a major IE protein. *Proc. Natl. Acad. Sci. USA* **2013**, *110*, 13126–13131. [CrossRef]

10. Scherer, M.; Schilling, E.M.; Stamminger, T. The human CMV IE1 protein: An offender of PML nuclear bodies. *Adv. Anat. Embryol. Cell Biol.* **2017**, *223*, 77–94. [CrossRef]

11. Scherer, M.; Stamminger, T. The human CMV IE1 protein: Past and present developments. *Future Virol.* **2014**, *9*, 415–430. [CrossRef]

12. Paulus, C.; Nevels, M. The human cytomegalovirus major immediate-early proteins as antagonists of intrinsic and innate antiviral host responses. *Viruses* **2009**, *1*, 760–779. [CrossRef] [PubMed]

13. Meier, J.L.; Stinski, M.F. Major immediate-early enhancer and its gene products. In *Cytomegaloviruses: From Molecular Pathogenesis to Intervention*; Reddehase, M.J., Ed.; Caister Academic Press: Norfolk, UK, 2013; Volume 1.

14. Stinski, M.F.; Petrik, D.T. Functional roles of the human cytomegalovirus essential IE86 protein. *Curr. Top. Microbiol. Immunol.* **2008**, *325*, 133–152. [PubMed]

15. Torres, L.; Tang, Q. Immediate-Early (IE) gene regulation of cytomegalovirus: IE1- and pp71-mediated viral strategies against cellular defenses. *Virol. Sin.* **2014**, *29*, 343–352. [CrossRef] [PubMed]

16. White, E.A.; Spector, D.H. Early viral gene expression and function. In *Human Herpesviruses: Biology, Therapy, and Immunoprophylaxis*; Arvin, A., Campadelli-Fiume, G., Mocarski, E., Moore, P.S., Roizman, B., Whitley, R., Yamanishi, K., Eds.; Cambridge University Press: Cambridge, UK, 2007.

17. Griffiths, P.D.; Boeckh, M. Antiviral therapy for human cytomegalovirus. In *Human Herpesviruses: Biology, Therapy, and Immunoprophylaxis*; Arvin, A., Campadelli-Fiume, G., Mocarski, E., Moore, P.S., Roizman, B., Whitley, R., Yamanishi, K., Eds.; Cambridge University Press: Cambridge, UK, 2007.

18. Gentry, B.G.; Bogner, E.; Drach, J.C. Targeting the terminase: An important step forward in the treatment and prophylaxis of human cytomegalovirus infections. *Antivir. Res.* **2019**, *161*, 116–124. [CrossRef]

19. Close, W.L.; Anderson, A.N.; Pellett, P.E. Betaherpesvirus virion assembly and egress. *Adv. Exp. Med. Biol.* **2018**, *1045*, 167–207. [CrossRef]

20. Britt, B. Maturation and egress. In *Human Herpesviruses: Biology, Therapy, and Immunoprophylaxis*; Arvin, A., Campadelli-Fiume, G., Mocarski, E., Moore, P.S., Roizman, B., Whitley, R., Yamanishi, K., Eds.; Cambridge University Press: Cambridge, UK, 2007.

21. Anders, D.G.; Kerry, J.A.; Pari, G.S. DNA synthesis and late viral gene expression. In *Human Herpesviruses: Biology, Therapy, and Immunoprophylaxis*; Arvin, A., Campadelli-Fiume, G., Mocarski, E., Moore, P.S., Roizman, B., Whitley, R., Yamanishi, K., Eds.; Cambridge University Press: Cambridge, UK, 2007.

22. Schwartz, M.; Stern-Ginossar, N. The transcriptome of latent human cytomegalovirus. *J. Virol.* **2019**, *93*, e00047-19. [CrossRef]

23. Collins-McMillen, D.; Buehler, J.; Peppenelli, M.; Goodrum, F. Molecular determinants and the regulation of human cytomegalovirus latency and reactivation. *Viruses* **2018**, *10*, 444. [CrossRef]

24. Gerna, G.; Kabanova, A.; Lilleri, D. Human cytomegalovirus cell tropism and host cell receptors. *Vaccines* **2019**, *7*, 70. [CrossRef]

25. Sinzger, C.; Jahn, G. Human cytomegalovirus cell tropism and pathogenesis. *Intervirology* **1996**, *39*, 302–319. [CrossRef]

26. Poole, E.; Reeves, M.; Sinclair, J.H. The use of primary human cells (fibroblasts, monocytes, and others) to assess human cytomegalovirus function. *Methods Mol. Biol.* **2014**, *1119*, 81–98. [CrossRef]

27. Boeckh, M.; Geballe, A.P. Cytomegalovirus: Pathogen, paradigm, and puzzle. *J. Clin. Investig.* **2011**, *121*, 1673–1680. [CrossRef] [PubMed]

28. Griffiths, P.; Baraniak, I.; Reeves, M. The pathogenesis of human cytomegalovirus. *J. Pathol.* **2015**, *235*, 288–297. [CrossRef] [PubMed]

29. Stern, L.; Withers, B.; Avdic, S.; Gottlieb, D.; Abendroth, A.; Blyth, E.; Slobedman, B. Human Cytomegalovirus latency and reactivation in allogeneic hematopoietic stem cell transplant recipients. *Front. Microbiol.* **2019**, *10*, 1186. [CrossRef]

30. Kenneson, A.; Cannon, M.J. Review and meta-analysis of the epidemiology of congenital cytomegalovirus (CMV) infection. *Rev. Med. Virol.* **2007**, *17*, 253–276. [CrossRef] [PubMed]

31. Dollard, S.C.; Grosse, S.D.; Ross, D.S. New estimates of the prevalence of neurological and sensory sequelae and mortality associated with congenital cytomegalovirus infection. *Rev. Med. Virol.* **2007**, *17*, 355–363. [CrossRef] [PubMed]

32. Marsico, C.; Kimberlin, D.W. Congenital Cytomegalovirus infection: Advances and challenges in diagnosis, prevention and treatment. *Ital. J. Pediatr.* **2017**, *43*, 38. [CrossRef]

33. de Vries, J.J.; Vossen, A.C.; Kroes, A.C.; van der Zeijst, B.A. Implementing neonatal screening for congenital cytomegalovirus: Addressing the deafness of policy makers. *Rev. Med. Virol.* **2011**, *21*, 54–61. [CrossRef]

34. Manicklal, S.; Emery, V.C.; Lazzarotto, T.; Boppana, S.B.; Gupta, R.K. The "silent" global burden of congenital cytomegalovirus. *Clin. Microbiol. Rev.* **2013**, *26*, 86–102. [CrossRef]

35. Cannon, M.J.; Schmid, D.S.; Hyde, T.B. Review of cytomegalovirus seroprevalence and demographic characteristics associated with infection. *Rev. Med. Virol.* **2010**, *20*, 202–213. [CrossRef]

36. Gane, E.; Saliba, F.; Valdecasas, G.J.; O'Grady, J.; Pescovitz, M.D.; Lyman, S.; Robinson, C.A. Randomised trial of efficacy and safety of oral ganciclovir in the prevention of cytomegalovirus disease in liver-transplant recipients. The Oral Ganciclovir International Transplantation Study Group [corrected]. *Lancet* **1997**, *350*, 1729–1733. [CrossRef]

37. Lowance, D.; Neumayer, H.H.; Legendre, C.M.; Squifflet, J.P.; Kovarik, J.; Brennan, P.J.; Norman, D.; Mendez, R.; Keating, M.R.; Coggon, G.L.; et al. Valacyclovir for the prevention of cytomegalovirus disease after renal transplantation. International Valacyclovir Cytomegalovirus Prophylaxis Transplantation Study Group. *N. Engl. J. Med.* **1999**, *340*, 1462–1470. [CrossRef] [PubMed]

38. Wreghitt, T.G.; Abel, S.J.; McNeil, K.; Parameshwar, J.; Stewart, S.; Cary, N.; Sharples, L.; Large, S.; Wallwork, J. Intravenous ganciclovir prophylaxis for cytomegalovirus in heart, heart-lung, and lung transplant recipients. *Transplant Int.* **1999**, *12*, 254–260. [CrossRef]

39. Brizic, I.; Lisnic, B.; Brune, W.; Hengel, H.; Jonjic, S. Cytomegalovirus infection: Mouse model. *Curr. Protoc. Immunol.* **2018**, e51. [CrossRef] [PubMed]

40. Reddehase, M.J.; Lemmermann, N.A.W. Mouse model of cytomegalovirus disease and immunotherapy in the immunocompromised host: Predictions for medical translation that survived the "test of time". *Viruses* **2018**, *10*, 693. [CrossRef]

41. Thackeray, R.; Wright, A.; Chipman, K. Congenital cytomegalovirus reference material: A content analysis of coverage and accuracy. *Matern. Child Health J.* **2014**, *18*, 584–591. [CrossRef]

42. Kimberlin, D.W.; Jester, P.M.; Sanchez, P.J.; Ahmed, A.; Arav-Boger, R.; Michaels, M.G.; Ashouri, N.; Englund, J.A.; Estrada, B.; Jacobs, R.F.; et al. Valganciclovir for symptomatic congenital cytomegalovirus disease. *N. Engl. J. Med.* **2015**, *372*, 933–943. [CrossRef]

43. Price, S.M.; Bonilla, E.; Zador, P.; Levis, D.M.; Kilgo, C.L.; Cannon, M.J. Educating women about congenital cytomegalovirus: Assessment of health education materials through a web-based survey. *BMC Womens Health* **2014**, *14*, 144. [CrossRef]

44. Harrison, G.J. Current controversies in diagnosis, management, and prevention of congenital cytomegalovirus: Updates for the pediatric practitioner. *Pediatr. Ann.* **2015**, *44*, e115–e125. [CrossRef]

45. Nigro, G.; Adler, S.P. Hyperimmunoglobulin for prevention of congenital cytomegalovirus disease. *Clin. Infect. Dis.* **2013**, *57* (Suppl. 4), S193–S195. [CrossRef]

46. Buxmann, H.; Stackelberg, O.M.; Schlosser, R.L.; Enders, G.; Gonser, M.; Meyer-Wittkopf, M.; Hamprecht, K.; Enders, M. Use of cytomegalovirus hyperimmunoglobulin for prevention of congenital cytomegalovirus disease: A retrospective analysis. *J. Perinat. Med.* **2012**, *40*, 439–446. [CrossRef]

47. Nigro, G.; Adler, S.P.; La Torre, R.; Best, A.M. Congenital Cytomegalovirus Collaborating, G. Passive immunization during pregnancy for congenital cytomegalovirus infection. *N. Engl. J. Med.* **2005**, *353*, 1350–1362. [CrossRef] [PubMed]

48. Nigro, G.; Adler, S.P.; Parruti, G.; Anceschi, M.M.; Coclite, E.; Pezone, I.; Di Renzo, G.C. Immunoglobulin therapy of fetal cytomegalovirus infection occurring in the first half of pregnancy—A case-control study of the outcome in children. *J. Infect. Dis.* **2012**, *205*, 215–227. [CrossRef] [PubMed]

49. Visentin, S.; Manara, R.; Milanese, L.; Da Roit, A.; Forner, G.; Salviato, E.; Citton, V.; Magno, F.M.; Orzan, E.; Morando, C.; et al. Early primary cytomegalovirus infection in pregnancy: Maternal hyperimmunoglobulin therapy improves outcomes among infants at 1 year of age. *Clin. Infect. Dis.* **2012**, *55*, 497–503. [CrossRef]

50. Stern, A.; Papanicolaou, G.A. CMV Prevention and Treatment in Transplantation: What's New in 2019. *Curr. Infect. Dis. Rep.* **2019**, *21*, 45. [CrossRef] [PubMed]

51. Gerna, G.; Lilleri, D. Human cytomegalovirus (HCMV) infection/re-infection: Development of a protective HCMV vaccine. *New Microbiol.* **2019**, *42*, 1–20. [PubMed]

52. Diamond, D.J.; La Rosa, C.; Chiuppesi, F.; Contreras, H.; Dadwal, S.; Wussow, F.; Bautista, S.; Nakamura, R.; Zaia, J.A. A fifty-year odyssey: Prospects for a cytomegalovirus vaccine in transplant and congenital infection. *Expert Rev. Vaccines* **2018**, *17*, 889–911. [CrossRef] [PubMed]

53. Plotkin, S.A.; Boppana, S.B. Vaccination against the human cytomegalovirus. *Vaccine* **2019**, *37*, 7437–7442. [CrossRef]

54. Fu, T.M.; An, Z.; Wang, D. Progress on pursuit of human cytomegalovirus vaccines for prevention of congenital infection and disease. *Vaccine* **2014**, *32*, 2525–2533. [CrossRef]

55. Wang, D.; Fu, T.M. Progress on human cytomegalovirus vaccines for prevention of congenital infection and disease. *Curr. Opin. Virol.* **2014**, *6*, 13–23. [CrossRef]

56. Lilja, A.E.; Mason, P.W. The next generation recombinant human cytomegalovirus vaccine candidates-beyond gB. *Vaccine* **2012**, *30*, 6980–6990. [CrossRef]

57. Riddell, S.R.; Greenberg, P.D. T-cell therapy of cytomegalovirus and human immunodeficiency virus infection. *J. Antimicrob. Chemother.* **2000**, *45* (Suppl. T3), 35–43. [CrossRef]

58. Peggs, K.S.; Verfuerth, S.; Pizzey, A.; Khan, N.; Guiver, M.; Moss, P.A.; Mackinnon, S. Adoptive cellular therapy for early cytomegalovirus infection after allogeneic stem-cell transplantation with virus-specific T-cell lines. *Lancet* **2003**, *362*, 1375–1377. [CrossRef]

59. Mui, T.S.; Kapp, M.; Einsele, H.; Grigoleit, G.U. T-cell therapy for cytomegalovirus infection. *Curr. Opin. Organ Transplant.* **2010**, *15*, 744–750. [CrossRef] [PubMed]

60. Britt, W.J.; Prichard, M.N. New therapies for human cytomegalovirus infections. *Antivir. Res.* **2018**, *159*, 153–174. [CrossRef] [PubMed]

61. Andrei, G.; De Clercq, E.; Snoeck, R. Novel inhibitors of human CMV. *Curr. Opin. Investig. Drugs* **2008**, *9*, 132–145. [PubMed]

62. Andrei, G.; De Clercq, E.; Snoeck, R. Drug targets in cytomegalovirus infection. *Infect. Disord. Drug Targets* **2009**, *9*, 201–222. [CrossRef]

63. Krishna, B.A.; Wills, M.R.; Sinclair, J.H. Advances in the treatment of cytomegalovirus. *Br. Med. Bull.* **2019**, *131*, 5–17. [CrossRef]

64. Al-Badr, A.A.; Ajarim, T.D.S. Ganciclovir. *Profiles Drug Subst. Excip. Relat. Methodol.* **2018**, *43*, 1–208. [CrossRef]

65. Shiraki, K. Antiviral drugs against alphaherpesvirus. *Adv. Exp. Med. Biol.* **2018**, *1045*, 103–122. [CrossRef]

66. El Helou, G.; Razonable, R.R. Safety considerations with current and emerging antiviral therapies for cytomegalovirus infection in transplantation. *Expert Opin. Drug Saf.* **2019**, *18*, 1017–1030. [CrossRef]

67. Razonable, R.R. Drug-resistant cytomegalovirus: Clinical implications of specific mutations. *Curr. Opin. Organ Transplant.* **2018**, *23*, 388–394. [CrossRef] [PubMed]

68. Chevillotte, M.; von Einem, J.; Meier, B.M.; Lin, F.M.; Kestler, H.A.; Mertens, T. A new tool linking human cytomegalovirus drug resistance mutations to resistance phenotypes. *Antivir. Res.* **2010**, *85*, 318–327. [CrossRef]

69. Frange, P.; Leruez-Ville, M. Maribavir, brincidofovir and letermovir: Efficacy and safety of new antiviral drugs for treating cytomegalovirus infections. *Med. Mal. Infect.* **2018**, *48*, 495–502. [CrossRef] [PubMed]

70. Gerna, G.; Lilleri, D.; Baldanti, F. An overview of letermovir: A cytomegalovirus prophylactic option. *Expert Opin. Pharmacother.* **2019**, *20*, 1429–1438. [CrossRef] [PubMed]

71. Fowler, K.B.; Boppana, S.B. Congenital cytomegalovirus infection. *Semin. Perinatol.* **2018**, *42*, 149–154. [CrossRef]

72. Dupont, L.; Reeves, M.B. Cytomegalovirus latency and reactivation: Recent insights into an age old problem. *Rev. Med. Virol.* **2016**, *26*, 75–89. [CrossRef]

73. Elder, E.; Sinclair, J. HCMV latency: What regulates the regulators? *Med. Microbiol. Immunol.* **2019**, *208*, 431–438. [CrossRef]

74. Stinski, M.F.; Isomura, H. Role of the cytomegalovirus major immediate early enhancer in acute infection and reactivation from latency. *Med. Microbiol. Immunol.* **2008**, *197*, 223–231. [CrossRef]

75. Tarrant-Elorza, M.; Rossetto, C.C.; Pari, G.S. Maintenance and Replication of the Human Cytomegalovirus Genome during Latency. *Cell Host Microbe* **2014**, *16*, 43–54. [CrossRef]

76. Arend, K.C.; Ziehr, B.; Vincent, H.A.; Moorman, N.J. Multiple transcripts encode full-length human cytomegalovirus IE1 and IE2 proteins during lytic infection. *J. Virol.* **2016**, *90*, 8855–8865. [CrossRef] [PubMed]

77. Puchtler, E.; Stamminger, T. An inducible promoter mediates abundant expression from the immediate-early 2 gene region of human cytomegalovirus at late times after infection. *J. Virol.* **1991**, *65*, 6301–6306. [CrossRef] [PubMed]

78. Stenberg, R.M.; Depto, A.S.; Fortney, J.; Nelson, J.A. Regulated expression of early and late RNAs and proteins from the human cytomegalovirus immediate-early gene region. *J. Virol.* **1989**, *63*, 2699–2708. [CrossRef] [PubMed]

79. Kondo, K.; Xu, J.; Mocarski, E.S. Human cytomegalovirus latent gene expression in granulocyte-macrophage progenitors in culture and in seropositive individuals. *Proc. Natl. Acad. Sci. USA* **1996**, *93*, 11137–11142. [CrossRef] [PubMed]

80. Yang, C.Q.; Miao, L.F.; Pan, X.; Wu, C.C.; Rayner, S.; Mocarski, E.S.; Ye, H.Q.; Luo, M.H. Natural antisense transcripts of UL123 packaged in human cytomegalovirus virions. *Arch. Virol.* **2014**, *159*, 147–151. [CrossRef] [PubMed]

81. Kondo, K.; Mocarski, E.S. Cytomegalovirus latency and latency-specific transcription in hematopoietic progenitors. *Scand. J. Infect. Dis. Suppl.* **1995**, *99*, 63–67.

82. Wilkinson, G.W.; Akrigg, A.; Greenaway, P.J. Transcription of the immediate early genes of human cytomegalovirus strain AD169. *Virus Res.* **1984**, *1*, 101–106. [CrossRef]

83. Stinski, M.F.; Thomsen, D.R.; Stenberg, R.M.; Goldstein, L.C. Organization and expression of the immediate early genes of human cytomegalovirus. *J. Virol.* **1983**, *46*, 1–14. [CrossRef]

84. Collins-McMillen, D.; Rak, M.; Buehler, J.C.; Igarashi-Hayes, S.; Kamil, J.P.; Moorman, N.J.; Goodrum, F. Alternative promoters drive human cytomegalovirus reactivation from latency. *Proc. Natl. Acad. Sci. USA* **2019**, *116*, 17492–17497. [CrossRef]

85. Isomura, H.; Tsurumi, T.; Stinski, M.F. Role of the proximal enhancer of the major immediate-early promoter in human cytomegalovirus replication. *J. Virol.* **2004**, *78*, 12788–12799. [CrossRef]

86. Keller, M.J.; Wu, A.W.; Andrews, J.I.; McGonagill, P.W.; Tibesar, E.E.; Meier, J.L. Reversal of human cytomegalovirus major immediate-early enhancer/promoter silencing in quiescently infected cells via the cyclic AMP signaling pathway. *J. Virol.* **2007**, *81*, 6669–6681. [CrossRef]

87. Lubon, H.; Ghazal, P.; Hennighausen, L.; Reynolds-Kohler, C.; Lockshin, C.; Nelson, J. Cell-specific activity of the modulator region in the human cytomegalovirus major immediate-early gene. *Mol. Cell. Biol.* **1989**, *9*, 1342–1345. [CrossRef] [PubMed]

88. Nelson, J.A.; Reynolds-Kohler, C.; Smith, B.A. Negative and positive regulation by a short segment in the 5′-flanking region of the human cytomegalovirus major immediate-early gene. *Mol. Cell. Biol.* **1987**, *7*, 4125–4129. [CrossRef] [PubMed]

89. Meier, J.L.; Stinski, M.F. Effect of a modulator deletion on transcription of the human cytomegalovirus major immediate-early genes in infected undifferentiated and differentiated cells. *J. Virol.* **1997**, *71*, 1246–1255. [CrossRef] [PubMed]

90. Ghazal, P.; Lubon, H.; Reynolds-Kohler, C.; Hennighausen, L.; Nelson, J.A. Interactions between cellular regulatory proteins and a unique sequence region in the human cytomegalovirus major immediate-early promoter. *Virology* **1990**, *174*, 18–25. [CrossRef]

91. Angulo, A.; Kerry, D.; Huang, H.; Borst, E.M.; Razinsky, A.; Wu, J.; Hobom, U.; Messerle, M.; Ghazal, P. Identification of a boundary domain adjacent to the potent human cytomegalovirus enhancer that represses transcription of the divergent UL127 promoter. *J. Virol.* **2000**, *74*, 2826–2839. [CrossRef] [PubMed]

92. Lashmit, P.E.; Lundquist, C.A.; Meier, J.L.; Stinski, M.F. Cellular repressor inhibits human cytomegalovirus transcription from the UL127 promoter. *J. Virol.* **2004**, *78*, 5113–5123. [CrossRef]

93. Lundquist, C.A.; Meier, J.L.; Stinski, M.F. A strong negative transcriptional regulatory region between the human cytomegalovirus UL127 gene and the major immediate-early enhancer. *J. Virol.* **1999**, *73*, 9039–9052. [CrossRef]

94. Lee, J.; Klase, Z.; Gao, X.; Caldwell, J.S.; Stinski, M.F.; Kashanchi, F.; Chao, S.H. Cellular homeoproteins, SATB1 and CDP, bind to the unique region between the human cytomegalovirus UL127 and major immediate-early genes. *Virology* **2007**, *366*, 117–125. [CrossRef]

95. Chao, S.H.; Harada, J.N.; Hyndman, F.; Gao, X.; Nelson, C.G.; Chanda, S.K.; Caldwell, J.S. PDX1, a cellular homeoprotein, binds to and regulates the activity of human cytomegalovirus immediate early promoter. *J. Biol. Chem.* **2004**, *279*, 16111–16120. [CrossRef]

96. Thomsen, D.R.; Stenberg, R.M.; Goins, W.F.; Stinski, M.F. Promoter-regulatory region of the major immediate early gene of human cytomegalovirus. *Proc. Natl. Acad. Sci. USA* **1984**, *81*, 659–663. [CrossRef]

97. Honess, R.W.; Gompels, U.A.; Barrell, B.G.; Craxton, M.; Cameron, K.R.; Staden, R.; Chang, Y.N.; Hayward, G.S. Deviations from expected frequencies of CpG dinucleotides in herpesvirus DNAs may be diagnostic of differences in the states of their latent genomes. *J. Gen. Virol.* **1989**, *70*, 837–855. [CrossRef] [PubMed]

98. Moritz, B.; Becker, P.B.; Göpfert, U. CMV promoter mutants with a reduced propensity to productivity loss in CHO cells. *Sci. Rep.* **2015**, *5*, 16952. [CrossRef] [PubMed]

99. Kong, Q.; Wu, M.; Huan, Y.; Zhang, L.; Liu, H.; Bou, G.; Luo, Y.; Mu, Y.; Liu, Z. Transgene expression is associated with copy number and cytomegalovirus promoter methylation in transgenic pigs. *PLoS ONE* **2009**, *4*, e6679. [CrossRef] [PubMed]

100. Osterlehner, A.; Simmeth, S.; Göpfert, U. Promoter methylation and transgene copy numbers predict unstable protein production in recombinant Chinese hamster ovary cell lines. *Biotechnol. Bioeng.* **2011**, *108*, 2670–2681. [CrossRef] [PubMed]

101. Yang, Y.; Mariati; Chusainow, J.; Yap, M.G. DNA methylation contributes to loss in productivity of monoclonal antibody-producing CHO cell lines. *J. Biotechnol.* **2010**, *147*, 180–185. [CrossRef]

102. Estekizadeh, A.; Landazuri, N.; Pantalone, M.R.; Davoudi, B.; Hu, L.F.; Nawaz, I.; Stragliotto, G.; Ekstrom, T.J.; Rahbar, A. 5-azacytidine treatment results in nuclear exclusion of DNA methyltransferase1, as well as reduced proliferation and invasion in human cytomegalovirusinfected glioblastoma cells. *Oncol. Rep.* **2019**, *41*, 2927–2936. [CrossRef] [PubMed]

103. Boom, R.; Geelen, J.L.; Sol, C.J.; Minnaar, R.P.; van der Noordaa, J. Resistance to methylation de novo of the human cytomegalovirus immediate early enhancer in a model for virus latency and reactivation in vitro. *J. Gen. Virol.* **1987**, *68*, 2839–2852. [CrossRef]

104. Hummel, M.; Yan, S.; Li, Z.; Varghese, T.K.; Abecassis, M. Transcriptional reactivation of murine cytomegalovirus ie gene expression by 5-aza-2'-deoxycytidine and trichostatin A in latently infected cells despite lack of methylation of the major immediate-early promoter. *J. Gen. Virol.* **2007**, *88*, 1097–1102. [CrossRef]

105. Nitzsche, A.; Paulus, C.; Nevels, M. Temporal dynamics of cytomegalovirus chromatin assembly in productively infected human cells. *J. Virol.* **2008**, *82*, 11167–11180. [CrossRef]

106. Knipe, D.M.; Lieberman, P.M.; Jung, J.U.; McBride, A.A.; Morris, K.V.; Ott, M.; Margolis, D.; Nieto, A.; Nevels, M.; Parks, R.J.; et al. Snapshots: Chromatin control of viral infection. *Virology* **2013**, *435*, 141–156. [CrossRef]

107. Liu, X.F.; Wang, X.; Yan, S.; Zhang, Z.; Abecassis, M.; Hummel, M. Epigenetic control of cytomegalovirus latency and reactivation. *Viruses* **2013**, *5*, 1325–1345. [CrossRef]

108. Paulus, C.; Nitzsche, A.; Nevels, M. Chromatinisation of herpesvirus genomes. *Rev. Med. Virol.* **2010**, *20*, 34–50. [CrossRef]

109. Liu, X.F.; Yan, S.; Abecassis, M.; Hummel, M. Establishment of murine cytomegalovirus latency in vivo is associated with changes in histone modifications and recruitment of transcriptional repressors to the major immediate-early promoter. *J. Virol.* **2008**, *82*, 10922–10931. [CrossRef] [PubMed]

110. Nitzsche, A.; Steinhäusser, C.; Mücke, K.; Paulus, C.; Nevels, M. Histone H3 lysine 4 methylation marks postreplicative human cytomegalovirus chromatin. *J. Virol.* **2012**, *86*, 9817–9827. [CrossRef] [PubMed]

111. Cuevas-Bennett, C.; Shenk, T. Dynamic histone H3 acetylation and methylation at human cytomegalovirus promoters during replication in fibroblasts. *J. Virol.* **2008**, *82*, 9525–9536. [CrossRef] [PubMed]

112. Groves, I.J.; Reeves, M.B.; Sinclair, J.H. Lytic infection of permissive cells with human cytomegalovirus is regulated by an intrinsic 'pre-immediate-early' repression of viral gene expression mediated by histone post-translational modification. *J. Gen. Virol.* **2009**, *90*, 2364–2374. [CrossRef] [PubMed]

113. Soo, B.P.C.; Tay, J.; Ng, S.; Ho, S.C.L.; Yang, Y.; Chao, S.H. Correlation between expression of recombinant proteins and abundance of H3K4Me3 on the enhancer of human cytomegalovirus major immediate-early promoter. *Mol. Biotechnol.* **2017**, *59*, 315–322. [CrossRef]

114. Moritz, B.; Woltering, L.; Becker, P.B.; Göpfert, U. High levels of histone H3 acetylation at the CMV promoter are predictive of stable expression in Chinese hamster ovary cells. *Biotechnol. Prog.* **2016**, *32*, 776–786. [CrossRef]

115. Ioudinkova, E.; Arcangeletti, M.C.; Rynditch, A.; De Conto, F.; Motta, F.; Covan, S.; Pinardi, F.; Razin, S.V.; Chezzi, C. Control of human cytomegalovirus gene expression by differential histone modifications during lytic and latent infection of a monocytic cell line. *Gene* **2006**, *384*, 120–128. [CrossRef]

116. Murphy, J.C.; Fischle, W.; Verdin, E.; Sinclair, J.H. Control of cytomegalovirus lytic gene expression by histone acetylation. *EMBO J.* **2002**, *21*, 1112–1120. [CrossRef]

117. Reeves, M.B.; MacAry, P.A.; Lehner, P.J.; Sissons, J.G.; Sinclair, J.H. Latency, chromatin remodeling, and reactivation of human cytomegalovirus in the dendritic cells of healthy carriers. *Proc. Natl. Acad. Sci. USA* **2005**, *102*, 4140–4145. [CrossRef] [PubMed]

118. Reeves, M.B.; Lehner, P.J.; Sissons, J.G.; Sinclair, J.H. An in vitro model for the regulation of human cytomegalovirus latency and reactivation in dendritic cells by chromatin remodelling. *J. Gen. Virol.* **2005**, *86*, 2949–2954. [CrossRef]

119. Rauwel, B.; Jang, S.M.; Cassano, M.; Kapopoulou, A.; Barde, I.; Trono, D. Release of human cytomegalovirus from latency by a KAP1/TRIM28 phosphorylation switch. *Elife* **2015**, *4*, 68. [CrossRef] [PubMed]

120. Rossetto, C.C.; Tarrant-Elorza, M.; Pari, G.S. Cis and trans acting factors involved in human cytomegalovirus experimental and natural latent infection of CD14 (+) monocytes and CD34 (+) Cells. *PLoS Pathog.* **2013**, *9*, e1003366. [CrossRef] [PubMed]

121. Abraham, C.G.; Kulesza, C.A. Polycomb repressive complex 2 silences human cytomegalovirus transcription in quiescent infection models. *J. Virol.* **2013**, *87*, 13193–13205. [CrossRef]

122. Terhune, S.S.; Moorman, N.J.; Cristea, I.M.; Savaryn, J.P.; Cuevas-Bennett, C.; Rout, M.P.; Chait, B.T.; Shenk, T. Human cytomegalovirus UL29/28 protein interacts with components of the NuRD complex which promote accumulation of immediate-early RNA. *PLoS Pathog.* **2010**, *6*, e1000965. [CrossRef]

123. Liang, Y.; Vogel, J.L.; Arbuckle, J.H.; Rai, G.; Jadhav, A.; Simeonov, A.; Maloney, D.J.; Kristie, T.M. Targeting the JMJD2 histone demethylases to epigenetically control herpesvirus infection and reactivation from latency. *Sci. Transl. Med.* **2013**, *5*, 167ra165. [CrossRef]

124. Nevels, M.; Paulus, C.; Shenk, T. Human cytomegalovirus immediate-early 1 protein facilitates viral replication by antagonizing histone deacetylation. *Proc. Natl. Acad. Sci. USA* **2004**, *101*, 17234–17239. [CrossRef]

125. Lee, S.H.; Albright, E.R.; Lee, J.H.; Jacobs, D.; Kalejta, R.F. Cellular defense against latent colonization foiled by human cytomegalovirus UL138 protein. *Sci. Adv.* **2015**, *1*, e1501164. [CrossRef]

126. Reeves, M.; Murphy, J.; Greaves, R.; Fairley, J.; Brehm, A.; Sinclair, J. Autorepression of the human cytomegalovirus major immediate-early promoter/enhancer at late times of infection is mediated by the recruitment of chromatin remodeling enzymes by IE86. *J. Virol.* **2006**, *80*, 9998–10009. [CrossRef]

127. Bigley, T.M.; Reitsma, J.M.; Mirza, S.P.; Terhune, S.S. Human cytomegalovirus pUL97 regulates the viral major immediate early promoter by phosphorylation-mediated disruption of histone deacetylase 1 binding. *J. Virol.* **2013**, *87*, 7393–7408. [CrossRef] [PubMed]

128. Kew, V.G.; Yuan, J.; Meier, J.; Reeves, M.B. Mitogen and stress activated kinases act co-operatively with CREB during the induction of human cytomegalovirus immediate-early gene expression from latency. *PLoS Pathog.* **2014**, *10*, e1004195. [CrossRef] [PubMed]

129. Yee, L.F.; Lin, P.L.; Stinski, M.F. Ectopic expression of HCMV IE72 and IE86 proteins is sufficient to induce early gene expression but not production of infectious virus in undifferentiated promonocytic THP-1 cells. *Virology* **2007**, *363*, 174–188. [CrossRef] [PubMed]

130. Reeves, M.B. Chromatin-mediated regulation of cytomegalovirus gene expression. *Virus Res.* **2010**, *157*, 134–143. [CrossRef]

131. Nevels, M.; Nitzsche, A.; Paulus, C. How to control an infectious bead string: Nucleosome-based regulation and targeting of herpesvirus chromatin. *Rev. Med. Virol.* **2011**, *21*, 154–180. [CrossRef]

132. Saffert, R.T.; Kalejta, R.F. Human cytomegalovirus gene expression is silenced by Daxx-mediated intrinsic immune defense in model latent infections established in vitro. *J. Virol.* **2007**, *81*, 9109–9120. [CrossRef]

133. Lallemand-Breitenbach, V.; de The, H. PML nuclear bodies: From architecture to function. *Curr. Opin. Cell Biol.* **2018**, *52*, 154–161. [CrossRef]

134. Full, F.; Ensser, A. Early nuclear events after herpesviral infection. *J. Clin. Med.* **2019**, *8*, 1408. [CrossRef]

135. Scherer, M.; Stamminger, T. Emerging role of PML nuclear bodies in innate immune signaling. *J. Virol.* **2016**, *90*, 5850–5854. [CrossRef]

136. Landolfo, S.; De Andrea, M.; Dell'Oste, V.; Gugliesi, F. Intrinsic host restriction factors of human cytomegalovirus replication and mechanisms of viral escape. *World J. Virol.* **2016**, *5*, 87–96. [CrossRef]

137. Tavalai, N.; Stamminger, T. Intrinsic cellular defense mechanisms targeting human cytomegalovirus. *Virus Res.* **2011**, *157*, 128–133. [CrossRef] [PubMed]

138. Geoffroy, M.C.; Chelbi-Alix, M.K. Role of promyelocytic leukemia protein in host antiviral defense. *J. Interferon Cytokine Res.* **2011**, *31*, 145–158. [CrossRef] [PubMed]

139. Dunphy, G.; Flannery, S.M.; Almine, J.F.; Connolly, D.J.; Paulus, C.; Jonsson, K.L.; Jakobsen, M.R.; Nevels, M.M.; Bowie, A.G.; Unterholzner, L. Non-canonical activation of the DNA sensing adaptor STING by ATM and IFI16 mediates NF-κB signaling after nuclear DNA damage. *Mol. Cell* **2018**, *71*, 745–760. [CrossRef] [PubMed]

140. Li, T.; Chen, J.; Cristea, I.M. Human cytomegalovirus tegument protein pUL83 inhibits IFI16-mediated DNA sensing for immune evasion. *Cell Host Microbe* **2013**, *14*, 591–599. [CrossRef] [PubMed]

141. Diner, B.A.; Lum, K.K.; Toettcher, J.E.; Cristea, I.M. Viral DNA sensors IFI16 and cyclic GMP-AMP synthase possess distinct functions in regulating viral gene expression, immune defenses, and apoptotic responses during herpesvirus infection. *MBio* **2016**, *7*, e01553-16. [CrossRef]

142. Gariano, G.R.; Dell'Oste, V.; Bronzini, M.; Gatti, D.; Luganini, A.; De Andrea, M.; Gribaudo, G.; Gariglio, M.; Landolfo, S. The intracellular DNA sensor IFI16 gene acts as restriction factor for human cytomegalovirus replication. *PLoS Pathog.* **2012**, *8*, e1002498. [CrossRef]

143. Cristea, I.M.; Moorman, N.J.; Terhune, S.S.; Cuevas, C.D.; O'Keefe, E.S.; Rout, M.P.; Chait, B.T.; Shenk, T. Human cytomegalovirus pUL83 stimulates activity of the viral immediate-early promoter through its interaction with the cellular IFI16 protein. *J. Virol.* **2010**, *84*, 7803–7814. [CrossRef]

144. Elder, E.G.; Krishna, B.A.; Williamson, J.; Lim, E.Y.; Poole, E.; Sedikides, G.X.; Wills, M.; O'Connor, C.M.; Lehner, P.J.; Sinclair, J. Interferon-responsive genes are targeted during the establishment of human cytomegalovirus latency. *MBio* **2019**, *10*, e02574-19. [CrossRef]

145. Wang, X.; Huong, S.M.; Chiu, M.L.; Raab-Traub, N.; Huang, E.S. Epidermal growth factor receptor is a cellular receptor for human cytomegalovirus. *Nature* **2003**, *424*, 456–461. [CrossRef]

146. Wu, K.; Oberstein, A.; Wang, W.; Shenk, T. Role of PDGF receptor-α during human cytomegalovirus entry into fibroblasts. *Proc. Natl. Acad. Sci. USA* **2018**, *115*, E9889–E9898. [CrossRef]

147. Soroceanu, L.; Akhavan, A.; Cobbs, C.S. Platelet-derived growth factor-α receptor activation is required for human cytomegalovirus infection. *Nature* **2008**, *455*, 391–395. [CrossRef] [PubMed]

148. Wang, D.; Shenk, T. Human cytomegalovirus virion protein complex required for epithelial and endothelial cell tropism. *Proc. Natl. Acad. Sci. USA* **2005**, *102*, 18153–18158. [CrossRef] [PubMed]

149. Nogalski, M.T.; Chan, G.C.; Stevenson, E.V.; Collins-McMillen, D.K.; Yurochko, A.D. The HCMV gH/gL/UL128-131 complex triggers the specific cellular activation required for efficient viral internalization into target monocytes. *PLoS Pathog.* **2013**, *9*, e1003463. [CrossRef] [PubMed]

150. Altman, A.M.; Mahmud, J.; Nikolovska-Coleska, Z.; Chan, G. HCMV modulation of cellular PI3K/AKT/mTOR signaling: New opportunities for therapeutic intervention? *Antivir. Res.* **2019**, *163*, 82–90. [CrossRef]

151. Buchkovich, N.J.; Yu, Y.; Zampieri, C.A.; Alwine, J.C. The TORrid affairs of viruses: Effects of mammalian DNA viruses on the PI3K-Akt-mTOR signalling pathway. *Nat. Rev. Microbiol.* **2008**, *6*, 266–275. [CrossRef]

152. Campadelli-Fiume, G.; Collins-McMillen, D.; Gianni, T.; Yurochko, A.D. Integrins as herpesvirus receptors and mediators of the host signalosome. *Annu. Rev. Virol.* **2016**, *3*, 215–236. [CrossRef]

153. Liu, P.; Cheng, H.; Roberts, T.M.; Zhao, J.J. Targeting the phosphoinositide 3-kinase pathway in cancer. *Nat. Rev. Drug Discov.* **2009**, *8*, 627–644. [CrossRef]

154. De Santis, M.C.; Gulluni, F.; Campa, C.C.; Martini, M.; Hirsch, E. Targeting PI3K signaling in cancer: Challenges and advances. *Biochim. Biophys. Acta Rev. Cancer* **2019**, *1871*, 361–366. [CrossRef]

155. Fruman, D.A.; Chiu, H.; Hopkins, B.D.; Bagrodia, S.; Cantley, L.C.; Abraham, R.T. The PI3K pathway in human disease. *Cell* **2017**, *170*, 605–635. [CrossRef]

156. Johnson, R.A.; Wang, X.; Ma, X.L.; Huong, S.M.; Huang, E.S. Human cytomegalovirus up-regulates the phosphatidylinositol 3-kinase (PI3-K) pathway: Inhibition of PI3-K activity inhibits viral replication and virus-induced signaling. *J. Virol.* **2001**, *75*, 6022–6032. [CrossRef]

157. Yu, Y.; Alwine, J.C. Human cytomegalovirus major immediate-early proteins and simian virus 40 large T antigen can inhibit apoptosis through activation of the phosphatidylinositide 3'-OH kinase pathway and the cellular kinase Akt. *J. Virol.* **2002**, *76*, 3731–3738. [CrossRef]

158. Cobbs, C.S.; Soroceanu, L.; Denham, S.; Zhang, W.; Kraus, M.H. Modulation of oncogenic phenotype in human glioma cells by cytomegalovirus IE1-mediated mitogenicity. *Cancer Res.* **2008**, *68*, 724–730. [CrossRef]

159. Kudchodkar, S.B.; Yu, Y.; Maguire, T.G.; Alwine, J.C. Human cytomegalovirus infection alters the substrate specificities and rapamycin sensitivities of raptor- and rictor-containing complexes. *Proc. Natl. Acad. Sci. USA* **2006**, *103*, 14182–14187. [CrossRef]

160. McFarlane, S.; Preston, C.M. Human cytomegalovirus immediate early gene expression in the osteosarcoma line U2OS is repressed by the cell protein ATRX. *Virus Res.* **2011**, *157*, 47–53. [CrossRef]

161. Chan, G.; Nogalski, M.T.; Yurochko, A.D. Activation of EGFR on monocytes is required for human cytomegalovirus entry and mediates cellular motility. *Proc. Natl. Acad. Sci. USA* **2009**, *106*, 22369–22374. [CrossRef]

162. Chan, G.; Nogalski, M.T.; Bentz, G.L.; Smith, M.S.; Parmater, A.; Yurochko, A.D. PI3K-dependent upregulation of Mcl-1 by human cytomegalovirus is mediated by epidermal growth factor receptor and inhibits apoptosis in short-lived monocytes. *J. Immunol.* **2010**, *184*, 3213–3222. [CrossRef]

163. Kim, J.H.; Collins-McMillen, D.; Buehler, J.C.; Goodrum, F.D.; Yurochko, A.D. Human cytomegalovirus requires epidermal growth factor receptor signaling to enter and initiate the early steps in the establishment of latency in CD34(+) human progenitor cells. *J. Virol.* **2017**, *91*, e01206-16. [CrossRef]

164. Rak, M.A.; Buehler, J.; Zeltzer, S.; Reitsma, J.; Molina, B.; Terhune, S.; Goodrum, F. Human cytomegalovirus UL135 interacts with host adaptor proteins to regulate epidermal growth factor receptor and reactivation from latency. *J. Virol.* **2018**, *92*, e00919-18. [CrossRef]

165. Mikell, I.; Crawford, L.B.; Hancock, M.H.; Mitchell, J.; Buehler, J.; Goodrum, F.; Nelson, J.A. HCMV miR-US22 down-regulation of EGR-1 regulates CD34+ hematopoietic progenitor cell proliferation and viral reactivation. *PLoS Pathog.* **2019**, *15*, e1007854. [CrossRef]

166. Buehler, J.; Zeltzer, S.; Reitsma, J.; Petrucelli, A.; Umashankar, M.; Rak, M.; Zagallo, P.; Schroeder, J.; Terhune, S.; Goodrum, F. Opposing regulation of the EGF receptor: A molecular switch controlling cytomegalovirus latency and replication. *PLoS Pathog.* **2016**, *12*, e1005655. [CrossRef]

167. Buehler, J.; Carpenter, E.; Zeltzer, S.; Igarashi, S.; Rak, M.; Mikell, I.; Nelson, J.A.; Goodrum, F. Host signaling and EGR1 transcriptional control of human cytomegalovirus replication and latency. *PLoS Pathog.* **2019**, *15*, e1008037. [CrossRef] [PubMed]

168. Chen, J.; Stinski, M.F. Role of regulatory elements and the MAPK/ERK or p38 MAPK pathways for activation of human cytomegalovirus gene expression. *J. Virol.* **2002**, *76*, 4873–4885. [CrossRef]

169. Kew, V.; Wills, M.; Reeves, M. HCMV activation of ERK-MAPK drives a multi-factorial response promoting the survival of infected myeloid progenitors. *J. Mol. Biochem.* **2017**, *6*, 13–25. [PubMed]

170. Rodems, S.M.; Spector, D.H. Extracellular signal-regulated kinase activity is sustained early during human cytomegalovirus infection. *J. Virol.* **1998**, *72*, 9173–9180. [CrossRef]

171. Reeves, M.B.; Breidenstein, A.; Compton, T. Human cytomegalovirus activation of ERK and myeloid cell leukemia-1 protein correlates with survival of latently infected cells. *Proc. Natl. Acad. Sci. USA* **2012**, *109*, 588–593. [CrossRef]

172. Michaelis, M.; Paulus, C.; Loschmann, N.; Dauth, S.; Stange, E.; Doerr, H.W.; Nevels, M.; Cinatl, J., Jr. The multi-targeted kinase inhibitor sorafenib inhibits human cytomegalovirus replication. *Cell. Mol. Life Sci.* **2011**, *68*, 1079–1090. [CrossRef]

173. Mukhopadhyay, R.; Venkatadri, R.; Katsnelson, J.; Arav-Boger, R. Digitoxin suppresses human cytomegalovirus replication via Na(+), K(+)/ATPase α1 subunit-dependent AMP-activated protein kinase and autophagy activation. *J. Virol.* **2018**, *92*, e01861-17. [CrossRef]

174. Dupont, L.; Du, L.; Poulter, M.; Choi, S.; McIntosh, M.; Reeves, M.B. Src family kinase activity drives cytomegalovirus reactivation by recruiting MOZ histone acetyltransferase activity to the viral promoter. *J. Biol. Chem.* **2019**, *294*, 12901–12910. [CrossRef]

175. Hutterer, C.; Eickhoff, J.; Milbradt, J.; Korn, K.; Zeittrager, I.; Bahsi, H.; Wagner, S.; Zischinsky, G.; Wolf, A.; Degenhart, C.; et al. A novel CDK7 inhibitor of the pyrazolotriazine class exerts broad-spectrum antiviral activity at nanomolar concentrations. *Antimicrob. Agents Chemother.* **2015**, *59*, 2062–2071. [CrossRef]

176. Yuan, J.; Li, M.; Torres, Y.R.; Galle, C.S.; Meier, J.L. Differentiation-coupled induction of human cytomegalovirus replication by union of the major enhancer retinoic acid, cyclic AMP, and NF-κB response elements. *J. Virol.* **2015**, *89*, 12284–12298. [CrossRef]

177. Kowalik, T.F.; Wing, B.; Haskill, J.S.; Azizkhan, J.C.; Baldwin, A.S., Jr.; Huang, E.S. Multiple mechanisms are implicated in the regulation of NF-κB activity during human cytomegalovirus infection. *Proc. Natl. Acad. Sci. USA* **1993**, *90*, 1107–1111. [CrossRef]

178. Yurochko, A.D.; Hwang, E.S.; Rasmussen, L.; Keay, S.; Pereira, L.; Huang, E.S. The human cytomegalovirus UL55 (gB) and UL75 (gH) glycoprotein ligands initiate the rapid activation of Sp1 and NF-κB during infection. *J. Virol.* **1997**, *71*, 5051–5059. [CrossRef]

179. DeMeritt, I.B.; Milford, L.E.; Yurochko, A.D. Activation of the NF-κB pathway in human cytomegalovirus-infected cells is necessary for efficient transactivation of the major immediate-early promoter. *J. Virol.* **2004**, *78*, 4498–4507. [CrossRef]

180. Sambucetti, L.C.; Cherrington, J.M.; Wilkinson, G.W.; Mocarski, E.S. NF-κB activation of the cytomegalovirus enhancer is mediated by a viral transactivator and by T cell stimulation. *EMBO J.* **1989**, *8*, 4251–4258. [CrossRef]

181. Cherrington, J.M.; Mocarski, E.S. Human cytomegalovirus ie1 transactivates the α promoter-enhancer via an 18-base-pair repeat element. *J. Virol.* **1989**, *63*, 1435–1440. [CrossRef]

182. Prösch, S.; Staak, K.; Stein, J.; Liebenthal, C.; Stamminger, T.; Volk, H.D.; Krüger, D.H. Stimulation of the human cytomegalovirus IE enhancer/promoter in HL-60 cells by TNFα is mediated via induction of NF-κB. *Virology* **1995**, *208*, 197–206. [CrossRef]

183. Roy, S.; Arav-Boger, R. New cell-signaling pathways for controlling cytomegalovirus replication. *Am. J. Transplant.* **2014**, *14*, 1249–1258. [CrossRef]

184. Murray, M.J.; Peters, N.E.; Reeves, M.B. Navigating the Host Cell Response during Entry into Sites of Latent Cytomegalovirus Infection. *Pathogens* **2018**, *7*, 30. [CrossRef]

185. Cohen, Y.; Stern-Ginossar, N. Manipulation of host pathways by human cytomegalovirus: Insights from genome-wide studies. *Semin. Immunopathol.* **2014**, *36*, 651–658. [CrossRef]

186. Hou, W.; Torres, L.; Cruz-Cosme, R.; Arroyo, F.; Irizarry, L.; Luciano, D.; Marquez, A.; Rivera, L.L.; Sala, A.L.; Luo, M.H.; et al. Two polypyrimidine tracts in intron 4 of the major immediate early gene are critical for gene expression switching from IE1 to IE2 and for replication of human cytomegalovirus. *J. Virol.* **2016**, *90*, 7339–7349. [CrossRef]

187. Lin, Y.T.; Prendergast, J.; Grey, F. The host ubiquitin-dependent segregase VCP/p97 is required for the onset of human cytomegalovirus replication. *PLoS Pathog.* **2017**, *13*, e1006329. [CrossRef]

188. Stamminger, T.; Puchtler, E.; Fleckenstein, B. Discordant expression of the immediate-early 1 and 2 gene regions of human cytomegalovirus at early times after infection involves posttranscriptional processing events. *J. Virol.* **1991**, *65*, 2273–2282. [CrossRef]

189. Abdalla, A.E.; Mahjoob, M.O.; Abosalif, K.O.A.; Ejaz, H.; Alameen, A.A.M.; Elsaman, T. Human cytomegalovirus-encoded MicroRNAs: A master regulator of latent infection. *Infect. Genet. Evol.* **2019**, *78*, 104119. [CrossRef]

190. Diggins, N.L.; Hancock, M.H. HCMV miRNA targets reveal important cellular pathways for viral replication, latency, and reactivation. *Noncodin. RNA* **2018**, *4*, 29. [CrossRef]

191. Hook, L.; Hancock, M.; Landais, I.; Grabski, R.; Britt, W.; Nelson, J.A. Cytomegalovirus microRNAs. *Curr. Opin. Virol.* **2014**, *7*, 40–46. [CrossRef]

192. Grey, F.; Meyers, H.; White, E.A.; Spector, D.H.; Nelson, J. A human cytomegalovirus-encoded microRNA regulates expression of multiple viral genes involved in replication. *PLoS Pathog.* **2007**, *3*, e163. [CrossRef]

193. Murphy, E.; Vanicek, J.; Robins, H.; Shenk, T.; Levine, A.J. Suppression of immediate-early viral gene expression by herpesvirus-coded microRNAs: Implications for latency. *Proc. Natl. Acad. Sci. USA* **2008**, *105*, 5453–5458. [CrossRef]

194. Lau, B.; Poole, E.; Van Damme, E.; Bunkens, L.; Sowash, M.; King, H.; Murphy, E.; Wills, M.; Van Loock, M.; Sinclair, J. Human cytomegalovirus miR-UL112-1 promotes the down-regulation of viral immediate early-gene expression during latency to prevent T-cell recognition of latently infected cells. *J. Gen. Virol.* **2016**, *97*, 2387–2398. [CrossRef]

195. Stern-Ginossar, N.; Saleh, N.; Goldberg, M.D.; Prichard, M.; Wolf, D.G.; Mandelboim, O. Analysis of human cytomegalovirus-encoded microRNA activity during infection. *J. Virol.* **2009**, *83*, 10684–10693. [CrossRef]

196. Scherer, M.; Klingl, S.; Sevvana, M.; Otto, V.; Schilling, E.M.; Stump, J.D.; Müller, R.; Reuter, N.; Sticht, H.; Müller, Y.A.; et al. Crystal structure of cytomegalovirus IE1 protein reveals targeting of TRIM family member PML via coiled-coil interactions. *PLoS Pathog.* **2014**, *10*, e1004512. [CrossRef]

197. Teng, M.W.; Bolovan-Fritts, C.; Dar, R.D.; Womack, A.; Simpson, M.L.; Shenk, T.; Weinberger, L.S. An endogenous accelerator for viral gene expression confers a fitness advantage. *Cell* **2012**, *151*, 1569–1580. [CrossRef] [PubMed]

198. Stump, J.D.; Sticht, H. Investigation of the dynamics of the viral immediate-early protein 1 in different conformations and oligomerization states. *J. Biomol. Struct. Dyn.* **2016**, *34*, 1029–1041. [CrossRef] [PubMed]

199. Waheed, I.; Chiou, C.J.; Ahn, J.H.; Hayward, G.S. Binding of the human cytomegalovirus 80-kDa immediate-early protein (IE2) to minor groove A/T-rich sequences bounded by CG dinucleotides is regulated by protein oligomerization and phosphorylation. *Virology* **1998**, *252*, 235–257. [CrossRef] [PubMed]

200. Ahn, J.H.; Chiou, C.J.; Hayward, G.S. Evaluation and mapping of the DNA binding and oligomerization domains of the IE2 regulatory protein of human cytomegalovirus using yeast one and two hybrid interaction assays. *Gene* **1998**, *210*, 25–36. [CrossRef]

201. Harel, N.Y.; Alwine, J.C. Phosphorylation of the human cytomegalovirus 86-kilodalton immediate-early protein IE2. *J. Virol.* **1998**, *72*, 5481–5492. [CrossRef]

202. Pajovic, S.; Wong, E.L.; Black, A.R.; Azizkhan, J.C. Identification of a viral kinase that phosphorylates specific E2Fs and pocket proteins. *Mol. Cell. Biol.* **1997**, *17*, 6459–6464. [CrossRef]

203. Müller, S.; Dejean, A. Viral immediate-early proteins abrogate the modification by SUMO-1 of PML and Sp100 proteins, correlating with nuclear body disruption. *J. Virol.* **1999**, *73*, 5137–5143. [CrossRef]

204. Spengler, M.L.; Kurapatwinski, K.; Black, A.R.; Azizkhan-Clifford, J. SUMO-1 modification of human cytomegalovirus IE1/IE72. *J. Virol.* **2002**, *76*, 2990–2996. [CrossRef]

205. Hofmann, H.; Floss, S.; Stamminger, T. Covalent modification of the transactivator protein IE2-p86 of human cytomegalovirus by conjugation to the ubiquitin-homologous proteins SUMO-1 and hSMT3b. *J. Virol.* **2000**, *74*, 2510–2524. [CrossRef]

206. Ahn, J.H.; Xu, Y.; Jang, W.J.; Matunis, M.J.; Hayward, G.S. Evaluation of interactions of human cytomegalovirus immediate-early IE2 regulatory protein with small ubiquitin-like modifiers and their conjugation enzyme Ubc9. *J. Virol.* **2001**, *75*, 3859–3872. [CrossRef]

207. Barrasa, M.I.; Harel, N.Y.; Alwine, J.C. The phosphorylation status of the serine-rich region of the human cytomegalovirus 86-kilodalton major immediate-early protein IE2/IEP86 affects temporal viral gene expression. *J. Virol.* **2005**, *79*, 1428–1437. [CrossRef] [PubMed]

208. Nevels, M.; Brune, W.; Shenk, T. SUMOylation of the human cytomegalovirus 72-kilodalton IE1 protein facilitates expression of the 86-kilodalton IE2 protein and promotes viral replication. *J. Virol.* **2004**, *78*, 7803–7812. [CrossRef] [PubMed]

209. Huh, Y.H.; Kim, Y.E.; Kim, E.T.; Park, J.J.; Song, M.J.; Zhu, H.; Hayward, G.S.; Ahn, J.H. Binding STAT2 by the acidic domain of human cytomegalovirus IE1 promotes viral growth and is negatively regulated by SUMO. *J. Virol.* **2008**, *82*, 10444–10454. [CrossRef]

210. Heider, J.A.; Yu, Y.; Shenk, T.; Alwine, J.C. Characterization of a human cytomegalovirus with phosphorylation site mutations in the immediate-early 2 protein. *J. Virol.* **2002**, *76*, 928–932. [CrossRef]

211. Sadanari, H.; Yamada, R.; Ohnishi, K.; Matsubara, K.; Tanaka, J. SUMO-1 modification of the major immediate-early (IE) 1 and 2 proteins of human cytomegalovirus is regulated by different mechanisms and modulates the intracellular localization of the IE1, but not IE2, protein. *Arch. Virol.* **2005**, *150*, 1763–1782. [CrossRef]

212. Kim, E.T.; Kim, Y.E.; Kim, Y.J.; Lee, M.K.; Hayward, G.S.; Ahn, J.H. Analysis of human cytomegalovirus-encoded SUMO targets and temporal regulation of SUMOylation of the immediate-early proteins IE1 and IE2 during infection. *PLoS ONE* **2014**, *9*, e103308. [CrossRef]

213. Reuter, N.; Reichel, A.; Stilp, A.C.; Scherer, M.; Stamminger, T. SUMOylation of IE2p86 is required for efficient autorepression of the human cytomegalovirus major immediate-early promoter. *J. Gen. Virol.* **2018**, *99*, 369–378. [CrossRef]

214. Berndt, A.; Hofmann-Winkler, H.; Tavalai, N.; Hahn, G.; Stamminger, T. Importance of covalent and noncovalent SUMO interactions with the major human cytomegalovirus transactivator IE2p86 for viral infection. *J. Virol.* **2009**, *83*, 12881–12894. [CrossRef]

215. Vardi, N.; Chaturvedi, S.; Weinberger, L.S. Feedback-mediated signal conversion promotes viral fitness. *Proc. Natl. Acad. Sci. USA* **2018**, *115*, E8803–E8810. [CrossRef]

216. Scherer, M.; Otto, V.; Stump, J.D.; Klingl, S.; Müller, R.; Reuter, N.; Müller, Y.A.; Sticht, H.; Stamminger, T. Characterization of recombinant human cytomegaloviruses encoding IE1 mutants L174P and 1-382 reveals that viral targeting of PML bodies perturbs both intrinsic and innate immune responses. *J. Virol.* **2015**, *90*, 1190–1205. [CrossRef]

217. Mocarski, E.S.; Kemble, G.W.; Lyle, J.M.; Greaves, R.F. A deletion mutant in the human cytomegalovirus gene encoding IE1(491aa) is replication defective due to a failure in autoregulation. *Proc. Natl. Acad. Sci. USA* **1996**, *93*, 11321–11326. [CrossRef] [PubMed]

218. Gawn, J.M.; Greaves, R.F. Absence of IE1 p72 protein function during low-multiplicity infection by human cytomegalovirus results in a broad block to viral delayed-early gene expression. *J. Virol.* **2002**, *76*, 4441–4455. [CrossRef] [PubMed]

219. Greaves, R.F.; Mocarski, E.S. Defective growth correlates with reduced accumulation of a viral DNA replication protein after low-multiplicity infection by a human cytomegalovirus ie1 mutant. *J. Virol.* **1998**, *72*, 366–379. [CrossRef] [PubMed]

220. Marchini, A.; Liu, H.; Zhu, H. Human cytomegalovirus with IE-2 (UL122) deleted fails to express early lytic genes. *J. Virol.* **2001**, *75*, 1870–1878. [CrossRef] [PubMed]

221. Heider, J.A.; Bresnahan, W.A.; Shenk, T.E. Construction of a rationally designed human cytomegalovirus variant encoding a temperature-sensitive immediate-early 2 protein. *Proc. Natl. Acad. Sci. USA* **2002**, *99*, 3141–3146. [CrossRef]

222. Harwardt, T.; Lukas, S.; Zenger, M.; Reitberger, T.; Danzer, D.; Übner, T.; Munday, D.C.; Nevels, M.; Paulus, C. Human cytomegalovirus immediate-early 1 protein rewires upstream STAT3 to downstream STAT1 signaling switching an IL6-type to an IFNγ-like response. *PLoS Pathog.* **2016**, *12*, e1005748. [CrossRef]

223. Knoblach, T.; Grandel, B.; Seiler, J.; Nevels, M.; Paulus, C. Human cytomegalovirus IE1 protein elicits a type II interferon-like host cell response that depends on activated STAT1 but not interferon-γ. *PLoS Pathog.* **2011**, *7*, e1002016. [CrossRef]

224. Reitsma, J.M.; Sato, H.; Nevels, M.; Terhune, S.S.; Paulus, C. Human cytomegalovirus IE1 protein disrupts interleukin-6 signaling by sequestering STAT3 in the nucleus. *J. Virol.* **2013**, *87*, 10763–10776. [CrossRef]

225. Paulus, C.; Krauss, S.; Nevels, M. A human cytomegalovirus antagonist of type I IFN-dependent signal transducer and activator of transcription signaling. *Proc. Natl. Acad. Sci. USA* **2006**, *103*, 3840–3845. [CrossRef]

226. Krauss, S.; Kaps, J.; Czech, N.; Paulus, C.; Nevels, M. Physical requirements and functional consequences of complex formation between the cytomegalovirus IE1 protein and human STAT2. *J. Virol.* **2009**, *83*, 12854–12870. [CrossRef]

227. Song, Y.J.; Stinski, M.F. Effect of the human cytomegalovirus IE86 protein on expression of E2F-responsive genes: A DNA microarray analysis. *Proc. Natl. Acad. Sci. USA* **2002**, *99*, 2836–2841. [CrossRef] [PubMed]

228. Spector, D.H. Human cytomegalovirus riding the cell cycle. *Med. Microbiol. Immunol.* **2015**, *204*, 409–419. [CrossRef]

229. Mücke, K.; Paulus, C.; Bernhardt, K.; Gerrer, K.; Schön, K.; Fink, A.; Sauer, E.M.; Asbach-Nitzsche, A.; Harwardt, T.; Kieninger, B.; et al. Human cytomegalovirus major immediate-early 1 protein targets host chromosomes by docking to the acidic pocket on the nucleosome surface. *J. Virol.* **2014**, *88*, 1228–1248. [CrossRef] [PubMed]

230. Fang, Q.; Chen, P.; Wang, M.; Fang, J.; Yang, N.; Li, G.; Xu, R.M. Human cytomegalovirus IE1 protein alters the higher-order chromatin structure by targeting the acidic patch of the nucleosome. *Elife* **2016**, *5*, 1911. [CrossRef]

231. Park, J.J.; Kim, Y.E.; Pham, H.T.; Kim, E.T.; Chung, Y.H.; Ahn, J.H. Functional interaction of the human cytomegalovirus IE2 protein with histone deacetylase 2 in infected human fibroblasts. *J. Gen. Virol.* **2007**, *88*, 3214–3223. [CrossRef]

232. Bryant, L.A.; Mixon, P.; Davidson, M.; Bannister, A.J.; Kouzarides, T.; Sinclair, J.H. The human cytomegalovirus 86-kilodalton major immediate-early protein interacts physically and functionally with histone acetyltransferase P/CAF. *J. Virol.* **2000**, *74*, 7230–7237. [CrossRef]

233. Hsu, C.H.; Chang, M.D.; Tai, K.Y.; Yang, Y.T.; Wang, P.S.; Chen, C.J.; Wang, Y.H.; Lee, S.C.; Wu, C.W.; Juan, L.J. HCMV IE2-mediated inhibition of HAT activity downregulates p53 function. *EMBO J.* **2004**, *23*, 2269–2280. [CrossRef]

234. Sinclair, J. Chromatin structure regulates human cytomegalovirus gene expression during latency, reactivation and lytic infection. *Biochim. Biophys. Acta* **2010**, *1799*, 286–295. [CrossRef]

235. Saffert, R.T.; Kalejta, R.F. Inactivating a cellular intrinsic immune defense mediated by Daxx is the mechanism through which the human cytomegalovirus pp71 protein stimulates viral immediate-early gene expression. *J. Virol.* **2006**, *80*, 3863–3871. [CrossRef]

236. Bieniasz, P.D. Intrinsic immunity: A front-line defense against viral attack. *Nat. Immunol.* **2004**, *5*, 1109–1115. [CrossRef]

237. Tavalai, N.; Stamminger, T. Interplay between herpesvirus infection and host defense by PML nuclear bodies. *Viruses* **2009**, *1*, 1240. [CrossRef]

238. Reeves, M.; Woodhall, D.; Compton, T.; Sinclair, J. Human cytomegalovirus IE72 protein interacts with the transcriptional repressor hDaxx to regulate LUNA gene expression during lytic infection. *J. Virol.* **2010**, *84*, 7185–7194. [CrossRef] [PubMed]

239. Hornig, J.; Choi, K.Y.; McGregor, A. The essential role of guinea pig cytomegalovirus (GPCMV) IE1 and IE2 homologs in viral replication and IE1-mediated ND10 targeting. *Virology* **2017**, *504*, 122–140. [CrossRef] [PubMed]

240. Tang, Q.; Maul, G.G. Mouse cytomegalovirus immediate-early protein 1 binds with host cell repressors to relieve suppressive effects on viral transcription and replication during lytic infection. *J. Virol.* **2003**, *77*, 1357–1367. [CrossRef] [PubMed]

241. Liu, X.J.; Yang, B.; Huang, S.N.; Wu, C.C.; Li, X.J.; Cheng, S.; Jiang, X.; Hu, F.; Ming, Y.Z.; Nevels, M.; et al. Human cytomegalovirus IE1 downregulates Hes1 in neural progenitor cells as a potential E3 ubiquitin ligase. *PLoS Pathog.* **2017**, *13*, e1006542. [CrossRef]

242. Kim, Y.E.; Lee, J.H.; Kim, E.T.; Shin, H.J.; Gu, S.Y.; Seol, H.S.; Ling, P.D.; Lee, C.H.; Ahn, J.H. Human cytomegalovirus infection causes degradation of Sp100 proteins that suppress viral gene expression. *J. Virol.* **2011**, *85*, 11928–11937. [CrossRef]

243. Khan, Z.; Yaiw, K.C.; Wilhelmi, V.; Lam, H.; Rahbar, A.; Stragliotto, G.; Soderberg-Naucler, C. Human cytomegalovirus immediate early proteins promote degradation of connexin 43 and disrupt gap junction communication: Implications for a role in gliomagenesis. *Carcinogenesis* **2014**, *35*, 145–154. [CrossRef]

244. Tavalai, N.; Adler, M.; Scherer, M.; Riedl, Y.; Stamminger, T. Evidence for a dual antiviral role of the major nuclear domain 10 component Sp100 during the immediate-early and late phases of the human cytomegalovirus replication cycle. *J. Virol.* **2011**, *85*, 9447–9458. [CrossRef]

245. Ahn, J.H.; Brignole, E.J., 3rd; Hayward, G.S. Disruption of PML subnuclear domains by the acidic IE1 protein of human cytomegalovirus is mediated through interaction with PML and may modulate a RING finger-dependent cryptic transactivator function of PML. *Mol. Cell. Biol.* **1998**, *18*, 4899–4913. [CrossRef]

246. Hou, W.; Cruz-Cosme, R.; Wen, F.; Ahn, J.H.; Reeves, I.; Luo, M.H.; Tang, Q. Expression of human cytomegalovirus IE1 leads to accumulation of mono-SUMOylated PML that is protected from degradation by herpes simplex virus 1 ICP0. *J. Virol.* **2018**, *92*, e01452-18. [CrossRef]

247. Kang, H.; Kim, E.T.; Lee, H.R.; Park, J.J.; Go, Y.Y.; Choi, C.Y.; Ahn, J.H. Inhibition of SUMO-independent PML oligomerization by the human cytomegalovirus IE1 protein. *J. Gen. Virol.* **2006**, *87*, 2181–2190. [CrossRef] [PubMed]

248. Schilling, E.M.; Scherer, M.; Reuter, N.; Schweininger, J.; Müller, Y.A.; Stamminger, T. The human cytomegalovirus IE1 protein antagonizes PML nuclear body-mediated intrinsic immunity via the inhibition of PML de novo SUMOylation. *J. Virol.* **2017**, *91*, e02049-16. [CrossRef] [PubMed]

249. Korioth, F.; Maul, G.G.; Plachter, B.; Stamminger, T.; Frey, J. The nuclear domain 10 (ND10) is disrupted by the human cytomegalovirus gene product IE1. *Exp. Cell Res.* **1996**, *229*, 155–158. [CrossRef] [PubMed]

250. Wilkinson, G.W.; Kelly, C.; Sinclair, J.H.; Rickards, C. Disruption of PML-associated nuclear bodies mediated by the human cytomegalovirus major immediate early gene product. *J. Gen. Virol.* **1998**, *79*, 1233–1245. [CrossRef]

251. Ahn, J.H.; Hayward, G.S. The major immediate-early proteins IE1 and IE2 of human cytomegalovirus colocalize with and disrupt PML-associated nuclear bodies at very early times in infected permissive cells. *J. Virol.* **1997**, *71*, 4599–4613. [CrossRef]

252. Ahn, J.H.; Hayward, G.S. Disruption of PML-associated nuclear bodies by IE1 correlates with efficient early stages of viral gene expression and DNA replication in human cytomegalovirus infection. *Virology* **2000**, *274*, 39–55. [CrossRef]

253. Sourvinos, G.; Tavalai, N.; Berndt, A.; Spandidos, D.A.; Stamminger, T. Recruitment of human cytomegalovirus immediate-early 2 protein onto parental viral genomes in association with ND10 in live-infected cells. *J. Virol.* **2007**, *81*, 10123–10136. [CrossRef]

254. Ishov, A.M.; Stenberg, R.M.; Maul, G.G. Human cytomegalovirus immediate early interaction with host nuclear structures: Definition of an immediate transcript environment. *J. Cell Biol.* **1997**, *138*, 5–16. [CrossRef]

255. Zhu, H.; Shen, Y.; Shenk, T. Human cytomegalovirus IE1 and IE2 proteins block apoptosis. *J. Virol.* **1995**, *69*, 7960–7970. [CrossRef]

256. Lukac, D.M.; Alwine, J.C. Effects of human cytomegalovirus major immediate-early proteins in controlling the cell cycle and inhibiting apoptosis: Studies with ts13 cells. *J. Virol.* **1999**, *73*, 2825–2831. [CrossRef]

257. Chiou, S.H.; Yang, Y.P.; Lin, J.C.; Hsu, C.H.; Jhang, H.C.; Yang, Y.T.; Lee, C.H.; Ho, L.L.; Hsu, W.M.; Ku, H.H.; et al. The immediate early 2 protein of human cytomegalovirus (HCMV) mediates the apoptotic control in HCMV retinitis through up-regulation of the cellular FLICE-inhibitory protein expression. *J. Immunol.* **2006**, *177*, 6199–6206. [CrossRef] [PubMed]

258. Tanaka, K.; Zou, J.P.; Takeda, K.; Ferrans, V.J.; Sandford, G.R.; Johnson, T.M.; Finkel, T.; Epstein, S.E. Effects of human cytomegalovirus immediate-early proteins on p53-mediated apoptosis in coronary artery smooth muscle cells. *Circulation* **1999**, *99*, 1656–1659. [CrossRef] [PubMed]

259. Bonin, L.R.; McDougall, J.K. Human cytomegalovirus IE2 86-kilodalton protein binds p53 but does not abrogate G1 checkpoint function. *J. Virol.* **1997**, *71*, 5861–5870. [CrossRef] [PubMed]

260. Speir, E.; Modali, R.; Huang, E.S.; Leon, M.B.; Shawl, F.; Finkel, T.; Epstein, S.E. Potential role of human cytomegalovirus and p53 interaction in coronary restenosis. *Science* **1994**, *265*, 391–394. [CrossRef] [PubMed]

261. Tsai, H.L.; Kou, G.H.; Chen, S.C.; Wu, C.W.; Lin, Y.S. Human cytomegalovirus immediate-early protein IE2 tethers a transcriptional repression domain to p53. *J. Biol. Chem.* **1996**, *271*, 3534–3540. [CrossRef]

262. Marques, M.; Ferreira, A.R.; Ribeiro, D. The interplay between human cytomegalovirus and pathogen recognition receptor signaling. *Viruses* **2018**, *10*, 514. [CrossRef]

263. DeFilippis, V.R. Induction and evasion of the type I interferon response by cytomegaloviruses. *Adv. Exp. Med. Biol.* **2007**, *598*, 309–324.

264. Biolatti, M.; Gugliesi, F.; Dell'Oste, V.; Landolfo, S. Modulation of the innate immune response by human cytomegalovirus. *Infect. Genet. Evol.* **2018**, *64*, 105–114. [CrossRef]

265. Bianco, C.; Mohr, I. Restriction of Human cytomegalovirus replication by ISG15, a host effector regulated by cGAS-STING double-stranded-DNA sensing. *J. Virol.* **2017**, *91*, e02483-16. [CrossRef]

266. Kim, Y.J.; Kim, E.T.; Kim, Y.E.; Lee, M.K.; Kwon, K.M.; Kim, K.I.; Stamminger, T.; Ahn, J.H. Consecutive inhibition of ISG15 expression and ISGylation by cytomegalovirus regulators. *PLoS Pathog.* **2016**, *12*, e1005850. [CrossRef]

267. Raghavan, B.; Cook, C.H.; Trgovcich, J. The carboxy terminal region of the human cytomegalovirus immediate early 1 (IE1) protein disrupts type II inteferon signaling. *Viruses* **2014**, *6*, 1502–1524. [CrossRef] [PubMed]

268. Jaworska, J.; Gravel, A.; Flamand, L. Divergent susceptibilities of human herpesvirus 6 variants to type I interferons. *Proc. Natl. Acad. Sci. USA* **2010**, *107*, 8369–8374. [CrossRef] [PubMed]

269. Kim, Y.E.; Ahn, J.H. Positive role of promyelocytic leukemia protein in type I interferon response and its regulation by human cytomegalovirus. *PLoS Pathog.* **2015**, *11*, e1004785. [CrossRef] [PubMed]

270. Taylor, R.T.; Bresnahan, W.A. Human cytomegalovirus immediate-early 2 gene expression blocks virus-induced beta interferon production. *J. Virol.* **2005**, *79*, 3873–3877. [CrossRef] [PubMed]

271. Taylor, R.T.; Bresnahan, W.A. Human cytomegalovirus immediate-early 2 protein IE86 blocks virus-induced chemokine expression. *J. Virol.* **2006**, *80*, 920–928. [CrossRef] [PubMed]

272. Botto, S.; Abraham, J.; Mizuno, N.; Pryke, K.; Gall, B.; Landais, I.; Streblow, D.N.; Früh, K.J.; DeFilippis, V.R. Human cytomegalovirus immediate early 86-kDa protein blocks transcription and induces degradation of the immature interleukin-1β protein during virion-mediated activation of the AIM2 inflammasome. *MBio* **2019**, *10*, e02510-18. [CrossRef]

273. Taylor, R.T.; Bresnahan, W.A. Human cytomegalovirus IE86 attenuates virus- and tumor necrosis factor α-induced NFκB-dependent gene expression. *J. Virol.* **2006**, *80*, 10763–10771. [CrossRef]

274. Kim, J.E.; Kim, Y.E.; Stinski, M.F.; Ahn, J.H.; Song, Y.J. Human cytomegalovirus IE2 86 kDa protein induces STING degradation and inhibits cGAMP-mediated IFN-β induction. *Front. Microbiol.* **2017**, *8*, 1854. [CrossRef]

275. Britt, W. Manifestations of human cytomegalovirus infection: Proposed mechanisms of acute and chronic disease. *Curr. Top. Microbiol. Immunol.* **2008**, *325*, 417–470.

276. Shanley, J.D.; Pesanti, E.L.; Nugent, K.M. The pathogenesis of pneumonitis due to murine cytomegalovirus. *J. Infect. Dis.* **1982**, *146*, 388–396. [CrossRef]

277. Grundy, J.E.; Shanley, J.D.; Griffiths, P.D. Is cytomegalovirus interstitial pneumonitis in transplant recipients an immunopathological condition? *Lancet* **1987**, *2*, 996–999. [CrossRef]

278. Grundy, J.E.; Shanley, J.D.; Shearer, G.M. Augmentation of graft-versus-host reaction by cytomegalovirus infection resulting in interstitial pneumonitis. *Transplantation* **1985**, *39*, 548–553. [CrossRef]

279. Scholz, M.; Doerr, H.W.; Cinatl, J. Inhibition of cytomegalovirus immediate early gene expression: A therapeutic option? *Antivir. Res.* **2001**, *49*, 129–145. [CrossRef]

280. Tanaka, K.; Koga, Y.; Lu, Y.Y.; Zhang, X.Y.; Wang, Y.; Kimura, G.; Nomoto, K. Murine cytomegalovirus-associated pneumonitis in the lungs free of the virus. *J. Clin. Investig.* **1994**, *94*, 1019–1025. [CrossRef]

281. Jacobson, M.A.; Zegans, M.; Pavan, P.R.; O'Donnell, J.J.; Sattler, F.; Rao, N.; Owens, S.; Pollard, R. Cytomegalovirus retinitis after initiation of highly active antiretroviral therapy. *Lancet* **1997**, *349*, 1443–1445. [CrossRef]

282. Wu, C.-C.; Jiang, X.; Wang, X.-Z.; Liu, X.-J.; Li, X.-J.; Yang, B.; Ye, H.-Q.; Harwardt, T.; Gan, L.; Jiang, M.; et al. Human cytomegalovirus immediate early 1 protein causes loss of SOX2 from neural progenitor cells by trapping unphosphorylated STAT3 in the nucleus. *J. Virol.* **2018**, *92*, e00340-18. [CrossRef]

283. Sylwester, A.W.; Mitchell, B.L.; Edgar, J.B.; Taormina, C.; Pelte, C.; Ruchti, F.; Sleath, P.R.; Grabstein, K.H.; Hosken, N.A.; Kern, F.; et al. Broadly targeted human cytomegalovirus-specific CD4+ and CD8+ T cells dominate the memory compartments of exposed subjects. *J. Exp. Med.* **2005**, *202*, 673–685. [CrossRef]

284. Braendstrup, P.; Mortensen, B.K.; Justesen, S.; Osterby, T.; Rasmussen, M.; Hansen, A.M.; Christiansen, C.B.; Hansen, M.B.; Nielsen, M.; Vindelov, L.; et al. Identification and HLA-tetramer-validation of human CD4+ and CD8+ T cell responses against HCMV proteins IE1 and IE2. *PLoS ONE* **2014**, *9*, e94892. [CrossRef]

285. Elkington, R.; Walker, S.; Crough, T.; Menzies, M.; Tellam, J.; Bharadwaj, M.; Khanna, R. Ex vivo profiling of CD8+-T-cell responses to human cytomegalovirus reveals broad and multispecific reactivities in healthy virus carriers. *J. Virol.* **2003**, *77*, 5226–5240. [CrossRef]

286. Link, E.K.; Brandmüller, C.; Suezer, Y.; Ameres, S.; Volz, A.; Moosmann, A.; Sutter, G.; Lehmann, M.H. A synthetic human cytomegalovirus pp65-IE1 fusion antigen efficiently induces and expands virus specific T cells. *Vaccine* **2017**, *35*, 5131–5139. [CrossRef]

287. Banas, B.; Boger, C.A.; Luckhoff, G.; Kruger, B.; Barabas, S.; Batzilla, J.; Schemmerer, M.; Kostler, J.; Bendfeldt, H.; Rascle, A.; et al. Validation of T-Track CMV to assess the functionality of cytomegalovirus-reactive cell-mediated immunity in hemodialysis patients. *BMC Immunol.* **2017**, *18*, 15. [CrossRef] [PubMed]

288. Barabas, S.; Spindler, T.; Kiener, R.; Tonar, C.; Lugner, T.; Batzilla, J.; Bendfeldt, H.; Rascle, A.; Asbach, B.; Wagner, R.; et al. An optimized IFN-γ ELISpot assay for the sensitive and standardized monitoring of CMV protein-reactive effector cells of cell-mediated immunity. *BMC Immunol.* **2017**, *18*, 14. [CrossRef] [PubMed]

289. Glass, M.; Busche, A.; Wagner, K.; Messerle, M.; Borst, E.M. Conditional and reversible disruption of essential herpesvirus proteins. *Nat. Methods* **2009**, *6*, 577–579. [CrossRef] [PubMed]

290. Sanders, R.L.; Clark, C.L.; Morello, C.S.; Spector, D.H. Development of cell lines that provide tightly controlled temporal translation of the human cytomegalovirus IE2 proteins for complementation and functional analyses of growth-impaired and nonviable IE2 mutant viruses. *J. Virol.* **2008**, *82*, 7059–7077. [CrossRef]

291. Busche, A.; Angulo, A.; Kay-Jackson, P.; Ghazal, P.; Messerle, M. Phenotypes of major immediate-early gene mutants of mouse cytomegalovirus. *Med. Microbiol. Immunol.* **2008**, *197*, 233–240. [CrossRef]

292. Busche, A.; Marquardt, A.; Bleich, A.; Ghazal, P.; Angulo, A.; Messerle, M. The mouse cytomegalovirus immediate-early 1 gene is not required for establishment of latency or for reactivation in the lungs. *J. Virol.* **2009**, *83*, 4030–4038. [CrossRef]

293. Sandford, G.R.; Schumacher, U.; Ettinger, J.; Brune, W.; Hayward, G.S.; Burns, W.H.; Voigt, S. Deletion of the rat cytomegalovirus immediate-early 1 gene results in a virus capable of establishing latency, but with lower levels of acute virus replication and latency that compromise reactivation efficiency. *J. Gen. Virol.* **2010**, *91*, 616–621. [CrossRef]

294. Gümbel, H.; Cinatl, J., Jr.; Vogel, J.U.; Scholz, M.; Hoffmann, F.; Rabenau, H.F. CMV retinitis: Clinical experience with the metal chelator desferroxamine. In *CMV-Related Immunopathology, Monogr Virol*; Scholz, M., Rabenau, H.F., Doerr, H.W., Cinatl, J., Jr., Eds.; Karger: Basel, Switzerland, 1998; Volume 21, pp. 173–179.

295. Azad, R.F.; Driver, V.B.; Tanaka, K.; Crooke, R.M.; Anderson, K.P. Antiviral activity of a phosphorothioate oligonucleotide complementary to RNA of the human cytomegalovirus major immediate-early region. *Antimicrob. Agents Chemother.* **1993**, *37*, 1945–1954. [CrossRef]

296. Detrick, B.; Nagineni, C.N.; Grillone, L.R.; Anderson, K.P.; Henry, S.P.; Hooks, J.J. Inhibition of human cytomegalovirus replication in a human retinal epithelial cell model by antisense oligonucleotides. *Investig. Ophthalmol. Vis. Sci.* **2001**, *42*, 163–169.

297. Anderson, K.P.; Fox, M.C.; Brown-Driver, V.; Martin, M.J.; Azad, R.F. Inhibition of human cytomegalovirus immediate-early gene expression by an antisense oligonucleotide complementary to immediate-early RNA. *Antimicrob. Agents Chemother.* **1996**, *40*, 2004–2011. [CrossRef]

298. Mulamba, G.B.; Hu, A.; Azad, R.F.; Anderson, K.P.; Coen, D.M. Human cytomegalovirus mutant with sequence-dependent resistance to the phosphorothioate oligonucleotide fomivirsen (ISIS 2922). *Antimicrob. Agents Chemother.* **1998**, *42*, 971–973. [CrossRef] [PubMed]

299. Stein, C.A.; Castanotto, D. FDA-approved oligonucleotide therapies in 2017. *Mol. Ther.* **2017**, *25*, 1069–1075. [CrossRef] [PubMed]

300. Vitravene Study, G. Randomized dose-comparison studies of intravitreous fomivirsen for treatment of cytomegalovirus retinitis that has reactivated or is persistently active despite other therapies in patients with AIDS. *Am. J. Ophtalmol.* **2002**, *133*, 475–483.

301. Vitravene Study, G. A randomized controlled clinical trial of intravitreous fomivirsen for treatment of newly diagnosed peripheral cytomegalovirus retinitis in patients with AIDS. *Am. J. Ophtalmol.* **2002**, *133*, 467–474.

302. Jabs, D.A.; Griffiths, P.D. Fomivirsen for the treatment of cytomegalovirus retinitis. *Am. J. Ophtalmol.* **2002**, *133*, 552–556. [CrossRef]

303. Trang, P.; Lee, M.; Nepomuceno, E.; Kim, J.; Zhu, H.; Liu, F. Effective inhibition of human cytomegalovirus gene expression and replication by a ribozyme derived from the catalytic RNA subunit of RNase P from Escherichia coli. *Proc. Natl. Acad. Sci. USA* **2000**, *97*, 5812–5817. [CrossRef]

304. Trang, P.; Hsu, A.; Zhou, T.; Lee, J.; Kilani, A.F.; Nepomuceno, E.; Liu, F. Engineered RNase P ribozymes inhibit gene expression and growth of cytomegalovirus by increasing rate of cleavage and substrate binding. *J. Mol. Biol.* **2002**, *315*, 573–586. [CrossRef]

305. Zou, H.; Lee, J.; Umamoto, S.; Kilani, A.F.; Kim, J.; Trang, P.; Zhou, T.; Liu, F. Engineered RNase P ribozymes are efficient in cleaving a human cytomegalovirus mRNA in vitro and are effective in inhibiting viral gene expression and growth in human cells. *J. Biol. Chem.* **2003**, *278*, 37265–37274. [CrossRef]

306. Sun, X.; Chen, W.J.; He, L.L.; Sheng, J.X.; Liu, Y.J.; Vu, G.P.; Yang, Z.; Li, W.; Trang, P.; Wang, Y.; et al. Inhibition of human cytomegalovirus immediate early gene expression and growth by a novel RNase P ribozyme variant. *PLoS ONE* **2017**, *12*, 6791. [CrossRef]

307. Davies-Sala, C.; Soler-Bistue, A.; Bonomo, R.A.; Zorreguieta, A.; Tolmasky, M.E. External guide sequence technology: A path to development of novel antimicrobial therapeutics. *Ann. N. Y. Acad. Sci.* **2015**, *1354*, 98–110. [CrossRef]

308. Derksen, M.; Mertens, V.; Pruijn, G.J.M. RNase P-mediated sequence-specific cleavage of RNA by engineered external guide sequences. *Biomolecules* **2015**, *5*, 3029–3050. [CrossRef] [PubMed]

309. Kilani, A.F.; Trang, P.; Jo, S.; Hsu, A.; Kim, J.; Nepomuceno, E.; Liou, K.; Liu, F. RNase P ribozymes selected in vitro to cleave a viral mRNA effectively inhibit its expression in cell culture. *J. Biol. Chem.* **2000**, *275*, 10611–10622. [CrossRef] [PubMed]

310. Xiaofei, E.; Stadler, B.M.; Debatis, M.; Wang, S.; Lu, S.; Kowalik, T.F. RNA interference-mediated targeting of human cytomegalovirus immediate-early or early gene products inhibits viral replication with differential effects on cellular functions. *J. Virol.* **2012**, *86*, 5660–5673. [CrossRef]

311. Hamilton, S.T.; Milbradt, J.; Marschall, M.; Rawlinson, W.D. Human cytomegalovirus replication is strictly inhibited by siRNAs targeting UL54, UL97 or UL122/123 gene transcripts. *PLoS ONE* **2014**, *9*, e97231. [CrossRef]

312. Wiebusch, L.; Truss, M.; Hagemeier, C. Inhibition of human cytomegalovirus replication by small interfering RNAs. *J. Gen. Virol.* **2004**, *85*, 179–184. [CrossRef]

313. Bai, Z.Q.; Li, L.; Wang, B.; Liu, Z.J.; Liu, H.Y.; Jiang, G.Y.; Wang, H.T.; Yan, Z.Y.; Qian, D.M.; Ding, S.Y.; et al. Inhibition of human cytomegalovirus infection by IE86-specific short hairpin RNA-mediated RNA interference. *Biosci. Biotechnol. Biochem.* **2010**, *74*, 1368–1372. [CrossRef]

314. Gergen, J.; Coulon, F.; Creneguy, A.; Elain-Duret, N.; Gutierrez, A.; Pinkenburg, O.; Verhoeyen, E.; Anegon, I.; Nguyen, T.H.; Halary, F.A.; et al. Multiplex CRISPR/Cas9 system impairs HCMV replication by excising an essential viral gene. *PLoS ONE* **2018**, *13*, e0192602. [CrossRef]

315. Chin, W.X.; Ang, S.K.; Chu, J.J. Recent advances in therapeutic recruitment of mammalian RNAi and bacterial CRISPR-Cas DNA interference pathways as emerging antiviral strategies. *Drug Discov. Today* **2017**, *22*, 17–30. [CrossRef]

316. Badia, R.; Ballana, E.; Este, J.A.; Riveira-Munoz, E. Antiviral treatment strategies based on gene silencing and genome editing. *Curr. Opin. Virol.* **2017**, *24*, 46–54. [CrossRef]

317. Efferth, T.; Romero, M.R.; Wolf, D.G.; Stamminger, T.; Marin, J.J.; Marschall, M. The antiviral activities of artemisinin and artesunate. *Clin. Infect. Dis.* **2008**, *47*, 804–811. [CrossRef]

318. Ho, W.E.; Peh, H.Y.; Chan, T.K.; Wong, W.S. Artemisinins: Pharmacological actions beyond anti-malarial. *Pharmacol. Ther.* **2014**, *142*, 126–139. [CrossRef] [PubMed]

319. Efferth, T.; Marschall, M.; Wang, X.; Huong, S.M.; Hauber, I.; Olbrich, A.; Kronschnabl, M.; Stamminger, T.; Huang, E.S. Antiviral activity of artesunate towards wild-type, recombinant, and ganciclovir-resistant human cytomegaloviruses. *J. Mol. Med.* **2002**, *80*, 233–242. [CrossRef] [PubMed]

320. Arav-Boger, R.; He, R.; Chiou, C.J.; Liu, J.; Woodard, L.; Rosenthal, A.; Jones-Brando, L.; Forman, M.; Posner, G. Artemisinin-derived dimers have greatly improved anti-cytomegalovirus activity compared to artemisinin monomers. *PLoS ONE* **2010**, *5*, e10370. [CrossRef] [PubMed]

321. Chou, S.; Marousek, G.; Auerochs, S.; Stamminger, T.; Milbradt, J.; Marschall, M. The unique antiviral activity of artesunate is broadly effective against human cytomegaloviruses including therapy-resistant mutants. *Antivir. Res.* **2011**, *92*, 364–368. [CrossRef]

322. Schnepf, N.; Corvo, J.; Pors, M.J.; Mazeron, M.C. Antiviral activity of ganciclovir and artesunate towards human cytomegalovirus in astrocytoma cells. *Antivir. Res.* **2011**, *89*, 186–188. [CrossRef] [PubMed]

323. Morere, L.; Andouard, D.; Labrousse, F.; Saade, F.; Calliste, C.A.; Cotin, S.; Aubard, Y.; Rawlinson, W.D.; Esclaire, F.; Hantz, S.; et al. Ex vivo model of congenital cytomegalovirus infection and new combination therapies. *Placenta* **2015**, *36*, 41–47. [CrossRef]

324. Reiter, C.; Fröhlich, T.; Gruber, L.; Hutterer, C.; Marschall, M.; Voigtlander, C.; Friedrich, O.; Kappes, B.; Efferth, T.; Tsogoeva, S.B. Highly potent artemisinin-derived dimers and trimers: Synthesis and evaluation of their antimalarial, antileukemia and antiviral activities. *Bioorg. Med. Chem.* **2015**, *23*, 5452–5458. [CrossRef]

325. Reiter, C.; Fröhlich, T.; Zeino, M.; Marschall, M.; Bahsi, H.; Leidenberger, M.; Friedrich, O.; Kappes, B.; Hampel, F.; Efferth, T.; et al. New efficient artemisinin derived agents against human leukemia cells, human cytomegalovirus and plasmodium falciparum: 2nd generation 1,2,4-trioxane-ferrocene hybrids. *Eur. J. Med. Chem.* **2015**, *97*, 164–172. [CrossRef]

326. Hutterer, C.; Niemann, I.; Milbradt, J.; Fröhlich, T.; Reiter, C.; Kadioglu, O.; Bahsi, H.; Zeittrager, I.; Wagner, S.; Einsiedel, J.; et al. The broad-spectrum antiinfective drug artesunate interferes with the canonical nuclear factor kappa B (NF-κB) pathway by targeting RelA/p65. *Antivir. Res.* **2015**, *124*, 101–109. [CrossRef]

327. Drouot, E.; Piret, J.; Boivin, G. Artesunate demonstrates in vitro synergism with several antiviral agents against human cytomegalovirus. *Antivir. Ther.* **2016**, *21*, 535–539. [CrossRef]

328. Oiknine-Djian, E.; Weisblum, Y.; Panet, A.; Wong, H.N.; Haynes, R.K.; Wolf, D.G. The artemisinin derivative artemisone is a potent inhibitor of human cytomegalovirus replication. *Antimicrob. Agents Chemother.* **2018**, *62*, e00288-18. [CrossRef] [PubMed]

329. Sonntag, E.; Hahn, F.; Bertzbach, L.D.; Seyler, L.; Wangen, C.; Müller, R.; Tannig, P.; Grau, B.; Baumann, M.; Zent, E.; et al. In vivo proof-of-concept for two experimental antiviral drugs, both directed to cellular targets, using a murine cytomegalovirus model. *Antivir. Res.* **2019**, *161*, 63–69. [CrossRef] [PubMed]

330. He, R.; Mott, B.T.; Rosenthal, A.S.; Genna, D.T.; Posner, G.H.; Arav-Boger, R. An artemisinin-derived dimer has highly potent anti-cytomegalovirus (CMV) and anti-cancer activities. *PLoS ONE* **2011**, *6*, e24334. [CrossRef] [PubMed]

331. Hahn, F.; Fröhlich, T.; Frank, T.; Bertzbach, L.D.; Kohrt, S.; Kaufer, B.B.; Stamminger, T.; Tsogoeva, S.B.; Marschall, M. Artesunate-derived monomeric, dimeric and trimeric experimental drugs—Their unique mechanistic basis and pronounced antiherpesviral activity. *Antivir. Res.* **2018**, *152*, 104–110. [CrossRef]

332. He, R.; Park, K.; Cai, H.; Kapoor, A.; Forman, M.; Mott, B.; Posner, G.H.; Arav-Boger, R. Artemisinin-derived dimer diphenyl phosphate is an irreversible inhibitor of human cytomegalovirus replication. *Antimicrob. Agents Chemother.* **2012**, *56*, 3508–3515. [CrossRef]

333. He, R.; Forman, M.; Mott, B.T.; Venkatadri, R.; Posner, G.H.; Arav-Boger, R. Unique and highly selective anticytomegalovirus activities of artemisinin-derived dimer diphenyl phosphate stem from combination of dimer unit and a diphenyl phosphate moiety. *Antimicrob. Agents Chemother.* **2013**, *57*, 4208–4214. [CrossRef]

334. Held, F.E.; Guryev, A.A.; Fröhlich, T.; Hampel, F.; Kahnt, A.; Hutterer, C.; Steingruber, M.; Bahsi, H.; von Bojnicic-Kninski, C.; Mattes, D.S.; et al. Facile access to potent antiviral quinazoline heterocycles with fluorescence properties via merging metal-free domino reactions. *Nat. Commun.* **2017**, *8*, 15071. [CrossRef]

335. Fröhlich, T.; Reiter, C.; Ibrahim, M.M.; Beutel, J.; Hutterer, C.; Zeittrager, I.; Bahsi, H.; Leidenberger, M.; Friedrich, O.; Kappes, B.; et al. Synthesis of novel hybrids of quinazoline and artemisinin with high activities against plasmodium falciparum, human cytomegalovirus, and leukemia cells. *ACS Omega* **2017**, *2*, 2422–2431. [CrossRef]

336. Capci Karagoz, A.; Reiter, C.; Seo, E.J.; Gruber, L.; Hahn, F.; Leidenberger, M.; Klein, V.; Hampel, F.; Friedrich, O.; Marschall, M.; et al. Access to new highly potent antileukemia, antiviral and antimalarial agents via hybridization of natural products (homo)egonol, thymoquinone and artemisinin. *Bioorg. Med. Chem.* **2018**, *26*, 3610–3618. [CrossRef]

337. Fröhlich, T.; Reiter, C.; Saeed, M.E.M.; Hutterer, C.; Hahn, F.; Leidenberger, M.; Friedrich, O.; Kappes, B.; Marschall, M.; Efferth, T.; et al. Synthesis of thymoquinone-artemisinin hybrids: New potent antileukemia, antiviral, and antimalarial agents. *ACS Med. Chem. Lett.* **2018**, *9*, 534–539. [CrossRef]

338. DeMeritt, I.B.; Podduturi, J.P.; Tilley, A.M.; Nogalski, M.T.; Yurochko, A.D. Prolonged activation of NF-κB by human cytomegalovirus promotes efficient viral replication and late gene expression. *Virology* **2006**, *346*, 15–31. [CrossRef] [PubMed]

339. Hancock, M.H.; Nelson, J.A. Modulation of the NFκb signalling pathway by human cytomegalovirus. *Virology* **2017**, *1*, 104. [PubMed]

340. Roy, S.; He, R.; Kapoor, A.; Forman, M.; Mazzone, J.R.; Posner, G.H.; Arav-Boger, R. Inhibition of human cytomegalovirus replication by artemisinins: Effects mediated through cell cycle modulation. *Antimicrob. Agents Chemother.* **2015**, *59*, 3870–3879. [CrossRef] [PubMed]

341. Kaptein, S.J.; Efferth, T.; Leis, M.; Rechter, S.; Auerochs, S.; Kalmer, M.; Bruggeman, C.A.; Vink, C.; Stamminger, T.; Marschall, M. The anti-malaria drug artesunate inhibits replication of cytomegalovirus in vitro and in vivo. *Antivir. Res.* **2006**, *69*, 60–69. [CrossRef] [PubMed]

342. Shapira, M.Y.; Resnick, I.B.; Chou, S.; Neumann, A.U.; Lurain, N.S.; Stamminger, T.; Caplan, O.; Saleh, N.; Efferth, T.; Marschall, M.; et al. Artesunate as a potent antiviral agent in a patient with late drug-resistant cytomegalovirus infection after hematopoietic stem cell transplantation. *Clin. Infect. Dis.* **2008**, *46*, 1455–1457. [CrossRef] [PubMed]

343. Wolf, D.G.; Shimoni, A.; Resnick, I.B.; Stamminger, T.; Neumann, A.U.; Chou, S.; Efferth, T.; Caplan, O.; Rose, J.; Nagler, A.; et al. Human cytomegalovirus kinetics following institution of artesunate after hematopoietic stem cell transplantation. *Antivir. Res.* **2011**, *90*, 183–186. [CrossRef] [PubMed]

344. Lau, P.K.; Woods, M.L.; Ratanjee, S.K.; John, G.T. Artesunate is ineffective in controlling valganciclovir-resistant cytomegalovirus infection. *Clin. Infect. Dis.* **2011**, *52*, 279. [CrossRef]

345. Germi, R.; Mariette, C.; Alain, S.; Lupo, J.; Thiebaut, A.; Brion, J.P.; Epaulard, O.; Saint Raymond, C.; Malvezzi, P.; Morand, P. Success and failure of artesunate treatment in five transplant recipients with disease caused by drug-resistant cytomegalovirus. *Antivir. Res.* **2014**, *101*, 57–61. [CrossRef]

346. Gilmore, T.D.; Herscovitch, M. Inhibitors of NF-κB signaling: 785 and counting. *Oncogene* **2006**, *25*, 6887–6899. [CrossRef]

347. Herrington, F.D.; Carmody, R.J.; Goodyear, C.S. Modulation of NF-κB signaling as a therapeutic target in autoimmunity. *J. Biomol. Screen* **2016**, *21*, 223–242. [CrossRef]

348. Frelin, C.; Imbert, V.; Griessinger, E.; Loubat, A.; Dreano, M.; Peyron, J.F. AS602868. a pharmacological inhibitor of IKK2. reveals the apoptotic potential of TNF-α in Jurkat leukemic cells. *Oncogene* **2003**, *22*, 8187–8194. [CrossRef] [PubMed]

349. Frelin, C.; Imbert, V.; Griessinger, E.; Peyron, A.C.; Rochet, N.; Philip, P.; Dageville, C.; Sirvent, A.; Hummelsberger, M.; Berard, E.; et al. Targeting NF-κB activation via pharmacologic inhibition of IKK2-induced apoptosis of human acute myeloid leukemia cells. *Blood* **2005**, *105*, 804–811. [CrossRef] [PubMed]

350. Caposio, P.; Musso, T.; Luganini, A.; Inoue, H.; Gariglio, M.; Landolfo, S.; Gribaudo, G. Targeting the NF-κB pathway through pharmacological inhibition of IKK2 prevents human cytomegalovirus replication and virus-induced inflammatory response in infected endothelial cells. *Antivir. Res.* **2007**, *73*, 175–184. [CrossRef] [PubMed]

351. Basha, W.; Kitagawa, R.; Uhara, M.; Imazu, H.; Uechi, K.; Tanaka, J. Geldanamycin, a potent and specific inhibitor of Hsp90, inhibits gene expression and replication of human cytomegalovirus. *Antivir. Chem. Chemother.* **2005**, *16*, 135–146. [CrossRef]

352. Evers, D.L.; Chao, C.F.; Zhang, Z.; Huang, E.S. 17-allylamino-17-(demethoxy)geldanamycin (17-AAG) is a potent and effective inhibitor of human cytomegalovirus replication in primary fibroblast cells. *Arch. Virol.* **2012**, *157*, 1971–1974. [CrossRef]

353. Wu, P.; Nielsen, T.E.; Clausen, M.H. FDA-approved small-molecule kinase inhibitors. *Trends Pharmacol. Sci.* **2015**, *36*, 422–439. [CrossRef]

354. Zhang, H.; Niu, X.; Qian, Z.; Qian, J.; Xuan, B. The c-Jun N-terminal kinase inhibitor SP600125 inhibits human cytomegalovirus replication. *J. Med. Virol.* **2015**, *87*, 2135–2144. [CrossRef]

355. Beelontally, R.; Wilkie, G.S.; Lau, B.; Goodmaker, C.J.; Ho, C.M.K.; Swanson, C.M.; Deng, X.; Wang, J.; Gray, N.S.; Davison, A.J.; et al. Identification of compounds with anti-human cytomegalovirus activity that inhibit production of IE2 proteins. *Antivir. Res.* **2017**, *138*, 61–67. [CrossRef]

356. Strang, B.L. RO0504985 is an inhibitor of CMGC kinase proteins and has anti-human cytomegalovirus activity. *Antivir. Res.* **2017**, *144*, 21–26. [CrossRef]

357. Khan, A.S.; Murray, M.J.; Ho, C.M.K.; Zuercher, W.J.; Reeves, M.B.; Strang, B.L. High-throughput screening of a GlaxoSmithKline protein kinase inhibitor set identifies an inhibitor of human cytomegalovirus replication that prevents CREB and histone H3 post-translational modification. *J. Gen. Virol.* **2017**, *98*, 754–768. [CrossRef]

358. Nehme, Z.; Pasquereau, S.; Herbein, G. Control of viral infections by epigenetic-targeted therapy. *Clin. Epigenet.* **2019**, *11*, 55. [CrossRef] [PubMed]

359. Liang, Y.; Quenelle, D.; Vogel, J.L.; Mascaro, C.; Ortega, A.; Kristie, T.M. A novel selective LSD1/KDM1A inhibitor epigenetically blocks herpes simplex virus lytic replication and reactivation from latency. *MBio* **2013**, *4*, e00558-12. [CrossRef] [PubMed]

360. Rai, G.; Kawamura, A.; Tumber, A.; Liang, Y.; Vogel, J.L.; Arbuckle, J.H.; Rose, N.R.; Dexheimer, T.S.; Foley, T.L.; King, O.N.; et al. Discovery of ML324, a JMJD2 demethylase inhibitor with demonstrated antiviral activity. In *Probe Reports from the NIH Molecular Libraries Program*; National Center for Biotechnology Information: Bethesda, MD, USA, 2010.

361. Fukui, Y.; Shindoh, K.; Yamamoto, Y.; Koyano, S.; Kosugi, I.; Yamaguchi, T.; Kurane, I.; Inoue, N. Establishment of a cell-based assay for screening of compounds inhibiting very early events in the cytomegalovirus replication cycle and characterization of a compound identified using the assay. *Antimicrob. Agents Chemother.* **2008**, *52*, 2420–2427. [CrossRef] [PubMed]

362. Gardner, T.J.; Cohen, T.; Redmann, V.; Lau, Z.; Felsenfeld, D.; Tortorella, D. Development of a high-content screen for the identification of inhibitors directed against the early steps of the cytomegalovirus infectious cycle. *Antivir. Res.* **2015**, *113*, 49–61. [CrossRef]

363. Cohen, T.; Williams, J.D.; Opperman, T.J.; Sanchez, R.; Lurain, N.S.; Tortorella, D. Convallatoxin-induced reduction of methionine import effectively inhibits human cytomegalovirus infection and replication. *J. Virol.* **2016**, *90*, 10715–10727. [CrossRef]

364. Mercorelli, B.; Luganini, A.; Nannetti, G.; Tabarrini, O.; Palu, G.; Gribaudo, G.; Loregian, A. drug repurposing approach identifies inhibitors of the prototypic viral transcription factor IE2 that block human cytomegalovirus replication. *Cell Chem. Biol.* **2016**, *23*, 340–351. [CrossRef]

365. Kapoor, A.; Cai, H.; Forman, M.; He, R.; Shamay, M.; Arav-Boger, R. Human cytomegalovirus inhibition by cardiac glycosides: Evidence for involvement of the HERG gene. *Antimicrob. Agents Chemother.* **2012**, *56*, 4891–4899. [CrossRef]

366. Cai, H.Y.; Wang, H.Y.L.; Venkatadri, R.; Fu, D.X.; Forman, M.; Bajaj, S.O.; Li, H.Y.; O'Doherty, G.A.; Arav-Boger, R. Digitoxin analogues with improved anticytomegalovirus activity. *ACS Med. Chem. Lett.* **2014**, *5*, 395–399. [CrossRef]

367. Hartley, C.; Hartley, M.; Pardoe, I.; Knight, A. Ionic Contra-Viral Therapy (ICVT); a new approach to the treatment of DNA virus infections. *Arch. Virol.* **2006**, *151*, 2495–2501. [CrossRef]

368. Whayne, T.F., Jr. Clinical Use of Digitalis: A State of the Art Review. *Am. J. Cardiovasc. Drugs* **2018**, *18*, 427–440. [CrossRef]

369. Cai, H.; Kapoor, A.; He, R.; Venkatadri, R.; Forman, M.; Posner, G.H.; Arav-Boger, R. In vitro combination of anti-cytomegalovirus compounds acting through different targets: Role of the slope parameter and insights into mechanisms of Action. *Antimicrob. Agents Chemother.* **2014**, *58*, 986–994. [CrossRef] [PubMed]

370. Orlov, S.N.; Klimanova, E.A.; Tverskoi, A.M.; Vladychenskaya, E.A.; Smolyaninova, L.V.; Lopina, O.D. Na(+)i,K(+)i-dependent and -independent signaling triggered by cardiotonic steroids: Facts and artifacts. *Molecules* **2017**, *22*, 635. [CrossRef] [PubMed]

371. Amarelle, L.; Lecuona, E. The antiviral effects of Na,K-ATPase inhibition: A minireview. *Int. J. Mol. Sci.* **2018**, *19*, 2154. [CrossRef] [PubMed]

372. Yamada, K.H.; Majima, R.; Yamaguchi, T.; Inoue, N. Characterization of phenyl pyrimidine derivatives that inhibit cytomegalovirus immediate-early gene expression. *Antivir. Chem. Chemother.* **2018**, *26*, 3193. [CrossRef]

373. Vasou, A.; Paulus, C.; Narloch, J.; Gage, Z.O.; Rameix-Welti, M.A.; Eleouet, J.F.; Nevels, M.; Randall, R.E.; Adamson, C.S. Modular cell-based platform for high throughput identification of compounds that inhibit a viral interferon antagonist of choice. *Antivir. Res.* **2018**, *150*, 79–92. [CrossRef]

374. Loregian, A.; Mercorelli, B.; Muratore, G.; Sinigalia, E.; Pagni, S.; Massari, S.; Gribaudo, G.; Gatto, B.; Palumbo, M.; Tabarrini, O.; et al. The 6-aminoquinolone WC5 inhibits human cytomegalovirus replication at an early stage by interfering with the transactivating activity of viral immediate-early 2 protein. *Antimicrob. Agents Chemother.* **2010**, *54*, 1930–1940. [CrossRef]

375. Stevens, M.; Balzarini, J.; Tabarrini, O.; Andrei, G.; Snoeck, R.; Cecchetti, V.; Fravolini, A.; De Clercq, E.; Pannecouque, C. Cell-dependent interference of a series of new 6-aminoquinolone derivatives with viral (HIV/CMV) transactivation. *J. Antimicrob. Chemother.* **2005**, *56*, 847–855. [CrossRef]

376. Mercorelli, B.; Muratore, G.; Sinigalia, E.; Tabarrini, O.; Biasolo, M.A.; Cecchetti, V.; Palu, G.; Loregian, A. A 6-aminoquinolone compound, WC5, with potent and selective anti-human cytomegalovirus activity. *Antimicrob. Agents Chemother.* **2009**, *53*, 312–315. [CrossRef]

377. Cecchetti, V.; Parolin, C.; Moro, S.; Pecere, T.; Filipponi, E.; Calistri, A.; Tabarrini, O.; Gatto, B.; Palumbo, M.; Fravolini, A.; et al. 6-Aminoquinolones as new potential anti-HIV agents. *J. Med. Chem.* **2000**, *43*, 3799–3802. [CrossRef]

378. Mercorelli, B.; Luganini, A.; Muratore, G.; Massari, S.; Terlizzi, M.E.; Tabarrini, O.; Gribaudo, G.; Palu, G.; Loregian, A. The 6-Aminoquinolone WC5 inhibits different functions of the immediate-early 2 (IE2) protein of human cytomegalovirus that are essential for viral replication. *Antimicrob. Agents Chemother.* **2014**, *58*, 6615–6626. [CrossRef]

379. Massari, S.; Mercorelli, B.; Sancineto, L.; Sabatini, S.; Cecchetti, V.; Gribaudo, G.; Palu, G.; Pannecouque, C.; Loregian, A.; Tabarrini, O. Design, synthesis, and evaluation of WC5 analogues as inhibitors of human cytomegalovirus Immediate-Early 2 protein, a promising target for anti-HCMV treatment. *ChemMedChem* **2013**, *8*, 1403–1414. [CrossRef]

380. Luganini, A.; Caposio, P.; Mondini, M.; Landolfo, S.; Gribaudo, G. New cell-based indicator assays for the detection of human cytomegalovirus infection and screening of inhibitors of viral immediate-early 2 protein activity. *J. Appl. Microbiol.* **2008**, *105*, 1791–1801. [CrossRef]

381. Mercorelli, B.; Luganini, A.; Celegato, M.; Palu, G.; Gribaudo, G.; Loregian, A. Repurposing the clinically approved calcium antagonist manidipine dihydrochloride as a new early inhibitor of human cytomegalovirus targeting the Immediate-Early 2 (IE2) protein. *Antivir. Res.* **2018**, *150*, 130–136. [CrossRef]

382. Luganini, A.; Mercorelli, B.; Messa, L.; Palu, G.; Gribaudo, G.; Loregian, A. The isoquinoline alkaloid berberine inhibits human cytomegalovirus replication by interfering with the viral Immediate Early-2 (IE2) protein transactivating activity. *Antivir. Res.* **2019**, *164*, 52–60. [CrossRef]

383. Mercorelli, B.; Gribaudo, G.; Palu, G.; Loregian, A. Approaches for the generation of new anti-cytomegalovirus agents: Identification of protein-protein interaction inhibitors and compounds against the HCMV IE2 protein. *Methods Mol. Biol.* **2014**, *1119*, 349–363. [CrossRef]

Review

Sphingolipids as Potential Therapeutic Targets against Enveloped Human RNA Viruses

Eric J. Yager [1] and Kouacou V. Konan [2,*]

[1]   Department of Basic and Clinical Sciences, Albany College of Pharmacy and Health Sciences, Albany, NY 12208, USA; Eric.Yager@acphs.edu

[2]   Department of Immunology and Microbial Disease, Albany Medical College, Albany, NY 12208-3479, USA

*   Correspondence: KonanK@amc.edu; Tel.: +1-518-262-9040

Received: 7 August 2019; Accepted: 28 September 2019; Published: 1 October 2019

Abstract: Several notable human diseases are caused by enveloped RNA viruses: Influenza, AIDS, hepatitis C, dengue hemorrhagic fever, microcephaly, and Guillain–Barré Syndrome. Being enveloped, the life cycle of this group of viruses is critically dependent on host lipid biosynthesis. Viral binding and entry involve interactions between viral envelope glycoproteins and cellular receptors localized to lipid-rich regions of the plasma membrane. Subsequent infection by these viruses leads to reorganization of cellular membranes and lipid metabolism to support the production of new viral particles. Recent work has focused on defining the involvement of specific lipid classes in the entry, genome replication assembly, and viral particle formation of these viruses in hopes of identifying potential therapeutic targets for the treatment or prevention of disease. In this review, we will highlight the role of host sphingolipids in the lifecycle of several medically important enveloped RNA viruses.

Keywords: sphingolipids; glycosphingolipids; viruses; lipid biosynthesis; antiviral

---

1. Introduction

Human viruses come in various shapes and sizes with DNA or RNA as their genetic material. The focal point of this review is on enveloped RNA virus particles, which consist of a lipid bilayer typically surrounding the genomic-RNA-protecting shell or capsid. While the lipid composition of the envelope varies between these RNA viruses, it is often enriched in phospholipids, cholesterol, and sphingolipids. In many instances, the integrity of the virus envelope is crucial for viral infectivity. For example, specific phospholipids in the envelope of some members of the Flaviviridae family of viruses are reported to facilitate virus attachment to host cells, a key step in viral entry [1–3].

Enveloped RNA viruses are further divided into two classes based on the polarity of the genome. For example, most positive-stranded RNA viruses replicate exclusively in the cytoplasm of the infected cell and in intimate contact with intracellular membranes. This strategy enables viral and host factors to concentrate in distinct cellular locations to optimize a new virus particle's formation and evade innate immune responses [4–8]. By contrast, the replication cycles of some negative-stranded RNA viruses (e.g. Influenza virus), and human immunodeficiency virus, occurs in the nucleus [9,10]. Hence, positive and negative-stranded RNA viruses require a distinct set of host membranes, and the lipids present therein, for successful virus propagation. This review will feature a few medically important enveloped RNA viruses, such as hepatitis C virus (HCV), dengue virus (DENV), Zika virus (ZIKV), human immunodeficiency virus (HIV), and influenza virus (IAV), and highlight the role of sphingolipids in their replication cycle and pathogenesis.

Sphingolipids are important biomolecules found in all eukaryotic membranes. They regulate membrane trafficking, cell signaling, and play a crucial role in influenza virus particles' release or cell surface binding of HIV-1 glycoprotein gp120 [11,12]. They are also major constituents of lipid rafts, which are integral components of the HCV replication complex [13,14]. Sphingolipid biosynthesis starts with the conversion of palmitoyl-CoA and serine into ceramide, a sphingolipid byproduct of the endoplasmic reticulum (ER) resident enzyme—serine palmitoyltransferase, or SPT [15–18] (Figure 1). Ceramide can be carried by ceramide transport protein (CERT) [19] to the trans-Golgi where it is converted into another sphingolipid called sphingomyelin. Alternatively, four-phosphate adaptor protein 2, or FAPP2 [20], carries ceramide to the cis-Golgi where glucosylceramide synthase (GCS) produces the first glycosphingolipid called glucosylceramide (GlcCer; Figure 1). Glucosylceramide is subsequently converted into more complex glycosphingolipids including lactosylceramide (LacCer), globosides (e.g., Gb3) and gangliosides (e.g., GM3, GM1, and GA1; Figure 1). GCS is the rate-limiting enzyme in glycosphingolipid biosynthesis. The insufficiency or overproduction of glycosphingolipids has been associated with disease in humans. Consequently, efforts were made to inhibit GCS activity to reduce glucosylceramide accumulation in patients. One such GCS inhibitor, Genz-112638 [21], has been approved for treating Gaucher disease linked to defective glucosylceramide catabolism [22–25] (Figure 1). Sphingolipids and glycosphingolipids are found in distinct internal membranes as well as the plasma membrane. Additionally, glycosphingolipids are highly enriched in neurons, skin epithelial cells and might contribute to the tropism, replication and pathogenicity of viruses targeting related organs. Traditional methods to detect sphingolipids and glycosphingolipids include thin liquid chromatography (TLC) [20], high pressure thin liquid chromatography (HPTLC) [26], immunocytochemistry, and enzyme-linked immunosorbent assay (ELISA) [27,28]. Recently, sphingolipids and glycosphingolipids have been detected with state-of-the-art liquid chromatography coupled with mass spectrometry (e.g., LC-MS/MS system). This approach has enabled investigators to accurately determine the levels of sphingolipid and glycosphingolipid species in cells or tissues [28,29] and the impact of viral infection on levels of these lipids [30].

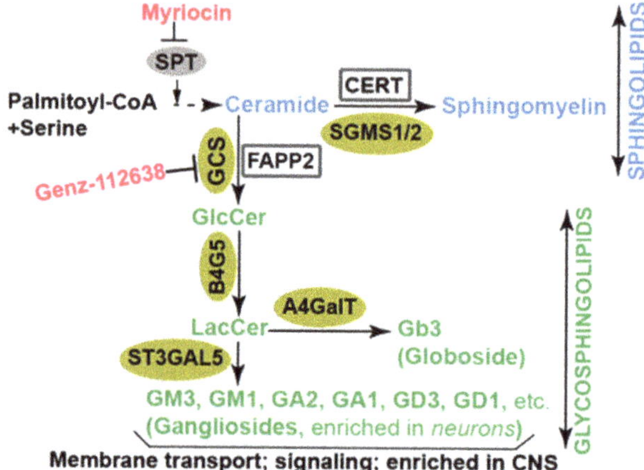

Figure 1. Diagram of sphingolipid biosynthetic pathways in mammalian cells. The initial step in the de novo biosynthesis of sphingolipids is the conversion of serine and palmitoyl CoA to ceramide. Following that, ceramide is subjected to conversion to sphingomyelin or to various glycosphingolipid intermediates on their way to becoming complex glycosphingolipids. The enzymes involved in the synthesis of sphingolipids and glycosphingolipids are denoted in gold. Chemical inhibitors of key enzymes are indicated in red. SGMS1/2: sphingomyelin synthase; GCS: glucosylceramide synthase; B4G5: lactosylceramide synthase; ST3GAL5: lactosylceramide alpha-2,3-sialyltransferase or GM3 synthase; A4GalT: alpha 1,4-galactosyltransferase or Gb3 synthase; SPT: serine palmitoyl transferase; CERT: ceramide transfer protein; FAPP2: four-phosphate adaptor protein 2; GlcCer: glucosylceramide; LacCer: lactosylceramide.

2. Hepatitis C Virus Propagation and Sphingolipids

Hepatitis C virus (HCV) is responsible for chronic liver disease in 60–90 million people worldwide [31] and is a member of the Flaviviridae family of viruses that encompass dengue virus, West Nile virus, and Zika virus. HCV is an enveloped virus with a positive-stranded RNA genome encoding structural proteins (Core, E1, and E2) (Figure 2) required for infectious HCV particles formation. The nonstructural proteins (NS2, NS3, NS4A, NS4B, NS5A, and NS5B) are involved in the replication of HCV genome and the packaging of the newly generated genome [32,33]. Major insights in the molecular and structural biology of HCV have led to the development of direct acting antivirals (DAAs) targeting HCV NS34/A protease (e.g., Simeprivir and Voxilaprevir), NS5A (e.g., Ledipasvir and Pibrentasvir), and NS5B RNA-dependent RNA polymerase or RDRp (e.g., Sofosbuvir and Dasabuvir) [34,35]. However, the high error rate (2.5×10^{-5} per nucleotide per genome replication) [36] of the HCV RdRp has resulted in the presence of resistance-associated variants [37,38]. Thus, novel antivirals targeting other HCV proteins, or host factors, are needed.

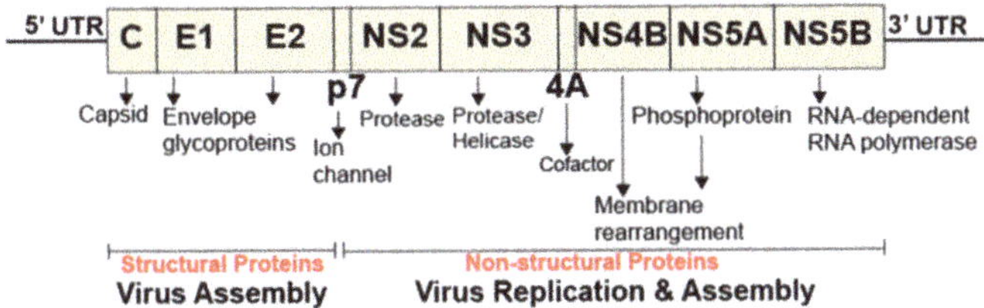

Figure 2. Diagram of hepatitis C virus genome. The HCV genome consists of a single open reading frame (ORF) flanked by the 5′ and 3′ untranslated regions (UTRs). The ORF is translated into a single polyprotein, which is further processed into individual proteins. The 5′ and 3′ UTRs are critical for internal ribosome binding, translation, and HCV genome replication.

2.1. Sphingolipids and HCV Entry

There is strong evidence in the literature implying that sphingolipids and glycosphingolipids play an intimate role in HCV replication in liver cells. First, Merz et al. utilized lipid mass spectrometry to demonstrate that affinity-purified HCV particles are enriched in sphingomyelin [39], suggesting that sphingolipids are integral components of HCV envelope. This study did not address the role of sphingomyelin in HCV propagation, but another study by Aizaki et al. [40] demonstrated that sphingomyelin facilitates HCV internalization, perhaps via fusion of the virus envelope with endocytic membrane to release HCV genome into the cytoplasm.

2.2. Sphingolipids, Glycosphingolipids, and HCV Genome Replication

Sphingolipids are also known to facilitate HCV genome replication. In one study led by Hirata et al. [14], the authors showed that HCV infection stimulates sphingomyelin production, leading to sphingomyelin enrichment in the HCV replication complex and sphingomyelin-induced stimulation of HCV RNA-dependent RNA polymerase to synthesize more viral RNA. During infection, HCV induces the formation of a distinct membrane structure in the cell. This platform called the membranous web [4,41,42] recruits viral and host factors to foster HCV genome amplification. In another study from our laboratory, we found that HCV redirects the glycosphingolipid carrier protein, FAPP2 (Figure 1), to the membranous web to facilitate HCV genome replication [20]. It is likely that FAPP2 transports glycosphingolipids to the virus replication platform, as FAPP2 knockdown impedes HCV replication, whereas providing glycosphingolipids to the FAPP2 knockdown cells rescues HCV genome replication [20]. Finally, FAPP2 knockdown was found to disrupt the membranous web and alter the colocalization of HCV replicase proteins, implying that FAPP2 and/or glycosphingolipids contribute to the formation or maintenance of the HCV replication platform.

3. Flavivirus Propagation and Sphingolipids

Dengue virus (DENV), West Nile virus (WNV), and Zika virus (ZIKV) are archetype flaviviruses transmitted by mosquitoes, mainly *Aedes aegypti* and *Aedes albopictus*. According to the world health organization, DENV is responsible for 50–100 million infections each year with mild complications resembling flu-like symptoms and major complications, including deadly dengue hemorrhagic fever. WNV causes flu-like symptoms, neuroinvasive disease, and death in many countries in the world. WNV was first introduced in the United States (US) in 1999 from infected Israeli birds imported into New York state. The virus has now spread to most states and is responsible for many deaths in birds, humans and horses. By contrast, ZIKV only emerged as a global health concern since 2016, due its association with neurological disorders, such as microcephaly in newborns and Guillain–Barré syndrome in adults [27,43]. Like hepatitis C virus (HCV), DENV, WNV, and ZIKV are enveloped viruses. Unlike HCV, the positive-stranded RNA genome of these flaviviruses encodes a slightly different set of structural proteins (Core, E, and prM) (Figure 3) required for virus particle formation, and nonstructural proteins (NS1, NS2A, NS2B, NS3, NS4A, NS4B, NS5) involved in viral genome replication, packaging, and pathogenesis [44–46]. No effective vaccine or specific antiviral treatments are currently available for DENV, WNV, or ZIKV infection.

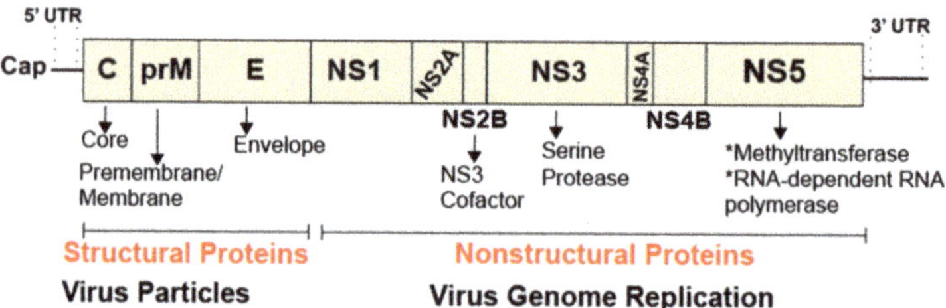

Figure 3. Organization of a flavivirus genome. The genome of flaviviruses, such as dengue and Zika, is ~11 kb in size and encodes a single, large polyprotein, which is proteolytically processed into three structural proteins and seven nonstructural proteins. The 5′ end of the genome contains a cap structure critical for the initiation of translation. RNA structures present in the 5′ and 3′ untranslated regions (UTR) are critical for capping and genome replication.

3.1. DENV Propagation and Sphingolipids

There is evidence that sphingolipids and glycosphingolipids are also required for the replication of some flaviviruses. For example, DENV has been reported to upregulate the expression of sphingolipids (ceramide and sphingomyelin) in mosquito cells and cause an accumulation of these lipids in a membrane fraction enriched in the viral replication complex [30]. That study did not directly address the role of sphingolipids in DENV replication. In a different study, Wang et al. [47] exploited mouse melanoma WT cells (B16) and a mutant counterpart (GM95) to demonstrate that the glycosphingolipid GM3 is required for DENV genome replication. The authors found higher levels of GM3 in DENV-infected cells and relocalization of GM3 to sites where DENV replicates its genome. Importantly, the authors found that inhibition of GM3 synthesis with soyasaponin I increases the survival rate of DENV-infected mice [47]. While the exact role of GM3 in DENV genome replication is currently unknown, these in vivo findings raise the prospect that pharmacological inhibitors targeting GM3 synthesis can serve as a foundation for new antiviral therapy.

3.2. WNV Propagation and Sphingolipids

3.2.1. Sphingolipids and WNV Entry and Genome Replication

The role of sphingomyelin in WNV replication is well documented. A study by Martin-Acebes and colleagues [48] showed that WNV replicates at a much higher level in mice deficient in acid sphingomyelinase (unable to catabolize sphingomyelin), or cells derived from Niemann–Pick disease type A patients (NPA; accumulate sphingomyelin) relative to the the their wild type controls. This suggested that sphingomyelin accumulation enhances WNV infectivity. Consistent with these findings, adding sphingomyelin to infected fibroblast cells markedly increased WNV infectivity. Further analysis showed that sphingomyelin colocalizes with WNV dsRNA at cytoplasmic foci, implying that sphingomyelin plays a role in the formation of the WNV replication platform [48]. Interestingly, pharmacological inhibitors of sphingomyelin synthesis (DS609 and SPK-601) markedly reduced the infectivity of WNV released from infected cells, but had little impact on the amount of released viral genome [48]. These findings imply that sphingomyelin is also required for WNV attachment, internalization, and/or virus–endosome fusion.

3.2.2. Sphingolipids and WNV Particle Formation

In an earlier report, Martin-Acebes and colleagues [49] also showed that WNV particles were enriched in sphingomyelin. Surprisingly, pharmacological inhibition of neutral sphingomyelinase (converts sphingomyelin into ceramide and phosphorylcholine) reduced WNV release from infected cells, implying perhaps that ceramide generated from sphingomyelin catabolism is critical for the infectious WNV particle. Subsequent analysis showed that inhibition of neutral sphingomyelinase activity reduces the budding of the immature WNV particles, a crucial step in infectious WNV particle formation [49]. While this study appears to be at odds with the putative role of sphingomyelin in WNV entry, it also highlights the role of sphingomyelin in two distinct steps of the WNV replication cycle.

3.3. ZIKV Replication and Sphingolipids

The role of sphingolipids and glycosphingolipids in ZIKV replication is not well understood. This is crucial because ZIKV patients with Guillain–Barré Syndrome, have elevated levels of antibodies targeting gangliosides GM2, GM1, GA1, and GD1 [27], hence highlighting the need to define the role of glucosylceramide-derived glycosphingolipids in ZIKV infectivity. Our group has evidence that ZIKV particles are enriched in sphingomyelin, ceramide, and glucosylecramide (unpublished data). Our current data suggest that glycosphingolipids leading to gangliosides biosynthesis (Figure 1) are required for ZIKV particle assembly (unpublished data). Current efforts are focused on understanding how sphingolipids and glycosphingolipids regulate ZIKV infectivity.

4. Human Immunodeficiency Virus's Propagation and Sphingolipids

Human immunodeficiency virus (HIV) is responsible for acquired immunodeficiency syndrome (AIDS) worldwide. In 2017 alone, approximately one-million people died of AIDS-related illnesses, highlighting the need for alternative or complementary remedies to eradicate HIV infection. HIV is an enveloped virus with two identical single-stranded RNAs which serve as a template for reverse transcription into a single double-stranded DNA intermediate that can integrate into the host genome. This DNA is the template for RNA that codes for new genomes or for viral proteins. Full-length RNA codes for a structural polyprotein that includes matrix (MA), capsid (CA), and nucleocapsid (NC) (Figure 4). Inefficient supression of a stop codon between the gag and the downstream pol reading frame produces a polyprotein that is extended to include the pol proteins protease (PR), reverse transcriptase (RT), and integrase (IN). All are cut into their constituent parts by PR. A singly spliced mRNA encodes the env glycoproteins gp41 (TM, transmembrane, a fusion protein) and gp120 (SU, surface glycoprotein). Other more extensively spliced messages, encode the regulatory proteins Tat, a transactivator of transcription, and Rev, an HIV-specific RNA exporter. Yet other messages code for accessory proteins Vif and Vpr, which facilitate the degradation of cellular defense proteins, and Vpu and Nef, which remove cellular proteins from the cell surface [50,51]. Knowledge of HIV biology and pathogenesis has lead to the identification of several pharmacological targets for antiretroviral drug therapy. Following approval of the first anti-HIV drug viral azidothymidine (AZT; nucleoside reverse transcriptase inhibitor), additional classes of drugs targeting viral attachment/fusion, viral genome replication, integrase activity, and the viral protease have been developed, some of which were subsequently included in combinational antiretroviral drug therapy, resulting in effective clinical management of HIV infection and AIDS-related illnesses [52]. Despite the efficacy and availability of these drugs, there exist several challenges that necessitate continued research in the identification of new viral targets and anti-HIV agents [53,54]. There is a continual threat of drug resistance due to HIV's high mutation rate. Each class of currently available drug has the potential to cause acute and chronic toxicities in patients (e.g., cardiovascular and metabolic abnormalities). Lastly, side-effects of the drugs adversely impact patient compliance.

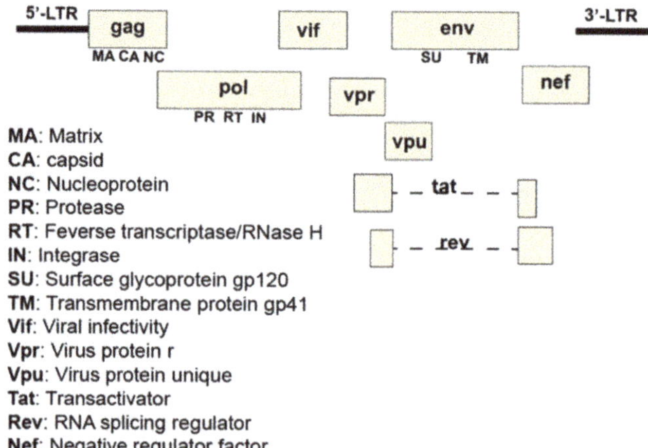

Figure 4. Organization of the HIV-1 genome. The HIV-1 genome consists of two identical copies of noncovalently linked, linear, positive-sense, single-stranded RNA molecules. Each identical copy contains nine genes that encode fifteen proteins. Many of the proteins are synthesized as precursor polyproteins which are proteolytically processed by host or viral proteases into individual proteins with roles in viral architecture, replication, regulation of cellular functions, and evasion of the host defenses. The gag gene encodes viral proteins involved in the structure of the virus. The pol gene encodes viral proteins critical for replication and integration of provirus into the host genome. The env gene encodes proteins needed for viral attachment and fusion with target cells.

4.1. Glycosphingolipids and HIV Entry

Glycosphingolipids are also key players in HIV infection. HIV-1 infection of susceptible cells involves fusion of the HIV membrane with the host cell membrane. This process involves interactions between the viral gp120 and gp 41 envelope glycoproteins and CD4 and chemokine coreceptors (CXCR4 or CCR5) on the host cell. Several studies have implicated glycosphingolipids in the fusion process. Hug et al. demonstrated that glucosylceramide-derived glycosphingolipids found on the target cell membrane are involved in the organization of gp120–gp41, CD4, and chemokine receptors into a membrane fusion complex [55]. Rawat et al. reported that expression levels of the membrane ganglioside GM3 impact CD4-dependent viral fusion and infection [56]. Furthermore, several glycosphingolipids (galactosylceramide, GD3, Gb3, and GM3) have been reported to bind to the HIV gp120 envelope protein and contribute in some cases to the HIV infection of CD4 negative cells [57–60]. Specifically, HIV-1 entry was impaired in human colonic cells (HT29; CD4 negative cells) with a synthetic analog or a monoclonal antibody to galactosylceramide (GalCer) [60,61], implying that GalCer is an alternative receptor for HIV replication entry. Similarly, antibodies to GalCer reduced HIV infectivity andinternalization in two CD4-negative neural cell lines, U373-MG and SK-N-MC [58]. Altogether, these findings imply that glycosphingolipids contribute to HIV membrane fusion and can serve as alternative receptors for HIV entry in some cells.

4.2. Glycosphingolipids and HIV Budding

HIV particle assembly and budding takes place in the plasma membrane's lipid rafts which contain high levels of cholesterol, sphingolipid (sphingomyelin), and glycosphingolipids. Notably, genetic analysis and lipid mass spectrometry showed that the HIV particle is enriched in sphingolipids (sphingomyelin and dihydro sphingomyelin) and glycosphingolipids (hexosylceramide, GM3, GM2, GM1) [62–65], implying that sphingolipids contribute to the biogenesis of infectious HIV particle and are acquired during HIV budding. There is also increasing evidence that glycosphinolipids help facilitate HIV's infection of macrophages. Indeed, multiple studies have shown that GM3, GM2, and GM1 in HIV's envelope interact with Siglec-1, a molecule that specifically binds to the sialyllactose moiety in glycosphinolipids [64–66]. This interaction between glycosphinolipids and Siglec-1 captures HIVs on macrophages, which might exist as a reservoir for HIV's transmission to T cells.

5. Influenza Virus Propagation and Sphingolipids

Seasonal influenza epidemics are responsible for over 200,000 hospitalizations in the United States and up to 500,000 deaths worldwide each year. Influenza A (IAV) is an enveloped virus possessing a negative sense, single-stranded, segmented RNA genome. Eight RNA segments encode 11 different viral proteins (Figure 5). The envelope spike glycoproteins HA and NA mediate viral entry and release, respectively, and serve as antigenic determinants of the virus. M2 is proton-selective ion channel involved in uncoating once the virus has entered the cell. Matrix protein (M1) forms a matrix under the viral envelope which is critical for maintaining the integrity and shape of the intact viral particle. Each segment of viral RNA is encapsulated with nucleocapsid protein (NP) and associated with the trimeric polymerase complex (PB1, PB2, and PA), forming what are referred to as a viral ribonucleoprotein complex (vRNP). Non-structural protein 1 (NS1) plays a critical role in evasion of host immunity. Non-structural protein 2 (NS2) is involved in the export of newly synthesized vRNPs from the nucleus to the cytoplasm for packaging.

The two classes of flu antiviral drugs include M2 ion channel blockers (e.g., amantadine and rimantadine) and neuraminidase inhibitors (e.g., zanamivir, oseltamivir). Early administration of these antivirals reduces disease symptoms, shortens the duration of illness, reduces hospitalization rates, and reduces viral transmission [67,68]. However, replication of the IAV genome involves a high error rate (10^{-3} to 10^{-4} substititution per genome) [69,70], resulting in the frequent accumulation of amino acid changes in IAV proteins. These changes enable IAV to evade host immunity acquired by prior exposure or vaccination and is the reason why IAV vaccines must be reformulated annually. Additionally, these amino acid changes may allow the virus to develop resistance against currently available antiviral agents that target the activity of the flu NA protein. As such, novel antivirals against influenza are critically needed.

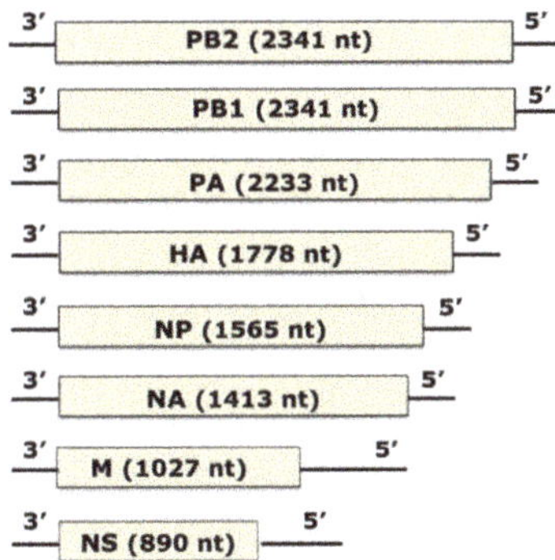

Figure 5. Organization of the Influenza A virus's genome. The influenza A virus's (IAV's) genome consists of eight segments of negative-sense, single-stranded RNA. Encoded by the genome are three polymerase proteins (PA, PB1, and PB2), nucleoprotein (NP), and two envelope proteins (HA and NA). The M and NS mRNAs can be alternatively spliced to yield M1 and M2, and NS1 and NS2 proteins, respectively. Boxes indicate coding regions, sized relative to the other gene segments. Black lines at each end represent the 3' and 5' untranslated regions. The total length of each segment (conding and non-coding regions) in nucleotides (nt) is indicated. Figure adapted from Bouvier and Palese [71].

5.1. Sphingolipids and IAV Entry

The influenza virus's envelope, derived from the host cell's plasma membrane, consists of a lipid bilayer decorated with the viral hemagglutinin (HA), neuraminidase (NA), and M2 proteins. The lipidome of purfied IAV particles consists of glycerophospholipids and sterols (primarily cholesterol) [72,73]. Further, almost all sphingolipid classes were detected in the viral envelope [72]. Attachment and entry into host cells requires interactions between the viral HA, concentrated in microdomains on the viral envelope, and sialic acid residues present on the cell surface. These microdomains, paralleling the lipids rafts present on the cell surface, are enriched with cholesterol and various sphingolipids, including sphingomyelin and glycosphingolipids [74–76]. Residues in the transmembrane domain and cytoplasmic tail of IAV HA are important for its association with lipid microdomains [75,76]. Disruption of these microdomains has an adverse effect on viral attachment. Sun and Whittaker showed that pretreatment of IAV virions with methly-β-cyclodextrin to deplete envelope cholesterol resulted in reduced viral fusion and infectivity [77]. Though not examined in their study, the authors suggested that cholesterol depletion perturbed the organization of HA in envelope lipid microdomains. Viral infectivity has been found to be reduced by approximately a thousandfold when HA fails to associate with lipid microdomains [75]. Similarly, it is expected that depletion of the IAV envelope sphingolipds would adversely impact IAV binding and infectivity.

5.2. Sphingolipids and IAV Replication

Following binding to host target cells, influenza virus enters the cytoplasm via receptor-mediated endocytosis. The viral M2 ion channel protein allows the influx of protons into the virion, triggering the release of viral ribonucleoproteins (vRNPs) into the cytoplasm. The vRNPs, made up of viral negative-stranded RNA, viral NP, and the viral RNA polymerase complex, then travel to the nucleus for transcription and replication of the influenza virus genome. Research suggests that products derived from sphingolipids are involved in influenza virus's genome replication. Seo et al. [78] found that cells infected with influenza virus possessed increased levels of the enzyme sphingosine kinase (SK1), which converts sphingosine into sphingosine 1-phosphate (S1P). The authors further showed that inhibition of SK1 impaired viral RNA synthesis and the subsequent nuclear export of newly generated vRNPs. Similarly, it was demonstrated that SK1 is critical for the nuclear export of viral proteins (NP, NS2, and M1) involved in transporting vRNPs from the nucleus to the cytoplasm [79].

5.3. Sphingolipids and IAV Egress

Like several other enveloped viruses, influenza uses "raft-like" microdomains on the cell surface as platforms for viral assembly. Newly synthesized HA and NA concentrate in microdomains enriched for sphingomyelin and cholesterol [76,80–82]. Tafesse et al. [12] showed that perturbation of host sphingomyelin biosynthesis adversely impacted the trafficking of influenza virus HA and NA to the cell surface, which in turn impaired viral maturation, budding, and release. Additionally, though abbreviated treatment with methyl-β-cyclodextrin at the late stages of infection was found to increase the release of viral particles from infected MDCK cells, the infectivity of the released particles was significantly reduced, as their envelope possessed lower contents of cholesterol and disrupted raft microdomains [83].

6. Conclusions

The role of sphingolipids and glycosphingolipids has been overlooked for many years in viral infection due to the difficulty of detecting or measuring these lipids. However, with the advent of lipid mass spectrometry, it is now possible to accurately determine the level of sphingolipids and glycosphingolipids in virus-infected cells and virus particles. Recent findings clearly imply that many enveloped RNA viruses have evolved to leverage sphingolipids and/or glycosphingolipids to enter target cells, replicate their genome, or form new virus particles enriched with these lipids (Figure 6).

Despite differences in the viral proteins and corresponding host cell receptor(s), the presence of sphingolipids in the envelopes of HCV, flaviviruses, HIV, and IAV appears critical for the proper organization of viral envelope proteins within microdomains to facilitate viral entry. Correspondingly, cell surface receptors required for viral adsorption and subsequent entry are concentrated in sphingolipid-rich microdomains present on host cell membranes, with sphingolipid itself serving as an alternative receptor for some viruses (e.g., HIV). Several of the viruses (e.g., HCV and DENV) discussed utilize specialized sites for genome replication. The trafficking of viral genomes and required replication platforms appear dependent on vesicular networks which involve lipid moieties, including sphingolipids. Lastly, the morphogeneses and egressions of the viruses discussed require the trafficking and assembly of components to budding sites on host membranes enriched for sphingolipids. Continued research is needed to ascertain the role of sphingolipids in the pathogenesis of these viruses and other enveloped RNA viruses.

Sphingolipids and glycosphingolipids are critical for membrane integrity and depleting them can have deleterious impact on distinct tissues or organs. However, minor changes in host glycosphingolipids can have a much greater impact on viruses that need them for successful infection. Hence, there is a need to develop more pharmacological inhibitors targeting sphingolipid and glycosphingolipid metabolic pathways. These inhibitors have the potential for a broad-spectrum antiviral activity. In addition, they can be combined with existing antivirals to increase their effectiveness and reduce the cost of these drugs.

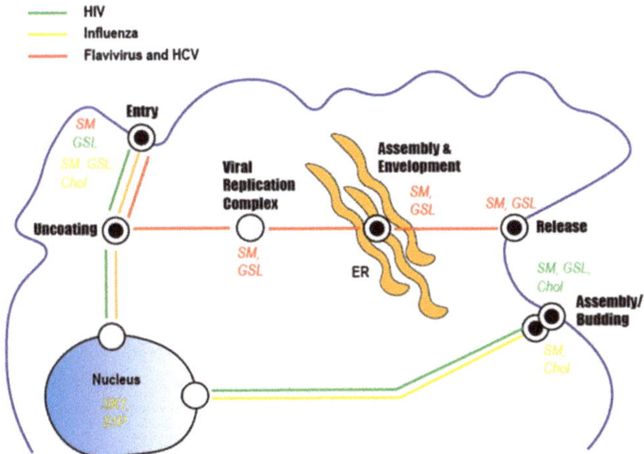

Figure 6. Sphingolipid involvement in the infection cycle of clinically important viruses. The three lines indicate the steps involved in the replication of HCV and flaviviruses (red), HIV (green), and influenza virus (yellow). Reported roles for lipids (sphingomyelin [SM], glycosphingolipids [GSL], and/or cholesterol (Chol)) in the lifecycle of each virus are indicated.

Author Contributions: K.V.K. and E.J.Y. contributed equally to the final manuscript.

Funding: K.V.K. is supported by the Albany Medical College Bridge Grant. E.J.Y. is supported by an Albany College of Pharmacy and Health Sciences Scholarship of Discovery Grant.

Acknowledgments: The authors thank Tammy Garren for her input on the design of figures and Carlos de Noronha for his assistance with the editing of the manuscript.

Conflicts of Interest: The authors declare no conflict of interest.

References

1. Nowakowski, T.J.; Pollen, A.A.; Di Lullo, E.; Sandoval-Espinosa, C.; Bershteyn, M.; Kriegstein, A.R. Expression Analysis Highlights AXL as a Candidate Zika Virus Entry Receptor in Neural Stem Cells. *Cell Stem Cell* **2016**, *18*, 591–596. [CrossRef] [PubMed]

2. Richard, A.S.; Shim, B.-S.; Kwon, Y.-C.; Zhang, R.; Otsuka, Y.; Schmitt, K.; Berri, F.; Diamond, M.S.; Choe, H. AXL-dependent infection of human fetal endothelial cells distinguishes Zika virus from other pathogenic flaviviruses. *Proc. Natl. Acad. Sci. USA* **2017**, *114*, 2024–2029. [CrossRef] [PubMed]

3. Meertens, L.; Carnec, X.; Lecoin, M.P.; Ramdasi, R.; Guivel-Benhassine, F.; Lew, E.; Lemke, G.; Schwartz, O.; Amara, A. The TIM and TAM families of phosphatidylserine receptors mediate Dengue virus entry. *Cell Host Microbe* **2012**, *12*, 544–557. [CrossRef] [PubMed]

4. Aligo, J.; Jia, S.; Manna, D.; Konan, K. V Formation and function of hepatitis C virus replication complexes require residues in the carboxy-terminal domain of NS4B protein. *Virology* **2009**, *393*, 68–83. [CrossRef] [PubMed]

5. Stone, M.; Jia, S.; Heo, W.D.; Meyer, T.; Konan, K.V. Participation of Rab5, an early endosome protein, in Hepatitis C virus RNA replication machinery. *J. Virol.* **2007**, *81*, 4551–4563. [CrossRef] [PubMed]

6. Mackenzie, J.M.; Jones, M.K.; Westaway, E.G. Markers for trans-Golgi membranes and the intermediate compartment localize to induced membranes with distinct replication functions in flavivirus-infected cells. *J. Virol.* **1999**, *73*, 9555–9567.

7. Westaway, E.G.; Mackenzie, J.M.; Kenney, M.T.; Jones, M.K.; Khromykh, A.A. Ultrastructure of Kunjin virus-infected cells: Colocalization of NS1 and NS3 with double-stranded RNA, and of NS2B with NS3, in virus-induced membrane structures. *J. Virol.* **1997**, *71*, 6650–6661.

8. Miyanari, Y.; Atsuzawa, K.; Usuda, N.; Watashi, K.; Hishiki, T.; Zayas, M.; Bartenschlager, R.; Wakita, T.; Hijikata, M.; Shimotohno, K. The lipid droplet is an important organelle for hepatitis C virus production. *Nat. Cell Biol.* **2007**, *9*, 1089–1097. [CrossRef]

9. Jackson, D.A.; Caton, A.J.; McCready, S.J.; Cook, P.R. Influenza virus RNA is synthesized at fixed sites in the nucleus. *Nature* **1982**, *296*, 366–368. [CrossRef]

10. Bukrinsky, M. A hard way to the nucleus. *Mol. Med.* **2004**, *10*, 1–5. [CrossRef]

11. Mahfoud, R.; Garmy, N.; Maresca, M.; Yahi, N.; Puigserver, A.; Fantini, J. Identification of a common sphingolipid-binding domain in Alzheimer, prion, and HIV-1 proteins. *J. Biol. Chem.* **2002**, *277*, 11292–11296. [CrossRef] [PubMed]

12. Tafesse, F.G.; Sanyal, S.; Ashour, J.; Guimaraes, C.P.; Hermansson, M.; Somerharju, P.; Ploegh, H.L. Intact sphingomyelin biosynthetic pathway is essential for intracellular transport of influenza virus glycoproteins. *Proc. Natl. Acad. Sci. USA* **2013**, *110*, 6406–6411. [CrossRef] [PubMed]

13. Aizaki, H.; Lee, K.-J.; Sung, V.M.-H.; Ishiko, H.; Lai, M.M. Characterization of the hepatitis C virus RNA replication complex associated with lipid rafts. *Virology* **2004**, *324*, 450–461. [CrossRef] [PubMed]

14. Hirata, Y.; Ikeda, K.; Sudoh, M.; Tokunaga, Y.; Suzuki, A.; Weng, L.; Ohta, M.; Tobita, Y.; Okano, K.; Ozeki, K.; et al. Self-enhancement of Hepatitis C virus replication by promotion of specific sphingolipid biosynthesis. *PLoS Pathog.* **2012**, *8*, e1002860. [CrossRef] [PubMed]

15. Bejaoui, K.; Uchida, Y.; Yasuda, S.; Ho, M.; Nishijima, M.; Brown, R.H.; Holleran, W.M.; Hanada, K. Hereditary sensory neuropathy type 1 mutations confer dominant negative effects on serine palmitoyltransferase, critical for sphingolipid synthesis. *J. Clin. Investig.* **2002**, *110*, 1301–1308. [CrossRef]

16. Hornemann, T.; Wei, Y.; von Eckardstein, A. Is the mammalian serine palmitoyltransferase a high-molecular-mass complex? *Biochem. J.* **2007**, *405*, 157–164. [CrossRef] [PubMed]

17. Rotthier, A.; Auer-Grumbach, M.; Janssens, K.; Baets, J.; Penno, A.; Almeida-Souza, L.; Van Hoof, K.; Jacobs, A.; De Vriendt, E.; Schlotter-Weigel, B.; et al. Mutations in the SPTLC2 subunit of serine palmitoyltransferase cause hereditary sensory and autonomic neuropathy type I. *Am. J. Hum. Genet.* **2010**, *87*, 513–522. [CrossRef]

18. Yasuda, S.; Nishijima, M.; Hanada, K. Localization, topology, and function of the LCB1 subunit of serine palmitoyltransferase in mammalian cells. *J. Biol. Chem.* **2003**, *278*, 4176–4183. [CrossRef]

19. Hanada, K.; Kumagai, K.; Tomishige, N.; Yamaji, T. CERT-mediated trafficking of ceramide. *Biochim. Biophys. Acta Mol. Cell Biol. Lipids* **2009**, *1791*, 684–691. [CrossRef]

20. Khan, I.; Katikaneni, D.S.; Han, Q.; Sanchez-Felipe, L.; Hanada, K.; Ambrose, R.L.; Mackenzie, J.M.; Konan, K.V. Modulation of hepatitis C virus genome replication by glycosphingolipids and four-phosphate adaptor protein 2. *J. Virol.* **2014**, *88*, 12276–12295. [CrossRef]

21. Cox, T.M. Eliglustat tartrate, an orally active glucocerebroside synthase inhibitor for the potential treatment of Gaucher disease and other lysosomal storage diseases. *Curr. Opin. Investig. Drugs* **2010**, *11*, 1169–1181. [PubMed]

22. Canals, D.; Perry, D.M.; Jenkins, R.W.; Hannun, Y.A. Drug targeting of sphingolipid metabolism: Sphingomyelinases and ceramidases. *Br. J. Pharmacol.* **2011**, *163*, 694–712. [CrossRef] [PubMed]

23. Fuller, M. Sphingolipids: The nexus between Gaucher disease and insulin resistance. *Lipids Health Dis.* **2010**, *9*, 113. [CrossRef] [PubMed]

24. Merrill, A.H., Jr.; Sullards, M.C.; Allegood, J.C.; Kelly, S.; Wang, E. Sphingolipidomics: High-throughput, structure-specific, and quantitative analysis of sphingolipids by liquid chromatography tandem mass spectrometry. *Methods* **2005**, *36*, 207–224. [CrossRef] [PubMed]

25. Patwardhan, G.A.; Liu, Y.-Y. Sphingolipids and expression regulation of genes in cancer. *Prog. Lipid Res.* **2011**, *50*, 104–114. [CrossRef] [PubMed]

26. Yamashita, T.; Wada, R.; Sasaki, T.; Deng, C.; Bierfreund, U.; Sandhoff, K.; Proia, R.L. A vital role for glycosphingolipid synthesis during development and differentiation. *Proc. Natl. Acad. Sci. USA* **1999**, *96*, 9142–9147. [CrossRef] [PubMed]

27. Cao-Lormeau, V.-M.; Blake, A.; Mons, S.; Lastère, S.; Roche, C.; Vanhomwegen, J.; Dub, T.; Baudouin, L.; Teissier, A.; Larre, P.; et al. Guillain-Barré Syndrome outbreak associated with Zika virus infection in French Polynesia: A case-control study. *Lancet* **2016**, *387*, 1531–1539. [CrossRef]

28. Hulkova, H.; Ledvinova, J.; Kuchar, L.; Smid, F.; Honzikova, J.; Elleder, M. Glycosphingolipid profile of the apical pole of human placental capillaries: The relevancy of the observed data to Fabry disease. *Glycobiology* **2012**, *22*, 725–732. [CrossRef] [PubMed]

29. Sullards, M.C.; Liu, Y.; Chen, Y.; Merrill, A.H. Analysis of mammalian sphingolipids by liquid chromatography tandem mass spectrometry (LC-MS/MS) and tissue imaging mass spectrometry (TIMS). *Biochim. Biophys. Acta Mol. Cell Biol. Lipids* **2011**, *1811*, 838–853. [CrossRef]

30. Perera, R.; Riley, C.; Isaac, G.; Hopf-Jannasch, A.S.; Moore, R.J.; Weitz, K.W.; Pasa-Tolic, L.; Metz, T.O.; Adamec, J.; Kuhn, R.J. Dengue virus infection perturbs lipid homeostasis in infected mosquito cells. *PLoS Pathog.* **2012**, *8*, e1002584. [CrossRef]

31. Manns, M.P.; Buti, M.; Gane, E.; Pawlotsky, J.-M.; Razavi, H.; Terrault, N.; Younossi, Z. Hepatitis C virus infection. *Nat. Rev. Dis. Primer* **2017**, *3*, 17006. [CrossRef] [PubMed]

32. Moradpour, D.; Penin, F.; Rice, C.M. Replication of hepatitis C virus. *Nat. Rev. Microbiol.* **2007**, *5*, 453–463. [CrossRef] [PubMed]

33. Paul, D.; Madan, V.; Bartenschlager, R. Hepatitis C virus RNA replication and assembly: Living on the fat of the land. *Cell Host Microbe* **2014**, *16*, 569–579. [CrossRef] [PubMed]

34. Li, D.K.; Chung, R.T. Overview of Direct-Acting Antiviral Drugs and Drug Resistance of Hepatitis C Virus. *Methods Mol. Biol.* **2019**, *1911*, 3–32. [PubMed]

35. Asselah, T.; Marcellin, P.; Schinazi, R.F. Treatment of hepatitis C virus infection with direct-acting antiviral agents: 100% cure? *Liver Int.* **2018**, *38*, 7–13. [CrossRef] [PubMed]

36. Ribeiro, R.M.; Li, H.; Wang, S.; Stoddard, M.B.; Learn, G.H.; Korber, B.T.; Bhattacharya, T.; Guedj, J.; Parrish, E.H.; Hahn, B.H.; et al. Quantifying the Diversification of Hepatitis C Virus (HCV) during Primary Infection: Estimates of the In Vivo Mutation Rate. *PLoS Pathog.* **2012**, *8*, e1002881. [CrossRef]

37. Hayes, C.N.; Imamura, M.; Chayama, K. Management of HCV patients in cases of direct-acting antiviral failure. *Expert Rev. Gastroenterol. Hepatol.* **2019**, *13*, 839–848. [CrossRef] [PubMed]

38. Lontok, E.; Harrington, P.; Howe, A.; Kieffer, T.; Lennerstrand, J.; Lenz, O.; McPhee, F.; Mo, H.; Parkin, N.; Pilot-Matias, T.; et al. Hepatitis C virus drug resistance-associated substitutions: State of the art summary. *Hepatology* **2015**, *62*, 1623–1632. [CrossRef]

39. Merz, A.; Long, G.; Hiet, M.-S.; Brügger, B.; Chlanda, P.; Andre, P.; Wieland, F.; Krijnse-Locker, J.; Bartenschlager, R. Biochemical and morphological properties of Hepatitis C virus particles and determination of their lipidome. *J. Biol. Chem.* **2011**, *286*, 3018–3032. [CrossRef]

40. Aizaki, H.; Morikawa, K.; Fukasawa, M.; Hara, H.; Inoue, Y.; Tani, H.; Saito, K.; Nishijima, M.; Hanada, K.; Matsuura, Y.; et al. Critical role of virion-associated cholesterol and sphingolipid in Hepatitis C virus infection. *J. Virol.* **2008**, *82*, 5715–5724. [CrossRef]

41. Egger, D.; Wölk, B.; Gosert, R.; Bianchi, L.; Blum, H.E.; Moradpour, D.; Bienz, K. Expression of hepatitis C virus proteins induces distinct membrane alterations including a candidate viral replication complex. *J. Virol.* **2002**, *76*, 5974–5984. [CrossRef] [PubMed]

42. Konan, K.V.; Giddings, T.H.; Ikeda, M.; Li, K.; Lemon, S.M.; Kirkegaard, K. Nonstructural protein precursor NS4A/B from hepatitis C virus alters function and ultrastructure of host secretory apparatus. *J. Virol.* **2003**, *77*, 7843–7855. [CrossRef] [PubMed]

43. Paploski, I.A.D.; Prates, A.P.P.B.; Cardoso, C.W.; Kikuti, M.; Silva, M.M.O.; Waller, L.A.; Reis, M.G.; Kitron, U.; Ribeiro, G.S. Time lags between exanthematous illness attributed to Zika virus, Guillain-Barré syndrome, and microcephaly, Salvador, Brazil. *Emerg. Infect. Dis.* **2016**, *22*, 1438–1444. [CrossRef] [PubMed]

44. Yoon, K.-J.; Song, G.; Qian, X.; Pan, J.; Xu, D.; Rho, H.-S.; Kim, N.-S.; Habela, C.; Zheng, L.; Jacob, F.; et al. Zika-virus-encoded NS2A disrupts mammalian cortical neurogenesis by degrading adherens junction proteins. *Cell Stem Cell* **2017**, *21*, 349–358. [CrossRef] [PubMed]

45. Liang, Q.; Luo, Z.; Zeng, J.; Chen, W.; Foo, S.-S.; Lee, S.-A.; Ge, J.; Wang, S.; Goldman, S.A.; Zlokovic, B.V.; et al. Zika virus NS4A and NS4B proteins deregulate Akt-mTOR signaling in human fetal neural stem cells to inhibit neurogenesis and induce autophagy. *Cell Stem Cell* **2016**, *19*, 663–671. [CrossRef]

46. Yuan, L.; Huang, X.-Y.; Liu, Z.-Y.; Zhang, F.; Zhu, X.-L.; Yu, J.-Y.; Ji, X.; Xu, Y.-P.; Li, G.; Li, C.; et al. A single mutation in the prM protein of Zika virus contributes to fetal microcephaly. *Science* **2017**, *358*, 933–936. [CrossRef]

47. Wang, K.; Wang, J.; Sun, T.; Bian, G.; Pan, W.; Feng, T.; Wang, P.; Li, Y.; Dai, J. Glycosphingolipid GM3 is indispensable for Dengue virus genome replication. *Int. J. Biol. Sci.* **2016**, *12*, 872–883. [CrossRef]

48. Martín-Acebes, M.A.; Gabandé-Rodríguez, E.; García-Cabrero, A.M.; Sánchez, M.P.; Ledesma, M.D.; Sobrino, F.; Saiz, J.-C. Host sphingomyelin increases West Nile virus infection in vivo. *J. Lipid Res.* **2016**, *57*, 422–432. [CrossRef]

49. Martin-Acebes, M.A.; Merino-Ramos, T.; Blazquez, A.-B.; Casas, J.; Escribano-Romero, E.; Sobrino, F.; Saiz, J.-C. The composition of West Nile virus lipid envelope unveils a role of sphingolipid metabolism in Flavivirus biogenesis. *J. Virol.* **2014**, *88*, 12041–12054. [CrossRef]

50. Freed, E.O.; Martin, M. A HIVs and their replication. In *Fields Virology*, 4th ed.; Lippincott Williams & Wilkins: Philadelphia, PA, USA, 2001; ISBN 9780781760607.

51. Li, G.; De Clercq, E. HIV Genome-Wide Protein Associations: A Review of 30 Years of Research. *Microbiol. Mol. Biol. Rev.* **2016**, *80*, 679–731. [CrossRef]

52. Levy, J.A. HIV pathogenesis: 25 years of progress and persistent challenges. *AIDS* **2009**, *23*, 147–160. [CrossRef] [PubMed]

53. Desai, M.; Iyer, G.; Dikshit, R.K. Antiretroviral drugs: Critical issues and recent advances. *Indian J. Pharmacol.* **2012**, *44*, 288–298. [CrossRef] [PubMed]

54. Turpin, J.A. The next generation of HIV/AIDS drugs: Novel and developmental antiHIV drugs and targets. *Expert Rev. Anti Infect. Ther.* **2003**, *1*, 97–128. [CrossRef] [PubMed]

55. Hug, P.; Lin, H.M.; Korte, T.; Xiao, X.; Dimitrov, D.S.; Wang, J.M.; Puri, A.; Blumenthal, R. Glycosphingolipids promote entry of a broad range of human immunodeficiency virus type 1 isolates into cell lines expressing CD4, CXCR4, and/or CCR5. *J. Virol.* **2000**, *74*, 6377–6385. [CrossRef] [PubMed]

56. Rawat, S.S.; Gallo, S.A.; Eaton, J.; Martin, T.D.; Ablan, S.; KewalRamani, V.N.; Wang, J.M.; Blumenthal, R.; Puri, A. Elevated Expression of GM3 in Receptor-Bearing Targets Confers Resistance to Human Immunodeficiency Virus Type 1 Fusion. *J. Virol.* **2004**, *78*, 7360–7368. [CrossRef] [PubMed]

57. Bhat, S.; Spitalnik, S.L.; Gonzalez-Scarano, F.; Silberberg, D.H. Galactosyl ceramide or a derivative is an essential component of the neural receptor for human immunodeficiency virus type 1 envelope glycoprotein gp120. *Proc. Natl. Acad. Sci. USA* **1991**, *88*, 7131–7134. [CrossRef] [PubMed]

58. Harouse, J.; Bhat, S.; Spitalnik, S.; Laughlin, M.; Stefano, K.; Silberberg, D.; Gonzalez-Scarano, F. Inhibition of entry of HIV-1 in neural cell lines by antibodies against galactosyl ceramide. *Science* **1991**, *253*, 320–323. [CrossRef] [PubMed]

59. Hammache, D.; Yahi, N.; Piéroni, G.; Ariasi, F.; Tamalet, C.; Fantini, J. Sequential interaction of CD4 and HIV-1 gp120 with a reconstituted membrane patch of ganglioside GM3: Implications for the role of glycolipids as potential HIV-1 fusion cofactors. *Biochem. Biophys. Res. Commun.* **1998**, *246*, 117–122. [CrossRef] [PubMed]

60. Yahi, N.; Baghdiguian, S.; Moreau, H.; Fantini, J. Galactosyl ceramide (or a closely related molecule) is the receptor for human immunodeficiency virus type 1 on human colon epithelial HT29 cells. *J. Virol.* **1992**, *66*, 4848–4854.

61. Fantini, J.; Hammache, D.; Delézay, O.; Yahi, N.; André-Barrès, C.; Rico-Lattes, I.; Lattes, A. Synthetic soluble analogs of galactosylceramide (GalCer) bind to the V3 domain of HIV-1 gp120 and inhibit HIV-1-induced fusion and entry. *J. Biol. Chem.* **1997**, *272*, 7245–7252. [CrossRef] [PubMed]

62. Brugger, B.; Glass, B.; Haberkant, P.; Leibrecht, I.; Wieland, F.T.; Krausslich, H.-G. The HIV lipidome: A raft with an unusual composition. *Proc. Natl. Acad. Sci. USA* **2006**, *103*, 2641–2646. [CrossRef]

63. Lorizate, M.; Sachsenheimer, T.; Glass, B.; Habermann, A.; Gerl, M.J.; Kräusslich, H.-G.; Brügger, B. Comparative lipidomics analysis of HIV-1 particles and their producer cell membrane in different cell lines. *Cell. Microbiol.* **2013**, *15*, 292–304. [CrossRef]

64. Izquierdo-Useros, N.; Lorizate, M.; Contreras, F.-X.; Rodriguez-Plata, M.T.; Glass, B.; Erkizia, I.; Prado, J.G.; Casas, J.; Fabriàs, G.; Kräusslich, H.-G.; et al. Sialyllactose in viral membrane gangliosides is a novel molecular recognition pattern for mature dendritic cell capture of HIV-1. *PLoS Biol.* **2012**, *10*, e1001315. [CrossRef] [PubMed]

65. Puryear, W.B.; Yu, X.; Ramirez, N.P.; Reinhard, B.M.; Gummuluru, S. HIV-1 incorporation of host-cell-derived glycosphingolipid GM3 allows for capture by mature dendritic cells. *Proc. Natl. Acad. Sci. USA* **2012**, *109*, 7475–7480. [CrossRef] [PubMed]

66. Hammonds, J.E.; Beeman, N.; Ding, L.; Takushi, S.; Francis, A.C.; Wang, J.-J.; Melikyan, G.B.; Spearman, P. Siglec-1 initiates formation of the virus-containing compartment and enhances macrophage-to-T cell transmission of HIV-1. *PLoS Pathog.* **2017**, *13*, e1006181. [CrossRef] [PubMed]

67. McClellan, K.; Perry, C.M. Oseltamivir. *Drugs* **2001**, *61*, 263–283. [CrossRef] [PubMed]

68. Burch, J.; Corbett, M.; Stock, C.; Nicholson, K.; Elliot, A.J.; Duffy, S.; Westwood, M.; Palmer, S.; Stewart, L. Prescription of anti-influenza drugs for healthy adults: A systematic review and meta-analysis. *Lancet Infect. Dis.* **2009**, *9*, 537–545. [CrossRef]

69. Shao, W.; Li, X.; Goraya, M.U.; Wang, S.; Chen, J.-L. Evolution of Influenza A Virus by Mutation and Re-Assortment. *Int. J. Mol. Sci.* **2017**, *18*, 1650. [CrossRef]

70. Ahlquist, P. RNA-Dependent RNA Polymerases, Viruses, and RNA Silencing. *Science* **2002**, *296*, 1270–1273. [CrossRef]

71. Bouvier, N.M.; Palese, P. The biology of influenza viruses. *Vaccine* **2008**, *26* (Suppl. 4), D49–D53. [CrossRef]

72. Gerl, M.J.; Sampaio, J.L.; Urban, S.; Kalvodova, L.; Verbavatz, J.-M.; Binnington, B.; Lindemann, D.; Lingwood, C.A.; Shevchenko, A.; Schroeder, C.; et al. Quantitative analysis of the lipidomes of the influenza virus envelope and MDCK cell apical membrane. *J. Cell Biol.* **2012**, *196*, 213–221. [CrossRef] [PubMed]

73. Tanner, L.B.; Chng, C.; Guan, X.L.; Lei, Z.; Rozen, S.G.; Wenk, M.R. Lipidomics identifies a requirement for peroxisomal function during influenza virus replication. *J. Lipid Res.* **2014**, *55*, 1357–1365. [CrossRef] [PubMed]

74. Skibbens, J.E.; Roth, M.G.; Matlin, K.S. Differential extractability of influenza virus hemagglutinin during intracellular transport in polarized epithelial cells and nonpolar fibroblasts. *J. Cell Biol.* **1989**, *108*, 821–832. [CrossRef] [PubMed]

75. Takeda, M.; Leser, G.P.; Russell, C.J.; Lamb, R.A. Influenza virus hemagglutinin concentrates in lipid raft microdomains for efficient viral fusion. *Proc. Natl. Acad. Sci. USA* **2003**, *100*, 14610–14617. [CrossRef] [PubMed]

76. Scheiffele, P.; Roth, M.G.; Simons, K. Interaction of influenza virus haemagglutinin with sphingolipid-cholesterol membrane domains via its transmembrane domain. *EMBO J.* **1997**, *16*, 5501–5508. [CrossRef] [PubMed]

77. Sun, X.; Whittaker, G.R. Role for influenza virus envelope cholesterol in virus entry and infection. *J. Virol.* **2003**, *77*, 12543–12551. [CrossRef]

78. Seo, Y.-J.; Pritzl, C.J.; Vijayan, M.; Bomb, K.; McClain, M.E.; Alexander, S.; Hahm, B. Sphingosine kinase 1 serves as a pro-viral factor by regulating viral RNA synthesis and nuclear export of viral ribonucleoprotein complex upon Influenza virus infection. *PLoS ONE* **2013**, *8*, e75005. [CrossRef]

79. Shimizu, T.; Takizawa, N.; Watanabe, K.; Nagata, K.; Kobayashi, N. Crucial role of the influenza virus NS2 (NEP) C-terminal domain in M1 binding and nuclear export of vRNP. *FEBS Lett.* **2011**, *585*, 41–46. [CrossRef]

80. Leser, G.P.; Lamb, R.A. Influenza virus assembly and budding in raft-derived microdomains: A quantitative analysis of the surface distribution of HA, NA and M2 proteins. *Virology* **2005**, *342*, 215–227. [CrossRef]

81. Barman, S.; Krylov, P.S.; Turner, J.C.; Franks, J.; Webster, R.G.; Husain, M.; Webby, R.J. Manipulation of neuraminidase packaging signals and hemagglutinin residues improves the growth of A/Anhui/1/2013 (H7N9) influenza vaccine virus yield in eggs. *Vaccine* **2017**, *35*, 1424–1430. [CrossRef]

82. Nayak, D.P.; Balogun, R.A.; Yamada, H.; Zhou, Z.H.; Barman, S. Influenza virus morphogenesis and budding. *Virus Res.* **2009**, *143*, 147–161. [CrossRef] [PubMed]

83. Barman, S.; Nayak, D.P. Lipid raft disruption by cholesterol depletion enhances Influenza A virus budding from MDCK cells. *J. Virol.* **2007**, *81*, 12169–12178. [CrossRef] [PubMed]

Review

Human Antimicrobial Peptides as Therapeutics for Viral Infections

Aslaa Ahmed [1], Gavriella Siman-Tov [1], Grant Hall [2], Nishank Bhalla [1] and Aarthi Narayanan [1,*]

[1] National Center for Biodefense and Infectious Disease, School of Systems Biology, George Mason University, Manassas, VA 20110, USA
[2] United States Military Academy, West Point, NY 10996, USA
* Correspondence: anaraya1@gmu.edu

Received: 28 May 2019; Accepted: 30 July 2019; Published: 1 August 2019

Abstract: Successful in vivo infection following pathogen entry requires the evasion and subversion of multiple immunological barriers. Antimicrobial peptides (AMPs) are one of the first immune pathways upregulated during infection by multiple pathogens, in multiple organs in vivo. In humans, there are many classes of AMPs exhibiting broad antimicrobial activities, with defensins and the human cathelicidin LL-37 being the best studied examples. Whereas historically the efficacy and therapeutic potential of AMPs against bacterial infection has been the primary focus of research, recent studies have begun to elucidate the antiviral properties of AMPs as well as their role in regulation of inflammation and chemoattraction. AMPs as therapeutic tools seem especially promising against emerging infectious viral pathogens for which no approved vaccines or treatments are currently available, such as dengue virus (DENV) and Zika virus (ZIKV). In this review, we summarize recent studies elucidating the efficacy and diverse mechanisms of action of various classes of AMPs against multiple viral pathogens, as well as the potential use of human AMPs in novel antiviral therapeutic strategies.

Keywords: human antimicrobial peptides; antiviral strategies; defensins; cathelicidins; hepcidins; transferrins

---

1. Introduction

Found in virtually all organisms, antimicrobial peptides (AMPs) are short, positively-charged oligopeptides that exhibit a diversity of structures and functions. AMPs are a fundamental component of the innate immune system and play a vital role in the initial immune response generated against both injury and infections. AMP-mediated immune responses are rapidly activated following infections as AMPs are primarily synthesized and stored in cells of myeloid origin and epithelial cells, among the first responders to infections. AMPs are expressed in a wide variety of tissues including skin, eyes, oral cavity, ears, airway, lung, female reproductive tract, cervical-vaginal fluid, intestines, and urinary tract [1,2]. The majority of AMPs are synthesized as large polyprotein precursors, the proteolytic processing of which releases active peptide segments which can be present alone or in multiple copies. Removal of signal peptide may be a post-translation or a co-translational process. Processed functional peptides have been characterized into many classes in mammals; in humans, they include defensins, cathelicidins, transferrins, hepcidin, human antimicrobial proteins, dermcidin, histones, AMPs derived from known proteins, chemokines, and AMPs from immune cells, antimicrobial neuropeptides, and Beta-amyloid peptides [3]. While all of the AMPs classes have been shown to possess antimicrobial activity, only a few classes have demonstrated antiviral properties.

A defining feature of AMPs is their rapid response to infections of bacteria, viruses, fungi, or protozoa [1,4]. Exhibiting inhibitory and immunomodulatory properties, AMPs have been intensively studied as alternatives to antibiotics in bacterial infections and in recent years have gained substantial attention as viral therapeutics [5]. Here we report on the application of human AMPs in the treatment of viral infections.

2. Defensins

2.1. Expression

Highly abundant and widely distributed, defensins modulate immune responses thereby playing a central role in innate immunity [6–8]. Defensins are classified into three subgroups: α, β and θ. Although humans do not produce functional members of the θ-defensin family of AMPs, expression of θ-defensin mRNA has been observed in humans. The θ-defensin mRNA contains a pre-mature stop codon which prevents translation; however, functional θ-defensins are present in non-human primates [9]. To date, six α-defensin and 31 β-defensin peptides have been identified in various species [9]. Originally isolated from neutrophils, four of the six distinct α-defensins are termed human neutrophil peptides (HNP-1 through 4). They are also produced by myeloid-lineage cells such as macrophages, natural killer (NK) cells and some classes of T and B-cells. α-defensins 5 (HD5) and 6 (HD6) are expressed in epithelial cells in the small intestine [6,8–10]. The β-defensin family of AMPs is commonly expressed in birds and mammals. In humans, three β-defensins (HBD-1 through 3) have been fully characterized and a fourth, HBD-4, was recently identified. β-defensins are primarily expressed by epithelial cells and keratinocytes, but can also be produced by neutrophils, macrophages, mast cells, NK cells, dendritic cells, and lymphocytes [6–8,10]. Current data suggest a functional redundancy when comparing the efficacy of α and β defensins against various pathogens [11].

Defensins are defined by the presence of a conserved spacing pattern comprised of cysteine residues, which is critical for the efficacy of their cationic antimicrobial properties [9,12]. Human α-defensins are composed of 29 to 34 amino acids with an overall positive charge [9,12,13]. Defensins exhibit a characteristic β-sheet structure with a distinctive six-cysteine motif for which stabilization is a consequence of the presence of three intramolecular disulfide bonds. The α-defensins are synthesized as pre-propeptides consisting of a N-terminal signal sequence, an anionic pro-peptide, and a C-terminal mature peptide comprised of approximately 30 amino acids. HNP1, HNP2, and HNP3 are synthesized by promyelocytes and stored in primary neutrophil granules as mature peptides [10]. In contrast, β-defensins have a short N-terminal pro-region and can retain antimicrobial activity in full-length form, and; therefore, do not require N-terminal processing to be fully active [14]. They are synthesized in epithelial compartments and can range from 38 to 42 amino acids in length.

2.2. Antiviral Activity of Defensins

The antiviral activity of defensins was first reported in 1986 [10]. Since then, defensins have demonstrated protection against human immunodeficiency virus (HIV), influenza A virus (IAV), human adenovirus (HAdV), severe acute respiratory syndrome coronavirus (SARSC), papillomavirus (HPV), respiratory syncytial virus (RSV), and herpes simplex virus (HSV) [5,10,15–18]. Recent studies have focused on elucidating the multiple mechanisms associated with defensins' antiviral activity (Table 1). Defensins can block viral infection through direct action on virus particles or interfere indirectly at various stages of the viral life cycle [10,18]. Available data suggest antiviral activity occurs predominantly at viral entry steps; however, antiviral effects at other stages of infection have also been reported, particularly affecting viral trafficking within infected cells [19]. Defensins can also modify the innate immune response to viral infections, including: modulation of T-cells, macrophage and dendritic cells recruitment to sites of infection, wound healing and angiogenesis, differentiation and maturation of dendritic cells, induction of the production of pro-inflammatory cytokines by macrophages, mast cells, and keratinocytes, and regulation of cell death pathways [9]. For example, HBD-3 can suppress activation of the caspase cascade to prevent apoptosis in infected cells [20]. Similarly, the concentration of HNPs released into the microenvironment upon activation of neutrophils during inflammation exerts a differential effect on cytokine production in activated monocytes [19]. HNP concentrations of 1 to 10 nM can upregulate the expression of tumor necrosis factor α (TNF-α) and interleukin-1β (IL-1β), whereas concentrations of 10 to 100 μM are cytotoxic to monocytes.

2.3. Adenovirus

Human adenovirus (HAdV) is a non-enveloped double-stranded DNA virus that is capable of infecting the respiratory, gastrointestinal, ocular, and excretory systems in humans. There are approximately 80 recognized HAdV serotypes, subdivided into species A–G [21,22]. Currently there are only a limited number of HAdV therapeutic strategies and vaccines available to treat HAdV infections. Alpha defensins have demonstrated an ability to hinder HAdV infections in vitro [21,23]. HD5 reduces HAdV replication by 95% when cells are exposed to the peptide (IC$_{50}$ = 3–4 μM) prior to infection, and by 50% when peptide is added 30–60 min post inoculation, suggesting that inhibition occurs at an early stage during viral infection [21]. Additional studies have shown that direct binding of HD5 (10 μM) to HAdV particles prior to infection prevents the release of internalized viral particles from endosomes [24]. Subsequently, viral particles appear to colocalize with lysosomes indicating altered viral trafficking following infection as a consequence of HD5 binding [24]. These findings suggest defensins' antiviral activity against HAdV results in blockage of HAdV uncoating and genome exposure [24]. In addition, HD5 antiviral activity is species specific; pre-treatment with 15μM HD5 decreased HAdV infectivity of subspecies A–C and E, while infectivity of HAdV subspecies D and F demonstrated no change [23]. The cause of species specificity of HD5 activity is yet to be determined.

2.4. Influenza A Virus

During the early infiltrate in influenza A virus (IAV), neutrophils predominate in infected airways, highlighting their importance in initiating immune responses against IAV [13]. Defensins are; thus, likely to interact with IAV. Neutrophil extracellular traps (NETs) displaying HNPs are formed in vivo and in vitro in response to IAV infection [13]. Cells incubated with defensins pre- or post-infection demonstrated minimal inhibitory activity against IAV, whereas incubation of HNPs with virions prior to infection is necessary for the antiviral activity of these AMPs against IAV (e.g., HNP1 IC$_{50}$ < 2 μg/mL [25]) [13,26]. In addition, the binding activity of defensins against IAV is increased by formation of multi-molecular assemblies of defensins, which may be responsible for pore formation in the IAV envelope, thereby destabilizing virions prior to receptor binding and cellular entry [13].

Expression of β-defensins HBD1, HBD2, and HBD3 has been reported in various epithelial cell tissues, with each β-defensin demonstrating a unique expression induction profile in response to IAV infection [13]. However, HBD1 and HBD2 have also been detected in monocytes, macrophages and monocyte-derived dendritic cells (DCs), and possess strong neutralizing activity against multiple IAV strains [10,13]. HBDs exhibit low potency as direct inhibitors of IAV virions, but are speculated to play important immunomodulatory roles by limiting inflammation during IAV infection [13]. While the exact sequence of immunomodulatory events is yet to be determined, it has been reported that deletion of the HBD1 analog in mice resulted in a more serious inflammatory reaction to IAV [13,27]. In addition, HBD3 has demonstrated strong anti-inflammatory effects in cells stimulated with 50 ng/mL lipopolysaccharide (LPS), confirmed by the inhibition of expression of inflammatory mediators such as (TNF-α) [28]. Conversely, α-defensins inhibit IAV replication in infected cells. Pre-incubation of virions with HNP-1 (25 μg/mL) is capable of reducing the replication of IAV strain H1N1 by 10- to 1000-fold in multiple cell lines when compared to replication in untreated cells [29]. Similarly, HNP-1 and 2 can reduce infectious virus of the Phil82 strain of IAV by 85% to 90% in various cell lines [26].

2.5. Human Immunodeficiency Virus

Defensins demonstrate antiviral activity against human immunodeficiency virus (HIV), mediated by direct virus–peptide interaction and/or inhibition of viral genome replication. Inhibition mediated by the direct binding of defensins with HIV virions is attributed to interactions between positively-charged HNPs and negatively-charged moieties of the HIV envelope glycoprotein gp120. HNP1, HNP2, and

HNP3 function as lectins by directly blocking the interaction of gp120 and the HIV receptor CD4. However, the exact mechanism of this interaction is not well characterized [10,13]. HNP-1 can also interfere with critical steps in the HIV replication cycle [30]. HNP-1 inhibits protein kinase C signaling, which is important for the transcription and nuclear import of the HIV genome [31]. HD5 exhibits a robust dose-dependent (IC_{50} = 400 nM) suppression of HIV-1 replication in absence of serum when pre-incubated with virions [32]. HD5 also blocks HIV-1 infection at a step prior to viral entry [32]. HD5 competitively binds to the CD4 receptor in a dose-dependent manner against HIV, thereby blocking HIV entry into target cells [32]. Interestingly, in contrast to HNP1, HNP2, HNP3, and HD5, HNP4 does not interact with CD4 or HIV gp120 [13]. HNP4 inhibits HIV replication with greater effectiveness than HNP1, HNP2, and HNP3, but it is unclear whether efficacy of HNP4 is mediated only through a direct effect on virions or also on host processes that ultimately affect viral replication.

β-defensins can also exert antiviral activities against HIV. Expression of HBD2 and HBD3 can be induced by microbial products such as endotoxins, viruses, bacteria, and pro-inflammatory cytokines such as TNF and IL-1β [13]. Expression of HBD2 and HBD3, but not HBD1, mRNA can be induced by HIV in human oral epithelial cells. HBD2 inhibits the formation of early HIV transcript products but does not affect cell–cell fusion [10].

2.6. Herpes Simplex Virus

Alpha and β-defensins can exhibit anti-HSV properties [33]. HNP-1-4 and HD6 inhibit HSV binding to its target receptor by directly interacting with either the HSV glycoprotein or by binding to heparan sulfate (HS), thereby preventing viral entry [8,33]. HBD3 binds either the HSV receptor or the HSV glycoprotein, thereby eliciting a stronger inhibition of viral entry. Furthermore, treatment of infected cells with defensins post-infection results in substantial reduction in viral replication, indicating that these peptides can exhibit post entry antiviral effects [33]. In addition, studies exploring HNP-1-3 have demonstrated their ability to reduce intracellular HSV protein transport and expression during infection [34]. HNP-1 (100 µg/mL) exhibits the greatest antiviral potential against HSV, reducing HSV titers up to 100,000-fold upon a combination of pre- and post-treatment of infected cells as compared to HNP-2 and HNP-3 (100 µg/mL), which can reduce titers by up to 100-fold following treatment [33].

2.7. Respiratory Syncytial Virus

The antiviral effects of human defensins against RSV are relatively unexplored. A study assessing leukotriene B4 (LTB4) stimulation of nasal neutrophil activity highlighted α-defensins as a possible source of antiviral activity against RSV [35]. Cells pre-treated with HBD-2 (4 µg/mL), can reduce RSV viral titers 100-fold following infection [36]. Electron microscopy images revealed damage to the lipid envelope of RSV following HBD-2 treatment, suggesting that defensins destabilize RSV virion envelopes, thereby inhibiting viral cellular entry [36]. The use of defensins in anti-RSV therapies may also limit viral evolutionary strategies that counters antiviral activities, due to the difficulty of changing the viral envelope lipid composition [37]. The evolutionary longevity of defensins suggests this to be a favorable strategy, making defensins an attractive therapeutic candidate for the treatment of RSV infections [16].

2.8. Human Papilloma Virus

Most α-defensins possess some level of anti-human papilloma virus (HPV) activity, with HD-6 being a notable exception [38]. Due to their low toxicity and high efficacy, HNP-1 and HD-5 have been most frequently tested as anti-HPV candidates. Recent studies have focused solely on HD5 as it is secreted by epithelial cells in the genitourinary tract. The antiviral activity of HNP-1 and HD5 (5 µg/mL) against HPV is time-independent, with robust inhibition even when peptides are introduced to cells six hours post infection in vitro [38]. Employing immunofluorescent confocal microscopy, the authors demonstrated that the peptides do not inhibit viral entry but rather prevent virion escape from cytoplasmic vesicles [38]. During the course of HPV entry, cleavage of HPV L2 capsid protein by furin, a cellular protease, is required for successful infection [39]. HD5 directly disrupts this proteolytic processing step, thereby preventing HPV genome escape from endosomes [39]. Interestingly, the use of a furin-cleaved HPV does not result in abrogation of HD5 activity. HD5 can still block HPV infection by preventing viral capsid dissociation from the genome, and by reducing viral trafficking within the host cells [40]. These results indicate that α-defensins, particularly HD5, demonstrate robust anti-HPV activity by targeting multiple steps during viral life cycle.

3. Cathelicidin, LL-37

3.1. Expression

Cathelicidins are peptides with a conserved 100-amino-acid cathelin domain, a protein sequence first identified in porcine leukocytes that is capable of inhibiting the protease cathepsin-L [9]. Cathelicidins are typically linear peptides that fold into amphipathic α-helical structures which are frequently cleaved from the highly variable C-terminal antimicrobial domain [12]. In humans, cathelicidin is produced by a vitamin D-dependent antimicrobial pathway [41]. Although there are multiple cathelicidins found in nature, humans only express a single cathelicidin, known as human cationic antimicrobial peptide 18 (hCAMP-18), hCAP18, or LL-37. LL-37 was identified and isolated in 1995 from neutrophils [1,42]. Like α-defensins, cathelicidins are synthesized as pre-propeptides; following proteolytic removal of the signal peptide, the inactive propeptide is tagged for storage in neutrophil granules. The active cationic molecule is generated by cleavage of the C-terminus end of the hCAP18 precursor protein yielding a linear 37-amino-acid-long peptide [1]. The name hCAP18 alludes to the molecular weight of the polypeptide (18 kDa) and the cationic character of the structure, whereas LL-37 refers to the 37 amino acid length of the peptide along with a Leu-Leu motif located at the N-terminus [1]. The peptide can also be produced in epithelial cells and may play an important role in the initial immune response to various pathogens [10,11]. In epithelial skin cells, LL-37 is further cleaved into shorter segments exhibiting potent antimicrobial activity [43]. LL-37 is also produced by monocytes, NK cells, mast cells, B cells, colon enterocytes, and keratinocytes [1], and has been detected in numerous tissues and biological fluids such as sweat, breast milk, wound fluid, vernix, tracheal aspirates of newborns, and seminal plasma [9]. The concentration of LL-37 and its precursor in tissues and body fluids can range between 2 and 5 μg/mL (0.4–1 μM); however, concentrations can increase up to 20 μg/mL (2.2 μM) during infections in bronchoalveolar fluid [44]. In nasal secretions LL-37 concentrations can vary from 1.2–80 μg/mL [44]. The expression of LL-37 is regulated by a number of endogenous factors including pro-inflammatory cytokines and growth factors such as the active form of vitamin D [1]. LL-37 functions as a chemoattractant for neutrophils, monocytes, dendritic and T-cells and is rapidly released by epithelial cells and leukocytes following infection [11,45]. LL-37 can stimulate IL-6 production in human dendritic cells and may act as both an anti and pro-inflammatory factor during the immune response to various infections [45]. Individuals with cathelicidin deficient neutrophils display an increased susceptibility to infection [11].

3.2. Antiviral Activity

The antiviral activity of LL-37 has been reported against a number of viruses including HIV-1, IAV, RSV, rhinovirus (HRV), vaccinia virus (VACV), HSV, ZIKV, and hepatitis C virus (HCV), mediated primarily by its interaction with the virus outer envelope (Table 1) [18,41,43,46]. LL-37 is proposed to remove the outer membrane of viruses in a single event during an antimicrobial attack rather than a gradual piece-by-piece removal [37]. This suggests a carpet model of antimicrobial peptide action, wherein a susceptible membrane remains intact until a threshold concentration of peptide is reached, following which a rapid disintegration of the targeted membrane occurs [47,48].

3.3. Influenza A virus

LL-37 therapeutic activity against influenza type A virus has been demonstrated in vivo and in vitro. It is likely that in vivo, IAV encounters LL-37 in the respiratory tract following innate immune responses against the virus and is secreted from neutrophils, macrophages, and epithelial cells [44,49]. Early studies assessed the antiviral activity of LL-37 in vivo using a mouse IAV strain [50]. Mice were nebulized with LL-37 (500 µg/mL) a day prior to infection with a lethal dose of IVA PR/8 mouse strain and survival and weight loss were monitored for 14 days following infection [50]. Initially, all mice exhibited weight loss, but weight loss ceased at day seven in mice treated with LL-37 or the IAV antiviral zanamivir. Mice treated with LL-37 and zanamivir exhibited 60% survival compared to the untreated group which succumbed to infection by day nine suggesting that therapeutic use of LL-37 reduces IAV infection severity in a manner comparable to zanamivir [50]. LL-37 also decreased expression of inflammatory cytokines particularly IL-1β, granulocyte-macrophage colony-stimulating factor (GM-CSF), keratinocytes chemoattractant (KC), and the chemotactic cytokine known as regulated on activation normal t-cell expressed and secreted (RANTES), in bronchoalveolar lavage fluid in mice infected with PR/8 at two days following LL-37 treatment as determined by immunoassay demonstrating the immunomodulatory properties of LL-37 [50]. In vitro plaque assays demonstrated one log inhibition of PR/8 when virus was pre-incubated with LL-37 (50 µg/mL) in Madin-Darby canine kidney (MDCK) cells [50].

During IAV infection, in vitro LL-37 treatment did not prevent viral uptake, cause viral aggregation, and was not associated with blocking of hemagglutinin (HA). Interestingly LL-37 inhibits IAV replication at post-entry steps prior to viral RNA or protein synthesis [44]. A reduction in viral load, direct antiviral effects in epithelial cells, and inflammatory cytokine production have all been linked to LL-37 activity [49]. LL-37 inhibits the NY01 strain of IAV with a significant reduction in uptake of virus into cells, and in a manner dependent on dosage [44,49]. For optimal anti-IAV activity the central helix of LL-37 is required, as evident by fragments of LL-37 containing the complete central sequence of the peptide demonstrating more robust antiviral responses as compared to fragments with shorter central fragments. At a concentration of <2 µM, NY01 was only partially inhibited; however, this inhibition was surprisingly lost at higher concentrations of LL-37 [44]. A strain containing only the pandemic HA (Mex 1:7); however, was inhibited by LL-37 at all concentrations tested (up to 10 µM). Consistent with previous studies, these results suggest the antiviral effects of LL-37 are not determined by direct interaction of LL-37 with the viral HA [44].

Experimental data also demonstrates the participation of LL-37 in host defenses against IAV through modulation of innate immune cells, particularly neutrophils. IAV infection induces a respiratory burst response in neutrophils, and this response is noticeably up-regulated by pre-incubation of LL-37 with IAV [49]. LL-37 alone does not stimulate this respiratory burst response; however, optimal enhancement of this antiviral response is achieved only when the virus is pre-incubated with LL-37 [49]. Furthermore, there is growing evidence that NET formations play an important role during IAV infections [49]. Recently, in vivo studies have provided evidence of NET formation in the lungs of IAV-infected mice [51]. On the other hand, in vitro evidence of binding of IVA to NETs suggests LL-37 induces an increase in NET formation in response to IAV, which may promote viral clearance in vivo [49,51]. Data also suggest that significant protection against IAV may be provided by therapeutic treatment of influenza infected individuals with LL-37 or by increasing natural cathelicidin expression in the IAV-infected lung [11]. Hence, to maximize anti-IAV functions, approaches, such as therapeutic administration of naturally-occurring cathelicidins, as well as increasing vitamin D levels to boost endogenous cathelicidin have been proposed [11].

3.4. Human Immunodeficiency Virus

Earlier studies provided evidence of LL-37 ability to protect against HIV-1 infection given epithelial expression of LL-37, including in peripheral blood mononuclear cells such as CD4+ T-cells in vitro [52]. LL-37 directly inhibits the activity of HIV-1 reverse transcriptase via a protein–protein interaction in a dose-dependent manner (IC_{50} = 15 μM) [9,53]. Inhibition of HIV-1 protease activity with LL-37 has also been reported; however, this activity is less potent when compared to inhibition of HIV-1 reverse transcriptase (20%–30% inhibition at 100 μM). In addition, the plasma levels of LL-37 in HIV positive individuals undergoing antiretroviral therapy (ART) are much higher than in patients who are not, corresponding with an increased susceptibility to secondary infections in patients not undergoing ART [54].

3.5. Dengue Virus

To date, little research has been performed to characterize the antiviral activity of LL-37 against DENV. However, a recent study demonstrated that treatment of dengue virus 2 (DENV-2) with LL-37 inhibits viral infection in green monkey kidney (Vero) cells. Incubation of virus with LL-37 (10-15 μM) prior to infection inhibits production of viral particles, whereas pre-treatment of cells with LL-37 demonstrates no effect on viral replication [43]. Molecular docking studies of DENV-2 E protein have revealed the direct binding of LL-37 with E2 protein moieties, further demonstrating the peptide's ability to act as an entry inhibitor [43]. A more recent study assessed truncated and full length variants of LL-37 against other serotypes of DENV which revealed the inhibitory properties of LL-37 required the full length peptide [55]. In comparison to DENV-2, DENV-4 required higher concentrations of LL-37 for inhibition with effect being not as potent as with the former. DENV-1 and DENV-3 inhibition however, was prominent at lower concentrations of LL-37 [55]. A recent study using DENV-2 infected keratinocytes has demonstrated the production of AMPs by infected cells as well as bystander cells [56]. Pre-incubation of cells with LL-37 prior to DENV-2 infection results in a significant decrease in viral titers and replication in infected keratinocytes, whereas HBD2 and HBD3 demonstrate minimal inhibition [56]. Additionally, as vitamin D is an inducer of LL-37 expression, supplementation of populations with vitamin D prior to DENV outbreak seasons has been suggested as a possible preventative strategy to control the virus at initial stages of infection [43].

3.6. Respiratory Syncytial Virus

A few studies have demonstrated the efficacy of LL-37 against RSV [11,57]. Cells pre-incubated with LL-37 (>10 μg/mL) are protected against RSV infection whereas addition of LL-37 two hours post-infection results in decreased antiviral activity [57]. Additionally, LL-37 can limit viral-induced cell death in infected cell cultures indicating that the peptide's activity is not limited to prophylactic treatment. Treatment of epithelial cells with LL-37 prior to infection results in peptide internalization and retention, which provides antiviral protection for several hours post-treatment [57]. Furthermore, RSV infection induces the production of cytokines and chemokines in lungs. LL-37 (50 μg/mL) can impact the expression of chemokines as well as viral load when pre-incubated with RSV [11]. While the exact mechanism of the antiviral activity of LL-37 against RSV is not well established, it is speculated that the peptide directly interacts with the virus prior to infection due to its dose-dependent early effects on RSV infection. Interestingly, children with lower cathelicidin levels are more susceptible to RSV infection and display an increase in the severity of RSV-associated bronchitis [58].

3.7. Human Rhinovirus

Human rhinoviruses (HRVs) are causative agents of the common cold and most viral respiratory tract infections. As respiratory epithelial cells are the primary targets of HRV infection, studies evaluating the efficacy of LL-37 on HRV have utilized airway epithelial cells. LL-37 (50 μg/mL) demonstrates direct antiviral activity against HRV when added as a pre-treatment by acting on viral particles, and when added post infection by acting on the host cell [59]. LL-37 can induce a significant reduction in the metabolic activity of infected cells, as measured by mitochondrial metabolic potential [59]. Studies evaluating HRV in cystic fibrosis cells have revealed that expression of LL-37 decreases HRV viral load in vivo [60]. Thus, LL-37 reduces HRV infections in respiratory cells as well as in cystic fibrosis cells.

3.8. Vaccinia Virus

Vaccinia virus (VACV) is a DNA virus that can infect many types of mammalian cells. LL-37 limits VACV replication and can alter viral membranes [61]. VACV gene expression and viral titers are reduced in a dose-dependent manner in cells pre-incubated with LL-37(25–50 µM) [61]. Transmission electron microscopy images have shown a disruption in the integrity of VACV viral membrane after 24 h incubation with LL-37. Whereas murine LL-37 has demonstrated great efficacy and protection against VACV during infection, the efficacy of human LL-37 against VACV is unknown [61].

3.9. Herpes Simplex Virus

Few studies evaluating the efficacy of LL-37 against HSV-1 have been performed, all of which assessed LL-37 inhibition of HSV-1 in the context of a corneal infection [62]. LL-37 (500 µg/mL) can inhibit HSV-1 infection when pre-incubated with virions in vitro [62]. LL-37 reduces viral titers in corneal epithelial cells by more than 100-fold when compared to a scrambled LL-37 control [62]. Another study evaluating the anti-HSV-1 activity on corneal implants assessed the release of LL-37 delivered through corneal implant-incorporated nanoparticles [63]. Whereas LL-37 did not clear viruses from infected cells, it blocked HSV-1 infection in corneal epithelial cells by preventing viral-cell attachment [63]. These studies reinforce the mechanism of LL-37 antiviral activity as entry inhibition. Interestingly, LL-37 released from HSV-1 infected keratinocytes can also enhance HIV-1 infection [64]. This study measured the susceptibility of Langerhans cells (LC) to HIV-1 and implicates LL-37 in increasing HIV-1 cell receptor counts, resulting in increased HIV-1 infection [64]. While there are numerous studies linking a decrease in HIV-1 infection as a consequence of LL-37 or defensin treatment, the difference in this activity of the AMPs is possibly attributed to different cell targets.

3.10. Zika Virus

Zika virus (ZIKV) is a positive-sense, single-stranded RNA virus that can cause fever, headaches, rashes, joint pain, and myalgia in children and adults, and "microcephaly, ventriculomegaly, intracranial calcifications, abnormalities of the corpus callosum, retinal lesions, craniofacial disorder, hearing loss, and dysphagia" in neonates [65]. The emergence of ZIKV is a global concern since it is the first major infectious disease that has been associated with birth defects in over five decades [66]. Currently, no vaccines or treatments are available to prevent ZIKV infection [66]. He et al. [46] conducted a study to determine whether LL-37 and synthetic derivatives can be used to treat ZIKV infection in primary human fetal astrocytes [46]. Whereas LL-37 is toxic to these cells (EC_{50} = 20 M), an LL-37 derivative, GF-17, can be safely used due to its lower toxicity (EC_{50} > 50 µM) [46]. Treatment of primary human fetal astrocytes with 10 µM of GF-17 24 h after ZIKV infection results in a seven-fold decrease in the number of ZIKV plaque forming units [46]. Pre-incubation of ZIKV between 1 and 4 h with GF-17 (10 µM), results in at least a 95% decrease in the number of active zika virions [46]. In addition to the possibility of GF-17 directly interacting with ZIKV virions as a mechanism of antiviral activity, GF-17 increases interferon-α2 (IFN-α2) expression in a dose-dependent manner, which further impacts the ability of ZIKV to infect primary human fetal astrocytes [46]. The study suggests that GF-17 may be a possible option for the prevention and treatment of ZIKV infections [46].

3.11. Hepatitis C Virus

Hepatitis C virus (HCV) is a major worldwide health concern with possible severe outcomes including cirrhosis, liver cancer, and even death if an infection is left untreated [67]. Whereas effective antivirals against HCV infections exist, there is an unmet need for novel anti-HCV treatments that can overcome current treatment barriers such as cost and access to healthcare [41,67]. LL-37 has demonstrated anti-HCV properties in cell culture. HCV titers are significantly reduced when HCV is pre-incubated with LL-37 and subsequently used to infect Huh-7 cells [41]. Although different strains of HCV were utilized in this study, the antiviral effects of LL-37 are not associated or dependent on a specific HCV strain. Decrease in viral replication occurs in a dose-dependent manner [41]. Furthermore, of LL-37 primarily acts against HCV extracellularly, consistent with the activity of LL-37 against other enveloped viruses.

3.12. Venezuelan Equine Encephalitis Virus

Venezuelan equine encephalitis virus (VEEV) is an alphavirus that has been categorized as biothreat agent due to ease of aerosolization and high retention of infectivity in the aerosol form [68,69]. Currently there are no FDA approved therapeutics to combat VEEV infections. LL-37 has recently demonstrated anti-VEEV activity in vitro [70]. A significant decrease in intracellular VEEV genomic RNA copies was observed upon pre-incubation of LL-37 (10 µg/mL) and VEEV. Microscopy data revealed the extracellular aggregation of VEEV virions, suggesting the mechanism of action of LL-37 against VEEV is through direct interaction with viral particles, thereby inhibiting entry [70]. Pre-treatment of human microglial cells with LL-37 prior to infection also resulted in a significant reduction in VEEV titers, suggesting entry prevention is not the only mechanism of LL-37-mediated inhibition. Indeed, LL-37 increased the expression of type I interferon (IFNβ), possibly inducing an antiviral state [70]. LL-37 could prove as a potent therapeutic candidate against VEEV and possibly other alphaviruses.

4. Transferrins

4.1. Expression

The most notable AMP exhibiting antiviral activity of the transferrin family of iron-binding proteins is lactoferrin (LF), a multifunctional 80 KDa glycoprotein [71,72]. Originally discovered in bovine milk, LF is highly conserved and can be found in humans, mice, and porcine species. It is expressed in most biological fluids such as exocrine secretions (milk, saliva, fluids of digestive tract, and tears) and in neutrophil granules [72,73]. LF serves as a key component of innate immune defenses and demonstrates antimicrobial activity against a wide range of bacteria and viruses through direct action on pathogen membrane and target host cell moieties as well as through modulation of inflammation (Table 1) [72,74].

LF expression may be induced under hormonal control by epithelial cells in mammary glands or at well-defined stages of cell cycle such as neutrophil differentiation [74,75]. LF structure consists of a polypeptide chain that is characterized by a highly basic and positively-charged N-terminal region [74,76]. The LF chain folds into two globular lobes linked by a three-turn alpha helix, with each lobe containing an iron-binding site [77]. There is a strong interaction between the two lobes when iron is bound, which renders LF resistant to proteolysis in this holo form as compared to an open apo form [77]. In the stomach, acidic pepsin hydrolysis of the N-terminal of LF yields lactoferricin (Lfcin), a 25-amino-acid peptide with multiple hydrophobic, positively-charged residues [74,76]. Lfcin retains the properties of LF and demonstrates potent antiviral activity.

4.2. Respiratory Syncytial Virus

Early studies have demonstrated the direct anti-RSV activity of LF in vitro [78–80]. In HEp-2 cells, LF (100 µg/mL) decreased the production of IL-8 induced by RSV infection when pre-incubated with virions. This LF-mediated inhibition was dependent on LF–RSV interactions as LF inhibited RSV entry into HEp-2 cells [79]. Pre-incubation of LF (100 µg/mL) with RSV resulted in decreased viral entry. LF was found to inhibit RSV entry by directly binding to the F_1 subunit of the RSV F protein, which mediates fusion of the virion with the cell membrane [79].

4.3. Influenza Virus and Parainfluenza Virus

Human and bovine LF have demonstrated anti-IAV activity. Bovine LF inhibits IAV-mediated programmed cell death and directly binds viral hemagglutinin leading to inhibition of viral hemagglutination [81–84]. Only a few studies; however, have evaluated the activity of human

LF in the context of IAV infection. In Madin-Darby canine kidney cells, human LF (20-80 μg/mL) demonstrated enhanced antiviral activity against avian IAV in a dose-dependent manner [85]. In addition, bovine LF inhibited parainfluenza virus replication in vitro and decreased viral adsorption onto cells, thereby preventing viral entry [86]. Together these results highlight the possible role of LF in preventing influenza and parainfluenza virus infections.

4.4. Adenovirus

Lactoferrin-mediated anti-HAdV activity occurs at multiple stages during infection. In one study LF inhibited HAdV replication during different phases of infection; when the peptide was introduced before infection, after viral adsorption, and when the peptide was present throughout the experiment, indicating more than one mechanism of action may be involved [87]. Bovine LF has also demonstrated anti-HAdV properties in a dose-dependent manner similar to human LF [87–89]. Bovine LF demonstrated the greatest inhibition when virus was pre-incubated with LF (1 mg/mL), which was validated by electron microscopy imaging the binding of LF to HAdV, thus suggesting direct LF inhibition as a mechanism of action [89], In contrast, human LF in tear fluids was implicated in promoting HAdV binding to epithelial cells, whereby HAdV hijacks human LF and utilizes the protein in order to bind to host cells [90].

4.5. Herpes Simplex Viruses 1 and 2

Studies assessing inhibition of HSV-1 have revealed the potency of LF at inhibiting replication of the viruses. LF (0.5–1 mg/mL) inhibited HSV-1 replication in human embryo lung cells and prevented virus adsorption and entry [91]. Marchetti et al. also demonstrated the ability of bovine and human LF to inhibit HSV-1 replication and adsorption in Vero cells independent of iron-binding [92,93]. Furthermore, intracellular trafficking of virions that gained entry was delayed [94]. Similar inhibitory properties were displayed by Lfcin when the peptide was tested against HSV-1 [94]. In addition, cell-to-cell viral spread was inhibited by LF as well as Lfcin [95]. The data presented by these studies demonstrate LF's antiviral properties against HSV.

4.6. Hepatitis C and B Virus

Lactoferrin has demonstrated efficacy against HCV and HBV. Bovine LF can prevent HCV infection in human hepatocytes and in turn was tested as a measure to control HCV viremia in chronic hepatitis C patients [96]. However, LF only reduced HCV RNA in patients with previously low HCV RNA serum concentration. Conversely, a randomized trial of bovine LF in patients with chronic hepatitis C reported that orally-administered LF did not demonstrate significant efficacy against HCV when compared to a placebo control [97]. Combination therapy using the antiviral molecule ribavirin, LF, and interferon therapy, on the other hand, suggested that LF contributed to decreased HCV RNA titers [98]. These contradictory results suggest that while orally-administered LF may not be a promising stand-alone therapeutic, it can enhance the effectiveness of currently used therapeutics. Studies on the interaction of HCV and LF reported the direct interaction and binding of human and bovine LF with the HCV envelope proteins E1 and E2 [99]. A more recent study concluded that LFs of various species directly prevented HCV cell entry by binding to virions. Furthermore, pre-incubation of virions or post-treatment of infected cells with human, bovine, sheep, or camel LF inhibited HCV replication in human hepatoma (HepG2) cells, suggesting that LF may be used prophylactically as well as therapeutically, particularly in combined therapies [100].

Pre-incubation of HBV with LF had no effect on viral replication; however, pre-treatment of cells with human or bovine LF prevented infection [101]. Bovine LF did not interact with HBV but indirectly protected cells from HBV infection [102]. In addition, iron or zin-saturated LF on HBV-infected HepG2 cells inhibited HBV-DNA amplification suggesting its use as a possible candidate for the treatment of HBV infections [103].

4.7. HIV

Circulating LF plasma levels in HIV-1-infected patients was significantly decreased compared to non-infected patients [104], suggesting an important role for LF during HIV-1 infection and disease progression. Experiments assessing the role of LF in HIV-1 infection in peripheral blood mononuclear cells revealed LF (IC_{50} = 9.6 μM) as a potent inhibitor of six clinical isolates of HIV-1 in a dose-dependent manner upon treatment of cells prior to infection [105]. Additionally, synergistic effects of human and bovine LF were assessed in combination with zidovudine, which also demonstrated a dose-dependent inhibition [105]. Bovine LF inhibited the HIV-1 entry process by binding to the viral envelope, and HIV-1 variants resistant to LF exhibit mutations in the viral envelope protein [106]. Bovine LF was found to be a more potent inhibitor than human LF in preventing dendritic cell-mediated HIV-1 transmission between cells by directly and strongly binding to DC-SIGN, a receptor molecule on dendritic cells that mediates HIV-1 internalization [106].

In pediatric HIV-1 infections, LF enhanced responses to antiretroviral therapy by decreasing plasma viral load and modulating the immune system [107,108]. LF treatment alone increased CD4+ cell counts, but a more significant increase was observed when LF was combined with antiretroviral therapy [107]. Result suggested that no HIV-1-related symptoms were evident for the duration of the experiments. However, large-scale studies are required in order to further assess LF's therapeutic potential against HIV.

4.8. Hantavirus

Hantavirus is a zoonotic RNA virus that causes severe human disease characterized by widespread and extensive hemorrhaging [109]. In vitro studies have demonstrated that pre-treatment of Vero cells with LF (ED_{50} = 39 μg/mL) significantly reduces Hantavirus foci number [110]. A combination of LF and ribavirin also significantly reduced the number of viral foci where pre-treatment of cells with LF (400 μg/mL), and treatment with ribavirin (100 μg/mL) post infection completely abolished viral foci [109]. In additional synthesis of viral glycoprotein, G2, and nucleocapsid protein (NP) was delayed in LF pretreated cells. In vivo testing revealed that prophylactic LF treatment (160 mg/kg) can significantly improve survival rates following hantavirus infection in suckling mice by up to 70% when compared with non-treated mice [109].

4.9. Human Papillomavirus

Human and bovine LF and Lfcin have been demonstrated to inhibit the internalization of HPV-16 particles, as visualized using an HPV virus-like particle (VLP) that fluoresces after cellular internalization [111]. This inhibition occurred in a dose-dependent manner on HaCaT cells [111]. Bovine LF-mediated inhibition was more potent than human LF. However, both LFs inhibited HPV-5 and HPV-16 [112]. Different synthetic derivatives or variants of human and bovine Lfcin demonstrated selective inhibition of HPV, such that a human variant of Lfcin with amino acids 1 to 49 displayed antiviral activity as well as inhibition of attachment of both HPV strains while the bovine variant (17–42) only inhibited HPV-5 infection [112]. These results suggest that different domains of Lfcin contain selective and specific antiviral properties which can be engineered to target specific infections and viral strains.

4.10. Rotavirus

Rotavirus is one of the world's major causes of gastroenteritis in children. As an effort to develop anti-rotavirus strategies, milk proteins including apo-LF (25 μM) and Fe-LF were assessed for inhibitory properties against rotavirus [113]. Results demonstrated apo-LF to be the most potent inhibitor of rotavirus by binding to viral particles and hindering virus attachment as well as preventing rotavirus hemagglutination [113]. A randomized trial; however, concluded that orally-administered LF did not provide protection against rotavirus infection [114]. This serves as an example that the pronounced inhibition in vitro does not necessarily recapitulate the effects that occurs in vivo.

4.11. Other Viruses

Lactoferrin antiviral activity against other viruses including poliovirus, alphaviruses, DENV, and Japanese encephalitis virus (JEV) has been studied. In Vero cells, iron-, manganese-, and zinc-saturated LFs have demonstrated anti-poliovirus properties in a dose-dependent manner when LFs were present either during the entire viral life cycle or during viral adsorption [115]. Against alphaviruses, human LF (200 µg/mL) inhibited infection of BHK-21 cells by Sindbis virus (SINV) and Semliki Forest viruses (SFV) adapted to bind heparan sulfate (HS) upon pre-treatment of cells prior to infection [116]. The LF-mediated inhibition was thought to occur by binding of LF to heparan sulfate moieties, thereby preventing virus–receptor interactions [116]. Bovine LF was tested against JEV and DENV, and was capable of inhibiting replication of both viruses. In JEV, similar to alphaviruses, HS-adapted JEV strains were inhibited by bovine LF, while wild-type strains were not [117]. Bovine LF inhibited binding of DENV to HS by directly interacting with HS. Additionally, morbidity of DENV-infected suckling mice was reduced upon administration of a mixture of DENV and LF as compared to DENV-only-infected mice [118].

5. Human Antimicrobial Proteins–Eosinophil Proteins

5.1. Expression

Eosinophils contain large cytoplasmic granules that play a critical role in innate immune responses. These granules are storage hotspots for major cationic antimicrobial proteins including eosinophil-derived neurotoxin (EDN) and eosinophil cationic protein (ECP). Both EDN and ECP are human antimicrobial proteins that are active ribonucleases and members of the human RNase A superfamily [119,120]. The antimicrobial properties of RNases were mapped to the N-terminal domain that is conserved among ribonucleases [3]. While both proteins contain characteristic RNase A superfamily structures and catalytic residues, they exhibit antimicrobial activity against both bacteria and viruses (Table 1).

5.2. Respiratory Syncytial Virus

Eosinophil-derived neurotoxin (EDN), also known as RNase 2, has demonstrated inhibitory activity against RSV in several studies. In one study, RSV infectivity decreased in a dose-dependent manner when eosinophils were introduced to a suspension of RSV [121]. This property was abolished upon addition of a ribonuclease inhibitor suggesting that an RNase was responsible for the antiviral activity. At 50 nM, a plasmid-constructed recombinant human EDN exhibited a 40-fold reduction in RSV infectivity [121]. A decrease in RSV genomic RNA copies as quantified by RT-PCR upon incubation of EDN with virions suggests EDN ribonuclease activity directly targets extracellular viral particles. EDN may contain unique features that help mediate its antiviral activity as ribonuclease A alone has no effect on RSV infectivity.

Additionally, ECP, which is also known as RNase 3, has been detected in infants with acute bronchiolitis following RSV infection [122]. ECP levels were significantly higher in infants who developed persistent wheezing compared to those that do not after a five-year follow-up [122]. Additionally, while ECP was able to reduce RSV infectivity in the same suspension as EDN, ECP is less effective when used to inhibit RSV alone, demonstrating only a six-fold reduction as compared to EDN at a concentration of 50 nM [121]. Hence while ECP exhibits some inhibitory activity against RSV, it is not sufficient in controlling infection as RSV infection itself induces the production of ECP with no evident antiviral effects. Instead, the production of ECP in infants with bronchiolitis can be used as tool in prediction the risk of wheezing cough development in those infants as higher ECP levels correlate with wheezing cough development.

5.3. HIV

In earlier studies, EDN did not demonstrate direct activity against HIV; However, it indirectly induces the production of HIV-inhibiting molecules. Soluble factors produced by hosts such as chemokines, interferons, RNases, and alloantigen-stimulated factors (ASF) have been reported to inhibit HIV-1 replication or binding [123]. A test to determine the soluble HIV-1 inhibitory activity utilized supernatants from mixed lymphocytes of healthy individuals that exhibit anti-HIV-1 activity. While the supernatant exhibited anti-HIV activity, this activity was blocked with antibodies specific for EDN by 58% [123]. Antibodies to RNase A had no effect on the anti-HIV-1 activity, thus ruling out its involvement. In addition, an RNase inhibitor significantly blocked the inhibitory activity indicating that EDN or a possible closely-related RNase is responsible for the HIV-1 inhibitory activity [123]. Later studies reported EDN and a recombinant EDN, containing a four amino acid extension of the N-terminus of END, both exhibit anti-HIV-1 activity independent of time of addition, before, during, or 2 h post infection [124].

6. AMPs from Immune Cells

6.1. Expression

Immune cells have the ability to produce AMPs such as the protease inhibitors elafin and secretory leukocyte protease inhibitor (SLPI) [3]. Studies investigating AMPs produced by gamma-delta T cells ($\gamma\delta$ T cells), recorded production of elafin [125]. Elafin is expressed in tissues such as skin, placenta, genital and gastrointestinal tracts, as well as by cells including neutrophils, epithelial cells, macrophages, and keratinocytes and thus can be found at many surfaces and in secretions [126]. Elafin is a 57-amino-acid single-polypeptide chain produced by the proteolytic cleavage of a precursor molecule called trappin-2 [127,128]. Similarly, SLPI is expressed by neutrophils, macrophages, epithelial cells of renal tubules and respiratory/alimentary tracts [129]. This peptide can be found in saliva, cervical mucus, breast milk, cerebral spinal fluid, tears, and secretions from nose and bronchi [129]. SLPI is a 107-amino-acid, cysteine-rich polypeptide chain that is highly basic [129]. The C-terminals of elafin and SLPI exhibit both high homology and similar protease activity [127] and have demonstrated antimicrobial as well as anti-inflammatory activities (Table 1) [129,130].

6.2. Herpes Simplex Virus

Elafin and its precursor trappin-2 (Tr) both exhibit anti-HSV activity. Drannik and colleagues (2013) investigated the efficacy of elafin and Tr on HSV-2 infection using a Tr-expressing, replication-deficient adenovirus vector as well as recombinant TR/elafin proteins. Cells were either infected with the adenovirus vector or treated with the recombinant proteins prior to HSV-2 infection. When endometrial and endocervical epithelial cells were pre-treated with these molecules (TR $IC_{50}\sim0.07$ µg/mL, E $IC_{50}\sim0.01$ µg/mL), viral titers following HSV-2 infection were significantly reduced, with the activity of Tr being more potent at inhibiting HSV-2 than elafin [126]. A decrease in viral titers correlated with a decrease in production of pro-inflammatory cytokines, whereas the antiviral IFN-β response was increased. Furthermore, the recombinant molecules decreased viral attachment to cells [126]. In vivo, Tr-transgenic mice, mice generated to express the full elafin/Tr human gene, demonstrated reduced central nervous system viral load and TNF-α expression upon intravaginal infection [126]. The data provides evidence for elafin and its precursor's potential as anti-HSV therapeutics.

6.3. HIV

Both proteases, elafin and SLPI, can provide protection against HIV-1 in vitro. SLPI in saliva has demonstrated anti-HIV-1 properties in vitro [131]. Genital secretions of females that are HIV-resistant contained significantly higher levels of elafin/Tr as compared to uninfected females [132–134]. In addition, elafin/Tr is found in and produced by epithelial cells of uterine, fallopian tubes, and cervix [132]. Ghosh et al. (2010) [132] investigated the ability of elafin/Tr in the reproductive tract to inhibit HIV-1and demonstrated HIV-1 inhibition was dose-dependent, particularly when HIV-1 was pre-incubated with elafin/Tr (10 ng/mL), suggesting the HIV-1 inhibition is mediated by direct interaction of elafin/Tr with the virus [132]. However, a recent report on multiple groups of HIV-infected women found no correlation between anti-HIV activity of mediators such as elafin and HIV susceptibility [135].

7. Hepcidin

7.1. Expression

Hepcidin, also known as human liver expressed antimicrobial peptide-1 (LEAP-1), is a 25-amino acids long AMP that is predominately expressed by liver hepatocytes [136]. It was first isolated in human blood filtrates. Hepcidin is synthesized as a pre-propeptide with two cleavage sites. The first cleavage site releases an N-terminal endoplasmic reticulum signal sequence while the second cleavage site releases the mature hepcidin peptide from a prodomain [136]. Hepcidin plays a major role in iron regulation and in systemic iron homeostasis in hepatocytes and other cells [136–138]. Hepcidin regulates iron uptake and efflux by directly binding to the iron exporter ferroportin, resulting in degradation of this transporter [137]. Consequently, high levels of hepcidin result in inhibition of iron uptake and sequestration of blood iron in macrophages, whereas low levels of hepcidin result in excessive uptake and toxic accumulation of dietary iron [137,138]. Hence, increased serum iron levels increase hepcidin expression. The vital role of hepcidin in iron homeostasis can determine the outcome of infections by limiting iron availability to invading pathogens.

Relatively little information is available on the effects of hepcidin on the pathogenesis of viral infections; however, hepcidin induction has been reported following a number of human and murine viral infections [137,138]. Hepcidin mRNA levels were significantly increased in mice during influenza virus PR8 infection [139]. This induction appeared to be IL-6 dependent as IL-6 knockout mice did not induce the expression of hepcidin mRNA. In addition, in primary hepatocytes that were stimulated with pathogen-associated molecular patterns (PAMPs), hepcidin was inhibited by the addition of IL-6 neutralizing antibodies [139]. These results indicate that hepcidin expression is increased following viral infections and infection-induced inflammatory responses, particularly IL-6 due to the central role this cytokine plays in hepcidin production.

7.2. Antiviral Activity

Liver injury as well as chronic liver infections, such as hepatitis B and C, can result in abnormal hepcidin expression. Hepcidin expression was found to be highly elevated in chronic hepatitis B and C patients, as reported by Wang and colleagues (2013). In patients with increased HBV DNA, hepcidin and IL-6 levels were elevated; however, an exact correlation between hepcidin and IL-6 was not determined [140]. In contrast, in studies assessing hepcidin levels in HBV and HCV acute infections during primary viremic phases, hepcidin levels were not upregulated and hypoferremia was not evident [141]. This difference could be attributed to the different stages of disease where hepcidin may be induced at later stages of infection when iron levels are elevated. In the same study Armitage et al. also measured hepcidin levels in HIV-1 positive plasma donors of both acute infections and patients transitioning to chronic infections. In the former group, hepcidin levels were increased and peaked with peak viral load and iron levels decreased due to retention in cells by hepcidin activity; whereas in the latter group, hepcidin levels remained elevated even in individuals undergoing ART [141]. The increase in hepcidin levels coincided with an increase in inflammatory cytokine expression [141]. HIV replication is iron-dependent and the hepcidin-induced sequestering of iron in cells such as lymphocytes and macrophages is a highly favorable condition for HIV pathogenesis.

8. Antimicrobial Neuropeptides

8.1. Expression

Neuropeptides (NPs) are low molecular weight, cationic, amphipathic AMPs. Traditionally NPs are thought to transmit and modulate signals in central and peripheral nervous systems. However, in recent years, NPs have demonstrated antimicrobial activities [142,143]. The NP, α-melanocyte-stimulating hormone (α-MSH) exhibits both antiviral and anti-inflammatory activity (Table 1). α-MSH is a 13-amino acid peptide that is a product of posttranslational processing of the precursor molecule pro-opiomelanocortin (POMC) [144]. α-MSH is widely expressed in peripheral brain tissue and by numerous cells including monocytes, dendritic cells, melanocytes, pituitary cells as well as T lymphocytes [142,144]. The active component of α-MSH resides in the C-terminal tripeptide (11-13, KPV) [144].

8.2. HIV

In HIV-infected monocytes, α-MSH exhibited inhibitory properties against HIV-1. Previously, α-MSH inhibited pro-inflammatory cytokine production such as TNFα and IL-1β in blood from HIV-infected individuals [145]. Barcellini et al. [55] aimed to determine if α-MSH exhibits anti-HIV properties by utilizing chronically infected monocytic U1 cells and acutely infected monocyte-derived macrophages. Not only did U1 cells consistently produce α-MSH, but α-MSH and the tripeptide (KPV) both (10 µM) inhibited HIV expression at low concentrations comparable to physiological concentrations of α-MSH, with inhibition more pronounced at higher concentrations [146]. In addition, both peptides inhibited nuclear factor-kB (NF-kB) activation which is known to enhance HIV viral protein expression.

9. Therapeutic Potential and Challenges of AMPs in Clinical Applications

In recent years the antiviral properties of AMPs, both naturally occurring and synthetic, and their therapeutic potential have generated increasing interest within the academic and pharmaceutical communities. Many of the AMPs originally thought to demonstrate only antibacterial activity have been found to exhibit antiviral and immune modulatory capabilities. While to date few AMP-based therapies have been clinically approved, so far, the suitability and efficacy of only a few AMPs as therapeutic agents has been studied in clinical trials as treatments for infections as well as agents for immune modulations [147]. Clinical studies utilizing AMPs include defensin-mimetic compounds (CTIX 1278) for the treatment of drug-resistant *Klebsiella pneumoniae*, an alpha-helical AMP (PL-5) for gram negative and positive bacterial skin infections, and a cationic peptide (DPK 060) for skin dermatitis [148]. AMPs are promising therapeutics due to their relatively low toxicity and tolerance in vivo. AMPs can be delivered in a variety of ways including intravenous injections, oral administration, as topical ointments, and via inhalation. In respiratory tract infections, supplementation and treatment with AMPs such as defensins and LL-37 may provide protection in the lungs. For example, nebulizing mice with LL-37 prior to infection with IAV can reduce disease severity and result in increased survival rates [50]. The same concept could be applied against other respiratory pathogens such RSV and rhinovirus with the inclusion of AMPs such as lactoferrin and EDN, which have both previously demonstrated efficacy against RSV [79,121].

AMPs can also be applied as ointments, gels, or creams. For instance, a gel for the treatment of vaginal candidiasis currently in clinical trials is derived from human α-MSH [147]. As α-MSH has previously demonstrated antiviral activity against HIV [146], this form of treatment could be developed into a cheap and effective strategy to reduce HIV transmission. The treatment can be expanded to other sexually transmitted diseases, such as HSV, and include many of the other classes of AMPs demonstrating antiviral activity against those particular viruses. Equally, oral formulations or intravenous injections are more feasible for viral infections affecting internal organs. Lactoferrin has been used as an oral supplement to antiviral therapy against HCV infections and greatly enhances the effectiveness of treatment [98]. While most studies utilizing oral LF against other pathogens alone have not provided promising results, different AMPs may prove more effective against different pathogens. An intravenous treatment of human LF is currently in clinical trials for the treatment of bacteremia and fungal infections [147]. Interestingly, the protease inhibitor Telaprevir, which is marketed for use against HCV suggests that eosinophil peptides that have dual functionality as protease inhibitors and AMPs may possess anti-HCV properties. In addition, AMPs that directly target viruses can serve as alternatives to antivirals for which resistance has become an issue as it is a challenge for viruses to change their envelope lipid compositions. Nonetheless, more research and investigation are required to assess the full potential of AMPs as novel therapeutics.

However, despite their appeal and several successful in vitro outcomes following the use of AMPs as antiviral therapeutics, their greater application as therapeutics faces many hurdles. Little clinical data on the use, toxicity, efficacy, methods of delivery, or host uptake and the in vivo response to AMPs is currently available to guide the development of AMPs into therapeutic products [147,149]. It is unclear whether and to what extent laboratory findings will translate into antiviral activity in vivo. Additionally, the relationship between peptide concentrations used in vitro experiments and physiologically effective concentrations required for in vivo activity is unknown for many AMPs. As an example, the anti-bacterial activities of AMPs appear to heavily depend on the environmental conditions used during experiments [147]. Several AMPs have been reported to control group A Streptococcus at physiological concentrations which are ineffective in vitro [150], suggested to be partly due to the physiological ionic environment not being recapitulated fully in laboratory experiments [151]. Moreover, AMPs minimally active in in vitro assays have been shown to efficiently control infections upon in vivo administration [152,153], highlighting a lack of clear correlation between efficacy and dose-dependence of AMPs derived from laboratory experiments and animal models of infection. Further research into the correlation between antiviral activity of AMPs in vitro and the translation of this activity into in vivo efficacy will be needed for further development of AMPs as a new class of therapeutics.

Another important challenge is isolation and large scale manufacturing of AMPs. The high cost of producing AMPs impedes the applicability for clinical use as a commercial-scale manufacturing platform does not yet exists. The complex secondary structures of AMPs also serve as an obstacle for large-scale production as their activity is structure-based. The use of expression systems such as bacterial systems can generate "correctly folded peptides" in large quantities; however, the peptides are susceptible to proteolytic degradation in these systems [17]. To turn AMPs into viable therapeutics for clinical use, these production disadvantages must be overcome. Meanwhile, as an alternative, non-human AMPs are being used. For example, numerous studies are examining the activity of LF utilize bovine instead of human LF due to ease of producing bovine LF in bulk. While bovine LF demonstrates equivalent and at times enhanced efficacy, as compared to the human form, other AMPs may not function similarly or may contain sequence variations that limit or abolish their properties. Synthetic peptides may prove to be a practical alternative; however, they are expensive to generate, and future use in therapies will require development of novel synthesis methods [154,155]. Several systems have demonstrated promise in mass-production of AMPs including mammalian cells and transgenic plants. While mammalian cell culture systems are preferred to produce human AMPs, they are not very cost-effective. Plant bioreactors; however, are a cost-effective method for commercial-scale AMPs production that have demonstrated applicability with other compounds [17]. Another interesting approach that could possibly counteract these challenges that has recently gained attention is the employment of agents to induce or increase expression of AMPs. For example, vitamin D induces LL-37 production that can be used to boost the natural response to infection and has been suggested for use prior to potential DENV outbreaks [43].

Furthermore, the metabolic stability of AMPs may limit their future clinical application [156–158]. There is a lack of data assessing the pharmacokinetics of AMPs to address issues such as the physiological half-life of peptides, the necessary dosing regiments, and the aggregation of peptides in vivo. For example, AMPs in oral and topical formulations will be subjected to enzymatic degradation, changes in pH and ionic concentrations, and face hurdles upon delivery, such as penetration of and uptake from the skin, intestinal, or nasal mucosa [159]. Intravenous administered AMPs will also face proteolytic degradation from enzymes in plasma and tissues, although the relatively shorter half-life of AMPs when compared to small-molecule drugs may be considered beneficial in some contexts due to lack of accumulation in tissues [159]. In addition, some AMPs are sensitive to salt or serum, a factor that substantially reduces their activity further limiting their applicability in clinical use.

The delivery of sufficient amounts of AMPs to an infection site may also present challenges and require careful calibration of dosing. For example, topical application of AMPs to the skin may require large-scale application of ointments, raising issues about sensitization and development of allergies against applied AMPs. Similarly, AMP activity in the context of natural innate immunity occurs in the presence numerous feedback pathways that collectively calibrate the anti-pathogen response to an appropriate level. It remains to be determined how the introduction of potentially high doses of AMPs during infections will interact with these pathways and modulate the activity of AMP therapeutics, especially as synthetic AMPs appear to elicit weaker immune responses upon administration when compared to the response produced by naturally secreted peptides during infection [160].

While these challenges may considerably impede clinical development of AMPs, the incorporation of innovative formulations can speed the process of bringing AMP products to clinical phases. The use of nanoparticles (NPs) as delivery vehicles may greatly improve the metabolic stability of AMPs. Nanoparticles can provide unique advantages, such as protection of AMPs from proteolytic degradation and control over the rate of peptide release. Several nanoparticles have been evaluated for delivery of AMPs. For example, self-assembling hyaluronic acid (HA) nanogels have been used to successfully deliver LLLKKK18, an analog of LL-37 via intra-tracheal administration in tuberculosis-infected mice [161]. Additionally, the HA nanocarrier was able to both stabilize the peptide and decrease peptide cytotoxicity in macrophages in vitro [161]. Poly lactic-co-glycolic acid (PGLA) NPs were also successfully utilized to deliver cationic AMPS [162]. Although these formulations have only been investigated in experimental animal models and in vitro applications, the outcomes for developing future AMP-based therapeutics appear promising.

10. Conclusions

Progress has been made in the last decade to elucidate the mechanisms of action of various AMPs. The primary mechanism of AMP-mediated antiviral activity has been attributed to direct interference with, and destabilization of, viral envelopes. However, AMPs have also demonstrated selective immune modulation. Antiviral activity against both enveloped and non-enveloped viruses has been reported with the latter hinting at the presence of undiscovered activities of AMPs, in addition to the known direct interaction with viral envelopes. Indeed, antiviral activity has also been reported at post entry steps affecting later stages in the viral life cycle, such as genome replication and viral protein trafficking. Additionally, studies have demonstrated that AMP treatment prior to viral infection results in peptide retention and internalization by cells which may reflect a more robust response to viruses compared to other potential therapeutics in development. Additionally, post-infection treatments have been reported to exhibit antiviral activity; however, to a lesser extent to that of treatments prior to infection. Nonetheless, these treatments play a role in altering viral replication and assembly as well as a role in accelerating immune activation or suppression. Hence, AMPs can directly impact viral infections or can modulate host processes that ultimately impact viral replication negatively. AMPs have been reported to drive interferon β (IFNβ) signaling, contributing to the induction of an antiviral state in susceptible cells. This dual functionality of AMPs is advantageous as they can be used as a prophylactic and/or as part of post-exposure antiviral measures. In vulnerable individuals, prophylactic expression of AMPs has the potential to become a preventative strategy against viral infections, especially during emerging pandemics. In addition, the simplicity of AMPs makes the development of synthetic peptide analogues a cost-effective measure to treat established viral infections.

AMPs and their synthetic derivatives are a promising avenue to yield new strategies to control and treat a wide range of viral diseases but their application is still at the preliminary stages. Therefore, further research is warranted to understand AMP antiviral activity both in vivo and in vitro and to determine underlying mechanisms involved in AMP-mediated immune modulation for clinical applications.

Table 1. Mechanisms of actions of antiviral AMPs.

AMP Family	Target	Proposed Mechanism of Action	References
Defensins	HAdV HIV HSV RSV HPV	Direct interaction with virions; reduction of cell trafficking; direct binding to cell receptor blocking entry (HS); inhibition of protein kinase C signaling; release inhibition of viral components from endosomes; decrease in proinflammaktory cytoine production.	[8,10,13,16,21–28,30–41]
Cathelicidin (LL-37)	HIV DENV RSV HRV VACV HSV ZIKV HCV VEEV	Direct interaction with virions; Increase in type I IFN expression; decrease in proinflammatory cytokine production.	[9,11,37,41,43,44,46–67,70,163]
Transferrin	RSV IAV HPIV HAdV HSV HCV HBV HIV Hantavirus HPV Rotavirus JEV SFV SINV DENV	Direct interaction with virions; inhibition of viral attachment/absorption; delay in viral protein synthesis; Inhibition of cellular trafficking; direct binding to cell receptor blocking entry (HS and DC-SIGN).	[71–118,164]
Eosinophil proteins	RSV HV	Direct interaction with virions	[119–124]
AMPS from Immune cells	HSV HIV	Direct interaction with virions; increase in type I IFN expression	[125–135]
Hepcidin	HBV HCV HIV	Sequester iron from pathogens	[136–141]
Antimicrobial Neuropeptides	HIV	Inhibition of NF-kB and cytokine production	[142–146]

Author Contributions: Conceptualization, A.A., G.S.-T. and A.N.; manuscript drafting, A.A., G.S-T. and N.B.; literature search, A.A., G.S.-T., N.B. and G.H.; Editing, N.B. and A.A.

Funding: This work was supported by the Defense Threat Reduction Agency (HDTRA11810041).

Conflicts of Interest: The authors declare no conflicts of interest.

References

1. Agier, J.; Efenberger, M.; Brzezinska-Blaszczyk, E. Cathelicidin impact on inflammatory cells. *Cent. Eur. J. Immunol.* **2015**, *40*, 225–235. [CrossRef] [PubMed]
2. De la Fuente-Nunez, C.; Silva, O.N.; Lu, T.K.; Franco, O.L. Antimicrobial peptides: Role in human disease and potential as immunotherapies. *Pharmacol. Ther.* **2017**, *178*, 132–140. [CrossRef] [PubMed]
3. Wang, G. Human antimicrobial peptides and proteins. *Pharmaceuticals* **2014**, *7*, 545–594. [CrossRef] [PubMed]
4. Bahar, A.A.; Ren, D. Antimicrobial peptides. *Pharmaceuticals* **2013**, *6*, 1543–1575. [CrossRef] [PubMed]
5. Mulder, K.C.; Lima, L.A.; Miranda, V.J.; Dias, S.C.; Franco, O.L. Current scenario of peptide-based drugs: The key roles of cationic antitumor and antiviral peptides. *Front. Microbiol.* **2013**, *4*, 321. [CrossRef] [PubMed]

6. Holly, M.K.; Diaz, K.; Smith, J.G. Defensins in Viral Infection and Pathogenesis. *Annu. Rev. Virol.* **2017**, *4*, 369–391. [CrossRef] [PubMed]

7. Oppenheim, J.J.; Biragyn, A.; Kwak, L.W.; Yang, D. Roles of antimicrobial peptides such as defensins in innate and adaptive immunity. *Ann. Rheum. Dis.* **2003**, *62*, ii17–ii21. [CrossRef]

8. Ding, J.; Chou, Y.Y.; Chang, T.L. Defensins in viral infections. *J. Innate Immun.* **2009**, *1*, 413–420. [CrossRef]

9. Findlay, F.; Proudfoot, L.; Stevens, C.; Barlow, P.G. Cationic host defense peptides; novel antimicrobial therapeutics against Category A pathogens and emerging infections. *Pathog. Glob. Health* **2016**, *110*, 137–147. [CrossRef]

10. Klotman, M.E.; Chang, T.L. Defensins in innate antiviral immunity. *Nat. Rev. Immunol.* **2006**, *6*, 447–456. [CrossRef]

11. Harcourt, J.L.; McDonald, M.; Svoboda, P.; Pohl, J.; Tatti, K.; Haynes, L.M. Human cathelicidin, LL-37, inhibits respiratory syncytial virus infection in polarized airway epithelial cells. *BMC Res. Notes* **2016**, *9*, 11. [CrossRef] [PubMed]

12. Ganz, T. The role of antimicrobial peptides in innate immunity. *Integr. Comp. Biol.* **2003**, *43*, 300–304. [CrossRef] [PubMed]

13. Hsieh, I.N.; Hartshorn, K.L. The Role of Antimicrobial Peptides in Influenza Virus Infection and Their Potential as Antiviral and Immunomodulatory Therapy. *Pharmaceuticals* **2016**, *9*, 53. [CrossRef] [PubMed]

14. Wilson, C.L.; Schmidt, A.P.; Pirila, E.; Valore, E.V.; Ferri, N.; Sorsa, T.; Ganz, T.; Parks, W.C. Differential Processing of {alpha}- and {beta}-Defensin Precursors by Matrix Metalloproteinase-7 (MMP-7). *J. Biol. Chem.* **2009**, *284*, 8301–8311. [CrossRef] [PubMed]

15. Jiang, Y.; Yang, D.; Li, W.; Wang, B.; Jiang, Z.; Li, M. Antiviral activity of recombinant mouse beta-defensin 3 against influenza A virus in vitro and in vivo. *Antivir. Chem. Chemother.* **2012**, *22*, 255–262. [CrossRef] [PubMed]

16. Park, M.S.; Kim, J.I.; Lee, I.; Park, S.; Bae, J.Y.; Park, M.S. Towards the Application of Human Defensins as Antivirals. *Biomol. Ther.* **2018**, *26*, 242–254. [CrossRef] [PubMed]

17. Wilson, S.S.; Wiens, M.E.; Holly, M.K.; Smith, J.G. Defensins at the Mucosal Surface: Latest Insights into Defensin-Virus Interactions. *J. Virol.* **2016**, *90*, 5216–5218. [CrossRef] [PubMed]

18. Pachon-Ibanez, M.E.; Smani, Y.; Pachon, J.; Sanchez-Cespedes, J. Perspectives for clinical use of engineered human host defense antimicrobial peptides. *FEMS Microbiol. Rev.* **2017**, *41*, 323–342. [CrossRef]

19. Castaneda-Sanchez, J.I.; Dominguez-Martinez, D.A.; Olivar-Espinosa, N.; Garcia-Perez, B.E.; Lorono-Pino, M.A.; Luna-Herrera, J.; Salazar, M.I. Expression of Antimicrobial Peptides in Human Monocytic Cells and Neutrophils in Response to Dengue Virus Type 2. *Intervirology* **2016**, *59*, 8–19. [CrossRef]

20. Nagaoka, I.; Niyonsaba, F.; Tsutsumi-Ishii, Y.; Tamura, H.; Hirata, M. Evaluation of the effect of human beta-defensins on neutrophil apoptosis. *Int. Immunol.* **2008**, *20*, 543–553. [CrossRef]

21. Smith, J.G.; Nemerow, G.R. Mechanism of adenovirus neutralization by Human alpha-defensins. *Cell Host Microbe* **2008**, *3*, 11–19. [CrossRef]

22. Hiwarkar, P.; Kosulin, K.; Cesaro, S.; Mikulska, M.; Styczynski, J.; Wynn, R.; Lion, T. Management of adenovirus infection in patients after haematopoietic stem cell transplantation: State-of-the-art and real-life current approach: A position statement on behalf of the Infectious Diseases Working Party of the European Society of Blood and Marrow Transplantation. *Rev. Med. Virol.* **2018**, *28*, e1980. [CrossRef]

23. Smith, J.G.; Silvestry, M.; Lindert, S.; Lu, W.; Nemerow, G.R.; Stewart, P.L. Insight into the mechanisms of adenovirus capsid disassembly from studies of defensin neutralization. *PLoS Pathog.* **2010**, *6*, e1000959. [CrossRef]

24. Nguyen, E.K.; Nemerow, G.R.; Smith, J.G. Direct evidence from single-cell analysis that human {alpha}-defensins block adenovirus uncoating to neutralize infection. *J. Virol.* **2010**, *84*, 4041–4049. [CrossRef]

25. Doss, M.; White, M.R.; Tecle, T.; Gantz, D.; Crouch, E.C.; Jung, G.; Ruchala, P.; Waring, A.J.; Lehrer, R.I.; Hartshorn, K.L. Interactions of alpha-, beta-, and theta-defensins with influenza A virus and surfactant protein D. *J. Immunol.* **2009**, *182*, 7878–7887. [CrossRef]

26. Hartshorn, K.L.; White, M.R.; Tecle, T.; Holmskov, U.; Crouch, E.C. Innate defense against influenza A virus: Activity of human neutrophil defensins and interactions of defensins with surfactant protein D. *J. Immunol.* **2006**, *176*, 6962–6972. [CrossRef]

27. Ryan, L.K.; Dai, J.; Yin, Z.; Megjugorac, N.; Uhlhorn, V.; Yim, S.; Schwartz, K.D.; Abrahams, J.M.; Diamond, G.; Fitzgerald-Bocarsly, P. Modulation of human beta-defensin-1 (hBD-1) in plasmacytoid dendritic cells (PDC), monocytes, and epithelial cells by influenza virus, Herpes simplex virus, and Sendai virus and its possible role in innate immunity. *J. Leukoc. Biol.* **2011**, *90*, 343–356. [CrossRef]

28. Semple, F.; Webb, S.; Li, H.N.; Patel, H.B.; Perretti, M.; Jackson, I.J.; Gray, M.; Davidson, D.J.; Dorin, J.R. Human beta-defensin 3 has immunosuppressive activity in vitro and in vivo. *Eur. J. Immunol.* **2010**, *40*, 1073–1078. [CrossRef]

29. Salvatore, M.; Garcia-Sastre, A.; Ruchala, P.; Lehrer, R.I.; Chang, T.; Klotman, M.E. alpha-Defensin inhibits influenza virus replication by cell-mediated mechanism(s). *J. Infect. Dis.* **2007**, *196*, 835–843. [CrossRef]

30. Pace, B.T.; Lackner, A.A.; Porter, E.; Pahar, B. The Role of Defensins in HIV Pathogenesis. *Mediat. Inflamm.* **2017**, *2017*, 5186904. [CrossRef] [PubMed]

31. Chang, T.L.; Vargas, J., Jr.; DelPortillo, A.; Klotman, M.E. Dual role of alpha-defensin-1 in anti-HIV-1 innate immunity. *J. Clin. Investig.* **2005**, *115*, 765–773. [CrossRef]

32. Furci, L.; Tolazzi, M.; Sironi, F.; Vassena, L.; Lusso, P. Inhibition of HIV-1 infection by human alpha-defensin-5, a natural antimicrobial peptide expressed in the genital and intestinal mucosae. *PLoS ONE* **2012**, *7*, e45208. [CrossRef]

33. Hazrati, E.; Galen, B.; Lu, W.; Wang, W.; Ouyang, Y.; Keller, M.J.; Lehrer, R.I.; Herold, B.C. Human alpha- and beta-defensins block multiple steps in herpes simplex virus infection. *J. Immunol.* **2006**, *177*, 8658–8666. [CrossRef]

34. Yasin, B.; Wang, W.; Pang, M.; Cheshenko, N.; Hong, T.; Waring, A.J.; Herold, B.C.; Wagar, E.A.; Lehrer, R.I. Theta defensins protect cells from infection by herpes simplex virus by inhibiting viral adhesion and entry. *J. Virol.* **2004**, *78*, 5147–5156. [CrossRef]

35. Widegren, H.; Andersson, M.; Borgeat, P.; Flamand, L.; Johnston, S.; Greiff, L. LTB4 increases nasal neutrophil activity and conditions neutrophils to exert antiviral effects. *Respir. Med.* **2011**, *105*, 997–1006. [CrossRef]

36. Kota, S.; Sabbah, A.; Chang, T.H.; Harnack, R.; Xiang, Y.; Meng, X.; Bose, S. Role of human beta-defensin-2 during tumor necrosis factor-alpha/NF-kappaB-mediated innate antiviral response against human respiratory syncytial virus. *J. Biol. Chem.* **2008**, *283*, 22417–22429. [CrossRef]

37. Mangoni, M.L.; McDermott, A.M.; Zasloff, M. Antimicrobial peptides and wound healing: Biological and therapeutic considerations. *Exp. Dermatol.* **2016**, *25*, 167–173. [CrossRef]

38. Buck, C.B.; Day, P.M.; Thompson, C.D.; Lubkowski, J.; Lu, W.; Lowy, D.R.; Schiller, J.T. Human alpha-defensins block papillomavirus infection. *Proc. Natl. Acad. Sci. USA* **2006**, *103*, 1516–1521. [CrossRef]

39. Wiens, M.E.; Smith, J.G. Alpha-defensin HD5 inhibits furin cleavage of human papillomavirus 16 L2 to block infection. *J. Virol.* **2015**, *89*, 2866–2874. [CrossRef]

40. Wiens, M.E.; Smith, J.G. Alpha-Defensin HD5 Inhibits Human Papillomavirus 16 Infection via Capsid Stabilization and Redirection to the Lysosome. *MBio* **2017**, *8*. [CrossRef]

41. Matsumura, T.; Sugiyama, N.; Murayama, A.; Yamada, N.; Shiina, M.; Asabe, S.; Wakita, T.; Imawari, M.; Kato, T. Antimicrobial peptide LL-37 attenuates infection of hepatitis C virus. *Hepatol. Res.* **2016**, *46*, 924–932. [CrossRef]

42. Agerberth, B.; Gunne, H.; Odeberg, J.; Kogner, P.; Boman, H.G.; Gudmundsson, G.H. FALL-39, a putative human peptide antibiotic, is cysteine-free and expressed in bone marrow and testis. *Proc. Natl. Acad. Sci. USA* **1995**, *92*, 195–199. [CrossRef]

43. Alagarasu, K.; Patil, P.S.; Shil, P.; Seervi, M.; Kakade, M.B.; Tillu, H.; Salunke, A. In-vitro effect of human cathelicidin antimicrobial peptide LL-37 on dengue virus type 2. *Peptides* **2017**, *92*, 23–30. [CrossRef]

44. Tripathi, S.; Wang, G.; White, M.; Qi, L.; Taubenberger, J.; Hartshorn, K.L. Antiviral Activity of the Human Cathelicidin, LL-37, and Derived Peptides on Seasonal and Pandemic Influenza A Viruses. *PLoS ONE* **2015**, *10*, e0124706. [CrossRef]

45. Bandurska, K.; Berdowska, A.; Barczynska-Felusiak, R.; Krupa, P. Unique features of human cathelicidin LL-37. *Biofactors* **2015**, *41*, 289–300. [CrossRef]

46. He, M.; Zhang, H.; Li, Y.; Wang, G.; Tang, B.; Zhao, J.; Huang, Y.; Zheng, J. Cathelicidin-Derived Antimicrobial Peptides Inhibit Zika Virus Through Direct Inactivation and Interferon Pathway. *Front. Immunol.* **2018**, *9*, 722. [CrossRef]

47. Dean, R.E.; O'Brien, L.M.; Thwaite, J.E.; Fox, M.A.; Atkins, H.; Ulaeto, D.O. A carpet-based mechanism for direct antimicrobial peptide activity against vaccinia virus membranes. *Peptides* **2010**, *31*, 1966–1972. [CrossRef]

48. Ulaeto, D.O.; Morris, C.J.; Fox, M.A.; Gumbleton, M.; Beck, K. Destabilization of alpha-Helical Structure in Solution Improves Bactericidal Activity of Antimicrobial Peptides: Opposite Effects on Bacterial and Viral Targets. *Antimicrob. Agents Chemother.* **2016**, *60*, 1984–1991. [CrossRef]

49. Tripathi, S.; Verma, A.; Kim, E.J.; White, M.R.; Hartshorn, K.L. LL-37 modulates human neutrophil responses to influenza A virus. *J. Leukoc. Biol.* **2014**, *96*, 931–938. [CrossRef]

50. Barlow, P.G.; Svoboda, P.; Mackellar, A.; Nash, A.A.; York, I.A.; Pohl, J.; Davidson, D.J.; Donis, R.O. Antiviral activity and increased host defense against influenza infection elicited by the human cathelicidin LL-37. *PLoS ONE* **2011**, *6*, e25333. [CrossRef]

51. LeMessurier, K.S.; Lin, Y.; McCullers, J.A.; Samarasinghe, A.E. Antimicrobial peptides alter early immune response to influenza A virus infection in C57BL/6 mice. *Antiviral Res.* **2016**, *133*, 208–217. [CrossRef]

52. Bergman, P.; Walter-Jallow, L.; Broliden, K.; Agerberth, B.; Soderlund, J. The antimicrobial peptide LL-37 inhibits HIV-1 replication. *Curr. HIV Res.* **2007**, *5*, 410–415. [CrossRef]

53. Wong, J.H.; Legowska, A.; Rolka, K.; Ng, T.B.; Hui, M.; Cho, C.H.; Lam, W.W.; Au, S.W.; Gu, O.W.; Wan, D.C. Effects of cathelicidin and its fragments on three key enzymes of HIV-1. *Peptides* **2011**, *32*, 1117–1122. [CrossRef]

54. Honda, J.R.; Connick, E.; MaWhinney, S.; Chan, E.D.; Flores, S.C. Plasma LL-37 correlates with vitamin D and is reduced in human immunodeficiency virus-1 infected individuals not receiving antiretroviral therapy. *J. Med. Microbiol.* **2014**, *63*, 997–1003. [CrossRef]

55. Jadhav, N.J.; Patil, P.S.; Alagarasu, K. Effect of full-length and truncated variants of LL-37 on dengue virus infection and immunomodulatory effects of LL-37 in dengue virus infected U937-DC-SIGN cells. *Int. J. Pept. Res. Ther.* **2019**, 1–9. [CrossRef]

56. Lopez-Gonzalez, M.; Meza-Sanchez, D.; Garcia-Cordero, J.; Bustos-Arriaga, J.; Velez-Del Valle, C.; Marsch-Moreno, M.; Castro-Jimenez, T.; Flores-Romo, L.; Santos-Argumedo, L.; Gutierrez-Castaneda, B.; et al. Human keratinocyte cultures (HaCaT) can be infected by DENV, triggering innate immune responses that include IFNlambda and LL37. *Immunobiology* **2018**, *223*, 608–617. [CrossRef]

57. Currie, S.M.; Findlay, E.G.; McHugh, B.J.; Mackellar, A.; Man, T.; Macmillan, D.; Wang, H.; Fitch, P.M.; Schwarze, J.; Davidson, D.J. The human cathelicidin LL-37 has antiviral activity against respiratory syncytial virus. *PLoS ONE* **2013**, *8*, e73659. [CrossRef]

58. Mansbach, J.M.; Piedra, P.A.; Borregaard, N.; Martineau, A.R.; Neuman, M.I.; Espinola, J.A.; Camargo, C.A., Jr. Serum cathelicidin level is associated with viral etiology and severity of bronchiolitis. *J. Allergy Clin. Immunol.* **2012**, *130*, 1007–1008. [CrossRef]

59. Sousa, F.H.; Casanova, V.; Findlay, F.; Stevens, C.; Svoboda, P.; Pohl, J.; Proudfoot, L.; Barlow, P.G. Cathelicidins display conserved direct antiviral activity towards rhinovirus. *Peptides* **2017**, *95*, 76–83. [CrossRef]

60. Schogler, A.; Muster, R.J.; Kieninger, E.; Casaulta, C.; Tapparel, C.; Jung, A.; Moeller, A.; Geiser, T.; Regamey, N.; Alves, M.P. Vitamin D represses rhinovirus replication in cystic fibrosis cells by inducing LL-37. *Eur. Respir. J.* **2016**, *47*, 520–530. [CrossRef]

61. Howell, M.D.; Jones, J.F.; Kisich, K.O.; Streib, J.E.; Gallo, R.L.; Leung, D.Y. Selective killing of vaccinia virus by LL-37: Implications for eczema vaccinatum. *J. Immunol.* **2004**, *172*, 1763–1767. [CrossRef]

62. Gordon, Y.J.; Huang, L.C.; Romanowski, E.G.; Yates, K.A.; Proske, R.J.; McDermott, A.M. Human cathelicidin (LL-37), a multifunctional peptide, is expressed by ocular surface epithelia and has potent antibacterial and antiviral activity. *Curr. Eye Res.* **2005**, *30*, 385–394. [CrossRef]

63. Lee, C.J.; Buznyk, O.; Kuffova, L.; Rajendran, V.; Forrester, J.V.; Phopase, J.; Islam, M.M.; Skog, M.; Ahlqvist, J.; Griffith, M. Cathelicidin LL-37 and HSV-1 Corneal Infection: Peptide Versus Gene Therapy. *Transl. Vis. Sci. Technol.* **2014**, *3*, 4. [CrossRef]

64. Ogawa, Y.; Kawamura, T.; Matsuzawa, T.; Aoki, R.; Gee, P.; Yamashita, A.; Moriishi, K.; Yamasaki, K.; Koyanagi, Y.; Blauvelt, A.; et al. Antimicrobial peptide LL-37 produced by HSV-2-infected keratinocytes enhances HIV infection of Langerhans cells. *Cell Host Microbe* **2013**, *13*, 77–86. [CrossRef]

65. Mittal, R.; Nguyen, D.; Debs, L.H.; Patel, A.P.; Liu, G.; Jhaveri, V.M.; SI, S.K.; Mittal, J.; Bandstra, E.S.; Younis, R.T.; et al. Zika Virus: An Emerging Global Health Threat. *Front. Cell Infect. Microbiol.* **2017**, *7*, 486. [CrossRef]

66. Petersen, L.R.; Jamieson, D.J.; Powers, A.M.; Honein, M.A. Zika Virus. *N. Engl. J. Med.* **2016**, *374*, 1552–1563. [CrossRef]

67. Yost, S.A.; Wang, Y.; Marcotrigiano, J. Hepatitis C Virus Envelope Glycoproteins: A Balancing Act of Order and Disorder. *Front. Immunol.* **2018**, *9*, 1917. [CrossRef]

68. Weaver, S.C.; Ferro, C.; Barrera, R.; Boshell, J.; Navarro, J.C. Venezuelan equine encephalitis. *Annu. Rev. Entomol.* **2004**, *49*, 141–174. [CrossRef]

69. Hawley, R.J.; Eitzen, E.M., Jr. Biological weapons—A primer for microbiologists. *Annu. Rev. Microbiol.* **2001**, *55*, 235–253. [CrossRef]

70. Ahmed, A.; Siman-Tov, G.; Keck, F.; Kortchak, S.; Bakovic, A.; Risner, K.; Lu, T.K.; Bhalla, N.; de la Fuente-Nunez, C.; Narayanan, A. Human cathelicidin peptide LL-37 as a therapeutic antiviral targeting Venezuelan equine encephalitis virus infections. *Antiviral Res.* **2019**, *164*, 61–69. [CrossRef]

71. Berlutti, F.; Pantanella, F.; Natalizi, T.; Frioni, A.; Paesano, R.; Polimeni, A.; Valenti, P. Antiviral properties of lactoferrin—A natural immunity molecule. *Molecules* **2011**, *16*, 6992–7018. [CrossRef]

72. Gonzalez-Chavez, S.A.; Arevalo-Gallegos, S.; Rascon-Cruz, Q. Lactoferrin: Structure, function and applications. *Int. J. Antimicrob. Agents* **2009**, *33*, 301.e1–301.e8. [CrossRef]

73. Wakabayashi, H.; Oda, H.; Yamauchi, K.; Abe, F. Lactoferrin for prevention of common viral infections. *J. Infect. Chemother.* **2014**, *20*, 666–671. [CrossRef]

74. Legrand, D.; Pierce, A.; Elass, E.; Carpentier, M.; Mariller, C.; Mazurier, J. Lactoferrin structure and functions. *Adv. Exp. Med. Biol.* **2008**, *606*, 163–194. [CrossRef]

75. Teng, C.T. Lactoferrin gene expression and regulation: An overview. *Biochem. Cell Biol.* **2002**, *80*, 7–16. [CrossRef]

76. Gifford, J.L.; Hunter, H.N.; Vogel, H.J. Lactoferricin: A lactoferrin-derived peptide with antimicrobial, antiviral, antitumor and immunological properties. *Cell. Mol. Life Sci.* **2005**, *62*, 2588–2598. [CrossRef]

77. Roseanu, A.; Brock, J.H. What are the structure and the biological function of lactoferrin in human breast milk? *IUBMB Life* **2006**, *58*, 235–237. [CrossRef]

78. Portelli, J.; Gordon, A.; May, J.T. Effect of compounds with antibacterial activities in human milk on respiratory syncytial virus and cytomegalovirus in vitro. *J. Med. Microbiol.* **1998**, *47*, 1015–1018. [CrossRef]

79. Sano, H.; Nagai, K.; Tsutsumi, H.; Kuroki, Y. Lactoferrin and surfactant protein A exhibit distinct binding specificity to F protein and differently modulate respiratory syncytial virus infection. *Eur. J. Immunol.* **2003**, *33*, 2894–2902. [CrossRef]

80. Grover, M.; Giouzeppos, O.; Schnagl, R.D.; May, J.T. Effect of human milk prostaglandins and lactoferrin on respiratory syncytial virus and rotavirus. *Acta Paediatr.* **1997**, *86*, 315–316. [CrossRef]

81. Ammendolia, M.G.; Agamennone, M.; Pietrantoni, A.; Lannutti, F.; Siciliano, R.A.; De Giulio, B.; Amici, C.; Superti, F. Bovine lactoferrin-derived peptides as novel broad-spectrum inhibitors of influenza virus. *Pathog. Glob. Health* **2012**, *106*, 12–19. [CrossRef]

82. Pietrantoni, A.; Dofrelli, E.; Tinari, A.; Ammendolia, M.G.; Puzelli, S.; Fabiani, C.; Donatelli, I.; Superti, F. Bovine lactoferrin inhibits influenza A virus induced programmed cell death in vitro. *Biometals* **2010**, *23*, 465–475. [CrossRef]

83. Pietrantoni, A.; Ammendolia, M.G.; Superti, F. Bovine lactoferrin: Involvement of metal saturation and carbohydrates in the inhibition of influenza virus infection. *Biochem. Cell Biol.* **2012**, *90*, 442–448. [CrossRef]

84. Scala, M.C.; Sala, M.; Pietrantoni, A.; Spensiero, A.; Di Micco, S.; Agamennone, M.; Bertamino, A.; Novellino, E.; Bifulco, G.; Gomez-Monterrey, I.M.; et al. Lactoferrin-derived Peptides Active towards Influenza: Identification of Three Potent Tetrapeptide Inhibitors. *Sci. Rep.* **2017**, *7*, 10593. [CrossRef]

85. Taha, S.H.; Mehrez, M.A.; Sitohy, M.Z.; Abou Dawood, A.G.; Abd-El Hamid, M.M.; Kilany, W.H. Effectiveness of esterified whey proteins fractions against Egyptian Lethal Avian Influenza A (H5N1). *Virol. J.* **2010**, *7*, 330. [CrossRef]

86. Yamamoto, H.; Ura, Y.; Tanemura, M.; Koyama, A.; Takano, S.; Uematsu, J.; Kawano, M.; Tsurudome, M.; O'Brian, M.; Komada, H. Inhibitory Effect of Bovine Lactoferrin on Human Parainfluenza Virus Type 2 Infection. *J. Health Sci.* **2010**, *56*, 613–617. [CrossRef]

87. Arnold, D.; Di Biase, A.M.; Marchetti, M.; Pietrantoni, A.; Valenti, P.; Seganti, L.; Superti, F. Antiadenovirus activity of milk proteins: Lactoferrin prevents viral infection. *Antivir. Res.* **2002**, *53*, 153–158. [CrossRef]

88. Di Biase, A.M.; Pietrantoni, A.; Tinari, A.; Siciliano, R.; Valenti, P.; Antonini, G.; Seganti, L.; Superti, F. Heparin-interacting sites of bovine lactoferrin are involved in anti-adenovirus activity. *J. Med. Virol.* **2003**, *69*, 495–502. [CrossRef]

89. Pietrantoni, A.; Di Biase, A.M.; Tinari, A.; Marchetti, M.; Valenti, P.; Seganti, L.; Superti, F. Bovine lactoferrin inhibits adenovirus infection by interacting with viral structural polypeptides. *Antimicrob. Agents Chemother.* **2003**, *47*, 2688–2691. [CrossRef]

90. Johansson, C.; Jonsson, M.; Marttila, M.; Persson, D.; Fan, X.L.; Skog, J.; Frangsmyr, L.; Wadell, G.; Arnberg, N. Adenoviruses use lactoferrin as a bridge for CAR-independent binding to and infection of epithelial cells. *J. Virol.* **2007**, *81*, 954–963. [CrossRef]

91. Hasegawa, K.; Motsuchi, W.; Tanaka, S.; Dosako, S. Inhibition with lactoferrin of in vitro infection with human herpes virus. *Jpn. J. Med. Sci. Biol.* **1994**, *47*, 73–85. [CrossRef]

92. Marchetti, M.; Longhi, C.; Conte, M.P.; Pisani, S.; Valenti, P.; Seganti, L. Lactoferrin inhibits herpes simplex virus type 1 adsorption to Vero cells. *Antivir. Res.* **1996**, *29*, 221–231. [CrossRef]

93. Marchetti, M.; Pisani, S.; Antonini, G.; Valenti, P.; Seganti, L.; Orsi, N. Metal complexes of bovine lactoferrin inhibit in vitro replication of herpes simplex virus type 1 and 2. *Biometals* **1998**, *11*, 89–94. [CrossRef]

94. Marr, A.K.; Jenssen, H.; Moniri, M.R.; Hancock, R.E.; Pante, N. Bovine lactoferrin and lactoferricin interfere with intracellular trafficking of Herpes simplex virus-1. *Biochimie* **2009**, *91*, 160–164. [CrossRef]

95. Jenssen, H.; Sandvik, K.; Andersen, J.H.; Hancock, R.E.; Gutteberg, T.J. Inhibition of HSV cell-to-cell spread by lactoferrin and lactoferricin. *Antivir. Res.* **2008**, *79*, 192–198. [CrossRef]

96. Tanaka, K.; Ikeda, M.; Nozaki, A.; Kato, N.; Tsuda, H.; Saito, S.; Sekihara, H. Lactoferrin inhibits hepatitis C virus viremia in patients with chronic hepatitis C: A pilot study. *Jpn. J. Cancer Res.* **1999**, *90*, 367–371. [CrossRef]

97. Ueno, H.; Sato, T.; Yamamoto, S.; Tanaka, K.; Ohkawa, S.; Takagi, H.; Yokosuka, O.; Furuse, J.; Saito, H.; Sawaki, A.; et al. Randomized, double-blind, placebo-controlled trial of bovine lactoferrin in patients with chronic hepatitis C. *Cancer Sci.* **2006**, *97*, 1105–1110. [CrossRef]

98. Kaito, M.; Iwasa, M.; Fujita, N.; Kobayashi, Y.; Kojima, Y.; Ikoma, J.; Imoto, I.; Adachi, Y.; Hamano, H.; Yamauchi, K. Effect of lactoferrin in patients with chronic hepatitis C: Combination therapy with interferon and ribavirin. *J. Gastroenterol. Hepatol.* **2007**, *22*, 1894–1897. [CrossRef]

99. Yi, M.; Kaneko, S.; Yu, D.Y.; Murakami, S. Hepatitis C virus envelope proteins bind lactoferrin. *J. Virol.* **1997**, *71*, 5997–6002.

100. El-Fakharany, E.M.; Sanchez, L.; Al-Mehdar, H.A.; Redwan, E.M. Effectiveness of human, camel, bovine and sheep lactoferrin on the hepatitis C virus cellular infectivity: Comparison study. *Virol. J.* **2013**, *10*, 199. [CrossRef]

101. Hara, K.; Ikeda, M.; Saito, S.; Matsumoto, S.; Numata, K.; Kato, N.; Tanaka, K.; Sekihara, H. Lactoferrin inhibits hepatitis B virus infection in cultured human hepatocytes. *Hepatol. Res.* **2002**, *24*, 228. [CrossRef]

102. Li, S.; Huang, G.; Zhou, H.; Liu, N. Study of inhibition effect of bovine lactoferrin in vitro on hepatitis B surface antigen. *Wei Sheng Yan Jiu* **2008**, *37*, 196–198.

103. Li, S.; Zhou, H.; Huang, G.; Liu, N. Inhibition of HBV infection by bovine lactoferrin and iron-, zinc-saturated lactoferrin. *Med. Microbiol. Immunol.* **2009**, *198*, 19–25. [CrossRef]

104. Defer, M.C.; Dugas, B.; Picard, O.; Damais, C. Impairment of circulating lactoferrin in HIV-1 infection. *Cell. Mol. Biol.* **1995**, *41*, 417–421.

105. Viani, R.M.; Gutteberg, T.J.; Lathey, J.L.; Spector, S.A. Lactoferrin inhibits HIV-1 replication in vitro and exhibits synergy when combined with zidovudine. *AIDS* **1999**, *13*, 1273–1274. [CrossRef]

106. Berkhout, B.; van Wamel, J.L.; Beljaars, L.; Meijer, D.K.; Visser, S.; Floris, R. Characterization of the anti-HIV effects of native lactoferrin and other milk proteins and protein-derived peptides. *Antivir. Res.* **2002**, *55*, 341–355. [CrossRef]

107. Zuccotti, G.V.; Salvini, F.; Riva, E.; Agostoni, C. Oral lactoferrin in HIV-1 vertically infected children: An observational follow-up of plasma viral load and immune parameters. *J. Int. Med. Res.* **2006**, *34*, 88–94. [CrossRef]

108. Salvini, F.; Gemmerllaro, L.; Bettiga, C.; Ruscitto, A.; Zuccotti, G.V.; Giovannini, M. 321 Immunological and Virological Effects of Bovine Lactoferrin in HIV-1 Vertically Infected Children. *Pediatric Res.* **2005**, *58*, 409. [CrossRef]

109. Murphy, M.E.; Kariwa, H.; Mizutani, T.; Tanabe, H.; Yoshimatsu, K.; Arikawa, J.; Takashima, I. Characterization of in vitro and in vivo antiviral activity of lactoferrin and ribavirin upon hantavirus. *J. Vet. Med. Sci.* **2001**, *63*, 637–645. [CrossRef]

110. Murphy, M.E.; Kariwa, H.; Mizutani, T.; Yoshimatsu, K.; Arikawa, J.; Takashima, I. In vitro antiviral activity of lactoferrin and ribavirin upon hantavirus. *Arch. Virol.* **2000**, *145*, 1571–1582. [CrossRef]

111. Drobni, P.; Naslund, J.; Evander, M. Lactoferrin inhibits human papillomavirus binding and uptake in vitro. *Antivir. Res.* **2004**, *64*, 63–68. [CrossRef]

112. Mistry, N.; Drobni, P.; Naslund, J.; Sunkari, V.G.; Jenssen, H.; Evander, M. The anti-papillomavirus activity of human and bovine lactoferricin. *Antivir. Res.* **2007**, *75*, 258–265. [CrossRef]

113. Superti, F.; Ammendolia, M.G.; Valenti, P.; Seganti, L. Antirotaviral activity of milk proteins: Lactoferrin prevents rotavirus infection in the enterocyte-like cell line HT-29. *Med. Microbiol. Immunol.* **1997**, *186*, 83–91. [CrossRef]

114. Yen, M.H.; Chiu, C.H.; Huang, Y.C.; Lin, T.Y. Effects of lactoferrin-containing formula in the prevention of enterovirus and rotavirus infection and impact on serum cytokine levels: A randomized trial. *Chang. Gung Med. J.* **2011**, *34*, 395–402.

115. Marchetti, M.; Superti, F.; Ammendolia, M.G.; Rossi, P.; Valenti, P.; Seganti, L. Inhibition of poliovirus type 1 infection by iron-, manganese- and zinc-saturated lactoferrin. *Med. Microbiol. Immunol.* **1999**, *187*, 199–204. [CrossRef]

116. Waarts, B.L.; Aneke, O.J.; Smit, J.M.; Kimata, K.; Bittman, R.; Meijer, D.K.; Wilschut, J. Antiviral activity of human lactoferrin: Inhibition of alphavirus interaction with heparan sulfate. *Virology* **2005**, *333*, 284–292. [CrossRef]

117. Chien, Y.J.; Chen, W.J.; Hsu, W.L.; Chiou, S.S. Bovine lactoferrin inhibits Japanese encephalitis virus by binding to heparan sulfate and receptor for low density lipoprotein. *Virology* **2008**, *379*, 143–151. [CrossRef]

118. Chen, J.M.; Fan, Y.C.; Lin, J.W.; Chen, Y.Y.; Hsu, W.L.; Chiou, S.S. Bovine Lactoferrin Inhibits Dengue Virus Infectivity by Interacting with Heparan Sulfate, Low-Density Lipoprotein Receptor, and DC-SIGN. *Int. J. Mol. Sci.* **2017**, *18*, 1957. [CrossRef]

119. Rosenberg, H.F. Eosinophil-derived neurotoxin / RNase 2: Connecting the past, the present and the future. *Curr. Pharm. Biotechnol.* **2008**, *9*, 135–140. [CrossRef]

120. Rosenberg, H.F.; Domachowske, J.B. Eosinophils, eosinophil ribonucleases, and their role in host defense against respiratory virus pathogens. *J. Leukoc. Biol.* **2001**, *70*, 691–698.

121. Domachowske, J.B.; Dyer, K.D.; Bonville, C.A.; Rosenberg, H.F. Recombinant human eosinophil-derived neurotoxin/RNase 2 functions as an effective antiviral agent against respiratory syncytial virus. *J. Infect. Dis.* **1998**, *177*, 1458–1464. [CrossRef]

122. Pifferi, M.; Ragazzo, V.; Caramella, D.; Baldini, G. Eosinophil cationic protein in infants with respiratory syncytial virus bronchiolitis: Predictive value for subsequent development of persistent wheezing. *Pediatr. Pulmonol.* **2001**, *31*, 419–424. [CrossRef]

123. Rugeles, M.T.; Trubey, C.M.; Bedoya, V.I.; Pinto, L.A.; Oppenheim, J.J.; Rybak, S.M.; Shearer, G.M. Ribonuclease is partly responsible for the HIV-1 inhibitory effect activated by HLA alloantigen recognition. *AIDS* **2003**, *17*, 481–486. [CrossRef]

124. Bedoya, V.I.; Boasso, A.; Hardy, A.W.; Rybak, S.; Shearer, G.M.; Rugeles, M.T. Ribonucleases in HIV type 1 inhibition: Effect of recombinant RNases on infection of primary T cells and immune activation-induced RNase gene and protein expression. *AIDS Res. Hum. Retrovir.* **2006**, *22*, 897–907. [CrossRef]

125. Marischen, L.; Wesch, D.; Schroder, J.M.; Wiedow, O.; Kabelitz, D. Human gammadelta T cells produce the protease inhibitor and antimicrobial peptide elafin. *Scand. J. Immunol.* **2009**, *70*, 547–552. [CrossRef]

126. Drannik, A.G.; Nag, K.; Sallenave, J.M.; Rosenthal, K.L. Antiviral activity of trappin-2 and elafin in vitro and in vivo against genital herpes. *J. Virol.* **2013**, *87*, 7526–7538. [CrossRef]

127. Ying, Q.L.; Simon, S.R. Kinetics of the inhibition of human leukocyte elastase by elafin, a 6-kilodalton elastase-specific inhibitor from human skin. *Biochemistry* **1993**, *32*, 1866–1874. [CrossRef]

128. Guyot, N.; Butler, M.W.; McNally, P.; Weldon, S.; Greene, C.M.; Levine, R.L.; O'Neill, S.J.; Taggart, C.C.; McElvaney, N.G. Elafin, an elastase-specific inhibitor, is cleaved by its cognate enzyme neutrophil elastase in sputum from individuals with cystic fibrosis. *J. Biol. Chem.* **2008**, *283*, 32377–32385. [CrossRef]

129. Doumas, S.; Kolokotronis, A.; Stefanopoulos, P. Anti-inflammatory and antimicrobial roles of secretory leukocyte protease inhibitor. *Infect. Immun.* **2005**, *73*, 1271–1274. [CrossRef]

130. King, A.E.; Wheelhouse, N.; Cameron, S.; McDonald, S.E.; Lee, K.F.; Entrican, G.; Critchley, H.O.; Horne, A.W. Expression of secretory leukocyte protease inhibitor and elafin in human fallopian tube and in an in-vitro model of Chlamydia trachomatis infection. *Hum. Reprod.* **2009**, *24*, 679–686. [CrossRef]

131. Skott, P.; Lucht, E.; Ehnlund, M.; Bjorling, E. Inhibitory function of secretory leukocyte proteinase inhibitor (SLPI) in human saliva is HIV-1 specific and varies with virus tropism. *Oral Dis.* **2002**, *8*, 160–167. [CrossRef]

132. Ghosh, M.; Shen, Z.; Fahey, J.V.; Cu-Uvin, S.; Mayer, K.; Wira, C.R. Trappin-2/Elafin: A novel innate anti-human immunodeficiency virus-1 molecule of the human female reproductive tract. *Immunology* **2010**, *129*, 207–219. [CrossRef]

133. Iqbal, S.M.; Ball, T.B.; Levinson, P.; Maranan, L.; Jaoko, W.; Wachihi, C.; Pak, B.J.; Podust, V.N.; Broliden, K.; Hirbod, T.; et al. Elevated elafin/trappin-2 in the female genital tract is associated with protection against HIV acquisition. *AIDS* **2009**, *23*, 1669–1677. [CrossRef]

134. Gonzalez, S.M.; Taborda, N.A.; Feria, M.G.; Arcia, D.; Aguilar-Jimenez, W.; Zapata, W.; Rugeles, M.T. High Expression of Antiviral Proteins in Mucosa from Individuals Exhibiting Resistance to Human Immunodeficiency Virus. *PLoS ONE* **2015**, *10*, e0131139. [CrossRef]

135. Ghosh, M.; Daniels, J.; Pyra, M.; Juzumaite, M.; Jais, M.; Murphy, K.; Taylor, T.N.; Kassaye, S.; Benning, L.; Cohen, M.; et al. Impact of chronic sexual abuse and depression on inflammation and wound healing in the female reproductive tract of HIV-uninfected and HIV-infected women. *PLoS ONE* **2018**, *13*, e0198412. [CrossRef]

136. Walker, A.P.; Partridge, J.; Srai, S.K.; Dooley, J.S. Hepcidin: What every gastroenterologist should know. *Gut* **2004**, *53*, 624–627. [CrossRef]

137. Armitage, A.E.; Eddowes, L.A.; Gileadi, U.; Cole, S.; Spottiswoode, N.; Selvakumar, T.A.; Ho, L.P.; Townsend, A.R.; Drakesmith, H. Hepcidin regulation by innate immune and infectious stimuli. *Blood* **2011**, *118*, 4129–4139. [CrossRef]

138. Michels, K.; Nemeth, E.; Ganz, T.; Mehrad, B. Hepcidin and Host Defense against Infectious Diseases. *PLoS Pathog.* **2015**, *11*, e1004998. [CrossRef]

139. Rodriguez, R.; Jung, C.L.; Gabayan, V.; Deng, J.C.; Ganz, T.; Nemeth, E.; Bulut, Y. Hepcidin induction by pathogens and pathogen-derived molecules is strongly dependent on interleukin-6. *Infect. Immun.* **2014**, *82*, 745–752. [CrossRef]

140. Wang, X.H.; Cheng, P.P.; Jiang, F.; Jiao, X.Y. The effect of hepatitis B virus infection on hepcidin expression in hepatitis B patients. *Ann. Clin. Lab. Sci.* **2013**, *43*, 126–134.

141. Armitage, A.E.; Stacey, A.R.; Giannoulatou, E.; Marshall, E.; Sturges, P.; Chatha, K.; Smith, N.M.; Huang, X.; Xu, X.; Pasricha, S.R.; et al. Distinct patterns of hepcidin and iron regulation during HIV-1, HBV, and HCV infections. *Proc. Natl. Acad. Sci. USA* **2014**, *111*, 12187–12192. [CrossRef]

142. Augustyniak, D.; Nowak, J.; Lundy, F.T. Direct and indirect antimicrobial activities of neuropeptides and their therapeutic potential. *Curr. Protein Pept. Sci.* **2012**, *13*, 723–738. [CrossRef]

143. Schluesener, H.J.; Su, Y.; Ebrahimi, A.; Pouladsaz, D. Antimicrobial peptides in the brain: Neuropeptides and amyloid. *Front. Biosci.* **2012**, *4*, 1375–1380. [CrossRef]

144. Cutuli, M.; Cristiani, S.; Lipton, J.M.; Catania, A. Antimicrobial effects of alpha-MSH peptides. *J. Leukoc. Biol.* **2000**, *67*, 233–239. [CrossRef]

145. Catania, A.; Garofalo, L.; Cutuli, M.; Gringeri, A.; Santagostino, E.; Lipton, J.M. Melanocortin peptides inhibit production of proinflammatory cytokines in blood of HIV-infected patients. *Peptides* **1998**, *19*, 1099–1104. [CrossRef]

146. Barcellini, W.; Colombo, G.; La Maestra, L.; Clerici, G.; Garofalo, L.; Brini, A.T.; Lipton, J.M.; Catania, A. Alpha-melanocyte-stimulating hormone peptides inhibit HIV-1 expression in chronically infected promonocytic U1 cells and in acutely infected monocytes. *J. Leukoc. Biol.* **2000**, *68*, 693–699.

147. Mahlapuu, M.; Hakansson, J.; Ringstad, L.; Bjorn, C. Antimicrobial Peptides: An Emerging Category of Therapeutic Agents. *Front. Cell. Infect. Microbiol.* **2016**, *6*, 194. [CrossRef]

148. Naafs, M.A.B. The Antimicrobial Peptides: Ready for Clinical Trials? *Biomed. J. Sci. Tech. Res.* **2018**, *7*. [CrossRef]

149. Bjorn, C.; Noppa, L.; Naslund Salomonsson, E.; Johansson, A.L.; Nilsson, E.; Mahlapuu, M.; Hakansson, J. Efficacy and safety profile of the novel antimicrobial peptide PXL150 in a mouse model of infected burn wounds. *Int. J. Antimicrob. Agents* **2015**, *45*, 519–524. [CrossRef]

150. Dorschner, R.A.; Pestonjamasp, V.K.; Tamakuwala, S.; Ohtake, T.; Rudisill, J.; Nizet, V.; Agerberth, B.; Gudmundsson, G.H.; Gallo, R.L. Cutaneous injury induces the release of cathelicidin anti-microbial peptides active against group A Streptococcus. *J. Investig. Dermatol.* **2001**, *117*, 91–97. [CrossRef]
151. Dorschner, R.A.; Lopez-Garcia, B.; Peschel, A.; Kraus, D.; Morikawa, K.; Nizet, V.; Gallo, R.L. The mammalian ionic environment dictates microbial susceptibility to antimicrobial defense peptides. *FASEB J.* **2006**, *20*, 35–42. [CrossRef]
152. Myhrman, E.; Hakansson, J.; Lindgren, K.; Bjorn, C.; Sjostrand, V.; Mahlapuu, M. The novel antimicrobial peptide PXL150 in the local treatment of skin and soft tissue infections. *Appl. Microbiol. Biotechnol.* **2013**, *97*, 3085–3096. [CrossRef]
153. Chennupati, S.K.; Chiu, A.G.; Tamashiro, E.; Banks, C.A.; Cohen, M.B.; Bleier, B.S.; Kofonow, J.M.; Tam, E.; Cohen, N.A. Effects of an LL-37-derived antimicrobial peptide in an animal model of biofilm Pseudomonas sinusitis. *Am. J. Rhinol. Allergy* **2009**, *23*, 46–51. [CrossRef]
154. Bray, B.L. Large-scale manufacture of peptide therapeutics by chemical synthesis. *Nat. Rev. Drug Discov.* **2003**, *2*, 587–593. [CrossRef]
155. Biswaro, L.S.; da Costa Sousa, M.G.; Rezende, T.M.B.; Dias, S.C.; Franco, O.L. Antimicrobial Peptides and Nanotechnology, Recent Advances and Challenges. *Front. Microbiol.* **2018**, *9*, 855. [CrossRef]
156. Mohammadi-Samani, S.; Taghipour, B. PLGA micro and nanoparticles in delivery of peptides and proteins; problems and approaches. *Pharm. Dev. Technol.* **2015**, *20*, 385–393. [CrossRef]
157. Gordon, Y.J.; Romanowski, E.G.; McDermott, A.M. A review of antimicrobial peptides and their therapeutic potential as anti-infective drugs. *Curr. Eye Res.* **2005**, *30*, 505–515. [CrossRef]
158. Marr, A.K.; Gooderham, W.J.; Hancock, R.E. Antibacterial peptides for therapeutic use: Obstacles and realistic outlook. *Curr. Opin. Pharmacol.* **2006**, *6*, 468–472. [CrossRef]
159. Vlieghe, P.; Lisowski, V.; Martinez, J.; Khrestchatisky, M. Synthetic therapeutic peptides: Science and market. *Drug Discov. Today* **2010**, *15*, 40–56. [CrossRef]
160. McGregor, D.P. Discovering and improving novel peptide therapeutics. *Curr. Opin. Pharmacol.* **2008**, *8*, 616–619. [CrossRef]
161. Silva, J.P.; Goncalves, C.; Costa, C.; Sousa, J.; Silva-Gomes, R.; Castro, A.G.; Pedrosa, J.; Appelberg, R.; Gama, F.M. Delivery of LLKKK18 loaded into self-assembling hyaluronic acid nanogel for tuberculosis treatment. *J. Control. Release* **2016**, *235*, 112–124. [CrossRef]
162. D'angelo, I.; Casciaro, B.; Miro, A.; Quaglia, F.; Mangoni, M.L.; Ungaro, F. Overcoming barriers in Pseudomonas aeruginosa lung infections: Engineered nanoparticles for local delivery of a cationic antimicrobial peptide. *Colloids Surf. B Biointerfaces* **2015**, *135*, 717–725. [CrossRef]
163. Tripathi, S.; Tecle, T.; Verma, A.; Crouch, E.; White, M.; Hartshorn, K.L. The human cathelicidin LL-37 inhibits influenza A viruses through a mechanism distinct from that of surfactant protein D or defensins. *J. Gen. Virol.* **2013**, *94*, 40–49. [CrossRef]
164. Li, F.; Wang, Y.; Yu, L.; Cao, S.; Wang, K.; Yuan, J.; Wang, C.; Wang, K.; Cui, M.; Fu, Z.F. Viral Infection of the Central Nervous System and Neuroinflammation Precede Blood-Brain Barrier Disruption during Japanese Encephalitis Virus Infection. *J. Virol.* **2015**, *89*, 5602–5614. [CrossRef]

Article

An Oleanolic Acid Derivative Inhibits Hemagglutinin-Mediated Entry of Influenza A Virus

Mengdie Ye [1,2,†], Yixian Liao [3,†], Li Wu [2], Wenbao Qi [2], Namrta Choudhry [1,2], Yahong Liu [1,2], Weisan Chen [4], Gaopeng Song [3,*] and Jianxin Chen [1,2,*]

1   Guangdong Provincial Key Laboratory of Veterinary Pharmaceutics Development and Safety Evaluation, Guangzhou 510642, China; mengdieye@stu.scau.edu.cn (M.Y.); choudharynamrta@gmail.com (N.C.); gale@scau.edu.cn (Y.L.)
2   College of Veterinary Medicine, South China Agricultural University, Guangzhou 510642, China; 13503033623@139.com (L.W.); qiwenbao@scau.edu.cn (W.Q.)
3   College of Materials and Energy, South China Agricultural University, Guangzhou 510642, China; cjjwcf@163.com
4   Department of Biochemistry and Genetics, La Trobe Institute for Molecular Science, La Trobe University, Melbourne, Victoria 3086, Australia; weisan.chen@latrobe.edu.au
*   Correspondence: songgp1021@scau.edu.cn (G.S.); jxchen@scau.edu.cn (J.C.); Tel./Fax: +86-20-8528-0293 (G.S.); +86-20-8528-0234 (J.C.)
†   These authors contributed equally to this work.

Received: 22 September 2019; Accepted: 14 February 2020; Published: 18 February 2020

Abstract: Influenza A viruses (IAV) have been a major public health threat worldwide, and options for antiviral therapy become increasingly limited with the emergence of drug-resisting virus strains. New and effective anti-IAV drugs, especially for highly pathogenic influenza, with different modes of action, are urgently needed. The influenza virus glycoprotein hemagglutinin (HA) plays critical roles in the early stage of virus infection, including receptor binding and membrane fusion, making it a potential target for the development of anti-influenza drugs. In this study, we show that OA-10, a newly synthesized triterpene out of 11 oleanane-type derivatives, exhibited significant antiviral activity against four different subtypes of IAV (H1N1, H5N1, H9N2 and H3N2) replications in A549 cell cultures with EC_{50} ranging from 6.7 to 19.6 μM and a negligible cytotoxicity (CC_{50} > 640 μM). It inhibited acid-induced hemolysis in a dose-dependent manner, with an IC_{50} of 26 μM, and had a weak inhibition on the adsorption of H5 HA to chicken erythrocytes at higher concentrations (≥40 μM). Surface plasmon resonance (SPR) analysis showed that OA-10 interacted with HA in a dose-dependent manner with the equilibrium dissociation constants (KD) of the interaction of 2.98×10^{-12} M. Computer-aided molecular docking analysis suggested that OA-10 might bind to the cavity in HA stem region which is known to undergo significant rearrangement during membrane fusion. Our results demonstrate that OA-10 inhibits H5N1 IAV replication mainly by blocking the conformational changes of HA2 subunit required for virus fusion with endosomal membrane. These findings suggest that OA-10 could serve as a lead for further development of novel virus entry inhibitors to prevent and treat IAV infections.

Keywords: oleanane-type derivatives; influenza A virus (IAV); virus entry inhibitors; hemagglutinin (HA)

1. Introduction

Influenza A virus (IAV), the causative agent of influenza, is a major pathogen that causes public health problem and socioeconomic burden world-wide. With waterfowl as the primary reservoir, the virus is able to infect a wide variety of birds and mammals, including humans. Due to this trait, zoonotic spillovers occur occasionally and can lead to pandemics with severe consequences for the human population [1,2]. The avian-origin H5N1 and H7N9 subtypes of influenza viruses are recent examples of animal viruses that acquired the potential to infect and cause disease in humans. H5N1 IAV are often highly pathogenic (HP) avian influenza viruses. In 1997 in Hongkong, six people died out of 18 confirmed human cases with HP H5N1 virus infection [3]. In 2003, novel H5N1 IAV genetic variants circulated in Southeast Asian countries, which led innumerable poultry to death and caused sporadic human infections in the following years. By the end of 2017, 860 laboratory confirmed cases of H5N1 IAV infection from 16 different countries, resulting in 454 deaths, had been reported to the World Health Organization (WHO) [2]. Influenza A viruses mutate easily because of their segmented RNA genome, making it difficult to produce a timely and sufficiently effective vaccine to prevent the potential epidemic outbreaks. Therefore, using anti-influenza agents could be a more efficient approach for prevention and treatment at the beginning of outbreak. To date, two types of anti-influenza drugs have been approved. One type pertains to the M2 ion channel inhibitors, including amantadine and rimantadine, which block the release of viral RNA into the cytoplasm [4]. The others are neuraminidase (NA) inhibitors, including zanamivir, oseltamivir and peramivir, which prevent progeny virus from being released by their host cells [5,6]. However, it has been reported that most of the circulating IAV strains are resistant to M2 inhibitors and may rapidly develop resistance to NA inhibitors, which limits the use of those licensed drugs [7–9]. Therefore, it is urgently required to develop novel anti-influenza agents preferentially with novel mechanisms of action to combat influenza, especially the HP influenza.

The first and most critical step of IAV infection is viral entry mediated by the interaction of viral envelope protein hemagglutinin (HA) and its receptor on host cell surface [10]. HA forms a trimer including three HA1 and three HA2 subunits. Each HA2 contains a fusion peptide, a soluble ectodomain (SE) and a transmembrane domain [11]. After binding to receptor through HA1, IAV is endocytosed. Within the endosome, increasing acidity, pH 5–6, induces HA2 to undergo an irreversible conformational change, leading to the fusion peptide extruding toward the endosomal membrane, and ultimately fusion of the viral and endosomal membranes [10]. HA2, as a major component of the stem region of HA, is a highly conserved subunit. Consequently, blocking HA2's conformational change could abolish viral membrane fusion and infection. Hence, HA2 is considered an attractive antiviral target [12,13]. The first discovered compound of this class was tertiary butylhydroquinone [14]. Later, several compounds that act in a similar way were described: BMY-27709, CL-385319, stachyflin and 4c [15–18]. However, these small molecular compounds were reported to display inhibitory effects on H1, H2 and H3 IAV subtypes rather than H5 subtype. Recently, guided by structural knowledge on the interactions of HA and anti-stem broadly neutralizing antibodies, Maria and colleagues successfully developed a small molecule JNJ4796 with the ability to inhibit HA-mediated fusion and protect mice against lethal and sublethal influenza challenges [19]. The potent anti-IAV activity of JNJ4796 further reinforces that HA2 could be a promising drug target for blocking IAV infection.

Pentacyclic triterpenoids (PTs), an abundant natural product in plants, have received considerable attention due to their wide spectrum of antiviral activities against various viruses, such as HIV, IAV and HCV [20–23], and some are already marketed as therapeutic agents or dietary supplements. For example, glycyrrhizin (glycyrrhizic acid) has been used for treating chronic hepatic diseases for over 40 years in China and Japan clinically.

Oleanane acid (OA) triterpene is probably the most well-known member of PTs with noticeable antiviral activities. Echinocystic acid (EA), an oleanane-type triterpene, was reported to have substantial inhibitory activity on HCV entry with EC_{50} at nanomolar range [24,25]. Zhou reported that a different chemical modification to the C-17-COOH of OA led to Y3 (an OA-acetyl galactose conjugate), which showed strong antiviral activity against H1N1 and H3N2 IAV infections in vitro [13]. Further study indicated that Y3 was able to bind tightly to HA protein, thereby disrupting the interaction of HA with its sialic acid receptor [13]. In our previous study, we used an efficient HIV-based pseudotyping system to screen a semisynthetic saponin library, and discovered that OA saponins with β-chacotriosyl residue at the C-3-OH of OA exhibited excellent inhibitory activity against H5N1

IAV entry [26]. Based on this finding, we further designed, synthesized and evaluated a series of 3-O-β-chacotriosyl oleanolic acid analogs as H5N1 IAV entry inhibitors. We found that the introduction of a disubstituted amide structure at the 17-COOH of OA could significantly improve the selective index while maintaining their antiviral activities in vitro [27]. However, as the previous antiviral evaluation of 3-O-β-chacotriosyl oleanolic acid analogs was based on a pseudotyping system, their inhibitory effects on IAV and mechanisms of action are unconfirmed. In the present study, antiviral activities of OA and its eleven analogs, including four newly synthesized derivatives, against H5N1 virus in A549 cells, were investigated, and their mechanisms of action and antiviral activities against other IAV subtypes based on a representative novel compound OA-10 were further investigated. To our knowledge, this is the first report on anti H5N1 IAV activity of OA derivatives.

2. Materials and Methods

2.1. Compounds

In total, 12 oleanane-type triterpenoid compounds (Figure S1 and Figure 1A), including oleanolic acid and its 11 derivatives, and their antiviral activities against H5N1 IAV infection in A549 cells were studied. Oleanolic acid was purchased from Chengdu Gelipu Biotechnology Co., Ltd. (Chengdu, China). Seven derivatives named OA-1, OA-2, OA-3, OA-4, OA-5, OA-6 and OA-7 were synthesized as described previously [27]. Four new derivatives named OA-8, OA-9, OA-10 and OA-11, which were confirmed by data from nuclear magnetic resonance (NMR) and high-resolution electrospray ionization mass spectrometry (HRESIMS) analysis (shown in the supplementary material), were synthesized by a method similar to that described previously [27]. Briefly, the known saponin 3-(β-D-glucopyranosyl)-oleanolic acid benzyl ester or 3-(α-D-mannopyranosyl)-oleanolic acid benzyl ester was subjected to PivCl in the presence of pyridine to selectively protect the 3,6-OHs of the β-glucopyranosyl or α-mannopyranosyl residue, and subsequent glycosylation of the 2, 4-OHs in sugars with 2,3,4-tri-O-acetyl-L-rhamnopyranosyl trichloroacetimidate using TMSOTf as the promoter, followed by deprotection of the acyl groups with NaOH to produce two tridesmosidic oleanolic acid benzyl ester saponins, respectively. Hydrogenolysis of the above oleanolic acid benzyl ester saponins over palladium/carbon in THF-MeOH led to OA-9 and 3-O-β-chacotriosyl oleanolic acid, was carried out to furnish OA-10 or OA-11 by four consecutive acetylation, acyl chlorination, amide condensation reaction and alkaline hydrolysis, respectively. The purities of all oleanane-type triterpenoids were ≥98% by HPLC analysis. Ribavirin hydrochloride was purchased from Guangdong Starlake Bioscience Co. Ltd. (Zhaoqing, China) with purities ≥ 99%. Peramivir in sterile 0.9% NaCl solution (0.3 g/100 mL) was purchased from Guangzhou Nucien Pharmaceutical Co., Ltd. (Guanzhou, China). Arbidol hydrochloride was purchased from Dalian Melun Biological Technology Co., Ltd. (Dalian, China) with purities ≥98%. Oleanane-type triterpenoids were dissolved in dimethyl sulfoxide (DMSO) and diluted with PBS to <0.4% DMSO for in vitro experiments. Ribavirin hydrochloride was dissolved in PBS for in vitro experiments.

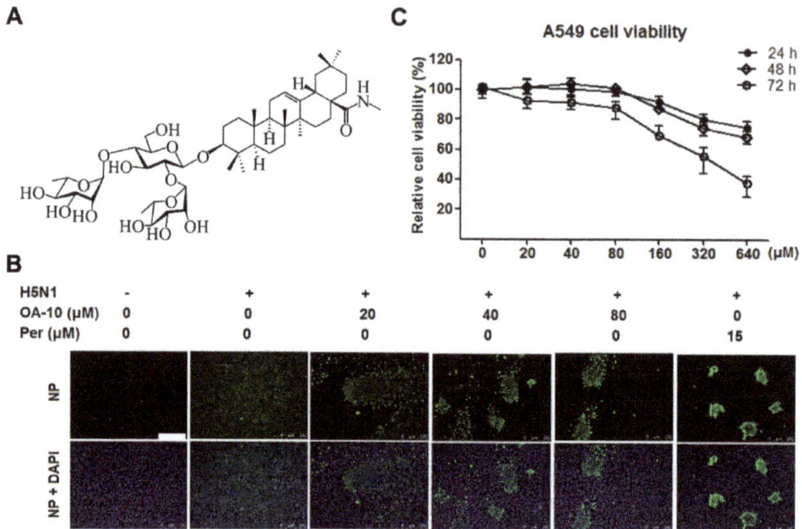

Figure 1. OA-10 inhibited H5N1 IAV replication in A549 cells with minimal cytotoxicity. (**A**) Chemical structure of OA-10. (**B**) Antiviral activity of OA-10 against H5N1 IAV infection. Confluent A549 cells grown in 96-well plates were infected with H5N1 IAV (0.1 MOI) for 1 h at 37 °C and then cultured in fresh medium containing various concentrations of OA-10 or 15 μM peramivir (Per). At 24 hpi, the cells were fixed with paraformaldehyde and the viral NP expression was detected by indirect immunofluorescence assay (IFA). Representative IFA images from three independent experiments are shown. Scale bar: 250 μm. (**C**) Cytotoxicity of OA-10 in A549 cells. The results are expressed as percentages (%) of mock treated cells.

2.2. Cell Lines and Influenza Virus

A549 (human lung carcinoma) cells and MDCK (Madin-Darby canine kidney) cells used for IAV infection were cultured in Dulbecco's modified Eagle's medium (DMEM, Gibco, UT, USA) supplemented with 10% fetal bovine serum (FBS, Biological Industries, Kibbutz Beit Haemek, Israel), 100 U/mL of penicillin and 100 μg/mL streptomycin in a humidified atmosphere with 5% CO_2. All cells were obtained from the Center of Cellular Resource, Chinese Academy of Sciences (Shanghai, China).

Avian influenza A virus strain A/Duck/Guangdong/99(H5N1 IAV) and A/Chicken/Guangdong/16/ 1996 (H9N2) were kindly provided by the Veterinary Technology Center of South China Agricultural University (Guangzhou, China). The H1N1 IAV strain A/Puerto Rico/8/34 (PR8) and H3N2 IAV strain A/Guangdong/Dongguan/1100/2006 viruses were obtained from the Chinese Center for Disease Control and Prevention (Beijing, China). Virus stocks were passaged in 10-day old embryonated chicken eggs for 48 h. The allantoid fluid was harvested and aliquots were stored at −80 °C until required. Experiments involving H5N1 virus were conducted in a physical containment level three (PC3) laboratory.

2.3. Cytotoxicity Assay

The cytotoxicity of tested compounds was evaluated using a MTT assay as described previously by Luo [28]. Briefly, cells were grown in 96-well plates for 24 h. The medium was replaced with fresh one containing serially diluted compounds and the cells were further incubated for 24 h, 48 h or 72 h. The culture medium was removed and replaced with 100 μL 3-(4,5-dimethylthiozol-2-yl)-3,5-dipheryl tetrazolium bromide (MTT; Sigma-Aldrich, MA, USA) solution (1 mg/mL in PBS) and incubated at 37 °C for 4 h. After removal of the supernatant, 150 μL of DMSO was added to all of the wells to dissolve the formazan crystals for 10 min at 37 °C. Cell viability was then measured as the absorbance at 490 nm with a microplate reader (Thermo fisher scientific, MA, USA) and expressed as a percentage of the control level. The mean optical density (OD) values from six replicated wells per treatment were used as the cell viability index. The 50% cytotoxic concentration (CC_{50}) was calculated by GraphPad Prism 7.0 (GraphPad Software, San Diego, CA, USA).

2.4. Indirect Immunofluorescence Assay (IFA)

Indirect immunofluorescence assay was used to rapidly evaluate antiviral activities of compounds against H5N1 IAV infection. For immunostaining, the H5N1-infected or control cells were fixed with 4% paraformaldehyde for 10 min, then permeabilized with 0.25% Triton X-100 for 10 min at room temperature (RT). Cells were blocked with 1% bovine serum albumin (BSA) for 60 min at RT and then incubated with a mouse monoclonal antibody against IAV nucleoprotein (NP protein) (1:500 dilution, Sino Biological, Beijing, China) at 4 °C overnight. After three washes with PBS, the cells were incubated for 1 h at RT with an anti-mouse IgG antibody conjugated with Alexa Fluor® 488 (green) (Cell Signaling Technology, MA, USA) at 1:1000 dilution. Nuclei were counterstained with 50 µL of 4,6-diamidino-2-phenylindole (DAPI, 300 nM; Sigma-Aldrich, MA, USA). Immunofluorescence was captured using the Leica DMI 4000B fluorescence microscope (Leica, Wetzlar, Germany). Blue and green fluorescence spots were counted as the total and IAV-infected cell numbers respectively in every IFA image.

Relative NP protein level (%) of each image was calculated based on the fluorescence optical density (OD) using Software Image J. Results from compound-treated samples were compared to those from corresponding DMSO-treated control groups (set as 100%). The EC_{50} value (the concentration required to protect 50% cells from IAV infection) was determined by plotting the relative NP protein level as a function of compound concentration and calculated using GraphPad Prism 7.0 software (GraphPad Software, San Diego, CA, USA).

2.5. In Vitro Virus Growth Inhibition Assay

A549 monolayers were infected with the virus for 1 h. Supernatants were removed, and cells were then incubated with DMEM containing serial concentrations of test compound. Cells and supernatants were collected at indicated time points post-infection and in total subjected to three freeze-thaw cycles at −80 °C and 4 °C to ensure maximal release of cellular virions [29]. Final viral titers in the supernatants were determined by an end point dilution assay using MDCK cells and expressed as $\log_{10}TCID_{50}$ /0.1 mL.

2.6. Real-Time Reverse-Transcription PCR (RT-PCR)

Total RNA was extracted from cells or mixtures of cells and supernatants using the total RNA rapid extraction kit (Fastagen, Shanghai, China) following the manufacturer's instructions. RNA was reverse-transcribed into first-strand cDNA using a reverse transcription kit (TaKaRa, Japan). PCR amplification was performed using 1 µL of reverse-transcribed product with primers designed for IAV-NP and GAPDH (glyceraldehyde-3-phosphate dehydrogenase, used as the endogenous control). The primers used for PCR amplification are listed in Table 1 [30]. Real-time PCR was performed using 2×RealStar Green Power Mixture (containing SYBR Green I Dye) (Genstar, Beijing, China) on a CFX96 Real-time PCR system (Bio-Rad, Hercules, CA, USA). Relative mRNA expression was calculated by $2^{-\Delta\Delta CT}$ method using DMSO-treated infected cells as reference samples for determining IAV-NP [31,32]. To assess the effect of OA-10 on transcriptional activation of NP in IAV infected cells, the relative fold change of each NP gene expression was calculated and compared between OA-10-treated virus-infected and virus-infected cells.

Table 1. Real-time PCR primer sequences.

Name [a]	Sequences 5′ to 3′
NP-F	5′-GGATTTGGCGTCAAGCGAACA-3′ - 3′ACCAGAAGATKTGTCMTTCCAGGG -3′
NP-R	5′-GTCCCTACCCCCTTTACTGC-3′ TACTCCTCCGCATTGTCTCCGAAG -3′
GAPDH-F	5′-GCACCGTCAAGGCTGAGAAC-3′
GAPDH-R	5′-TGGTGAAGACGCCAGTGGA-3′

[a] F: forward primer; R: reverse primer

2.7. Time Course Inhibition Assay

To estimate the influence of OA-10 on the IAV replication cycle, A549 cells were grown in 24-well plates to confluence and then infected with H5N1 IAV (0.1 MOI) for 1 h at 37 °C. OA-10 or ribavirin or peramivir was added before, during or after IAV infection. For pretreatment, cells were incubated with the indicated compound for 2 h at 37 °C, followed by three washes with PBS and then infected with H5N1 IAV for 1 h. For co-treatment, cells were simultaneously incubated with H5N1 IAV and the compound. After 1 h, the virus–drug mixture was removed, and the cells were washed three times with PBS before fresh medium was added. For post-treatment, cells were first infected with H5N1 IAV for 1 h followed by three washes with PBS and then incubated in the fresh medium containing the compound. At 24 hpi, progeny viruses in the supernatants were determined by an end point dilution assay and the extent of virus infection in the cells was assessed by IFA for NP protein, respectively.

To determine the specific stage(s) of the viral life cycle affected by OA-10, a time-of-addition assay was performed as described by Luo et al. [28]. Briefly, confluent monolayers of A549 cells grown in 24-well plates were infected with H5N1 IAV (1.0 MOI) for 1 h at 37 °C. Cells were washed three times with PBS to remove unbound viruses and incubated in fresh medium. In total, 80 µM of OA-10 or ribavirin or peramivir was then added for 0 to 2, 2 to 4, 4 to 6, 6 to 8 h or 0 to 8 hpi. After each incubation period, the cells were washed three times with PBS and incubated with fresh medium at 37 °C. At 8 hpi, the cells were subjected to viral NP protein analysis using IFA and viral mRNA analysis using RT-PCR, respectively.

2.8. Hemagglutination Inhibition Assay (HAI)

The inhibitory activity of OA-10 on HA-mediated hemagglutination of chicken red blood cells (RBCs) was assessed by an HAI assay. Briefly, 25 µL of A/Duck/Guangdong/99(H5N1) (4 hemagglutination units) was incubated with 25 µL H5 standard antiserum (H5 antiserum were provided by the Veterinary Technology Center of South China Agricultural University, Guangzhou, China) or OA-10 at indicated concentrations for 30 min at RT. Then, 50 µL chicken RBCs (0.5%) in saline solution were added to each well and incubated at 37 °C for 30 min. The plates were taken images at 45 degree inclination for recording hemagglutination and the fluidity of deposited RBCs.

2.9. Hemolysis Inhibition Assay (HIA)

The inhibitory effects of compounds on virus-induced hemolysis at low pH were determined by a procedure described previously by Basu et al. [33]. Briefly, chicken RBCs were washed twice with PBS and resuspended to 2% (vol/vol) in PBS and stored at 4 °C until use; 100 µL of compound diluted in PBS was then mixed with an equal volume of H5N1 IAV (10^8 PFU/mL) in a 96-well plate. After incubating the virus-compound mixture at room temperature for 30 min, 200 µL of 2% chicken RBCs (pre-warmed at 37 °C) was added. The mixture was incubated at 37 °C for another 30 min. To trigger hemolysis, 100 µL of sodium acetate-acetic acid buffer solution (0.5 M, pH 5.2) was added and mixed with the RBC suspension. The mixture was incubated at 37 °C for another 30 min for HA acidification and hemolysis. To separate unlysed RBC from the lysed ones, plates were centrifuged at the end of incubation at 1200 rpm for 6 min; 300 μL of the supernatant was transferred to another flat-bottom 96-well plate. The OD_{540} was read on a microtiter plate reader. IC_{50} is defined as the compound concentration that generates 50% protection on chicken RBCs lysed by virus.

2.10. Surface Plasmon Resonance (SPR) Analysis

Interactions between the influenza HA and the compounds were analyzed using the Berthold bScreen LB 991 (LabXMedia group, Midland, Canada) at 4 °C. Recombinant influenza HA protein (Abcam, Cambridge, England) from virus strain A/Vietnam/1203/2004 (H5N1) was immobilized on a sensor chip (Photo-cross-linker SensorCHIP™) using an amine coupling kit (GE Healthcare, Buckinghamshire, UK). Subsequently, compounds were injected as analytes at various concentrations, and we used PBST (10 mM phosphate buffer with 0.1% Tween 20, pH 5.0) as running buffer. For binding studies, analytes were applied at corresponding concentrations in running buffer at a flow rate of 30 μL/min with a contact time of 600 s and a dissociation time of 360 s. Chip platforms were regenerated with regeneration buffer (Glycine-HCl, pH = 2.0) after each test cycle. The processing and analyses of association and dissociation rate constants (Ka/Kon and Kd/Koff respectively) and the equilibrium dissociation constant (KD, kd/ka) were performed using the data analysis software of the bScreen LB 991 unlabelled microarray system according to a single-site binding model (1:1 Langmuir binding) with mass transfer limitations for binding kinetics determination.

2.11. Molecular Docking

The protein structure of hemagglutinin (PDB ID: 6CFG) was used for the docking study. All calculations were performed using Discovery Studio 2017. The 3D structures of OA-10 were constructed using the Discovery Studio small molecule window, and energy was minimized by CHARMm force field in a two-step method: steepest descent with RMS gradient convergent to 0.1, and the final step was conjugate gradient with RMS gradient convergent to 0.0001. Prior to the docking procedure, all bound water molecules were removed from the protein crystal structure. A site sphere radius was set to 13.39 Å in which the other parameters were set as default. The docking program CDOCKER was used to perform the automated molecular simulation in which the top hits were set as 10, and the random conformations were set as 20. The top compounds were ranked by the corresponding values of -CDOCKER interaction energy.

2.12. Statistical Analysis

All values are expressed as means ± SDs from at least three independent experiments. Statistical significance was determined by Student's *t*-test when only two groups were compared or by one-way analysis of variance (ANOVA) when more than two groups were compared. Statistical analyses were performed using GraphPad Prism 7 (GraphPad Software, San Diego, CA, USA). * $P < 0.05$, ** $P < 0.01$ and *** $P < 0.001$ were considered to be statistically significant at different levels.

3. Results

3.1. Compound OA-10 Inhibits IAV Infections in A549 Cells with Minimal Cytotoxicity

Chemical structures of oleanane acid (OA-0) and its 11 derivatives, including four new derivatives, are shown in Figure S1 of the Supplementary Material. The cytotoxicity of OA-0 and its derivatives on A549 cells was first evaluated using the 3-(4,5-dimethylthiozol-2-yl)-3,5-dipheryl tetrazolium bromide (MTT) assay. For each compound, CC_{50} value, the concentration required to reduce normal cell viability by 50% after 24 h of compound treatment, was determined, as shown in Table 2. Derivatives OA-1, OA-2, OA-4, OA-5, OA-7, OA-8 and OA-9 exhibited greater cytotoxicity on A549 cells with $CC_{50} \leq$ 20.5 μM, while other derivatives and oleanolic acid exhibited less cytotoxicity with $CC_{50} \geq 31.1$ μM. Noticeably, OA-10 showed the least cytotoxicity with $CC_{50} > 640$ μM (Figure 1C). DMSO (0.4%, used as solvent) did not exhibit detectable cytotoxicity on A549 cells. No obvious cytotoxicity was observed for OA-10 at concentrations $\leq 80\ \mu M$ after 24, 48 or 72 h of treatment, as shown in Figure 1C. Thus, $80\ \mu M$ of OA-10 was selected as the maximum concentration for further studies.

Table 2. Cellular toxicity and inhibitory activity of oleanane-type triterpenoid derivatives against H5N1 influenza A virus (IAV) replication in A549 cells.

Compound	MW (g/mol)	CC_{50} [a] (μM)	EC_{50} [b] (μM)	SI [c]
OA-0	456	67.4 ± 4.8	14.2 ± 1.9	4.7
OA-1	911	15.1 ± 1.3	2.30 ± 0.4	6.6
OA-2	996	20.5 ± 1.2	5.22 ± 0.6	3.9
OA-3	939	31.1 ± 2.5	2.87 ± 0.4	11
OA-4	966	12.1 ± 1.0	4.45 ± 0.7	16
OA-5	897	7.55 ± 0.8	4.24 ± 0.6	1.8
OA-6	924	>640	>86.6	-
OA-7	924	9.88 ± 1.1	7.79 ± 1.1	1.3
OA-8	924	15.5 ± 1.5	3.03 ± 0.5	5.1
OA-9	911	10.8 ± 0.9	4.61 ± 0.8	2.3
OA-10	924	>640	14.0 ± 2.3	>45
OA-11	1000	60.0 ± 4.5	6.7 ± 0.9	9.0

[a] CC_{50}, the concentration required to reduce normal, noninfected cell viability by 50%; [b] EC_{50}, the concentration required to protect 50% cells from H5N1 IAV infection; [c] SI (selectivity index) is the ratio of CC_{50} to EC_{50}. Data were presented as means $\pm$ SDs of results from three independent experiments.

The antiviral activities of synthesized OA derivatives against H5N1 IAV were evaluated via relative nucleoprotein (NP) expression level (%) detected by immunofluorescence microscopy at 24 hpi. The initial results showed that OA-0 and the 11 derivatives exhibited various antiviral activities against H5N1 IAV infection (Table 2). It was found that OA-0 and all other derivatives except OA-6 significantly reduced H5N1 IAV infection in A549 cells with $EC_{50} \leq 14.2\ \mu M$, while OA-6 showed the lowest antiviral activity with $EC_{50} > 86.6\ \mu M$. However, most of the effective derivatives, including OA-1, OA-2, OA-3, OA-4, OA-5, OA-7, OA-8, OA-9 and OA-11, exhibited high cytotoxicity on A549 cells with $CC_{50} \leq 31.1\ \mu M$, while OA-0, OA-10 and OA-11 showed relatively lower cytotoxicity on A549 cells with $CC_{50} \geq 60.0\ \mu M$. Most interestingly, OA-10 showed significant inhibition on H5N1 IAV replication in a dose-dependent manner in concentrations ranging from 20 to 80 μM (EC_{50}: 14.0 μM) with very low cytotoxicity, as shown in Figure 1B,C. Among the 12 evaluated compounds, OA-10 showed the highest select index (SI > 45). Thus, OA-10 was selected for further studies.

To explore whether OA-10 possesses a broad inhibitory effect on different IAV subtypes, three other IAV strains (PR8, H9N2 and H3N2) were evaluated using IFA and A549 cells at 24 hpi. As shown in Figure S2A and Table 3, OA-10 also exhibited significant inhibitions on PR8 (EC_{50}: 6.7μM), H9N2 (EC_{50}: 15.3 μM) and H3N2 (EC_{50}: 19.6 μM) replications.

Table 3. Inhibitory activity of OA-10 against influenza A virus replication in A549 cells.

IAV Strain	H5N1	PR8 (H1N1)	H9N2	H3N2
$EC_{50}\ (\mu M)$	14.0 ± 2.3	6.7 ± 1.4	15.3 ± 2.5	19.6 ± 3.7

Data are presented as means $\pm$ SDs from three independent experiments.

To more accurately assess OA-10's inhibitory role, we further examined its antiviral effects against H5N1 IAV infection using virus titration and RT-PCR at 48 hpi. As shown in Figure 2A, treatment with OA-10 resulted in a significant reduction of progeny virus titer in a dose-dependent manner. Treatment with 80 μM of OA-10 led to a 1.8 log reduction in progeny virus production compared to that in DMSO-treated control. In fact, OA-10 at concentrations from 20 to 80 μM significantly inhibited H5N1 IAV NP RNA levels in A549 cells in a dose-dependent manner (Figure 2B). We further studied the viral inhibition kinetics by OA-10 at 80 μM. In the H5N1 IAV-infected control, the levels of virus titer and virus mRNA expression increased continuously from 24 to 72 hpi (Figure 2C,D). The addition of 80 µM OA-10 significantly reduced progeny virus titers and viral RNA levels at all time-points, as shown in Figure 2C,D. Simultaneously, treatment with OA-10 also reduced progeny virus titer on PR8, H9N2 and H3N2 infection, respectively, in a dose-dependent manner at 48 hpi, as shown in Figure S2B. Peramivir, a well-known neuraminidase inhibitor, was used as a positive antiviral drug control in this study. Our results showed that 15 µM of peramivir exhibited a significant inhibition on IAV infections in the same assays.

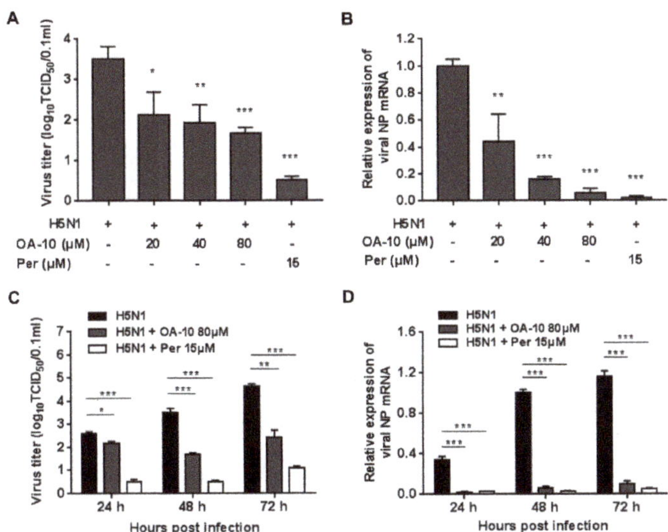

Figure 2. OA-10 inhibited H5N1 IAV infection in A549 cells. A549 cells grown in 24-well plates were infected with H5N1 IAV (0.1 MOI) for 1 h, and then cultured in fresh medium containing various concentrations of OA-10 or 15 µM peramivir. At 48 hpi (**A,B**) or indicated time-points post infection (**C,D**), cells and supernatants of each well were mixed and subjected to viral titer or RT-PCR analysis. H5N1 IAV expression of GAPDH was used as a loading control, and a DMSO-treated sample (H5N1 IAV infected and non-drug treated) at 48 hpi was used as treatment control (set as 1). Data are presented as means ± SDs of results from three independent experiments. *$P < 0.05$, ** $P < 0.01$ and *** $P < 0.001$ compared to the respective virus control.

3.2. OA-10 Interferes with Virus Entry

To identify the OA-10 affected stage(s) during an influenza infection cycle, we first performed a virus binding (attachment) assay by coculture A549 cells with IAV at 4 °C to permit attachment yet avoid viral entry in the presence or absence of OA-10. The co-incubation of 80 µM OA-10 with H5N1 IAV at 4 °C for 1 h followed by removal of excess virus and 37 °C culture did not affect IAV infection (Figure S3), indicating that OA-10 does not affect IVA binding to cells and also does not directly inactivate IAV particles. We next performed time course studies for the inhibitory effects of OA-10. A549 cells were treated with OA-10 for 2 h prior to virus infection (pre-treatment), or for 1 h during the viral infection (co-treatment), or for 24 h after 1 h virus infection and removal (post-treatment), as shown in Figure 3A. Results in Figure 3B–D show that pre-treatment of 80 µM OA-10 did not reduce progeny virus yield and viral NP production. This result indicates that OA-10 does not impair the susceptibility of A549 cells to IAV infection. When cells were incubated with H5N1 IAV in the presence of 80 µM OA-10 for 1 h (co-treatment), a mild but significant reduction in progeny virus yield as well as NP protein expression was observed. This result indicates that OA-10 likely interacts with the virus early during the infection cycle. When cells were treated with 80 µM OA-10 for 24 h post H5N1 IAV infection (post-treatment), one log reduction in progeny virus yield was observed (Figure 3B), which was also reflected by decreased NP production (Figure 3C,D). This result is consistent with the above results and further indicates that OA-10 likely exerts its antiviral effects during virus entry. In these assays, 15 µM peramivir exhibited more significant inhibition on H5N1 IAV's replication in both co-treatment and post-treatment modes, while 30 µM ribavirin exhibited inhibition only in the post-treatment mode, and its inhibition was significantly weaker than that of peramivir. In addition, the combination of 40 µM OA-10 and 15 µM ribavirin treatment showed a synergistic effect on reducing progeny virus titer and NP production, suggesting their complimentary inhibiting mechanisms against IAV replication (Figure 3B–D).

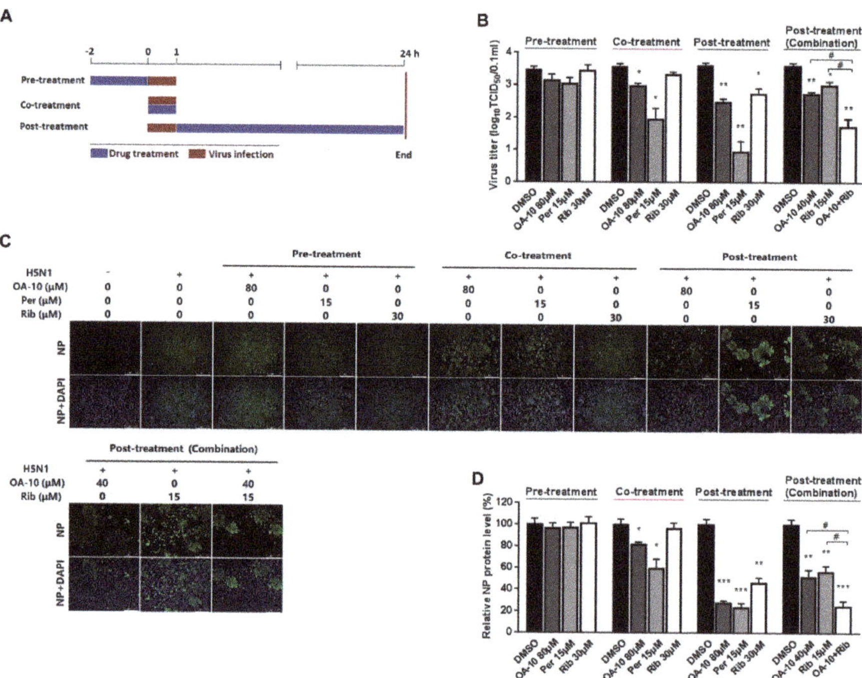

Figure 3. OA-10 exhibited inhibition on H5N1 IAV replication in co- and post-treatment modes. A549 cells grown in 24-well plates were treated with indicated compounds for 2 h prior to virus infection (pre-treatment), or for 1 h during viral infection (co-treatment), or for 24 h after 1 h virus infection and removal (post-treatment) (**A**). For three treatment models, 0.1 MOI of H5N1 IAV was used for infecting cells for 1 h. At 24 hpi, supernatants were collected for determining virus titer using the end point dilution assay (**B**), and the cells were subjected to viral NP protein analysis using IFA (**C,D**). Results shown in (**D**) are normalized NP protein levels based on the fluorescence optical densities (OD) of the images from three independent experiments. Software Image J was used to digitize image OD. Results from OA-10 or peramivir or ribavirin treated samples were compared to those from corresponding DMSO-treated control groups (set as 100%) (**D**). Representative IFA images of the three independent experiments are shown in C. Scale bar: 250 µm. * $P < 0.05$, ** $P < 0.01$ and *** $P < 0.001$ compared to the respective virus (DMSO-treated) control.

It was estimated that from IAV entering into a cell to producing its progeny takes on average 6–8 h, depending on cell type [34]. To identify the exact stage(s) affected by OA-10 during such a IAV replication cycle, we treated the infected cells with OA-10 at four separate 2 h time intervals (0–2, 2–4, 4–6 or 6–8, and 0–8 h as a control) and monitored viral NP RNA and protein expression (Figure 4A). As shown in Figure 4B, a more significant inhibition of viral replication, as represented by decreased viral NP RNA level, was observed when OA-10 was added to the A549 cells during 0–2 and 2–4 hpi. A similar profile was also observed in viral NP expression (Figure 4C,D). These results again indicate that OA-10 exerts its effect during the early stages of IAV infection; i.e., virus internalization, endosome fusion, RNA release and replication. Ribavirin inhibits viral RNA synthesis. As expected, it exerted better IAV inhibition when added to the infected A549 cells before 4 hpi and showed no inhibition when added after 4 hpi. On the other hand, peramivir blocks progeny virus release from infected cells, and did not exhibit any inhibition of either viral RNA synthesis or NP protein expression during a single infection cycle in this assay, although its antiviral activity was the most remarkable, as shown in Figure 3, when multiple infection cycles were studied.

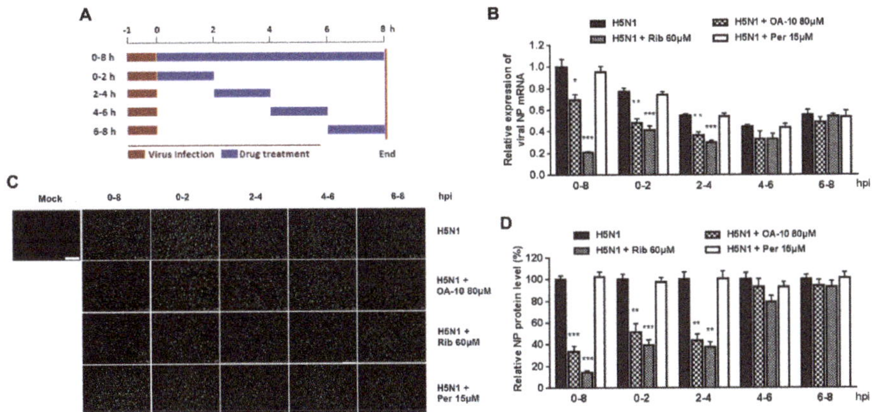

Figure 4. OA-10 inhibits H5N1 IAV replication by targeting the earlier stage(s) of the viral infection cycle. A549 cells grown in 24-well plates were infected with 1.0 MOI of H5N1 IAV for 1 h. After two washes with PBS, OA-10 or ribavirin or peramivir was added at the indicated time points and removed after 2 h or 8 h. After each incubation period, the cells were incubated in fresh medium, harvested at 8 hpi and NP mRNA and protein analyzed by RT-PCR and IFA, respectively. (**A**) The experimental design of time-of-addition assay. (**B**) Relative virus NP mRNA level. (**C,D**) Viral NP protein expression, representative IFA images (**C**) and digitized NP expression (**D**) from three independent experiments. Scale bar: 250 μm. * $P < 0.05$, ** $P < 0.01$ and *** $P < 0.001$ compared to the respective virus control.

3.3. OA-10 Partly Blocks IAV HA Adsorption to Chicken RBCs at High Concentrations

IAV is able to adsorb to chicken RBC, resulting in hemagglutination through the interaction of the receptor-binding domain in viral HA1 subunits with the sialic acid receptors on RBC membrane. To see whether HA is the potential target of OA-10, we questioned whether OA-10 could inhibit hemagglutination by interfering with H5 HA adsorption to RBC. OA-10 did not inhibit HA adsorption to chicken RBCs at concentrations of 10 and 20 μM. At the concentrations of OA-10 ≥ 40 μM, it was observed that a portion of RBCs settled to the bottom of the wells but the settled RBCs were not able to flow, as shown in Figure 5A. Meanwhile, OA-10 treatment did not affect RBC properties in the absence of H5 virus, and the positive control antisera against H5 hemagglutinin (Anti-H5) could effectively inhibit H5 HA adsorption to chicken RBCs with a titer at 1:16, and the settled RBCs could move when the plate was tilted (Figure 5A). These results indicated that OA-10 might have a weak interaction with the receptor binding domain of H5 HA1, which needs further demonstration.

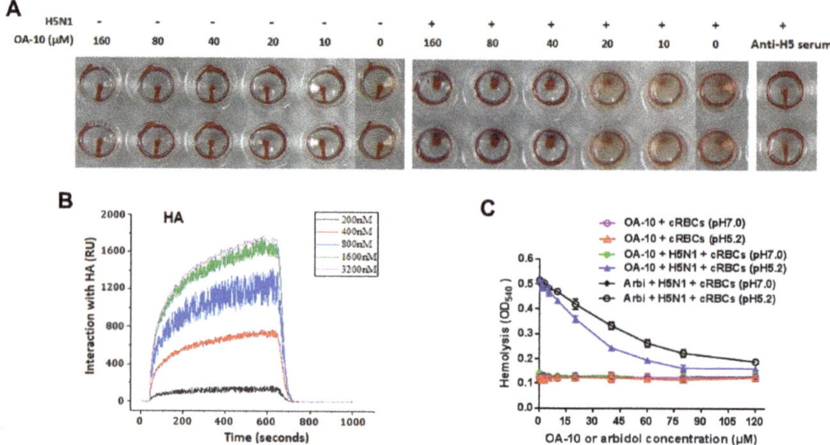

Figure 5. HA is a potential target of OA-10. (**A**) Comparisons of the behaviors of OA-10 vs. antisera against H5 HA in inhibition of H5N1 IAV-induced aggregation of chicken RBCs. OA-10 partially blocked H5N1 virus adsorption to chicken RBCs at concentrations ≥40 μM. (**B**) Characterization of the affinity between OA-10 and HA protein using surface plasmon resonance (SPR) analysis. (**C**) Inhibition of OA-10 on HA-mediated chicken RBCs hemolysis using arbidol as the positive control.

3.4. OA-10 Exhibits a Strong Interaction with HA Protein

To investigate the interaction of OA-10 with HA, we studied the affinities of OA-10 with HA by surface plasmon resonance (SPR). A recombinant HA protein from virus strain A/Vietnam/1203/2004 (H5N1) was used as representative HA of H5N1 IAV strains, because the amino acid sequence identity of the full-length HA proteins between A/Vietnam/1203/2004 strain and A/Duck/Guangdong/99(H5N1) strain used in above antiviral assay was 97.0% (Supplementary Material 5) [35]. As shown in Figure 5B, OA-10 interacted with HA in a dose-dependent manner, and the equilibrium dissociation constant (KD) of the interaction was 2.98×10^{-12} M, indicating a strong affinity between OA-10 with HA. OA-10 did not interact with negative control protein FKBP12 (FK506 binding protein 1A, 12 kDa), indicating that interactions of HA with OA-10 were specific [36]. This result indicates that HA is most likely the target of OA-10 by which it inhibits virus entry and subsequent replication.

3.5. OA-10 Inhibits H5N1 IAV-Mediated Hemolysis at Low pH

We next investigated whether OA-10 was able to inhibit HA-induced chicken RBC hemolysis at a low pH [33]. To trigger hemolysis, the virus-cell suspension was acidified (pH 5.2) to initiate HA subunit HA2 conformational changes that lead to the lysis of chicken RBCs and their hemoglobin release. Arbidol, a licensed antiviral agent against IAV infection in Russia and China, was demonstrated to inhibit IAV replication by binding to HA2 in acidic condition and used as a positive hemolysis inhibitor control in this study [37]. As shown in Figure 5C, as expected, arbidol showed hemolysis inhibition with an IC_{50} of 51 μM. Interestingly, OA-10 more potently inhibited HA-induced hemolysis in a dose-dependent manner, with an IC_{50} of 26 μM. Taken together, these results show that OA-10 interacts with HA2.

3.6. Molecular Model of HA-OA-10 Binding

In order to further understand the molecular basis of OA-10 interacting with HA, we performed docking experiments using CDOCKER protocol to predict the binding modes of OA-10 into the CR6261 antibody binding site which is a highly conserved hydrophobic groove at the HA1-HA2 interface in the HA stem region [19]. The docked conformation of OA-10 was determined based on the minimum

CDOCKER interaction energy. As shown in Figure 6, the modeled structure of HA complexed with OA-10 indicated that C2-OH and C3-OH of the L-rhamnose moiety linked to C2-OH of D-glucose were positioned within appropriate distance (2.46 Å, 1.99 Å and 2.18 Å) to make three strong hydrogen bonds with Gly20 and Val18 in HA2, respectively. Furthermore, C3-OH of D-glucose and the C2-OH of the L-rhamnose moiety linked to C4-OH of D-glucose formed hydrogen bonds with Asp19 and Gln42 with the distances 2.69 Å and 2.78 Å, respectively. In addition, oleanane scaffold also form numerous hydrophobic contacts with the HA1 residue (His38) and HA2 residues (Trp21, Lys38 and Ile45). According to docked conformation of OA-10, 3-O-β-chacotriosyl moiety obviously plays an important role in binding residues in HA2 with higher docking score.

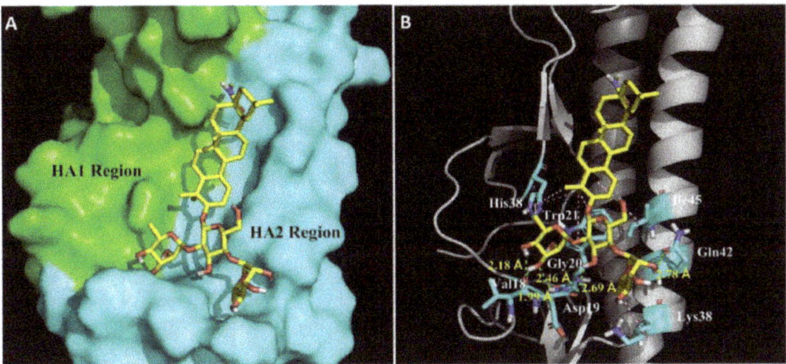

Figure 6. Structural model of OA-10 binding site in H5 HA. (**A**) CR6261 antibody binding site surface. The modeling was based on the published crystal structure of the A/Vietnam/1203/2004 (H5N1) (PDB: 6CFG). HA1 and HA2 regions are highlighted in green and blue, respectively. Compound OA-10 is shown as yellow sticks. (**B**) Modeled structure of HA complexed with OA-10. Green lines represent hydrogen bond interaction, with the distances indicated; and pink lines represent hydrophobic interaction between OA-10 and HA.

4. Discussion

In the present study, OA-10, a newly synthesized oleanane-type triterpene, exhibits significant antiviral activity against highly pathogenic H5N1 IAV replication with an EC_{50} of 14.0 µM in A549 cell cultures. The in vitro cytotoxicity of OA-10 is quite low, with a CC_{50} of more than 640 µM and a selective index of more than 45. OA-10 also exhibits similar inhibitory effect on other IAV subtypes including PR8, H3N2 and H9N2, with EC_{50} values ranging from 6.7 to 19.6 µM. The time course inhibition study indicates that OA-10 exerts its antiviral effect during early stages of viral infection post binding to cell surface receptors. SPR analyses demonstrate that OA-10 interacts with HA protein strongly. The inhibition of OA-10 on H5N1 IAV induced hemolysis of chicken RBCs at low pH confirms interaction of OA-10 with HA subunit HA2. Furthermore, computer-aided molecular docking analysis suggested that OA-10 might bind to the interface of HA1 and HA2 in the HA stem region, which was known to undergo significant rearrangement during membrane fusion [19].

One IAV replication cycle is orderly composed of virus binding, internalization, RNA replication and viral protein synthesis, assembly, budding and release from infected cells [38]. Through the time course inhibition experiments, we found that OA-10 inhibited IAV replication in the co- and post-treatment modes (Figure 3), but was unable to block IAV binding to A549 cells (Figure S3). To identify the exact stage(s) of IAV replication cycle affected by OA-10, we investigated the time course inhibition within one IAV lifecycle and found that OA-10 exerted its effect during the early stages of IAV infection (Figure 4). During this stage, IAV attachment to target cells is mediated by HA1 via sialic acid-receptor binding, and subsequent virus-endosome membrane fusion is mediated by rearrangement of HA2 at low pH. Our results showed that OA-10 did not block IAV binding to cells at 4 °C (Figure S3). In addition, OA-10 did not inhibit IAV adsorption to chicken RBCs at concentrations of 10 and 20 µM, in spite of the fact that non-classical hemagglutination inhibitions of OA-10 at 40 and 80 µM were observed. These results suggest that the HA1 receptor binding domain is likely not a main target. Given that OA-10 exhibited antiviral activity during early stage(s) of IAV infection cycle, we speculated that OA-10 might target the membrane fusion step mediated by the more conserved hemagglutinin transmembrane subunit HA2. This hypothesis was confirmed by the activity of blocking hemolysis of OA-10 in a low pH environment in a dose-dependent manner (Figure 5C), as hemolysis is mediated by HA2 rather than HA1 [39].

To investigate the binding intensity of OA-10 with HA, SPR analyses were conducted. SPR data showed that OA-10 interacted with HA strongly with KD of 2.98×10^{-12} M (Figure 5B), which was consistent with its potent hemolysis inhibition at low pH. To study the possible binding site, docking simulation analyses were performed, by which a highly conserved hydrophobic cavity at the HA1-HA2 interface in the HA stem region was identified as the possible binding site of OA-10. The binding cavity is formed by residues of the two HA subunits, including Gly20, Val18, Trp21, Lys38 and Ile45 in HA2, and His38 in HA1, which was previously reported to be one of the critical regions responsible for the conformational changes in HA2 at low pH [19]. In addition, this region is recognized by an antibody with a broad-spectrum neutralizing ability to avian and human influenza A viruses [19]; thus, it could serve as a potential drug target for developing IAV entry inhibitors. Indeed, our data indicated that OA-10 might bind to this cavity through hydrogen bonds and hydrophobic interaction.

Influenza A viruses have been classified into 18 hemagglutinin subtypes (H1 to H18), which can be divided phylogenetically into two groups (1 and 2). H1, H5 and H9 belong to Group 1 HA of IAV, while H3 belongs to Group 2 HA of IAV [40]. It is known that the amino acid sequence identities of the HA2 portion between different HA subtypes are much higher than those of the full-length HA proteins [40], which makes HA2 an attractive target for developing broad-spectrum therapeutic antibodies and antiviral drugs. In fact, several broadly neutralizing antibodies (bnAbs) against IAV conserved HA stem have been developed, and their broad protection against IAV infection has been demonstrated in clinical trials. One such examples is the antibody CR6261 for most group 1 IAV subtypes [41,42]. Recently, guided by structural knowledge on the interactions of HA and anti-stem bnAb CR6261, a small molecule JNJ4796 that mimics the bnAb functionality with the ability to inhibit HA-mediated fusion was successfully developed. Importantly, this compound demonstrated potent antiviral activities against H1 and H5 strains in vitro and in vivo [19], but not group 2 IAVs such as H3 and H7 subtypes [19]. An OA derivative Y3 was reported to have significant antiviral activity in vitro against H3 and H1, suggesting broad-spectrum inhibitions of OA derivatives against both Group 1 and Group 2 IAVs.[13]. Further, HR2 in influenza HA2 was recently shown to be the target domain for Y3 [43]. In the present study, we show that OA-10 has promising antiviral activities against four IAV subtypes, including H1N1, H5N1, H9N2 (Group 1 IAVs) and H3N2 (Group 2 IAV), further demonstrating the potential of OA derivatives as broad-spectrum IAV entry inhibitors. To our knowledge, this is the first reported antiviral activity of OA derivatives against H5N1 IAV infection. Further in vivo studies will be carried out to clarify the efficacy of OA-10 as an IAV entry inhibitor in animals.

Another potential advantage of OA-10's antiviral activity is its synergistic inhibition on viral replication when used together with ribavirin, a broad-spectrum virus RNA polymerase inhibitor (Figure 3). Such synergy was likely the result of the simultaneous disruption of HA-mediated viral entry by OA-10 (earlier step) and polymerase-mediated RNA replication by ribavirin (later step). It is expected that using an inhibitor like OA-10 together with ribavirin or NA inhibitors would not only enhance antiviral effects against IAV infection, but also prevent or significantly delay drug resistance.

In summary, we demonstrate that OA-10, a novel synthesized oleanane-type triterpenoid, inhibits four different IAV subtypes, including Group 1 H1N1 (PR8), H5N1 and H9N2, and Group 2 H3N2 infections with potent activity and negligible cytotoxicity in A549 cells. Mechanically, OA-10 blocks the conformational changes of the HA2 subunit at the low pH required for IAV to fuse with an endosomal membrane. These effects are attributed to a conserved hydrophobic cavity in the HA stem region as the likely binding site of OA-10. It could serve as a lead for optimization in order to design novel compounds with improved antiviral potency. OA-10 and its derivatives hold promise to be developed as broad-spectrum anti-influenza drugs.

Supplementary Materials: The following are available online at http://www.mdpi.com/1999-4915/12/2/225/s1: Figure S1: Chemical structures of oleanane acid (OA-0) and its 11 derivatives. OA-8, OA-9, OA-10 and OA-11 were newly synthesized derivatives. Figure S2: OA-10 inhibits IAV PR8 (H1N1), H9N2 and H3N2 replications in A549 cells. Figure S3: OA-10 does not block IAV binding to A549 cells. Figure S4: Amino acid sequence alignment of the full-length HA proteins between strains of A/Vietnam/1203/2004 (H5N1) and A/Duck/Guangdong/99(H5N1).

Author Contributions: J.C. and G.S. designed the study and wrote the paper. M.Y. and Y.L. (Yixian Liao) performed the experiments. L.W. and N.C. participated in the establishment of the methodology and data analysis. W.Q. and Y.L. (Yahong Liu) provided some critical reagents. W.C. thoroughly revised the manuscript and participated in study design. All authors have read and agreed to the published version of the manuscript.

Funding: This research was funded by the National Key Research and Development Program of China (grant number 2017YFD0501404) and the National Natural Science Foundation of China (grant numbers 31572565 and 31872521) to Jianxin Chen.

Conflicts of Interest: The authors declare no conflict of interest. The funders had no role in the design of the study; in the collection, analyses, or interpretation of data; in the writing of the manuscript, or in the decision to publish the results.

References

1. Yen, H.-L.; Webster, R.G. Pandemic influenza as a current threat. *Curr. Top. Microbiol. Immunol.* **2009**, *333*, 3–24.

2. World Health Organization. Influenza (Avian and Other Zoonotic). Available online: https://www.who.int/en/news-room/fact-sheets/detail/influenza-(avian-and-other-zoonotic) (accessed on 13 November 2018).

3. Bender, C.; Hall, H.; Huang, J.; Klimov, A.; Cox, N.; Hay, A.; Gregory, V.; Cameron, K.; Lim, W.; Subbarao, K. Characterization of the surface proteins of influenza A (H5N1) viruses isolated from humans in 1997-1998. *Virology* **1999**, *254*, 115–123. [CrossRef]

4. Hsieh, H.-P.; Hsu, J.T.A. Strategies of development of antiviral agents directed against influenza virus replication. *Curr. Pharm. Des.* **2007**, *13*, 3531–3542. [CrossRef]

5. Xie, Y.; Xu, D.; Huang, B.; Ma, X.; Qi, W.; Shi, F.; Liu, X.; Zhang, Y.; Xu, W. Discovery of N-Substituted Oseltamivir Derivatives as Potent and Selective Inhibitors of H5N1 Influenza Neuraminidase. *J. Med. Chem.* **2014**, *57*, 8445–8458. [CrossRef]

6. Kelly, H.; Cowling, B.J. Influenza: The rational use of oseltamivir. *Lancet* **2015**, *385*, 1700–1702. [CrossRef]

7. Gupta, R.K.; Nguyen-Van-Tam, J.S. Oseltamivir resistance in influenza A (H5N1) infection. *New Engl. J. Med.* **2006**, *354*, 1423–1424.

8. Le, Q.M.; Kiso, M.; Someya, K.; Sakai, Y.T.; Nguyen, T.H.; Nguyen, K.H.L.; Pham, N.D.; Ngyen, H.H.; Yamada, S.; Muramoto, Y.; et al. Avian flu—Isolation of drug-resistant H5N1 virus. *Nature* **2005**, *437*, 1108. [CrossRef] [PubMed]

9. Moscona, A. Global Transmission of Oseltamivir-Resistant Influenza. *New Engl. J. Med.* **2009**, *360*, 953–956. [CrossRef] [PubMed]

10. Liu, S.; Li, R.; Zhang, R.; Chan, C.C.S.; Xi, B.; Zhu, Z.; Yang, J.; Poon, V.K.M.; Zhou, J.; Chen, M.; et al. CL-385319 inhibits H5N1 avian influenza A virus infection by blocking viral entry. *Eur. J. Pharmacol.* **2011**, *660*, 460–467. [CrossRef] [PubMed]

11. Ranaweera, A.; Ratnayake, P.U.; Weliky, D.P. The Stabilities of the Soluble Ectodomain and Fusion Peptide Hairpins of the Influenza Virus Hemagglutinin Subunit II Protein Are Positively Correlated with Membrane Fusion. *Biochemistry* **2018**, *57*, 5480–5493. [CrossRef]

12. Vanderlinden, E.; Naesens, L. Emerging Antiviral Strategies to Interfere with Influenza Virus Entry. *Med. Res. Rev.* **2014**, *34*, 301–339. [CrossRef] [PubMed]

13. Yu, M.; Si, L.; Wang, Y.; Wu, Y.; Yu, F.; Jiao, P.; Shi, Y.; Wang, H.; Xiao, S.; Fu, G.; et al. Discovery of pentacyclic triterpenoids as potential entry inhibitors of influenza viruses. *J. Med. Chem.* **2014**, *57*, 10058–10071. [CrossRef]

14. Bodian, D.L.; Yamasaki, R.B.; Buswell, R.L.; Stearns, J.F.; White, J.M.; Kuntz, I.D. Inhibition of the fusion-inducing conformational change of influenza hemagglutinin by benzoquinones and hydroquinones. *Biochemistry* **1993**, *32*, 2967–2978. [CrossRef] [PubMed]

15. Luo, G.; Torri, A.; Harte, W.E.; Danetz, S.; Cianci, C.; Tiley, L.; Day, S.; Mullaney, D.; Yu, K.-L.; Ouellet, C.; et al. Molecular mechanism underlying the action of a novel fusion inhibitor of influenza A virus. *J. Virol.* **1997**, *71*, 4062–4070. [CrossRef] [PubMed]

16. Plotch, S.J.; O'Hara, B.; Morin, J.; Palant, O.; Larocque, J.; Bloom, J.D.; Lang, S.A., Jr.; Digrandi, M.J.; Bradley, M.; Nilakantan, R.; et al. Inhibition of influenza A virus replication by compounds interfering with the fusogenic function of the viral hemagglutinin. *J. Virol.* **1999**, *73*, 140–151. [CrossRef]

17. Yoshimoto, J.; Kakui, M.; Iwasaki, H.; Fujiwara, T.; Sugimoto, H.; Hattori, N. Identification of a novel HA conformational change inhibitor of human influenza virus. *Arch. Virol.* **1999**, *144*, 865–878. [CrossRef]

18. Vanderlinden, E.; Goktas, F.; Cesur, Z.; Froeyen, M.; Reed, M.L.; Russell, C.J.; Cesur, N.; Naesens, L. Novel Inhibitors of Influenza Virus Fusion: Structure-Activity Relationship and Interaction with the Viral Hemagglutinin. *J. Virol.* **2010**, *84*, 4277–4288. [CrossRef]

19. Van Dongen, M.J.P.; Kadam, R.U.; Juraszek, J.; Lawson, E.; Brandenburg, B.; Schmitz, F.; Schepens, W.B.G.; Stoops, B.; van Diepen, H.A.; Jongeneelen, M.; et al. A small-molecule fusion inhibitor of influenza virus is orally active in mice. *Science* **2019**, *363*, 1056. [CrossRef]

20. Wang, C.; Lu, L.; Na, H.; Li, X.; Wang, Q.; Jiang, X.; Xu, X.; Yu, F.; Zhang, T.; Li, J.; et al. Conjugation of a Nonspecific Antiviral Sapogenin with a Specific HIV Fusion Inhibitor: A Promising Strategy for Discovering New Antiviral Therapeutics. *J. Med. Chem.* **2014**, *57*, 7342–7354. [CrossRef]

21. Xiao, S.; Tian, Z.; Wang, Y.; Si, L.; Zhang, L.; Zhou, D. Recent progress in the antiviral activity and mechanism study of pentacyclic triterpenoids and their derivatives. *Med. Res. Rev.* **2018**, *38*, 951–976. [CrossRef]

22. Grishko, V.V.; Galaiko, N.V.; Tolmacheva, I.A.; Kucherov, I.I.; Eremin, V.F.; Boreko, E.I.; Savinova, O.V.; Slepukhin, P.A. Functionalization, cyclization and antiviral activity of A-secotriterpenoids. *Eur. J. Med. Chem.* **2014**, *83*, 601–608. [CrossRef] [PubMed]

23. Dang, Z.; Ho, P.; Zhu, L.; Qian, K.; Lee, K.-H.; Huang, L.; Chen, C.-H. New Betulinic Acid Derivatives for Bevirimat-Resistant Human Immunodeficiency Virus Type-1. *J. Med. Chem.* **2013**, *56*, 2029–2037. [CrossRef] [PubMed]

24. Yu, F.; Wang, Q.; Zhang, Z.; Peng, Y.; Qiu, Y.; Shi, Y.; Zheng, Y.; Xiao, S.; Wang, H.; Huang, X.; et al. Development of Oleanane-Type Triterpenes as a New Class of HCV Entry Inhibitors. *J. Med. Chem.* **2013**, *56*, 4300–4319. [CrossRef] [PubMed]

25. Yu, F.; Peng, Y.; Wang, Q.; Shi, Y.; Si, L.; Wang, H.; Zheng, Y.; Lee, E.; Xiao, S.; Yu, M.; et al. Development of bivalent oleanane-type triterpenes as potent HCV entry inhibitors. *Eur. J. Med. Chem.* **2014**, *77*, 258–268. [CrossRef] [PubMed]

26. Song, G.; Yang, S.; Zhang, W.; Cao, Y.; Wang, P.; Ding, N.; Zhang, Z.; Guo, Y.; Li, Y. Discovery of the First Series of Small Molecule H5N1 Entry Inhibitors. *J. Med. Chem.* **2009**, *52*, 7368–7371. [CrossRef] [PubMed]

27. Song, G.; Shen, X.; Li, S.; Li, Y.; Si, H.; Fan, J.; Li, J.; Gao, E.; Liu, S. Structure-activity relationships of 3-O-beta-chacotriosyl oleanane-type triterpenoids as potential H5N1 entry inhibitors. *Eur. J. Med. Chem.* **2016**, *119*, 109–121. [CrossRef]

28. Luo, L.; Han, W.; Du, J.; Yang, X.; Duan, M.; Xu, C.; Zeng, Z.; Chen, W.; Chen, J. Chenodeoxycholic Acid from Bile Inhibits Influenza A Virus Replication via Blocking Nuclear Export of Viral Ribonucleoprotein Complexes. *Molecules* **2018**, *23*, 3315. [CrossRef]

29. Granados, A.; Petrich, A.; McGeer, A.; Gubbay, J.B. Measuring influenza RNA quantity after prolonged storage or multiple freeze/thaw cycles. *J. Virol. Methods* **2017**, *247*, 45–50. [CrossRef]

30. Cai, W.; Li, Y.; Chen, S.; Wang, M.; Zhang, A.; Zhou, H.; Chen, H.; Jin, M. 14-Deoxy-11,12-dehydroandrographolide exerts anti-influenza A virus activity and inhibits replication of H5N1 virus by restraining nuclear export of viral ribonucleoprotein complexes. *Antivir. Res.* **2015**, *118*, 82–92. [CrossRef]

31. Rao, X.; Huang, X.; Zhou, Z.; Lin, X. An improvement of the 2^(-delta delta CT) method for quantitative real-time polymerase chain reaction data analysis. *Biostat. Bioinform. Biomath.* **2013**, *3*, 71–85.

32. Schmittgen, T.D.; Livak, K.J. Analyzing real-time PCR data by the comparative C-T method. *Nat. Protoc.* **2008**, *3*, 1101–1108. [CrossRef] [PubMed]

33. Basu, A.; Antanasijevic, A.; Wang, M.; Li, B.; Mills, D.M.; Ames, J.A.; Nash, P.J.; Williams, J.D.; Peet, N.P.; Moir, D.T.; et al. New Small Molecule Entry Inhibitors Targeting Hemagglutinin-Mediated Influenza A Virus Fusion. *J. Virol.* **2014**, *88*, 1447–1460. [CrossRef] [PubMed]

34. Liao, Q.; Qian, Z.; Liu, R.; An, L.; Chen, X. Germacrone inhibits early stages of influenza virus infection. *Antivir. Res.* **2013**, *100*, 578–588. [CrossRef] [PubMed]

35. Senne, D.A.; Panigrahy, B.; Kawaoka, Y.; Pearson, J.E.; Suss, J.; Lipkind, M.; Kida, H.; Webster, R.G. Survey of the hemagglutinin (HA) cleavage site sequence of H5 and H7 avian influenza viruses: Amino acid sequence at the HA cleveage site as a marker of pathogenicity potential. *Avian Dis.* **1996**, *40*, 425–437. [CrossRef]

36. Ye, M.; College of Veterinary Medicine, South China Agricultural University, Guangzhou, China. Personal Communication, 2019.

37. Kadam, R.U.; Wilson, I.A. Structural basis of influenza virus fusion inhibition by the antiviral drug Arbidol. *Proc. Natl. Acad. Sci. USA* **2017**, *114*, 206–214. [CrossRef]

38. Edinger, T.O.; Pohl, M.O.; Stertz, S. Entry of influenza A virus: Host factors and antiviral targets. *J. Gen. Virol.* **2014**, *95*, 263–277. [CrossRef]

39. Lin, D.; Luo, Y.; Yang, G.; Li, F.; Xie, X.; Chen, D.; He, L.; Wang, J.; Ye, C.; Lu, S.; et al. Potent influenza A virus entry inhibitors targeting a conserved region of hemagglutinin. *Biochem. Pharmacol.* **2017**, *144*, 35–51. [CrossRef]

40. Russell, R.J.; Kerry, P.S.; Stevens, D.J.; Steinhauer, D.A.; Martin, S.R.; Gamblin, S.J.; Skehel, J.J. Structure of influenza hemagglutinin in complex with an inhibitor of membrane fusion. *Proc. Natl. Acad. Sci. USA* **2008**, *105*, 17736–17741. [CrossRef]

41. Ekiert, D.C.; Bhabha, G.; Elsliger, M.-A.; Friesen, R.H.E.; Jongeneelen, M.; Throsby, M.; Goudsmit, J.; Wilson, I.A. Antibody Recognition of a Highly Conserved Influenza Virus Epitope. *Science* **2009**, *324*, 246–251. [CrossRef]

42. Throsby, M.; van den Brink, E.; Jongeneelen, M.; Poon, L.L.M.; Alard, P.; Cornelissen, L.; Bakker, A.; Cox, F.; van Deventer, E.; Guan, Y.; et al. Heterosubtypic Neutralizing Monoclonal Antibodies Cross-Protective against H5N1 and H1N1 Recovered from Human IgM(+) Memory B Cells. *PLoS ONE* **2008**, *3*, e3942. [CrossRef]

43. Si, L.; Meng, K.; Tian, Z.; Sun, J.; Li, H.; Zhang, Z.; Soloveva, V.; Li, H.; Fu, G.; Xia, Q.; et al. Triterpenoids manipulate a broad range of virus-host fusion via wrapping the HR2 domain prevalent in viral envelopes. *Sci. Adv.* **2018**, *4*. [CrossRef] [PubMed]

Article

Brevilin A, a Sesquiterpene Lactone, Inhibits the Replication of Influenza A Virus In Vitro and In Vivo

Xiaoli Zhang [1,*,†]**, Yiping Xia** [1,†]**, Li Yang** [1]**, Jun He** [2]**, Yaolan Li** [3] **and Chuan Xia** [1,*]

1 Department of Biotechnology, College of Life Science and Technology, Jinan University, Guangzhou 510632, Guangdong, China
2 Institute of Laboratory Animal Science, Jinan University, Guangzhou 510632, Guangdong, China
3 Institute of Traditional Chinese Medicine and Natural Products, Jinan University, Guangzhou 510632, Guangdong, China
* Correspondence: xiaolizhang@jnu.edu.cn (X.Z.); xiachuan@jnu.edu.cn (C.X.)
† These authors contributed equally to this work.

Received: 8 August 2019; Accepted: 6 September 2019; Published: 8 September 2019

Abstract: With the emergence of drug-resistant strains of influenza A viruses (IAV), new antivirals are needed to supplement the existing counter measures against IAV infection. We have previously shown that brevilin A, a sesquiterpene lactone isolated from *C. minima*, suppresses the infection of influenza A/PR/8/34 (H1N1) in vitro. Here, we further investigate the antiviral activity and mode of action of brevilin A against different IAV subtypes. Brevilin A inhibited the replication of influenza A H1N1, H3N2, and H9N2 viruses in vitro. The suppression effect of brevilin A was observed as early as 4–8 hours post infection (hpi). Furthermore, we determined that brevilin A inhibited viral replication in three aspects, including viral RNA (vRNA) synthesis, expression of viral mRNA, and protein encoded from the M and NS segments, and nuclear export of viral ribonucleoproteins (vRNPs). The anti-IAV activity of brevilin A was further confirmed in mice. A delayed time-to-death with 50% surviving up to 14 days post infection was obtained with brevilin A (at a dose of 25 mg/kg) treated animals compared to the control cohorts. Together, these results are encouraging for the exploration of sesquiterpene lactones with similar structure to brevilin A as potential anti-influenza therapies.

Keywords: influenza A virus; brevilin A; antiviral; sesquiterpene lactone; replication

---

1. Introduction

Influenza A viruses (IAV) are a major cause of respiratory infection in humans and are responsible for significant morbidity and mortality worldwide. Annually, seasonal influenza epidemics affect approximately 5%–15% of the global population, resulting in 290,000–650,000 deaths [1,2]. Vaccination programs are important for preventing and controlling influenza. However, the efficacy of vaccination is typically only 40%–60% and can be lower than 20% during years of vaccine mismatch [3]. Thus, antiviral therapies are a vital option for the treatment of influenza.

Until now, three types of antivirals have been approved by the FDA for influenza prevention and therapies, including M2 ion channel inhibitors (i.e., adamantanes and rimantadine) [4], neuraminidase inhibitors (NAIs, i.e., oseltamivir and zanamivir) [5], and a cap-dependent endonuclease inhibitor [6,7]. However, the M2 inhibitors are no longer used clinically as currently almost all circulating IAV strains are resistant to adamantanes [8]. Moreover, the 2008–2009 seasonal H1N1 influenza virus strain in North America presented nearly complete resistance to oseltamivir [9]. Baloxavir marboxil was approved for treating influenza last year, but recent work has shown that viral resistance is still a concern [6,10,11]. Amidst concerns about drug resistance, the development of novel antivirals with distinct mechanisms of action is necessary.

Brevilin A (chemical structure shown in Figure 1) is a sesquiterpene lactone isolated from medicinal herb *Centipeda minima*. As a major constituent of *C. minima* [12], it has been reported that brevilin A displays multiple activities such as anti-tumor [13–17], anti-bacterial [18], and antiprotozoal [19]. We previously evaluated the antiviral activity of 16 sesquiterpene lactones isolated from *C. minima* against influenza A/PR/8/34 (H1N1) virus in vitro. Eight of them showed significant antiviral activity. Among them, brevilin A exhibited the strongest antiviral effect [20], but the mechanism of this antiviral effect was not extensively studied. Here, we extend our previous findings by investigating the antiviral effects of brevilin A against various IAVs and mode of actions in vitro at a noncytotoxic concentration. We found that brevilin A exhibits significant antiviral activities against all tested IAV strains, and it inhibits the vRNA synthesis and the expression of some viral proteins. Furthermore, the anti-IAV effect of this compound in vivo was also evaluated.

Figure 1. The chemical structure of brevilin A.

2. Materials and Methods

2.1. Compounds and Reagents

Brevilin A (purity >95% by HPLC) was isolated from the supercritical fluid extract of *C. minima*. Ribavirin was purchased from Sigma-Aldrich (St. Louis, MO, USA). Both compounds were dissolved in DMSO to prepare a solution with the concentration of 50 mM and stored at −20 °C for in vitro experiments. Brevilin A did not show cytotoxicity in Madin–Darby canine kidney (MDCK) epithelial cells up to 8 μM, which was used as the maximum concentration for in vitro antiviral assays. For in vivo experiments, brevilin A was dissolved in 10% Lipovenos containing 0.2% DMSO, 10% PEG300, and 2.5% glycerol, while oseltamivir carboxylate (Tamiflu, Roche, Basel, Switzerland), purchased from Guangzhou Overseas Chinese Hospital (Guangzhou, China), was dissolved in distilled water. Leptomycin B (LMB, a nuclear export inhibitor) solution was obtained from Beyotime Institute of Biotechnology, Shanghai, China.

Mouse anti-IAV NP (ab128193) and M2 (ab5416) antibodies, mouse anti-GAPDH antibody (ab181603), and donkey anti-mouse IgG (H + L) secondary antibody (ab150105) were purchased from Abcam Company Ltd., Shanghai, China. Mouse anti-IAV HA (GTX28262), NA (GTX629696) and M1 (GTX125928) antibodies, rabbit anti-IAV NS1 (GTX125990) and NS2 (GTX125953) antibodies were obtained from GeneTex, Alton Pkwy Irvine, CA, USA.

2.2. Cells and Viruses

Madin–Darby canine kidney (MDCK) epithelial cells (obtained from the Key Laboratory of Veterinary Vaccine Innovation of the Ministry of Agriculture, P. R. China) were maintained as monolayers in Dulbecco's modified Eagle's medium (DMEM, Gibco, Beijing, China) supplemented with 10% fetal bovine serum (FBS, Gemini, Calabasas, CA, USA), and 1% penicillin/streptomycin (Pen/Strep, Gibco) at 37 °C, 5% CO_2.

Influenza A/PR/8/34 H1N1 (PR8, a gift from the Key Laboratory of Veterinary Vaccine Innovation of the Ministry of Agriculture, P. R. China), A/FM/1/47 H1N1, A/Hong Kong/498/97 H3N2, and A/chicken/Guangdong/1996 H9N2 viruses were grown in 10-day-old chicken embryos, titrated on

MDCK cells, and then stored at −80 °C. For virus infection, cells were washed with PBS and infected with virus in PBS containing 0.3% BSA (Sigma) and 1% Pen/Strep for 1 h at 37 °C. The inoculum was aspirated, and cells were incubated in infection medium supplemented with DMEM, 0.3% BSA, 2 µg/mL TPCK-trypsin (Sigma-Aldrich), and 1% Pen/Strep.

2.3. Animals

Female BALB/c mice, six to eight-week-old (average weight, 16.0 ± 2.0 g), were obtained from Beijing Vital River Laboratory Animal Technology Co., Ltd. (Beijing, China). The mice were quarantined 2–3 days prior to the experimental manipulation and were fed standard rodent chow and had ad libitum access to water. Animal experiments were conducted under the guidance of both the Guangdong Provincial Center for Disease Control and Prevention's Institutional Animal Care and Use Committee. The protocols were approved by South China Agricultural University' committee on the Ethics of Animal Experiments of Animal Biosafety Level 3 (permit no. 2017A002).

2.4. Plaque Assay and Plaque Reduction Assay

Plaque assay was performed as described previously [21]. Briefly, MDCK cells (8×10^5/well) were seeded into 6-well plates and incubated overnight. The infected cells were overlaid with F12-DMEM (Gibco) containing 2% Oxoid agar, 0.2% BSA, DEAE dextran, and 2 µg/mL TPCK-treated trypsin, and further incubated for 48–72 h. The cells were fixed with 4% paraformaldehyde for 30 min and then stained with 0.1% crystal violet. Virus titers were determined by counting the PFU (plaques) for each sample and expressed as PFU/mL.

The plaque reduction assay was performed to determine EC_{50}, as described previously [22]. Briefly, monolayer MDCK cells were incubated with virus (~50 pfu/well) at 37 °C for 1 h, rocking every 15 min. Cells were washed twice with PBS and an agar overlay with or without brevilin A (0.5–8 µM) or ribavirin (5–30 µM) was added to each well. After 48–72 h incubation, the cells were fixed and plaques were counted by crystal violet staining. The concentration required to reduce the plaque number by 50% (EC_{50}) was calculated using the log (inhibitor) versus response logistic nonlinear regression equation in Graphpad Prime 6.0 software (LaJolla, CA, USA).

2.5. Immunofluorescence

At the indicated time points after infection, MDCK cells were washed with PBS three times and fixed in 4% paraformaldehyde for 15 min at 4 °C and permeabilized with 0.25% triton in PBS for 15 min at room temperature, and then incubated for 1 h with anti-NP (1:100) monoclonal antibody. After washing with PBS, the cells were incubated for 1 h with the secondary antibodies, Alexa Fluor 488-conjugated goat anti-mouse IgG (H + L) (1:1000; Thermo Fisher, Waltham, MA, USA) antibody. Nuclei were counterstained with DAPI. Cells were observed under the Leica confocal microscope.

2.6. Western Blot Assay

Confluent MDCK cell monolayers were first infected with PR8 (MOI = 1), and then treated with brevilin A or vehicle. At the indicated time of 4 or 8 hour-post-infection (hpi), proteins from total cell lysates were separated using SDS-PAGE and transferred onto a polyvinylidene difluoride membrane (Millipore, Burlington, MA, USA). The membrane was blocked with 5% nonfat milk or 5% BSA in PBS containing 0.1% Tween 20 and incubated overnight at 4 °C with primary mouse anti-HA (1:1000), mouse anti-NA (1:1000), mouse anti-M1 (1:3000), mouse anti-M2 (1:1000), rabbit anti-NS1 (1:1000), rabbit anti-NS2 (1:1000), and mouse anti-NP (1:1000) antibody, or mouse anti-GAPDH and mouse anti-β-actin antibody as control. The membrane was washed five times for 5 min with PBS-Tween and incubated for 1 h at room temperature with the respective secondary antibodies conjugated to horseradish peroxidase (HRP). After five washes for 5 min with PBS-Tween buffer, the chemiluminescence of the labeled proteins was visualized with HRP substrate and captured using the LICOR Odyssey imaging system. The relative densities of proteins were all determined by using ImageJ (NIH) v.1.46r (Wayen Rasband, US National Institutes of Health, Bethesda, MD, USA).

2.7. Real-Time Quantitative PCR (RT-qPCR)

Total RNAs were extracted from MDCK cells at 6, 12, or 24 hpi, using the RNAfast200 kit (Fastagen Biotech, Shanghai, China) according to the Manufacturer's instructions. Reverse transcription (RT) was carried out on 0.5 µg of total RNA by using PrimeScript RT reagent kit (Takara, Tokyo, Japan). Reverse transcription (RT) was conducted using specific oligonucleotides for vRNA (5′-AGC AAA AGC AGG-3′), cRNA (5′-AGT AGA AAC AAG G-3′), and mRNA [oligo(dT)]. A glyceraldehyde-3-phosphate dehydrogenase (GAPDH)-specific primer (5′-GAA GAT GGT GAT GGG ATT TC-3′) was also included in the RT reaction mixture for vRNA or cRNA analysis. The quantitative real-time PCR was performed in a 20-µL reaction mixture containing 50 nM forward and reverse primers (NP forward primer 5′-GAT TGG AAT TGG ACG AT-3′, reverse primer 5′-AGA GCA CCA TTC TCT CTA TT-3′; M1 forward primer 5′-AAG ACC AAT CCT GTC ACC TCT GA-3′, reverse primer 5′-CAA AGC GTC TAC GCT GCA GTC C-3′; M2 forward primer 5′- CCG AGG TCG AAA CGC CTA TC-3′, reverse primer 5′-CTT TGG CAC TCC TTC CGT AG-3′; NS1 forward primer 5′-CTT CGC CGA GAT CAG AAA TC-3′, reverse primer 5′-TGG ACC AGT CCC TTG ACA TT-3′; NS2 forward primer 5′-GTT GGC GAA ATT TCA CCA TTG CCT TCT CT-3′, reverse primer 5′-TTA AAT AAG CTG AAA TGA GAA AGT TCT-3′), 1 × SYBR green master mix (Takara Biotech, Dalian, China), and various amounts of template. To quantify the changes in gene expression, the change in threshold cycle (ΔC_T) method was used to calculate the relative changes normalized against the GAPDH gene (forward primer, 5′-AAT TCC ACG GCA CAG TCA AGG C-3′; reverse primer, 5′-AAC ATA CTC AGC ACC AGC ATC ACC-3′). The C_T was defined as the cycle at which fluorescence was determined to be significantly greater than the background. The ratio of viral RNA to the internal control was normalized to the control level 0 h after infection, which was arbitrarily set equal to 1.0.

2.8. In Vivo Experiments

Six to eight-week-old female BALB/c mice (average weight, 16.0 ± 2.0 g) were separated into 4 groups. Pretreatment was done by administering one dose of brevilin A (25 mg/kg or 10 mg/kg) or solvent (10% Lipovenos containing 0.2% DMSO, 10% PEG300 and 2.5% glycerol) intraperitoneally (i.p.) every other day in a volume of 0.1 mL for 6 days starting 1 hpi. The reference drug oseltamivir (20 mg/kg) was applied to mice orally via gavage once a day in a volume of 0.1 mL for 6 days starting 1 hpi. Mice were anesthetized with isoflurane (RWD R450, Shenzhen, China) and inoculated intranasally with 50 µL of a solution containing PR8 virus. Control animals were treated with distilled water. Body weight and survival were monitored daily for 14 days.

2.9. Statistical Analysis

Statistical significance was analyzed using GraphPad Prism 6 (GraphPad Software Inc.), and *, $p \leq 0.05$; **, $p \leq 0.01$; ***, $p \leq 0.001$ were considered statistically significant. For paired samples, a paired t test was performed; otherwise, an unpaired Student t test was used. Differences in group survival were analyzed using Log-rank (Mantel-Cox) test. Error bars represent means ± standard deviations (SD).

3. Results

3.1. Brevilin A Shows a Broad-Spectrum Antiviral Activity against IAV

In our previous work, brevilin A showed potent antiviral activity against PR8 virus assessed by cytopathogenic effect (CPE) reduction assay and the cell viability assay [20]. To further verify its anti-IAV activity, brevilin A was tested in a plaque reduction assay using several IAV strains including A/PR/8/34 H1N1, A/FM/1/47 H1N1, A/Hong Kong/498/97 H3N2, and A/chicken/Guangdong/1996

H9N2 viruses. Ribavirin served as a positive control. The concentration for 50% of maximal effect (EC_{50}) of brevilin A obtained with PR8 for viral plaque formation was calculated to be 2.96 ± 1.10 μM. This result concurs with the EC_{50} of 1.75 ± 0.59 μM that we evaluated in previous work. Comparable to PR8, the EC_{50} values of brevilin A obtained with H1N1 (FM1), H3N2, and H9N2 were 1.60 ± 1.14, 3.28 ± 1.09, and 2.07 ± 1.12 μM, respectively (Table 1). While the EC_{50} of ribavirin obtained with these four IAV strains were between 7.05 to 10.76 μM. These results indicate that brevilin A exhibits better anti-IAV activity than ribavirin, and the effects of both are not IAV type/subtype specific. In order to test whether brevilin A possesses antiviral activity against other RNA viruses, the effect of brevilin A on respiratory syncytial virus (RSV) was evaluated by a CPE reduction assay. However, brevilin A did not show inhibitory effect on RSV at a noncytotoxic concentration.

Table 1. Anti-IAV activities of brevilin A.

Comp.	IAV	EC_{50} (μM) [a]	SI [b]
brevilin A	A/PR/8/34 H1N1	2.96 ± 1.10	8
	A/FM/1/47 H1N1	1.60 ± 1.14	14
	A/Hong Kong/498/97 H3N2	3.28 ± 1.09	7
	A/chicken/Guangdong/1996 H9N2	2.07 ± 1.12	11
ribavirin	A/PR/8/34 H1N1	7.05 ± 1.10	>14
	A/FM/1/47 H1N1	9.19 ± 1.02	>20
	A/Hong Kong/498/97 H3N2	10.76 ± 1.07	>18
	A/chicken/Guangdong/1996 H9N2	10.35 ± 1.04	>18

[a] Effective concentration required for reducing virus-induced plaque number by 50%. [b] Selectivity index, CC_{50}/EC_{50}.

3.2. Brevilin A Inhibits Progeny Virus Production in Various Virus-To-Cell Ratios

To examine to what extent the anti-IAV activities of brevilin A is affected by virus-to-cell ratio, the cells were infected with PR8 at a MOI (MOI, defined as the ratio of input infectious viral particles per target cell) of 0.001 or 1 in the presence of either brevilin A (8 μM) or vehicle control (DMSO). Virus titers in the supernatants at the indicated time points were quantified by plaque assays. As shown in Figure 2A, after infection with virus at a MOI of 0.001, the amount of progeny virus in the supernatants increased over the incubation time and peaked at 48 hpi in vehicle control, while treatment with brevilin A could significantly reduce the production of infectious virus from cells at 24 or 48 hpi. Even when cells were infected with virus at a higher MOI (MOI = 1), treatment of brevilin A also significantly decreased virus production by about 10-fold at 8 and 12 hpi (Figure 2B). These findings imply that the treatment of brevilin A strongly suppresses the replication of IAV, of note, the inhibitory activity of brevilin A is still rather effective against a relatively higher dose of input virus.

Additionally, we also analyzed the impact of brevilin A on replication of H1N1 (FM1), H3N2, or H9N2 in MDCK cells over multiple replication cycles. As shown in Figure 2C–E, compared to the vehicle, virus titers at each time point were markedly reduced by treatment with brevilin A.

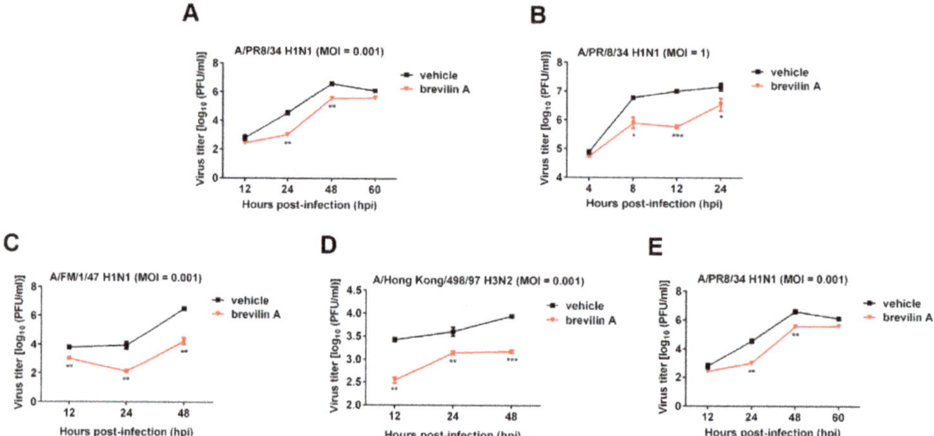

Figure 2. The inhibitory effect of brevilin A on the growth curves of various influenza A viruses (IAV) strains. Madin–Darby canine kidney (MDCK cells) were infected with influenza A/PR/8/34 H1N1 virus at a MOI of 0.001 (**A**) or 1 (**B**), or A/FM/1/47 H1N1 virus (**C**), A/Hong Kong/498/97 H3N2 virus (**D**), or A/chicken/Guangdong/1996 H9N2 virus (**E**) at a MOI of 0.001. Cells were then treated with 8 µM of brevilin A or vehicle. At the indicated time points after infection, virus titers in the supernatants were determined by a plaque assay. The data represent means ± SD. *, $p < 0.05$; **, $p < 0.01$; ***, $p < 0.001$ are considered statistically significant, compared to vehicle.

3.3. Brevilin A Is Effective at the Viral Genome Replication and Translation Stage

The life cycle of influenza virus is around 8–10 h and is divided into three steps: virus entry (0–2 h), viral genome replication and translation (2–8 h), and progeny virion release (8–10 h) [23]. To investigate the stage of viral cycle where brevilin A exhibits its activity against PR8 virus, we next performed time of addition experiments. MDCK cells were infected with PR8 virus, and brevilin A was added or removed at the indicated time points. The expression levels of influenza M2 protein in infected cells were measured at four time-intervals, 0–2, 2–4, 4–8 and 8–10 hpi. We found that the M2 level at the interval 4–8 hpi was reduced around 70%, compared to the vehicle. In contrast, no antiviral activity was detected for the remaining three intervals (0–2, 2–4, 8–10 hpi) (Figure 3A). These data indicate that brevilin A is effective at the viral genome replication and translation stage. No inhibitory effect was observed at virus entry stage or progeny virion assembly/release stage.

We then performed two other experiments to examine the mode of action of brevilin A. First, brevilin A was added to the IAV infected cells at 0, 4, 6, or 8 hpi, the supernatant was collected at 24 hpi, and the virus titers were determined by plaque assay. The virus titers were decreased when brevilin A was added at 0–4 hpi, compared to vehicle. However, similar titers were observed in the treatment and vehicle control when the compound was added 6 h after infection (Figure 3B). Moreover, the results obtained with an immunofluorescent assay showed that compared to the vehicle control, the expression of viral protein M2 was markedly reduced when brevilin A was added at 0 to 4 hpi, and addition of brevilin A at 6 hpi still had a minor impact on M2 expression (Figure 3C). These data suggest that brevilin A blocks the intermediate stage (s) of the influenza virus life cycle between approximately 4 to 6 hpi.

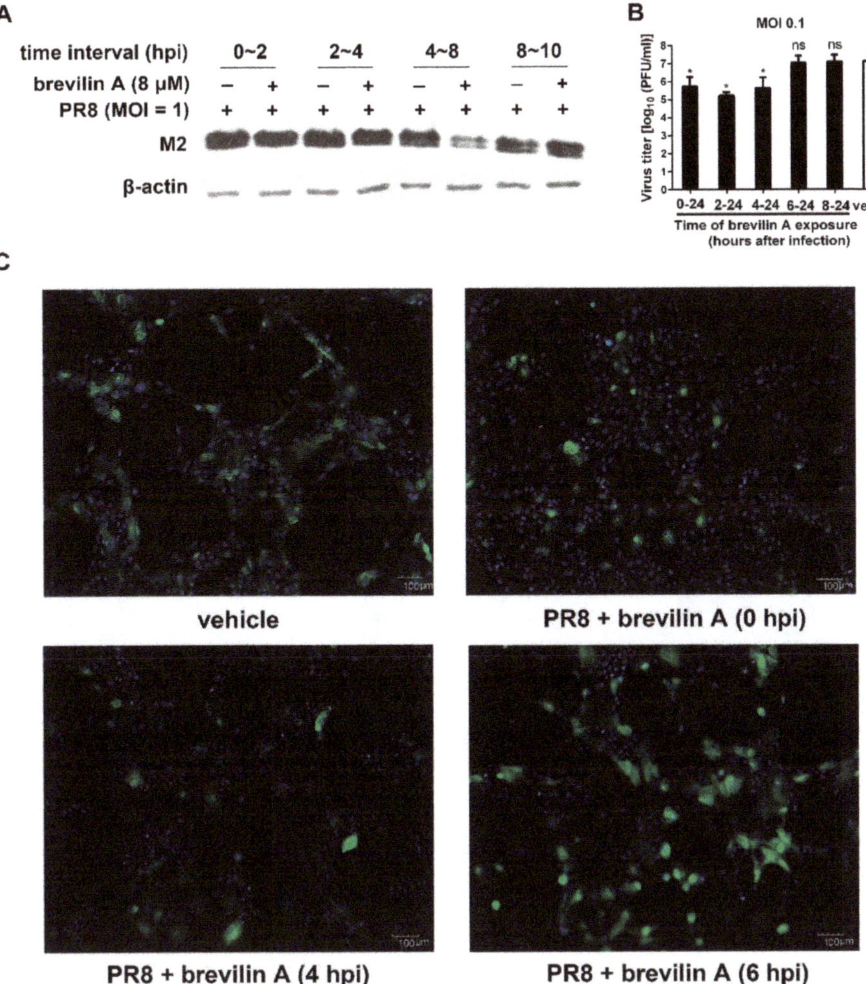

Figure 3. Influence of different treatment conditions of brevilin A on IAV replication. (**A**) MDCK cells were infected with PR8 at a MOI of 1, brevilin A or vehicle was present at four time-intervals, 0–2, 2–4, 4–8, and 8–10 hpi. At 12 hpi, cell lysates were analyzed by Western blot assay; (**B**) MDCK cells were infected with PR8 at a MOI of 0.1, then treated with brevilin A (8 µM) at 0, 2, 4, or 6 hpi. The virus titers in the supernatant were determined by plaque assay at 24 hpi; (**C**) MDCK cells infected with PR8 (MOI = 1) were treated with brevilin A at the indicated times. The M2 protein expression was determined by immunofluorescence assay at 12 hpi. The scale bar in the images is 100 µm. The data represent means ± SD. *, $p < 0.05$ is considered statistically significant compared to vehicle.

3.4. Brevilin A Inhibits Influenza Viral RNA Synthesis

Transcription of vRNA produces mRNA and complementary RNA (cRNA). The former serve as the template for synthesis of viral proteins, and the latter are templates for synthesis of more vRNA for production of new virions [24]. We next performed RT-qPCR to evaluate the effect of brevilin A on viral RNA synthesis. MDCK cells were infected with PR8 at 1 MOI, then treated with brevilin A or DMSO. Total RNA was extracted at 6, 12, or 24 hpi and reverse transcription was performed with specific primers for viral NP gene. At 6 hpi, when the viral RNA synthesis was already completed [25], the levels of vRNA in brevilin A-treated samples were significantly reduced in comparison to those in vehicle treated cells by 50%. In contrast, the levels of mRNA were only decreased by ~20% and cRNA was not altered upon the treatment with brevilin A (Figure 4A). We next tested several later time points post-infection to determine the inhibitory effects of brevilin A during multiple replication cycles of IAV. The effect of brevilin A on cRNA levels was not significant at 12 or 24 hpi (Figure 4B). However, the inhibition of mRNA levels became notable at 24 hpi with a reduction of about 45% (Figure 4C). These results indicate that the inhibitory effect of brevilin A is intimately involved with the vRNA synthesis.

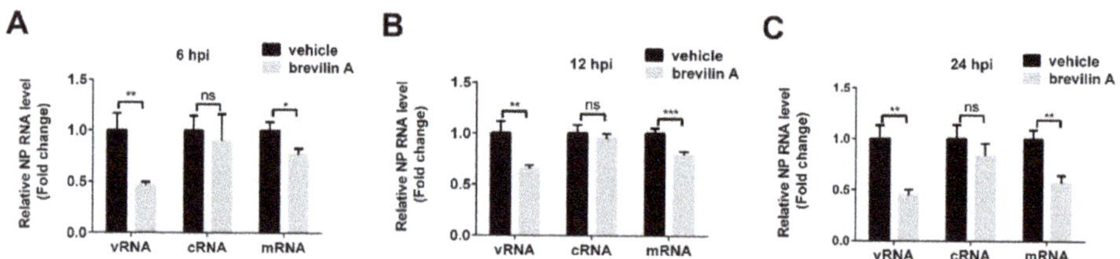

Figure 4. The inhibitory effects of brevilin A on the expression of virus RNA. MDCK cells were infected with PR8 at a MOI of 1, then treated with brevilin A or vehicle. Total RNA was extracted at 6 hpi (**A**), 12 hpi (**B**), or 24 hpi (**C**), and the vRNA, cRNA, and mRNA were quantified by RT-qPCR. *, $p < 0.05$; **, $p < 0.01$; ***, and $p < 0.001$ are considered statistically significant, compared to vehicle.

3.5. Brevilin A Decreases the Levels of Viral mRNA and Proteins Expressed from the M and NS Segments

Next, we investigated the expression pattern of viral proteins during IAV infection observed in the presence of brevilin A. For this purpose, MDCK cells were infected with PR8 (MOI = 1) and treated with brevilin A at a concentration of 8 µM. At the time point of 4 hpi, only the expressions of NA, M1, and NS1 were observed in the control cells. Treatment with brevilin A resulted in about 20% reduction of these proteins (Figure 5A). At 8 hpi, the expressions of viral NA, HA, and NP were not affected by brevilin A, and a modest reduction of M1 (~30%) was observed upon brevilin A treatment. However, the reduction of M2, NS1, and NS2 proteins was much more dramatic: about 70% in M2, 50% in NS1, and 70% in NS2, respectively (Figure 5A). The M2 protein is translated from M2 mRNA, which is produced from the alternative splicing of M1; the NS1 and NS2 proteins were respectively translated from NS1 and NS2 mRNA generated from the NS segment. We next performed RT-qPCR to determine the effect of brevilin A on the mRNA expressions of M1, M2, NS1, and NS2. We infected MDCK cells with PR8 virus and then treated cells with brevilin A. Whole-cell RNA was isolated from the infected cells at 6 hpi and quantified for mRNA using RT-qPCR assay. The M2 mRNA was markedly reduced following treatment with brevilin A, with a reduction of 55%–60%, whereas the reduction rates of M1, NS1, and NS2 mRNA levels were similar to that of NP (with a reduction of ~20%) compared to vehicle (Figure 5B–E). These results suggest that brevilin A reduces the mRNA and protein expressions of viral M and NS.

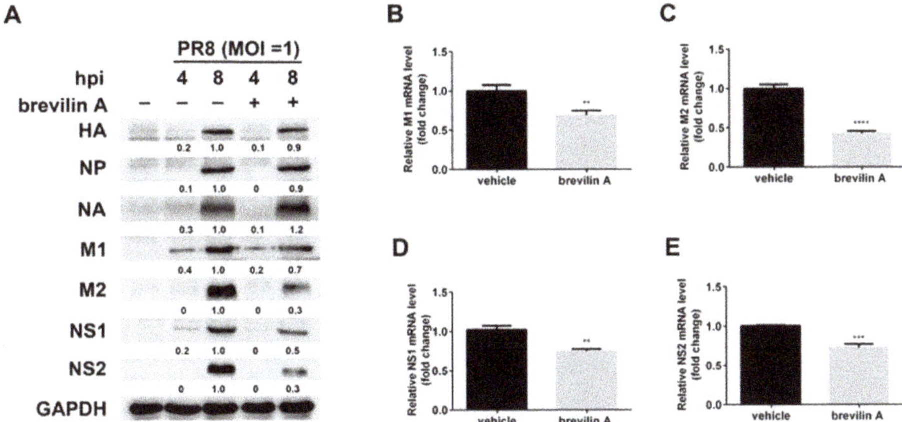

Figure 5. The inhibitory effects of brevilin A on the mRNA and protein expression of viral NS and M. (**A**) MDCK cells were infected with PR8 at a MOI of 1, and then treated with brevilin A or vehicle. At the indicated time points, cell lysates were analyzed by Western blot assay for the indicated antigens; (**B–E**) MDCK cells were infected with PR8 at a MOI of 1, and then treated with brevilin A (8 µM) at 1 hpi, total RNA was extracted from the infected cells at 6 hpi, and mRNA was quantified with RT-qPCR. **, $p < 0.01$; ***, $p < 0.001$; ****, and $p < 0.0001$ are considered statistically significant, compared with vehicle.

3.6. Brevilin A Induces Influenza Viral RNP Aggregation in the Nucleus

In contrast with most RNA viruses, the replication and transcription of IAV were carried out in the nucleus of the infected cells. After uncoating, the vRNP complex, which consists of the viral PB1-PB2-PA (3P) heterotrimeric RNA polymerase and NP protein, is imported into the nucleus for virus RNA replication and transcription and then exported to the cytoplasm for packaging into newly formed virions at the cellular plasma membranes [26]. The matrix protein M1, the viral nuclear export protein (NS2/NEP), and the M2 ion channel protein are essential proteins involved in viral trafficking, releasing into the cytoplasm, and budding [27]. To investigate the effects of brevilin A on the nucleocytoplasmic trafficking of vRNPs, MDCK cells were infected with PR8 virus for 1 h and then treated with DMSO, brevilin A, or LMB. The viral NP protein was detected by indirect immunofluorescence microscopy at 4, 8, and 11 hpi to determine the vRNP localization. As shown in Figure 6, in vehicle treated cells, the NP protein was detected exclusively in the nucleus at 4 hpi and shifted toward the cytoplasm at 8 hpi. By 11 hpi, vRNPs were mainly distributed in the cytoplasm. As a control, we used LMB, a potent and specific nuclear export inhibitor, which has been demonstrated to be able to cause the nuclear accumulation of newly generated vRNPs [28]. Consistent with the literature report, LMB prevented the export of vRNPs, even at the late stage of infection (11 hpi), whereas their nuclear import was not affected. Similar to LMB, upon treatment with brevilin A, the translocation of vRNPs to the nucleus was delayed. The aggregation of vRNPs was also observed at 11 hpi (Figure 6). These results suggest that brevilin A induces vRNPs aggregation in the nucleus.

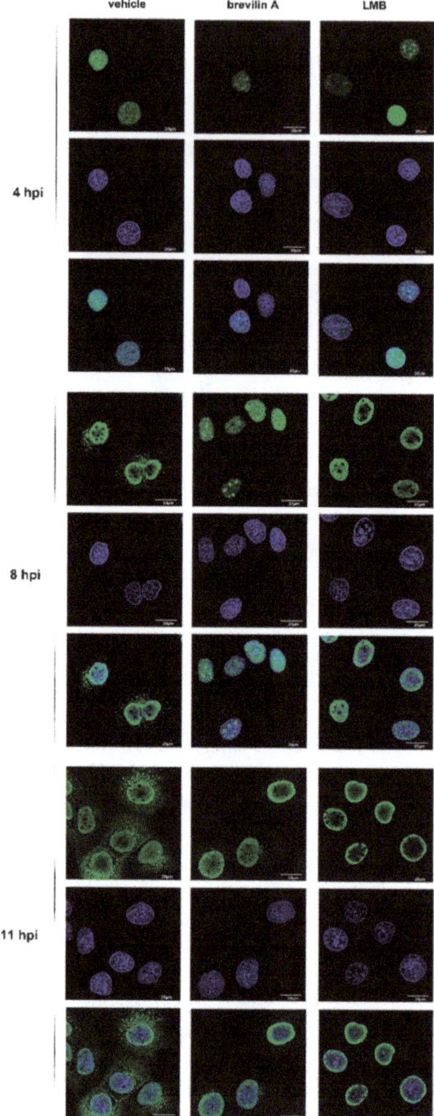

Figure 6. The effects of brevilin A on the viral ribonucleoproteins (vRNP) localization. MDCK cells were infected with PR8 at a MOI of 3 and then treated with brevilin A, LMB, or vehicle as indicated. Samples were fixed at 4, 8, or 11 hpi, and then stained with anti-NP body (green) and DAPI (blue). Immunofluorescence was observed with confocal microscopy. The scale bar in the images is 20 μm.

3.7. Brevilin A Protects Mice from IAV Pathogenesis

Since brevilin A showed antiviral activity against influenza A virus in vitro, we next examined whether brevilin A could also protect mice against influenza virus infection. Mice were infected intratracheally with PR8 at a dose of three 50% lethal doses (LD_{50}). Brevilin A was given once every other day intraperitoneally immediately after infection at two concentrations (10 mg/kg or 25 mg/kg). Oseltamivir was used as the positive control that it is commonly used for treating influenza virus infection in the clinic. The experiment was conducted following the scheme illustrated in Figure 7A. Morbidity and mortality were monitored daily by measuring the body weight and survival rate (Figure 7B,C). Animals falling below the threshold of 75% of their initial body weight were humanely euthanized. Vehicle-treated mice showed severe morbidity after infection with influenza virus, and 100% mortality at 10 days post-infection, while the uninfected group (normal) and oseltamivir-treated group (20 mg/kg/day) showed 100% survival during the entire experiment. Mice treated with brevilin A at 10 mg/kg did not show significant differences in terms of body weight loss or survival rate (Figure 7B,C). However, treatment of brevilin A at 25 mg/kg sustained the body weights of mice in comparison to vehicle-treated group (Figure 7B). Also, brevilin A-treated mice (25 mg/kg) showed a delayed time-to-death with 50% survival up to 14 days post-infection (Figure 7C). Thus, these results show that brevilin A protects mice from IAV pathogenesis.

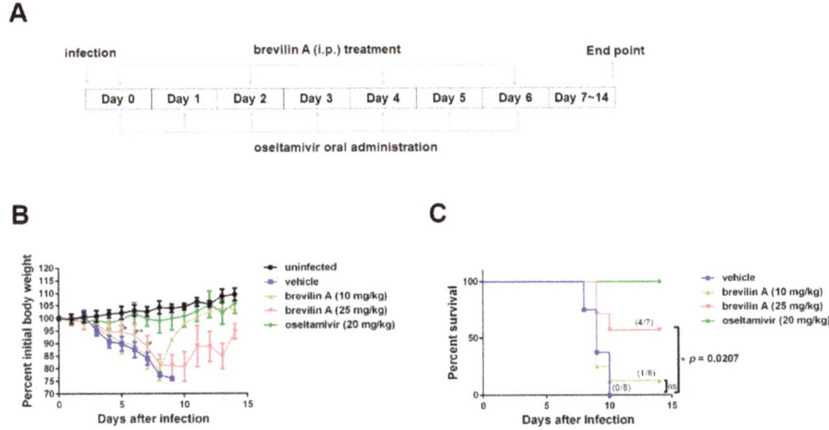

Figure 7. Brevilin A decreases the lethality observed in IAV-infected mice. (**A**) BALB/c mice were infected intranasally with three 50% lethal doses (LD_{50}) of PR8 virus. Brevilin A (25 mg/kg or 10 mg/kg) was intraperitoneally (i.p.) administered to mice 1 h after virus infection, and then once every other day for 6 days beginning on the day of infection (n = 8). Oseltamivir phosphate (20 mg/kg) used as a positive control was administered by oral gavage every day (n = 7). Vehicle (10% Lipovenos containing 0.2% DMSO, 10% PEG300 and 2.5% glycerol, n = 8) was used as a negative control. (**B**) The body weight of mice from each group was monitored daily from day 0 to day 14. The data represent means ± SD. (**C**) The survival rates of the mice were calculated. Animals falling below the threshold of 75% of their initial body weight were humanely euthanized. The p-value is shown (Log-rank (Mantel–Cox) test).

4. Discussion

The present study shows that brevilin A at a noncytotoxic concentration has a broad-spectrum antiviral activity against IAV, including H1N1, H3N2, and H9N2. Mode of mechanism studies demonstrate that brevilin A exhibits its antiviral activity by regulating the replication and translation stages of IAV life cycle. Brevilin A strongly decreased viral RNA level, reduced the expression of viral proteins expressed from the smaller segments (M and NS), and impaired the nuclear export of vRNPs. Furthermore, we showed that brevilin A reduced influenza-associated morbidity and mortality in vivo.

In the current study, we determined that the anti-IAV activity of brevilin A is not viral subtype specific as brevilin A displayed a broad-spectrum antiviral activity against many IAV types/subtypes. These effects were assessed by plaque reduction assay and generation of virus growth curves (Figure 2). Using a time-of-addition assay, we deduced that brevilin A acts at the replication and translation stages of infection (Figure 3), which could explain why the virus titers at 4 hpi were not reduced by treatment with brevilin A (Figure 2B). Further examination reveals that brevilin A preferentially regulates the synthesis of vRNA but not the complementary positive strand cRNA or the mRNA (Figure 4). Moreover, by analyzing the expression of viral protein, we determined that not all of the viral proteins are equally affected by brevilin A. Expression of M2, NS1, and NS2 proteins are more severely inhibited in comparison to other viral proteins (Figure 5A). It has been reported that some IAV genes (segments 1, 2, 3, 5, and unspliced 8) are preferentially expressed early and the others (segments 4, 6, unspliced 7, and two spliced transcripts) are expressed late during infection [23]. However, this could not explain our observation that at early time points post-infection, the proteins translated from the M and NS mRNAs were down-regulated by treatment with brevilin A, while the expression of HA and NA (supposed to be preferentially expressed early) were not affected. Considering that the M and NS segments are coincidentally the ones that produce spliced products, the mRNA levels of M1, M2, NS1, and NS2 were analyzed. Our results demonstrate that the production of M2 mRNA was strongly reduced following the treatment with brevilin A (Figure 5B–E), indicating that the alternative splicing of the M1 mRNA was affected. Splicing is a necessary step for influenza replication, while NS2 is required for nucleocytoplasmic transport of vRNPs [29], and M2 is an important factor in viral pathogenicity [30,31]. During the nuclear replication stage, numerous host-splicing factors, such as the spliceosome complex and host splicing regulators, are necessary to process the M and NS segments [32]. How brevilin A affects the alternative splicing of influenza A viruses needs to be further investigated.

Brevilin A's antiviral effects in vivo were also evaluated in a mouse model upon influenza virus infection. Often, a compound possessing potent inhibitory activity in vitro fails when tested in vivo. However, we found that treatment of IAV-infected mice with brevilin A markedly improved their survival, compared to vehicle control mice (Figure 7). Under our experimental settings, brevilin A (25 mg/kg) was delivered every other day for four times rather than every day for 6 days, since both delivery frequency showed similar protective effects in IAV-infected mice. Further exploration of the in vivo potential of brevilin A is warranted to assess its in vivo toxicity, and define the best conditions of treatment, including analysis of dosage and route(s) of inoculation. Besides, brevilin A's bioavailability and pharmacokinetics would also be considered in future research to get more detailed information about absorption, metabolism, and disposition.

In summary, the sesquiterpene lactone brevilin A is a promising candidate lead compound for development of antiviral agents that broadly inhibit IAV replication by impairing the vRNA synthesis and the viral protein translation. Further investigation is warranted of this and other similar inhibitors as potential therapeutic agents against influenza.

Author Contributions: Conceptualization, X.Z. and C.X.; methodology, X.Z. and Y.X.; investigation, X.Z., Y.X., L.Y., J.H., Y.L. and C.X.; data, X.Z., Y.X., L.Y. and C.X.; writing—original draft preparation, X.Z. and C.X.; writing—review and editing, X.Z. and C.X.; supervision, X.Z. and C.X.; funding acquisition, X.Z. and J.H.

Funding: This research was funded by National Natural Science Foundation of China (81703371, 81603165, 81672006), the Guangdong Innovative and Entrepreneurial Research Team Program (2014ZT05S136) and China Postdoctoral Science Foundation (2017M622924).

Acknowledgments: The authors thank Wenbao Qi from the Key Laboratory of Veterinary Vaccine Innovation of the Ministry of Agriculture, P. R. China for providing the influenza virus A/PR/8/34 (H1N1) and MDCK cells. We thank Fenyong Liu (University of California, Berkeley), Rong Hai (University of California, Riverside), Hua Zhu (Rutgers New Jersey Medical School), Jun Chen (Jinan University), and Ting Wang (Jinan University) for critical comments, reagents, and technical assistance.

Conflicts of Interest: The authors declare no conflicts of interests.

References

1. Petrova, V.N.; Russell, C.A. The evolution of seasonal influenza viruses. *Nat. Rev. Microbiol.* **2018**, *16*, 47–60. [CrossRef] [PubMed]

2. Iuliano, A.D.; Roguski, K.M.; Chang, H.H.; Muscatello, D.J.; Palekar, R.; Tempia, S.; Cohen, C.; Gran, J.M.; Schanzer, D.; Cowling, B.J.; et al. Global Seasonal Influenza-associated Mortality Collaborator, N., Estimates of global seasonal influenza-associated respiratory mortality: A modelling study. *Lancet* **2018**, *391*, 1285–1300. [CrossRef]

3.   Paules, C.I.; Sullivan, S.G.; Subbarao, K.; Fauci, A.S. Chasing Seasonal Influenza—The Need for a Universal Influenza Vaccine. *N. Engl. J. Med.* **2018**, *378*, 7–9. [CrossRef] [PubMed]

4.   Hu, Y.; Musharrafieh, R.; Ma, C.; Zhang, J.; Smee, D.F.; DeGrado, W.F.; Wang, J. An M2-V27A channel blocker demonstrates potent in vitro and in vivo antiviral activities against amantadine-sensitive and -resistant influenza A viruses. *Antivir. Res.* **2017**, *140*, 45–54. [CrossRef] [PubMed]

5.   Hussain, M.; Galvin, H.D.; Haw, T.Y.; Nutsford, A.N.; Husain, M. Drug resistance in influenza A virus: The epidemiology and management. *Infect. Drug Resist.* **2017**, *10*, 121–134. [CrossRef] [PubMed]

6.   Hayden, F.G.; Sugaya, N.; Hirotsu, N.; Lee, N.; de Jong, M.D.; Hurt, A.C.; Ishida, T.; Sekino, H.; Yamada, K.; Portsmouth, S.; et al. Baloxavir Marboxil for Uncomplicated Influenza in Adults and Adolescents. *N. Engl. J. Med.* **2018**, *379*, 913–923. [CrossRef] [PubMed]

7.   O'Hanlon, R.; Shaw, M.L. Baloxavir marboxil: The new influenza drug on the market. *Curr. Opin. Virol.* **2019**, *35*, 14–18. [CrossRef] [PubMed]

8.   Bright, R.A.; Medina, M.J.; Xu, X.; Perez-Oronoz, G.; Wallis, T.R.; Davis, X.M.; Povinelli, L.; Cox, N.J.; Klimov, A.I. Incidence of adamantane resistance among influenza A (H3N2) viruses isolated worldwide from 1994 to 2005: A cause for concern. *Lancet* **2005**, *366*, 1175–1181. [CrossRef]

9.   Moscona, A. Global transmission of oseltamivir-resistant influenza. *N. Engl. J. Med.* **2009**, *360*, 953–956. [CrossRef]

10. Taniguchi, K.; Ando, Y.; Nobori, H.; Toba, S.; Noshi, T.; Kobayashi, M.; Kawai, M.; Yoshida, R.; Sato, A.; Shishido, T.; et al. Inhibition of avian-origin influenza A(H7N9) virus by the novel cap-dependent endonuclease inhibitor baloxavir marboxil. *Sci. Rep.* **2019**, *9*, 3466. [CrossRef]

11. Omoto, S.; Speranzini, V.; Hashimoto, T.; Noshi, T.; Yamaguchi, H.; Kawai, M.; Kawaguchi, K.; Uehara, T.; Shishido, T.; Naito, A.; et al. Characterization of influenza virus variants induced by treatment with the endonuclease inhibitor baloxavir marboxil. *Sci. Rep.* **2018**, *8*, 9633. [CrossRef]

12. Chan, C.O.; Jin, D.P.; Dong, N.P.; Chen, S.B.; Mok, D.K. Qualitative and quantitative analysis of chemical constituents of Centipeda minima by HPLC-QTOF-MS & HPLC-DAD. *J. Pharm. Biomed. Anal.* **2016**, *125*, 400–407. [PubMed]

13. Liu, Y.; Chen, X.Q.; Liang, H.X.; Zhang, F.X.; Zhang, B.; Jin, J.; Chen, Y.L.; Cheng, Y.X.; Zhou, G.B. Small compound 6-O-angeloylplenolin induces mitotic arrest and exhibits therapeutic potentials in multiple myeloma. *PLoS ONE* **2011**, *6*, e21930. [CrossRef] [PubMed]

14. Wu, P.; Su, M.X.; Wang, Y.; Wang, G.C.; Ye, W.C.; Chung, H.Y.; Li, J.; Jiang, R.W.; Li, Y.L. Supercritical fluid extraction assisted isolation of sesquiterpene lactones with antiproliferative effects from Centipeda minima. *Phytochemistry* **2012**, *76*, 133–140. [CrossRef] [PubMed]

15. Chen, X.; Du, Y.; Nan, J.; Zhang, X.; Qin, X.; Wang, Y.; Hou, J.; Wang, Q.; Yang, J. Brevilin A, a novel natural product, inhibits janus kinase activity and blocks STAT3 signaling in cancer cells. *PLoS ONE* **2013**, *8*, e63697. [CrossRef] [PubMed]

16. Su, M.; Chung, H.Y.; Li, Y. 6-O-Angeloylenolin induced cell-cycle arrest and apoptosis in human nasopharyngeal cancer cells. *Chem. Biol. Interact.* **2011**, *189*, 167–176. [CrossRef]

17. You, P.; Wu, H.; Deng, M.; Peng, J.; Li, F.; Yang, Y. Brevilin A induces apoptosis and autophagy of colon adenocarcinoma cell CT26 via mitochondrial pathway and PI3K/AKT/mTOR inactivation. *Biomed. Pharmacother.* **2018**, *98*, 619–625. [CrossRef] [PubMed]

18. Taylor, R.S.; Towers, G.H. Antibacterial constituents of the Nepalese medicinal herb, Centipeda minima. *Phytochemistry* **1998**, *47*, 631–634. [CrossRef]

19. Yu, H.W.; Wright, C.W.; Cai, Y.; Yang, S.L.; Phillipson, J.D.; Kirby, G.C.; Warhurst, D.C. Antiprotozoal activities of Centipeda minima. *Phytother. Res.* **1994**, *8*, 436–438. [CrossRef]

20. Zhang, X.; He, J.; Huang, W.; Huang, H.; Zhang, Z.; Wang, J.; Yang, L.; Wang, G.; Wang, Y.; Li, Y. Antiviral Activity of the Sesquiterpene Lactones from Centipeda minima against Influenza A Virus in vitro. *Nat. Prod. Commun.* **2018**, *13*, 115–119. [CrossRef]

21. Nachbagauer, R.; Wohlbold, T.J.; Hirsh, A.; Hai, R.; Sjursen, H.; Palese, P.; Cox, R.J.; Krammer, F. Induction of broadly reactive anti-hemagglutinin stalk antibodies by an H5N1 vaccine in humans. *J. Virol.* **2014**, *88*, 13260–13268. [CrossRef] [PubMed]

22. Wang, T.T.; Tan, G.S.; Hai, R.; Pica, N.; Petersen, E.; Moran, T.M.; Palese, P. Broadly protective monoclonal antibodies against H3 influenza viruses following sequential immunization with different hemagglutinins. *PLoS Pathog.* **2010**, *6*, e1000796. [CrossRef] [PubMed]

23. Shapiro, G.I.; Gurney, T., Jr.; Krug, R.M. Influenza virus gene expression: Control mechanisms at early and late times of infection and nuclear-cytoplasmic transport of virus-specific RNAs. *J. Virol.* **1987**, *61*, 764–773. [PubMed]

24. Scull, M.A.; Rice, C.M. A big role for small RNAs in influenza virus replication. *Proc. Natl. Acad. Sci. USA* **2010**, *107*, 11153–11154. [CrossRef] [PubMed]

25. Vester, D.; Lagoda, A.; Hoffmann, D.; Seitz, C.; Heldt, S.; Bettenbrock, K.; Genzel, Y.; Reichl, U. Real-time RT-qPCR assay for the analysis of human influenza A virus transcription and replication dynamics. *J. Virol. Methods* **2010**, *168*, 63–71. [CrossRef] [PubMed]

26. Te Velthuis, A.J.; Fodor, E. Influenza virus RNA polymerase: Insights into the mechanisms of viral RNA synthesis. *Nat. Rev. Microbiol.* **2016**, *14*, 479–493. [CrossRef] [PubMed]

27. Eisfeld, A.J.; Neumann, G.; Kawaoka, Y. At the centre: Influenza A virus ribonucleoproteins. *Nat. Rev. Microbiol.* **2015**, *13*, 28–41. [CrossRef] [PubMed]

28. Watanabe, K.; Takizawa, N.; Katoh, M.; Hoshida, K.; Kobayashi, N.; Nagata, K. Inhibition of nuclear export of ribonucleoprotein complexes of influenza virus by leptomycin B. *Virus Res.* **2001**, *77*, 31–42. [CrossRef]

29. O'Neill, R.E.; Talon, J.; Palese, P. The influenza virus NEP (NS2 protein) mediates the nuclear export of viral ribonucleoproteins. *EMBO J.* **1998**, *17*, 288–296. [CrossRef]

30. Chiang, C.; Chen, G.W.; Shih, S.R. Mutations at Alternative 5′ Splice Sites of M1 mRNA Negatively Affect Influenza A Virus Viability and Growth Rate. *J. Virol.* **2008**, *82*, 10873–10886. [CrossRef]

31. Hutchinson, E.C.; Curran, M.D.; Read, E.K.; Gog, J.R.; Digard, P. Mutational analysis of cis-acting RNA signals in segment 7 of influenza A virus. *J. Virol.* **2008**, *82*, 11869–11879. [CrossRef] [PubMed]

32. Dubois, J.; Terrier, O.; Rosa-Calatrava, M. Influenza viruses and mRNA splicing: Doing more with less. *MBio* **2014**, *5*, e00070-14. [CrossRef] [PubMed]

Article

Anti-Respiratory Syncytial Virus Activity of *Plantago asiatica* and *Clerodendrum trichotomum* Extracts In Vitro and In Vivo

Kiramage Chathuranga [1], Myun Soo Kim [2], Hyun-Cheol Lee [1], Tae-Hwan Kim [1], Jae-Hoon Kim [1,3], W. A. Gayan Chathuranga [1], Pathum Ekanayaka [1], H. M. S. M. Wijerathne [1], Won-Kyung Cho [4], Hong Ik Kim [2], Jin Yeul Ma [4] and Jong-Soo Lee [1,*,†]

[1]   Microbiology Laboratory, Department of Preventive Veterinary Medicine, College of Veterinary Medicine, Chungnam National University, Daejeon 34134, Korea
[2]   Vitabio Corporation, Daejeon 34540, Korea
[3]   Laboratory Animal Resource Center, Korea Research Institute of Bioscience and Biotechnology, University of Science and Technology (UST), Daejeon 34141, Korea
[4]   Korean Medicine (KM) Application Center, Korea Institute of Oriental Medicine, Daegu 41062, Korea
*   Correspondence: jongsool@cnu.ac.kr; Tel.: +82-(42)-821-6753; Fax: +82-(42)-825-7910
†   Current address: College of Veterinary Medicine, Chungnam National University, 99 Daehak-ro, Yuseong-Gu, Daejeon 34134, Korea.

Received: 17 April 2019; Accepted: 1 July 2019; Published: 3 July 2019

Abstract: The herbs *Plantago asiatica* and *Clerodendrum trichotomum* have been commonly used for centuries in indigenous and folk medicine in tropical and subtropical regions of the world. In this study, we show that extracts from these herbs have antiviral effects against the respiratory syncytial virus (RSV) in vitro cell cultures and an in vivo mouse model. Treatment of HEp2 cells and A549 cells with a non-cytotoxic concentration of *Plantago asiatica* or *Clerodendrum trichotomum* extract significantly reduced RSV replication, RSV-induced cell death, RSV gene transcription, RSV protein synthesis, and also blocked syncytia formation. Interestingly, oral inoculation with each herb extract significantly improved viral clearance in the lungs of BALB/c mice. Based on reported information and a high-performance liquid chromatography (HPLC) analysis, the phenolic glycoside acteoside was identified as an active chemical component of both herb extracts. An effective dose of acteoside exhibited similar antiviral effects as each herb extract against RSV in vitro and in vivo. Collectively, these results suggest that extracts of *Plantago asiatica* and *Clerodendrum trichotomum* could provide a potent natural source of an antiviral drug candidate against RSV infection.

Keywords: *Plantago asiatica*; *Clerodendrum trichotomum*; RSV; therapeutic effects; acteoside

---

1. Introduction

Acute respiratory infections caused by viruses are the most common cause of morbidity and mortality in children worldwide. Respiratory syncytial virus (RSV) is one of the major causes of lower respiratory tract infections, which cause a huge global disease burden [1]. It is the most important viral agent of serious respiratory tract illness in infants and young children. Nearly all infants have been infected with RSV at least once by the age of two years [2]. RSV is also a major cause of acute respiratory illness in the elderly, and it can have a detrimental effect in immune-compromised individuals. Even though RSV infection generally occurs at an early age, individuals may be re-infected throughout their lifetime, because naturally acquired immunity does not provide persistent protection [3]. At present, RSV vaccines and antiviral drugs are in the preclinical and clinical stage of development; however, no RSV vaccines or antiviral drugs suitable for typical use are commercially available at this time [4]. Ribavirin and immunoglobulin preparations with high titers of RSV-specific neutralizing antibodies are currently approved to treat and prevent RSV infections [5]. However, neither of these options is cost-effective or convenient to administer. Due to the high infant morbidity and mortality rates, the lack of an effective vaccine, and the availability of just one antiviral agent (Ribavirin), which is used only in severe cases, novel therapies for RSV infection warrant investigation.

Fossil evidence has revealed that human use of plants as folk medicine dates back at least 60,000 years [6]. According to the World Health Organization (WHO), almost 65% of the world's population reports the use of natural compounds as medicinal agents [7]. Modern analytical technologies applied towards the active compounds found in plants have allowed for greater insights into plant-derived pharmaceutical compounds [8]. *Plantago asiatica* is a perennial belonging to the family Plantaginaceae and is commonly used as a folk medicine in Korea, China, and Japan [9]. *Plantago asiatica* extract (PAE) has been used to treat a variety of health conditions, such as wounds, cholesterolemia, diarrhea, bronchitis, and chronic constipation [10–12]. Moreover, it has been shown to inhibit cancer and leukemia growth and to enhance cell-mediated immunity [13]. *Clerodendrum trichotomum* is a deciduous shrub that belongs to the family Lamiaceae (formerly Verbenaceae) and is widely distributed in South Korea, China, Japan, and the Philippines. *Clerodendrum trichotomum* extract (CTE) possesses broad-spectrum anti-inflammatory [14,15], anti-hypertensive [16], anti-asthmatic [17], anti-oxidative [18] and immunotoxin [19] properties. However, the effects of PAE and CTE and their active components on RSV replication in vitro and in vivo have not been reported.

In this study, we evaluated the antiviral activities of *P. asiatica* and *C. trichotomum* aqueous extracts against RSV in vitro and in vivo. Furthermore, we identified and confirmed the antiviral function of acteoside (verbascoside), a phenolic glycoside presents in both aqueous extracts, as the potential active component with antiviral activity against RSV infection.

2. Materials and Methods

2.1. Plant Materials and Total Aqueous Extract Preparation

A water-soluble extract of *P. asiatica* and *C. trichotomum* was prepared by Herbal Medicine Improvement Research Center, Korea Institute of Oriental Medicine, Daejeon, Republic of Korea. Crude plant materials were purchased from a local store (Jaecheon Oriental Herbal Market) and verified by Professor Ki-Hwan Bae at the College of Pharmacy, Chungnam National University. In the proses, 100 g of the plant materials were placed in 1 L of distilled water and extracted by heating for 2.5 h at 105 °C using a medical heating plate. After the extraction proses, the extract was subjected to filtration using a filter paper (0.45 μm, Millex® , Darmstadt, Germany) and stored at 4 °C for 24 h. The extract was then centrifuged at 8000× *g* for 15 min. The supernatant was collected, and the pH was adjusted to 7.0. Following pH adjustment total successive aqueous extract was subjected to membrane syringe filtration (0.22 μm) and stored at −20 °C until further use.

2.2. Reagents, Chemicals and Antibodies

Verbascoside (Acteoside) was purchased from Sigma (V4015). Trypan blue solution was purchased from Gibco (Waltham, MA, USA). Cell cytotoxicity assay kit was purchased from Dojindo Molecular Technologies, INC (CK04: Cell Counting Kit-8, Japan). Antibodies used in the immunoblotting study were as follows: Anti-RSV Glycoprotein (RSV-G) (Abcam, #ab94966, Cambridge, UK), β-actin (Santa Cruz, SC 47778, Dallas, TX, USA), HRP-conjugated anti-mouse IgG (Gene Tex, GTX213111-01, Taichung, Taiwan), HRP-conjugated anti-rabbit IgG (Cell signaling technology, 7074P2, Danvers, MA, USA).

2.3. Cell Culture and Virus Infection

Human epithelial type 2: HEp2 cells with HeLa contaminant (ATCC CCL-23, Manassas, VA, USA) and A549 cells (ATCC CCL-185, Manassas, VA, USA) were maintained in Dulbecco's Modified Eagle's Medium (DMEM) (Invitrogen, Waltham, MA, USA) supplemented with 10% fetal bovine serum (FBS) (Hyclone, Australia) and 1% antibiotic/antimycotic solution (Gibco, Waltham, MA, USA) at 37 °C with a 5% CO_2 environment. The Green Fluorescent Protein fused Respiratory syncytial virus (RSV-GFP) from Dr. Jae U. Jung, Department of Molecular Microbiology and Immunology, University of Southern California, USA. RSV-GFP propagated on confluent HEp2 cells, and titer was determined by a standard plaque assay.

2.4. Antiviral Assays

RSV-GFP virus replication inhibition assay was performed in vitro using HEp2 cells, as described previously with minor modifications [20]. Briefly, HEp2 cells and A549 cells were seeded in 12 well cell culture plates with the cell number of 2.5×10^5 cells/well and incubated for 12 h. Medium was changed with DMEM containing 1% FBS and Cells were infected with RSV-GFP [multiplicity of infection (MOI) 0.1] for 2 h. Cells were washed with PBS and medium was replace with DMEM containing 10% FBS and cells were treated with indicated concentrations of PAE, CTE or acteoside. GFP expression was measured at 48 h post infection (hpi) with Glomax multidetection system following manufacturer's directions. Virus titer was determined in supernatant and cells by plaque assay in HEp2 cells or A549 cells [21]. Cell viability was evaluated using a trypan blue exclusion test as described previously [22].

2.5. Determination of Effective Concentration (EC$_{50}$) of Extracts and Acteoside

HEp2 cells were grown in 24-well cell culture plates (1.25×10^5 cells/well) and incubated at 37 °C in a 5% CO_2 atmosphere. After 12 h, the medium was replaced with DMEM containing 1% FBS and cells were infected with RSV-GFP (0.1 MOI) for 2 h. Then cells were washed with PBS once, and the medium was replaced with DMEM containing 10% FBS. Cells were treated with indicated concentrations of herb extracts or acteoside. The experiment was performed in triplicate. GFP expression was measured 48 hpi with the Glomax multi-detection system (Promega, Fitchburg, WI, USA) according to the manufacturer's instructions. The EC$_{50}$ values were then calculated as the extract concentration yielding 50% GFP expression.

2.6. Determination of Cytotoxic Concentration (CC$_{50}$) of Extracts and Acteoside

Cell cytotoxic concentration of herb extracts and acteoside was determined using Cell counting kit-8 Dojindo Molecular Technologies, as described previously [23]. Briefly, HEp2 cells were seeded into 96-well cell culture plates (2.5×10^4 cells/well) and incubated for 12 h. Cells were treated with indicated concentrations of herb extracts or acteoside. Next, at 48-h post treatment, 10 µL of CCK-8 solution was added to each well of the plate. Then, it was incubated for 1 h at 37 °C, and absorbance was measured at 450 nm using microplate reader (molecular devices).

2.7. Quantitative RT-PCR (qRT PCR)

Total RNA was extracted from cells, or 1 g of lung homogenate using the RNeasy Mini kit (Qiagen, Hiden, Germany) and cDNA synthesis was performed using the enzyme reverse transcriptase (TOYOBO). Next, qRT-PCR was performed using the Rotor Gene Q instrument (Qiagen, Hiden, Germany), with the QuantiTect SYBR Green PCR Master Mix (Qiagen, Hiden, Germany). The transcription level of mRNA was obtained by the $2^{-\Delta\Delta Ct}$ method as described previously [24] and expressed as fold induction. The RT-PCR primer sequences used as follows, RSV-G forward primer 5'-CCAAACAAACCC AATAATGATTT-3' reverse primer 5'-GCCCAGCAGGTTGGATTGT-3' Glyceraldehyde 3-phosphate dehydrogenase (GAPDH): Forward primer 5'-TGACCACAGTCCATGCCATC-3' reverse primer 5'-GACGGACACATTGGGGGTAG-3'.

2.8. Immunoblot Analysis

HEp2 cells were seeded in six well cell culture plates (5×10^5 cells/well) and incubated for 12 h. Medium was changed to DMEM containing 1% FBS and cells were infected with RSV-GFP (0.1 MOI) for 2 h. Cells were treated with PAE, CTE or acteoside once replacing the medium with DMEM containing

10% FBS. Cells were harvested at 0, 12, 24, 36, 48, hpi and subjected to immunoblot analysis. Briefly, harvested cells were lysed with lysis buffer containing 1% NP-40, 150 mM NaCl, 50 mM Tris-HCl pH 8.0 and a protease inhibitor (Sigma). Whole cell lysates were mixed with 10x sample buffer (Sigma, St. Louis, MO, USA) at 1:1 ratio, and the total protein was separated in 12% gel by SDS-PAGE and transferred to a PVDF membrane (BioRad, Hercules, CA, USA). The membrane was blocked in 5% bovine serum albumin (BSA, Sigma) and incubated with anti RSV-G antibody or anti-β-actin antibody with 5% BSA and TBST (Tris-buffered saline (LPS Solution) + Tween 20 (Life science, #0777-1L, Suwanee, GA, USA)) respectively. Proteins were detected by incubating with a secondary anti-rabbit IgG-HRP or anti-mouse IgG-HRP for 1 h at room temperature. The membrane was developed with ECL reagent mix (LPS solution, FEMTO-100) and images were captured with an Enhanced Chemiluminescence Detection (ECL) system (GE Life science, Pittsburgh, PA, USA), using Las-4000 mini lumino-image analyzer (GE Life Science). Band intensity was calculated using ImageQuant LAS 4000 control software (Pittsburgh, PA, USA).

2.9. Syncytium Formation Assay

The ability of the herb extract and acteoside to block cell to cell spread was evaluated using GFP expression in the cells. HEp2 cells were cultured in 12 well cell culture plates (2.5×10^5 cells/well) and incubate for 12 h. Medium was changed to DMEM containing 1% FBS and cells were infected with RSV-GFP (0.1 MOI) for 2 h. Cells were treated with PAE/CTE or acteoside once replacing the medium with DMEM containing 10% FBS. After 48 h incubation cells were washed with cooled PBS and cell images were taken under 400 magnifications. Syncytium formation was quantified with imageJ 1.48 program (https://imagej.net/ImageJ).

2.10. RSV-GFP Challenge Experiment in Mouse Model

Five-week-old female BALB/c mice were purchased from orient bio (South Korea) and acclimated for three days under experimental condition prior to use. Mice were separated into experimental groups as virus infected, and herb or acteoside treated group ($n = 5$), virus only infected group ($n = 5$) and un infected group ($n = 2$). Mice were anesthetized with ketamine for a short time period, and RSV-GFP 1×10^6 Plaque forming unit (PFU) per mice in the total volume 28 μL was infected intranasally. Mice were orally administered 0.5 mg/mL concentration of PAE or CTE at a total volume of 200 μL at 6, 12, 18 and 24 hpi. In the case of acteoside, 80 mg/Kg of body weight/mice was intraperitonially administered to mice at 6 hpi in the total volume of 100 μL. Lung tissues from euthanized mice were collected aseptically at 3- and 5-day post infection (dpi). Lung RSV titration was determined by RSV-G protein mRNA transcription fold quantification. RSV-G protein mRNA level was quantified as described before.

2.11. Ethical Approval

The animal study was conducted under appropriate conditions with the approval of the Institutional Animal Care and Use Committee of Chungnam National University (Reference number CNU-00816).

2.12. Identification of Acteoside through HPLC

A reversed-phase high-performance liquid chromatography (HPLC) was performed using Agilent technologies 1200 series HPLC system equipped with a DAD detector (Agilent technologies, Santa Clara, CA, USA). The binary mobile phase consisted of water containing 1% formic acid (solvent A) and acetonitrile (solvent B). All solvents were filtered through a 0.45 μm filter prior to use. The mobile phase consisted of 1% Formic acid (Solvent A) and Acetonitrile (Solvent B) in the gradient mode as follows: 0–20 min 0–40% B; 20–22 min 40–100% B; 22–25 min 100–0% B at flow rate of 1.0 mL/min at 30 °C.

2.13. Statistical Analysis

Data are presented as the means ± standard deviations (SD) and are representative of at least three independent experiments. Graphs and all Statistical analysis were performed using GraphPad Prism software version 6 (San Diego, CA, USA) for Windows. Differences between untreated and herb or acteoside treated groups were analyzed by Unpaired t-test. $p < 0.05$, $p < 0.01$ or $p < 0.001$ was regarded as significant.

3. Results

3.1. Antiviral Effects of PAE and CTE

A library of herb extracts was screened to detect antiviral activity against RSV. Among them, PAE and CTE were selected. The ability of the two herb extracts to inhibit the replication of GFP-tagged RSV (RSV-GFP) was further confirmed in a dose-dependent experiment. The expression of RSV-GFP, the virus titer, and the recovery of RSV-induced cell death were evaluated in HEp2 cells upon herb treatment. As shown in Figure 1A,B, HEp2 cells and A549 cells treated with PAE and CTE (10, 30, or 50 µg/mL) exhibited a marked reduction in GFP expression compared to untreated HEp2 cells and A549 cells. Moreover, all doses of the two-herb extract significantly reduced the RSV titer compared to the untreated group (Figure 1C). Interestingly, treatment with PAE and CTE significantly reduced RSV-induced HEp2 cell and A549 cell death at 48 hpi (Figure 1D). Therefore, both herb extracts could significantly reduce RSV replication in HEp2 cells and A549 cells. Since the 50 µg/mL concentration was the most effective at inhibiting viral replication and virus-induced cytotoxicity, this concentration was used for further in vitro experiments.

3.2. Therapeutic Effect of PAE and CTE against RSV Infection

The ability of PAE and CTE to inhibit virus replication after the infection was determined. HEp2 cells were infected with RSV-GFP 0.1 MOI, and herb extracts were added at the indicated time points. GFP expression was measured at 48 hpi. As expected, increased GFP expression at 48 hpi was observed as the amount of time between viral infection and herb treatment increased (Figure 2A,B). Similarly, RSV titers increased with increasing time between viral infection and herb extract treatment (Figure 2C,D). Next, to assess the effect of PAE and CTE on virus replication, an assay was performed, and GFP expression was measured at different times after virus infection. As shown in Figure 2E,F, a 50 µg/mL concentration of each herb extract significantly reduced the GFP expression at 36 hpi and 48 hpi but, interestingly, not at 12 hpi or 24 hpi.

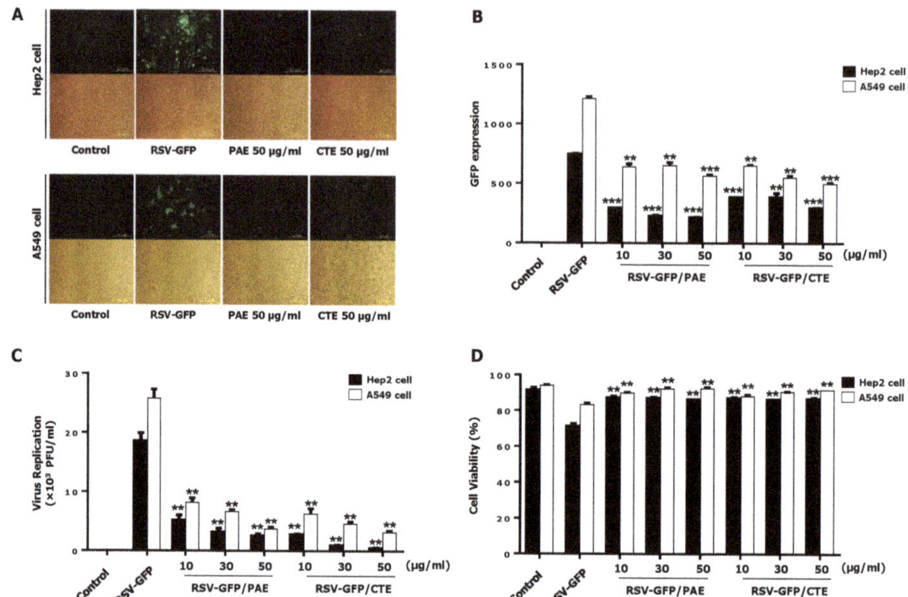

Figure 1. Antiviral activity of *Plantago asiatica* extract (PAE) and *Clerodendrum trichotomum* extract (CTE) in HEp2 cells and A549 cells. HEp2 cells and A549 cells were seeded into 12 well cell culture plates with the cell number of 2.5×10^5 cells/well. Twelve hours later, the medium was changed to 1% fetal bovine serum (FBS) containing Dulbecco's Modified Eagle's Medium (DMEM) and cells were infected with Green Fluorescent Protein fused Respiratory syncytial virus (RSV-GFP) 0.1multiplicity of infection (MOI) or kept uninfected. Two hours later, the medium was replaced with 10% FBS containing DMEM and cells were treated with 10, 30, 50 (µg/mL) PAE or CTE. Cells without any treatment regard as virus only. (**A**) After 48 h, images were obtained (200× magnification). (**B**) GFP absorbance levels were measured by Gloma multi-detection luminometer (Promega). (**C**) Virus titration was done from the cell supernatant and cells by standard plaque assay and expressed as plaque forming unit (PFU). (**D**) Cell viability was determined by trypan blue exclusion assay at 48hour post infection (hpi). GFP absorbance, cell viability and virus titer expressed as mean ± standard deviations (SD). Error bars indicate the range of values obtained from counting duplicate in three independent experiments (** $p < 0.01$ and *** $p < 0.001$ regarded as significant difference).

3.3. Synergistic Anti-RSV Effect of PAE and CTE in HEp2 Cells

Next, synergistic anti-RSV effect of PAE and CTE were determined in HEp2 cells. HEp2 cells were infected with RSV-GFP (0.1MOI) for 2 h in DMEM containing 1% FBS. Then cells were treated with PAE, CTE along or as a combination (1:1) in DMEM containing 10% FBS. At 48 hpi GFP absorbance were taken, and the virus titer was determined by standard plaque assay. GFP expression level was significantly reduced with the increasing dose either PAE, CTE along or combination treatment (Figure 3A). However, there was no significant difference observed in combined herb extract treatment compared to individual herb extract treatment at the same final dose treatment. Furthermore, a similar result was observed in the virus replication quantification by standard plaque assay (Figure 3B). Therefore, this data demonstrates that PAE or CTE does not enhance anti-RSV activity synergistically.

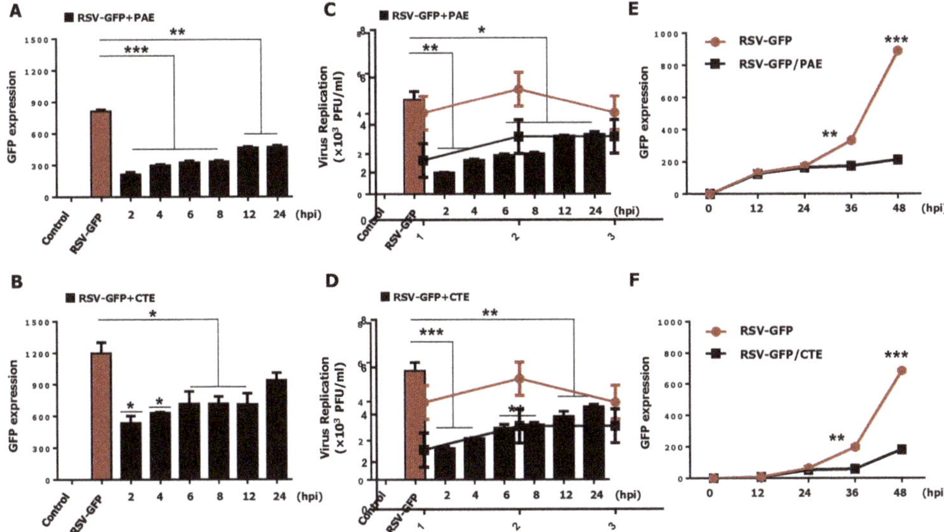

Figure 2. The therapeutic effect of PAE and CTE against RSV-GFP infection. HEp2 cells were seeded into 12 well cell culture plates and left for 12 h. Medium was changed with DMEM containing 1% FBS and cells were infected with RSV-GFP (0.1MOI) for 2 h. (**A,B**) RSV-GFP infected cells were treated with 50 µg/mL PAE or CTE at different times after post infection as indicated or left untreated, and GFP expression level was measured at 48 hpi. (**C,D**) Virus titer was measured from both supernatant and cells by standard plaque assay at 48 hpi and expressed as PFU. (**E,F**) Cells were treated with 50 µg/mL PAE or CTE, and GFP expression level was measured at different time after virus infection as indicated. GFP absorbance and virus titer expressed as mean ± SD. Error bars indicate the range of values obtained from counting duplicate in three independent experiments (* $p < 0.05$, ** $p < 0.01$ and *** $p < 0.001$ regarded as significant difference).

3.4. Determination of the Effective Concentration (EC$_{50}$) and Cytotoxic Concentration (CC$_{50}$) of PAE and CTE

EC$_{50}$ values of the herb extracts were determined against RSV on HEp2 cells. For this experiment, a GFP assay was performed with some modifications, as described previously [25,26]. Briefly, RSV-GFP virus was used, and a 50% reduction in GFP expression was considered equivalent to a 50% reduction in virus titer. As shown in Figure 3D,F, PAE and CTE inhibited RSV-GFP infection (MOI, 0.1) by 50% at concentrations of 39.82 µg/mL and 27.95 µg/mL, respectively. Next, we determined the CC$_{50}$ values of the two extracts based on a cell cytotoxicity assay using HEp2 cells. The assay showed CC$_{50}$ values of 938.43 µg/mL and 764.17 µg/mL for PAE and CTE, respectively (Figure 3F,G). Interestingly, the cell viability at the effective concentrations of both extracts was greater than 80%. The selectivity index (SI) indicates the safety of a crude extract against RSV infection [19]. The SIs of PAE and CTE were 23.5 and 27.3, respectively (Figure 3H). This data suggests that both PAE and CTE could be used safely as therapeutic agents against RSV infection.

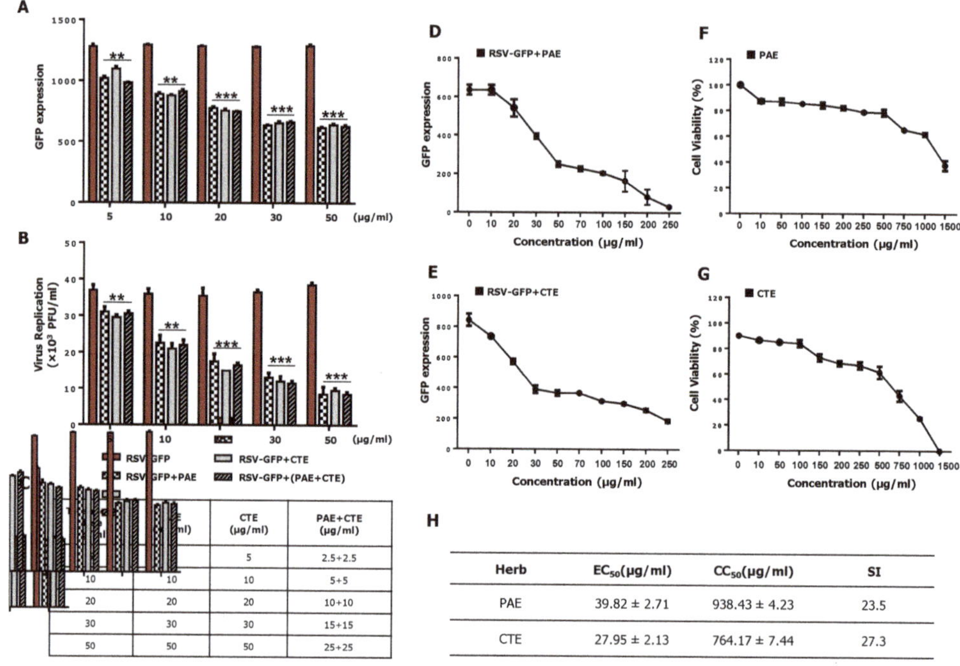

Figure 3. Synergistic effect and effective concentration (EC$_{50}$), cytotoxic concentration (CC$_{50}$) of PAE and CTE in HEp2 cells. (**A**) HEp2 cells were infected with RSV-GFP (0.1 MOI) for 2 h with DMEM containing 1% FBS. Cells were treated with PAE, CTE or combination of both at different concentrations with DMEM containing 10% FBS. At 48 hpi GFP expression level was determined. (**B**) Virus titer was measured from both supernatant and cells by standard plaque assay at 48 hpi and expressed as PFU. (**C**) Dose information of PAE and CTE single or combination treatment. (**D,E**) HEp2 cells were infected with RSV-GFP (O.1MOI) for 2 h with DMEM containing 1% FBS. Then, the medium was changed to DMEM containing 10% FBS and cells were treated with various concentrations of PAE (**D**) or CTE (**E**). 48 hpi GFP expression level was determined. (**F,G**) HEp2 cells were treated with various concentrations of PAE (**F**) or CTE (**G**), and cell viability was determined at 48 h post treatment (hpt) by cell cytotoxicity assay kit. (**H**) To calculate EC$_{50}$ value, 50% reduction of GFP expression was considered as equivalent to the 50% reduction in virus titer. The ratio between CC$_{50}$ and EC$_{50}$ was considered as Selectivity Index (SI). GFP absorbance and cell viability expressed as mean ± SD. Error bars indicate the range of values obtained from counting duplicate in three independent experiments. (** $p < 0.01$ and *** $p < 0.001$ regarded as significant difference).

3.5. Effect of PAE and CTE on the Production of Viral RNA and Protein in HEp2 Cells

Inhibition of intracellular viral RNA transcription and protein translation by herb extract treatment was evaluated in HEp2 cells. HEp2 cells were infected with RSV-GFP and treated with PAE and CTE (50 µg/mL) at 2 hpi, then cells were harvested at the indicated time points, and viral gene expression at the RNA and protein level was determined by qRT-PCR and immunoblot analysis, respectively. Interestingly, PAE treatment reduced the transcription of RSV-G mRNA level by 39-fold at 36 hpi and 22-fold at 48 hpi. Similarly, CTE treatment reduced the transcription of RSV-G mRNA level by 10-fold at 36 hpi and 11-fold at 48 hpi (Figure 4A). Furthermore, the reduction in viral gene transcription in HEp2 cells treated with herb extracts was associated with reduced RSV-G protein synthesis (Figure 4B,C). Thus, the reduction of viral gene transcription and protein synthesis by PAE and CTE correlated with their antiviral activity in vitro (Figure 2B,C).

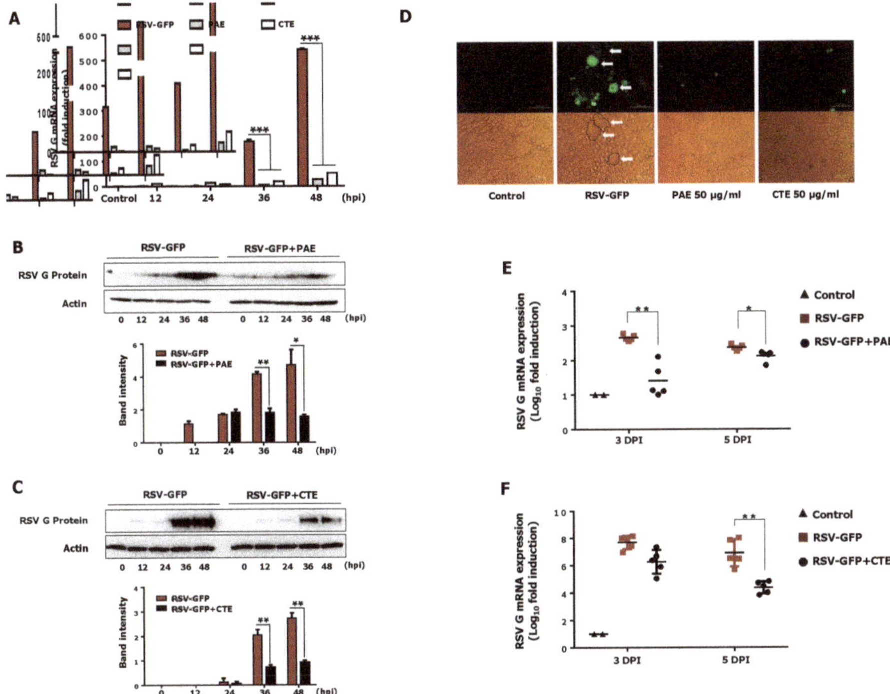

Figure 4. Reduction of RSV Glycoprotein (RSV-G) gene transcription, protein translation and syncytium formation in-vitro and inhibition of RSV replication *in-vivo* by PAE and CTE. HEp2 cells were seeded in six well cell culture plats and incubate for 12 h. Medium was changed into DMEM containing 1% FBS and RSV-GFP (0.1MOI) was infected for 2 h. Then cells were treated with 50 µg/mL PAE or CTE. (**A**) Cells were harvested at indicated time points, and RSV-G protein mRNA level was measured at 12, 24, 36, 48 hpi by qRT-PCR, Glyceraldehyde 3-phosphate dehydrogenase (GAPDH) was used for the normalization. (**B,C**) Immunoblot analysis was performed using cell lysates harvested at indicated time points to measure RSV-G protein and β-actin protein expression level time-dependently. The intensity of the RSV-G was quantified. (**D**) Cell and the GFP image were taken at 48 hpi to see the syncytial formation inhibition by PAE and CTE (400× magnification). (**E,F**) Five weeks old (16 g/mice) BALB/c mice (*n* = 5) were intranasally infected with RSV-GFP (1 × 10⁶ PFU/mice) in the total volume of 28 µL. PAE or CTE were orally administrated at a dose of 200 µL/mice (0.5 mg/mL) at 6, 12, 18 and 24 hpi. At 3 and 5-day post infection (dpi), lung tissues were collected, and the transcription level of RSV-G protein mRNA was determined by qRT-PCR. The arrow indicates the RSV syncytium formation in HEp2 cells. mRNA expression, band intensity expressed as mean ± SD. Error bars indicate the range of values obtained from three independent experiments. In vivo experiment was performed in duplicate. (* $p < 0.05$, ** $p < 0.01$ and *** $p < 0.001$ regarded as significant difference).

3.6. PAE and CTE Inhibits RSV Syncytium Formation

Syncytium formation by RSV is a well-known mechanism of cell-to-cell infection that contributes significantly to virus spread in vivo. Therefore, to determine whether PAE and CTE prevent the cell-to-cell spread of the virus after infection, a syncytium formation assay was performed on infected HEp2 cells. Monolayers of HEp2 cells were infected with RSV-GFP and incubated at 37 °C for 2 h. Cells were left untreated or treated with PAE or CTE (50 µg/mL) and examined for syncytium formation. Interestingly, in untreated HEp2 cells, large areas of syncytium formation were visible. By contrast, PAE and CTE treated HEp2 cells showed significantly reduced syncytium formation at 48 hpi (Figure 4D).

3.7. Oral Administration of PAE and CTE Enhance Protection against RSV Infection in BALB/c Mice

Next, we designed a mouse model to evaluate the therapeutic effect of PAE and CTE against RSV infection in vivo. BALB/c mice were intranasally infected with RSV-GFP (1×10^6 PFU/mouse) or left uninfected, and PBS, PAE, or CTE were orally administrated at 6, 12, 18, and 24 hpi. The RSV infection titer (1×10^6 PFU/mouse), inoculation time (6, 12, 18, and 24 hpi) and inoculation dose (200 μL (0.5 mg/mL) mice/time were chosen based on preliminary studies. Lungs were collected aseptically at 3 dpi and 5 dpi. RSV-G protein mRNA level in the lungs was determined by qRT-PCR. As shown in Figure 4E, PAE-treated mice showed significantly reduced viral mRNA in the respiratory tract at both 3 dpi and 5 dpi compared to the control (PBS)-treated group. Similarly, the transcription of RSV-G mRNA was reduced in the CTE-treated group compared to the PBS-treated group at 3 dpi and 5 dpi, and was significantly reduced at 5 dpi (Figure 4F). This data demonstrates that both herb extracts have the ability to inhibit viral replication in the mouse respiratory tract and protect against RSV infection in vivo.

3.8. Acteoside Inhibits RSV Replication at Non-Cytotoxic Concentrations in vitro

It has been previously reported that acteoside is an important phenolic glycoside in PAE [27] and CTE [15,17]. To investigate for the presence of this glycoside in PAE and CTE, we performed reverse-phase high-performance liquid chromatography (HPLC) on extracts from PAE and CTE. Interestingly, we found that both extracts contained acteoside as one of their major active components (Figure 5A). To assess the antiviral effect of acteoside, a monolayer of HEp2 cells was infected with RSV-GFP, and at 2 hpi, cells were treated with 10, 30, or 50 ng/mL acteoside and virus replication were monitored. As reported in Figure 5C,D, acteoside treatment significantly reduced the GFP expression compared to untreated HEp2 cells. Furthermore, virus titers were determined by plaque formation assay. Similar to the results of the GFP expression analysis, acteoside significantly reduced the RSV plaque titer (Figure 5E). Cell death induced by RSV infection was also reduced in acteoside-treated HEp2 cells compared with untreated cells (Figure 5F). Next, the EC_{50} and CC_{50} of acteoside were determined in HEp2 cells as 15.64 ± 1.07 ng/mL and 740.34 ± 8.23 ng/mL, respectively (Figure 5G,H). The SI of acteoside against RSV-GFP was 47.33 (Figure 5I). Based on this data, the reduced viral replication caused by actioside (Figure 5C–F) was due to its antiviral properties and not its cytotoxicity. Moreover, the mRNA and protein expression of viral genes in HEp2 cells treated with acteoside was determined, as with the herb extracts. Interestingly, acteoside-treated cells showed significantly reduced RSV-G mRNA compared with untreated cells (1.7-fold reduction at 36 hpi and 4-fold reduction at 48 hpi) (Figure 6A). Similarly, RSV-G protein synthesis was inhibited by acteoside treatment (Figure 6B). Therefore, our results demonstrate that acteoside can inhibit RSV replication in HEp2 cells.

3.9. Intraperitoneal Administration of Acteoside Inhibits RSV Infection In Vivo

To further evaluate the therapeutic effect of acteoside against RSV, we investigated its antiviral activity in BALB/c mice. Mice were intranasally infected with RSV-GFP (1×10^6 PFU/mouse), and acteoside was administrated intraperitoneally at 6 hpi at a dose of 80 mg/kg, as described previously [28]. Lungs from all mice were collected at 3 dpi and 5 dpi, and RSV-G mRNA was measured by qRT-PCR. As shown in Figure 6C, the level of viral mRNA was significantly lower in the acteoside-treated group than in the PBS control group at both 3 dpi and 5 dpi. When taken together, acteoside exhibited a strong in vivo antiviral effect and protected mice from RSV infection.

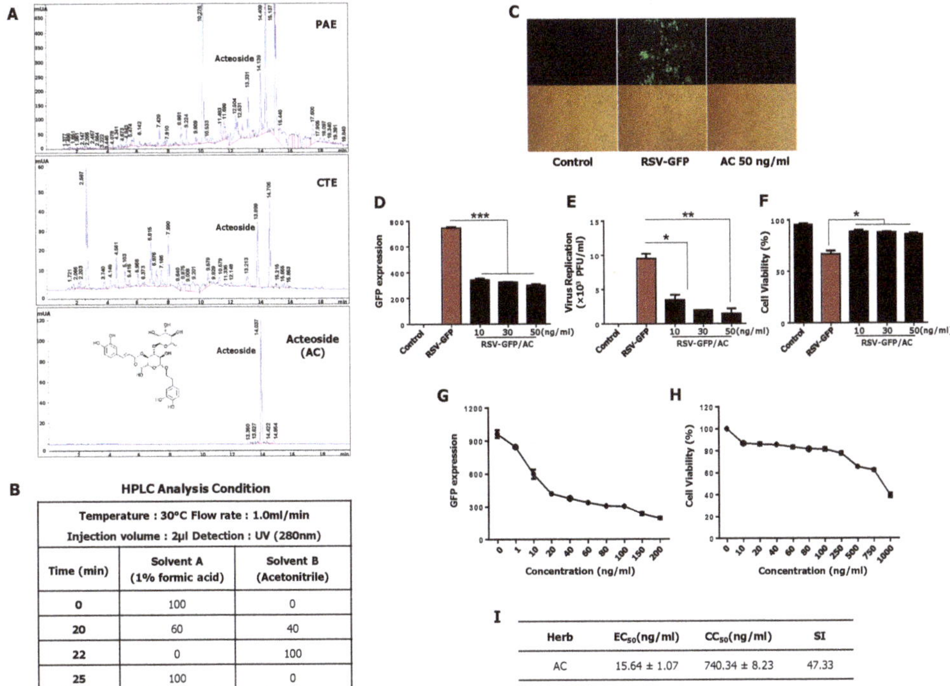

Figure 5. Identification and antiviral effect of acteoside (AC) *in-vitro*. (**A,B**) Chemical compounds in PAE and CTE were analyzed by the reversed phase HPLC. The monolayer of HEp2 cells was infected with RSV-GFP (0.1MOI) for 2 h with DMEM containing 1% FBS. Then, the medium was replaced with DMEM containing 10% FBS and cells were treated with 10, 30, 50 (ng/mL) AC. (**C**) After 48 h, images were obtained (200× magnification). (**D**) GFP absorbance levels were measured by Gloma multi-detection luminometer (Promega). (**E**) Viruses were titrated from the cell supernatant and cells by standard plaque assay. (**F**) Cell viability was determined by trypan blue exclusion assay at 48 hpi. (**A**) HEp2 cells were infected with RSV-GFP (O.1MOI) for 2 h with DMEM containing 1% FBS. Then, the medium was changed to DMEM containing 10% FBS and cells were treated with various concentrations AC. 48 hpi GFP expression level was determined. (**G**) HEp2 cells were treated with various concentrations of AC and cell viability was determined at 48 hpi by cell cytotoxicity assay kit. (**H**) To calculate EC_{50} value, 50% reduction of GFP expression was considered as equivalent to the 50% reduction in virus titer. (**I**) The ratio between CC_{50} and EC_{50} considered as Selectivity Index (SI). GFP absorbance, cell viability and virus titer expressed as mean ± SD. Error bars indicate the range of values obtained from counting duplicate in three independent experiments. In vivo experiment was performed in duplicate. (* $p < 0.05$, ** $p < 0.01$ and *** $p < 0.001$ regarded as significant difference).

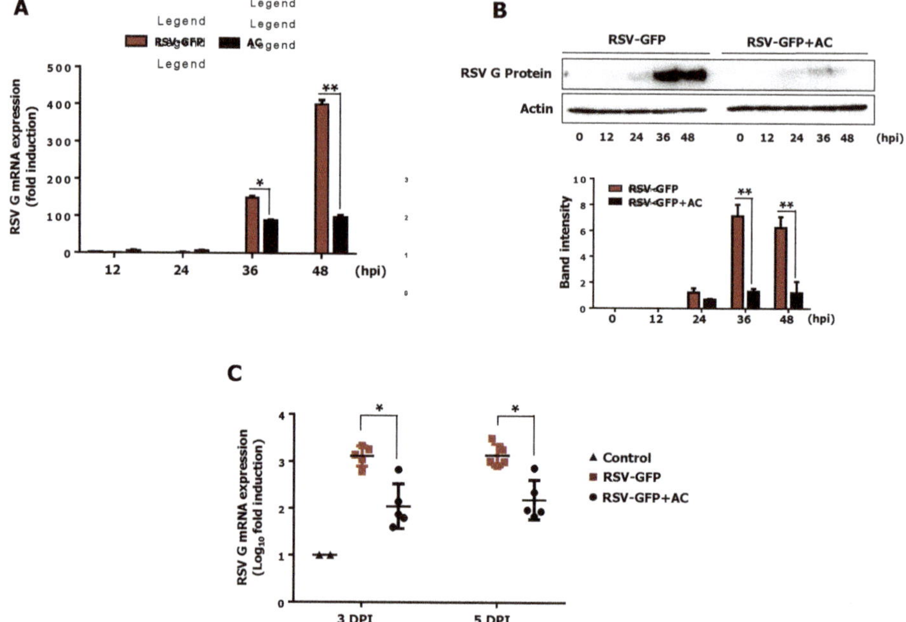

Figure 6. Antiviral effect of acteoside *in-vitro* and *in-vivo*. (**A**) RSV-GFP infected cells were treated with 50 ng/mL of AC at 2 hpi and cells were harvested at indicated time points. RSV-G protein mRNA transcription level was determined by qRT-PCR. GAPDH was used for normalization. (**B**) Infected cells were treated with 50 ng/mL concentration of AC at 2 hpi and cells were harvested at indicated time points. RSV-G protein expression was determined by immunoblotting with anti-RSV-G protein antibody, and the intensity of the RSV-G was quantified. (**C**) Five weeks old (16 g/mice) BALB/c mice (n = 5) were intranasally infected with RSV-GFP (1×10^6 PFU/mice) in the total volume of 28 µL. 6 hpi AC was intraperitoneally administrated at a dose of 80 mg/Kg body weight of mice. At 3 and 5 dpi, lung tissues were collected, and the transcription level of RSV-G protein mRNA was determined by qRT-PCR. mRNA expression and band intensity expressed as mean ± SD. Error bars indicate the range of values obtained from three independent experiments. In vivo experiment was performed in duplicate. (* $p < 0.05$ and ** $p < 0.01$ regarded as significant difference).

4. Discussion

Traditional medicines have been used as remedies against infectious diseases for thousands of years, due to their significant anti-inflammatory, anti-microbial activity and low rate of adverse effects [29,30]. These medicines have been gaining in popularity, due to concerns related to the side effects, high cost, and lack of efficacy of conventional Western medicines [31]. In the years 2001 and 2002, approximately one-quarter of the bestselling drugs worldwide were natural products or were derived from natural products [32]. Recent publications show that traditional Chinese medicinal herbs account for 10% of the prescription drugs in China. They are perceived as harmless and natural and are widely used in many parts of the world, individually or in combination [33]. In particular, medicinal plants have shown potential therapeutic effects against a wide range of respiratory tract-related viral infections, including Severe Acute Respiratory Syndrome (SARS) [34–36], Influenza [37,38], and RSV [26,27,39]. Among the thousands of promising medicinal herbs, *Plantago asiatica* and *Clerodendrum trichotomum* are well-known and commonly used in traditional medicine in China, Japan, and South Korea [9,14,15].

In the present study, we screened a library of herb extracts to identify novel therapeutic inhibitors of RSV infection. Interestingly, we identified both *Plantago asiatica* extract (PAE) and *Clerodendrum trichotomum* extract (CTE) as hits with potent antiviral effects against RSV infection in HEp2 cells (Supplementary Figures S1 and S2). Next, we confirmed the dose-dependent anti-RSV activity of both herbs in the HEp2 cell line in detail (Figures 1 and 2).

In addition, the CC_{50} values of PAE and CTE were several magnitudes higher than the EC_{50} values (Figure 3), which is consistent with a favorable safety profile. Furthermore, both herb extracts reduced intracellular viral gene transcription and protein synthesis in vitro, and oral administration of the herbs to infected mice significantly reduced viral gene transcription in the lungs (Figure 4). Finally, we found that acteoside, a common phenolic glycoside present in both herb extracts, is involved in the antiviral activity of the herbs against RSV infection (Figures 5 and 6).

PAE and CTE significantly reduced RSV replication and RSV-mediated syncytial formation in the HEp2 cell line in a dose-dependent manner (Figure 1C). Moreover, we were curious to evaluate whether PAE and CTE can work synergistically to enhance anti-RSV activity. However, herb extracts did not show synergistic anti-RSV effect compared to the individual treatment at the same dose (Figure 3A,B). Since, both extracts reduce the virus replication at the same level when treated alone, or together, at same dose (Ex: PAE, 50 µg/mL and CTE, 50 µg/mL or PAE, 25 µg/mL +CTE, 25 µg/mL), it's possible that both herb extracts undergo their own mechanism of action for anti-RSV function other than work synergistically.

However, PAE and CTE did not show any cytotoxic effect in the HEp2 cell line (Figure 3F,G). The CC_{50} values of PAE and CTE were 938.43 µg/mL and 764.17 µg/mL, respectively, which were several magnitudes higher than EC_{50} values of 39.82 µg/mL and 27.95 µg/mL, respectively. Even though both extracts were used at a concentration of 50 µg/mL for in vitro experiments, their high SI indicates a broad safety margin for therapeutic purposes.

To support the observation mentioned above, we investigated the intracellular RSV gene transcription in infected epithelial cells. Interestingly, PAE and CTE-treated HEp2 cells showed significantly reduced RSV-G mRNA at 36 and 48 hpi (Figure 4A). This reduction in viral mRNA transcription positively correlated with the low level of viral replication observed at the same time points (Figure 2A,B). Furthermore, PAE- and CTE-treated HEp2 cells showed significantly reduced RSV-G protein synthesis at late time points after virus infection (Figure 4B,C). Therefore, it is possible that PAE and CTE may affect the replication of RSV not only at the transcriptional level, but also at the posttranscriptional level. Syncytium formation by RSV is a well-known mechanism of cell-to-cell infection that contributes significantly to virus spread in vivo. PAE and CTE also significantly reduced the RSV-dependent formation of syncytia in HEp2 cells (Figure 4D). Since both herb extracts could inhibit RSV replication in vitro, we went on to test their antiviral potential in vivo. In vivo replication of RSV can also be accurately assessed by qRT-PCR [40].

Interestingly, our in vivo results (Figure 4E,F) revealed that oral administration of PAE and CTE significantly reduced the amount of RSV-G transcription in the lung at 3 and 5 dpi. These in vivo results are consistent with the low virus titers observed in HEp2 cells treated with PAE and CTE in vitro. Even though the exact underling therapeutic anti-RSV mechanisms of PAE and CTE are still under investigation, our results of the in vitro and in vitro antiviral assays show that both herb extracts significantly abolished cell to cell virus infection of RSV, and ultimate inhibition of virus spreading in the infected sites.

Phenylethanoid glycosides are widely found in edible plants and foodstuffs derived from plants [41,42], and these compounds have numerous biological properties, including anti-hepatotoxic [43], anti-inflammatory, anti-nociceptive [44], and anti-oxidant [45] effects. Phenylethanoid glycosides are one of the major bioactive constituents present in both *Plantago asiatica* [46] and *Clerodendrum trichotomum* [47]. HPLC analysis was conducted to confirm this finding (Figure 5A), and acteoside was identified as one of the major constituents of both PAE and CTE. Acteoside, also called kusagin or verbascoside [48], is a phenylethanoid glycoside isolated from many dicotyledons. Reportedly, acteoside has anti-oxidant and anti-inflammatory properties, and prevents cell apoptosis [39,49]. A recent study demonstrated that acteoside induced ERK activation and subsequent IFN-γ production; thus, showed antiviral activity against influenza and vesicular stomatitis virus (VSV) [28]. Based on these reported findings, we evaluated the antiviral effect of acteoside against RSV. Interestingly, acteoside reduced RSV replication and virus-induced cell death in HEp2 cells (Figure 5E,F) similar to PAE or CTE treatment. In addition, the SI of acteoside against RSV in vitro indicates a high safety margin for its therapeutic effect (Figure 5I). We also confirmed that acteoside reduced the level of RSV-G mRNA and RSV-G protein synthesis at 36 and 48 hpi (Figure 6A). Specifically, intraperitoneal treatment with acteoside showed an anti-RSV effect in a mouse model (Figure 6C). These results are similar to the antiviral activity of acteoside against Influenza or VSV [28]. However, further studies demonstrating a detailed mechanism of how acteoside inhibit RSV replication are needed.

In conclusion, the favorable safety profile and antiviral activity of PAE, CTE, and acteoside suggest that both herb extracts may be good candidates for antiviral therapy for RSV infection. Thus, oral administration of *Plantago asiatica* and *Clerodendrum trichotomum* or administration of acteoside could have potential therapeutic applications in both humans and livestock.

Supplementary Materials: The following are available online at http://www.mdpi.com/1999-4915/11/7/604/s1, Figure S1. Herb extract library screening results, Figure S2. Selected herb extracts second round screening results.

Author Contributions: K.C. designed and executed all virus infection and mouse experiments; M.S.K., H.-C.L., T.-H.K., J.-H.K., W.A.G.C., P.E., H.M.S.M.W., W.-K.C. performed cell biological experiments and mouse experiments, J.Y.M., H.I.K. analyzed the data. J.-S.L. designed the overall study and wrote the paper.

Acknowledgments: This work was supported by the Ministry for Food, Agriculture, Forestry and Fisheries, Republic of Korea (Grant No. 315044-3, 318039-3), the Small and Medium Business Administration (Grant No. S2130867, S2165234), the Korean Institute of Oriental Medicine by the Ministry of Education, Science and Technology (MEST) (Grant No. K16281), National Research Foundation (Grant No. 2018M3A9H4078703, 2019R1A2C2008283) and the research fund of Chungnam National University.

Conflicts of Interest: The authors declare no conflict of interest.

References

1. Piedimonte, G.; Perez, M.K. Alternative Mechanisms for Respiratory Syncytial Virus (RSV) Infection and Persistence: Could RSV Be Transmitted Through the Placenta And Persist Into Developing Fetal Lungs? *Curr. Opin. Pharmacol.* **2014**, *16*, 82–88. [CrossRef] [PubMed]

2. Mazur, N.I.; Martinón-Torres, F.; Baraldi, E.; Fauroux, B.; Greenough, A.; Heikkinen, T.; Manzoni, P.; Mejias, A.; Nair, H.; Papadopoulos, N.G.; et al. Lower Respiratory Tract Infection Caused by Respiratory Syncytial Virus: Current Management and New Therapeutics. *Lancet Respir. Med.* **2015**, *3*, 888–900. [CrossRef]

3. Walsh, E.; Falsey, A.R. Respiratory Syncytial Virus Infection in Adult Populations. *Infect. Disord.-Drug Targets* **2012**, *12*, 98–102. [CrossRef] [PubMed]

4. Le Nouen, C.; Brock, L.G.; Luongo, C.; McCarty, T.; Yang, L.; Mehedi, M.; Wimmer, E.; Mueller, S.; Collins, P.L.; Buchholz, U.J.; et al. Attenuation of Human Respiratory Syncytial Virus by Genome-Scale Codon-Pair Deoptimization. *Proc. Natl. Acad. Sci. USA* **2014**, *111*, 13169–13174. [CrossRef]

5. Kneyber, M.C.; Moll, H.A.; De Groot, R. Treatment and prevention of respiratory syncytial virus infection. *Eur. J. Pediatr.* **2000**, *159*, 399–411. [CrossRef]

6. Solecki, R.S.; Shanidar, I.V. A Neanderthal Flower Burial in Northern Iraq. *Science* **1975**, *190*, 880–881. [CrossRef]

7. Cowan, M.M. Plant Products as Antimicrobial Agents. *Clin. Microbiol. Rev.* **1999**, *12*, 564–582. [CrossRef]

8. Schuppan, D.; Afdhal, N.H. Liver Cirrhosis. *Lancet* **2008**, *371*, 838–851. [CrossRef]

9. Lin, C.C.; Kan, W.S. Medicinal Plants Used for the Treatment of Hepatitis in Taiwan. *Am. J. Chin. Med.* **1990**, *18*, 35–43. [CrossRef]

10. Marlett, J.A.; Fischer, M.H. The Active Fraction of Psyllium Seed Husk. *Proc. Nutr. Soc.* **2003**, *62*, 207–209. [CrossRef]

11. Ramkumar, D.; Rao, S.S. Efficacy and Safety of Traditional Medical Therapies for Chronic Constipation: Systematic Review. *Am. J. Gastroenterol.* **2005**, *100*, 936–971. [CrossRef] [PubMed]

12. Singh, B. Psyllium as Therapeutic and Drug Delivery Agent. *Int. J. Pharm.* **2007**, *334*, 1–14. [CrossRef] [PubMed]
13. Chiang, L.C.; Chiang, W.; Chang, M.Y.; Lin, C.C. In Vitro Cytotoxic, Antiviral and Immunomodulatory Effects of Plantago Major and *Plantago Asiatica. Am. J. Chin. Med.* **2003**, *31*, 225–234. [CrossRef] [PubMed]
14. Choi, J.H.; Whang, W.K.; Kim, H.J. Studies on the Anti-Inflammatory Effects Ofclerodendron Trichotomum Thunberg Leaves. *Arch. Pharmacal Res.* **2004**, *27*, 189–193. [CrossRef]
15. Kim, K.H.; Kim, S.; Jung, M.Y.; Ham, I.H.; Whang, W.K. Anti-Inflammatory Phenylpropanoid Glycosides from Clerodendron Trichotomum Leaves. *Arch. Pharmacal Res.* **2009**, *32*, 7–13. [CrossRef] [PubMed]
16. Kang, D.G.; Lee, Y.S.; Kim, H.J.; Lee, Y.M.; Lee, H.S. Angiotensin Converting Enzyme Inhibitory Phenylpropanoid Glycosides from Clerodendron Trichotomum. *J. Ethnopharmacol.* **2003**, *89*, 151–154. [CrossRef]
17. Lee, J.Y.; Lee, J.G.; Sim, S.S.; Whang, W.K.; Kim, C.J. Anti-Asthmatic Effects of Phenylpropanoid Glycosides from Clerodendron Trichotomum Leaves and Rumex Gmelini Herbes in Conscious Guinea-Pigs Challenged with Aerosolized Ovalbumin. *Phytomedicine* **2011**, *18*, 134–142. [CrossRef] [PubMed]
18. Chae, S.W.; Kang, K.A.; Kim, J.S.; Kim, H.K.; Lee, E.J.; Hyun, J.W.; Kang, S.S. Antioxidant activities of acetylmartynosides from Clerodendron trichotomum. *J. Appl. Biol. Chem.* **2007**, *50*, 270–274.
19. Lee, S.J.; Moon, H.I. Immunotoxicity activity of 2,6,10,15-tetrameheptadecane from the essential oils of *Clerodendron trichotomum* Thunb. against *Aedes aegypti* L. *Immunopharmacol. Immunotoxicol.* **2010**, *32*, 705–707. [CrossRef]
20. Moon, H.J.; Lee, J.S.; Choi, Y.K.; Park, J.Y.; Talactac, M.R.; Chowdhury, M.Y.; Poo, H.; Sung, M.H.; Lee, J.H.; Jung, J.H.; et al. Induction of Type I Interferon by High-Molecular Poly-Γ-Glutamate Protects B6.A2G-Mx1 Mice Against Influenza a Virus. *Antiviral Res.* **2012**, *94*, 98–102. [CrossRef]
21. Nguyen, D.T.; De Witte, L.; Ludlow, M.; Yüksel, S.; Wiesmüller, K.H.; Geijtenbeek, T.B.H.; Osterhaus, A.D.M.E.; De Swart, R.L. The Synthetic Bacterial Lipopeptide Pam3csk4 Modulates Respiratory Syncytial Virus Infection Independent of TLR Activation. *PLoS Pathog.* **2010**, *6*, e1001049. [CrossRef] [PubMed]
22. Strober, W. Trypan blue exclusion test of cell viability. *Curr. Protoc. Immunol.* **2001**. [CrossRef]
23. Kuhn, D.M.; Balkis, M.; Chandra, J.; Mukherjee, P.; Ghannoum, M. Uses and Limitations of the XTT Assay in Studies of Candida Growth and Metabolism. *J. Clin. Microbiol.* **2003**, *41*, 506–508. [CrossRef] [PubMed]
24. Livak, K.J.; Schmittgen, T.D. Analysis of relative gene expression data using real-time quantitative PCR and the $2^{-\Delta\Delta Ct}$ method. *Methods* **2001**, *25*, 402–408. [CrossRef] [PubMed]
25. Magadula, J.; Suleimani, H. Cytotoxic and Anti-HIV Activities of Some Tanzanian Garcinia Species. *Tanzan. J. Health Res.* **2010**, *12*, 144–149. [CrossRef]
26. Lee, B.H.; Chathuranga, K.; Uddin, M.B.; Weeratunga, P.; Kim, M.S.; Cho, W.K.; Kim, H.I.; Ma, J.Y.; Lee, J.S. Coptidis rhizoma extract inhibits replication of respiratory syncytial virus in vitro and in vivo by inducing antiviral state. *J. Microbiol.* **2017**, *55*, 488–498. [CrossRef] [PubMed]
27. Huang, D.F.; Tang, Y.F.; Nie, S.P.; Wan, Y.; Xie, M.Y.; Xie, X.M. Effect of Phenylethanoid Glycosides and Polysaccharides from The Seed of *Plantago Asiatica* L. On the Maturation of Murine Bone Marrow-Derived Dendritic Cells. *Eur. J. Pharmacol.* **2009**, *620*, 105–111. [CrossRef]
28. Song, X.; He, J.; Xu, H.; Hu, X.P.; Wu, X.L.; Wu, H.Q.; Liu, L.Z.; Liao, C.H.; Zeng, Y.; Li, Y. The Antiviral Effects of Acteoside and The Underlying IFN-γ-Inducing Action. *Food Funct.* **2016**, *7*, 3017–3030. [CrossRef]
29. Ma, S.C.; Du, J.; But, P.P.H.; Deng, X.L.; Zhang, Y.W.; Ooi, V.E.C.; Xu, H.X.; Lee, S.H.S.; Lee, S.F. Antiviral Chinese medicinal herbs against respiratory syncytial virus. *J. Ethnopharmacol.* **2002**, *79*, 205–211. [CrossRef]
30. Li, L.; Yu, C.H.; Ying, H.Z.; Yu, J.M. Antiviral effects of modified Dingchuan decoction against respiratory syncytial virus infection in vitro and in an immunosuppressive mouse model. *J. Ethnopharmacol.* **2013**, *147*, 238–244. [CrossRef]
31. Pan, S.Y.; Zhou, S.F.; Gao, S.H.; Yu, Z.L.; Zhang, S.; Tang, M.; Sun, J.; Ma, D.; Han, Y.; Fong, W.; et al. New Perspectives on How to Discover Drugs from Herbal Medicines: CAM's Outstanding Contribution to Modern Therapeutics. *Evid.-Based Complement. Altern. Med.* **2013**, *2013*, 627375. [CrossRef] [PubMed]
32. Butler, M.S. The role of natural product chemistry in drug discovery. *J. Nat. Prod.* **2004**, *67*, 2141–2153. [CrossRef] [PubMed]
33. Li, T.; Peng, T. Traditional Chinese Herbal Medicine as A Source of Molecules with Antiviral Activity. *Antivir. Res.* **2013**, *97*, 1–9. [CrossRef] [PubMed]

34. Lin, C.W.; Tsai, F.J.; Tsai, C.H.; Lai, C.C.; Wan, L.; Ho, T.Y.; Hsieh, C.; Chao, P. Anti-SARS Coronavirus 3C-Like Protease Effects of Isatis Indigotica Root and Plant-Derived Phenolic Compounds. *Antivir. Res.* **2005**, *68*, 36–42. [CrossRef] [PubMed]

35. Chen, C.J.; Michaelis, M.; Hsu, H.K.; Tsai, C.C.; Yang, K.D.; Wu, Y.C.; Cinatl, J.; Doerr, H. Toona Sinensis Roem Tender Leaf Extract Inhibits SARS Coronavirus Replication. *J. Ethnopharmacol.* **2008**, *120*, 108–111. [CrossRef] [PubMed]

36. Chen, F.; Chan, K.H.; Jiang, Y.; Kao, R.Y.; Lu, H.T.; Fan, K.W.; Cheng, V.; Tsui, W.; Hung, I.; Lee, T. In Vitro Susceptibility of 10 Clinical Isolates of SARS Coronavirus to Selected Antiviral Compounds. *J. Clin. Virol.* **2004**, *31*, 69–75. [CrossRef] [PubMed]

37. Haruyama, T.; Nagata, K. Anti-Influenza Virus Activity of Ginkgo Biloba Leaf Extracts. *J. Nat. Med.* **2012**, *67*, 636–642. [CrossRef]

38. Makau, J.N.; Watanabe, K.; Kobayashi, N. Anti-influenza activity of *Alchemilla mollis* extract: Possible virucidal activity against influenza virus particles. *Drug Discov. Ther.* **2013**, *7*, 18–195. [CrossRef]

39. Reinke, D.; Kritas, S.; Polychronopoulos, P.; Skaltsounis, A.L.; Aligiannis, N.; Tran, C.D. Herbal substance, acteoside, alleviates intestinal mucositis in mice. *Gastroenterol. Res. Pract.* **2015**, *2015*, 327872. [CrossRef]

40. Boukhvalova, M.S.; Yim, K.C.; Prince, G.A.; Blanco, J.C. Methods for monitoring dynamics of pulmonary RSV replication by viral culture and by real-time reverse transcription–PCR in vivo: Detection of abortive viral replication. *Curr. Protoc. Cell Biol.* **2010**, *46*, 26.6.1–26.6.19.

41. Li, L.; Tsao, R.; Liu, Z.; Liu, S.; Yang, R.; Young, J.C.; Zhu, H.; Deng, Z.; Xie, M.; Fu, Z. Isolation and purification of acteoside and isoacteoside from *Plantago psyllium* L. by high-speed counter-current chromatography. *J. Chromatogr. A* **2005**, *1063*, 161–169. [CrossRef]

42. Miyase, T.; Ishino, M.; Akahori, C.; Ueno, A.; Ohkawa, Y.; Tanizawa, H. Phenylethanoid Glycosides from *Plantago Asiatica*. *Phytochemistry* **1991**, *30*, 2015–2018. [CrossRef]

43. Xiong, Q.; Hase, K.; Tezuka, Y.; Tani, T.; Namba, T.; Kadota, S. Hepatoprotective Activity of Phenylethanoids Fromcistanche Deserticola. *Planta Medica* **1998**, *64*, 120–125. [CrossRef] [PubMed]

44. Schapoval, E.E.; Winter de Vargas, M.R.; Chaves, C.G.; Bridi, R.; Zuanazzi, J.A.; Henriques, A.T. Antiinflammatory and Antinociceptive Activities of Extracts and Isolated Compounds from Stachytarpheta Cayennensis. *J. Ethnopharmacol.* **1998**, *60*, 53–59. [CrossRef]

45. He, Z.D.; Lau, K.M.; Xu, H.X.; Li, P.C.; Pui-Hay, B. Antioxidant Activity of Phenylethanoid Glycosides from Brandisia Hancei. *J. Ethnopharmacol.* **2000**, *71*, 483–486. [CrossRef]

46. Sahpaz, S.; Garbacki, N.; Tits, M.; Bailleul, F. Isolation and Pharmacological Activity of Phenylpropanoid Esters from Marrubium Vulgare. *J. Ethnopharmacol.* **2002**, *79*, 389–392. [CrossRef]

47. Kim, H.J.; Woo, E.R.; Shin, C.G.; Hwang, D.J.; Park, H.; Lee, Y.S. HIV-1 Integrase Inhibitory Phenylpropanoid Glycosides Fromclerodendron Trichotomum. *Arch. Pharmacal Res.* **2001**, *24*, 286–291. [CrossRef]

48. Alipieva, K.; Korkina, L.; Orhan, I.E.; Georgiev, M.I. Verbascoside—A Review of its Occurrence, (Bio)Synthesis and Pharmacological Significance. *Biotechnol. Adv.* **2014**, *32*, 1065–1076. [CrossRef]

49. Aligiannis, N.; Mitaku, S.; Tsitsa-Tsardis, E.; Harvala, C.; Tsaknis, I.; Lalas, S.; Haroutounian, S. Methanolic Extract Ofverbascum Macrurumas a Source of Natural Preservatives Against Oxidative Rancidity. *J. Agric. Food Chem.* **2003**, *51*, 7308–7312. [CrossRef]

Article

Prevalence and Correlates of Pre-Treatment HIV Drug Resistance among HIV-Infected Children in Ethiopia

Birkneh Tilahun Tadesse [1,2,*], Olivia Tsai [3], Adugna Chala [4], Tolossa Eticha Chaka [5], Temesgen Eromo [6], Hope R. Lapointe [7], Bemuluyigza Baraki [3], Aniqa Shahid [3], Sintayehu Tadesse [4], Eyasu Makonnen [4,8], Zabrina L. Brumme [3,7], Eleni Aklillu [2] and Chanson J. Brumme [7,9,*]

[1] Department of Pediatrics, College of Medicine and Health Sciences, Hawassa University, Hawassa 1560, Ethiopia
[2] Division of Clinical Pharmacology, Department of Laboratory Medicine, Karolinska Institute, Karolinska University Hospital Huddinge, 141 86 Stockholm, Sweden; eleni.aklillu@ki.se
[3] Faculty of Health Sciences, Simon Fraser University, Burnaby, BC V5A 1S6, Canada; otsai@sfu.ca (O.T.); bbaraki@sfu.ca (B.B.); aniqa_shahid@sfu.ca (A.S.); zbrumme@sfu.ca (Z.L.B.)
[4] Department of Pharmacology, College of Health Sciences, Addis Ababa University, Addis Ababa 9086, Ethiopia; adugnadema@gmail.com (A.C.); ssintayehu2010@gmail.com (S.T.); eyasumakonnen@yahoo.com (E.M.)
[5] Adama General Hospital and Medical College, Adama 84, Ethiopia; tecb2006@gmail.com
[6] Department of Microbiology, SNNPR Reference Laboratory, Hawassa 149, Ethiopia; temeeromo@yahoo.com
[7] British Columbia Centre for Excellence in HIV/AIDS, Vancouver, BC V6Z 1Y6, Canada; hlapointe@alumni.ubc.ca
[8] Center for Innovative Drug Development and Therapeutic Trials for Africa (CDT Africa), College of Health Sciences, Addis Ababa University, Addis Ababa 9086, Ethiopia
[9] Division of Infectious Diseases, Department of Medicine, University of British Columbia, Vancouver, BC V5Z 1M9, Canada
* Correspondence: birknehtilahun@gmail.com (B.T.T.); cbrumme@cfenet.ubc.ca (C.J.B.); Tel.: +251-911551807 (B.T.T.); +1-604-682-2344 (ext. 63211) (C.J.B.)

Received: 1 August 2019; Accepted: 17 September 2019; Published: 19 September 2019

Abstract: Pediatric human immunodeficiency virus (HIV) care in resource-limited settings remains a major challenge to achieving global HIV treatment and virologic suppression targets, in part because the administration of combination antiretroviral therapies (cART) is inherently complex in this population and because viral load and drug resistance genotyping are not routinely available in these settings. Children may also be at elevated risk of transmission of drug-resistant HIV as a result of suboptimal antiretroviral administration for prevention of mother-to-child transmission. We investigated the prevalence and the correlates of pretreatment HIV drug resistance (PDR) among HIV-infected, cART-naive children in Ethiopia. We observed an overall PDR rate of 14%, where all cases featured resistance to non-nucleoside reverse transcriptase inhibitors (NNRTIs): ~9% of participants harbored resistance solely to NNRTIs while ~5% harbored resistance to both NNRTIs and nucleoside reverse transcriptase inhibitors (NRTIs). No resistance to protease inhibitors was observed. No sociodemographic or clinical parameters were significantly associated with PDR, though limited statistical power is noted. The relatively high (14%) rate of NNRTI resistance in cART-naive children supports the use of non-NNRTI-based regimens in first-line pediatric treatment in Ethiopia and underscores the urgent need for access to additional antiretroviral classes in resource-limited settings.

Keywords: HIV; pediatrics; Ethiopia; pre-treatment drug resistance; combination antiretroviral therapy (cART); dried plasma spots; dried blood spots

---

1. Introduction

Morbidity and mortality associated with human immunodeficiency virus (HIV) infection have substantially decreased with the introduction of effective combination antiretroviral therapy (cART) [1–3]. Towards realizing the Joint United Nations Programme on HIV/AIDS (UNAIDS) "90-90-90" targets to help end the acquired immunodeficiency syndrome (AIDS) epidemic [4], which outline ambitious goals for timely HIV diagnosis, sustained treatment, and maintenance of virologic suppression to limit onward viral transmission [5], the number of HIV infected individuals on cART is estimated to have reached 66% globally in 2018 [6]. Ensuring the sustained effectiveness of cART is key to maintaining these gains and is particularly critical in settings where cART options remain relatively limited.

Pediatric HIV treatment, especially in resource-limited settings, remains a major challenge to achieving and sustaining global treatment and virologic suppression targets [7]. This is in part due to the complexities of administering cART in children combined with the lack of routine availability of viral load and genotypic drug resistance testing to guide and monitor HIV treatment efficacy in resource-limited settings. Moreover, children in such settings may be particularly vulnerable to harboring pretreatment HIV drug resistance mutations, as antiretrovirals prescribed during pregnancy for prevention of mother-to-child transmission (PMTCT), if suboptimally administered, could lead to selection and transmission of HIV drug resistance mutations [8,9].

In Ethiopia, prevalence of HIV infection in the adult population is 1.4%, though there is substantial urban–rural variation [10]. The provision of cART to pregnant women for PMTCT in Ethiopia has been steadily increasing, and coverage reached 50–69% in 2017 [11]. This is partially the result of the 2013 nationwide implementation of PMTCT "option B+", which recommends initiating lifelong cART during pregnancy and providing nevirapine for six weeks for the neonate with exclusive breast feeding for the first six months and complementary feeding thereafter [12]. Surveillance for pretreatment HIV drug resistance (PDR), also called pretherapy drug resistance [13], is therefore critical to the care and the treatment of HIV-infected infants and children; however, no studies have assessed burden of PDR among HIV infected children in this region.

In Ethiopia, neither routine plasma viral load monitoring nor HIV drug resistance testing are readily available to guide individualized patient care. Empiric choice of cART is therefore the routine practice, and for this reason, understanding the overall burden and the pattern of PDR among HIV-infected children could help inform empiric treatment guidelines and practices. Towards this goal, the current study assessed the prevalence and the correlates of PDR among HIV-infected cART-naive children from a resource-limited setting in Ethiopia using dried plasma and blood samples (DPS and DBS, respectively).

2. Materials and Methods

2.1. Study Participants

The present study comprised children who were originally screened for eligibility for inclusion in the Efavirenz Pediatric Dose Optimization Study (EPDOS) cohort. EPDOS enrolled cART naïve HIV-infected children at seven HIV/AIDS treatment centers across two of the largest administrative regions in Ethiopia: Oromia and Southern Nations Nationalities and Peoples Region (SNNPR), whose combined populations exceed 55 million [14]. The centers in Oromia region were: Adama General Hospital, Asela Referral and Teaching Hospital, and Shashemene General Hospital. From SNNPR, they were: Hawassa University Comprehensive Specialized Hospital, Adare General Hospital, Otona Referral Hospital, and Arbaminch General Hospital. These regions were selected because of the high HIV prevalence and the large population. In Ethiopia, the estimated HIV pediatric population is 62,000 (38,000–86,000) [15]. Children who had tuberculosis and those who had previously been on combination antiretroviral therapy were not eligible to be enrolled in the parent EPDOS study, though children with PMTCT exposure would have been eligible if data were available. Note, however, that PMTCT history was not available for study participants.

Beginning in 2001, Ethiopia adopted PMTCT intervention Option A, under which eligible pregnant women with CD4 < 350 copies/mm^3 were initiated on cART. At this time, women who did not meet the CD4-based eligibility criteria were provided antepartum zidovudine (AZT) and intrapartum single dose nevirapine. In 2013, the guidelines were amended to recommend Option B+, and uptake has scaled up since then. By 2014, around 2500 health facilities had started providing PMTCT services. Currently, PMTCT coverage estimates in Ethiopia vary between 50–70% depending on region [11,16,17].

A total of 117 children were screened for EPDOS, of whom 111 were eventually enrolled [14]). The present study analyzed the baseline cross-section of a subset of the children originally screened for EPDOS; specifically, 93 (of 117; 79.5%) participants for whom a pre-cART sample was available or could newly be obtained. Participants were between the ages of 3–18 years, were cART naïve, and had no acute severe illnesses at enrollment. Since Ethiopia practices the "test and treat" strategy, participants were enrolled within 2–4 weeks of HIV diagnosis after the required pretreatment counseling was completed. Moreover, we could confirm that all children were indeed cART naive at time of enrollment, as HIV care and treatment in Ethiopia is managed under a centralized system.

Sociodemographic, clinical, and laboratory data were collected at baseline. Viral loads were determined using the RealTime HIV-1 viral load test (Abbott, Des Plaines, IL, USA).

2.2. Ethics Statement

Ethical approval for this study was obtained from the Institutional Ethics Review Boards of Addis Ababa University College of Health Sciences, Karolinska Institutet in Stockholm, Sweden, Simon Fraser University and Providence Health Care/University of British Columbia. The renewed National Research and Ethics Review Committee Ethics certificate is SHE/SM/14.3/0421/1/2019. Blood samples were collected after obtaining written informed consent in accordance with the tenets of the Declaration of Helsinki. For participants ≤ 12 years, written informed consent was obtained from their parent or guardian, while for participants > 12 years of age, consent was obtained from the parent or guardian and assent obtained from the participant. All informed consent documents were provided in the local language.

2.3. Specimen Collection, Handling and Storage

Up to 5 blood spots (DBS) for 22 HIV infected children and up to 5 plasma spots (DPS) for 71 HIV infected children were collected. Each spot contained approximately 50 μL. The DBS were prepared from participants by fingerprick on blood spot cards (Labmate, Cape Town, South Africa) and dried overnight at room temperature. The DPS samples were prepared by thawing previously-collected plasma samples stored at −80 °C, spotting ~50 μL aliquots on blood spot cards, and drying overnight at room temperature. Each card was individually packed in a plastic specimen bag with desiccant pack and shipped to Simon Fraser University (SFU) for HIV drug resistance genotyping. Spots were stored at room temperature until shipment to Simon Fraser University (SFU); upon receipt, they were stored at −80 °C until tested.

2.4. HIV Drug Resistance Genotyping and Phylogenetic Inference

Using a standard $\frac{1}{4}$" manual hole punch or a pair of metal forceps, two spots (plasma or blood, as provided) per participant were transferred into sterile tubes for nucleic acid extraction [18]. The hole punch was cleaned of residual material between participant cards by punching 10 holes into clean filter paper [18]; the forceps were cleaned using bleach. Total nucleic acids were extracted using the NucliSENS easyMAG System according to manufacturer's instructions (BioMerieux, Marcy-l'Étoile, France).

HIV Protease and a portion of reverse transcriptase (RT) spanning a minimum of codons 1–234 were amplified using an initial reverse transcriptase step (Expand Reverse Transcriptase; Roche, Basel, Switzerland) followed by nested PCR (Expand Hifi System; Roche) or alternatively by RT-PCR (using the SuperScript III One-Step RT-PCR System with Platinum Taq High Fidelity DNA Polymerase;

Invitrogen, Massachusetts, USA) followed by nested PCR (using the Expand HiFi System; Roche) [19]. Amplification was attempted using up to 4 oligonucleotide primer sets designed to amplify various HIV-1 group M subtypes (Table 1). If amplification failed using the primary set, amplification was attempted using the backup sets. If amplification failed again, fresh nucleic acid extracts were prepared from remaining DPS or DBS, and amplification was re-attempted as above. Amplicons were visualized on a 1% agarose gel and bulk (directly) sequenced on a 3130xl or 3730xl automated DNA sequencer (Applied Biosystems, Foster City, CA, USA). Chromatograms were analyzed using Sequencher version 5.0.1 (Gene Codes, Ann Arbor, MI, USA) or the automated basecalling software RECall [20], where nucleotide mixtures were called if the secondary peak exceeded 25% of the dominant peak height (Sequencher) or 17.5% of the dominant peak area (RECall).

Table 1. Primers used for human immunodeficiency virus (HIV)-1 protease and reverse transcriptase (RT) amplification.

Primer Set		First Round		Second Round	
		HXB2 Coordinates (Start/End)	Sequence (5′→3′)	HXB2 Coordinates (Start/End)	Sequence (5′→3′)
1 *	F	1979/2005	AAGAAGGGCACMT AGCCARAAAYTGYA	2011/2039	CCTAGGAAAAARGGC TGTTGGAARTGTGG
	R	3333/3301	CCACTAACTTCTGTATGT CATTGACAGTCCAGC	3280/3255	ATAGGCTGTACTG TCCATTTATCAGG
2	F	2008/2031	GCCCCTAGGAAA AAGGGCTGTTGG	2011/2039	CCTAGGAAAAARGGCTG TTGGAARTGTGG
	R	3361/3342	TAAATCTGA CTTGCCCART	3323/3303	CTGTATRTC ATTRACWGTCCA
3	F	1979/2005	AAGAAGGGCACMTA GCCARAAAYTGYA	2011/2039	CCTAGGAAAAARGG CTGTTGGAARTGTGG
	R	3859/3831	GCTCCTACTATGGG TTCTTTYTCYARYTG	3798/3777	CAAACTCCCAY TCAGGRATCCA
4	F	1992/2015	AGCCAGAAATTGC AGGGCCCCTAG	2074/2095	AGACAGGCTA ATTTTTTAGGGA
	R	3322/3303	TGTATRTCATT GACAGTCCA	3271/3252	ACTGTCCAT TTRTCAGGATG

* indicates the primary primer set. HXB2 is a HIV-1 subtype B reference strain.

HIV sequences were aligned using HIV Align (options: MAFFT, codon-alignment) [21] and manually inspected using AliView [22]. Maximum likelihood phylogenies were inferred from HIV sequence alignments using PhyML under a general time reversible model of nucleotide substitution [23]. Phylogenies were generated from full alignments as well as alignments stripped of all codons associated with HIV surveillance drug resistance mutations [24] to control for any potential effects of these mutations on tree topology [25]. Phylogenies were visualized using Figtree (version 1.4.4). HIV subtype determination was performed using the Recombinant Identification Program tool hosted by the Los Alamos HIV Sequence Database (LANL) [26]. HIV sequences were deposited into GenBank (accession numbers MN244083–MN244139).

2.5. Drug Resistance Genotype Interpretation

Drug resistance genotype interpretation was performed using the Calibrated Population Resistance (CPR) tool on the Stanford University HIV Drug Resistance Database [27–29], which identifies mutations conferring resistance to protease inhibitors (PIs), nucleoside reverse transcriptase inhibitors (NRTIs), and non-nucleoside reverse transcriptase inhibitors (NNRTIs) as defined by the World Health Organization (WHO) 2009 list of mutations for surveillance of pretreatment HIV drug resistance [24].

2.6. Statistical Analysis

Statistical analyses were conducted using GraphPad Prism (version 8, San Diego, CA, USA). The dependent variable was PDR (yes/no) for NRTI, NNRTI. or any resistance. Associations between PDR and sociodemographic, clinical, and laboratory variables were assessed using Fisher's exact test for categorical independent variables and the Wilcoxon Rank Sum test for continuous independent variables. p-values < 0.05 were considered statistically significant.

3. Results

3.1. Patient Characteristics

A total of 93 children were included in the study; their characteristics at enrollment and the number of participants for which data were available for each characteristic are summarized in Table 2. Their median age was 9 years [interquartile range (IQR): 5–12]; 48 (51.6%) were male. Upon clinical evaluation, 56/85 (65.9%) were identified as having symptoms that define WHO clinical stage 2 or above (Table 2). These symptoms included papular pruritic eruptions (21/85; 24.7%), mucocutaneous viral infections (24/85; 28.2%), chronic diarrhea (18/83; 21.2%), and features of fungal infection (15/85; 17.7%). The median CD4+ T-lymphocyte cell (T-cell) count was 319 (IQR: 141–615) cells per mm^3, and the median viral load was 4.3 log$_{10}$ copies/mL (IQR: 3.7–4.9 log$_{10}$ copies/mL) in the 41 and 83 children where CD4 and pVL data were available, respectively. There were two cases with hepatitis C virus (HCV) coinfection, but no hepatitis B virus (HBV) coinfection was reported. Regarding nutritional status at HIV diagnosis, 30/82 (36.6%), 23/82 (28.1%), and 20/53 (37.7%) participants were stunted, wasted, and underweight, defined as Z score < -2, respectively.

Table 2. Characteristics of children included in the study.

Variable	Summary Statistic	Total N
Age in years, median (IQR)	9.0 (5.0–12.0)	93
Male, N (%)	48 (51.6)	93
Symptoms at diagnosis, Yes N (%)	56 (65.9)	85
CD4 count, median (IQR) cells/mm^3	319 (141–615)	41
Plasma viral load in log$_{10}$ copies/mL of, median (IQR)	4.3 (3.7–4.9)	83
Weight for age Z-score, Median (IQR), Z-score	-1.2 (-2.7–(-0.7))	53 *
Height for age Z-score, Median (IQR)	-1.6 (-2.6–(-0.7))	82
Body mass index Z-score, Median (IQR)	-1.1 (-2.2–(-0.1))	82

IQR—interquartile range; normal weight, height, and body mass index Z scores range from -2 to $+2$. Z-scores between -2 and -3 indicate moderate undernutrition while Z-scores below -3 indicate severe malnutrition. * Weight-for-age Z score was not calculated for some children as the age and/or weight were out of range while using the World Health Organization (WHO) Anthro or AnthroPlus software (https://www.who.int/childgrowth/software/en/).

3.2. Prevalence of PDR and Detected Drug Resistance Mutation Types

Dried blood spots (DBS; $N = 22$) or dried plasma spots (DPS; $N = 71$) were collected from all 93 study participants. HIV drug resistance genotyping was successful for 57/93 (61.3%) of these samples, an overall success rate that is comparable to other studies utilizing dried blood products as starting material [30–32]. Success rates did not differ between DPS (14 of 22; 63.6%) and DBS (43 of 71; 60.6%) (Fisher's exact test, $p = 1$). Somewhat surprisingly, samples for which resistance genotyping was successful did not have significantly different viral loads from those samples where resistance was not successful [median 4.3 (IQR: 3.8–5.0) vs 4.4 (IQR: 3.2–4.9) $p = 0.47$]. Moreover, genotyping success rate was not significantly associated with any of the participant characteristics listed in Table 2, with the exception of weight-for-age Z-score. Participants for whom genotyping was successful exhibited a median Z-score of −1.9 [IQR −2.9–(−0.9) compared to −1 (IQR −1.7–(−0.1)] for whom genotyping was unsuccessful (Mann–Whitney U-test $p = 0.03$), though we acknowledge that this may be a chance finding. Despite using dried blood or plasma as starting material, some sequences nevertheless bore evidence of amplification of within-host sequence diversity. The number of nucleotide mixtures detected in successful genotypes ranged from 0 to 40 (median 0, IQR 0–14). One of the isolated sequences (EPDOS_9) was defective with an internal stop codon near the end of the sequence.

Consistent with the HIV epidemic in Ethiopia being predominately composed of subtype C [33–38], a total of 54 of 57 (94.7%) participants harbored HIV subtype C, one (1.7%) harbored an AG recombinant, one (1.7%) harbored an ACG recombinant, and one (1.7%) harbored subtype A (Figure 1). With the exception of one sibling pair whose viral sequences clustered closely together on the phylogeny, no other phylogenetically-linked infections were observed, as would be expected in a pediatric cohort. Tree topology was not substantially impacted by the presence of drug resistance codons).

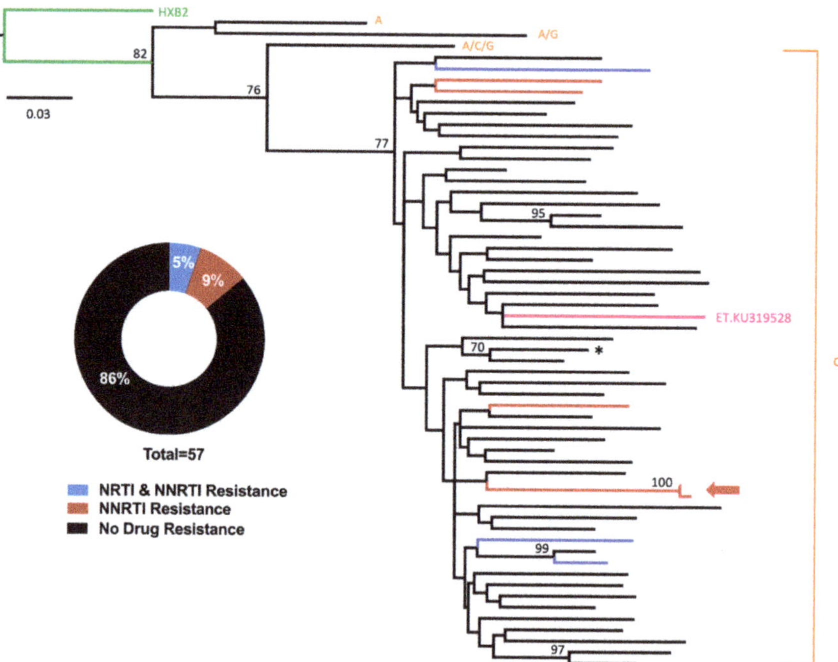

Figure 1. Prevalence of HIV-1 pretreatment HIV drug resistance (PDR) among combination antiretroviral therapies (cART)-naïve Ethiopian children. A maximum-likelihood phylogeny was inferred from the inclusive HIV consensus sequences of the 57 participants for whom genotyping was successful. Drug resistance codons were removed from the alignment prior to phylogenetic inference. Scale indicates expected substitutions per nucleotide site. Nodes with bootstrap values ≥ 70% are indicated on the tree. Colors indicate resistance genotype. HIV-1 subtypes are indicated at tree tips. Reference strains HXB2 (subtype B, green) and KU319528 (subtype C-Ethiopia, pink) are included. The arrow denotes a sibling pair harboring similar HIV sequences. The asterisk denotes a single participant harboring an E138A mutation in reverse transcriptase; this mutation is not on the list of WHO surveillance drug resistance mutations, and therefore this participant is classified in the "no drug resistance" category. However, this mutation confers low-level resistance to the non-nucleoside reverse transcriptase inhibitors (NNRTI) rilpivirine [36].

Overall, 8/57 successfully genotyped participants harbored HIV drug resistance mutations, yielding a total PDR prevalence of 14% (95% CI: 4.8–24.0%) in our study (Figure 1). All eight cases of PDR featured resistance to NNRTIs; specifically, five (62.5%) solely harbored (NNRTI) resistance mutations while three (37.5%) harbored both NRTI and NNRTI resistance mutations. No participant harbored protease inhibitor (PI) resistance mutations. The rates of drug resistance mutations did not differ significantly between the participants genotyped from DPS (5/43 successfully genotyped samples harbored resistance) and DBS (3/14 successfully genotyped samples harbored resistance) (Fisher's exact test, $p = 0.4$).

The mutation profiles observed in the eight PDR cases are listed in Table 3. The most commonly observed NNRTI resistance mutations were G190A and Y181C, each observed three times; K103N, K103S, and Y188L were also observed once each. NRTI resistance mutations observed included M184I, M184V, L210W, T215Y, and K219N. In general, children with PDR harbored a single mutation conferring resistance to one or more drugs in the class. Exceptions were EPDOS_8, who harbored NNRTI resistance with K103S and G190A, and EPDOS_53, who harbored dual-class (NNRTI/NRTI) resistance with three NRTI resistance mutations (M184V, L210W, T215Y).

Table 3. Mutational profiles of participants harboring HIV-1 antiretroviral drug resistance.

Sample ID	Dried Spot Type	NRTI Mutations	NNRTI Mutations
EPDOS_6	Plasma	None	G190G/A
EPDOS_8	Plasma	None	K103S, G190A
EPDOS_9	Plasma	None	K103N
EPDOS_22	Plasma	None	G190G/A
EPDOS_29	Plasma	None	Y181C
EPDOS_37	Blood	K219N	Y181C
EPDOS_39	Blood	M184I	Y188L
EPDOS_53	Blood	M184V, L210W, T215Y	Y181C

EPDOS—Efavirenz Pediatric Dose Optimization Study; NRTI—nucleoside reverse transcriptase inhibitors.

3.3. Correlates of PDR in Ethiopian Children

Even though the number of children with observed PDR was small, thereby potentially limiting the power to detect associations, we nevertheless wished to identify correlates of pretreatment HIV drug resistance mutations (Table 4). For this analysis, children were classified as having any (versus no) PDR; however, as all children with PDR harbored NNRTI resistance mutations, this can also be considered an analysis of correlates of NNRTI resistance mutations. Children with PDR were marginally, though not statistically significantly, younger at enrollment [median age 5 (IQR 0.3–10) years among participants with PDR versus 8 (5–12) years in those without PDR; $p = 0.06$]. Moreover, children with PDR had modestly lower albumin levels at HIV diagnosis as compared to their counterparts [2.9 (IQR: 2.5–3.4) mg/dL among participants with PDR versus 3.8 (3.2–4.2) mg/dL among participants with PDR; $p = 0.04$]. Even though a low albumin might suggest undernutrition in HIV-infected children who harbor HIV drug resistance mutations, anthropometric indicators did not show statistically significant differences between children with pretreatment HIV drug resistance mutations and those without. No significant associations were observed between other laboratory and clinical parameters and the presence of HIV drug resistance mutations. There were insufficient cases of NRTI resistance ($N = 3$, all of whom harbored both dual class resistance) to robustly evaluate this category separately; however, we observed that children with dual-class NNRTI/NRTI PDR tended to be younger than those who harbored single or no PDR 0.2 [(0.1–0.3) years for dual-class versus 8 (5–12) for single or none] ($p = 0.02$).

<table>
<tr><td rowspan="2" align="center">Variable</td><td rowspan="2" align="center">Number Missing</td><td colspan="2" align="center">Any Resistance</td><td rowspan="2" align="center">p Value</td></tr>
<tr><td align="center">Yes (N = 8)</td><td align="center">No (N = 49)</td></tr>
<tr><td>Age in years, median (IQR)</td><td align="center">0</td><td align="center">5 (0.3–10)</td><td align="center">8 (5–12)</td><td align="center">0.06</td></tr>
<tr><td>Sex (% Male)</td><td align="center">0</td><td align="center">4 (57.1)</td><td align="center">24 (51.1)</td><td align="center">0.54</td></tr>
<tr><td>WAZ, median (IQR)</td><td align="center">23</td><td align="center">−1.6 (−2.8–(−0.9))</td><td align="center">−1.9 (−2.9–(−0.9))</td><td align="center">0.91</td></tr>
<tr><td>HAZ, median (IQR)</td><td align="center">7</td><td align="center">−1.8 (−1.9–(−0.7))</td><td align="center">−1.6 (−3.0–(−0.7))</td><td align="center">0.62</td></tr>
<tr><td>BAZ, median (IQR)</td><td align="center">7</td><td align="center">−1.4 (−2.2–(−1.3))</td><td align="center">−1.2 (−2.2–(−0.5))</td><td align="center">0.41</td></tr>
<tr><td>CD4, median (IQR), cells/mm^3</td><td align="center">33</td><td align="center">267 (217–317)</td><td align="center">370 (161–788)</td><td align="center">0.53</td></tr>
<tr><td>Log10 pVL, median (IQR) copies/mL</td><td align="center">6</td><td align="center">4.3 (4.1–4.8)</td><td align="center">4.2 (3.8–5.0)</td><td align="center">0.47</td></tr>
<tr><td>WHO clinical stage, N (%)
1
2
3
4</td><td align="center">4</td><td align="center">1 (25)
2 (25.0)
4 (50)
0 (0)</td><td align="center">15 (30.6)
10 (20.4)
19 (38.8)
2 (4.1)</td><td align="center">0.71</td></tr>
</table>

Table 4. Factors associated with PDR among cART-naïve HIV infected children, Ethiopia, 2018–2019.

IQR—interquartile range; ALT—alanine aminotransferase; AST—aspartate aminotransferase; HCT—hematocrit; WAZ—weight-for-age Z score; HAZ—height-for-age Z score; BAZ—body mass index-for-age Z score; WHO—World Health Organization.

4. Discussion

Our study represents the first characterization of pretreatment HIV drug resistance (PDR) among children newly diagnosed with HIV in Ethiopia in 2017–2019. We found that 14% of participants for whom HIV PR-RT sequences could be obtained harbored PDR; of these, approximately two-thirds solely harbored (NNRTI) resistance mutations, and the remainder harbored dual-class (NRTI and NNRTI) resistance mutations. Importantly, all participants with NNRTI resistance had intermediate-to high-level resistance to efavirenz (EFV) and nevirapine (NVP), the NNRTI components of first line cART in Ethiopia. Resistance to PIs was not observed. One important limitation of this study is that we had no information on the PMTCT exposure of the participants. Noting that limitation, the level of HIV drug resistance observed in the present study is comparable to PMTCT-unexposed HIV infected children [39] and is threefold lower than among PMTCT-exposed treatment naïve HIV infected children [39,40] in sub Saharan Africa. A recent analysis of multiple African countries reported a high prevalence of PDR to any antiretroviral drug among infants newly diagnosed with HIV (54.1% overall; 53.0% for NNRTIs and 8.8% for NRTIs) [41]. The much higher prevalence of pediatric PDR in the latter study may be attributable to greater PMTCT exposure, which was reported to be as high as 75–85% in surveys recorded from 2011–2014.

Our findings also indicate that, in Ethiopia, the burden of PDR among children newly diagnosed with HIV is substantially higher than that in adults newly diagnosed with HIV, in which PDR prevalence has been estimated at 3.9% [42]. Despite the limited PDR data in Ethiopia for both of these patient populations, our observations are consistent with PDR prevalence among adults and children in other high prevalence settings [43]. The difference in the burden of PDR between adults and children could be explained by PMTCT exposure in these children [44], although the observation that the majority of the children studied were born prior to the 2013 implementation of Option B+ suggests only limited PMTCT exposure in this cohort.

Over the past two decades, rates of HIV PDR in many sub Saharan African regions have been increasing, in some cases to alarming levels [39,45]. Our observations further underscore HIV drug resistance as a major threat to HIV control in resource-limited settings. HIV infected children in Ethiopia are potentially at risk of poor treatment outcomes as a result of high HIV PDR levels [46].

Our findings may also have implications for future treatment practices. Currently, the WHO guidelines recommend against the use of an NNRTI-based regimen as a first line treatment if the prevalence of NNRTI PDR in the region exceeds 10% [47]. In Ethiopia, based on the WHO guidelines, PI-based regimens are recommended for children under three years of age, while for children older than three years, two NRTIs (lamivudine with either abacavir, tenofovir, or AZT) plus one NNRTI (either EFV or NVP) are recommended [48,49]. For children older than 10 years, a dolutegravir (DTG) based regimen is recommended [50]. Our observations that 14% of HIV-positive children evaluated had evidence of NNRTI PDR while no children harbored PI resistance supports the consideration of non-NNRTI-based firstline regimens for children of all ages in Ethiopia. Specifically, our findings may support the use of PI-based or DTG-based first regimens in children of all ages. It is acknowledged, however, that the bitter taste of certain pediatric PI-based regimens can be unpalatable, particularly for children, and therefore our findings underscore the urgent need to expand access to newer antiretrovirals and additional drug classes, particularly integrase inhibitors, in Ethiopia.

Some limitations of our study merit mention. The lack of PMTCT information in the EPDOS cohort precludes us from interpreting results in the context of prior antiretroviral exposure and complicates comparisons with other studies from the region. However, as the majority of the children studied were diagnosed at a relatively late age (median nine years), it is possible that these children were only tested subsequent to one or both parents' recent HIV diagnosis—a common clinical occurrence in Ethiopia. Moreover, most of these children were born before the 2013 scale up of PMTCT Option B+. Taken together, it is likely that most children were not exposed to PMTCT. However, it is important to note that, for those children with prior PMTCT exposure, later diagnosis and enrollment into EPDOS may have allowed resistance mutations associated with NVP exposure to revert to wild-type, leading to an underestimation of the burden of PDR in this cohort. Moreover, our use of Sanger sequencing could have limited our ability to detect low frequency mutations.

In conclusion, the overall prevalence of PDR reported in the current study (14%) is comparable to the prevalence of PDR among PMTCT-unexposed HIV infected children in sub Saharan Africa [39]. As the study sites enrolled HIV infected children with different ethnic and sociodemographic characteristics across two large administrative regions of central and southern Ethiopia, the findings reflect the burden of PDR in Ethiopia at large. The observation that all PDR cases featured mutations that would confer intermediate-to-high-level resistance to efavirenz or nevirapine, the NNRTIs currently available for pediatric first-line treatment in Ethiopia [48–50], supports the use of non-NNRTI-based first-line regimens for newly diagnosed HIV infected children in southern Ethiopia (including integrase inhibitor based regimens for eligible children) and calls for establishing a routine drug resistance surveillance in the setting. The relatively low prevalence of NRTI resistance (~5%) and complete lack of PI resistance supports the preferred use of these agents in firstline cART regimens for HIV infected children. Nevertheless, the observation of dual-class PDR, albeit in a minority (5%) of cases, also underscores the urgent need for expanded and affordable access to newer antiretrovirals and additional drug classes, particularly integrase inhibitors, in resource-limited settings.

Author Contributions: Conceptualization, B.T.T., Z.L.B., and E.A.; methodology, B.T.T., A.C., T.E.C., E.M.; Z.L.B., C.J.B.; software, B.T.T., Z.L.B., C.J.B.; validation, B.T.T., O.T., B.B., A.S., E.M., Z.L.B., E.A.,A.C., T.E.C., S.T.; formal analysis, B.T.T., O.T., Z.L.B., C.J.B., E.A., T.E., A.C., T.E.C., and S.T.; investigation, O.T., B.B., A.S., Z.L.B., A.C., H.R.L., S.T.; data curation, B.T.T., O.T.,T.E.C., T.E., H.R.L., B.B., A.S., S.T., and A.C.; visualization, B.T.T., O.T., Z.L.B., and C.J.B.; supervision, E.A., E.M., Z.L.B., and C.J.B.; project administration, B.T.T., A.C., T.E.C., Z.L.B., O.T., E.M. and E.A.; funding acquisition, E.A., and Z.L.B.; resources, E.A., Z.L.B., C.J.B.; writing—original draft preparation, B.T.T., O.T., and Z.L.B.; and writing—review and editing, A.C., T.E.C., T.E., H.R.L., B.B., A.S., S.T., E.M., E.A., C.J.B.

Funding: This research was funded by the Swedish Research Council (Vetenskapsrådet), grant number: 2015-03295 (to E Aklillu) and by the Canadian Institutes for Health Research, grant number PJT-148621 (to ZLB). ZLB is supported by a Scholar Award from the Michael Smith Foundation for Health Research.

Acknowledgments: We thank Natalie Kinloch, Getahun Hailemeskel, and Hanwei Sudderruddin for technical assistance and helpful discussions. We are grateful to the children and parents who participated in the study as well as the healthcare professionals who were involved with participant follow-up. We also thank Julio Montaner and the BC Centre for Excellence in HIV/AIDS for support.

Conflicts of Interest: The authors declare no conflict of interest.

References

1. Michaels, S.H.; Clark, R.; Kissinger, P. Declining morbidity and mortality among patients with advanced human immunodeficiency virus infection. *N. Engl. J. Med.* **1998**, *339*, 405–406. [CrossRef] [PubMed]
2. Weverling, G.J.; Mocroft, A.; Ledergerber, B.; Kirk, O.; Gonzales-Lahoz, J.; d'Arminio Monforte, A.; Proenca, R.; Phillips, A.N.; Lundgren, J.D.; Reiss, P. Discontinuation of Pneumocystis carinii pneumonia prophylaxis after start of highly active antiretroviral therapy in HIV-1 infection. EuroSIDA Study Group. *Lancet* **1999**, *353*, 1293–1298. [CrossRef]
3. Hogg, R.S.; Heath, K.V.; Yip, B.; Craib, K.J.; O'Shaughnessy, M.V.; Schechter, M.T.; Montaner, J.S. Improved survival among HIV-infected individuals following initiation of antiretroviral therapy. *JAMA* **1998**, *279*, 450–454. [CrossRef] [PubMed]
4. UNAIDS. 90-90-90 Targets. 2017. Available online: https://www.unaids.org/en/resources/documents/2017/90-90-90 (accessed on 1 June 2019).
5. Granich, R.M.; Gilks, C.F.; Dye, C.; De Cock, K.M.; Williams, B.G. Universal voluntary HIV testing with immediate antiretroviral therapy as a strategy for elimination of HIV transmission: A mathematical model. *Lancet* **2009**, *373*, 48–57. [CrossRef]
6. UNAIDS. UNAIDS Data 2018. Available online: https://www.unaids.org/sites/default/files/media_asset/unaids-data-2018_en.pdf (accessed on 21 June 2019).
7. Davies, M.A.; Pinto, J. Targeting 90-90-90-don't leave children and adolescents behind. *J. Int. AIDS Soc.* **2015**, *18*, 20745. [CrossRef] [PubMed]
8. Louis, F.J.; Segaren, N.; Desinor, O.; Beard, R.S.; Jean-Louis, R.; Chang, J.; Boisson, S.; Hulland, E.N.; Wagar, N.; DeVos, J.; et al. High Levels of HIV-1 Drug Resistance in Children Who Acquired HIV Infection Through Mother to Child Transmission in the Era of Option B+, Haiti, 2013 to 2014. *Pediatric Infect. Dis. J.* **2019**, *38*, 503–507. [CrossRef] [PubMed]
9. Inzaule, S.C.; Hamers, R.L.; Calis, J.; Boerma, R.; Sigaloff, K.; Zeh, C.; Mugyenyi, P.; Akanmu, S.; Rinke de Wit, T.F. When prevention of mother-to-child HIV transmission fails: Preventing pretreatment drug resistance in African children. *AIDS* **2018**, *32*, 143–147. [CrossRef]
10. African Health Observatory: Analytical Summary—HIV/AIDS. Available online: http://www.aho.afro.who.int/profiles_information/index.php/Ethiopia:Analytical_summary_-_HIV/AIDS (accessed on 2 September 2019).
11. Avert. Global Information and Education on HIV and AIDS: Prevention of Mother-to-Child Transmission (PMTCT) of HIV. 2019. Available online: https://www.avert.org/professionals/hiv-programming/prevention/prevention-mother-child (accessed on 20 June 2019).
12. FMOH. National Guidelines for Comprehensive HIV Prevention, Care and Treatment. 2017. Available online: https://www.medbox.org/ethiopia-national-guidelines-for-comprehensive-hiv-prevention-care-and-treatment-1/download.pdf (accessed on 20 June 2019).
13. Hamers, R.L.; Rinke de Wit, T.F.; Holmes, C.B. HIV drug resistance in low-income and middle-income countries. *Lancet HIV* **2018**, *5*, e588–e596. [CrossRef]
14. Tadesse, B.T.; Foster, B.A.; Chala, A.; Chaka, T.E.; Bizuayehu, T.; Ayalew, F.; H/Meskel, G.; Tadesse, S.; Jerene, D.; Makonnen, E.; et al. HIV and cART-Associated Dyslipidemia Among HIV-Infected Children. *J. Clin. Med.* **2019**, *8*, 430. [CrossRef]
15. UNAIDS. Ethiopia: Country Factsheets. 2017. Available online: https://www.unaids.org/en/regionscountries/countries/ethiopia (accessed on 20 June 2019).
16. Ministry of Health of Federal Democratic Republic of Ethiopia. *National Strategic Plan for Elimination of Mother to Child Transmission of HIV (e-MTCT of HIV) 2013–2015*; Ministry of Health of Federal Democratic Republic of Ethiopia: Addis Ababa, Ethiopia, 2013.
17. Federal Democratic Republic of Ethiopia HIV Prevention and Control Office. *Country Progress Report on the HIV Response*; Federal Democratic Republic of Ethiopia HIV Prevention and Control Office: Addis Ababa, Ethiopia, 2014.

18. Buckton, A.J.; Prabhu, D.P.; Cane, P.A.; Pillay, D. No evidence for cross-contamination of dried blood spots excised using an office hole-punch for HIV-1 drug resistance genotyping. *J. Antimicrob. Chemother.* **2009**, *63*, 615–616. [CrossRef]

19. Tadesse, B.T.; Kinloch, N.N.; Baraki, B.; Lapointe, H.R.; Cobarrubias, K.D.; Brockman, M.A.; Brumme, C.J.; Foster, B.A.; Jerene, D.; Makonnen, E.; et al. High Levels of Dual-Class Drug Resistance in HIV-Infected Children Failing First-Line Antiretroviral Therapy in Southern Ethiopia. *Viruses* **2018**, *10*, 60. [CrossRef] [PubMed]

20. Woods, C.K.; Brumme, C.J.; Liu, T.F.; Chui, C.K.; Chu, A.L.; Wynhoven, B.; Hall, T.A.; Trevino, C.; Shafer, R.W.; Harrigan, P.R. Automating HIV drug resistance genotyping with RECall, a freely accessible sequence analysis tool. *J. Clin. Microbiol.* **2012**, *50*, 1936–1942. [CrossRef] [PubMed]

21. Gaschen, B.; Kuiken, C.; Korber, B.; Foley, B. Retrieval and on-the-fly alignment of sequence fragments from the HIV database. *Bioinformatics* **2001**, *17*, 415–418. [CrossRef] [PubMed]

22. Larsson, A. AliView: A fast and lightweight alignment viewer and editor for large datasets. *Bioinformatics* **2014**, *30*, 3276–3278. [CrossRef] [PubMed]

23. Guindon, S.; Dufayard, J.F.; Lefort, V.; Anisimova, M.; Hordijk, W.; Gascuel, O. New algorithms and methods to estimate maximum-likelihood phylogenies: Assessing the performance of PhyML 3.0. *Syst. Biol.* **2010**, *59*, 307–321. [CrossRef] [PubMed]

24. Bennett, D.E.; Camacho, R.J.; Otelea, D.; Kuritzkes, D.R.; Fleury, H.; Kiuchi, M.; Heneine, W.; Kantor, R.; Jordan, M.R.; Schapiro, J.M.; et al. Drug resistance mutations for surveillance of transmitted HIV-1 drug-resistance: 2009 update. *PLoS ONE* **2009**, *4*, e4724. [CrossRef] [PubMed]

25. Shafer, R.W.; Rhee, S.Y.; Pillay, D.; Miller, V.; Sandstrom, P.; Schapiro, J.M.; Kuritzkes, D.R.; Bennett, D. HIV-1 protease and reverse transcriptase mutations for drug resistance surveillance. *AIDS* **2007**, *21*, 215–223. [CrossRef] [PubMed]

26. Siepel, A.C.; Halpern, A.L.; Macken, C.; Korber, B.T. A computer program designed to screen rapidly for HIV type 1 intersubtype recombinant sequences. *AIDS Res. Hum. Retrovir.* **1995**, *11*, 1413–1416. [CrossRef] [PubMed]

27. Gifford, R.J.; Liu, T.F.; Rhee, S.Y.; Kiuchi, M.; Hue, S.; Pillay, D.; Shafer, R.W. The calibrated population resistance tool: Standardized genotypic estimation of transmitted HIV-1 drug resistance. *Bioinformatics* **2009**, *25*, 1197–1198. [CrossRef]

28. Rhee, S.Y.; Gonzales, M.J.; Kantor, R.; Betts, B.J.; Ravela, J.; Shafer, R.W. Human immunodeficiency virus reverse transcriptase and protease sequence database. *Nucleic Acids Res.* **2003**, *31*, 298–303. [CrossRef]

29. Shafer, R.W. Rationale and uses of a public HIV drug-resistance database. *J. Infect. Dis.* **2006**, *194* (Suppl. 1), S51–S58. [CrossRef] [PubMed]

30. Calcagno, A.; Motta, I.; Milia, M.G.; Rostagno, R.; Simiele, M.; Libanore, V.; Fontana, S.; D'Avolio, A.; Ghisetti, V.; Di Perri, G.; et al. Dried plasma/blood spots for monitoring antiretroviral treatment efficacy and pharmacokinetics: A cross-sectional study in rural Burundi. *Br. J. Clin. Pharm.* **2015**, *79*, 801–808. [CrossRef] [PubMed]

31. Garrido, C.; Zahonero, N.; Fernandes, D.; Serrano, D.; Silva, A.R.; Ferraria, N.; Antunes, F.; Gonzalez-Lahoz, J.; Soriano, V.; de Mendoza, C. Subtype variability, virological response and drug resistance assessed on dried blood spots collected from HIV patients on antiretroviral therapy in Angola. *J. Antimicrob. Chemother.* **2008**, *61*, 694–698. [CrossRef] [PubMed]

32. Rodriguez-Auad, J.P.; Rojas-Montes, O.; Maldonado-Rodriguez, A.; Alvarez-Muñoz, M.T.; Muñoz, O.; Torres-Ibarra, R.; Vazquez-Rosales, G.; Lira, R. Use of Dried Plasma Spots for HIV-1 Viral Load Determination and Drug Resistance Genotyping in Mexican Patients. *BioMed Res. Int.* **2015**, *2015*, 240407. [CrossRef] [PubMed]

33. Delatorre, E.O.; Bello, G. Phylodynamics of HIV-1 subtype C epidemic in east Africa. *PLoS ONE* **2012**, *7*, e41904. [CrossRef] [PubMed]

34. Hemelaar, J.; Gouws, E.; Ghys, P.D.; Osmanov, S. WHO-UNAIDS Network for HIV Isolation and Characterisation. Global trends in molecular epidemiology of HIV-1 during 2000–2007. *AIDS* **2011**, *25*, 679–689. [CrossRef]

35. Kalu, A.W.; Telele, N.F.; Gebreselasie, S.; Fekade, D.; Abdurahman, S.; Marrone, G.; Sonnerborg, A. Monophylogenetic HIV-1C epidemic in Ethiopia is dominated by CCR5-tropic viruses-an analysis of a prospective country-wide cohort. *BMC Infect. Dis.* **2017**, *17*, 37. [CrossRef]

36. Kassu, A.; Fujino, M.; Matsuda, M.; Nishizawa, M.; Ota, F.; Sugiura, W. Molecular epidemiology of HIV type 1 in treatment-naive patients in north Ethiopia. *AIDS Res. Hum. Retrovir.* **2007**, *23*, 564–568. [CrossRef]

37. Mulu, A.; Lange, T.; Liebert, U.G.; Maier, M. Clade homogeneity and Pol gene polymorphisms in chronically HIV-1 infected antiretroviral treatment naive patients after the roll out of ART in Ethiopia. *BMC Infect. Dis.* **2014**, *14*, 158. [CrossRef]

38. Tully, D.C.; Wood, C. Chronology and evolution of the HIV-1 subtype C epidemic in Ethiopia. *AIDS* **2010**, *24*, 1577–1582. [CrossRef]

39. Boerma, R.S.; Sigaloff, K.C.; Akanmu, A.S.; Inzaule, S.; Boele van Hensbroek, M.; Rinke de Wit, T.F.; Calis, J.C. Alarming increase in pretreatment HIV drug resistance in children living in sub-Saharan Africa: A systematic review and meta-analysis. *J. Antimicrob. Chemother.* **2017**, *72*, 365–371. [CrossRef] [PubMed]

40. Ngo-Giang-Huong, N.; Huynh, T.H.K.; Dagnra, A.Y.; Toni, T.D.; Maiga, A.I.; Kania, D.; Eymard-Duvernay, S.; Peeters, M.; Soulie, C.; Peytavin, G.; et al. Prevalence of pretreatment HIV drug resistance in West African and Southeast Asian countries. *J. Antimicrob. Chemother.* **2019**, *74*, 462–467. [CrossRef] [PubMed]

41. Jordan, M.R.; Penazzato, M.; Cournil, A.; Vubil, A.; Jani, I.; Hunt, G.; Carmona, S.; Maphalala, G.; Mthethwa, N.; Watera, C.; et al. Human Immunodeficiency Virus (HIV) Drug Resistance in African Infants and Young Children Newly Diagnosed With HIV: A Multicountry Analysis. *Clin. Infect. Dis.* **2017**, *65*, 2018–2025. [CrossRef] [PubMed]

42. Telele, N.F.; Kalu, A.W.; Gebre-Selassie, S.; Fekade, D.; Abdurahman, S.; Marrone, G.; Neogi, U.; Tegbaru, B.; Sonnerborg, A. Pretreatment drug resistance in a large countrywide Ethiopian HIV-1C cohort: A comparison of Sanger and high-throughput sequencing. *Sci. Rep.* **2018**, *8*, 7556. [CrossRef] [PubMed]

43. WHO. HIV Drug Resistance Report 2017. Available online: https://apps.who.int/iris/bitstream/handle/10665/255896/9789241512831-eng.pdf (accessed on 7 July 2019).

44. Firdu, N.; Enquselassie, F.; Jerene, D. HIV-infected adolescents have low adherence to antiretroviral therapy: A cross-sectional study in Addis Ababa, Ethiopia. *Pan Afr. Med. J.* **2017**, *27*, 80. [CrossRef] [PubMed]

45. Chimukangara, B.; Lessells, R.J.; Rhee, S.Y.; Giandhari, J.; Kharsany, A.B.M.; Naidoo, K.; Lewis, L.; Cawood, C.; Khanyile, D.; Ayalew, K.A.; et al. Trends in Pretreatment HIV-1 Drug Resistance in Antiretroviral Therapy-naive Adults in South Africa, 2000–2016: A Pooled Sequence Analysis. *EClinicalMedicine* **2019**, *9*, 26–34. [CrossRef]

46. Ngo-Giang-Huong, N.; Aghokeng, A.F. HIV Drug Resistance in Resource-limited Countries: Threat for HIV Elimination. *EClinicalMedicine* **2019**, *9*, 3–4. [CrossRef] [PubMed]

47. WHO. Guidelines on the Public Health Response to Pretreatment HIV Drug Resistance. 2017. Available online: https://apps.who.int/iris/bitstream/handle/10665/255880/9789241550055-eng.pdf (accessed on 7 July 2019).

48. World Health Organization (WHO). *Consolidated Guidelines on the Use of Antiretroviral Drugs for Treating and Preventing HIV Infection: Recommendations for a Public Health Approach*; World Health Organization: Geneva, Switzerland, 2013.

49. Ministry of Health Federal Democratic Republic of Ethiopia. *National Guidelines for Comprehensive HIV Prevention, Care and Treatment*; Ethiopian Ministry of Health: Addis Ababa, Ethiopia, 2014.

50. Ministry of Health Federal Democratic Republic of Ethiopia. *Continuum of HIV Services Refers to a Comprehensive Package of HIV Prevention, Diagnostic, Treatment, Care and Support Services Provided for People at Risk of HIV Infection or Living with HIV and Their Families*; Ethiopian Ministry of Health: Addis Ababa, Ethiopia, 2018.

Article

An Engineered Microvirin Variant with Identical Structural Domains Potently Inhibits Human Immunodeficiency Virus and Hepatitis C Virus Cellular Entry

Munazza Shahid [1,†], **Amina Qadir** [1,†], **Jaewon Yang** [2], **Izaz Ahmad** [1], **Hina Zahid** [1], **Shaper Mirza** [1], **Marc P. Windisch** [2,3] **and Syed Shahzad-ul-Hussan** [1,*]

1   Department of Biology, Syed Babar Ali School of Science and Engineering, Lahore University of Management Sciences, Lahore 54792, Pakistan; 14130004@lums.edu.pk (M.S.); 16140029@lums.edu.pk (A.Q.); 18140018@lums.edu.pk (I.A.); 15140003@lums.edu.pk (H.Z.); shaper.mirza@lums.edu.pk (S.M.)
2   Applied Molecular Virology Laboratory, Discovery Biology Division, Institut Pasteur Korea, 696, Seongnam 13488, Korea; jaewon.yang@ip-korea.org (J.Y.); marc.windisch@ip-korea.org (M.P.W.)
3   Division of Bio-Medical Science and Technology, University of Science and Technology, Daejeon 34141, Korea
*   Correspondence: shahzad.hussan@lums.edu.pk; Tel.: +92-423-560-8351
†   These authors contribute equally to this work.

Received: 13 January 2020; Accepted: 22 January 2020; Published: 11 February 2020

Abstract: Microvirin (MVN) is one of the human immunodeficiency virus (HIV-1) entry inhibitor lectins, which consists of two structural domains sharing 35% sequence identity and contrary to many other antiviral lectins, it exists as a monomer. In this study, we engineered an MVN variant, LUMS1, consisting of two domains with 100% sequence identity, thereby reducing the chemical heterogeneity, which is a major factor in eliciting immunogenicity. We determined carbohydrate binding of LUMS1 through NMR chemical shift perturbation and tested its anti-HIV activity in single-round infectivity assay and its anti-hepatitis C virus (HCV) activity in three different assays including HCVcc, HCVpp, and replicon assays. We further investigated the effect of LUMS1 on the activation of T helper (T_h) and B cells through flow cytometry. LUMS1 showed binding to α(1-2)mannobiose, the minimum glycan epitope of MVN, potently inhibited HIV-1 and HCV with EC_{50} of 37.2 and 45.3 nM, respectively, and showed negligible cytotoxicity with $CC_{50} > 10$ μM against PBMCs, Huh-7.5 and HepG2 cells, and 4.9 μM against TZM-bl cells. LUMS1 did not activate T_h cells, and its stimulatory effect on B cells was markedly less as compared to MVN. Together, with these effects, LUMS1 represents a potential candidate for the development of antiviral therapies.

Keywords: microvirin; lectin; human immunodeficiency virus; hepatitis C virus; antiviral inhibitor; non-immunogenic; viral entry; protein drugs; LUMS1

---

1. Introduction

Human immunodeficiency virus (HIV-1) and hepatitis C virus (HCV) infections continue to be a healthcare challenge globally, accounting for an enormous disease burden [1–3]. An effective vaccine against these viruses remains to be developed. Within the past decade, advancement in the development of anti-viral regimens has improved the situation, particularly in controlling HCV infections [4,5]. However, the outcome of antiviral therapies could be limited by several factors, including the possible emergence of drug-resistant viral variants. This scenario, therefore, signifies the continuous efforts towards the development of new anti-viral therapies or preventive measures.

The common feature between HIV-1 and HCV is the presence of highly glycosylated outer envelope—the envelope glycoprotein 120 (gp120) of HIV-1 and E2 of HCV exhibits over 20 and 11

N-glycosylation sites, respectively [6–8]. This glycan shield decorating the surface of viruses has been exploited as a potential target for therapeutic or preventive interventions. In recent years, carbohydrate-binding agents, in particular, lectins have been identified which can inhibit cellular entry of the viruses by specifically binding to these viral surface glycans [9–12]. Given their potent nature, use of these lectins as topical microbicides has been suggested for the prevention of sexual transmission of HIV-1 [13–15]. As an example, griffithsin, a lectin from red algae, is currently in phase-1 clinical trials [16]. Two of the major challenges hampering the clinical application of these lectins are their potential cytotoxicity and immunogenicity. In general, chemical and structural heterogeneity of proteins is one of the primary factors responsible for their immunogenicity.

Microvirin (MVN) is one of the potent anti-HIV lectins, which was initially isolated from *Microcystis aeruginosa* and has been shown to have only minor cytotoxicity and mitogenic effects as compared to other antiviral lectins [17,18]. MVN has been reported to specifically recognize α(1-2)mannobiose present at the termini of branched high mannose type glycans on the viral surface. This 12 kDa lectin consists of two structural domains, which share 35% sequence identity, and unlike other anti-viral lectins, it exists as a monomer (Figure 1a). Moreover, there is a four residues long insertion in domain-A as compared to domain-B of MVN [19]. In this study, we engineered an MVN variant, LUMS1 (the name derived from Lahore University of Management Sciences), exhibiting 100% sequence identity between its two structural domains, thereby markedly decreasing the chemical heterogeneity. We investigated this protein for its potential to inhibit cellular entry of HIV and HCV, and studied its cytotoxicity, carbohydrate specificity, and preliminary effects on the activation of immune cell surface markers.

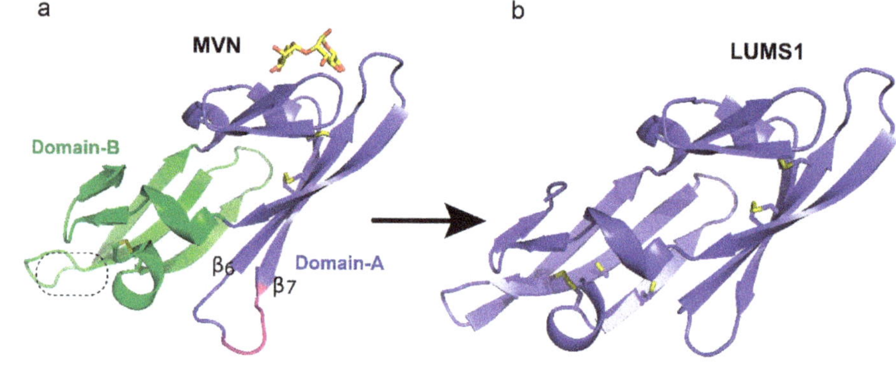

Figure 1. Description of the protein design: (**a**) microvirin (MVN) structure (PDB ID 2YHH) shown in cartoon presentation with two structural domains colored blue and green while bound glycan is colored yellow. Insertion of four amino acids in domain-A as compared to domain-B is indicated in magenta. The second putative carbohydrate binding site is indicated by a dotted circle. (**b**) The homology-modeled structure of LUMS1 was created through SWISS-MODEL online tools using MVN as a template. Qualitative model energy analysis (QMEAN) scoring function was used to access the quality of the model. Side chains of all cysteine residues in both proteins are shown in gold sticks. Alignment of amino acid sequence of two domains of MVN and LUMS1 is shown at the bottom of the respective protein structure. N, C indicates N- and C-termini of the protein sequences.

2. Materials and Methods

2.1. Protein Expression

For the recombinant expression of LUMS1, the gene encoding for LUMS1 amino acid sequence was synthesized through commercial facilities (Genscript, Piscataway, NJ, USA), sub-cloned into pET32a expression vector, subsequently expressed in a bacterial system (BL21 strain), and purified through different chromatographic techniques including nickel-affinity, size exclusion, and ion exchange chromatography. For the expression of the ^{15}N-labelled protein, the transformed bacteria were grown in minimal media supplemented with ^{15}N-ammonium chloride as the only source of nitrogen. The purified protein was transferred into PBS buffer of pH 7.4 for all biological assays, and into 20 mM phosphate buffer containing 50 mM NaCl for NMR experiments, through dialyses using dialysis membrane of 3.5 KDa cutoff (Slide-A-Lyzer™ MINI Dialysis Device, Thermo Fisher Scientific, Waltham, MA, USA) [19].

2.2. NMR Experiments

NMR experiments were performed on Bruker Avance Neo 600 MHz NMR spectrometer equipped with TXI triple resonance probe at 298 K. Two dimensional ^{15}NHSQC spectra were recorded with 16 scans and 256 data points in the indirect dimension. Topspin 4.0.5 software was used to acquire and process the NMR data [19].

2.3. HIV Inhibition Assay

HIV-1 entry inhibition by LUMS1 was studied by using pseud-typed virus-based single-round infectivity assay, according to a previously reported method [20]. In this regard, LUMS1 at varying concentration was mixed with HXB2 strain of HIV-1 pseudo-typed viral particles at 37 °C followed by the addition of TZM-bl cells (NIH AIDS reagent program) at a concentration of 1×10^4 cells/100 µL. After 48 h, cells were lysed and percent infection was measured through luciferase activity (BrightGlo, Promega, Maddison, WI, USA) for each dilution of inhibitor with respect to control containing no inhibitor. Similarly, the activity of LUMS1 against vesicular stomatitis virus (VSV) was also tested using virus pseudo-typed with VSV envelope and HIV-1 backbone.

2.4. HCV Infection Assay

The anti-HCV activity of LUMS1 was evaluated using cell culture-derived infectious HCV (HCVcc) expressing an NS5A-GFP fusion protein in the presence of inhibitors as previously described [21]. Briefly, Huh-7.5 cells were seeded in 384-well plates (2.5×10^3 cells/well). LUMS1 were serially diluted in complete DMEM, added to each well of the plates, inoculated with HCVcc and incubated at 37 °C for 3 days. On day 3 post-infection (p.i.), cultured cells were fixed with 2% paraformaldehyde in PBS containing 10 µg/mL Hoechst 33,342 (Life Technologies, Waltham, MA, USA) for 30 min. HCV replication was analyzed by determining the number of GFP-positive cells using automated confocal microscopy (Opera, PerkinElmer, Waltham, MA, USA). Cytotoxicity was assessed by counting cell nuclei stained with Hoechst 33342 and normalized to untreated control cells.

2.5. HCV Replication Assay

HCV subgenomic replicon cells were treated with inhibitors as described previously [21]. Briefly, replicon cells were seeded in 384-well plates and treated with inhibitors for 72 h. The inhibitory effect of the protein on HCV RNA replication was monitored together with RNA polymerase inhibitor (sofosbuvir) as a reference compound. Viral replication and cytotoxicity was assessed as described above.

2.6. HCV Pseudoparticle Assay

Viral entry was assessed using HCV pseudoparticle (HCVpp) expressing a luciferase reporter gene as described before [22]. In brief, HEK293 T cells seeded for 1 day in T-75 flasks were co-transfected with 5 µg of an HCV E1/E2 envelope protein expression vector (genotype 1a) and 15 µg of pNL4.3.Luc.R⁻ E⁻, HIV Gag-Pol expression packaging vector containing luciferase reporter gene using Lipofectamine 3000 (Invitrogen, Waltham, MA, USA). Supernatants containing the pseudoparticles were harvested at 72 h and used as HCVpp in the entry inhibition assay using Huh7.5 cells as described above in Section 2.3. EI-1, a known potent inhibitor of HCV entry [23] was used as a positive control at a single concentration of 10 µM.

2.7. Flow Cytometry Analysis of PBMCs

For the flow cytometry analysis blood from two healthy volunteers was collected after informed consent and PBMCs were isolated separately using density gradient centrifugation (Polymorphprep; Cosmo Bio, Tokyo, Japan) and washed with PBS (1% FBS). After treating PBMCs (10^6 cells/mL) with varying concentrations of LUMS1, MVN, and cyanovirin-N (CVN) for 72 h at 37 °C and 5% CO_2, cells were washed with PBS (2% FBS) and incubated with APC-conjugated anti-cluster of differentiation-4 (CD4), PE-conjugated anti-CD25, and percp cy5.5-conjugated anti-CD20 antibodies for 30 min at 4 °C. Respective isotype controls were used for compensation of backgrounds. Finally, cells were washed with PBS (2% FBS), fixed with 1% formaldehyde, and analyzed by FACS (Calibur; BD Biosciences, Franklin lakes, NJ, USA), using CellQuest software for data acquisition [18].

2.8. MTT Assay

TZM-bl cells and PBMCs were seeded in 96-well plates at optimized concentrations of 8×10^4 and 7×10^5 cells per 100 µL, respectively, with various concentrations of the LUMS1 protein (1.25, 2.5, 5, and 10 µM), and incubated for 48 h. MTT reagent was added at a final concentration of 0.5 mg/mL and incubated for 4 h as reported previously [24]. The plate containing PBMCs was centrifuged for 5 min at 300× g. After completely removing the media, 100 µL DMSO was added and thoroughly mixed to dissolve the crystals, and finally absorbance was measured at 570 nm.

3. Results

3.1. Designing of the LUMS1 Protein and Characterization of Its Carbohydrate Binding

A characteristic feature of HIV-1 entry inhibitor lectins is multivalent recognition through more than one carbohydrate-binding site to attain high avidity of interaction required for potent antiviral activity. MVN, however, contains one carbohydrate-binding site present in its domain-A (Figure 1a). In this design, we removed a four-residues long insertion between strands B6 and B7 in domain-A of MVN and changed the amino acid sequence of domain-B making it identical to that of domain-A and creating two carbohydrate-binding sites. The removal of the four residues corresponding to a long flexible loop could further minimize chemical heterogeneity and reduce the protein size. Since the domain-B of MVN has been reported to adopt the structure homologous to its domain-A without this insert [19], it was conceivable that the removal of these four residues may not disturb the protein folding. Subsequently, we built a homology model of the designed protein, LUMS1, using MVN as a template to obtain a preliminary idea about its structure. The resultant model exhibited a similar structure to MVN, but unlike MVN, each of the structural domains of LUMS1 contained a putative carbohydrate-binding site and two potential inter-strand disulfide linkages (Figure 1b). In order to experimentally investigate LUMS1, we produced the recombinant protein in two different forms, unlabeled and isotopically labeled with ^{15}N.

To find out if the purified protein was folded, we recorded a two-dimensional ^{15}NHSQC spectrum of isotopically labeled LUMS1. Well-dispersed ^{1}H-^{15}N correlation cross-peaks were observed indicating that protein was folded (Figure 2). In order to test the binding of LUMS1 to α(1-2)mannobiose, the minimum glycan epitope of MVN, we used the NMR chemical shift perturbation technique, as backbone ^{1}H-^{15}N resonances are sensitive to change in the environment resulting from the binding of a ligand [25]. In this regard, ^{15}NHSC spectra were acquired on a sample containing ^{15}N-lableled LUMS1 alone in solution and in the presence of increasing concentrations of α(1-2)mannobiose. On addition of α(1-2)mannobiose, cross-peaks of several amino acids in ^{1}H-^{15}N correlation spectra either broadened or underwent chemical shift changes, indicating the binding of the carbohydrate. Upon the addition of the carbohydrate at two equivalents of the protein molar concentration, titration appeared to be completed as no further changes in the spectra were observed upon addition of more quantities of the carbohydrate. ^{15}NHSQC spectra of LUMS1 free and in the presence of one equivalent of α(1-2)mannobiose is shown in Figure 2 with the expansion of a cross-peak showing a stepwise change in chemical shift with the addition of one and two equivalents of carbohydrate. At one equivalent of carbohydrate, the cross-peak in the expanded region representing free and carbohydrate-bound state of LUMS1 appeared, while in the presence of two equivalents of carbohydrate, the cross-peak representing the free state of LUMS1 completely disappeared. This indicated the binding stoichiometry as 2:1 and the presence of two carbohydrate-binding sites on LUMS1.

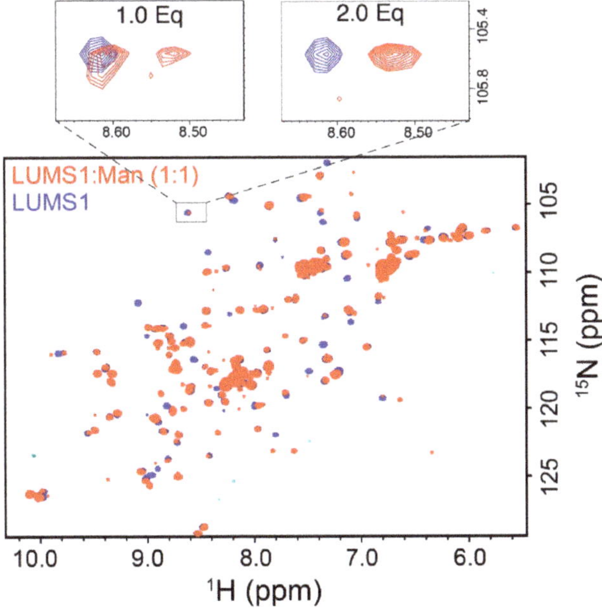

Figure 2. Carbohydrate binding of LUMS1: ^{15}NHSQC spectra of LUMS1 alone (blue) and in the presence of one equivalent of the α(1-2)mannobiose glycan (red), superimposed. Expansions of a region of spectrum containing single cross-peak in the absence (blue) and presence of one and two equivalents of α(1-2)mannobiose are shown at the top.

3.2. LUMS1 Inhibits HIV-1 Cellular Entry

We tested LUMS1 for its anti-HIV activity in pseud-typed virus-based single-round infection assay using the HXB2 strain of HIV-1 [19]. LUMS1 potently inhibited the HIV-1 entry with EC$_{50}$ of 37.2 ± 4.4 nM (Figure 3) measured from its dose-response curve. For comparison, the activity of MVN was also determined as positive control, and its EC$_{50}$ was measured as 8.0 ± 1.4 nM. To determine viral specificity, the activity of LUMS1 against an amphotropic virus, VSV, was tested and it was found that LUMS1 did not inhibit VSV at a concentration as high as 10 µM (Figure S2, Supplementary Material).

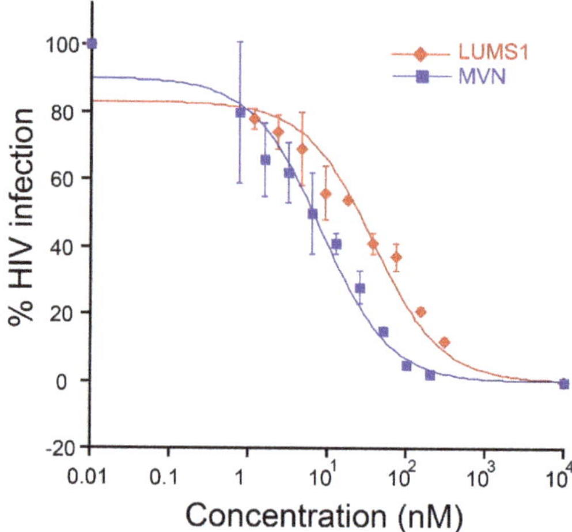

Figure 3. Human immunodeficiency virus (HIV-1) entry inhibition by LUMS1: dose-response curve showing inhibition of pseudo-typed virus, HIV-1 strain HXB2, by LUMS1 and MVN. The proteins at varying concentration was mixed with the virus at 37 °C followed by the addition of TZM-bl. After 48 h, cells were lysed, and percent infection was measured through luciferase activity. The assay was performed in triplicates.

3.3. LUMS1 Inhibits HCV Cellular Entry

Next, to evaluate the effect of LUMS1 on HCV infection, we used cell culture-derived infectious HCV (HCVcc) expressing an NS5A-GFP fusion protein [21]. In this assay, the infection of liver-derived cell line Huh-7.5 by HCVcc was analyzed by determining the number of GFP-positive cells using automated confocal microscopy. The expression of the NS5A-GFP fusion protein, which served as a marker for productive HCV infection, was inhibited by LUMS1 in a dose-dependent manner with a calculated EC_{50} of 45.3 ± 18.6 nM and CC_{50} > 10 μM (Figure 4a). Representative images are shown in Figure 4b. To investigate whether LUMS1 can also interfere with some post entry event of viral replication cycle, we tested the effect of LUMS1 using subgenomic HCV replicon cells [21]. We observed that HCV replication was not affected by LUMS1 at a concentration as high as 10 μM, while it was significantly inhibited by sofosbuvir, a known inhibitor of HCV replication (Figure 4c). This suggested that LUMS1 interfered with HCV entry rather than HCV replication. Results of the HCVcc and replicon assays are summarized in Table 1. To further confirm that LUMS1 interferes with HCV E1/E2-mediated viral entry, the HCV pseudoparticle (HCVpp) system was employed [22]. LUMS1 inhibited the HCVpp with EC_{50} of 142.1 ± 23 nM demonstrating that LUMS1 specifically inhibited viral entry (Figure 4d).

Table 1. Overview of antiviral activity of LUMS1 and compounds used as positive control.

Assay System	Inhibitor	EC_{50} (nM)
HCVcc	LUMS1	45.3 ± 18.6
	MVN [#]	35.6 ± 3.98
HCV replicon	LUMS1	n.d.
	Sofosbuvir	54.0 ± 32.8
HCVpp	LUMS1	142.1 ± 23.0
HIV-1 single	LUMS1	37.2 ± 4.4
round infectivity	MVN	8.0 ± 1.4

[#] Dose response curve is given in Figure S3, Supplementary Material.

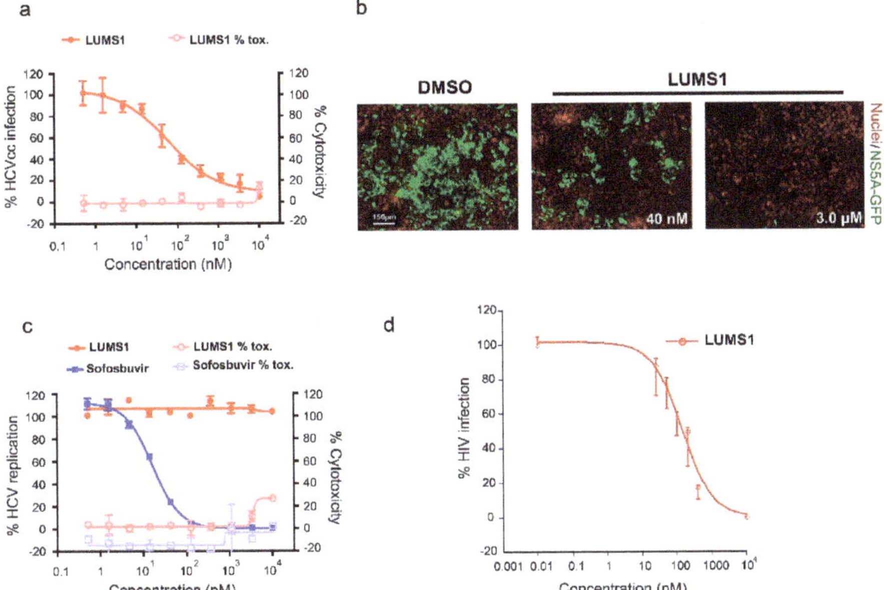

Figure 4. Hepatitis C virus (HCV) inhibition by LUMS1: (**a**) Anti-HCV activity in Huh-7.5 cells. Cells were pretreated with increasing concentrations of LUMS1 for 2 h followed by infection with HCVcc (JFH1) for 72 h in the presence of proteins. (**b**) HCV infectivity and total cell number were assessed by determining the number of GFP-positive cells (green) and nuclei (red), respectively, for 3 days in the presence of LUMS1). Images were acquired by confocal microscopy. (**c**) HCV subgenomic replicon cells were treated with LUMS1 and sofosbuvir. (**d**) HCVpp was mixed with different concentrations of LUMS1 and subsequently added to Huh-7.5 cells. After 72 h incubation, cells were lysed and percent infection was measured through luciferase activity (BrightGlo, Promega, USA) for each dilution of inhibitor with respect to control containing no inhibitor.

3.4. LUMS1 Does Not Stimulate Cellular Activation Markers and Shows Negligible Cytotoxic Effect

In the context of evaluating immunogenic effects of a protein in vitro, activation of T_h and B cells is considered an important marker of immunogenicity as these cells are involved in inducing monoclonal antibody-based immunogenicity [26,27]. In this regard the effect of LUMS1 and in parallel of MVN, was analyzed on the expression of CD4, CD25, and CD20 activation markers through flow cytometry (FACS) using freshly isolated PBMCs from healthy individuals. CVN was used as a positive control for the expression of CD4 and CD25 as its effect on these cellular activation markers have already been reported [28]. Both LUMS1 and MVN did not increase the population of CD4[+] and CD25[+] cells (T_h cells), while CVN in this case showed significantly high activation even at a concentration of 50 nM (Figure 5).

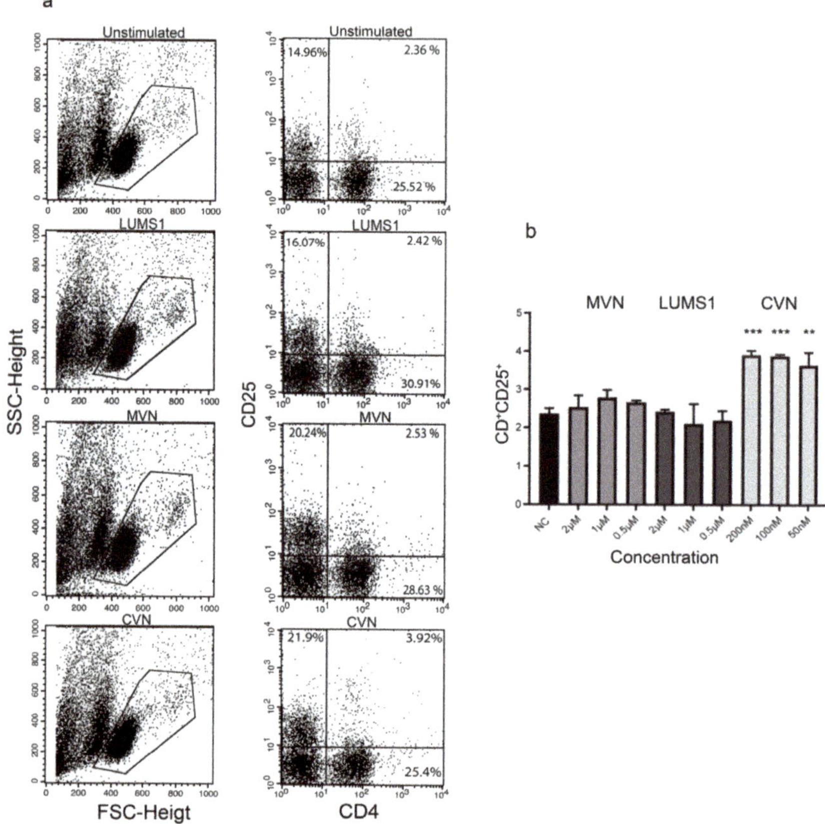

Figure 5. Effect of LUMS1 on the activation of T_h cells: (**a**) Flow cytometry analysis of PBMCs to determine the population of CD4$^+$ and CD25$^+$ cells in freshly isolated PBMCs in response to treatment with LUMS1, MVN, cyanovirin-N (CVN). After treating PBMCs (10^6 cells/mL) with varying concentrations of LUMS1, MVN, and CVN for 72 h at 37 °C and 5% CO_2, cells were washed with PBS and incubated with APC-conjugated anti-CD4 and PE-conjugated anti-CD25 antibodies for 30 min at 4 °C. Finally, cells were washed with PBS (2% FBS), fixed with 1% formaldehyde, and analyzed by FACS, using CellQuest software for data acquisition. Data were statistically analyzed using GraphPad Prism software. (**a**) Left panel, representative dot plots of forward scatter (FSC) and side scatter (SSC) indicating the subpopulation of cells in PBMCs; right panel, dot plots showing the relative population of cells with CD4 and CD25 activation markers. (**b**) Plot showing the percent population of CD4$^+$ and CD25$^+$ cells after treating with LUMS1, MVN, and CVN separately. The data represent the mean of three independent experiments and one-way ANOVA was used to compare different groups. ** $p \leq 0.01$; *** $p \leq 0.001$.

However, LUMS1 showed more a pronounced difference with MVN on the activation of CD20$^+$ cells. Treatment of LUMS1 at a concentration as high as 4 µM did not significantly increase the population of CD20$^+$ cells, while the effect of MVN in this regard was significantly high even at 2 µM concentration (Figure 6a). In addition to evaluating the cytotoxicity of LUMS1 against Huh7.5 cells (Figure 4a,c) we also determined the effect of LUMS1 on the viability of TZM-bl cells, PBMCs, and HepG2 cells using MTT assay. LUMS1 did not show a cytotoxic effect on Huh7.5 cells, HepG2 cells, and PBMCs at a concentration as high as 10 µM whereas its CC$_{50}$ value against TZM-bl cells was calculated as 4.9 ± 0.166 µM (Figure 6b and Table 2).

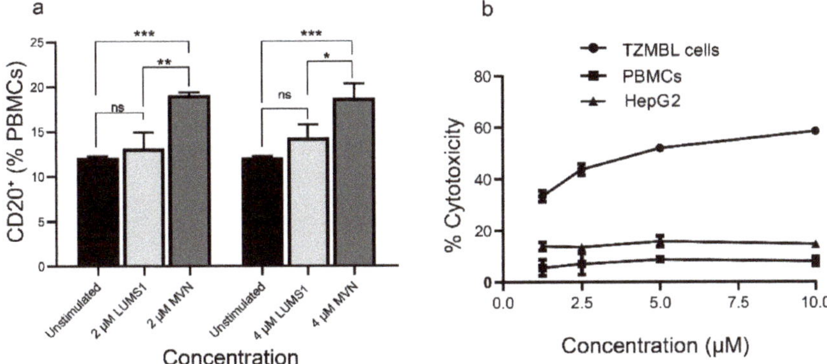

Figure 6. Evaluation of B cells activation by LUMS1 and its cytotoxicity: (**a**) graph presenting the flow cytometry analysis to determine the population of $CD20^+$ cells in freshly isolated PBMCs. PBMCs were isolated from freshly collected blood using density gradient centrifugation and washed with PBS. After treating PBMCs (10^6 cells/mL) with varying concentrations of LUMS1 and MVN for 72 h at 37 °C and 5% CO_2, cells were washed with PBS and incubated with percp cy5.5-conjugated anti-CD20 antibodies for 30 min at 4 °C. Finally, cells were washed with PBS (2% FBS), fixed with 1% formaldehyde and analyzed by FACS, using CellQuest software for data acquisition. Data were statistically analyzed using GraphPad Prism software. Analyses were performed in triplicate and one-way ANOVA with multiple comparisons was used to compare different groups. * $p \leq 0.05$, ** $p \leq 0.01$; *** $p \leq 0.001$. (**b**) The plot showing concentration dependent cytotoxic effect of LUMS1 on PBMCs, HepG2, and TZM-bl cells.

Table 2. Overview of the cytotoxicity of LUMS1 against different cell lines.

Cell Types	CC_{50} of LUMS1 (nM)
Huh7.5	>10,000
PBMCs	>10,000
TZMbl	4900 ± 166
HepG2	>10,000

4. Discussion

In this study, we engineered a lectin, LUMS1, by modifying MVN to incorporate two carbohydrate-binding sites and reduce chemical heterogeneity, a major factor in potential immunogenicity of a protein. The NMR analysis of the carbohydrate binding of LUMS1 suggested that it exhibited two carbohydrate-binding sites, and it has the same carbohydrate specificity as MVN—both recognize the α(1-2)mannobiose glycan as the minimum epitope.

MVN has been reported to inhibit HIV-1 entry with EC_{50} values ranging from 2 to 12 nM against HIV-1, and we reproduced the reported EC_{50} against HXB2 strain of HIV-1. LUMS1, however, inhibited the same strain with ≈4.5-fold lower potency (EC_{50} 37 nM). LUMS1 contains two carbohydrate-binding sites and is expected to exhibit higher avidity as compared to MVN by engaging more than one glycan or glycan branches at the surface of the virus. Lower potency of LUMS1 against HIV-1 as compared to MVN could be attributed to the different possible mechanism by which these lectins attain high avidity of interactions with the viral envelope; multivalent recognition in the case of LUMS1 and the bind-and-hop mechanism through single site interactions in the case of MVN [19], although these mechanisms remain to be experimentally validated. Binding studies of these lectins with the stabilized HIV-1 gp120 trimer, the form of the envelope protein exists on the viral spike, through isothermal titration calorimetry (ITC) or fluorescence resonance energy transfer (FRET) measuring microevents of binding could illustrate the detailed mechanism [29,30]. Moreover, multivalent recognition can be identified by solving the structure of the complex of lectin and gp120-trimer through X-ray or cryo electron microscopy. While comparing the carbohydrate binding of these lectins, the cross-peaks of carbohydrate-binding site amino acids of MVN in ^{1}H-^{15}N correlation NMR spectra experienced chemical shift changes on the addition of ligand suggesting slow exchange on the NMR time scale [19]. On the other hand, in the case of LUMS1, chemical shift perturbation in most of the cross-peaks were in the form of line broadening suggesting intermediate exchange on the NMR time scale. Slow exchange on the NMR time scale is related to higher binding affinity (lower K_D values) as compared to intermediate exchange [31]. The two lectins, therefore, demonstrate the difference in binding to carbohydrate in terms of affinity. The apparent lower carbohydrate-binding affinity of LUMS1 as compared to MVN indicates that its carbohydrate-binding sites may not be structurally optimal, which could be understood only after the structure of the complex of LUMS1 and carbohydrate is available. However, one of the significant aspects of LUMS1 was found to be its ability to potently inhibit HCV infection with an EC_{50} of 45 and 142.1 nM as determined in HCVcc and HCVpp assays, respectively. HCV inhibition by different oligomeric forms of MVN has been reported but only qualitatively [32]. By testing LUMS1 in HCVpp and replicon system, in addition to HCVcc assay, we clearly demonstrated that exclusively HCV E1/E2-mediated viral entry was inhibited. Many HIV-1 entry inhibitor lectins have been reported also to inhibit the entry of HCV [33–35], which could be attributed to likely similar glycan density on the surface of both viruses [8,36]. This apparent similarity potentiates the development of universal therapy against both viruses. Moreover, LUMS1 demonstrated its specificity for HIV-1 and HCV, as it did not inhibit an amphotropic virus VSV.

The major obstacle in the advancement towards clinical application of anti-viral lectins is their potential cytotoxicity and immunogenicity [17,27]. In cell viability assays using four different types of cells, LUMS1 demonstrated negligible cytotoxic effects with $CC_{50} > 10$ μM against Huh7.5 cells, HepG2 cells and PBMCs, and 4.9 ± 0.166 μM against the TZM-bl cells with a selectivity index (SI) value of 108, calculated by the ratio of the smallest CC_{50} value (4.9 μM) and the EC_{50} value (45.3 nM, against HCVcc) indicating its promising safety profile. As foreign peptides are prone to induce immunogenicity in patients, we tested LUMS1 for its effect on the activation of B and T_h cells in vitro and observed that LUMS1 demonstrated no significant increase in the expression of activation markers for these cells at a concentration as high as 4 μM. LUMS1 demonstrated significantly lesser effect in inducing the CD20 activation marker however its effect on the induction of CD4 and CD25 activation markers was comparable to MVN but slightly less. The detailed safety profile of LUMS1, however, remains to be investigated in animal models in the follow-up study. Taken together, LUMS1 represents an attractive potential therapeutic candidate against HIV-1 and HCV, as it potently inhibits both of these viruses, demonstrates lack of cytotoxicity and negligible activation of B and T_h cells. With the emerging trend of protein drugs, further optimization of LUMS1 to enhance its carbohydrate-binding affinity leading to increase anti-viral potency, and its detailed investigation in vitro and in vivo are the further aspects to be considered.

Supplementary Materials: The following are available online at http://www.mdpi.com/1999-4915/12/2/199/s1, Detailed methods about protein expression; Figure S1, Image containing overlaid structures of MVN and LUMS1; Figure S2, Effect of LUMS1 on VSV infection; and Figure S3, Anti-HCV activity of MVN.

Author Contributions: M.S., A.Q., J.Y., I.A., H.Z. and S.M. performed experiments. All authors contributed to the data analysis and interpretation, and M.P.W. and S.S.-u.-H. wrote the manuscript. All authors have read and agreed to the published version of the manuscript.

Funding: This study was supported by Higher Education Commission of Pakistan through NRPU grant (20-4437/NRPU/R&D/14), and the National Research Foundation of Korea (MSIT 2017M3A9G6068246 and NRF-2014R1A2A1A11052535) and the Gyeonggi Provincial Government.

Acknowledgments: We are thankful to Carole Bewley at NIDDK, National Institutes of Health, USA, for providing plasmids containing genes for MVN and CVN. We are also thankful to Punjab AIDS Control Program, Pakistan for providing a monthly stipend to Munazza Shahid.

Conflicts of Interest: The authors declare no conflict of interest.

References

1. Jefferies, M.; Rauff, B.; Rashid, H.; Lam, T.; Rafiq, S. Update on global epidemiology of viral hepatitis and preventive strategies. *World J. Clin. Cases* **2018**, *6*, 589–599. [CrossRef]
2. Mahy, M.; Marsh, K.; Sabin, K.; Wanyeki, I.; Daher, J.; Ghys, P.D. HIV estimates through 2018: Data for decision making. *Aids* 2019. [CrossRef]
3. Ashraf, M.U.; Iman, K.; Khalid, M.F.; Salman, H.M.; Shafi, T.; Rafi, M.; Javaid, N.; Hussain, R.; Ahmad, F.; Shahzad-Ul-Hussan, S.; et al. Evolution of efficacious pangenotypic hepatitis C virus therapies. *Med. Res. Rev.* **2019**, *39*, 1091–1136. [CrossRef] [PubMed]
4. Dhiman, R.K.; Grover, G.S.; Premkumar, M. Hepatitis C elimination: A Public Health Perspective. *Curr. Treat. Options Gastroenterol.* 2019. [CrossRef] [PubMed]
5. Kanters, S.; Vitoria, M.; Doherty, M.; Socias, M.E.; Ford, N.; Forrest, J.I.; Popoff, E.; Bansback, N.; Nsanzimana, S.; Thorlund, K.; et al. Comparative efficacy and safety of first-line antiretroviral therapy for the treatment of HIV infection: A systematic review and network meta-analysis. *Lancet HIV* **2016**, *3*, e510–e520. [CrossRef]
6. Goffard, A.; Dubuisson, J. Glycosylation of hepatitis C virus envelope proteins. *Biochimie* **2003**, *85*, 295–301. [CrossRef]
7. Doores, K.J.; Bonomelli, C.; Harvey, D.J.; Vasiljevic, S.; Dwek, R.A.; Burton, D.R.; Crispin, M.; Scanlan, C.N. Envelope glycans of immunodeficiency virions are almost entirely oligomannose antigens. *Proc. Natl. Acad. Sci. USA* **2010**, *107*, 13800–13805. [CrossRef]
8. Fenouillet, E.; Gluckman, J.C.; Jones, I.M. Functions of HIV envelope glycans. *Trends Biochem. Sci.* **1994**, *19*, 65–70. [CrossRef]
9. Shahzad-ul-Hussan, S.; Ghirlando, R.; Dogo-Isonagie, C.I.; Igarashi, Y.; Balzarini, J.; Bewley, C.A. Characterization and carbohydrate specificity of pradimicin S. *J. Am. Chem. Soc.* **2012**, *134*, 12346–12349. [CrossRef]
10. Balzarini, J. Carbohydrate-binding agents: A potential future cornerstone for the chemotherapy of enveloped viruses? *Antivir. Chem. Chemother.* **2007**, *18*, 1–11. [CrossRef]
11. Mori, T.; O'Keefe, B.R.; Sowder, R.C., 2nd; Bringans, S.; Gardella, R.; Berg, S.; Cochran, P.; Turpin, J.A.; Buckheit, R.W., Jr.; McMahon, J.B.; et al. Isolation and characterization of griffithsin, a novel HIV-inactivating protein, from the red alga Griffithsia sp. *J. Biol. Chem.* **2005**, *280*, 9345–9353. [CrossRef] [PubMed]
12. Akkouh, O.; Ng, T.B.; Singh, S.S.; Yin, C.; Dan, X.; Chan, Y.S.; Pan, W.; Cheung, R.C. Lectins with anti-HIV activity: A review. *Molecules* **2015**, *20*, 648–668. [CrossRef] [PubMed]
13. Yang, H.; Li, J.; Patel, S.K. Design of Poly(lactic-co-glycolic Acid) (PLGA) Nanoparticles for Vaginal Co-Delivery of Griffithsin and Dapivirine and Their Synergistic Effect for HIV Prophylaxis. *Pharmaceutics* **2019**, *11*, 184. [CrossRef] [PubMed]
14. O'Keefe, B.R.; Vojdani, F.; Buffa, V.; Shattock, R.J.; Montefiori, D.C.; Bakke, J.; Mirsalis, J.; d'Andrea, A.L.; Hume, S.D.; Bratcher, B.; et al. Scaleable manufacture of HIV-1 entry inhibitor griffithsin and validation of its safety and efficacy as a topical microbicide component. *Proc. Natl. Acad. Sci. USA* **2009**, *106*, 6099–6104. [CrossRef] [PubMed]
15. Huskens, D.; Schols, D. Algal lectins as potential HIV microbicide candidates. *Mar. Drugs* **2012**, *10*, 1476–1497. [CrossRef]
16. Griffithsin-Based Rectal Microbicide for PREvention of Viral ENTry (PREVENT). Available online: https://clinicaltrials.gov/ct2/show/NCT04032717 (accessed on 9 November 2019).
17. Kehr, J.C.; Zilliges, Y.; Springer, A.; Disney, M.D.; Ratner, D.D.; Bouchier, C.; Seeberger, P.H.; de Marsac, N.T.; Dittmann, E. A mannan binding lectin is involved in cell-cell attachment in a toxic strain of Microcystis aeruginosa. *Mol. Microbiol.* **2006**, *59*, 893–906. [CrossRef]
18. Huskens, D.; Ferir, G.; Vermeire, K.; Kehr, J.C.; Balzarini, J.; Dittmann, E.; Schols, D. Microvirin, a novel alpha(1,2)-mannose-specific lectin isolated from Microcystis aeruginosa, has anti-HIV-1 activity comparable with that of cyanovirin-N but a much higher safety profile. *J. Biol. Chem.* **2010**, *285*, 24845–24854. [CrossRef]
19. Shahzad-ul-Hussan, S.; Gustchina, E.; Ghirlando, R.; Clore, G.M.; Bewley, C.A. Solution structure of the monovalent lectin microvirin in complex with Man(alpha)(1-2)Man provides a basis for anti-HIV activity with low toxicity. *J. Biol. Chem.* **2011**, *286*, 20788–20796. [CrossRef]

20. Gustchina, E.; Louis, J.M.; Lam, S.N.; Bewley, C.A.; Clore, G.M. A monoclonal Fab derived from a human nonimmune phage library reveals a new epitope on gp41 and neutralizes diverse human immunodeficiency virus type 1 strains. *J. Virol.* **2007**, *81*, 12946–12953. [CrossRef]

21. Lee, M.; Yang, J.; Park, S.; Jo, E.; Kim, H.Y.; Bae, Y.S.; Windisch, M.P. Micrococcin P1, a naturally occurring macrocyclic peptide inhibiting hepatitis C virus entry in a pan-genotypic manner. *Antivir. Res.* **2016**, *132*, 287–295. [CrossRef]

22. Hsu, M.; Zhang, J.; Flint, M.; Logvinoff, C.; Cheng-Mayer, C.; Rice, C.M.; McKeating, J.A. Hepatitis C virus glycoproteins mediate pH-dependent cell entry of pseudotyped retroviral particles. *Proc. Natl. Acad. Sci. USA* **2003**, *100*, 7271–7276. [CrossRef] [PubMed]

23. Baldick, C.J.; Wichroski, M.J.; Pendri, A.; Walsh, A.W.; Fang, J.; Mazzucco, C.E.; Pokornowski, K.A.; Rose, R.E.; Eggers, B.J.; Hsu, M.; et al. A novel small molecule inhibitor of hepatitis C virus entry. *PLoS Pathog.* **2010**, *6*, e1001086. [CrossRef] [PubMed]

24. Kumar, P.; Nagarajan, A.; Uchil, P.D. Analysis of cell viability by the MTT assay. *Cold Spring Harb. Protoc.* **2018**, *2018*, pdbprot095505. [CrossRef] [PubMed]

25. Bewley, C.A.; Shahzad-Ul-Hussan, S. Characterizing carbohydrate-protein interactions by nuclear magnetic resonance spectroscopy. *Biopolymers* **2013**, *99*, 796–806. [CrossRef] [PubMed]

26. Baker, M.P.; Reynolds, H.M.; Lumicisi, B.; Bryson, C.J. Immunogenicity of protein therapeutics: The key causes, consequences and challenges. *Self Nonself* **2010**, *1*, 314–322. [CrossRef] [PubMed]

27. Ito, S.; Ikuno, T.; Mishima, M.; Yano, M.; Hara, T.; Kuramochi, T.; Sampei, Z.; Wakabayashi, T. In vitro human helper T-cell assay to screen antibody drug candidates for immunogenicity. *J. Immunotoxicol.* **2019**, *16*, 125–132. [CrossRef]

28. Huskens, D.; Vermeire, K.; Vandemeulebroucke, E.; Balzarini, J.; Schols, D. Safety concerns for the potential use of cyanovirin-N as a microbicidal anti-HIV agent. *Int. J. Biochem. Cell. Biol.* **2008**, *40*, 2802–2814. [CrossRef]

29. Lee, J.H.; Ozorowski, G.; Ward, A.B. Cryo-EM structure of a native, fully glycosylated, cleaved HIV-1 envelope trimer. *Science* **2016**, *351*, 1043–1048. [CrossRef]

30. Dam, T.K.; Brewer, C.F. Effects of clustered epitopes in multivalent ligand-receptor interactions. *Biochemistry* **2008**, *47*, 8470–8476. [CrossRef]

31. Li, Y.; Kang, C. Solution NMR Spectroscopy in Target-Based Drug Discovery. *Molecules* **2017**, *22*, 1399. [CrossRef]

32. Min, Y.Q.; Duan, X.C.; Zhou, Y.D.; Kulinich, A.; Meng, W.; Cai, Z.P.; Ma, H.Y.; Liu, L.; Zhang, X.L.; Voglmeir, J. Effects of microvirin monomers and oligomers on hepatitis C virus. *Biosci. Rep.* **2017**, *37*. [CrossRef] [PubMed]

33. Takebe, Y.; Saucedo, C.J.; Lund, G.; Uenishi, R.; Hase, S.; Tsuchiura, T.; Kneteman, N.; Ramessar, K.; Tyrrell, D.L.; Shirakura, M.; et al. Antiviral lectins from red and blue-green algae show potent in vitro and in vivo activity against hepatitis C virus. *PLoS ONE* **2013**, *8*, e64449. [CrossRef] [PubMed]

34. Kachko, A.; Loesgen, S.; Shahzad-Ul-Hussan, S.; Tan, W.; Zubkova, I.; Takeda, K.; Wells, F.; Rubin, S.; Bewley, C.A.; Major, M.E. Inhibition of hepatitis C virus by the cyanobacterial protein Microcystis viridis lectin: Mechanistic differences between the high-mannose specific lectins MVL, CV-N, and GNA. *Mol. Pharm.* **2013**, *10*, 4590–4602. [CrossRef] [PubMed]

35. Helle, F.; Wychowski, C.; Vu-Dac, N.; Gustafson, K.R.; Voisset, C.; Dubuisson, J. Cyanovirin-N inhibits hepatitis C virus entry by binding to envelope protein glycans. *J. Biol. Chem.* **2006**, *281*, 25177–25183. [CrossRef] [PubMed]

36. Goffard, A.; Callens, N.; Bartosch, B.; Wychowski, C.; Cosset, F.L.; Montpellier, C.; Dubuisson, J. Role of N-linked glycans in the functions of hepatitis C virus envelope glycoproteins. *J. Virol.* **2005**, *79*, 8400–8409. [CrossRef] [PubMed]

Article

Activity of Selected Nucleoside Analogue ProTides against Zika Virus in Human Neural Stem Cells

Jean A. Bernatchez [1,2,†], Michael Coste [3,†], Sungjun Beck [1,†], Grace A. Wells [3], Lucas A. Luna [3], Alex E. Clark [1,4], Zhe Zhu [5,6], David Hecht [3,7], Jeremy N. Rich [5,6], Christal D. Sohl [3,*], Byron W. Purse [3,8,*] and Jair L. Siqueira-Neto [1,2,*]

[1]   Skaggs School of Pharmacy and Pharmaceutical Sciences, University of California, San Diego, La Jolla, CA 92093, USA; jbernatchez@ucsd.edu (J.A.B.); sjbeck@ucsd.edu (S.B.); alexclark@ucsd.edu (A.E.C.)
[2]   Center for Discovery and Innovation in Parasitic Diseases, University of California, San Diego, La Jolla, CA 92093, USA
[3]   Department of Chemistry and Biochemistry, San Diego State University, San Diego, CA 92182, USA; michael.coste1@gmail.com (M.C.); graceawells@gmail.com (G.A.W.); luluna@ad.ucsd.edu (L.A.L.); dhecht@swccd.edu (D.H.)
[4]   Department of Cellular and Molecular Medicine, University of California, San Diego, La Jolla, CA 92093, USA
[5]   Sanford Consortium for Regenerative Medicine, La Jolla, CA 92093, USA; zhz504@ucsd.edu (Z.Z.); drjeremyrich@gmail.com (J.N.R.)
[6]   Department of Medicine, Division of Regenerative Medicine, School of Medicine, University of California, San Diego, La Jolla, CA 92093, USA
[7]   Department of Chemistry, Southwestern College, Chula Vista, CA 91910, USA
[8]   The Viral Information Institute, San Diego State University, San Diego, CA 92182, USA
[*]   Correspondence: csohl@sdsu.edu (C.D.S.); bpurse@sdsu.edu (B.W.P.); jairlage@ucsd.edu (J.L.S.-N.); Tel.: +1-619-594-2053 (C.D.S.); +1-619-594-6215 (B.W.P.); +1-858-822-5595 (J.L.S.-N.)
[†]   These authors contributed equally to this paper.

Received: 27 March 2019; Accepted: 18 April 2019; Published: 20 April 2019

Abstract: Zika virus (ZIKV), an emerging flavivirus that causes neurodevelopmental impairment to fetuses and has been linked to Guillain-Barré syndrome continues to threaten global health due to the absence of targeted prophylaxis or treatment. Nucleoside analogues are good examples of efficient anti-viral inhibitors, and prodrug strategies using phosphate masking groups (ProTides) have been employed to improve the bioavailability of ribonucleoside analogues. Here, we synthesized and tested a small library of 13 ProTides against ZIKV in human neural stem cells. Strong activity was observed for 2′-*C*-methyluridine and 2′-*C*-ethynyluridine ProTides with an aryloxyl phosphoramidate masking group. Substitution of a 2-(methylthio) ethyl phosphoramidate for the aryloxyl phosphoramidate ProTide group of 2′-*C*-methyluridine completely abolished antiviral activity of the compound. The aryloxyl phosphoramidate ProTide of 2′-*C*-methyluridine outperformed the hepatitis C virus (HCV) drug sofosbuvir in suppression of viral titers and protection from cytopathic effect, while the former compound's triphosphate active metabolite was better incorporated by purified ZIKV NS5 polymerase over time. These findings suggest both a nucleobase and ProTide group bias for the anti-ZIKV activity of nucleoside analogue ProTides in a disease-relevant cell model.

Keywords: Zika virus; nucleoside analogues; antiviral agents; NS5; prodrugs; ProTides; neural stem cells; RNA-dependent RNA polymerase

---

1. Introduction

The explosive spread of Zika virus (ZIKV) during the 2015–2016 epidemics in Latin America attracted worldwide attention to this previously neglected disease. The lack of effective vaccines or small molecules to prevent or treat this infection remains a cause for concern and emphasizes the urgent need for new therapeutic options [1]. ZIKV, an emerging flavivirus infection, causes several serious neurodevelopmental anomalies in the fetus, including microcephaly, congenital ZIKV syndrome (CZS), and even fetal demise [2]. While most cases of ZIKV infection are asymptomatic, reports of rash, conjunctivitis, pain in the muscles or joints, and fever have been recorded in clinical manifestations of the disease, and in rare cases, ZIKV infection has been linked to the neuroinflammatory Guillain-Barré syndrome [2].

Viral polymerases remain attractive drug targets for the development of selective antiviral therapies [3–5]. Generally, clinically-approved inhibitors that target these proteins fall into two broad classes [6,7]. The first class consists of nucleoside analogues that mimic the natural substrate of the enzyme. Upon analogue incorporation by the virally-encoded polymerase, DNA or RNA synthesis is abrogated by preventing further nucleotide incorporation (chain termination), thereby arresting viral replication. The second class is known as non-nucleoside inhibitors, which bind allosterically and arrest viral nucleic acid synthesis by distorting the polymerase active site geometry to interfere with nucleotide binding or nucleotide incorporation.

A common prodrug strategy used for antiviral ribonucleoside analogues involves the chemical synthesis of nucleoside analogue monophosphates with metabolically-removable masking groups [8–10]. These masking groups neutralize the negative charge of the phosphate on the nucleoside and allow for better membrane penetrance of these compounds. Chemical addition of the first phosphate group is crucial for improving intracellular levels of the active triphosphate metabolite form of ribonucleoside analogues. Metabolic addition of the first phosphate group is relatively slow for this class of compounds and represents a rate-limiting step for the triphosphorylation of ribonucleoside analogues to active drug molecules.

This "ProTide" strategy has been successfully deployed in the synthesis of bioactive ribonucleoside analogues [11,12], including the Food and Drug Administration-(FDA) approved hepatitis C virus (HCV) drug, sofosbuvir [13]. Sofosbuvir has recently been shown by multiple groups to also be active against ZIKV in vitro and in vivo [14–17], demonstrating that ribonucleoside analogue ProTides are an attractive avenue for the development of novel, selective antivirals against ZIKV.

Numerous nucleoside analogues have recently been explored as antiviral agents against ZIKV [18–23]. However, many of these compounds were tested in either cell-free systems or using cells that are not clinically relevant for the disease (such as Vero cells). Further, sofosbuvir is the only FDA approved inhibitor tested against ZIKV thus far in ProTide form. In addition, recent work by our group and others has suggested that sofosbuvir and ProTides in general have differential activity depending on the cell line used, which may be linked to the cell-specific metabolism of ProTides [17,24]. In this work, we chemically synthesized a library of 13 ProTides and tested them for activity against ZIKV in human neural stem cells, a disease-specific cell model of infection. Interestingly, the activity of 2′-C-modified aryloxyl phosphoramidate ProTides was strongly biased towards uridylate derivatives, with only modest activity observed for the 2′-C-modified adenylate and cytidylate aryloxyl phosphoramidate ProTides we tested. Changing the aryloxyl phosphoramidate masking group to a 2-(methylthio)ethyl phosphoramidate abolished the activity of the 2′-C-methyluridine ProTide, a representative hit from our library. In a head-to-head comparison between the 2′-C-methyluridine aryloxyl phosphoramidate ProTide and sofosbuvir, better suppression of viral titers was observed for the former compound. In addition, better protection from ZIKV-induced cytopathic effect was observed for the 2′-C-methyluridine aryloxyl phosphoramidate ProTide over sofosbuvir in a highly ZIKV-susceptible stem cell model of infection with strong phenotypic changes during ZIKV infection. These results represent the first broad study of nucleoside analogue ProTides against ZIKV.

2. Materials and Methods

2.1. Chemistry

All nucleoside analogue ProTides, including McGuigan aryloxy phosphoramidite ProTides and 2-(methylthio)ethyl tryptamine ProTides, were synthesized using established phosphorus chemistry [25,26]. The 2′-*C*-ethenyluridine ProTide was synthesized using a known ethynyl ribose precursor **1**, which was obtained by following a previously reported synthetic route (Scheme 1) [27]. The ethenyl group of compound **2** was obtained by hydrogenation of the alkyne using 5% Pd/BaSO$_4$ in (1:1) quinoline/benzene [28]. Ribosylation of uracil to compound **2** using *N,O*-bis(trimethylsilyl)acetamide (BSA) and SnCl$_4$ in ACN afforded benzoyl-protected 2′-*C*-ethenyluridine **3**. The deprotection of compound **3** was performed using NaOMe in MeOH to afford 2′-*C*-ethenyluridine **4**, which was then treated with *t*BuMgCl in THF followed by *N*-[(*S*)-(2,3,4,5,6-pentafluorophenoxy)phenoxyphosphinyl]-l-alanine 1-methylethyl ester and to afford 2′-*C*-ethenyluridine aryloxy phosphoramidate **5**.

2.1.1. General Synthetic Experimental

All reagents, chemicals, and nucleoside analogues precursors were purchased from Acros Organics, Carbosynth, Chem-Impex International, AK Scientific, and Fisher Chemical at ACS reagent grade or higher purity and used as received. The (3R,4R,5R)-5-(benzoyloxy)methyl)-3-ethynyltetrahydrofuran-2,3,4-triyl tribenzoate **1** was synthesized by following a previously reported procedure [27]. Synthetic details for aryloxy phosphoramidate ProTides other than for 2′-*C*-ethynyluridine and 2′-*C*-ethenyluridine were published previously by our group [24]. All reactions were carried out in either an oven-dried round bottom flask or a Schlenk tube under a nitrogen atmosphere using commercially available anhydrous solvents and monitored by thin-layer chromatography with detection by UV light. The ^{1}H NMR and the ^{31}P NMR spectra were acquired on a Varian 400-MHz spectrometer and recorded at 298 K. Chemical shifts were referenced to the residual protio solvent peak and are given in parts per million (ppm). Splitting patterns are denoted as s (singlet), d (doublet), dd (doublet of doublet), dq (doublet of quartet), ddd (doublet of doublet of doublet), ddt (doublet of doublet of triplet), t (triplet), td (triplet of doublet), tdd (triplet of doublet of doublet), q (quartet), and m (multiplet).

2′-*C*-Ethynyluridine Aryloxy Phosphoramidate

To a stirred suspension of 2′-*C*-ethynyluridine (0.09 mmol, dried under vacuum at 50 °C overnight) in dry THF (1 mL) was added a 2.0 M solution of tert-butyl magnesium chloride in THF (96 µL, 0.19 mmol). The mixture was stirred at 0 °C for 30 min and then allowed to warm to room temperature and stirred for an additional 30 min. The reaction mixture was then cooled to 0 °C and *N*-[(*S*)-(2,3,4,5,6-pentafluorophenoxy)phenoxyphosphinyl]-L-alanine 1-methylethyl ester (46 mg, 0.10 mmol) was added. The reaction mixture was stirred for 18 h as the temperature was allowed to warm to room temperature. The solvent was removed by rotary evaporation. The reaction mixture was purified first using flash chromatography (a gradient from 0 to 30% methanol in dichloromethane) and then using preparative, normal-phase HPLC (10 to 40% MeOH in dichloromethane gradient) to afford 25.5 mg of the product (54.3%) in ≥ 95% purity. ^{1}H NMR (400 MHz, CD$_3$OD): δ= 7.65 (d, 1H, *J* = 8.1 Hz), 7.40–7.34 (m, 2H), 7.27 (dq, 2H, *J* = 7.7, 1.2 Hz), 7.20 (ddd, 1H, *J* = 8.2, 7.1, 1.1 Hz), 6.04 (s, 1H), 5.61 (d, 1H, *J* = 8.1 Hz), 4.97 (m, 1H), 4.49 (ddd, 1H, *J* = 11.8, 6.0, 2.1 Hz), 4.36 (ddd, 1H, *J* = 11.8, 6.1, 3.7 Hz), 4.16 (d, 1H, *J* = 9.1 Hz), 4.08 (ddt, 1H, *J* = 9.0, 3.9, 2.1 Hz), 3.92 (dq, 1H, *J* = 9.9, 7.1 Hz), 3.08 (s, 1H), 1.35 (dd, 3H, *J* = 7.2, 1.0 Hz), 1.22 (dd, 6H, *J* = 6.3, 1.8 Hz). ^{31}P NMR (162 MHz, CD$_3$OD) δ= 3.78.

5-((Benzoyloxy)methyl)-3-vinyltetrahydrofuran-2,3,4-triyl tribenzoate 2

To a stirring solution of **1** (208.7 mg, 0.362 mmol, 1 equivalent) in benzene (2 mL) and ethanol (2 mL) under H$_2$ (g), we added 5% palladium on barium sulfate (20.8 mg, 10 wt%) followed by quinolone (22 µL) and stirred at room temperature for 2 h. The mixture was diluted in ethyl acetate, washed 3 times with water, and dried over anhydrous sodium sulfate. The reaction mixture was purified using flash chromatography (0 to 30% ethyl acetate in hexane gradient) to afford purified product; (162.7 mg, 0.281 mmol, 77.7%) 1H NMR (400 MHz, CDCl$_3$): δ = 8.23–8.09 (m, 4H), 8.07–8.03 (m, 2H), 7.92–7.88 (m, 2H), 7.69–7.39 (m, 10H), 7.18–7.12 (m, 2H), 6.46 (dd, 1H, J = 17.6, 11.2 Hz), 6.25 (d, 1H, J = 8.3 Hz), 4.54 (dd, 2H, J = 12.2, 4.8 Hz), 4.81 (ddd, 2H, J = 8.4, 4.7, 3.9 Hz), 4.73 (dd, 1H, J = 12.2, 3.9 Hz), 4.54 (dd, 1H, J = 12.2, 4.8 Hz)

5-((Benzoyloxy)Methyl)-2-(2,4-Dioxo-3,4-Hihydropyrimidin-1(2H)-yl)-3-Vinyltetrahydrofuran-3,4-diyl Benzoate 3

Uracil (63.0 mg, 0.562 mmol, 2 equivalent) and **2** (162.7 mg, 0.281 mmol, 1 equivalent) were dried under high vacuum in separate round bottom flasks for 2 h. Under N$_2$ (g) and stirring, we added dry acetonitrile (2 mL) to uracil followed by the addition of bis(trimethylsilyl)acetamide (550.1 µL, 2.250 mmol, 8 equivalent). The reaction mixture was refluxed at 80 °C for 1 h then cooled to 0 °C. Then, compound 2 in dry acetonitrile (2 mL) was added to the reaction mixture followed by tin (IV) chloride (229.9 µL, 1.968 mmol, 7 equivalent) and heated to 60 °C for 3 h. The reaction mixture was poured into a separatory funnel containing ice cold water, extracted 3 times with ethyl acetate, and the combined organic layer was dried over anhydrous sodium sulfate. The reaction mixture was purified using flash chromatography (0 to 100% ethyl acetate in hexane gradient) to afford purified product; (89.4 mg, 0.153 mmol, 54.6%) 1H NMR (400 MHz, CDCl$_3$): δ= 9.22 (s, 1H), 8.09 (m, 4H), 7.86–7.82 (m, 2H), 7.63–7.56 (m, 2H), 7.63–7.56 (m, 6H), 7.29–7.21 (m, 2H), 6.65 (s, 1H), 6.12 (dd, 1H, J = 17.5, 11.1 Hz), 6.04 (d, 1H, J = 5.2 Hz), 5.64 (dd, 1H, J = 8.2, 2.1 Hz), 5.46–5.40 (dd, 2H), 4.94 (dd, 1H, J = 12.3, 3.2 Hz), 4.81 (dd, 1H, J = 12.3, 5.7 Hz), 4.66 (td, 1H, J = 5.5, 3.2 Hz)

1-(3,4-dihydroxy-5-(hydromethyl)-3-vinyltetrahydrofuran2-yl)pyrimidine-2,4(1H,3H)-dione 4

Compound 3 (90.5 g, 0.155 mmol, 1 equivalent) was dried overnight on high vacuum. Under N$_2$ (g), methanol (1.5 mL) was added, then the reaction mixture was cooled to 0 °C followed by the dropwise addition of sodium methoxide (86.5 µL, 1.553 mmol, 10 equivalent). The reaction mixture was raised to room temperature and stirred for 1.5 h. The reaction mixture was cooled to 0 °C followed by the addition of formic acid until pH = 4. The reaction mixture was dried in vacuo then purified using flash chromatography (0 to 40% methanol in dicholoromethane gradient) to afford a purified product; (30.2 mg, 0.146 mmol, 93.8%) 1H NMR (400 MHz, CDCl$_3$): δ= 8.13 (d, 1H, J = 8.1 Hz), 5.95 (s, 1H), 5.74–5.65 (m, 2H), 5.44 (dd, 1H, J = 17.3, 1.3 Hz), 5.26 (dd, 1H, J = 10.8, 1.3 Hz), 4.22 (d, 1H, J = 9.2 Hz), 4.03–3.97 (m, 2H), 3.84–3.79 (m, 1H)

isopropyl((((5-(2,4-dioxo-3,4-dihydropyridimin-1(2H)-yl)-3,4-dihydroxy-3-inyltetrahydrofuran-2-yl)methoxy)(phenoxy)phosphoryl)-L-alaninate 5

To a stirring solution of **4** (36.4 mg, 0.176 mmol, 1 equivalent) in dry tetrahydrofuran (1 mL) at 0 °C was added tert-butyl magnesium chloride (184.5 µL, 0.369 mmol, 2.1 equivalent). The reaction mixture was raised to room temperature and allowed to react for 30 min. The reaction mixture was cooled to 0 °C, then N-[(S)-(2,3,4,5,6-pentafluorophenoxy)phenoxyphosphinyl]-L-alanine 1-methylethyl ester (95.5 mg, 0.211 mmol, 1.2 equivalent) was added and gradually warmed to room temperature overnight. The reaction mixture was purified using flash chromatography (0 to 30% methanol [MeOH] in dichloromethane gradient) to afford the purified product; (42.5 mg, 0.0788 mmol, 44.8%) 1H NMR (400 MHz, CD$_3$OD): δ= 7.76 (d, 1H, J = 8.1 Hz), 7.38 (dd, 2H, J = 8.6, 7.2 Hz), 7.31–7.25 (m, 2H), 7.25–7.16 (m, 2H), 5.94 (s, 1H), 5.68 (dd, 1H, J = 17.3, 10.8 Hz), 5.60 (d, 1H, J = 8.1 Hz), 5.48 (d, 1H, J = 1.4 Hz), 5.44 (d, 1, J = 1.4 Hz), 5.27 (dd, 1H, J = 10.8, 1.4 Hz), 4.96 (1H, m), 4.58–4.49 (m, 1H), 4.45–4.38 (m, 1H), 4.17 (s, 2H), 4.00–3.87 (m, 1H), 1.38 (dd, 3H, J = 7.1, 1.0 Hz), 1.21 (d, 6H, J = 6.3 Hz). ^{31}P NMR (162 MHz, CD$_3$OD) δ = 3.78.

(5-(2,4-dioxo-3,4-dihydropyrimidin-1(2H)-yl)-3,4-dihydroxy-4-methyltetrahydrofuran-2-yl)methyl (2-(methylthio)ethyl) (2-(1H-indol-3-yl)ethyl)phosphoramidate (2′-*C*-methyluridine methylthioethyl tryptamine ProTide)

The 2′-*C*-methyluridine (45.3 mg, 0.1790 mmol, 1 equivalent) and the triethylammonium 2-(methylthio)ethyl phosphonate (92.1 mg, 0.3579 mmol, 1.5 equivalent) were added to an oven-dried schlenk tube and dried overnight on the high vacuum. To the flask under N_2 (g) was added dry pyridine (2 mL) followed by the dropwise addition of trimethylacetyl chloride (60.6 µL, 0.4923 mmol, 2.75 equivalent), and the reaction was stirred for 3 h at room temperature. The reaction was quenched by saturated sodium bicarbonate and extracted 3 times with dicholormethane. The organic layer was dried over anhydrous sodium sulfate and concentrated in vacuo at room temperature and dried for 2 h on high vacuum. The dried residue was dissolved in pyridine (2 mL) under N_2 (g). To the reaction mixture was simultaneously added trimethylamine (49.9 µL, 0.3579 mmol, 2 equivalent), tryptamine (114.7 mg, 0.7158 mmol, 4 equivalent, dissolved in dry pyridine 2 mL), and carbon tetrachloride (26.0 µL, 0.2685 mmol, 1.5 equivalent) and stirred for 25 min at room temperature. The reaction mixture was diluted with methanol and purified using flash chromatography (0 to 30% methanol in dichloromethane gradient) to afford a purified product as a mixture of two epimers at phosphorus; (8.0 mg, 0.01443 mmol, 7.3%) ^{1}H NMR (400 MHz, CD_3OD): δ= 7.72 (dd, 1H, J = 21.4, 8.2 Hz), 7.55–7.50 (m, 1H), 7.33–7.29 (m, 1H), 7.06 (t, 2H, J = 7.3 Hz), 6.97 (tdd, 1H, J = 7.5, 4.6, 2.4 Hz), 5.95 (d, 1H, J = 5.2 Hz), 5.64 (dd, 1H, J = 8.1, 3.8 Hz), 4.37–4.29 (m, 1H), 4.17 (ddd, 1H, J = 11.7, 5.0, 3.1 Hz), 4.04 (dt, 1H, J = 10.4, 6.8 Hz), 4.00–3.90 (m, 1H), 3.77 (dd, 1H, J = 12.4, 9.3 Hz), 3.21 (td, 2H, J = 10.9, 5.6 Hz), 2.96 (dt, 2H, J = 7.3, 4.4 Hz), 2.66 (td, 2H, J = 6.7, 2.3 Hz), 2.07 (d, 3H, J = 1.4 Hz), 1.13 (s, 3H). ^{31}P NMR (162 MHz, CD_3OD) δ= 10.52 and 10.41.

Scheme 1. Synthesis of 2′-*C*-ethenyluridine ProTide. (i) H_2 (g), Pd/BaSO$_4$, quinoline, benzene, 78%; (ii) bovine serum albumin (BSA), uracil, SnCl$_4$, ACN, 60 °C to 80 °C, 55%; (iii) NaOMe, MeOH, 0 °C to reflux, 94%; (iv) tBuMgCl, *N*-[(*S*)-(2,3,4,5,6-pentafluorophenoxy)phenoxyphosphynyl]-*L*-alanine 1-methylethyl ester, THF, −70 °C to room temperature (RT), 45%.

2.2. Cell Culture

Human fetal NSCs were purchased from Clontech (human neural cortex; catalog number Y40050). Maintenance of hNSCs was performed in Neurobasal medium without phenol red (Thermo Fisher, Waltham, MA, USA) and in the presence of B-27 supplement (1:100; catalog number 12587010; Thermo Fisher), N-2 supplement (1:200; catalog number 17502-048; Invitrogen, San Diego, CA, USA), 20 ng/mL fibroblast growth factor (catalog number 4114-TC-01M; R&D Systems, Minneapolis, MN, USA, 20 ng/mL epidermal growth factor (catalog number 236-EG-01M; R&D Systems), GlutaMAX (catalog number 35050061; Thermo Fisher), and sodium pyruvate. Cells were prepared in 6-well plates precoated with laminin (10 mg/mL; catalog number L2020; Sigma-Aldrich, St. Louis, MO, USA) and were grown to near confluence (80 to 90%) prior to passage. For passaging, cells were rinsed gently with phosphate-buffered saline (PBS) (without calcium and magnesium) and then treated with TryPLE for 5 min at 37 °C for cell detachment (catalog number 12563029; Thermo Fisher). The GSC 387 cell line was prepared and maintained similarly to what has been previously described using the media and detachment conditions used for hNSCs described above [29]. Vero cells (ATCC CCL-81) were purchased from ATCC and cultured in Dulbecco's Modified Eagle Medium (DMEM) (Gibco, Waltham, MA, USA) with high glucose (4.5 g/L), 10% fetal calf serum (FCS) (Sigma), and 1% Penicillin-Streptomycin (PS) (Sigma-Aldrich, St. Louis, MO, USA) at 37 °C and 5% CO_2.

2.3. Viruses

Viral strains used in this study were provided through BEI Resources, NIAID, NIH: Zika virus H/PAN/2016/BEI-259634, NR-50210 (GenBank accession number KX198135), Zika virus PRVABC59/Human/2015/Puerto Rico, NR-50240 (GenBank accession number KU501215).

Vero cells were used to expand viral cultures for 2 to 3 serial passages in order to amplify the titers. Centrifugation of infected cell supernatants was performed to remove cell debris, and then viral stocks were concentrated through use of a sucrose cushion. Neural maintenance medium base [50% DMEM–F-12 medium–GlutaMAX, 50% Neurobasal-A medium, 1× N-2 supplement, 1× B-27 supplement (Life Technologies)] supplemented with 1% DMSO (Sigma) and 5% fetal bovine serum (FBS) (Gibco) was used to resuspend concentrated viral stocks, and aliquots were stored at −80 °C. Plaque assay on Vero cells was used to determine viral titers, which were greater than or equal to 2×10^8 PFU/mL.

2.4. CellTiter-Glo Activity Assay

NSCs (5000 cells/well) were seeded into 1536-well black, clear bottom plates containing 20-point, 2-fold serial dilutions of ProTides or DMSO vehicle controls and were either mock-infected, infected with ZIKV H/PAN/2016/BEI-259634 at the multiplicity of infection (MOI) of 10, or infected with ZIKV PRVABC59/Human/2015/Puerto Rico at MOI of 10. Cell viability was measured using the CellTiter-Glo assay (Promega, Madison, WI, USA) 72 h post-infection. An EnVision plate reader (PerkinElmer) was used for readouts of luminescence intensity. All data were normalized to DMSO vehicle controls and were expressed as % relative luminescence intensity. Normalized activity and toxicity data were plotted against ProTide concentration and fit to a sigmoidal dose-response curve with variable slope using Prism 6 (GraphPad Software) to generate EC_{50} and CC_{50} curves.

2.5. Bright-Field Microscopy

GSC 387 cells (5000 cells/well) were seeded into 1536-well black, clear bottom plates with wells containing either 10 or 30 µM sofosbuvir, 10 or 30 µM 2′-C-methyluridine aryloxyl phosphoramidate ProTide, or DMSO vehicle controls, and were either mock-infected or infected with ZIKV H/PAN/2016/BEI-259634 at MOI of 10. Cell foci images were acquired 72 h post-infection in bright-field using an ImageXpress Micro automated microscope (Molecular Devices, San Jose, CA, USA).

2.6. Plaque Assay

Vero cells were used for the titering experiments for drug-treated and vehicle-treated hNSC cultures infected with ZIKV H/PAN/2016/BEI-259634. Vero cells were seeded at a density of 30,000 cells per well in 96-well plates and incubated in 5% CO_2 at 37 °C in high-glucose DMEM with 1% FBS and 1% penicillin/streptomycin for 24 h before the addition of the viral inoculum from ZIKV-infected hNSCs. Inocula were collected from ZIKV H/PAN/2016/BEI-259634-infected NSCs (MOI 0.1) in the presence of 10 µM or 30 µM 2′-C-methyluridine aryloxyl phosphoramidate ProTide, 10 µM or 30 µM sofosbuvir, mock-infected cultures, infected cultures without compound treatment, and infected cultures with the

DMSO vehicle at 48 h post-infection. Collected inocula were diluted 10-fold in high-glucose DMEM containing 1% FBS and 1% penicillin-streptomycin in 96-well plates. These inocula were added to the pre-prepared Vero cells for 1 h in a volume of 100 µL and at a temperature of 37 °C, covered with an agarose overlay, and lastly incubated for 72 h at 37 °C and 5% CO_2. Fixation with 4% formaldehyde (final concentration) for 24 h was then performed for the samples. Following this, the overlay was removed, and crystal violet staining was performed to count the ZIKV plaques.

2.7. ZIKV RdRP Expression and Purification

The RdRP domain of ZIKV NS5 (amino acids 2772-3423 of accession number AMB18850) was fused to an N-terminal 6×-histidine tag containing a TEV cleavage site, and the cDNA was codon-optimized and synthesized by IDT (San Diego, CA, USA). This construct was then inserted into a pET-17b vector (Invitrogen) and heterologously expressed in the Rosetta strain of *Escherichia coli*. Cultures transfected with the RdRP construct were grown at 37 °C in LB media containing 1% ethanol and 100 µg/mL amplicillin until an OD_{600} of 0.6–0.7 was reached. Protein expression was then induced with 100 µM IPTG prior to an additional 5 h incubation at 20 °C and lower speed shaking (130 RPM). Cells were then harvested by centrifugation at 9000 RPM for 30 min at 4 °C, and pellets were stored at −80 °C. For purification, pellets were resuspended in lysis buffer (50 mM HEPES at pH 7.5, 20 mM NaCl, 20% glycerol, 0.1% octyl-glucoside, 5 mM β-mercaptoethanol) supplemented with a protease inhibitor tablet (Thermo Fisher). Cells were lysed using a microfluidizer (Microfluidics International Corporation, Newton, MA, USA) and clarified via centrifugation at 12,000 RPM for 30 min at 4 °C. The supernatant was loaded onto TALON cobalt affinity beads (Clontech, Mountain View, CA, USA) at 4 °C. The unbound protein was washed with 50 mL each of the following buffers: wash buffer (50 mM HEPES at pH 7.5, 5 mM β-mercaptoethanol) containing 100 mM NaCl and 10 mM imidazole, wash buffer containing 500 mM NaCl and 20 mM imidazole, and finally wash buffer containing 200 mM NaCl and 30 mM imidazole. The protein was eluted in wash buffer containing 200 mM NaCl and a linear imidazole gradient of 30–300 mM. Protein fractions were pooled and dialyzed overnight against 50 mM HEPES at pH 7.0, 100 mM NaCl, 10% glycerol, 0.01% CHAPS, and 5 mM dithiothreitol at 4 °C. The protein was concentrated to ≥6 µM and stored in single-use aliquots at −80 °C.

2.8. Single Nucleotide Incorporation Assay

To prepare the double-stranded RNA substrate, primer (5′-AGUUGUUGAUC3′) was ^{32}P-labeled at the 5′ position upon incubation with [γ-^{32}P] ATP (PerkinElmer, Waltham, MA, USA) and T4 polynucleotide kinase (NEB, Ipswich, MA, USA) at 37 °C for 30 min. Excess [γ-^{32}P] ATP was removed with a Bio-Spin 6 column (Bio-Rad, Hercules, CA, USA). After heat inactivation, the 5′-^{32}P labeled primer was combined with an excess template (5′-AUUCACUCAGAUCAACAACU-3′). The primer and the template were annealed via step incubation at 95 °C (5 min), 55 °C (15 min), and 37 °C (10 min) to generate the primer/template substrate (P/T substrate), where the next correct nucleotide to be incorporated was UTP.

Single nucleotide incorroration assays were modeled from previous work [19]. Briefly, ZIKV RdRP (1 µM) and P/T substrate (200 nM) were pre-incubated at 37 °C for 20 min. Reactions were initiated at 37 °C with the addition of 150 µM UTP or analog in 50 mM HEPES pH 7.5, 15% glycerol, 15 mM NaCl, 7.5 mM $MnCl_2$, and 10 mM dithiothreitol. At specified time points, reaction aliquots were quenched with ethylenediamine tetra-acetic acid (EDTA) to a final concentration of 0.3 M.

Prior to loading equi-volume samples on a 20% polyacrylamide, 8 M urea, TBE (89 mM Tris-Borate, 2 mM EDTA) denaturing gel, the quenched samples were combined with dye (formamide, 0.01% xylene cyanol, 0.01% bromophenol blue) and 1 µL of 5 µM D10 (5′-AGTTGTTGAT-3′) trap, followed by incubation at 95 °C for 5 min. The trap was used to bind to RNA template to reduce gel smearing [30]. Oligos were separated via electrophoresis at a maximum of 1900 V and 75 W for 3–6 h. Phosphorimaging (FLA 9500 imager, GE, Marlborough, MA, USA) was used to visualize the bands, and bands were quantified using ImageQuant software (GE). Plots of percent incorporation versus time were fit to a single exponential equation using Prism (Graphpad Software, San Diego, CA, USA) to compare relative incorporation. While single-turnover conditions were used, non-physiologically relevant rates were obtained, as incorporation occurred during a steady-state time scale. This has been a common issue in the field when measuring ZIKV RdRP incorporation [19,20,31] due to measuring incorporation into a double-stranded template instead of physiologically preferred de novo RNA replication (an issue seen in HCV incorporation assays [32]), and perhaps the presence of purification tags [33]. However, since here we were interested in comparing relative incorporation efficiency rather than obtaining rate constants, we and others [19,31] have found this experimental set-up ideal and better suited to scaling up the number of analogues to be assessed.

2.9. Structural Alignment

The following model and structures were aligned to HCV RdRP complexed with RNA, Mn^{2+}, and sofosbuvir-DP (PDB 4WTD [34]) using Pymol [35]: 1) a model of ZIKV RdRP in complex with RNA, Mn^{2+} ions, and an incoming ADP modified to an ATP (gc-o3 [36]) generated from ZIKV RdRP stucture PDB 5TFR [37] (160 atoms aligned, RMS = 1.650); 2) a crystal structure of apo ZIKV RdRP (PDB 5WZ3 [38]) (154 atoms aligned, RMS = 2.860); and 3) a crystal structure of apo HCV RdRP (PDB 3MWV [39]) (530 atoms aligned, RMS = 0.874).

3. Results

3.1. ProTide Library Design

To investigate the activity of a set of nucleotide analogue ProTides against ZIKV and obtain basic structure-activity relationships, we selected a set of nucleotides designed to inhibit viral RdRPs by three different mechanisms: obligate chain termination, nonobligate chain termination, and lethal mutagenesis (Figure 1) [40,41]. Past work—and the successful results of drug development against HCV—shows that nucleosides alone lack adequate potency owing to slow phosphorylation in the target cells [42]. McGuigan's nucleotide aryloxy phosphoramidates include amino acid esters and are well established; sofosbuvir includes this functional group, and we chose it as our starting point [9,43]. One potential drawback of the McGuigan design is that unmasking is initiated by the action of carboxyesterases, which are highly active in the hepatocytes targeted by HCV but less active in the neural cell targets of ZIKV [42]. For that reason, we chose to investigate a recently developed methylthioethyl tryptamine ProTide from Wagner, which is unmasked by chemical cleavage of methylthioethanol followed by tryptamine removal by HINT1 [26]. HINT1 also catalyzes P-N bond cleavage in the classic McGuigan ProTide. Our hypothesis was that the carboxyesterase-independent unmasking of the methylthioethyl tryptamine ProTide would provide greater potency against ZIKV in human neural stem cells.

The selection of nucleoside analogues for testing focused on ribose modifications known to induce chain termination in vitro when used as a nucleoside triphosphate analogue with the viral RdRP. Some of the selected compounds, e.g., the 3′-deoxy and the 3′-O-methyl, are obligate chain terminators, but phosphorylation of this class of compounds in cells is often inefficient [18,19]. Better success in antiviral drug development has been found when using nonobligate terminators, especially 2′ modifications that alter the nucleoside conformation and prevent extension after incorporation during RNA synthesis [14,31]. We included sofosbuvir in this set as a point of reference, along with ProTides of the 2′-C-methyl ribosides, which have known potency against the Flaviviridae family [19,31,44]. Also included were new ProTides of 2′-C-ethynyl- and 2′-C-ethenyluridine. The 2′-C-ethynyl ribosyl triphosphates are known sub-µM inhibitors of flavivirus RdRPs [31,45], and we prepared the new 2′-C-ethenyluridine ProTide to probe the active site of the RdRP and the metabolic consequences of further modification of the 2′-C-β position. A 5-fluorouridine ProTide was included in our set of analogues to test inhibition by lethal mutagenesis, although toxicity of mutagenic nucleoside analogues is typically prohibitive of clinical applications.

R^4 = classic McGuigan aryloxy phosphoramidate ProTide

R^4 = methylthioethyl tryptamine ProTide

base

obligate chain terminators
3'-deoxy (R^1 = H, R^2 = OH, R^3 = H)
3'-O-methyl (R^1 = H, R^2 = OH, R^3 = OCH$_3$)
3'-deoxy-3'-fluoro (R^1 = H, R^2 = OH, R^3 = F)
nonobligate chain terminators
2'-C-methyl (R^1 = CH$_3$, R^2 = OH, R^3 = OH)
2'-C-ethynyl (R^1 = C≡CH, R^2 = OH, R^3 = OH)
2'-C-ethenyl (R^1 = CH=CH$_2$, R^2 = OH, R^3 = OH)
2'-O-methyl (R^1 = H, R^2 = OCH$_3$, R^3 = OH)
2'-deoxy-2'-fluoro-2'-methyl (R^1 = CH$_3$, R^2 = F, R^3 = OH)
mutagens
fluorinated nucleobase (R^1 = H, R^2 = OH, R^3 = OH)

Figure 1. Design and structure of nucleotide analogue inhibitors tested against Zika virus (ZIKV) in this study.

3.2. Activity against ZIKV and Toxicity of Nucleoside Analogue ProTides in Human Neural Stem Cells

To profile the activity and toxicity profiles of our library of ProTides, we screened compounds in a 20-point, two-fold dose response with a highest concentration of 50 μM against ZIKV PRVABC59/Human/2015/Puerto Rico and ZIKV H/PAN/2016/BEI-259634 in neural stem cells (MOI of 10) using a luminescent cell-viability assay (CellTiter Glo) and determined EC$_{50}$ (effective concentration 50) and CC$_{50}$ (cytotoxic concentration 50) values from these data (Table 1). We observed that the compounds with the best selectivity indexes (SIs) were the 2'-C-methyluridine aryloxyl phosphoramidate ProTide (EC$_{50}$PRVABC59: 1 μM, EC$_{50}$H/PAN: 2 μM, CC$_{50}$: 47 μM, SIs: >47, >23, respectively) and the 2'-C-ethynyluridine aryloxyl phosphoramidate ProTide (EC$_{50}$PRVABC59: 0.3 μM, EC$_{50}$H/PAN: 0.8 μM, CC$_{50}$: >50 μM, SIs: >166, >62, respectively). These outperformed sofosbuvir in terms of anti-ZIKV activity (EC$_{50}$PRVABC59: 35 μM, EC$_{50}$H/PAN: 46 μM, CC$_{50}$: >50 μM, SIs: >1.4, >1.1, respectively). Activity was observed for the 2'-C-methyladenosine aryloxyl phosphoramidate ProTide (EC$_{50}$PRVABC59: 10 μM, EC$_{50}$H/PAN: 18 μM, CC$_{50}$: 42 μM, SIs: >4, >2, respectively) and 2'-C-methylcytidine aryloxyl phosphoramidate ProTide against PRVABC59 strain (EC$_{50}$PRVABC59: 48 μM, CC$_{50}$PRVABC59: >50 μM, SI: >1), albeit at lower levels. None of the other ProTides tested displayed activity. Interestingly, the 2'-C-ethenyluridine aryloxyl phosphoramidate ProTide was completely inactive and quite toxic to the host cells, suggesting that this functional group may be more promiscuous in terms of target binding or reactive with cellular components [46]. Replacement of the classic McGuigan ProTide masking group with methylthioethyl tryptamine phosphoramidate for the 2'-C-methyluridine ProTide completely abolished activity. This result was not expected because the classic McGuigan ProTides are unmasked with the involvement of both a carboxyesterase and HINT1, whereas the methylthioethyl tryptamine phosphoramidate requires only HINT1 in addition to spontaneous, chemical steps [26]. Clearly, cell line specificity for ProTide strategies must be considered in the design of nucleotide analogue inhibitors of ZIKV.

Table 1. Anti-ZIKV activity and toxicity of nucleoside analogue ProTides in neural stem cells.

Nucleoside Analogue ProTide	EC$_{50}$ PRVABC59 ZIKV, μM	EC$_{50}$ H/PAN ZIKV, μM	CC$_{50}$, μM	SI [1] PRVABC59 ZIKV	SI [1] H/PAN ZIKV
2'-C-Methylcytidine aryloxyl phosphoramidate	48 ± 1	>50	>50	>1	>1
3'-Deoxyadenosine aryloxyl phosphoramidate	>50	>50	>50	>1	>1
5'-Fluorocytidine aryloxyl phosphoramidate	>50	>50	10 ± 1	>0.2	>0.2

Table 1. *Cont.*

Nucleoside Analogue ProTide	EC_{50} PRVABC59 ZIKV, μM	EC_{50} H/PAN ZIKV, μM	CC_{50}, μM	SI [1] PRVABC59 ZIKV	SI [1] H/PAN ZIKV
3'-O-Methyluridine aryloxyl phosphoramidate	>50	>50	>50	>1	>1
2'-O-Methylcytidine aryloxyl phosphoramidate	>50	>50	>50	>1	>1
2'-C-Methyluridinearyloxyl phosphoramidate	1 ± 1	2 ± 1	47 ± 1	>47	>23
2'-C-Methyladenosine aryloxyl phosphoramidate	10 ± 1	18 ± 1	42 ± 1	>4	>2
3'-Deoxycytidine aryloxyl phosphoramidate	>50	>50	>50	>1	>1
3'-Deoxy-3'-Fluoroguanosine aryloxyl phosphoramidate	>50	>50	>50	>1	>1
2'-C-Ethenyluridine aryloxyl phosphoramidate	>50	>50	10 ± 1	>0.1	>0.1
2'-C-Ethynyluridine aryloxyl phosphoramidate	0.3 ± 1	0.8 ± 1	>50	>166	>62
2'-C-Methyluridine 2-(methylthio)ethyl phosphoramidate	>50	>50	>50	>1	>1
Sofosbuvir	35 ± 1	46 ± 1	>50	>1.4	>1.1

[1] SI = selectivity index: CC_{50}/EC_{50} for a given viral strain. Data values are the represented ± standard error (SE), $n = 4$.

3.3. Protection from ZIKV-Induced Cytopathic Effect by 2'-C-Methyluridine Aryloxyl Phosphoramidate ProTide and Sofosbuvir in a ZIKV-Sensitive Glioblastoma Stem Cell Line

We and others have previously shown that certain lines of brain tumors with stem cell-like genetic programs, such as glioblastomas, are highly susceptible to ZIKV infection [29,47–49]. These cells display a striking phenotypic change upon infection. To highlight the ability of a representative ProTide hit from our library, 2'-C-methyluridine aryloxyl phosphoramidate, to protect against ZIKV-induced cytopathic effect, we treated the glioblastoma stem cell line GSC 387 with either 10 or 30 μM 2'-C-methyluridine aryloxyl phosphoramidate and compared this to 10 or 30 μM sofosbuvir or the DMSO vehicle control treatment. Samples were prepared in both ZIKV-infected (MOI of 10) and uninfected conditions. After 72 h post-infection, bright-field microscopy images were obtained for each sample and are shown in Figure 2. While both compounds were able to protect against the cytopathic effect at 30 μM (intact cell foci were seen), the 2'-C-methyluridine aryloxyl phosphoramidate ProTide was better able to protect against ZIKV-induced cytopathic effect at the lower concentration of 10 μM than sofosbuvir, which is in line with our EC_{50} data in neural stem cells. These data suggest that monosubstituted 2'-C-modified uridylate aryloxyl phosphoramidate ProTides may be an attractive chemical series to pursue for more potent and selective ZIKV polymerase inhibitors than sofosbuvir.

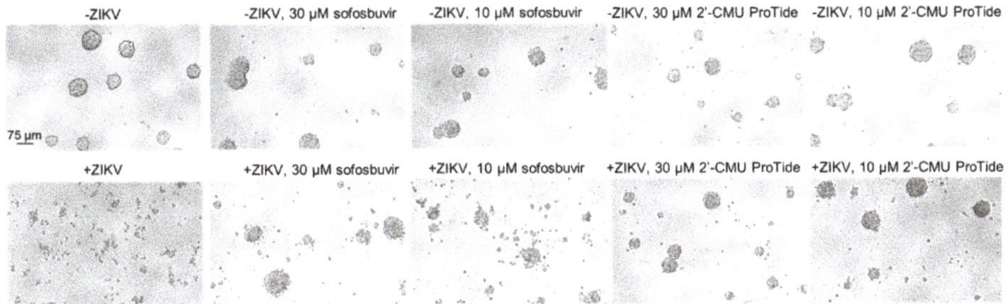

Figure 2. Protection from ZIKV-induced cytopathic effect by 2'-C-methyluridine aryloxyl phosphoramidate ProTide and sofosbuvir in a ZIKV-sensitive glioblastoma stem cell model. DMSO-treated GSC 387 cells [-ZIKV and +ZIKV, H/PAN multiplicity of infection (MOI) 10], 10 or 30 µM sofosbuvir-treated cells or 10 or 30 µM 2'-C-methyluridine aryloxyl phosphoramidate ProTide (2'-CMU ProTide)-treated cells were incubated for 72 h at 37 °C and 5% CO_2, and cell foci were subsequently imaged in bright-field.

3.4. Repression of ZIKV Titers by 2'-C-Methyluridine Aryloxyl Phosphoramidate ProTide and Sofosbuvir in Neural Stem Cells

To further assess the antiviral activity of the 2'-C-methyluridine aryloxyl phosphoramidate ProTide against ZIKV, we performed plaque assays at 48 h post-infection to titer ZIKV (H/PAN, MOI of 0.1) levels during treatment with either 10 or 30 µM of 2'-C-methyluridine aryloxyl phosphoramidate ProTide or 10 or 30 µM of sofosbuvir in human fetal neural stem cells (Figure 3). At 30 µM compound, both 2'-C-methyluridine aryloxyl phosphoramidate ProTide and sofosbuvir were able to repress viral titers to undetectable levels. However, only 2'-C-methyluridine aryloxyl phosphoramidate ProTide was able to reduce ZIKV titers to undetectable levels at the 10 µM compound, again highlighting its superior antiviral activity compared to sofosbuvir.

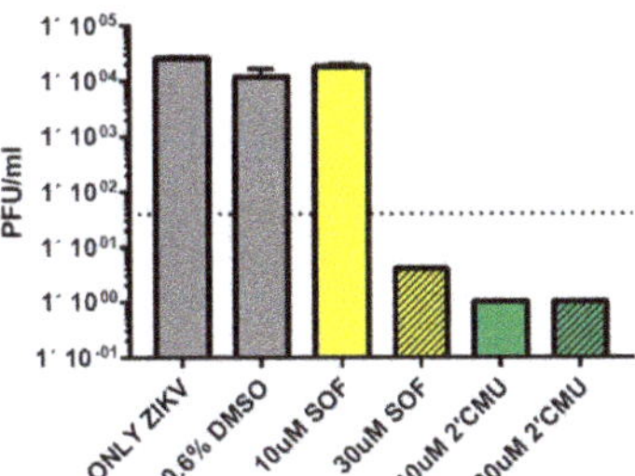

Figure 3. Reduction in ZIKV titers by 2'-C-methyluridine aryloxyl phosphoramidate ProTide and sofosbuvir in neural stem cells as observed by plaque assay. Viral infections were conducted in human fetal neural stem cells (H/PAN ZIKV, MOI 0.1) using vehicle-untreated (ONLY ZIKV), DMSO vehicle-treated, 10 or 30 µM sofosbuvir (SOF)-treated, and 10 or 30 µM 2'-C-methyluridine aryloxyl phosphoramidate ProTide (2'CMU)-treated samples. The dotted line represents the limit of detection of the assay. Virus was titered as described in the methods section and plaque forming units (PFU) per ml were calculated for each sample, ± standard error (SE), $n = 4$.

3.5. Enzymatic Incorporation of the Active Triphosphate Metabolites of 2'-C-Methyluridine and Sofosbuvir Over Time Reveals Higher Levels of Incorporation for the Former Compound

Single nucleotide incorporation assays were used to compare the relative preference of ZIKV RdRP for UTP versus the UTP analogs 2'-C-methyluridine triphosphate and 2'-fluoro-2'-C-methyluridine triphosphate (the active form of sofosbuvir) (Figure 4). Both analogues appeared to serve as chain terminators of ZIKV RdRP polymerization, as previously reported [19,31]. Unsurprisingly, UTP was most efficiently incorporated, with 50% incorporation by 5 min and full incorporation by ~30 min, followed by misincorporating extension. The 2′-fluoro-2′-*C*-methyluridine triphosphate had significantly less efficient incorporation by ZIKV RdRP, as previously reported [19,31,50]. Incorporation plateaued at 15%, and half maximal incorporation was only reached after nearly 90 min. In contrast, 2′-*C*-methyluridine triphosphate reached half of its eventual maximal incorporation of 35% after ~45 min, showing improved incorporation over the active form of sofosbuvir. These trends are supported by previous work focusing on biochemical characterization of 2′-*C*-methyluridine triphosphate and 2′-fluoro-2′-*C*-methyluridine triphosphate incorporation [19,20]. Importantly, here we verify the preference of 2′-*C*-methyluridine over sofosbuvir in both biochemical and cell-based assays, suggesting a robust inhibitor testing pipeline that supports the hypothesis that cellular efficacy occurs as a result of RdRP inhibition by nucleoside analogs.

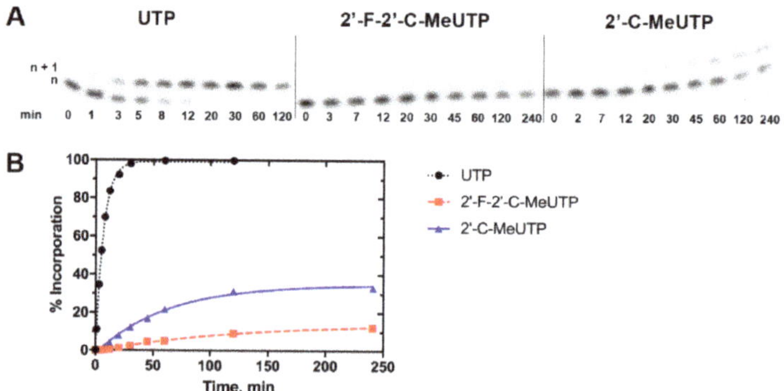

Figure 4. Kinetic characterization of nucleotide and nucleotide analogue incorporation by ZIKV RdRP. (**A**) ZIKV RdRP and primer/template (P/T) substrate (n) was mixed with excess UTP, 2′-F-2′-*C*-methyluridine triphosphate (2′-F-2′-*C*-MeUTP), or 2-*C*-methyluridine triphosphate (2′-*C*-MeUTP), and the reaction was quenched at the indicated timepoints. The substrate and the single nucleotide extension product (n + 1) were separated by using a 20% polyacrylamide denaturing gel, and bands were quantified to determine percent incorporation and plotted against time. (**B**) A single exponential fit was used to determine an observed rate of incorporation, k_{obs} (% incorporation min^{-1}) of 0.151 ± 0.004, 0.0085 ± 0.001, and 0.016 ± 0.001, for UTP (dotted black line), 2′-F-2′-*C*-MeUTP (dashed red line), and 2′-*C*-MeUTP (solid blue line), respectively. SE is reported as deviation from the fit, $n = 2$.

3.6. Molecular Superpositioning of the ZIKV NS5 Active Site: Consequences for 2′-*C*-Derivatitized Nucleoside Analogues

A previously prepared model (gc-o3 [36]) of ZIKV RdRP in complex with RNA and ATP generated from PDB 5TFR [37] and PDB 4WTD [34], an apo structure of ZIKV RdRP (PDB 5WZ3 [38]), and an apo structure of HCV RdRP (PDB 3MWV [39]) were aligned with a crystal structure of HCV RdRP complexed with RNA and sofosbuvir diphosphate (PDB 4WTG [34]), shown in Figure 5. ZIKV RdRP is shown in magenta (apo form) or in red (in complex with ligands), and HCV RdRP is shown in dark blue (apo form) or in cyan (in complex). The crystal structure of inhibited HCV RdRP (cyan) shows that residue D225 moved upward to accommodate sofosbuvir diphosphate, while in the ZIKV model (red) [36], the corresponding residue D1141 in ZIKV RdRP pointed down in the opposite direction to accommodate the nucleotide (Figure 5). While the position of the aspartic acid in both the apo forms of the viral RdRPs was similar, it is interesting that the nucleoside ligands were accommodated in different ways, with ZIKV RdRP displaying a larger change in positioning in this residue. While modeling was used to generate the holo-structure of ZIKV RdRP, this suggests some evidence that the ZIKV binding pocket is poised to accommodate a variety of nucleoside analogues.

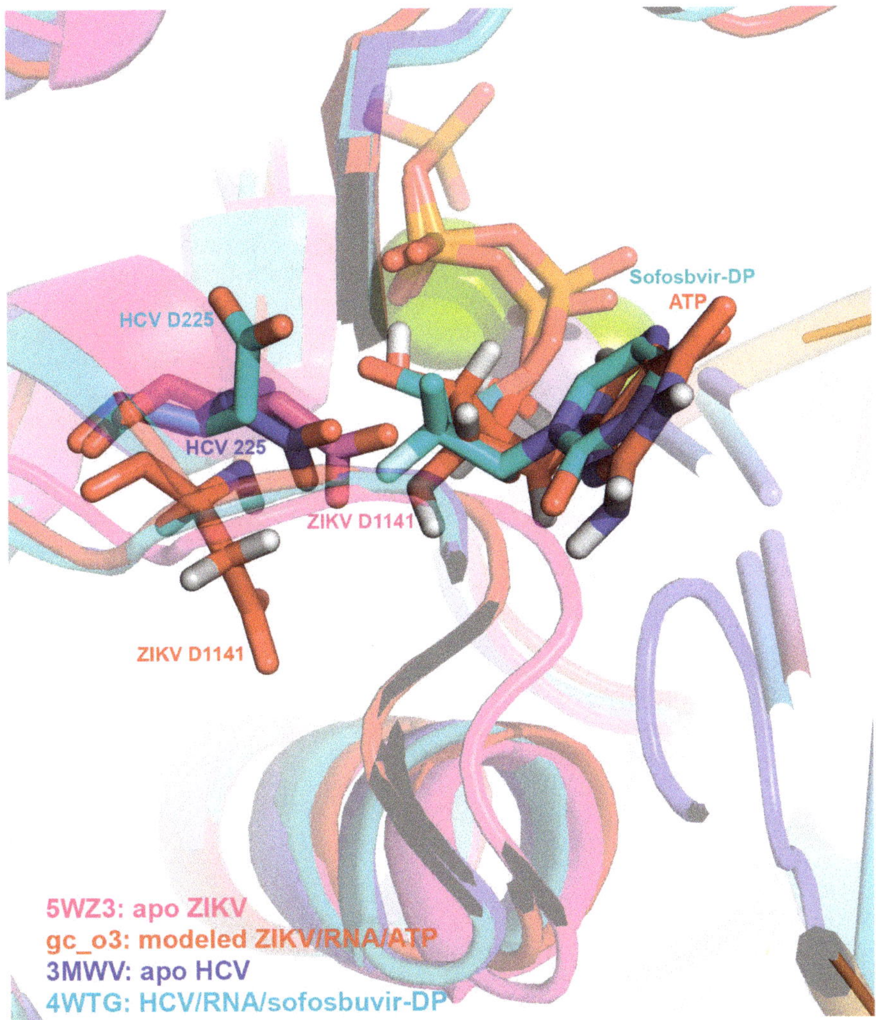

Figure 5. Structural superpositioning of apo and bound ZIKV RdRP and hepatitis C virus (HCV) RdRP. The ZIKV RdRP model gc-o3 [34] in complex with two manganese ions (green), ATP (red/CPK coloring), and RNA (orange/CPK coloring) is shown in red, and a crystal structure of apo ZIKV RdRP is shown in magenta (PDB 5WZ3 [38]). A crystal structure of HCV RdRP complexed with sofosbuvir diphosphate (cyan/CPK coloring), RNA (orange/CPK coloring), and a manganese ion (purple) is shown in cyan (PDB 4WTG [36]), and an apo structure of HCV RdRP is shown in blue (PDB 3MWV [39]). A comparison of the homologous residue D225 (HCV RdRP) and D1141 (ZIKV RdRP) is highlighted in stick form.

4. Discussion

In this work, we demonstrated that ProTides other than sofosbuvir have promise as potential antiviral drugs to treat ZIKV infections. Interestingly, an apparent bias towards 2'-*C*-modified uridylate ProTides (with the exception of the ethenyl derivative) was observed, suggesting a possible nucleobase bias to compound activity. Recent biochemical data examining levels of off-target incorporation of nucleoside analogue triphosphates by the human mitochondrial RNA polymerase may help explain the phenotypic relationship we observed; both 2′-*C*-methyluridine triphosphate and 2′-*C*-methyl-2′-*C*-fluorouridine triphosphate (the active metabolite of sofosbuvir) show much lower levels of incorporation by this host enzyme than 2′-*C*-modified analogues with other nucleobases [51]. We propose that this may account for the better safety profile observed for sofosbuvir compared to other 2′-*C*-modified compounds that were tested in clinical trials. This may also suggest that there is greater selectivity for 2′-*C*-modified uridylate incorporation by flavivirus polymerases over host polymerases, though the exact mechanism by which this occurs should be examined in future studies.

We also observed that changing the ProTide functionality of 2′-*C*-methyluridine from an aryloxyl phosphoramidate masking group to a 2-(methylthio)ethyl phosphoramidate resulted in a complete loss of activity, suggesting that the local metabolic environment of the host cell is critical for processing of the ProTide functionality. Further studies examining the relationship between different ProTide functionalities and the cell type dependence of their antiviral activity as well as testing the impact of these effects using in vivo animal model testing are planned in our research groups.

We have demonstrated here for the first time that ProTide technology can be broadly applied to anti-ZIKV drug development with nucleobase and ProTide group selectivity observed. These data will guide future work to design highly selective, safe, and bioavailable compounds for the treatment of ZIKV infections.

Author Contributions: J.A.B., M.C., S.B., G.A.W., L.A.L., A.E.C., Z.Z., J.N.R., C.D.S., B.W.P., and J.L.S.-N. conceived and designed the experiments; J.A.B., M.C., S.B., G.A.W., L.A.L., D.H., A.E.C., and Z.Z. performed experiments; J.A.B., M.C., S.B., G.A.W., C.D.S., B.W.P., and J.L.S.N. analyzed the data; J.A.B., M.C., S.B., G.A.W., C.D.S., B.W.P., and J.L.S.-N. wrote the original draft of manuscript; J.A.B., M.C., S.B., G.A.W., L.A.L., A.E.C., Z.Z., D.H., J.N.R., C.D.S., B.W.P., and J.L.S.-N. reviewed and edited the manuscript.

Funding: This research was supported by Clinical and Translational (CTRI) pilot grant UL1TR001442 (to J.L.S.-N.), and a CSUPERB New Investigator grant (to C.D.S). B.W.P., M.C., G.A.W., L.A.L., and C.D.S. thank San Diego State University and the California Metabolic Research Foundation for financial support.

Acknowledgments: The authors acknowledge the access and use of the UCSD Screening Core to perform part of the experiments presented in this manuscript.

Conflicts of Interest: The authors declare no conflict of interest. The funders had no role in the design of the study; in the collection, analyses, or interpretation of data; in the writing of the manuscript, or in the decision to publish the results.

References

1. De Araújo, T.V.B.; de Alencar Ximenes, R.A.; de Barros Miranda-Filho, D.; Souza, W.V.; Montarroyos, U.R.; de Melo, A.P.L.; Valongueiro, S.; Braga, C.; Brandão Filho, S.P.; Cordeiro, M.T.; et al. Association between microcephaly, Zika virus infection, and other risk factors in Brazil: Final report of a case-control study. *Lancet Infect. Dis.* **2018**, *18*, 328–336.

2. Pierson, T.C.; Diamond, M.S. The emergence of Zika virus and its new clinical syndromes. *Nature* **2018**, *560*, 573–581. [CrossRef]

3. Ferrero, D.; Ferrer-Orta, C.; Verdaguer, N. Viral RNA-Dependent RNA Polymerases: A Structural Overview. *Subcell. Biochem.* **2018**, *88*, 39–71. [PubMed]

4. Zarrouk, K.; Piret, J.; Boivin, G. Herpesvirus DNA polymerases: Structures, functions and inhibitors. *Virus Res.* **2017**, *234*, 177–192. [CrossRef]

5. Menéndez-Arias, L.; Sebastián-Martín, A.; Álvarez, M. Viral reverse transcriptases. *Virus Res.* **2017**, *234*, 153–176. [CrossRef] [PubMed]

6. Mayberry, J.; Lee, W.M. The revolution in treatment of hepatitis C. *Med. Clin. N. Am.* **2019**, *103*, 43–55. [CrossRef]

7. Sluis-Cremer, N.; Wainberg, M.A.; Schinazi, R.F. Resistance to reverse transcriptase inhibitors used in the treatment and prevention of HIV-1 infection. *Future Microbiol.* **2015**, *10*, 1773–1782. [CrossRef] [PubMed]

8. McGuigan, C.; Pathirana, R.N.; Mahmood, N.; Devine, K.G.; Hay, A.J. Aryl phosphate derivatives of AZT retain activity against HIV1 in cell lines which are resistant to the action of AZT. *Antivir. Res.* **1992**, *17*, 311–321. [CrossRef]

9.   Cahard, D.; McGuigan, C.; Balzarini, J. Aryloxy phosphoramidate triesters as pro-tides. *Mini. Rev. Med. Chem.* **2004**, *4*, 371–381. [CrossRef]

10.   Mehellou, Y.; Balzarini, J.; McGuigan, C. Aryloxy phosphoramidate triesters: A technology for delivering monophosphorylated nucleosides and sugars into cells. *Chem. Med. Chem.* **2009**, *4*, 1779–1791. [CrossRef] [PubMed]

11.   Ovadia, R.; Khalil, A.; Li, H.; De Schutter, C.; Mengshetti, S.; Zhou, S.; Bassit, L.; Coats, S.J.; Amblard, F.; Schinazi, R.F. Synthesis and anti-HCV activity of β-D -2′-deoxy-2′-α-chloro-2′-β-fluoro and β-d-2′-deoxy-2′-α-bromo-2′-β-fluoro nucleosides and their phosphoramidate prodrugs. *Bioorg. Med. Chem.* **2019**, *27*, 664–676. [CrossRef] [PubMed]

12.   McGuigan, C.; Harris, S.A.; Daluge, S.M.; Gudmundsson, K.S.; McLean, E.W.; Burnette, T.C.; Marr, H.; Hazen, R.; Condreay, L.D.; Johnson, L.; et al. Application of phosphoramidate pronucleotide technology to abacavir leads to a significant enhancement of antiviral potency. *J. Med. Chem.* **2005**, *48*, 3504–3515. [CrossRef] [PubMed]

13.   Murakami, E.; Tolstykh, T.; Bao, H.; Niu, C.; Steuer, H.M.M.; Bao, D.; Chang, W.; Espiritu, C.; Bansal, S.; Lam, A.M.; et al. Mechanism of activation of PSI-7851 and its diastereoisomer PSI-7977. *J. Biol. Chem.* **2010**, *285*, 34337–34347. [CrossRef]

14.   Bullard-Feibelman, K.M.; Govero, J.; Zhu, Z.; Salazar, V.; Veselinovic, M.; Diamond, M.S.; Geiss, B.J. The FDA-approved drug sofosbuvir inhibits Zika virus infection. *Antiviral Res.* **2017**, *137*, 134–140. [CrossRef] [PubMed]

15.   Ferreira, A.C.; Zaverucha-do-Valle, C.; Reis, P.A.; Barbosa-Lima, G.; Vieira, Y.R.; Mattos, M.; de Paiva Silva, P; Sacramento, C.; de Castro Faria Neto, H.C.; Campanati, L.; et al. Sofosbuvir protects Zika virus-infected mice from mortality, preventing short- and long-term sequelae. *Sci. Rep.* **2017**, *7*, 9409. [CrossRef]

16.   Mesci, P.; Macia, A.; Moore, S.M.; Shiryaev, S.A.; Pinto, A.; Huang, C.-T.; Tejwani, L.; Fernandes, I.R.; Suarez, N.A.; Kolar, M.J.; et al. Blocking Zika virus vertical transmission. *Sci. Rep.* **2018**, *8*, 1218. [CrossRef] [PubMed]

17.   Mumtaz, N.; Jimmerson, L.C.; Bushman, L.R.; Kiser, J.J.; Aron, G.; Reusken, C.B.E.M.; Koopmans, M.P.G.; van Kampen, J.J.A. Cell-line dependent antiviral activity of sofosbuvir against Zika virus. *Antiviral Res.* **2017**, *146*, 161–163. [CrossRef]

18.   Eyer, L.; Nencka, R.; Huvarová, I.; Palus, M.; Joao Alves, M.; Gould, E.A.; De Clercq, E.; Růžek, D. Nucleoside inhibitors of zika virus. *J. Infect. Dis.* **2016**, *214*, 707–711. [CrossRef]

19.   Lu, G.; Bluemling, G.R.; Collop, P.; Hager, M.; Kuiper, D.; Gurale, B.P.; Painter, G.R.; De La Rosa, A.; Kolykhalov, A.A. Analysis of Ribonucleotide 5′-Triphosphate Analogs as Potential Inhibitors of Zika Virus RNA-Dependent RNA Polymerase by Using Nonradioactive Polymerase Assays. *Antimicrob. Agents Chemother.* **2017**, *61*. [CrossRef] [PubMed]

20.   Hercík, K.; Kozak, J.; Šála, M.; Dejmek, M.; Hřebabecký, H.; Zborníková, E.; Smola, M.; Ruzek, D.; Nencka, R.; Boura, E. Adenosine triphosphate analogs can efficiently inhibit the Zika virus RNA-dependent RNA polymerase. *Antiviral Res.* **2017**, *137*, 131–133. [CrossRef] [PubMed]

21.   Deng, Y.-Q.; Zhang, N.-N.; Li, C.-F.; Tian, M.; Hao, J.-N.; Xie, X.-P.; Shi, P.-Y.; Qin, C.-F. Adenosine analog NITD008 is a potent inhibitor of zika virus. *Open Forum Infect. Dis.* **2016**, *3*, ofw175. [CrossRef]

22.   Zmurko, J.; Marques, R.E.; Schols, D.; Verbeken, E.; Kaptein, S.J.F.; Neyts, J. The Viral Polymerase Inhibitor 7-Deaza-2′-C-Methyladenosine Is a Potent Inhibitor of In Vitro Zika Virus Replication and Delays Disease Progression in a Robust Mouse Infection Model. *PLoS Negl. Trop. Dis.* **2016**, *10*, e0004695. [CrossRef] [PubMed]

23.   Kamiyama, N.; Soma, R.; Hidano, S.; Watanabe, K.; Umekita, H.; Fukuda, C.; Noguchi, K.; Gendo, Y.; Ozaki, T.; Sonoda, A.; et al. Ribavirin inhibits Zika virus (ZIKV) replication in vitro and suppresses viremia in ZIKV-infected STAT1-deficient mice. *Antiviral Res.* **2017**, *146*, 1–11. [CrossRef] [PubMed]

24.   Bernatchez, J.A.; Yang, Z.; Coste, M.; Li, J.; Beck, S.; Liu, Y.; Clark, A.E.; Zhu, Z.; Luna, L.A.; Sohl, C.D.; et al. Development and validation of a phenotypic high-content imaging assay for assessing the antiviral activity of small-molecule inhibitors targeting the Zika virus. *Antimicrob. Agents Chemother.* **2018**, *62*. [CrossRef]

25.   Ross, B.S.; Reddy, P.G.; Zhang, H.-R.; Rachakonda, S.; Sofia, M.J. Synthesis of diastereomerically pure nucleotide phosphoramidates. *J. Org. Chem.* **2011**, *76*, 8311–8319. [CrossRef]

26. Okon, A.; Han, J.; Dawadi, S.; Demosthenous, C.; Aldrich, C.C.; Gupta, M.; Wagner, C.R. Anchimerically Activated ProTides as Inhibitors of Cap-Dependent Translation and Inducers of Chemosensitization in Mantle Cell Lymphoma. *J. Med. Chem.* **2017**, *60*, 8131–8144. [CrossRef] [PubMed]

27. Anandan, S.K.; Aulakh, V.S.; Fenaux, M.; Lin, X.; Mao, L.; Saunders, O.; Sweeney, Z.K.; Yokokawa, F.; Zhong, W. 2′-Ethynyl Nucleoside Derivatives for Treatment of Viral Infections 2017. Available online: https://app.dimensions.ai/details/patent/US-9814739-B2 (accessed on 1 March 2019).

28. Itoh, T.; Melik-Ohanjanian, R.G.; Ishikawa, I.; Kawahara, N.; Mizuno, Y.; Honma, Y.; Hozumi, M.; Ogura, H. Improved procedures for the syntheses of pyrido-and pyrrolo(2,3-d) pyrimidines, and ribosides thereof. *Chem. Pharm. Bull.* **1989**, *37*, 3184–3190. [CrossRef]

29. Zhu, Z.; Gorman, M.J.; McKenzie, L.D.; Chai, J.N.; Hubert, C.G.; Prager, B.C.; Fernandez, E.; Richner, J.M.; Zhang, R.; Shan, C.; et al. Zika virus has oncolytic activity against glioblastoma stem cells. *J. Exp. Med.* **2017**, *214*, 2843–2857. [CrossRef]

30. Arnold, J.J.B.; Ghosh, S.K.; Bevilacqua, P.C.; Cameron, C.E. Single-nucleotide resolution of RNA strands in the presence of their RNA complements. *BioTechniques* **1999**, *27*, 450–456.

31. Potisopon, S.; Ferron, F.; Fattorini, V.; Selisko, B.; Canard, B. Substrate selectivity of Dengue and Zika virus NS5 polymerase towards 2′-modified nucleotide analogues. *Antiviral Res.* **2017**, *140*, 25–36. [CrossRef]

32. Jin, Z.; Leveque, V.; Ma, H.; Johnson, K.A.; Klumpp, K. Assembly, purification, and pre-steady-state kinetic analysis of active RNA-dependent RNA polymerase elongation complex. *J. Biol. Chem.* **2012**, *287*, 10674–10683. [CrossRef] [PubMed]

33. Kamkaew, M.; Chimnaronk, S. Characterization of soluble RNA-dependent RNA polymerase from dengue virus serotype 2: The polyhistidine tag compromises the polymerase activity. *Protein Expr. Purif.* **2015**, *112*, 43–49. [CrossRef] [PubMed]

34. Appleby, T.C.; Perry, J.K.; Murakami, E.; Barauskas, O.; Feng, J.; Cho, A.; Fox, D.; Wetmore, D.R.; McGrath, M.E.; Ray, A.S.; et al. Viral replication. Structural basis for RNA replication by the hepatitis C virus polymerase. *Science* **2015**, *347*, 771–775. [CrossRef] [PubMed]

35. Schrodinger PyMOL: The PyMOL Molecular Graphics System, Version 1.8, Schrödinger, LLC. Available online: http://www.sciepub.com/reference/159710 (accessed on 1 March 2019).

36. Šebera, J.; Dubankova, A.; Sychrovský, V.; Ruzek, D.; Boura, E.; Nencka, R. The structural model of Zika virus RNA-dependent RNA polymerase in complex with RNA for rational design of novel nucleotide inhibitors. *Sci. Rep.* **2018**, *8*, 11132. [CrossRef] [PubMed]

37. Upadhyay, A.K.; Cyr, M.; Longenecker, K.; Tripathi, R.; Sun, C.; Kempf, D.J. Crystal structure of full-length Zika virus NS5 protein reveals a conformation similar to Japanese encephalitis virus NS5. *Acta Crystallogr. Sect. F Struct. Biol. Commun.* **2017**, *73*, 116–122. [CrossRef]

38. Duan, W.; Song, H.; Wang, H.; Chai, Y.; Su, C.; Qi, J.; Shi, Y.; Gao, G.F. The crystal structure of Zika virus NS5 reveals conserved drug targets. *EMBO J.* **2017**, *36*, 919–933. [CrossRef]

39. LaPlante, S.R.; Gillard, J.R.; Jakalian, A.; Aubry, N.; Coulombe, R.; Brochu, C.; Tsantrizos, Y.S.; Poirier, M.; Kukolj, G.; Beaulieu, P.L. Importance of ligand bioactive conformation in the discovery of potent indole-diamide inhibitors of the hepatitis C virus NS5B. *J. Am. Chem. Soc.* **2010**, *132*, 15204–15212. [CrossRef]

40. De Clercq, E.; Neyts, J. Antiviral agents acting as DNA or RNA chain terminators. *Handb. Exp. Pharmacol.* **2009**, *189*, 53–84.

41. Bonnac, L.F.; Mansky, L.M.; Patterson, S.E. Structure-activity relationships and design of viral mutagens and application to lethal mutagenesis. *J. Med. Chem.* **2013**, *56*, 9403–9414. [CrossRef]

42. Sofia, M.J. Nucleotide prodrugs for the treatment of HCV infection. *Adv. Pharmacol.* **2013**, *67*, 39–73.

43. Mehellou, Y.; Rattan, H.S.; Balzarini, J. The protide prodrug technology: From the concept to the clinic. *J. Med. Chem.* **2018**, *61*, 2211–2226. [CrossRef] [PubMed]

44. Eldrup, A.B.; Allerson, C.R.; Bennett, C.F.; Bera, S.; Bhat, B.; Bhat, N.; Bosserman, M.R.; Brooks, J.; Burlein, C.; Carroll, S.S.; et al. Structure-activity relationship of purine ribonucleosides for inhibition of hepatitis C virus RNA-dependent RNA polymerase. *J. Med. Chem.* **2004**, *47*, 2283–2295. [CrossRef] [PubMed]

45. Wang, G.; Lim, S.P.; Chen, Y.-L.; Hunziker, J.; Rao, R.; Gu, F.; Seh, C.C.; Ghafar, N.A.; Xu, H.; Chan, K.; et al. Structure-activity relationship of uridine-based nucleoside phosphoramidate prodrugs for inhibition of dengue virus RNA-dependent RNA polymerase. *Bioorg. Med. Chem. Lett.* **2018**, *28*, 2324–2327. [CrossRef] [PubMed]

46. Jin, Z.; Kinkade, A.; Behera, I.; Chaudhuri, S.; Tucker, K.; Dyatkina, N.; Rajwanshi, V.K.; Wang, G.; Jekle, A.; Smith, D.B.; et al. Structure-activity relationship analysis of mitochondrial toxicity caused by antiviral ribonucleoside analogs. *Antiviral Res.* **2017**, *143*, 151–161. [CrossRef] [PubMed]

47. Kaid, C.; Goulart, E.; Caires-Júnior, L.C.; Araujo, B.H.S.; Soares-Schanoski, A.; Bueno, H.M.S.; Telles-Silva, K.A.; Astray, R.M.; Assoni, A.F.; Júnior, A.F.R.; et al. Zika virus selectively kills aggressive human embryonal CNS tumor cells in vitro and in vivo. *Cancer Res.* **2018**, *78*, 3363–3374. [CrossRef]

48. Mazar, J.; Li, Y.; Rosado, A.; Phelan, P.; Kedarinath, K.; Parks, G.D.; Alexander, K.A.; Westmoreland, T.J. Zika virus as an oncolytic treatment of human neuroblastoma cells requires CD24. *PLoS ONE* **2018**, *13*, e0200358. [CrossRef]

49. Chen, Q.; Wu, J.; Ye, Q.; Ma, F.; Zhu, Q.; Wu, Y.; Shan, C.; Xie, X.; Li, D.; Zhan, X.; et al. Treatment of Human Glioblastoma with a Live Attenuated Zika Virus Vaccine Candidate. *MBio* **2018**, *9*. [CrossRef]

50. Xu, H.-T.; Hassounah, S.A.; Colby-Germinario, S.P.; Oliveira, M.; Fogarty, C.; Quan, Y.; Han, Y.; Golubkov, O.; Ibanescu, I.; Brenner, B.; et al. Purification of Zika virus RNA-dependent RNA polymerase and its use to identify small-molecule Zika inhibitors. *J. Antimicrob. Chemother.* **2017**, *72*, 727–734. [CrossRef] [PubMed]

51. Ehteshami, M.; Zhou, L.; Amiralaei, S.; Shelton, J.R.; Cho, J.H.; Zhang, H.; Li, H.; Lu, X.; Ozturk, T.; Stanton, R.; et al. Nucleotide Substrate Specificity of Anti-Hepatitis C Virus Nucleoside Analogs for Human Mitochondrial RNA Polymerase. *Antimicrob. Agents Chemother.* **2017**, *61*. [CrossRef]

Article

Decanoyl-Arg-Val-Lys-Arg-Chloromethylketone: An Antiviral Compound That Acts against Flaviviruses through the Inhibition of Furin-Mediated prM Cleavage

Muhammad Imran [1,2,3,4], **Muhammad Kashif Saleemi** [4], **Zheng Chen** [1,2,3], **Xugang Wang** [1,2,3], **Dengyuan Zhou** [1,2,3], **Yunchuan Li** [1,2,3], **Zikai Zhao** [1,2,3], **Bohan Zheng** [1,2,3], **Qiuyan Li** [1,2,3], **Shengbo Cao** [1,2,3,*] and **Jing Ye** [1,2,3,*]

[1]   State Key Laboratory of Agricultural Microbiology, Huazhong Agricultural University, Wuhan 430070, Hubei, China; dr.mimran@uaf.edu.pk (M.I.); chenzheng19860227@163.com (Z.C.); wangxugang@webmail.hzau.edu.cn (X.W.); zhoudy6@webmail.hzau.edu.cn (D.Z.); liyunchuan@webmail.hzau.edu.cn (Y.L.); zikaizhao@hotmail.com (Z.Z.); zhengbohan@webmail.hzau.edu.cn (B.Z.); lqylqy6@webmail.hzau.edu.cn (Q.L.)

[2]   Key Laboratory of Preventive Veterinary Medicine in Hubei Province, College of Veterinary Medicine, Huazhong Agricultural University, Wuhan 430070, Hubei, China

[3]   The Cooperative Innovation Center for Sustainable Pig Production, Huazhong Agricultural University, Wuhan 430070, Hubei, China

[4]   Department of Pathology, Faculty of Veterinary Science, University of Agriculture, Faisalabad 38040, Pakistan; drkashif313@gmail.com

*   Correspondence: sbcao@mail.hzau.edu.cn (S.C.); yej@mail.hzau.edu.cn (J.Y.)

Received: 19 September 2019; Accepted: 30 October 2019; Published: 31 October 2019

Abstract: Flaviviruses, such as Zika virus (ZIKV), Japanese encephalitis virus (JEV), Dengue virus (DENV), and West Nile virus (WNV), are important arthropod-borne pathogens that present an immense global health problem. Their unpredictable disease severity, unusual clinical features, and severe neurological manifestations underscore an urgent need for antiviral interventions. Furin, a host proprotein convertase, is a key contender in processing flavivirus prM protein to M protein, turning the inert virus to an infectious particle. For this reason, the current study was planned to evaluate the antiviral activity of decanoyl-Arg-Val-Lys-Arg-chloromethylketone, a specific furin inhibitor, against flaviviruses, including ZIKV and JEV. Analysis of viral proteins revealed a significant increase in the prM/E index of ZIKV or JEV in dec-RVKR-cmk-treated Vero cells compared to DMSO-treated control cells, indicating dec-RVKR-cmk inhibits prM cleavage. Plaque assay, qRT-PCR, and immunofluorescence assay revealed a strong antiviral activity of dec-RVKR-cmk against ZIKV and JEV in terms of the reduction in virus progeny titer and in viral RNA and protein production in both mammalian cells and mosquito cells. Time-of-drug addition assay revealed that the maximum reduction of virus titer was observed in post-infection treatment. Furthermore, our results showed that dec-RVKR-cmk exerts its inhibitory action on the virus release and next round infectivity but not on viral RNA replication. Taken together, our study highlights an interesting antiviral activity of dec-RVKR-cmk against flaviviruses.

Keywords: flavivirus; Zika virus; Japanese encephalitis virus; furin inhibitor; precursor membrane protein

---

1. Introduction

Flaviviruses are arthropod-borne pathogens (arboviruses) that pose a serious threat to global health and cause millions of infections annually. Regardless of extensive research and public health issues, currently there are no specific antiviral treatments in clinical use for flavivirus infections and, despite licensed vaccines, outbreaks still occur, highlighting challenges in executing effective control measures [1]. Some flaviviruses, like Japanese encephalitis virus (JEV), West Nile virus (WNV), and tick-borne encephalitis virus (TBEV) are associated with severe encephalitis; dengue virus (DENV) and yellow fever (YFV), cause hemorrhagic fever; and the newly emerging Zika virus may cause microcephaly in neonates and Guillain–Barré syndrome in adults [2,3]. Flaviviruses are enveloped, icosahedral with positive sense single-stranded RNA viruses that enter host cells by receptor-mediated endocytosis and transport to endosomes, where an acidic condition triggers conformational changes in the envelope (E) protein that prompt virus and host cell membrane fusion [4]. The released genomic RNA is translated to polyprotein precursor of about 3.4 k amino acids in length. This polypeptide gives rise to three structural (core (C), precursor membrane (prM), and envelope (E)) and seven nonstructural (NS) (NS1, NS2A, NS2B, NS3, NS4A, NS4B, and NS5) proteins by host cell signalases and virus-encoded proteases [5]. NS proteins are mainly involved in virus replication and evasion from host immune response while structural proteins are responsible for virus assembly and successful viral entrance into and exit from host cells [6,7].

Immature virions assemble in the endoplasmic reticulum (ER). These non-infectious, immature viral particles contain heterodimers of E and prM proteins [8,9]. Maturation of virions occurs after the exocytic pathway in the trans-Golgi network (TGN) through the host cell endoprotease, called furin [10,11]. This calcium-dependent endoprotease cleaves prM to M protein in an acidic environment of the TGN after recognition of the sequence R-X-R/K-R in all flaviviruses. Mature particles then become infectious and are released in the extracellular environment by exocytosis [12]. This furin-mediated cleavage of prM is a critical step for flavivirus assembly and maturation [13]. Inhibition of furin functioning during the viral life cycle may debilitate flavivirus infectivity and pathogenicity.

Decanoyl-Arg-Val-Lys-Arg-chloromethylketone (dec-RVKR-cmk) and hexa-D-arginine (D6R) are small synthetic furin inhibitors that are suitable for clinical purposes. CMK has now been used by many groups as a reference inhibitor to study the effect of furin and related proprotein convertases. It also significantly inhibits viral infection because of its capacity to irreversibly block furin [14,15]. As reported previously, CMK is more effective than D6R in the reduction of HBV replication by inhibiting furin-mediated processing of the hepatitis B e antigen (HBeAg) precursor into mature HBeAg [16]. Dec-RVKR-cmk is a small, synthetic, irreversible, and cell-permeable competitive inhibitor of all proprotein convertases (PC1, PC2, PC4, PACE 4, PC5, PC7, and furin). This peptidyl chloromethylketone is reported to inhibit furin-mediated cleavage and fusion activity of viral glycoproteins, and acts as an antiviral agent against different viruses, including human immunodeficiency virus [17], Chikungunya virus [18], chronic hepatitis B virus, influenza A, Ebola virus infection [16], duck hepatitis B virus [19], and papilloma virus [20]. The structural-activity relationship of dec-RVKR-cmk was studied by Becker and colleagues by replacing the P1-arginine group with 4-amidinobenzylamide and the N-terminal deconyl group with phenyl acetyl group, and they derived a new compound, named phenylacetyl-Arg-Val-Arg-4-amidinobenzylamide, that exhibited more potency to inhibit the cleavage of hemagglutinin of fowl plague virus compared to dec-RVKR-cmk [15]. Smith et al. and Steinmetzer et al. also patented a peptidomimetic furin inhibitor by modifying the C-terminal of dec-RVKR-cmk with decarboxylated arginine mimetics, resulting in highly potent furin inhibitors [21].

There has been skepticism about whether a wide-range inhibitor of furin and other proprotein convertases (PCs) would interfere with normal cellular processes [22,23]. In the case of the PCs family, the concept of redundancy was observed, which constitutes an advantage over the other protease families. According to this concept, it is assumed that the inhibitory effect of a PC inhibitor in normal cells would be minimized by the redundant actions of other co-expressed PCs. Although the mechanism of the redundancy concept is still not well defined [23], this concept has been extensively validated in vitro, and recently has been explored in an animal study. It was observed that liver-specific furin knockout mice showed no obvious adverse effects, thus suggesting that the redundancy effects of other PCs can compensate for the molecular ablation of furin in normal cellular process [24]. Recent developments in medicinal chemistry have explored whether a peptide-based inhibitor has overcome the particular issues associated to their use, bioavailability, and toxicity [23].

The furin-mediated flavivirus maturation encouraged us to evaluate the therapeutic potential of dec-RVKR-cmk against flaviviruses. Our results highlight the efficacy of dec-RVKR-cmk as an interesting anti-flavivirus agent with significant antiviral activities at a non-cytotoxic concentration, suggesting dec-RVKR-cmk as a potential candidate for the treatment of flavivirus.

2. Materials and Methods

2.1. Cell Culture and Virus

African green monkey kidney cells (Vero, ATCC-CCL-81) and baby hamster kidney fibroblast cells (BHK-21, ATCC-CCL-10) were purchased from the American Type Culture Collection (ATCC) and cultured and maintained in Dulbecco's modified Eagle's medium (DMEM) supplemented by 10% fetal bovine serum (FBS), 100 U/mL penicillin, and 100 mg/mL streptomycin in a 5% CO_2 incubator at 37 °C. *Aedes albopictus* C6/36 cells (ATCC CRL-1660) were cultured and maintained in Roswell Park Memorial Institute (RPMI) 1640 medium supplemented with 10% fetal bovine serum (FBS), 100 U/mL penicillin, and 100 mg/mL streptomycin in a 5% CO_2 incubator at 27 °C. The JEV P3 strain (GenBank: U47032.1) was stored in our laboratory and was propagated and titrated on BHK-21 cells. ZIKV-MR-766 strain (GenBank: AY632535.2) was kindly provided by Dr. Xiaowu Pang (College of Dentistry, Howard University, USA) and was propagated and titrated on Vero cells.

2.2. Reagents

Dec-RVKR-cmk was purchased from Cayman Chemical (Ann Arbor, Michigan, USA). A stock solution was prepared in dimethyl sulfoxide (DMSO) with a solubility of 33 mg/mL. Further dilutions of this stock solution were made in DMEM prior to performing biological experiments. The structure of dec-RVKR-cmk is shown in Figure 1A. Antibodies against ZIKV prM were purchased from GeneTex (2456 Alton Pkwy Irvine, CA 92606 USA). The monoclonal antibodies against ZIKV (E, NS5) and JEV (prM, E, NS5) were generated in our laboratory. Anti-mouse and anti-rabbit IgG secondary antibodies conjugated with horse reddish peroxidase were purchased from Boster (Wuhan, China).

2.3. Cell Viability Assay and Efficacy Study of dec-RVKR-cmk

The cytotoxic concentration 50 (CC50) of dec-RVKR-cmk was determined using the CellTiter-GLO One Solution Assay kit (Promega). This assay was used to detect the viability of cultured cells on the basis of ATP quantification of cells. Briefly, Vero and C6/36 cells were seeded (10,000 cells per well) in a 96-well plate, 24 h before compound treatment. Culture supernatants were replaced with different concentrations of dec-RVKR-cmk or DMSO. Each concentration was tested in triplicate. After 72 h, cells were washed with phosphate-buffered saline (PBS) and an equal volume of (100 μL) CellTiter-GLO reagent was added to each well. For appropriate cell lysis, cells were agitated in a shaker for 2 min and then incubated for 10 min at room temperature. A multimode plate reader was used to quantitate luminescence signals in each condition and then the luminescence value was compared with its corresponding DMSO control. The efficacy of dec-RVKR-cmk against ZIKV (0.2 MOI) and JEV (0.2 MOI) was studied by using different concentrations (1, 10, 50, and 100 μM). The inhibitory concentration 50 (IC50) of dec-RVKR-cmk was determined by counting visible plaques produced by ZIKV or JEV. Both CC50 and IC50 were calculated by non-linear regression model using GraphPad prism7.

2.4. Immunofluorescence Assay (IFA)

Vero and C6/36 cells were infected with ZIKV or JEV-P3 at a multiplicity of infection as indicated in the results section for 1 h and the media were replaced with different concentrations of dec-RVKR-cmk or DMSO. Cells were fixed at various time points with ice-cold methanol for 10 min and then washed with PBS. Afterwards, cells were blocked with 10% bovine serum albumin (BSA) in PBS for 30 min at room temperature. Later, cells were incubated with mouse polyclonal anti-NS5 primary antibody of ZIKV or JEV for 1 h at room temperature. After washing with PBS, cells were stained with a second antibody (Alexa Fluor 488) for 30 min at room temperature. Cell nuclei were stained with 6-diamidino-2-phenyl indole (DAPI; Invitrogen). The cells were observed under fluorescence microscope (Zeiss).

2.5. Plaque Assay

Viral titers in cell culture supernatants were assessed as described in our previous study [25]. Briefly, virus-infected Vero and C6/36 cells were treated with dec-RVKR-cmk and DMSO. After different time points, as indicated in the results section, virus-containing cell culture supernatants were removed, serially diluted in DMEM, and adsorbed on a Vero (ZIKV) or BHK-21 (JEV) monolayer for 1 h. Afterwards, unbound viral particles were washed and overlaid with 2% carboxymethyl cellulose (CMC). After 5 days of incubation, cells were fixed with 10% formaldehyde for 12 h and then stained with 0.1% crystal violet for 6 h. The visible plaques were counted and viral loads were measured as plaque-forming unit (PFU) per ml of supernatant. All data are expressed as the means of triplicate samples.

2.6. Time-of-Drug Addition Assay

ZIKV- (0.2 MOI) and JEV- (0.2 MOI) infected Vero cells were treated with dec-RVKR-cmk under the following conditions: 1 h prior to infection, at the time of infection, or 1, 6, and 12 h post infection (hpi). Regardless of treatment time, cells were infected for 1 h. After 1 h, infectious media were replaced with fresh media and dec-RVKR-cmk added at the above time points. Supernatants were collected at 36 hpi to determine viral titer by plaque assay while cells were used to quantify viral genome copies by qRT-PCR.

2.7. Western Blot Analysis

Cells were lysed using radioimmunoprecipitation assay (RIPA) buffer (sigma) containing protease inhibitor (Roche). Samples were mixed with loading buffer and heated at 95 °C for 10 min and then fractionated by SDS-PAGE. Proteins were transferred to a polyvinylidine fluoride membrane (Millipore) using Mini Trans-Blot Cell (Bio-Rad) and blocked with 1% bovine serum albumin. Blots were probed with relevant primary and secondary antibodies and proteins were detected by enhanced chemiluminescent reagent (Thermo Scientific).

2.8. RNA Extraction and Quantitative Real-Time PCR

Trizol reagent (Invitrogen) was used to extract intracellular and extracellular RNA and transcribed into cDNA using a cDNA synthesis kit (Toyobo) according to the manufacturer's instructions. Quantitative real-time PCR was performed using Applied Biosystems 7500 real-time PCR system and TaqMan real-time PCR mix (NovoStart, China). The ZIKV MR-766 E gene or JEV-P3 E gene were used to generate the standard curve for the quantification of viral copy numbers. Each sample was analyzed in triplicate. ZIKV or JEV copy numbers were extrapolated from the generated standard curve using the Applied Biosystems protocol. Primers were as follows: ZIKV, 5′-CCGCTGCCCAACACAAG-3′ (forward) and 5′-CCACTAACGTTCTTTTGCAGACAT-3′ (reverse); ZIKV probe, 5′-AGCCTAACCTTGACAAGCAATCAGACACTCAA-3′; JEV, 5′-TGGTTTCATGACCTCGCTCTC-3′ (forward) and 5′- CCATGAGGAGTTCTCTGTTTCT-3′ (reverse); and JEV probe, 5′-CCTGGACGCCCCCTTCGAGCACAGCGT-3′.

2.9. Statistical Analysis

All experiments were performed at least three times with similar conditions. GraphPad Prism, version 7, was used for data analyses of outcomes. Results are presented as the mean ± standard error (SEM). CC50 and IC50 were calculated by non-linear regression. Viral titers are expressed as the medians. Statistical differences were determined by independent t-test or one-way analysis of variance (ANOVA), with Dunnett's multiple comparison test, and a *p* value < 0.05 was considered as significant.

3. Results

3.1. Cytotoxicity of dec-RVKR-cmk in Vero Cells

First, we examined the cytotoxicity of dec-RVKR-cmk in Vero cells by using luminescence-based cell viability assay. Viable cells were determined on the basis of ATP quantification of cells, which indicates the presence of metabolically active cells. Different concentrations of dec-RVKR-cmk were used against Vero cells and the results revealed that up to a 100 μM concentration dec-RVKR-cmk exhibited no cytotoxic effect, while 500 and 1000 μM concentrations were significantly toxic to cells (Figure 1b). The CC50 of dec-RVKR-cmk, which is the concentration that results in 50% cell viability, was determined to be 712.9 μM (log CC50 = 2.853) (Figure 1c).

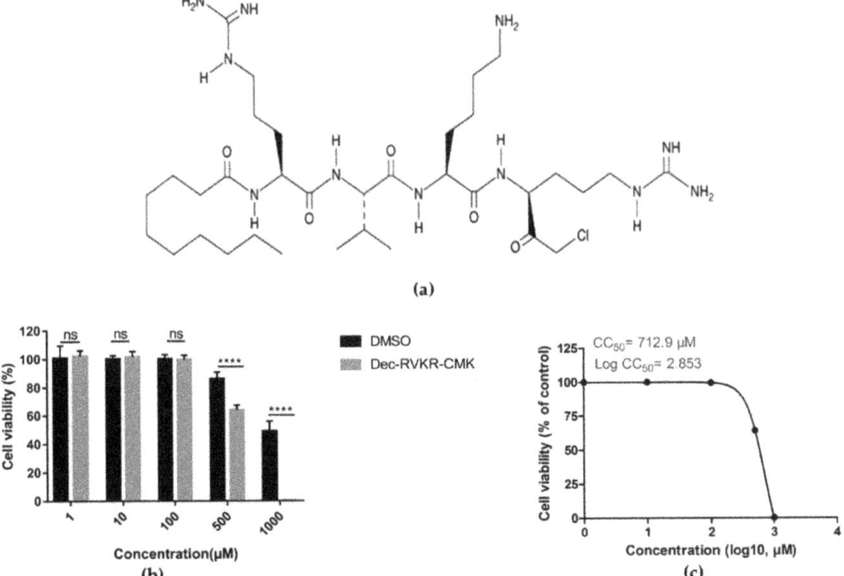

Figure 1. Determination of cytotoxicity of dec-Arg-Val-Lys-Arg-cmk on Vero cells. (**a**) Chemical structure of dec-RVKR-cmk. (**b**) Cytotoxicity of dec-RVKR-cmk on Vero cells determined by CellTiter-GLO One Solution Assay kit (Promega). (**c**) The CC50 value was calculated from GraphPad Prism using non-linear regression analysis. Data are presented as mean ± SEM from three independent experiments.

3.2. Dec-RVKR-cmk Inhibits prM Cleavage during ZIKV and JEV Infection

Flavivirus maturation is associated with the status of the M protein of viral particles, where the prM precursor protein is cleaved in TGN by the host proprotein convertase furin protease. Therefore, cleavage of the prM protein was analyzed by Western blotting (WB) at 36 hpi from Vero cells infected with JEV or ZIKV and treated or untreated with dec-RVKR-cmk. As expected, both dec-RVKR-cmk treated and untreated viral immune complexes showed almost identical bands of E and NS5 proteins, while a prominent thicker band of prM was detected in the dec-RVKR-cmk-treated cells when compared to untreated cells (Figure 2a,b), indicating a larger amount of prM protein accumulated in the treatment group of cells. After that, the relative quantification of detected signals of protein bands E and prM were analyzed by image J software. The proportion of prM was determined by dividing the prM adjusted signal over the E adjusted signal, and then we made a comparison of the prM/E ratio between the ZIKV or JEV treated sample and untreated control. Results revealed a significant increase in the prM/E index in the treated group compared to the untreated control (Figure 2c,d). Taken together, the data suggest that dec-RVKR-cmk exerts its inhibitory action on prM cleavage.

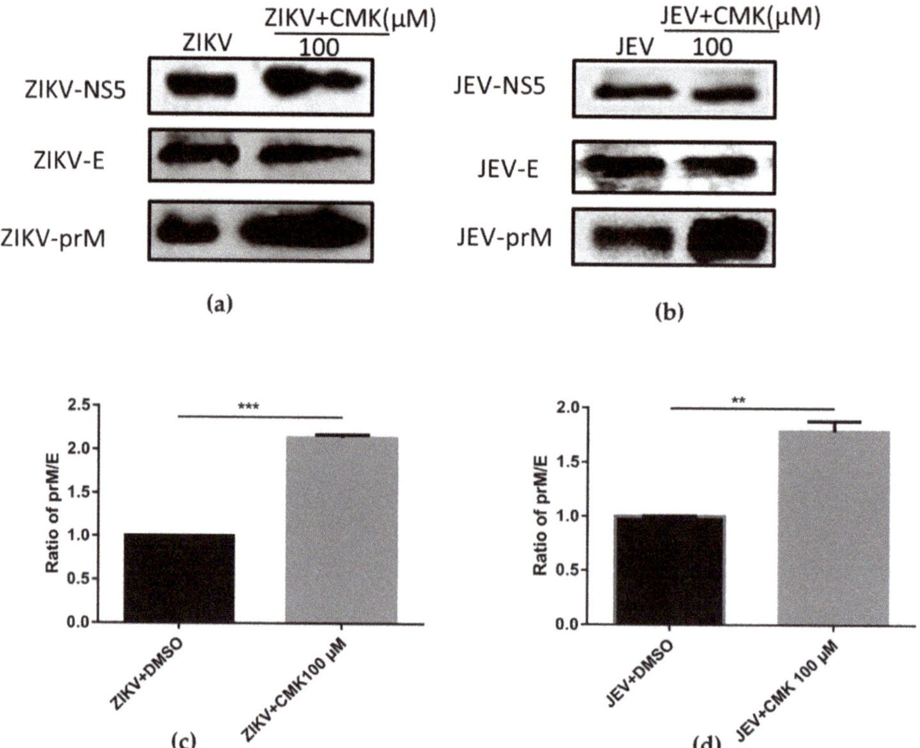

Figure 2. Dec-RVKR-cmk inhibits viral maturation process by preventing prM cleavage. Vero cells were infected with ZIKV-0.2 MOI and JEV-0.2 MOI followed by dec-RVKR-cmk treatment using 100 μM concentration. (**a**) ZIKV and (**b**) JEV viral proteins were analyzed using SDS-PAGE at 36 hpi from the peptidyl CMK-treated and untreated infected cells and then detected by Western blotting (WB) using relevant antibodies. Quantification of E and prM proteins were analyzed by image J software and the ratio of prM/E between the dec-RVKR-cmk-treated and untreated control group was compared for both (**c**) ZIKV and (**d**) JEV.

3.3. Dec-RVKR-cmk Inhibits ZIKV and JEV Infection at Different Times of Drug Administration

The ZIKV and JEV life cycle has multiple vulnerable points where suitable therapeutics can potentially be developed. For this purpose, a time-of-addition assay was performed using 100 μM dec-RVKR-cmk added to ZIKV and JEV infected cells 1 h prior to infection, at the time of infection, and 1, 6, and 12 hpi, followed by virus titer determination by plaque assay and quantification of viral genome copies inside the cells by qRT-PCR at 36 hpi. The data indicated that the maximum reduction in both ZIKV and JEV viral titer (Figure 3a,b) and genome copies (Figure 3c,d) were observed when dec-RVKR-cmk was added post infection. Interestingly, the infectious JEV and ZIKV particles released into the media from infected cells and inside the cells were still reduced even when peptidyl cmk was added at the time of infection.

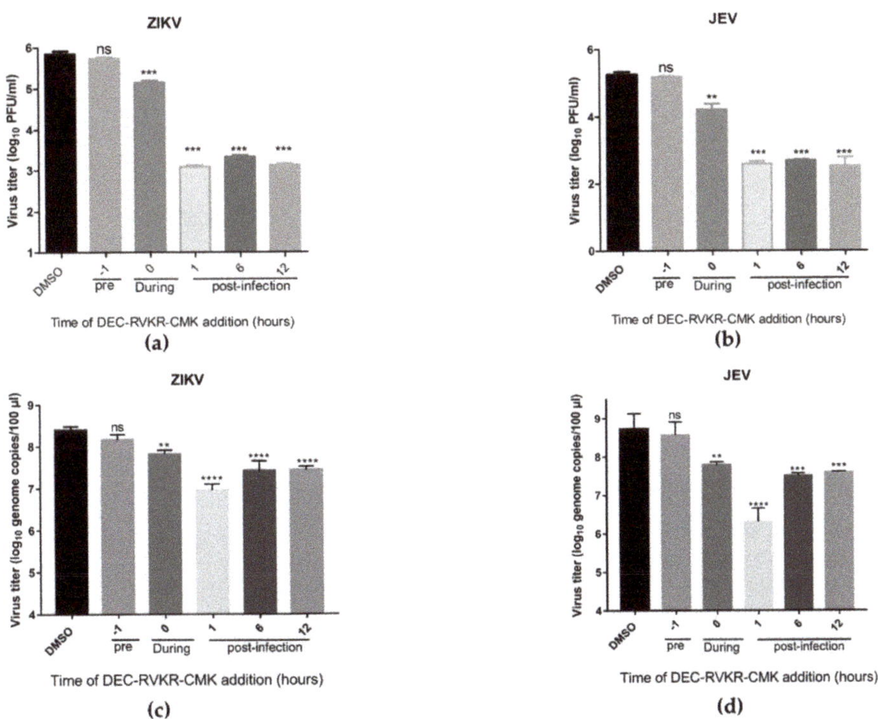

Figure 3. Time-of-addition assay of dec-RVKR-cmk against ZIKV and JEV infection. ZIKV (0.2 MOI) and JEV (0.2 MOI) infected Vero cells were treated with dec-RVKR-cmk (100 μM) at the indicated condition. Time-of-addition study revealed the effect of dec-RVKR-cmk against (**a,c**) ZIKV and (**b,d**) JEV viral titer production and intracellular genome copies, respectively. Data are presented as the mean ± SEM from three independent experiments.

3.4. Dec-RVKR-cmk Suppresses ZIKV and JEV Propagation in a Dose-Dependent Manner

Next we investigated the antiviral activity of dec-RVKR-cmk against ZIKV and JEV. Vero cells were infected with ZIKV (0.2 MOI) or JEV (0.2 MOI) in the presence of increasing concentrations of dec-RVKR-cmk (1, 10, 50, 100 μM), followed by viral titer determination in supernatant by plaque assay. Dec-RVKR-cmk inhibited ZIKV and JEV in a dose-dependent manner in the viral titer reduction assay. In the case of ZIKV infection, a 1.48 log10 and 2.44 log10 decrease in virus titer was observed with 50 and 100 μM dec-RVKR-cmk treatment, respectively (Figure 4a, left panel). In the case of JEV infection, treatment with 50 μM dec-RVKR-cmk led to a 1.22 log10 decrease in virus titer, while treatment with 100 μM led to a 2.53 log10 decrease in virus titter (Figure 4b, left panel). No significant inhibition of ZIKV and JEV was observed at 1 and 10 μM concentrations of dec-RVKR-cmk. The IC50 of dec-RVKR-cmk, i.e., the concentration at which 50% virus inhibition occurred, was determined to be 18.59 and 19.91 μM against ZIKV and JEV, respectively (Figure 4a,b, right panel).

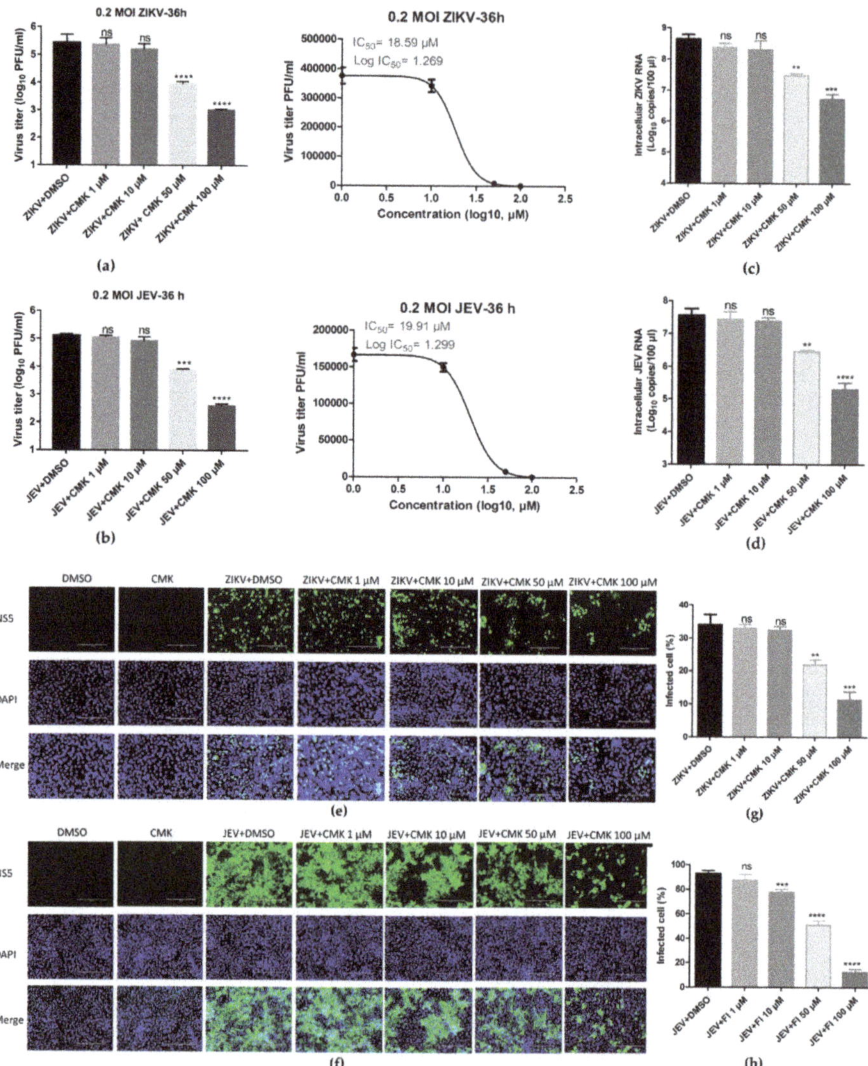

Figure 4. Antiviral assessment of dec-RVKR-cmk against ZIKV and JEV in a dose-dependent manner. Vero cells were infected with ZIKV or JEV at 0.2 MOI followed by dec-RVKR-cmk treatment using the indicated concentration. Right panels (**a**,**b**) indicate virus titer while left panels indicate IC50 of dec-RVKR-cmk against the indicated MOI of ZIKV and JEV. (**c**) ZIKV-0.2 MOI and (**d**) JEV-0.2 MOI infected Vero cells were treated by dec-RVKR-cmk in a dose-dependent manner and then analyzed by qRT-PCR for absolute genome copies using standard curve of in vitro transcribed ZIKV and JEV RNA at 36 hpi. Meanwhile, infected Vero cells with similar condition of RNA analysis were fixed to analyze virus spreading from infected cells to neighboring cells. Immunofluorescence images of (**e**) ZIKV- and (**f**) JEV-infected Vero cells were acquired at 36 hpi, and quantified (**g**) ZIKV and (**h**) JEV immunoreactive positive cells to visualize the inhibition of infection in a dose-dependent manner. Data are presented as the mean ± SEM from three independent experiments.

To obtain a detailed insight into the efficacy of dec-RVKR-cmk against ZIKV and JEV, RT-qPCR was performed. Reduced ZIKV (0.2 MOI) and JEV (0.2 MOI) genome copies were observed in a dose-dependent treatment in Vero cells at 36 hpi. A significant decline in the viral RNA of ZIKV (~2 log10) and JEV (2.2 log10) was observed in the 100 μM treatment of dec-RVKR-cmk. While at the 50 μM treatment, a 1 log10 decrease in ZIKV RNA and a 1.16 log10 decrease in JEV RNA were observed (Figure 4c,d). Treatment with 1 and 10 μM concentrations of dec-RVKR-cmk did not cause significant inhibition of ZIKV and JEV RNA.

Meanwhile, the effect of dec-RVKR-cmk was observed on virus spreading from infected to bystander cells by IFA. Immunofluorescence images of ZIKV- (0.2 MOI) and JEV- (0.2 MOI) infected Vero cells were taken after 36 hpi using different concentrations of dec-RVKR-cmk (Figure 4e,f). The fluorescence signals revealed that a significant ~22.67-fold inhibition of infection was found when ZIKV-infected cells were treated with 100 μM dec-RVKR-cmk and, to a lesser extent, with 50 μM (~12-fold) and with 10 μM (~1.7-fold), as compared to the control (Figure 4g). Similarly, in the case of JEV-infected cells, the counting of immunoreactive cells showed a strong decrease of viral spreading in a dose-dependent manner (Figure 4h). Taken together, the data suggest that dec-RVKR-cmk significantly reduces intracellular and extracellular virus particles along with the inhibition of ZIKV and JEV spreading from infected cells to bystanders in a dose-dependent manner.

3.5. Dec-RVKR-cmk Inhibits ZIKV and JEV Propagation at Various Time Points

We next studied the anti-viral efficacy of dec-RVKR-cmk against both ZIKV (0.2 MOI) and JEV (0.2 MOI) in Vero cells at different time points, i.e., 24, 36, and 48 hpi, using the 100 μM concentration of dec-RVKR-cmk. Cell culture supernatant was used to determine the virus progeny titer by plaque assay. Figure 5a indicates that significant inhibition of ZIKV progeny titer was seen at 24, 36, and 48 hpi (1.6 log10, 2.25 log10, and 2.23 log10, respectively). Similarly, in the case of JEV, significant inhibition was observed at 24, 36, and 48 hpi (1.08 log10, 2.37 log10, and 2.72 log10, respectively) (Figure 5b). For IFA, infected Vero cells were fixed from both the dec-RVKR-cmk-treated and DMSO-treated control groups to see the extent of infection by counting immunoreactive cells. No significant difference in fluorescence signal of ZIKV-NS5 (Figure 5c) and JEV-NS5 (Figure 5d) were observed at 24 hpi between the dec-RVKR-cmk-treated and DMSO-treated control groups. Significant ZIKV (Figure 5c,e) and JEV (Figure 5d,f) inhibition was seen at 36 and 48 hpi, in terms of reduction in the percentage of immunoreactive cells, in the dec-RVKR-cmk-treated group as compared to the control group for ZIKV (14% and 52% inhibition, respectively) and JEV (70% and 23% inhibition, respectively). Thus, the data suggest that the maximum effects of dec-RVKR-cmk were observed at 36 and 48 hpi in terms of reduction of virus progeny titer and inhibition of ZIKV and JEV spreading to neighboring bystander cells.

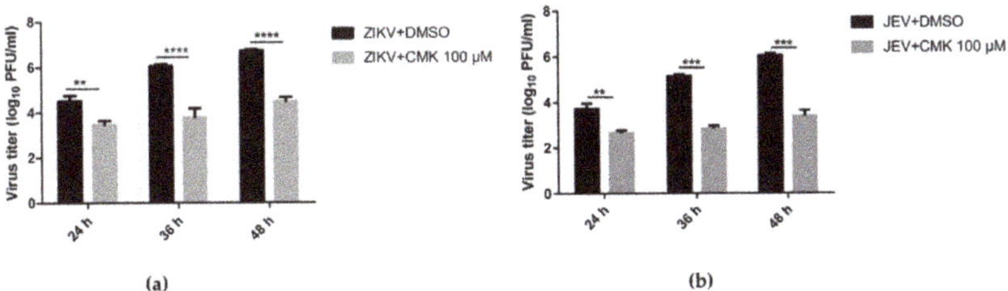

Figure 5. *Cont.*

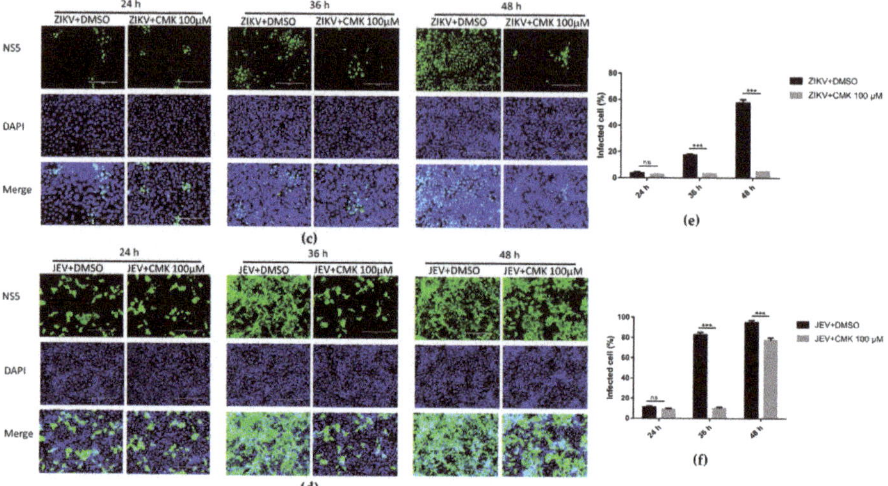

Figure 5. Dec-RVKR-cmk inhibited ZIKV and JEV infection at various time points. Vero cells were infected with ZIKV-0.2 MOI and JEV-0.2 MOI, followed by dec-RVKR-cmk treatment using the 100 μM concentration. Cell supernatant used to determine (**a**) ZIKV and (**b**) JEV viral titer by plaque assay at indicated time points. While immunofluorescence images of (**c**) ZIKV- and (**d**) JEV-infected Vero cells were acquired at different time points in both control (virus + DMSO) and treated (virus + CMK) Vero cells and quantified (**e**) ZIKV and (**f**) JEV immunoreactive positive cells to see the extent of infection. Data are presented as the mean ± SEM from three independent experiments.

We conducted an experiment to assess virus replication in more detail with the one-step growth cycle. We treated the cells using a 100 μM concentration of dec-RVKR-cmk or DMSO 1 hpi. Vero cells were infected with ZIKV at an MOI of 5 for 1 h and media were replaced with dec-RVKR-cmk or DMSO after washing the cells. The culture medium and cells were collected immediately following infection (0 h) and then every 3 h through 21 hpi. Virus titer was determined in both supernatants and cells by plaque assay on Vero cells. The data indicated that infectious viral particles began to produce between 9 and 12 h post infection in DMSO-treated cells (Figure 6a). Following initial amplification, the subsequent rise was observed at 15, 18, and 21 hpi. Similar findings were observed in the extracellular compartment (Figure 6b). In dec-RVKR-cmk-treated cells, a significant reduction in both intracellular (>1 $\log_{10}$ at 15, 18, and 21 hpi) and extracellular (>2 $\log_{10}$ at 15, 18, and 21 hpi) virus titers were observed. Taken together, the data suggests that dec-RVKR-cmk inhibits ZIKV and JEV propagation in the one-step growth cycle. Furthermore, the greater reduction in extracellular virus titer than that in intracellular virus titer indicates that dec-RVKR-cmk might inhibit virus release rather than replication.

3.6. Dec-RVKR-cmk Inhibits ZIKV Release and Next Round Infectivity

To elaborate on this phenomena, both ZIKV- (1 MOI) infected cells and their supernatants from treatment (dec-RVKR-cmk 100 μM) and control (DMSO) group were subjected to RT-qPCR analysis. Meanwhile, half the volume of supernatant was used to determine viral titer through plaque assay. According to the one-step growth cycle results (Figure 6), 12–21 h may reflect a single round of replication [26]. To clarify at which stage in virus life cycle dec-RVKR-cmk works, both ZIKV- and JEV-infected cells and their supernatants from the treatment (dec-RVKR-cmk 100 μM) and control (DMSO) groups were subjected to RT-qPCR analysis and plaque assay at 12, 16, and 20 hpi. At the intracellular level, there was no difference in genome copies in both the treated and control groups at different time points (Figure 7a). Extracellular viral titer, in terms of genome copies (Figure 7b) and plaque forming unit (Figure 7c), showed a significant difference between the control and treatment groups. Afterwards, we calculated the infectivity of the virus by comparing the viral titer of ZIKV, determined by plaque assay and RT-qPCR in supernatant, for both the control and treated groups. Figure 7d indicates no significant difference in virus infectivity at 12 and 16 hpi, but at 20 hpi a 20% reduction was observed in the released virus. Taken together, these data validate the results, seen in Figure 6, which show that no significant difference in genomic RNA copies was observed in dec-RVKR-cmk-treated and untreated cells at 12, 16 and 20 hpi. Thus, dec-RVKR-cmk cannot inhibit virus replication in the first round of infection but exerts its inhibitory action on virus release and next round infectivity.

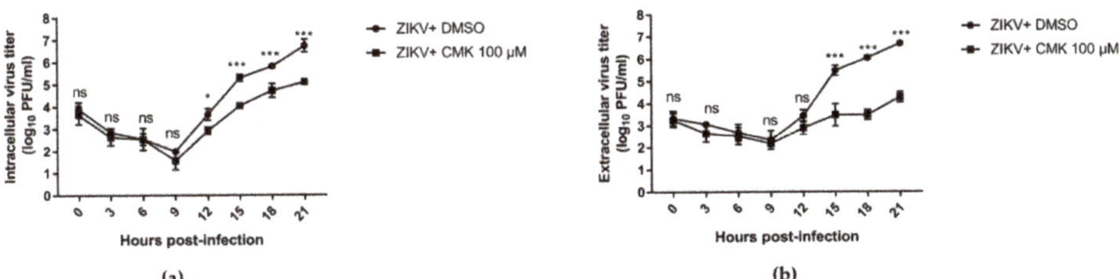

Figure 6. Dec-RVKR-cmk inhibits ZIKV infection in the one-step growth cycle. Vero cells were infected with ZIKV at an MOI of 5 for 1 h followed by dec-RVKR-cmk or DMSO treatment after washing the cells. At the indicated time points after infection, culture medium was collected and cells were subjected to three freeze-thaw cycles to liberate cell-associated viruses. (**a**) Intracellular and (**b**) extracellular virus titer was determined by plaque assay. Data are presented as the mean ± SEM from three independent experiments.

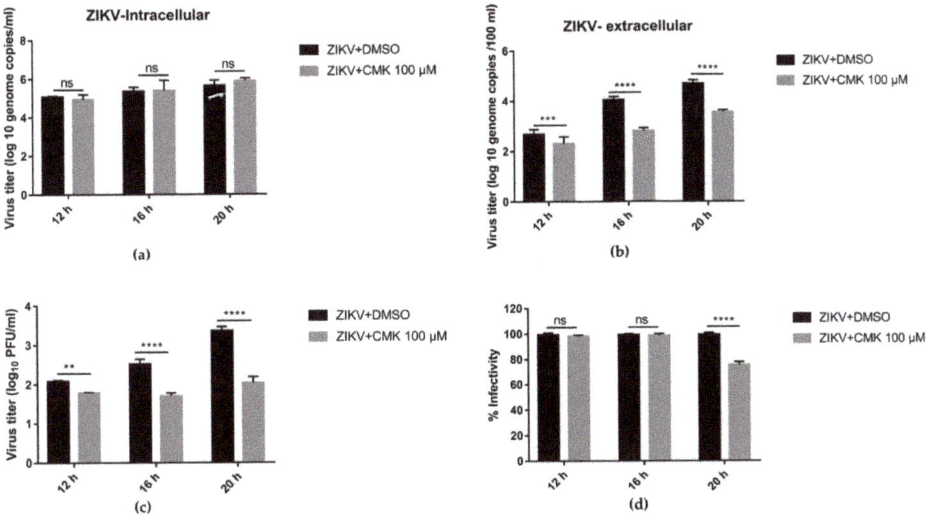

Figure 7. Dec-RVKR-cmk inhibits ZIKV release and next round infectivity. Vero cells were infected with ZIKV (1 MOI) followed by dec-RVKR-cmk treatment using the 100 µM concentration. (**a**) Infected cells were analyzed by qRT-PCR for absolute genome copies using the standard curve of in vitro transcribed ZIKV for the indicated time points. Cell supernatant was used to determine ZIKV viral titer by (**b**) qRT-PCR and also by (**c**) plaque assay. (**d**) The percentage infectivity was calculated by dividing the viral titer determined by plaque assay by the viral titer determined by qRT-PCR in supernatant, for both the control and treated groups, and then the value of the control group was normalized to 100 and compared with the treated group. Data are presented as the mean ± SEM from three independent experiments.

3.7. Dec-RVKR-cmk Inhibits ZIKV and JEV Infection in Mosquito Cells

It is interesting to know whether dec-RVKR-cmk inhibits ZIKV and JEV infection only in mammalian cells or inhibits infection in a mosquito cell line. To address this question, firstly we examined the cytotoxicity of dec-RVKR-cmk to C6/36 cells by using luminescence-based cell viability assay. Likewise, Vero cells treated with dec-RVKR-cmk exhibited no cytotoxicity to C6/36 cell line at the concentrations of 1, 10, 50, and 100 µM (Figure 8a). Afterwards, C6/36 cells were infected with ZIKV (1 MOI) and then treated with 100 µM of dec-RVKR-cmk. Cell supernatant was harvested at 24 hpi for virus titer determination through plaque assay and cells were fixed for immunofluorescence imaging. Figure 8b indicates that a significant inhibition of 2.23 log10 ZIKV progeny titer was observed when compared to the DMSO-treated control. The fluorescence image (Figure 8c) revealed a ~39% reduction in immunonoreactive positive cells in the dec-RVKR-cmk treatment group as compared to the control (Figure 8d). These results suggest that dec-RVKR-cmk could inhibit flavivirus propagation in both mammalian cells and mosquito cells.

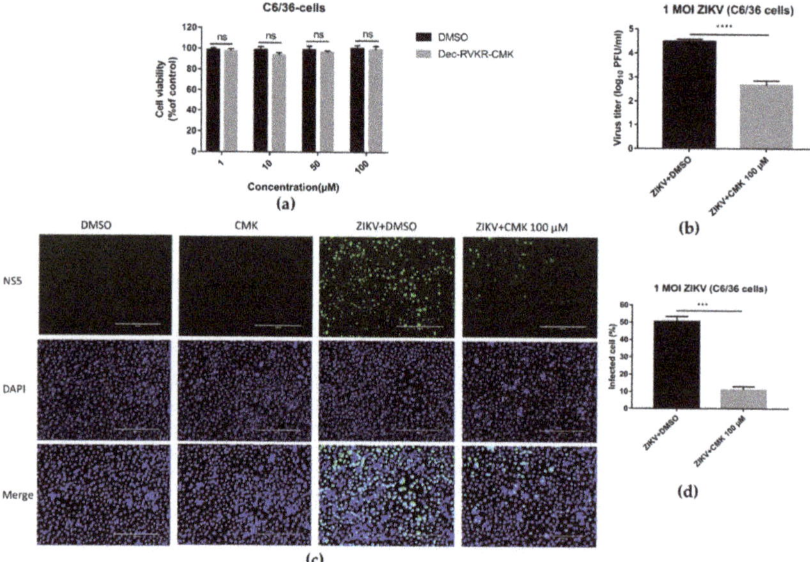

Figure 8. Dec-RVKR-cmk inhibits ZIKV and JEV infection in mosquito cell line C6/36. (**a**) Cytotoxicity of dec-RVKR-cmk in C6/36 cell line was determined by CellTiter-GLO One Solution Assay kit (Promega). (**b**) C6/36 cells were infected with ZIKV (1 MOI) followed by dec-RVKR-cmk treatment using the 100 µM concentration. Cell supernatant was used to determine ZIKV viral titer by plaque assay at 24 hpi. (**c**) Immunofluorescence images of ZIKV-infected c6/36 cells were acquired to quantify (**d**) ZIKV immunoreactive positive cells. Data are presented as the mean ± SEM from three independent experiments.

4. Discussion

Encephalitis, hemorrhagic disease, biphasic fever, jaundice, and flaccid paralysis are typical manifestations of flaviviruses in human beings [2]. Among mosquito-borne flaviviruses, ZIKA and JEV are medically important pathogens. JEV is known to be neurotropic and cause encephalitis, while ZIKV is considered to cause febrile illness. However, in a recent worldwide outbreak, ZIKV has been associated with neurological manifestations, including Guillain–Barré syndrome in adults and fetal microcephaly [27]. Continued rise in flavivirus infections across the world highlights an urgent need for antivirals to combat these challenges.

The objective of this study was to evaluate the antiviral activity of dec-RVKR-cmk against flavivirus. Processing of viral proteins by host cellular protease is a characteristic feature to achieve virus maturation in different viruses of various families. Dec-RVKR-cmk is reported to inhibit cleavage and fusion activity of various glycoproteins mediated by furin and to play a key role in activating several bacteria and viruses, including anthrax, botulinum, influenza A, measles, Ebola, HIV, HBV, and CHIKV [16,28–33]. Among flaviviruses, an important step in the production of infectious virions is the processing of prM protein to the anchored membrane M stump and the "pr" peptide that takes place in the TGN by the host proprotein convertase furin protease, prior to release from infected cell [18,34]. Therefore, we assessed furin inhibition activity of dec-RVKR-cmk against ZIKV and JEV. In this respect, our observations confirmed the critical role of furin in flavivirus maturation. We found that peptidyl CMK inhibited ZIKV and JEV more significantly in the later stage of their life cycle by preventing efficient cleavage of prM protein through the host proprotein convertase furin protease resulting in the effective arrest of subsequent viral infection.

The time-of-drug addition assay revealed that dec-RVKR-cmk worked more efficiently post infection but, interestingly, inhibition was also observed when dec-RVKR-cmk added at the time of infection. This antiviral activity of dec-RVKR-cmk was still observed even when the drug was added 12 hpi. Collectively, this suggests that dec-RVKR-cmk is more effective when added post infection. A dose-related inhibition of ZIKV and JEV in Vero cells was observed in terms of extracellular viral progeny titer, intracellular viral genome copies, and viral spreading from infected to bystander cells. These findings were consistent with the results of dec-RVKR-cmk against Chikungunya infection in muscle cells [18]. The effectiveness of dec-RVKR-cmk was observed against both ZIKV and JEV infections at different time points (24, 36, and 48 hpi) in terms of lowering viral titer and viral spreading from infected to neighboring cells. Maximum antiviral activity of dec-RVKR-cmk was perceived at 36 and 48 hpi. In the case of flavivirus, a single round of replication was reported to continue for 16 h [26]. Our study suggests that dec-RVKR-cmk could inhibit virus propagation in a one-step growth cycle, and a mechanism study suggests that it cannot inhibit virus replication, but exerts its inhibitory action on the virus release and next round infectivity. This might be due to inhibition of prM cleavage, which would affect viral packaging or result in accumulation of immature viral particles, while virus release and next round infectivity would be diminished.

Hepatitis C virus (HCV) is also an important member of Flaviviridae but has been placed in a genus separate from the other flaviviruses and it is not clear to what extent HCV envelope proteins behave like those of other members of the family Flaviviridae. As previously reported, proprotein convertases furin was responsible for proteolytic cleavage of pro-TGF-β1 into its bioactive form in HCV-infected cells that positively regulates HCV RNA replication [34]. Based on this study, dec-RVKR-cmk might be active against HCV but with a different mechanism of action.

The phenomenon of redundancy existing between furin and other proprotein convertases to overcome the side effects PCs inhibitor [23,24] suggesting dec-RVKR-cmk could not interfere with the normal cellular function of furin. These findings, together with the observed antiviral activity of dec-RVKR-cmk at a non-cytotoxic concentration in this study, support the possibility of the therapeutic application of the furin inhibitor against flavivirus infection, and our future study of this furin inhibitor will enable verification of therapeutic efficacy in an animal model.

In summary, this is the first study that shows the antiviral activity of dec-RVKR-cmk against flaviviruses (ZIKV and JEV). The observed IC50 and cytotoxicity profile together with time-of-addition and molecular mechanism data suggest dec-RVKR-cmk as a potential candidate for treatment.

Author Contributions: Conceptualization, M.I., J.Y., and S.C.; methodology, M.I.; software, M.K.S. and M.I.; validation, Z.C., X.W., Z.Z., Y.L., D.Z., B.Z. and Q.L.; writing—original draft preparation, M.I.; editing, J.Y. and S.C.; supervision, S.C. and J.Y.; project administration, S.C.; funding acquisition, S.C.

Funding: This work was supported by National Key Research and Development Program of China (2016YFD0500407), National Natural Science Foundation of China (31825025, 31572517, 31972721), Fundamental Research Funds for the Central Universities (2662018QD025), and Natural Science Foundation of Hubei Province (2019CFA010).

Acknowledgments: We acknowledge Huazhong Agricultural University for the provision of the biosafety laboratory. We are also grateful to Xiaowu Pang (College of Dentistry, Howard University, USA) for kindly providing the ZIKV-MR-766 strain.

Conflicts of Interest: The authors declare no conflict of interest.

References

1. Panayiotou, C.; Lindqvist, R.; Kurhade, C.; Vonderstein, K.; Pasto, J.; Edlund, K.; Upadhyay, A.S.; Overby, A.K. Viperin Restricts Zika Virus and Tick-Borne Encephalitis Virus Replication by Targeting NS3 for Proteasomal Degradation. *J. Virol.* **2018**, *92*, e02054-17. [CrossRef] [PubMed]

2. Gould, E.; Solomon, T. Pathogenic flaviviruses. *Lancet* **2008**, *371*, 500–509. [CrossRef]

3. Puerta-Guardo, H.; Glasner, D.R.; Espinosa, D.A.; Biering, S.B.; Patana, M.; Ratnasiri, K.; Wang, C.; Beatty, P.R.; Harris, E. Flavivirus NS1 Triggers Tissue-Specific Vascular Endothelial Dysfunction Reflecting Disease Tropism. *Cell Rep.* **2019**, *26*, 1598–1613. [CrossRef] [PubMed]

4. Kim, S.; Li, B.; Linhardt, R. Pathogenesis and inhibition of flaviviruses from a carbohydrate perspective. *Pharmaceuticals* **2017**, *10*, 44. [CrossRef]

5. Heinz, F.X.; Stiasny, K. Flaviviruses and flavivirus vaccines. *Vaccine* **2012**, *30*, 4301–4306. [CrossRef] [PubMed]

6. Wu, Y.; Liu, Q.; Zhou, J.; Xie, W.; Chen, C.; Wang, Z.; Yang, H.; Cui, J. Zika virus evades interferon-mediated antiviral response through the co-operation of multiple nonstructural proteins in vitro. *Cell Discov.* **2017**, *3*, 17006. [CrossRef] [PubMed]

7. Nambala, P.; Su, W.-C. Role of Zika Virus prM protein in viral pathogenicity and use in vaccine development. *Front. Microbiol.* **2018**, *9*, 1797. [CrossRef] [PubMed]

8. Lorenz, I.C.; Kartenbeck, J.; Mezzacasa, A.; Allison, S.L.; Heinz, F.X.; Helenius, A. Intracellular assembly and secretion of recombinant subviral particles from tick-borne encephalitis virus. *J. Virol.* **2003**, *77*, 4370–4382. [CrossRef] [PubMed]

9. Mackenzie, J.M.; Westaway, E.G. Assembly and maturation of the flavivirus Kunjin virus appear to occur in the rough endoplasmic reticulum and along the secretory pathway, respectively. *J. Virol.* **2001**, *75*, 10787–10799. [CrossRef] [PubMed]

10. Li, L.; Lok, S.-M.; Yu, I.-M.; Zhang, Y.; Kuhn, R.J.; Chen, J.; Rossmann, M.G. The flavivirus precursor membrane-envelope protein complex: Structure and maturation. *Science* **2008**, *319*, 1830–1834. [CrossRef]

11. Yu, I.-M.; Zhang, W.; Holdaway, H.A.; Li, L.; Kostyuchenko, V.A.; Chipman, P.R.; Kuhn, R.J.; Rossmann, M.G.; Chen, J. Structure of the immature dengue virus at low pH primes proteolytic maturation. *Science* **2008**, *319*, 1834–1837. [CrossRef] [PubMed]

12. Elshuber, S.; Mandl, C.W. Resuscitating mutations in a furin cleavage-deficient mutant of the flavivirus tick-borne encephalitis virus. *J. Virol.* **2005**, *79*, 11813–11823. [CrossRef] [PubMed]

13. Yoshii, K.; Igarashi, M.; Ichii, O.; Yokozawa, K.; Ito, K.; Kariwa, H.; Takashima, I. A conserved region in the prM protein is a critical determinant in the assembly of flavivirus particles. *J. Gen. Virol.* **2012**, *93*, 27–38. [CrossRef] [PubMed]

14. Garten, W.; Hallenberger, S.; Ortmann, D.; Schafer, W.; Vey, M.; Angliker, H.; Shaw, E.; Klenk, H.D. Processing of viral glycoproteins by the subtilisin-like endoprotease furin and its inhibition by specific peptidylchloroalkylketones. *Biochimie* **1994**, *76*, 217–225. [CrossRef]

15. Becker, G.L.; Sielaff, F.; Than, M.E.; Lindberg, I.; Routhier, S.; Day, R.; Lu, Y.; Garten, W.; Steinmetzer, T. Potent inhibitors of furin and furin-like proprotein convertases containing decarboxylated P1 arginine mimetics. *J. Med. Chem.* **2010**, *53*, 1067–1075. [CrossRef]

16. Pang, Y.J.; Tan, X.J.; Li, D.M.; Zheng, Z.H.; Lei, R.X.; Peng, X.M. Therapeutic potential of furin inhibitors for the chronic infection of hepatitis B virus. *Liver Int.* **2013**, *33*, 1230–1238. [CrossRef]

17. Hallenberger, S.; Bosch, V.; Angliker, H.; Shaw, E.; Klenk, H.-D.; Garten, W. Inhibition of furin-mediated cleavage activation of HIV-1 glycoprotein gp160. *Nature* **1992**, *360*, 358. [CrossRef]

18. Ozden, S.; Lucas-Hourani, M.; Ceccaldi, P.-E.; Basak, A.; Valentine, M.; Benjannet, S.; Hamelin, J.; Jacob, Y.; Mamchaoui, K.; Mouly, V. Inhibition of Chikungunya virus infection in cultured human muscle cells by furin inhibitors impairment of the maturation of the E2 surface glycoprotein. *J. Biol. Chem.* **2008**, *283*, 21899–21908. [CrossRef]

19. Tong, Y.; Tong, S.; Zhao, X.; Wang, J.; Jun, J.; Park, J.; Wands, J.; Li, J. Initiation of duck hepatitis B virus infection requires cleavage by a furin-like protease. *J. Virol.* **2010**, *84*, 4569–4578. [CrossRef]
20. Day, P.M.; Schiller, J.T. The role of furin in papillomavirus infection. *Future Microbiol.* **2009**, *4*, 1255–1262. [CrossRef]
21. Couture, F.; Kwiatkowska, A.; Dory, Y.L.; Day, R. Therapeutic uses of furin and its inhibitors: A patent review. *Expert Opin. Ther. Pat.* **2015**, *25*, 379–396. [CrossRef] [PubMed]
22. Creemers, J.W.; Khatib, A.M. Knock-out mouse models of proprotein convertases: Unique functions or redundancy? *Front. Biosci. J. Virtual Libr.* **2008**, *13*, 4960–4971. [CrossRef] [PubMed]
23. Couture, F.; D'Anjou, F.; Day, R. On the cutting edge of proprotein convertase pharmacology: From molecular concepts to clinical applications. *Biomol. Concepts* **2011**, *2*, 421–438. [CrossRef] [PubMed]
24. Roebroek, A.J.; Taylor, N.A.; Louagie, E.; Pauli, I.; Smeijers, L.; Snellinx, A.; Lauwers, A.; Van de Ven, W.J.; Hartmann, D.; Creemers, J.W. Limited redundancy of the proprotein convertase furin in mouse liver. *J. Biol. Chem.* **2004**, *279*, 53442–53450. [CrossRef] [PubMed]
25. Chen, Z.; Ye, J.; Ashraf, U.; Li, Y.; Wei, S.; Wan, S.; Zohaib, A.; Song, Y.; Chen, H.; Cao, S. MicroRNA-33a-5p modulates Japanese Encephalitis Virus replication by targeting eukaryotic translation elongation factor 1A1. *J. Virol.* **2016**, *90*, 3722–3734. [CrossRef]
26. Mukherjee, S.; Lin, T.Y.; Dowd, K.A.; Manhart, C.J.; Pierson, T.C. The infectivity of prM-containing partially mature West Nile virus does not require the activity of cellular furin-like proteases. *J. Virol.* **2011**, *85*, 12067–12072. [CrossRef]
27. Kovanich, D.; Saisawang, C.; Sittipaisankul, P.; Ramphan, S.; Kalpongnukul, N.; Somparn, P.; Pisitkun, T.; Smith, D.R. Analysis of the Zika and Japanese encephalitis virus NS5 interactomes. *J. Proteome Res.* **2019**. [CrossRef]
28. Becker, G.L.; Lu, Y.; Hardes, K.; Strehlow, B.; Levesque, C.; Lindberg, I.; Sandvig, K.; Bakowsky, U.; Day, R.; Garten, W.; et al. Highly potent inhibitors of proprotein convertase furin as potential drugs for treatment of infectious diseases. *J. Biol. Chem.* **2012**, *287*, 21992–22003. [CrossRef]
29. Peng, M.; Watanabe, S.; Chan, K.W.K.; He, Q.; Zhao, Y.; Zhang, Z.; Lai, X.; Luo, D.; Vasudevan, S.G.; Li, G. Luteolin restricts dengue virus replication through inhibition of the proprotein convertase furin. *Antivir. Res.* **2017**, *143*, 176–185. [CrossRef]
30. Thomas, G. Furin at the cutting edge: From protein traffic to embryogenesis and disease. *Nat. Rev. Mol. Cell Biol.* **2002**, *3*, 753–766. [CrossRef]
31. Decroly, E.; Vandenbranden, M.; Ruysschaert, J.M.; Cogniaux, J.; Jacob, G.S.; Howard, S.C.; Marshall, G.; Kompelli, A.; Basak, A.; Jean, F.; et al. The convertases furin and PC1 can both cleave the human immunodeficiency virus (HIV)-1 envelope glycoprotein gp160 into gp120 (HIV-1 SU) and gp41 (HIV-I TM). *J. Biol. Chem.* **1994**, *269*, 12240–12247. [PubMed]
32. Bolt, G.; Pedersen, L.O.; Birkeslund, H.H. Cleavage of the respiratory syncytial virus fusion protein is required for its surface expression: Role of furin. *Virus Res.* **2000**, *68*, 25–33. [CrossRef]
33. Molloy, S.S.; Anderson, E.D.; Jean, F.; Thomas, G. Bi-cycling the furin pathway: From TGN localization to pathogen activation and embryogenesis. *Trends Cell Biol.* **1999**, *9*, 28–35. [CrossRef]
34. Presser, L.D.; Haskett, A.; Waris, G. Hepatitis C virus-induced furin and thrombospondin-1 activate TGF-beta1: Role of TGF-beta1 in HCV replication. *Virology* **2011**, *412*, 284–296. [CrossRef] [PubMed]

Communication

5,6-Dichloro-2-Phenyl-Benzotriazoles: New Potent Inhibitors of Orthohantavirus

Giuseppina Sanna [1,*,†], Sandra Piras [2,†], Silvia Madeddu [1], Bernardetta Busonera [1], Boris Klempa [3,4], Paola Corona [2], Roberta Ibba [2], Gabriele Murineddu [2], Antonio Carta [2,*] and Roberta Loddo [1]

1   Department of Biomedical Sciences, University of Cagliari, Cittadella Universitaria, 09042 Monserrato, Cagliari, Italy; silvia.madeddu@unica.it (S.M.); iurru99@hotmail.com (B.B.); rloddo@unica.it (R.L.)
2   Department of Chemistry and Pharmacy, University of Sassari, Via Muroni 23/A, 07100 Sassari, Italy; piras@uniss.it (S.P.); pcorona@uniss.it (P.C.); ribba@uniss.it (R.I.); muri@uniss.it (G.M.)
3   Institute of Virology, Biomedical Research Center Slovak Academy of Sciences, 845 05 Bratislava, Slovakia; boris.klempa@savba.sk
4   Institute of Virology, Helmut-Ruska-Haus, Charité School of Medicine, 10117 Berlin, Germany
*   Correspondence: g.sanna@unica.it (G.S.); acarta@uniss.it (A.C.);
    Tel.: +39-070 6754161 (G.S.); +39-079 228722 (A.C.)
†   These authors contributed equally to this work.

Received: 6 November 2019; Accepted: 15 January 2020; Published: 20 January 2020

Abstract: Orthohantaviruses, previously known as hantaviruses (family Hantaviridae, order Bunyavirales), are emerging zoonoses hosted by different rodent and insectivore species. Orthohantaviruses are transmitted by aerosolized excreta (urine, saliva and feces) of their reservoir hosts. When transmitted to humans, they cause hemorrhagic fever with renal syndrome (HFRS) in Asia and Europe and hantavirus (cardio) pulmonary syndrome (HPS) in the Americas. Clinical studies have shown that early treatments of HFRS patients with ribavirin (RBV) improve prognosis. Nevertheless, there is the need for urgent development of specific antiviral drugs. In the search for new RNA virus inhibitors, we recently identified a series of variously substituted 5,6-dichloro-1(2)-phenyl-1(2)H-benzo[d][1,2,3]triazole derivatives active against the human respiratory syncytial virus (HRSV). Interestingly, several 2-phenyl-benzotriazoles resulted in fairly potent inhibitors of the Hantaan virus in a chemiluminescence focus reduction assay (C-FRA) showing an EC_{50} = 4–5 µM, ten-fold more active than ribavirin. Currently, there are no FDA approved drugs for the treatment of orthohantavirus infections. Antiviral activities and cytotoxicity profiles suggest that 5,6-dichloro-1(2)-phenyl-1(2)H-benzo[d][1,2,3]triazoles could be promising candidates for further investigation as a potential treatment of hantaviral diseases.

Keywords: orthohantavirus; phenyl-benzotriazoles; antiviral activity; C-FRA

---

1. Introduction

Orthohantaviruses are classified as emerging viruses that cause two life-threatening diseases: hemorrhagic fever with renal syndrome (HFRS) and orthohantaviruses pulmonary syndrome (HPS), also known as hantavirus cardiopulmonary syndrome (HCPS) [1]. Small mammals are natural hosts of orthohantavirus, mainly rodents but recently reptiles and fishes [2] have also been discovered as carriers of these viruses that are transmitted to humans through the aerosol route. They are responsible for persistent infections without evident illness signs in their hosts [3]. The two diseases that are orthohantaviruses-related both induce an impressive rise in blood vessel permeability, strong immune responses and inflammation and viruses such as the Hantaan virus (HTNV) and Sin Nombre virus

(SNV) are the causative agents. Although orthohantaviruses are distributed worldwide, HFRS and HCPS occur generally in Eurasia and the Americas, respectively [4].

Orthohantaviruses are members of the Hantaviridae family, order Bunyavirales. Their tripartite, single-stranded, negative sense RNA genome codes for four proteins. The L segment, S segment, and M segment encode an RNA-dependent RNA polymerase (RdRp), a nucleocapsid protein (N protein) and Gn and Gc glycoproteins, respectively. The two surface glycoproteins, before being exposed on the viral surface, are post-translationally processed via the endoplasmic reticulum and Golgi apparatus. These proteins interact with integrin receptors allowing viruses to enter new host cells [5].

The three genomic RNA molecules form a complex within the virion with N protein and, most probably, with RdRp. The viral RdRp mediates the genomic and anti-genomic viral RNAs and the transcription of viral mRNAs exclusively in the cytoplasm [6].

Over the last few years, the search for an effective treatment for orthohantaviruses infections has undergone a considerable increase [7]. Ribavirin (RBV), a broad-spectrum inhibitor, is the only antiviral with recognized in vitro and in vivo activity on hantavirus replication [8,9]. In China [10] and Russia [11], clinical trials have been conducted for the treatment of HFRS using post-exposure, intravenous RBV but while significant results have been obtained in the first case, the second resulted ineffective, as well as the trial conducted in patients with HPS. Furthermore, the use of RBV is limited by its myelosuppression and toxicity [12]. Besides RBV, the use of some nucleoside analogues resulted in being highly inhibitory in animal models [13]. A few other candidates have been evaluated: Favipiravir, a pyrazine derivative endowed of anti-influenza properties and Vandetanib, a tyrosine-kinase inhibitor. The use of corticosteroids, unfortunately, does not determine benefit and immunotherapy although it has given encouraging results; in preventing and treating human hantavirus infections, it remains challenging [14].

Nowadays, there are no U.S. Food and Drug Administration (FDA) granted antivirals [14], vaccines, or immunotherapeutic for the treatment of HFRS or HPS, and consequently, therapeutic approaches are usually based on supportive care. It is precisely for that reason that an effort to develop potential therapeutic agents is strongly desirable. At the present time, a very limited number of antivirals have been tested for orthohantavirus [14].

In recent years, our research group has published several 1(2)H-benzo[d][1,2,3]triazole, usually called benzotriazole, derivatives that have shown marked antiviral activity against many viruses [15–18]. The versatile biological behavior of benzotriazole and its derivatives have recently been described in an in-depth review [19]. Among these benzotriazole derivatives, the 5,6-dichloro1(2)phenyl-benzotriazole scaffold turned out to be endowed with high activity against several different viruses.

In recent times, we described a series of 5,6-dichloro1(2)phenyl-benzotriazole derivatives that are active against te human respiratory syncytial virus (HRSV) in low micromolar range with low cytotoxicity (CC_{50} >100 µM) and very high selectivity when analyzed on a wide panel of positive- and negative-sense single-stranded RNA, double-stranded RNA, and DNA viruses [20]. In order to further investigate if these derivatives are able to inhibit viral infection processes of other negative sense RNA virus families, we subjected them (Figure 1) to a broad antiviral screening including the Hantaan virus (HTNV), a segmented RNA virus of the family Hantaviridae, for which there is currently no antiviral or approved vaccine.

In general terms, these molecules showed little or no activity except for HTNV. Series 2 derivatives (Figure 1) showed most interesting EC_{50} values, so we performed the same broad antiviral screening on a series of 2-phenyl-benzotriazole from our library (3k-n and 4k-n) [21] or newly synthetized (3f, j and 4f, j), all showed in Figure 2, in order to better understand the role of substituent in position C5 and C6 on the benzotriazole moiety and describe structure-activity relationships (SARs).

1a-c,e-j

a) NH_2
b) $NHCOCH_3$
c) $NHCOCH_2CH_3$
d) $N(COCH_2CH_3)_2$
e) $N(COCH_2CH_2CH_3)_2$
f) NHCO-4-CH_3-Ph
g) NHCO-4-Cl-Ph
h) NHCO-4-NO_2-Ph
i) NHCO-4-OCH_3-Ph
j) NHCO-3,4,5-OCH_3-Ph
k) $NHCONHCH_2CH_3$
l) $NHCONH(CH_2)_2CH_3$
m) $NHCONH(CH_2)_3CH_3$
n) NHCONH-cyclohexyl

Figure 1. 5,6-dichloro-1-phenyl-1*H*-benzo[*d*][1,2,3]triazoles (1a-c, e-j) and 5,6-dichloro-2-phenyl-2*H*-benzo[*d*][1,2,3]triazoles (2a-n).

3f, j-n   **4f, j-n**

f) NHCO-4-CH_3-Ph
j) NHCO-3,4,5-OCH_3-Ph
k) $NHCONHCH_2CH_3$
l) $NHCONH(CH_2)_2CH_3$
m) $NHCONH(CH_2)_3CH_3$
n) NHCONH-cyclohexyl

Figure 2. 2-phenyl-2*H*-benzo[*d*][1,2,3]triazoles (3f, j-n) and 5,6-dimethyl-2-phenyl-2*H*-benzo[*d*][1,2,3] triazoles (4f, j-n).

2. Materials and methods

2.1. Chemistry

5,6-dichloro-1-phenyl-1*H*-benzotriazoles (1a-c, e-j), 5,6-dichloro-2-phenyl-2*H*-benzotriazoles (2a-n), 1-(4-(2*H*-benzotriazol-2-yl)phenyl)-3-alkylureas (3k-n) and 1-(4-(5,6-dimethyl-2*H*-benzotriazol -2-yl)phenyl)-3-alkylureas (4k-n) were obtain as recently described [20,21].

2.2. General Procedure for Preparation of Derivatives 3f, j and 4f, j.

Scheme 1 represents the synthetic route for obtaining derivatives 3f, j and 4f, j.

To a stirred solution of 1.47 mmol of 4-(2*H*-benzo[*d*][1,2,3]triazol-2-yl)aniline (3a) or 4-(5,6-dimethyl-2*H*-benzo[*d*][1,2,3]triazol-2-yl)aniline (4a) in anhydrous *N,N*-dimethylformamide (8 mL), 4.98 mmol (ratio 1:3 for compound 3j) or 1.76 mmol (ratio 1:1 + 20% for compounds 3f, 4f, 4j) of proper benzoyl-chloride (5f, 5j) in anhydrous N,N-dimethylformamide (DMFa) were added. The mixture was stirred at 80 °C for 2 h (3j) or 24 h (3f, 4f, 4j). At the end, reaction mixture was cooled to room temperature and the solids were obtained by filtration. Mother liquor was dried by evaporation obtaining further solid. The crude solids were purified by flash chromatography, using a mixture of petroleum spirit/ethyl acetate in a ratio 7/3 as eluent, followed by crystallization from ethanol.

3a R= H
4a R= CH$_3$

5f R$_1$= 4-CH$_3$
5j R$_1$= 3,4,5-OCH$_3$

3f,j and 4f,j

Scheme 1. Synthesis the amide derivatives (3f, j and 4f, j).

2.2.1. *N*-(4-(2H-benzo[d][1,2,3]triazol-2-yl)phenyl)-4-methylbenzamide (3f)

Compound was obtained in 80% of total yield; m.p. 266–268 °C; TLC (petroleum spirit/ethyl acetate 1/1), Rf 0,90. ^{1}H-NMR (DMSO-d_6) δ: (1H, s, NHCO), 8.32 (2H, d, J = 8.4 Hz, H-2′, 6′), 8.09 (2H, d, J = 8.4 Hz, H-3′, 5′), 8.05-8.00 (2H, m, H-4, 7), 7.92 (2H, d, J = 7.6 Hz, H-2″, 6″), 7.53–7.50 (2H, m, H-5, 6), 7.37 (2H, d, J = 7.6 Hz, H-3″, 5″). ^{13}C-NMR (DMSO-d_6) δ: 165.60 (CO), 144.34 (C), 141.86 (2C), 140.36 (C), 134.90 (C), 131.72 (C), 128.94 (2CH), 127.77 (2CH), 127.43 (2CH), 120.89 (2CH), 120.68 (2CH), 118.04 (2CH), 21.00 (CH$_3$). Anal. Calcd for C$_{20}$H$_{16}$N$_4$O, C, 73.15; H, 4.91; N, 17.06. Found C, 73.23; H, 4.94; N, 16.98. MW 328.37, LC/MS: 329 (M + H).

2.2.2. *N*-(4-(2H-benzo[d][1,2,3]triazol-2-yl)phenyl)-3,4,5-trimethoxybenzamide (3j)

Compound was obtained in 30% of total yield; m.p. 227–229 °C; TLC (petroleum spirit/ethyl acetate 7/3), Rf 0,29. ^{1}H-NMR (DMSO-d_6) δ: 10.44 (1H, s, NHCO), 8.34 (2H, d, J = 8.4 Hz, CH-2′, 6′), 8.04–8.00 (4H, m, CH-4, 7, 3′, 5′), 7.53–7.50 (2H, m, CH-5, 6), 7.33-7.28 (2H, m, CH-2″, 6″). ^{13}C-NMR (DMSO-d_6) δ: 165.17 (CO), 152.65 (2C), 144.35 (2C), 140.36 (C), 135.02 (C), 129.71 (2C), 127.48 (2CH), 121.15 (2CH), 120.72 (2CH), 118.03 (2CH), 105.41 (2CH), 60.12 (OCH$_3$), 56.12 (2OCH$_3$). Anal. Calcd for C$_{22}$H$_{20}$N$_4$O$_4$, C, 65.34; H, 4.98, N, 13.85. Found C, 65.46; H, 5.20; N, 13.58. MW 404.42, LC/MS: 405 (M + H).

2.2.3. *N*-(4-(5,6-dimethyl-2H-benzo[d][1,2,3]triazol-2-yl)phenyl)-4-methylbenzamide (4f)

This compound was obtained in 46% of total yield; m.p. 295–296 °C; TLC (petroleum spirit/ethyl acetate 7/3), Rf 0,53. ^{1}H-NMR (DMSO-d_6) δ: 10.46 (1H, s, NH), 8.25 (2H, d, J = 9.2 Hz, H-2′, 6′), 8.05 (2H, d, J = 9.2 Hz, H-3′, 5′), 7.91 (2H, d, J = 8.4 Hz, H-2″, 6″), 7.76 (2H, s, H-4, 7), 7.37 (2H, d, J = 8.4 Hz, H-3″, 5″), 2.51 (3H, s, CH$_3$), 2.40 (6H, s, 2CH$_3$). ^{13}C-NMR (DMSO-d_6) δ: 165.57 (CO), 143.79 (2C), 141.85 (C), 139.88 (C), 137.86 (2C), 135.06 (C), 131.73 (C), 128.95 (2CH), 127.74 (2CH), 121.40 (2CH), 120.90 (2CH), 116.17 (2CH), 29.96 (CH$_3$), 20.99 (2CH$_3$). Anal. Calcd for C$_{22}$H$_{20}$N$_4$O, C, 74.14; H, 5.66; N, 15.72. Found C, 74.18; H, 5.82; N, 15.67. MW 356.42; LC/MS: 457 (M + H).

2.2.4. *N*-(4-(5,6-dimethyl-2H-benzo[d][1,2,3]triazol-2-yl)phenyl)-3,4,5-trimethoxybenzamide (4j)

This compound was obtained in 20% of total yield; m.p. 272–273 °C; TLC (petroleum spirit/ethyl acetate 6/4), Rf 0,34. ^{1}H-NMR (DMSO-d_6) δ: 10.40 (1H, s, NH), 8.28 (2H, d, J = 8.8 Hz, CH-2′, 6′), 8.02 (2H, d, J = 9.2 Hz, CH-3′, 5′), 7.76 (2H, s, CH-2″, 6″), 7.32 (2H, s, CH-4, 7), 3.90 (6H, s, 2OCH$_3$), 3.75 (3H, s, OCH$_3$), 2.41 (6H, s, 2CH$_3$). ^{13}C-NMR (DMSO-d_6) δ: 165.11 (CO), 152.64 (2C), 143.8 (2C), 140.47 (C), 139.68 (C), 137.88 (2C), 135.18 (C), 129.73 (C), 121.14 (2CH), 120.32 (2CH), 116.26 (2CH), 105.38 (2CH), 60.12 (OCH$_3$), 56.12 (2OCH$_3$), 20.41 (2CH$_3$). Anal. Calcd for C$_{24}$H$_{24}$N$_4$O$_4$, C, 66.65; H, 5.59; N, 12.96. Found C, 69.46; H, 5.82; N, 12.58. MW 432.47; LC/MS: 433 (M + H).

2.3. Cells, Viruses, Reagents

All experimental work involving viruses was performed in a biosafety level 3 (BSL3) containment laboratory. Hantaan (HTNV) (strain 76–118) (kindly provided by Prof. Dr. D. H. Krüger and Dr. Boris Klempa, Institute of Virology, Charitè University of Berlin) was used for all experiments. Vero E6 cells (ATCC CRL 1586) were maintained in complete EMEM (minimum essential medium with Earle's salt, 25mM Hepes supplemented with 10% fetal bovine serum, 1% glutamine, 1% sodium pyruvate (NaPy), 1% non-essential amino acids NEAA, 0.1% gentamicyn). RBV was purchased from Sigma–Aldrich Co. (St. Louis, MO, USA). The primary antibody, rabbit anti-Malacky Ab, was obtained from Dr. Boris Klempa, Berlin. The secondary antibody was goat anti-rabbit IgG (H+L) HRP conjugate (Cat. No. 170-6515, Bio-Rad, Inc., CA, USA). SuperSignal West Pico Chemiluminescent Substrate (Cat. No. 34080, Thermo Scientific, MA, USA)

2.4. Cytotoxicity Assays

Vero E6 cells were seeded at an initial density of 4×10^5 cells/mL in 6-well plates, in culture medium (EMEM 25mM HEPES buffer) supplemented with 1% L-glutamine, 10% fetal bovine serum (FBS), 1% NaPy, 1% NEAA, and 0.1% gentamycin. Cell cultures were then incubated at 37 °C in a humidified, 5% CO$_2$ atmosphere in the absence or presence of serial dilutions of test compounds. Cell viability was determined after 7 days at 37 °C by the crystal violet staining method.

2.5. Orthohantavirus Antiviral Screening Assay

Vero-E6 cells were seeded in 6-well plates at an initial density of 4×10^5 cells/mL. Cell cultures were then infected and incubated at 37 °C in a humidified, 5% CO$_2$ atmosphere for 7 days in the absence or presence of serial dilutions of test compounds. The antiviral activity was determined by chemiluminescence focus reduction assay (C-FRA), as described previously [22] with small modifications. Briefly, after 7 days of incubation, overlay medium were discarded and cell monolayers were fixed with ice-cold methanol for 8 min at room temperature. After fixing, the methanol was removed, and the cells let dry. Cells were rinsed twice with washing buffer, and then a rabbit anti-Malacky antibody was added at 1:1000 dilution for 1 h at 37 °C in a humidified, 5% CO$_2$ atmosphere. Cells were then washed five times with washing buffer and a secondary goat anti-rabbit IgG antibody conjugated to horseradish peroxidase (HRP) was added for an additional hour in 5% CO$_2$ atmosphere. After washing five times with washing buffer cells were incubated with chemiluminescent substrate immediately before the detection. Infected cell foci were detected with the CCD camera for 3 min (or Chemidoc for 30 sec).

2.6. Yield Reduction assay

Vero E6 cells were inoculated with HTNV at an m.o.i. of 0.05 in maintenance medium and tested compounds at non-cytotoxic concentrations. Following a 1-h adsorption period at 37 °C and 5% CO$_2$ on a rocking platform, the inoculum was removed and replaced with fresh medium containing 20 µM concentration of compounds (2j, 2l, 2n). After 3 days, the cell supernatant was harvested and centrifuged (at 3000 rpm, at 4 °C, for 10 min) to remove debris and measured for the presence of infectious orthohantavirus hantavirus. The infectious progeny virus in the cell supernatant was evaluated by chemiluminescence focus reduction assay, as described above [22]. RBV was used as the reference compound.

2.7. Statistical Analysis

Data are represented as mean ± standard deviation (SD) unless otherwise stated. For the yield reduction assay, statistical comparisons were performed using the unpaired *t*-test, and *p*-values less than 0.05 were considered to be statistically significant. All analyses were performed with GraphPad Prism 6.

3. Results and Discussion

In Figure 3, we report the effects of compounds 5,6-dichloro-2-phenyl-2H-benzo[d][1,2,3]triazoles (2n, 2l, 2j, 2f, 2k) on the replication of HTNV in comparison to RBV, a known broad-spectrum antiviral agent. Series 1, 3, and 4 derivatives turned out as completely inactive and are not reported here (see Table S1).

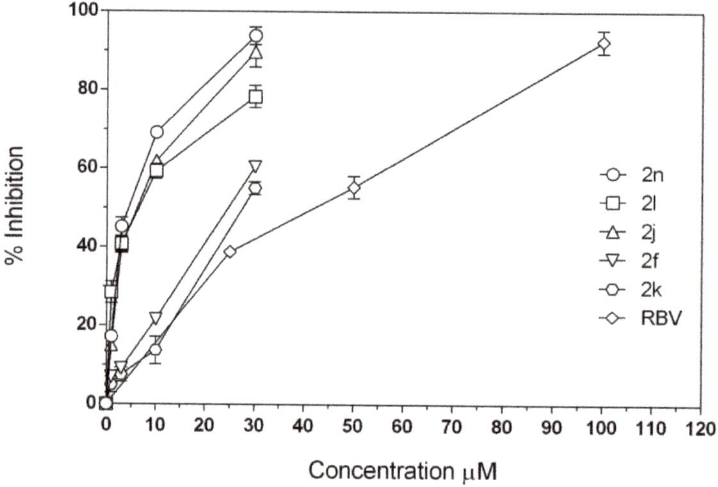

Figure 3. Anti-hantaviral effect of 2n, 2l, 2j, 2f, 2k, and RBV in vitro. Each symbol shows mean values for three independent determinations + SD.

Compounds 1a-c, 1e-j, 2a-n, 3f, 3j-n, and 4f, 4j-n, were evaluated for their potential inhibitory activity against HTNV using a chemiluminescence focus reduction assay (C-FRA). Three compounds (2j, 2l, 2n) showed interesting inhibitory activity (i.e., $\leq$5 μM; Figure 3). In parallel, moderate to low cytotoxicity was detected for almost all compounds, with CC_{50} values mostly in the high micromolar range (>30 μM) against all tested cell lines (see Table S1).

From a structure–activity relationship perspective, the most relevant results concerned the potent and selective activity of two urea derivatives (2l and 2n) and one amide derivative (2j), all belonging to the series of the 5,6-dichloro-2-phenyl-derivatives. By contrast, none of the molecules of the series 5,6-dichloro-1-phenyl-derivatives exhibited anti-HTNV activity, with only the exception of the compound 1h (EC_{50} = 21 μM). Regarding the presence of the two chlorine atoms in positions C5 and C6 of benzotriazole, their elimination (3f, j-n) or substitution with methyl groups (4f, j-n) results in total loss of antiviral activity. Moreover, we can highlight that derivative 2l (R= NHCONH-propyl) showed an EC_{50} value of 5 μM, but when decreasing or increasing the steric hindrance of the side chain with ethyl-group (2k) or butyl-group (2m), it turned out in a total loss of antiviral activity.

Derivative 2n, recently described by our research group as an interesting inhibitor of HRSV entry [16], showed a remarkable EC_{50} value of 4 μM against HTNV, resulting in being ten-fold more potent than RBV (EC_{50} = 37 μM); Figure 4.

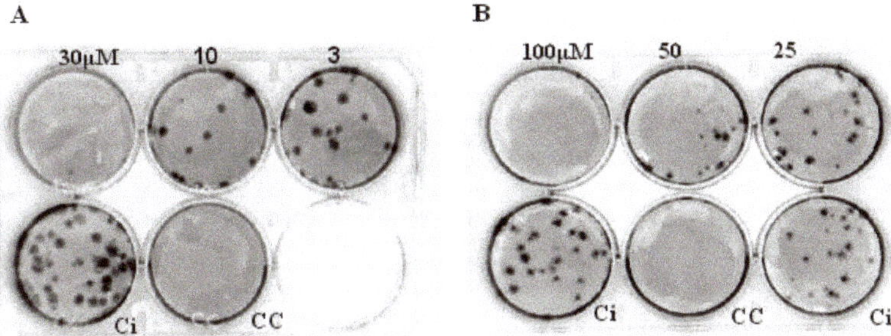

Figure 4. Focus reduction assay for 2n and ribavirin. VeroE6 cells were infected with HTNV (m.o.i. 0.05). The infected cultures were treated with 2n (panel **A**) and RBV (panel **B**; used as reference) at indicated doses. Ci = untreated control; CC = uninfected control. Antiviral activity was determined by focus reduction assay at day 7 post-infection.

2n also resulted in being more active in in vitro assays against HTNV than two orthohantavirus candidate drugs: ETAR (1-beta-d-ribofuranosyl-3-ethynyl-[1,2,4] triazole) (EC_{50} = 10 μM) and Favipiravir (EC_{50} =150.8 μM), respectively 2- and 37-fold more potent [13,23].

Accordingly, to obtain a detailed insight on the efficacy of 2j, 2l, and 2n against HTNV, a yield reduction assay was performed. The reduction of virus titer in the presence of the active compound during a single round of viral infection was determined. Treatment with non-cytotoxic 20 μM concentration of 2j, 2l and 2n caused a significant reduction of viral titer (Figure 5). A significant decline in viral titer of HTNV was also observed at 50 μM treatment of RBV (* *p*-value <0.05, unpaired *t*-test).

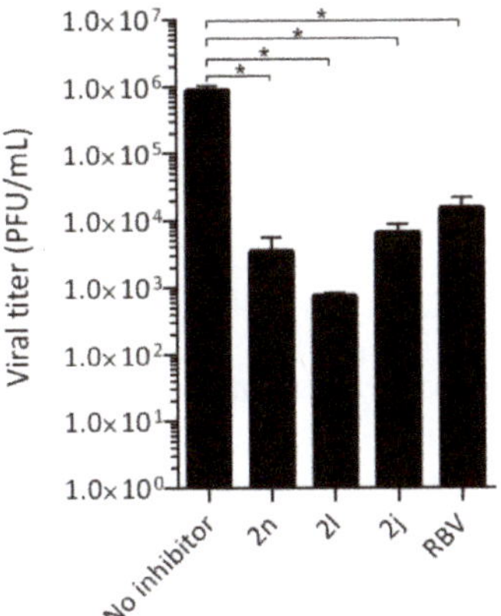

Figure 5. Yield of infectious Hantaan (HTNV) viruses produced in infected VeroE6 cells treated with selected benzotriazoles (2j, 2l, 2n) (20 μM) and RBV (50 μM). Results are expressed as means ± standard deviations from 3 separate experiments done in triplicate. * indicates the *p*-value <0.05.

VeroE6 cells were infected with HTNV (m.o.i. 0.05). The infected cultures were treated with 2j, 2l, 2n at indicated doses, and RBV was used as a reference compound. Viral yields in the culture supernatant were determined by viral titer reduction assay at day 3 post-infection.

In our assay, benzotriazole derivatives 2n, 2l, 2j (20 μM) determined a very interesting reduction of viral titer compared to control cells infected in the absence of inhibitors. The production of the virus was significantly reduced upon treatment with 2l (3,08 log10). The same trend of reduction in viral loads was detected for 2n (2,41 log10) and 2j (2.14 log10). Conversely, in order to obtain a titer decrease, less than 2 logs (1,78 log10) if compared to untreated control, the reference molecule ribavirin needed to be added at 50 μM concentration.

4. Conclusions

Over the past 50 years, hantavirus infections have caused severe diseases with serious clinical effects on human health. Recently, this has become more evident with the emergence of new outbreaks. As mentioned, there are no FDA approved antivirals, vaccines, or immunotherapeutic agents available for the treatment of HFRS or HPS. The objective of this study was to evaluate the antiviral activity of 2-phenyl-benzotriazoles against HTNV.

All compounds were generally endowed with low cytotoxicity against cell monolayers employed in our assays to support the replication of HTNV but also against a panel of cell cultures, as reported in SM2. We identified three promising lead compounds, derivatives 2j, 2l and 2n, characterized by interesting inhibitory activity against HTNV, in vitro ten-fold more potent than the nucleoside analogue RBV, currently the only antiviral with recognized in vitro and in vivo activity. This is the first report detailing benzotriazole anti-hantavirus activity. In conclusion, the therapeutic potential of these derivatives could be considered as a good starting point for the development of second-generation effective candidates for treatment against orthohantavirus infections.

Supplementary Materials: The following are available online at http://www.mdpi.com/1999-4915/12/1/122/s1, Table S1: Cytotoxicity and antiviral activity of phenyl-benzotriazoles against HTNV.

Author Contributions: G.S., R.L., S.P. and A.C. conceived and designed the experiments. S.P., P.C., R.I. and G.M. synthesized the compounds. G.S., B.K., S.M. and B.B. performed virus-related experiments. All authors analyzed data and contributed new reagents analytic tools. G.S., R.L., S.P., and A.C. wrote the paper. All authors reviewed and approved the manuscript.

Funding: This research was funded by MIUR PRIN 2015, grant number PRIN 2015C7PCYZ_002" and Assessorato della Programmazione, Bilancio, Credito e Assetto del territorio, della Regione Autonoma della Sardegna" (Italy), Grant number Legge Regionale 7 agosto 2007: CRP1_574". The APC was funded by PRIN 2015C7PCYZ_002".

Acknowledgments: The authors are highly grateful to D. H. Krüger, Institute of Virology, Charitè University of Berlin for providing Hantaan virus strain.

Conflicts of Interest: The authors declare that they have no competing interests.

References

1. Schmaljohn, C.; Hooper, J.W. Bunyaviridae: The viruses and their replication. In *Fields Virology*, 4th ed.; Knipe, D.M., Howley, P.M., Griffin, D.E., Lamb, R.A., Martin, M.A., Roizman, B., Straus, S.E., Eds.; Lippincott Williams & Wilkins: Philadelphia, PA, USA, 2001; pp. 1581–1602.
2. Shi, M.; Lin, X.D.; Chen, X.; Tian, J.H.; Chen, L.J.; Li, K.; Wang, W.; Eden, J.S.; Shen, J.J.; Liu, L.; et al. The evolutionary history of vertebrate RNA viruses. *Nature* **2018**, *556*, 197–202. [CrossRef] [PubMed]
3. Jonsson, C.B.; Figueiredo, L.T.M.; Vapalahti, O. A Global Perspective on Hantavirus Ecology, Epidemiology, and Disease. *Clin. Microbiol. Rev.* **2010**, *23*, 412–441. [CrossRef]
4. Hjelle, B.; Torres-Pérez, F. Hantaviruses in the Americas and Their Role as Emerging Pathogens. *Viruses* **2010**, *2*, 2559–2586. [CrossRef] [PubMed]
5. Gavrilovskaya, I.N.; Brown, E.J.; Ginsberg, M.H.; Mackow, E.R. Cellular entry of hantaviruses which cause hemorrhagic fever with renal syndrome is mediated by β3 integrins. *J. Virol.* **1999**, *73*, 3951–3959. [CrossRef]
6. Jonsson, C.B.; Schmaljohn, C.S. Replication of hantaviruses. *Curr. Top. Microbiol. Immunol.* **2001**, *256*, 15–32.
7. Sidwell, R.W.; Smee, D.F. Viruses of the Bunya- and Togaviridae families: Potential as bioterrorism agents and means of control. *Antivir. Res.* **2003**, *57*, 101–111. [CrossRef]
8. Ogg, M.; Jonsson, C.B.; Camp, J.V.; Hooper, J.W. Ribavirin Protects Syrian Hamsters against Lethal Hantavirus Pulmonary Syndrome—After Intranasal Exposure to Andes Virus. *Viruses* **2013**, *5*, 2704–2720. [CrossRef] [PubMed]
9. Krüger, D.H.; Schönrich, G.; Klempa, B. Human pathogenic hantaviruses and prevention of infection. *Hum. Vaccin.* **2011**, *7*, 685–693. [CrossRef]

10. Huggins, J.W.; Hsiang, C.M.; Cosgri, T.M.; Guang, M.Y.; Smith, J.I.; Wu, Z.O.; LeDuc, J.W.; Zheng, Z.M.; Meegan, J.M.; Wang, Q.N.; et al. Prospective, double-blind, concurrent, placebo-controlled clinical trial of intravenous ribavirin therapy of hemorrhagic fever with renal syndrome. *J. Infect. Dis.* **1991**, *164*, 1119–1127. [CrossRef]

11. Malinin, O.V.; Platonov, A.E. Insufficient efficacy and safety of intravenous ribavirin in treatment of haemorrhagic fever with renal syndrome caused by puumala virus. *Infect. Dis.* **2017**, *49*, 514–520. [CrossRef]

12. Chapman, L.E.; Mertz, G.J.; Peters, C.J. Intravenous ribavirin for hantavirus pulmonary syndrome: Safety and tolerance during 1 year of open-label experience. Ribavirin Study Group. *Antivir. Ther.* **1999**, *4*, 211–219. [PubMed]

13. Chung, D.H.; Kumarapperuma, S.C.; Sun, Y.; Li, Q.; Chu, Y.K.; Arterburn, J.B.; Parker, W.B.; Smith, J.; Spik, K.; Ramanathan, H.N.; et al. Synthesis of 1-beta-D-ribofuranosyl-3-ethynyl-[1,2,4]triazole and its in vitro and in vivo efficacy against Hantavirus. *Antivir. Res.* **2008**, *79*, 19–27. [CrossRef] [PubMed]

14. Brocato, R.L.; Hooper, J.W. Progress on the Prevention and Treatment of Hantavirus Disease. *Viruses* **2019**, *11*, 610. [CrossRef] [PubMed]

15. Carta, A.; Loriga, G.; Piras, S.; Paglietti, G.; Ferrone, M.; Fermeglia, M.; Pricl, S.; La Colla, P.; Secci, B.; Collu, G.; et al. Synthesis and in vitro evaluation of the anti-viral activity of N-[4-(1H(2H) benzotriazol-1(2)-yl)phenyl]alkylcarboxamides. *Med. Chem.* **2006**, *2*, 577–589. [CrossRef] [PubMed]

16. Carta, A.; Loriga, M.; Piras, S.; Paglietti, G.; Ferrone, M.; Fermeglia, M.; Pricl, S.; La Colla, P.; Collu, G.; Sanna, T.; et al. Synthesis and anti-picornaviridae in vitro activity of a new class of helicaseinhibitors the N,N'-bis[4 -(1H(2H)-benzotriazol-1(2)-yl)phenyl] alkyldicarboxamides. *Med. Chem.* **2007**, *3*, 520–532. [CrossRef]

17. Carta, A.; Pricl, S.; Piras, S.; Fermeglia, M.; La Colla, P.; Loddo, R. Activity and molecular modeling of a new small molecule active against NNRTI-resistant HIV-1 mutants. *Eur. J. Med. Chem.* **2009**, *44*, 5117–5122. [CrossRef]

18. Carta, A.; Briguglio, I.; Piras, S.; Corona, P.; Boatto, G.; Nieddu, M.; Giunchedi, P.; Marongiu, M.E.; Giliberti, G.; Iuliano, F.; et al. Quinolinetricyclic derivatives. Design, synthesis and evaluation of the antiviral activity of three new classes of RNA-dependent RNA polymerase inhibitors. *Bioorg. Med. Chem.* **2011**, *19*, 7070–7084. [CrossRef]

19. Briguglio, I.; Piras, S.; Corona, P.; Gavini, E.; Nieddu, M.; Boatto, G.; Carta, A. Benzotriazole: An overview on its versatile biological behavior. *Eur. J. Med. Chem.* **2015**, *97*, 612–648. [CrossRef]

20. Piras, S.; Sanna, G.; Carta, A.; Corona, P.; Ibba, R.; Loddo, R.; Madeddu, S.; Caria, P.; Aulic, S.; Laurini, E.; et al. Dichloro-phenyl-benzotriazoles: A new selective class of humanrespiratory syncytial virus entry inhibitors. *Front. Chem.* **2019**, *7*, 247. [CrossRef]

21. Piras, S.; Corona, P.; Ibba, R.; Riu, F.; Sanna, G.; Madeddu, S.; Delogu, I.; Loddo, R.; Carta, A. Novel Benzotriazole Derivatives as Selective Inhibitors of Coxsackies B5 virus: 1-[4-(5,6-dimethyl(H)-1H(2H)-benzotriazol-1(2)-yl)phenyl]-3-alkyl(aryl)urea. *Med. Chem.* **2020**, *16*, 1. [CrossRef]

22. Heider, H.; Ziaja, B.; Priemer, C.; Lundkvist, A.; Neyts, J.; Krüger, D.H.; Ulrich, R. A chemiluminescence detection method of hantaviral antigens in neutralisation assays and inhibitor studies. *J. Virol. Methods* **2001**, *96*, 17–23. [CrossRef]

23. Ye, C.; Wang, D.; Liu, H.; Ma, H.; Dong, Y.; Yao, M.; Wang, Y.; Zhang, H.; Zhang, L.; Cheng, L.; et al. An Improved Enzyme-Linked Focus Formation Assay Revealed Baloxavir Acid as a Potential Antiviral Therapeutic Against Hantavirus Infection. *Front. Pharmacol.* **2019**, *10*, 1203. [CrossRef] [PubMed]

Article

Genistein Has Antiviral Activity against Herpes B Virus and Acts Synergistically with Antiviral Treatments to Reduce Effective Dose

Julia C. LeCher [1,2,*], Nga Diep [2], Peter W. Krug [2] and Julia K. Hilliard [2]

[1] Department of Molecular and Cellular Biology, Kennesaw State University, Kennesaw, GA 30189, USA
[2] Viral Immunology Center, Department of Biology, Georgia State University, Atlanta, GA 30303, USA;
    diepnga103@gmail.com (N.D.); pkrug@gsu.edu (P.W.K.); jhilliard@gsu.edu (J.K.H.)
* Correspondence: jwand@kennesaw.edu

Received: 9 May 2019; Accepted: 30 May 2019; Published: 31 May 2019

Abstract: Herpes B virus is a deadly zoonotic agent that can be transmitted to humans from the macaque monkey, an animal widely used in biomedical research. Currently, there is no cure for human B virus infection and treatments require a life-long daily regimen of antivirals, namely acyclovir and ganciclovir. Long-term antiviral treatments have been associated with significant debilitating side effects, thus, there is an ongoing search for alternative efficacious antiviral treatment. In this study, the antiviral activity of genistein was quantified against B virus in a primary cell culture model system. Genistein prevented plaque formation of B virus and reduced virus production with an IC_{50} value of 33 and 46 µM for human and macaque fibroblasts, respectively. Genistein did not interfere directly with viral entry, but instead targeted an event post-viral replication. Finally, we showed that genistein could be used at its IC_{50} concentration in conjunction with both acyclovir and ganciclovir to reduce their effective dose against B virus with a 93% and 99% reduction in IC_{50} values, respectively. The results presented here illuminate the therapeutic potential of genistein as an effective antiviral agent against B virus when used alone or in combination with current antiviral therapies.

Keywords: herpes B virus; macacine herpesvirus-1; genistein; flavonoids; acyclovir; ganciclovir; antiviral agents

---

1. Introduction

Herpes B virus (Macacine herpesvirus-1) is an alphaherpesvirus endemic in macaque monkeys, an animal widely used in biomedical research [1]. B virus is of significant interest because it is capable of potentially deadly human zoonosis and thus poses a major occupational hazard to researchers, veterinarians, and animal handlers. B virus can be spread to humans via direct contact with infectious macaque body fluids and tissues during bites, scratches or splashes. Pathogenesis typically occurs within two weeks post-infection, with symptoms ranging from influenza-like illness, headache, vesicular lesions, ataxia and/or ascending paralysis. In humans, B virus is highly neuroinvasive and can lead to severe neurologic complications, such as transverse myelitis, encephalitis and potentially death. Without timely diagnosis and proper treatment, the mortality rate for infected individuals is as high as 70% [1–5]. Currently, there is no cure for human B virus infection, only life-long antiviral treatments, and patients live under the constant threat of potential virus reactivation.

The survival rate following human B virus infection is 80% when antiviral treatments are started early in infection [4]. Acyclovir (ACV) or ganciclovir (GCV) are typically the first drugs given in most clinical cases of B virus infection [1]. Following infection, antivirals are administered intravenously to put the virus into a latent state; then patients will remain on an oral antiviral regimen throughout their lifetime to prevent virus reactivation. ACV and GCV are attractive antivirals against B virus because they enter the host cell in an inactive form and must be phosphorylated intracellularly by virus-encoded thymidine kinase to become active [6–8]. However, these antivirals have several limitations. ACV, for example, has poor solubility in water, a short plasma half-life, and low blood-brain barrier passage. Following oral administration of ACV, there is a 20% bioavailability of the drug in the plasma and even lower levels in the cerebrospinal fluid [9,10]. In addition, in vitro studies have shown that B virus is approximately 10-fold less sensitive to ACV than the closely related human herpessimplex virus-1 (HSV-1) [8]. Thus, ACV must be administered in significantly greater concentration to control the disease, especially when intravenous administration is required. Increased doses and long-term administration comes with increased risk of negative side-effects. ACV can form crystals in the renal tubules during clearance from the kidney and, consequently, patients on ACV treatment often suffer from neurotoxicity associated with impaired renal function [11–13]. Further, there are reports of resistance of HSV-1 to ACV when patients are infected with virus isolates that carry mutations in either the DNA polymerase and/or the thymidine kinase gene [14]. It is anticipated that similar problems will ultimately arise when using ACV as a prophylactic treatment for B virus infection. Clearly, it is critical that new drugs be identified to provide a safer and more effective intervention against primary and reactivated B virus infections.

Genistein is a flavonoid found in plants of the Fabaceae (bean) family and in soy-based foods. In plants, flavonoids serve many functions critical to plant physiology. In humans, they possess many pharmacologic properties and therapeutic potential, such as antioxidant, anti-microbial, anti-carcinogenic, and anti-inflammatory activity, and have been summarized in several reviews [15,16]. In 1999, the Food and Drug Administration approved the food labeling of health claims of soy-based foods in reducing the risk of coronary heart diseases [17]. Since 1998, there have been 65 Phase I-III clinical trials registered with the National Institute of Health assessing the biological benefits of genistein and its derivatives [18].

Genistein is a general protein kinase inhibitor with selectivity towards tyrosine residues. Tyrosine phosphorylation regulates many cellular processes, including cell proliferation and angiogenesis. Therefore, genistein has been widely recognized as an anticancer agent, though the idea of flavonoids acting as antivirals is far from novel. Genistein has been shown to have antiviral properties against several herpesviruses including HSV-1, cytomegalovirus, and bovine herpesvirus-1, as well as non-herpesviruses, including SV40, human papilloma virus, porcine reproductive and respiratory syndrome virus, African swine fever virus, and human immunodeficiency virus [19–35]. Given the similarities of B virus to other herpesvirus against which genistein exhibits antiviral activities, we hypothesized that genistein would exhibit antiviral activity against B virus and, further, that genistein could be used in conjunction with current antiviral treatments to reduce their effective dose and enhance their antiviral properties.

2. Materials and Methods

2.1. Cells, Media, and Virus

Human foreskin fibroblasts (HFFs; ATCC® CCD-1112SK™, Manassas, VA, USA), Rhesus macaque foreskin fibroblasts (RMF; isolated in house), and African Green Monkey kidney cells (Vero; ATCC® CCL-81™, Manassas, VA, USA) were used in this study. The media compositions employed were (1) HFF: minimum essential medium (MEM), 10% inactivated fetal bovine serum (ΔFBS; Atlas Biologicals F-0050-A, Fort Collins, CO, USA), 100 U/mL penicillin-streptomycin (pen-strep; Mediatech/Corning #30-002-CL Tweaksbury, MA, USA), 1% sodium pyruvate (Mediatech/Corning #25-000-CL Tweaksbury, MA, USA), 1% glutaMAX™ (Gibco # 35050-061), and 1% non-essential amino acids (Mediatech/Corning #25-025-CL Tweaksbury, MA, USA); (2) RMF: Dulbecco's modified essential medium (DMEM; Sigma-Aldrich #D-6046, St. LouisMOUSA), 18% ΔFBS, and 100 U/mL pen-strep; and (3) Vero: DMEM, 10% ΔFBS, 100 U/mL pen-strep. For all experiments, cells were grown at 37 °C in a 95% O_2, 5% CO_2 incubator. B virus (laboratory strain E2490) was propagated in Vero cells. Virus stock titer was determined by standard plaque assay in Vero cells. Propagation and harvesting of B virus isolates were performed in the Georgia State University (GSU) BSL4 Laboratory in accordance with the guidelines of the 5th edition of Biosafety in Microbiological and Biomedical Laboratories.

2.2. Chemicals and Reagents

Stock solutions of acyclovir (ACV) and genistein were prepared at 100 mM in dimethyl sulfoxide (DMSO) and ganciclovir (GCV) were prepared at 40 mM in sterile water. Citrate buffer was prepared with 50 mM sodium citrate and 4 mM potassium chloride in sterile water, pH 3. Methylcellulose was prepared as a 1% solution in MEM. Crystal violet was prepared as a 0.5% solution in diH_2O with 20% ethanol.

2.3. Trypan Blue Dye Exclusion Assay

Cells were grown to ~75% confluency in 48-well plates. For experiments using synchronized cells, cells were serum-starved for ~18 h prior to treatment. Cells were then treated with 0–100 µM genistein or 0.1% DMSO (control) and grown for 48 h. Cells were collected by trypsinization (0.25% trypsin-EDTA; Gibco #25200-056), stained with trypan blue dye (Gibco #15250-061) and counted on a hemocytometer. All experiments were run in triplicate. Cell viability was calculated as the percent of live cells divided by the total number of cells.

2.4. Cell Proliferation Assay

Cell proliferation was measured in two ways. First, cells were grown, treated, and collected as above in the Trypan Blue Dye Exclusion Assay and then counted to obtain the total number of cells. This number was compared to the starting number of cells at the time of seeding. Secondly, cells were assayed by MTS assay (Promega-CellTiter 96® AQueous one solution cell proliferation assay). Cells were seeded into 96-well plates at ~75% confluency, allowed to adhere for ~6 h, then serum-starved for ~18 h before genistein treatment (0–100 µM) or 0.1% DMSO (control) was added to cells in fresh serum-containing medium. Cells were grown for an additional 48–72 h, then assayed for metabolic activity following the manufacturer's protocol. Metabolic activity was calculated by subtracting background EU values (cells with no MTS reagent) from total EU values, then setting the untreated group to 100% and comparing the treated and untreated groups. All experiments were run in triplicate.

2.5. Bromodeoxyuridine (BrdU) Incorporation Assay

Cells were grown to ~90% confluency on 8-well chamber well slides (NUNC™ Lab-Tek™), then serum-starved for ~18 h, after which genistein treatment (0–100 µM) or 0.1% DMSO (diluent control) was added to cells in serum-containing medium with BrdU labeling reagent (Invitrogen 00-0103, Carlsbad, CA, USA). Cells were incubated for 4 h, then fixed and stained for BrdU following the manufacturer's protocol (ThermoFisher, Waltham, MA, USA). Region of interest (ROI)analysis was performed using Zeiss AxioVision rel 4.8 software.

2.6. DNA Isolation and RT-PCR

DNA was isolated from cells using the manufacturer's protocol with slight modification. Viral DNA was detected by real-time PCR using a 6-carboxyfluorescein/tetramethylrhodamine (FAM/TAMRA)-labeled probe with primers to glycoprotein G (F 5′ 3′ R 5′ 3′). DNA isolated from Vero-infected cells was used as a positive control. Average C_T values were calculated from replicate groups then compared to the uninfected control group.

2.7. Indirect Virus Yield Assay

HFFs or RMFs were grown to confluency in 12-well plates, then infected with 150 PFU of B virus for 1 h in the presence of 0–100 µM genistein or DMSO control. After absorption, the media was removed and fresh media added. After 48 h, cells/supernatants were collected, serially diluted, and added to Veros grown to confluency in 12-well plates for standard plaque assay.

2.8. Plaque Reduction Assay

Cells were grown to confluency in 12-well plates then infected with 150 PFU of B virus for 1 h in the presence of 0–100 µM genistein or DMSO control. After absorption, the media was removed and cells overlaid with 0.5% methylcellulose with the indicated concentration of genistein. To assay the effects of the drugs on virus replication, the old methylcellulose was replaced with new methylcellulose containing 50 µM genistein at 0, 2, 4, 6, and 8 h post-infection. For combination plaque assays, infected cells were treated with 50 µM genistein and 0–100 µM ACV or 0–40 µM GCV, both during and after the absorption period. In all cases, at 48 h post-infection (hpi), cells were fixed with 100% methanol and plaques were visualized and counted either by immunohistochemistry using BV+ macaque sera as the primary antibody and HRP-conjugated goat-antihuman as the secondary antibody with DAB detection or by staining with crystal violet. IC_{50} values were calculated using logarithmic regression line generated from a semi-log dose–response curve.

2.9. Direct Inactivation Using Modified Plaque Reduction Assay

150 PFU of B virus were incubated with 0 µM or 50 µM genistein at 37 °C for 15, 30, 60, 90, and 120 min. At the end of the pre-incubation period, the virus inoculum was removed, diluted 10-fold, then used to infect Veros. All treatments were carried out in duplicate. After 1 h adsorption, cells were overlaid with methylcellulose and incubated at 37 °C for 48 h. Plaques were then counted and the titer calculated.

2.10. Virus Yield Assay

Vero cells were grown to confluency in 6-well plates then infected with B virus at a multiplicity of infection (MOI) of 1 and incubated for 1 h to allow for viral attachment. Cells were washed with 1 mL of 50 mM citrate buffer for 2 min to remove unabsorbed viruses, then rinsed twice with sterile PBS. Next, 0–100 µM of genistein or 0.1% DMSO was added to the infected cells and infection was continued for 24 h. The infected cells were harvested by scraping and lysed by three cycles of alternating freezing and thawing. The total infectious virus yields in the presence of each drug concentrations tested were determined by titration on Vero using a standard plaque assay.

2.11. Cell-Based ELISA Assay

HFFs and RMFs were grown to confluency in 12-well plates, then infected with 150 PFU of B virus for 1 h in the presence of 0–100 µM genistein or DMSO control. After 1 h absorption, the media was removed and fresh media added. After 48 h, cells/supernatants were collected, serially diluted, and added to Vero grown to confluency in 96-well plates. A standard curve was generated from 10-fold serial dilutions of B virus (1:100–10^9). After 1 h absorption, the inoculum was removed, fresh media added and infection was continued for 24 h. Next, cells were fixed with 85% acetone, washed three times with borate-buffered salt solution containing 0.05% Tween 20 (BBST), blocked for one hour in ELISA-Blotto (BBST + 2.5% dry milk), and incubated with rabbit anti-gB sera (generated from injection of rabbits with a recombinant B virus gB protein, as previously described [36]) at 1:50 dilution overnight at 4 °C. The primary antibody was removed, cells were washed three times in BBST, incubated with goat anti-rabbit IgG-alkaline phosphatase for 1 h and washed again three times in BBST. For detection, 200 µL of p-nitrophenyl phosphate (pNPP) substrate was added to each well and incubated for 30 min at room temperature. OD values were generated by reading absorbance on an ELISA microplate reader at 405 nm and 490 nm wavelengths. The average OD values were calculated from quadruplicates per experiment. Net OD values were obtained by subtracting the average OD values of the background control (ODbkg) from the average OD values of the test samples (ODts). A titration curve was constructed by plotting the net OD values of the virus standard versus the experimental virus dilution. The titer of the virus was determined from the curve using a cutoff valued that is determined from the background controls. We used the average of the background control OD values plus three standard deviations as the cutoff. The titer of the experimental samples was calculated from the standard curve.

2.12. Statistical Analysis

All analyses were performed either in Excel (*t*-test) or using GraphPad Prism 6 software (One-way ANOVA with correction for multiple comparisons—Dunnet's correction when comparing all groups and Tukey's correction when comparing all groups to the control group). Deviation from the mean was calculated using the standard deviation (Excel) or the standard error of the mean (GraphPad Prism 6 software).

3. Results

3.1. Genistein Does Not Exhibit Cytotoxic Effects on Primary Fibroblasts

Primary B virus infection occurs at the epithelial layer of the skin and/or mucosa. Thus, we chose to test genistein's efficacy in primary human and macaque fibroblasts (HFF and RMF respectively). Genistein, in certain cell types, has been reported to inhibit cellular proliferation via cell cycle arrest, so we first verified that genistein would not have any cytotoxic effects on primary fibroblasts at doses used in this study. We tested the effects of genistein on cell viability, proliferation, metabolic activity, and DNA synthesis. Cell viability was measured at 48 h post-genistein treatment via trypan blue dye exclusion assay. Genistein had no notable effect on cell viability at 48 h post-treatment (Figure 1A, \B). Next we asked if genistein affected cell proliferation. Initially, we assayed cellular division in an asynchronistic population of cells and saw no effect (Figure 1C,D). We then tested if genistein's effect on cell doubling was masked in the asynchronistic population. Cells were synced in G_0/G_1 phase via serum-starvation, then released from cell cycle arrest by the reintroduction of serum with or without the addition of genistein. At all doses, the tested cells showed similar doubling times to the controls (Figure 1C,D). An alternative method of measuring cellular proliferation was performed using an MTS assay, which measures cellular metabolic activity via the cell's ability to breakdown tetrazolium dye or MTT (3-(4,5-dimethylthiazol-2-yl)-2,5-diphenyltetrazolium bromide) to insoluble formazan. Cells were serum-starved overnight then released from cell cycle arrest with the reintroduction of serum in the presence or absence of genistein. Cells were kept for either 48 or 72 h and then assayed for their ability to reduce MTT. Genistein did not reduce cellular metabolic activity in HFFs; in fact, an increase in metabolic activity was noted, particularly at median doses, which was most apparent after 72 h (Figure 1E). A slight, though insignificant increase in metabolic activity was also repeatedly noted in both cell lines following treatment with DMSO (vehicle control). In RMFs, genistein had a modest inhibitory effect at high doses with a 24% and 36% reduction in metaboloic activity at 50 µM and 100 µM, respectively, compared with the DMSO vehicle control (Figure 1F). When compared to untreated cells, the effect was less dramatic, with a 10% reduction in metabolic activity at 50 µM and 26% reduction at 100 µM. Finally, we assayed for the ability of cells to undergo DNA synthesis in the presence of genistein. Cells were brought into cell cycle arrest via serum-starvation, after which serum was added back in the presence or absence of genistein along with a thymidine analog, bromodeoxyuridine (BrdU), that is incorporated into newly synthesized DNA. There was no notable difference in the pattern of BrdU uptake in cells at any genistein concentration tested versus the controls [37]. Collectively, these data support that genistein is non-toxic to primary fibroblasts at the concentrations used in this study.

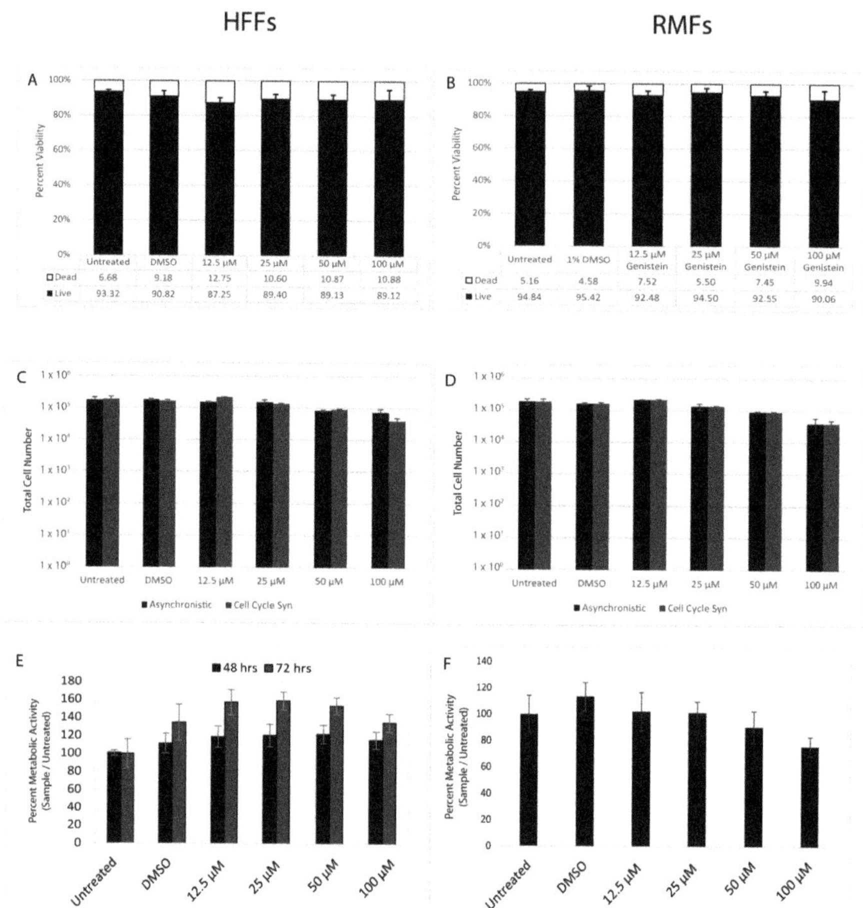

Figure 1. Cytotoxicity of genistein in human foreskin fibroblasts (HFFs) and Rhesus macaque foreskin fibroblasts (RMFs). Cells were left untreated (negative control) or treated with either 0.1% dimethyl sulfoxide (DMSO) (experimental control) or increasing concentrations of genistein. After 48 h, HFFs (**A**) and RMFs (**B**) were collected and assayed for live vs. dead cells via trypan blue dye exclusion. HFFs (**C**) and RMFs (**D**) were kept as an asychronous population or growth was synchronized via overnight serum-starvation prior to genistein treatment. Cells were collected and the total cell numbers counted after 48 h. HFFs (**E**) and RMFs (**F**) were serum-starved overnight, then genistein-treatments were added and cells were incubated for 48 h (**E**) or 72 h (**E**,**F**), and then assayed for metabolic activity via MTS assay. For A–F, $n = 9$, with the statistical analysis performed via one-way ANOVA with Dunnet's correction for multiple comparisons.

3.2. Genistein Reduces B Virus Spread and Replication in a Dose-Dependent Manner

We used three methods to assay for antiviral properties of genistein against B virus. Initially, we performed high input infections (MOI 5) in the presence or absence of genistein for 24 h, then collected cells/supernatants and back-titered these suspensions on Vero cells via a standard plaque assay. As shown in Figure 2A, genistein significantly reduced plaque formation in a dose-dependent manner. Since this assay required passaging the virus through a second cell type, we next asked if similar results could be obtained by assaying the virus directly into fibroblasts using either cell-ELISA or plaque reduction assays. Cell-ELISA also revealed a clear trend of reduced virus replication with genistein in a dose-dependent manner with an IC_{50} value of 33 μM in HFF (Figure 2B) and 46 μM in RMF (Figure 2C). Next, we performed direct plaque reduction assays. Genistein reduced plaque formation and plaque size in a dose-dependent manner (Figure 2C). Of note, at high doses, virus antigen was rarely detected in the cytoplasm of cells, instead localized to the nucleus. These data suggest that genistein can reduce both productive virus replication and spread, not only in cell-to-cell scenarios but also within the infected cell. Further, our findings support that genistein is effective in restricting B virus infection in both human and macaque cell lines.

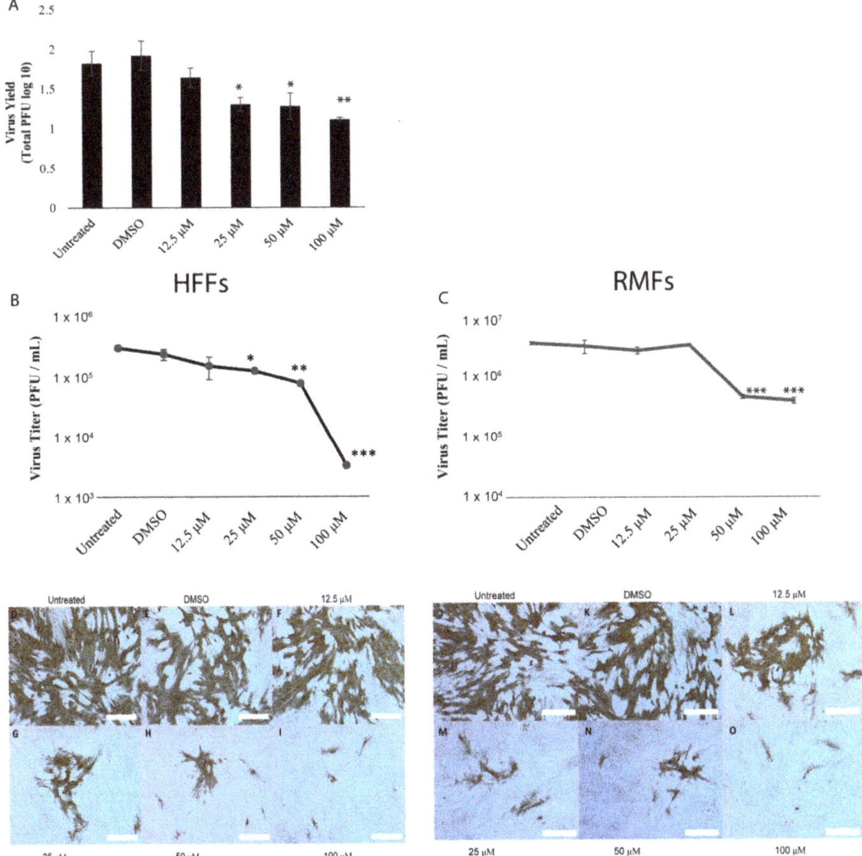

Figure 2. Genistein has antiviral activity against B virus in human and macaque fibroblasts. HFFs and RMFs were left untreated (negative control) or treated with either 1% DMSO (experimental control) or increasing concentrations of genistein and infected with 150 PFU of B virus for 48 h. (**A**) Indirect virus yield assay in HFFs. After 48 h, cells/supernatants were harvested and back-titered on Veros. Data shows total PFU counted in Veros. Cell-based Elisa in HFFs (**B**) and RMFs (**C**). After 48 h, cells were fixed with 100% methanol and assayed via ELISA for amount of virus antigen. An in-assay standard curve was generated to calculate virus titer and IC_{50} values were calculated using a logarithmic regression line. Statistical analysis performed via one-way ANOVA with Dunnet's correction for multiple comparisons. N = 9. Error bars show standard error of the mean. * p = <0.05, ** p < 0.01, *** p =< 0.001. Direct virus yield assay in HFFs (**D–I**) and RMFs (**J–O**). After 48 h, cells were fixed with 100% methanol and virus antigen was visualized via IHC with DAB detection. Scale bars shown at 200 μM.3.3. Genistein Targets B Virus Post-Viral Entry and Genomic Replication.

Antiviral agents may act directly to inactivate viral particles or, following virus entry into cells, there are several steps that antivirals can target: Immediate-early and early gene synthesis, DNA replication, late gene synthesis, virus assembly and virus budding. Direct plaque reduction assays are an effective means of calculating the reduction in virus titer; however, as shown in Figure 2C, B virus does not produce a clear plaque in primary fibroblasts, making it difficult to obtain reliable quantitative data using this methodology. For these reasons, we performed plaque reduction assays in Veros, as this cell line is sensitive to B virus infection and B virus produces clear plaques in these cells. We verified that genistein could effectively reduce B virus replication in a dose-dependent manner in Veros. Genistein inhibited plaque formation with an IC_{50} value of 56 µM for B virus (data not shown). To examine if genistein could directly inactivate B virus, 50 µM of genistein was pre-incubated with the virus for 15, 30, 60, 90 or 120 min, and then cells were infected with the virus/drug mix. To verify that the effect of genistein was only on the virus and not the cell, the mix was diluted to obtain a 5 µM final concentration of genistein prior to cellular infection. Our results showed that pre-incubation of B virus with 50 µM genistein for up to 2 h prior to infection had no effect on plaque formation, suggesting that genistein does not directly inactivate the virus and its antiviral activity is due to interference in post-infection processes (Figure 3A).

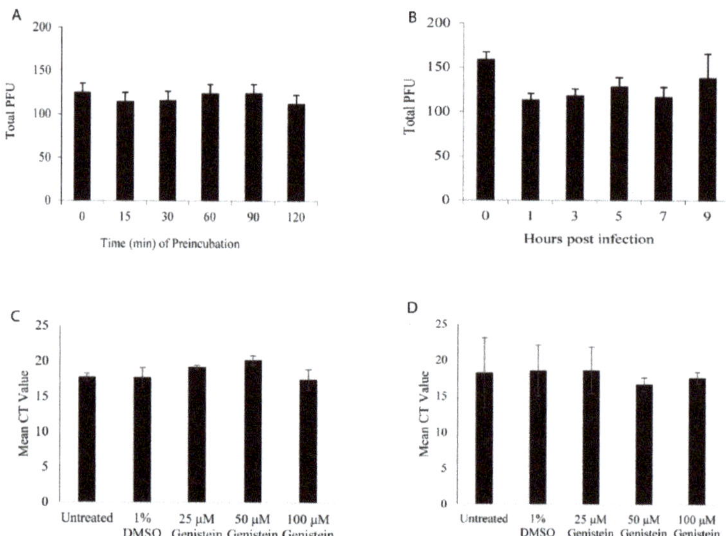

Figure 3. Genistein does not directly inactivate B virus; instead, it inhibits the virus after virus replication. (**A**) Direct inactivation assay. Veros were infected with 150 PFU of B virus following pre-incubation of B virus with 50 µM genistein for 0–120 min. A standard plaque assay was performed and the total PFU counted. (**B**) Plaque reduction assay. Veros were infected with 150 PFU of B virus and 50 µM of genistein was added at various times post-infection. A standard plaque assay was performed and the total PFU counted. RT-PCR results for total genomic virus post-48 h infection of HFFs (**C**) or RMFs (**D**) following B virus infection in the presence of increasing concentrations of genistein or 1% DMSO (control).

B virus' life cycle is ordered into three distinct phases of virus gene transcription/translation: Immediate-early (IE), early (E), and late (L), wherein each phase is dependent on the previous. The full replicative cycle is ~12 h in Vero, with IE genes expressed by 2 hpi, E genes expressed between 2–4 hpi and followed by genomic replication, and L gene synthesis up to 8 hpi. From 8–12 hpi, virions are assembled, glycoproteins undergo glycosylation, and virions are released from the infected cell. To look at the effect of genistein at all phases of the viral life cycle, we added 50 µM genistein at 1, 3, 5, 7, and

9 hpi. There was no difference in virus replication at any time point (Figure 3B). Further, there was no statistically significant difference when genistein was added at the same time as B virus (data not shown; Figure 4C), suggesting that genestein does not impact viral entry. To verify that HFF and RMF would behave in a similar fashion, DNA was isolated from cells infected with increasing concentrations of genistein. No significant difference in the total viral DNA levels in infected cells at any concentration was found (Figure 3C,D). Collectively, these data suggest genistein targets a viral event post-genomic replication and L gene synthesis and, further, that this is not a species-dependent mechanism.

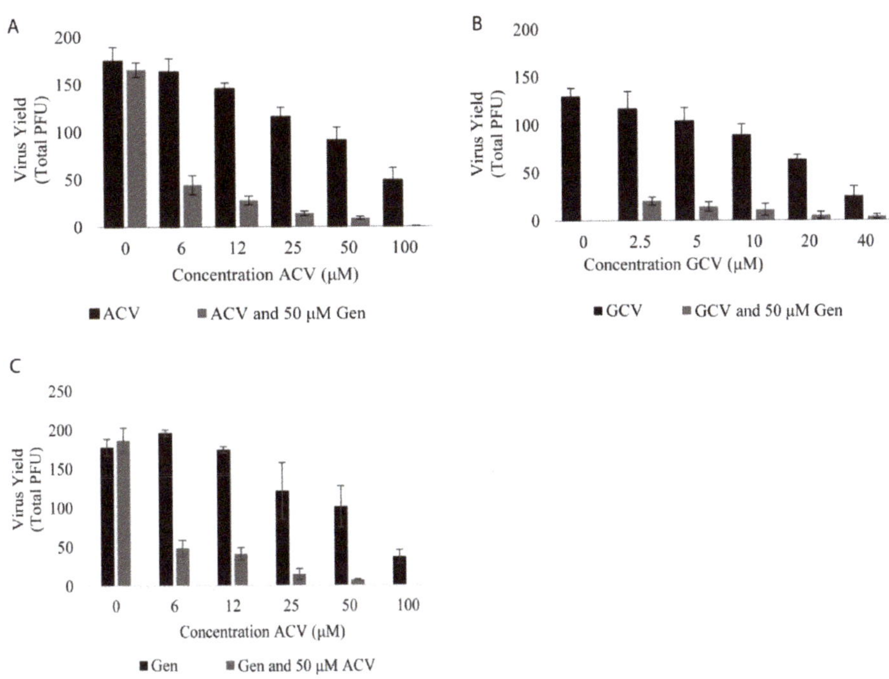

Figure 4. Genistein acts synergistically with antiviral agents to reduce effective dose. (**A**) and (**B**) show the results of virus yield assays, wherein Vero cells were infected with 150 PFU of B virus and 50 μM of genistein with increasing concentrations of ACV (**A**) or GCV (**B**). (**C**) Vero cells were infected with 150 PFU of B virus and 50 μM ACV with increasing concentrations of genistein. Total PFU was determined by standard plaque assay.

3.3. Genistein Reduces Effective Dose of Acyclovir and Ganciclovir against B Virus

Acyclovir (AVC) is typically the first antiviral drug administered following human B virus infection; however, if the patient shows symptoms of CNS involvement, ganciclovir (GCV) is typically administered. Like ACV, GCV has significant side effects, particular at high doses and with long-term usage. We examined whether genistein, when administered at 50 μM in combination with ACV or GCV, could enhance inhibition of B virus. First, we tested the IC_{50} dose of genistein with various concentrations of ACV or GCV. Infected cells receiving a combined genistein–ACV/GCV treatment showed marked reduction in plaque formation compared to cells treated with genistein, ACV or GCV alone (Figure 4A,B; Table 1). For example, whereas a 6 μM dose of ACV had little effect on B virus, a co-treatment with genistein at its IC_{50} value of 50 μM resulted in a 75% decrease in plaque formation (Figure 4A; Table 1). GCV is more effective against B virus than ACV, with an IC_{50} value of 16 μM versus 55 μM in Vero (Table 1), and, thus, was tested at a smaller range of concentration. Genistein exhibited the same additive effect when given with GCV (Figure 4B; Table 1). The similar level of plaque reduction was expected since these two nucleoside analogs share similar antiviral actions. The data also showed that increasing the concentration of ACV or GCV produced little changes to the reduction by genistein, suggesting that genistein and these two nucleoside analogs are not necessarily interacting with each other chemically, but may work via independent mechanisms to prevent viral infection. Conversely, we tested the IC_{50} dose of ACV with varying concentrations of genistein and inhibition in plaque formation was virtually identical as that achieved by a cotreatment of ACV and 50 µM genistein (Figure 4C; Table 1). Thus, neither drug can be concluded as an enhancer, but rather that genistein and ACV interact in a synergistic manner resulting in an enhanced antiviral response.

Table 1. Calculated IC_{50} Values against B Virus.

Agent	Concentration (µM)
Genistein	56
Acyclovir	55
Acyclovir + Genistein (50 µM)	3.9
Ganciclovir	16
Ganciclovir + Genistein (50 µM)	0.002

4. Discussion

This study tested the hypothesis that the plant flavonoid genistein has antiviral activity against B virus. Using plaque reduction assays and plaque titration assays, we not only showed that genistein can reduce virus replication and spread in a dose-dependent manner, but also that it targets a point in the virus life cycle after DNA replication. Further, we showed that genistein could act synergistically with current antiviral therapies to reduce their effective dose. Finally, we showed that efficacious doses had no cytotoxic effects on primary human and macaque cell lines.

Genistein had IC_{50} values against B virus of 33, 46, and 55 µM in HFF, RMF, and Vero cell lines, respectively. While the antiviral activity of genistein in HFF and RMF cells has not been previously reported on, our data for Vero is congruent with values reported from other groups. The effective concentration of genistein ranges between 12–50 µM [27–30,32], while toxicity is observed between 25–250 µM, depending on the technique and cell type. The highest value in the toxicity range was reported from Lyu et al. via MTT assay [21]. In our study, cellular toxicity was measured using four complimentary methods and no cytotoxicity was noted in primary fibroblasts at any of the tested doses (12.5–100 µM), however, toxicity was noted above 50 µM in Veros (data not shown). Animal studies of genistein in mice and pigs have shown no adverse effects on growth, reproduction or development [19,31]. Therefore, genistein is an attractive candidate for antiviral therapy.

Our data suggest that genistein is targeting a late-phase event in the life cycle of B virus to block new virion production and prevent viral spread. In our current study, we found that pre-exposure of cells to genistein had no enhanced effect on virus reduction or reduction in viral DNA production, suggesting that genistein targets an intracellular event during the viral life cycle. Productive virus replication proceeds with the organized and temporal expression of three gene family groups—IE, E, and L—with viral DNA replication occurring between E and L phases. Numerous phosphorylation events are involved during each of these phases, which could provide potential targets for a protein kinase inhibitor such as genistein. Some studies have already demonstrated that genistein interferes with late infection by inhibiting phosphorylation of glycoproteins [20,27,30]. In this study, immunohistochemical (IHC) staining for viral antigen revealed a very restricted intracellular distribution with increasing concentrations of genistein. This observation supports the possibility that genistein is interfering with late phase virion assembly and/or trafficking through the endomembrane system. Currently, we are investigating the specific mechanism by which genistein inhibits virion production and viral spread.

The macaque is widely used in the biomedical research community owing, in part, to its high degree of genetic homology to the human. The threat of zoonosis has led to the establishment of specific pathogen-free (SPF) breeding colonies. In the macaque, B virus presents with minimal pathology and macaques in these colonies often show sero-conversion to B virus as they reach sexual maturity [38]. Sporadic sero-conversion, coupled with asymptomatic shedding from infected animals, produces a potentially dangerous situation whereby animal handlers, veterinarians, and researchers can be at risk of exposure. In this study, genistein was equally effective at limiting B virus infection in macaque fibroblasts as human fibroblasts. These observations raise the interesting possibility of using genistein to control outbreaks within these SPF colonies.

Genistein has antiviral activity against several DNA and RNA viruses, though this is the first report of its antiviral activity against B virus in primary human and macaque cell lines [19–32]. Our data suggest that genistein may have therapeutic potential as an augmenter to current antiviral drugs for B virus. Combination therapy has been found to be useful against HSV-1 [39–43]. In most cases, combination therapy against virus infection requires a lower dose of either drug to achieve the same or greater antiviral response. If genistein can synergize with ACV or GCV to reduce the effective dose, then such a combination could greatly reduce negative side effects seen with long-term administration to B virus-infected patients. In a clinical setting, genistein can be provided as a topical ointment to be applied at the site of exposure. This method can enhance the absorption of genistein to the infected site since it bypasses the rapid hepatic metabolism and low bioavailability associated with oral consumption of flavonoids. Furthermore, it can limit toxicity only to the infected area. Topical delivery has been suggested to work well for other flavonoids with demonstrated antiviral activity [44,45]

In conclusion, genistein is a promising new antiviral treatment of human B virus infection and may serve as an augmenter to current, more toxic antiviral treatments.

Author Contributions: J.C.L., N.D., P.W.K. and J.K.H. all contributed to the conceptualization of this project. J.C.L. and N.D. performed data curation. J.C.L., N.D. and J.K.H. performed formal analysis of the data. J.C.L., N.D. and P.W.K. performed experimental investigations. J.C.L., N.D., P.W.K. and J.K.H. developed methodologies. J.C.L. and N.D. performed validations of the findings. J.C.L., N.T. and J.K.H. contributed to writing and editing. J.C.L. and J.K.H. contributed to writing the final draft. P.W.K. and J.K.H. provided supervision and project administration. J.K.H. provided resources and funding acquisition.

Funding: This research received no external funding.

Acknowledgments: This work was supported by the B Virus Resource Center in the Viral Immunology Center at Georgia State University and the Georgia Research Alliance. We thank Mugdha Vasireddi for her contribution to experimental design, advice, and trouble shooting. We thank Chadi Filfili for his support and training in the BSL4. We thank the resource medical technologists and administration specialists from the National B Virus Resource Center for their help, support, and thoughtful conversations. We thank Martin L. Hudson for his support and his assistance in manuscript revision.

Conflicts of Interest: The authors declare no conflict of interest.

References

1. Cohen, J.I.; Davenport, D.S.; Stewart, J.A.; Deitchman, S.; Hilliard, J.K.; Chapman, L.E.; the B Virus Working Group. Recommendations for Prevention of and Therapy for Exposure to B Virus (Cercopithecine Herpesvirus1). *Clin. Infect. Dis.* **2002**, *35*, 1191–1203. [CrossRef] [PubMed]

2. Eberle, R.; Hilliard, J. The simian herpesviruses. *Infect Agents Dis.* **1995**, *4*, 55–70. [PubMed]

3. Hilliard, J.K.; Whitley, R.J. Cercopithecine Herpesvirus (B Virus). In *Fields Virology*; Whitley, R.J., Griffen, D., Howely, P., Knipe, D., Lamb, R., Martin, M., Philiadelphia, Lippincott Williams and Wilkins, Eds.; Philiadelphia: Linppincott Williams and Wilkins: New York, NY, USA, 2006; pp. 2835–2847.

4. E Palmer, A. B virus, Herpesvirus simiae: historical perspective. *J. Med Primatol.* **1987**, *16*, 99–130.

5. Weigler, B.J. Biology of B Virus in Macaque and Human Hosts: A Review. *Clin. Infect. Dis.* **1992**, *14*, 555–567. [CrossRef]

6. De Clercq, E.; Field, H.J. Antiviral prodrugs–the development of successful prodrug strategies for antiviral chemotherapy. *Br. J. Pharmacol.* **2006**, *147*, 1–11. [CrossRef] [PubMed]

7. Elion, G.B.; Furman, P.A.; Fyfe, J.A.; Miranda, P.D.; Beauchamp, L.; Schaeffer, H.J. Selectivity of action of an antiherpetic agent, 9-(2-hydroxyethoxymethyl)guanine. *Proc. Natl. Acad. Sci. USA* **1977**, *74*, 5716–5720. [CrossRef]

8. Zwartouw, H.T.; Humphreys, C.R.; Collins, P. Oral chemotherapy of fatal B virus (herpesvirus simiae) infection. *Antivir. Res.* **1989**, *11*, 275–283. [CrossRef]

9. De Miranda, P.; Whitley, R.J.; Blum, M.R.; Keeney, R.E.; Barton, N.; Cocchetto, D.M.; Good, S.; Hemstreet, G.P., 3rd; Kirk, L.E.; Page, D.A.; et al. Acyclovir kinetics after intravenous infusion. *Clin. Pharmacol. Ther.* **1979**, *26*, 718–728. [CrossRef] [PubMed]

10. Svennerholm, B.; Appelgren, L.; Lycke, J.; Andersen, O.; Dahlöf, C. Acyclovir concentrations in serum and cerebrospinal fluid at steady state. *J. Antimicrob. Chemother.* **1989**, *24*, 947–954.

11. Brigden, D.; Rosling, A.E.; Woods, N.C. Renal function after acyclovir intravenous injection. *Am. J. Med.* **1982**, *73*, 182–185. [CrossRef]

12. Focher, F.; Lossani, A.; Verri, A.; Spadari, S.; Maioli, A.; Gambino, J.J.; Wright, G.E.; Eberle, R.; Black, D.H.; Medveczky, P.; et al. Sensitivity of Monkey B Virus (Cercopithecine herpesvirus 1) to Antiviral Drugs: Role of Thymidine Kinase in Antiviral Activities of Substrate Analogs and Acyclonucleosides. *Antimicrob. Agents Chemother.* **2007**, *51*, 2028–2034. [CrossRef] [PubMed]

13. Smith, R.; Morroni, J.; Wilcox, C. Lack of effect of treatment with penciclovir or acyclovir on the establishment of latent HSV-1 in primary sensory neurons in culture. *Antivir. Res.* **2001**, *52*, 19–24. [CrossRef]

14. Morfin, F.; Thouvenot, D. Herpes simplex virus resistance to antiviral drugs. *J. Clin. Virol.* **2003**, *26*, 29–37. [CrossRef]

15. Martin, J.; Crotty, S.; Warren, P.; Nelson, P. Does an apple a day keep the doctor away because a phytoestrogen a day keeps the virus at bay? A review of the anti-viral properties of phytoestrogens. *Phytochemistry* **2007**, *68*, 266–274. [CrossRef] [PubMed]

16. Middleton, E.; Kandaswami, C.; Theoharides, T.C. The effects of plant flavonoids on mammalian cells: implications for inflammation, heart disease, and cancer. *Pharmacol. Rev.* **2000**, *52*, 673–751. [PubMed]

17. Food and Drug Administration. FDA Approves New Health Claim for Soy Protein and Coronary Heart Disease. FDA Talk Paper. Available online: http://www.fda.gov/bbs/topics/ANSWERS/ANS00980.html.1999 (accessed on 30 May 2019).

18. U.S. National Library of Medicine. Available online: www.clinicaltrials.gov (accessed on 28 April 2019).

19. Greiner, L.L.; Stahly, T.S.; Stabel, T.J. The effect of dietary soy genistein on pig growth and viral replication during a viral challenge. *J. Sci.* **2001**, *79*, 1272. [CrossRef]

20. Akula, S.M.; Hurley, D.J.; Wixon, R.L.; Wang, C.; Chase, C.C. Effect of genistein on replication of bovine herpesvirus type 1. *Am. J.Vet. Res.* **2002**, *63*, 1124–1128. [CrossRef]

21. Lyu, S.Y.; Rhim, J.Y.; Park, W.B. Antiherpetic activities of flavonoids against herpes simplex virus type 1 (HSV-1) and type 2 (HSV-2) in vitro. *Arch. Pharmacal Res.* **2005**, *28*, 1293–1301. [CrossRef]

22. Evers, D.L.; Chao, C.F.; Wang, X.; Zhang, Z.; Huong, S.M.; Huang, E.S. Human cytomegalovirus-inhibitory flavonoids: studies on antiviral activity and mechanism of action. *Antivir. Res.* **2005**, *68*, 124–134. [CrossRef]

23. Engel, S.; Heger, T.; Mancini, R.; Herzog, F.; Kartenbeck, J.; Hayer, A.; Helenius, A. Role of endosomes in simian virus 40 entry and infection. *J. Virol.* **2011**, *85*, 4198–4211. [CrossRef] [PubMed]

24. Stantchev, T.S.; Markovic, I.; Telford, W.G.; Clouse, K.A.; Broder, C.C. The tyrosine kinase inhibitor genistein blocks HIV-1 infection in primary human macrophages. *Virus Res.* **2007**, *123*, 178–189. [CrossRef]

25. Arabyan, E.; Hakobyan, A.; Kotsinyan, A.; Karalyan, Z.; Arakelov, V.; Arakelov, G.; Nazaryan, K.; Simonyan, A.; Aroutiounian, R.; Ferreira, F.; Zakaryan, H. Genistein inhibits African swine fever virus replication in vitro by disrupting viral DNA synthesis. *Antivir. Res.* **2018**, *156*, 128–137. [CrossRef] [PubMed]

26. Özçelik, B.; Kartal, M.; Orhan, I. Cytotoxicity, antiviral and antimicrobial activities of alkaloids, flavonoids, and phenolic acids. *Pharm. Boil.* **2011**, *49*, 396–402. [CrossRef] [PubMed]

27. Kubo, Y.; Ishimoto, A.; Amanuma, H. Genistein, a protein tyrosine kinase inhibitor, suppresses the fusogenicity of Moloney murine leukemia virus envelope protein in XC cells. *Arch. Virol.* **2003**, *148*, 1899–1914. [CrossRef]

28. Querbes, W.; Benmerah, A.; Tosoni, D.; Di Fiore, P.P.; Atwood, W.J. A JC virus-induced signal is required for infection of glial cells by a clathrin- and eps15-dependent pathway. *J. Virol.* **2004**, *78*, 250–256. [CrossRef] [PubMed]

29. Shaw, A.; Braun, L.; Frew, T.; Hurley, D.; Rowland, R.; Chase, C.; Chase, C. A Role for Bovine Herpesvirus 1 (BHV-1) Glycoprotein E (gE) Tyrosine Phosphorylation in Replication of BHV-1 Wild-Type Virus but Not BHV-1 gE Deletion Mutant Virus. *Virology* **2000**, *268*, 159–166. [CrossRef] [PubMed]

30. Yura, Y.; Yoshida, H.; Sato, M. Inhibition of herpes simplex virus replication by genistein, an inhibitor of protein-tyrosine kinase. *Arch. Virol.* **1993**, *132*, 451–461. [CrossRef] [PubMed]

Viruses **2019**, *11*, 499

31. Ullmann, U.; Bendik, I.; Flühmann, B. Bonistein (synthetic genistein), a food component in development for a bone health nutraceutical. *J. Physiol. Pharmacol.* **2005**, *56*, 79–95. [PubMed]

32. Hirasawa, K.; Kim, A.; Han, H.-S.; Han, J.; Jun, H.-S.; Yoon, J.-W. Effect of p38 Mitogen-Activated Protein Kinase on the Replication of Encephalomyocarditis Virus. *J. Virol.* **2003**, *77*, 5649–5656. [CrossRef] [PubMed]

33. Dangoria, N.S.; Breau, W.C.; Anderson, H.A.; Cishek, D.M.; Norkin, L.C. Extracellular Simian Virus 40 Induces an ERK/MAP Kinase-independent Signalling Pathway that Activates Primary Response Genes and Promotes Virus Entry. *J. Virol.* **1996**, *77*, 2173–2182. [CrossRef] [PubMed]

34. Vela, E.M.; Bowick, G.C.; Herzog, N.K.; Aronson, J.F. Genistein treatment of cells inhibits arenavirus infection. *Antivir. Res.* **2008**, *77*, 153–156. [CrossRef]

35. Vella, C.; Festenstein, H. Coxsackievirus B4 infection of the mouse pancreas: the role of natural killer cells in the control of virus replication and resistance to infection. *J. Virol.* **1992**, *73*, 1379–1386. [CrossRef]

36. Perelygina, L.; Patrusheva, I.; Hombaiah, S.; Zurkuhlen, H.; Wildes, M.J.; Patrushev, N.; Hilliard, J. Production of Herpes B Virus Recombinant Glycoproteins and Evaluation of Their Diagnostic Potential. *J. Clin. Microbiol.* **2005**, *43*, 620–628. [CrossRef] [PubMed]

37. LeCher, J.C. Georgia State University: Atlanta, GA, USA; Analysis of BrdU incorporation into human and macaque fibroblasts post treatment with increasing concentrations of genestein. Unpublished data. 2015.

38. Elmore, D.; Eberle, R. Monkey B Virus (Cercopithecine herpesvirus 1). *Comp. Med.* **2008**, *58*, 11–21. [PubMed]

39. Hayashi, K.; Hayashi, T.; Otsuka, H.; Takeda, Y. Antiviral activity of 5,6,7-trimethoxyflavone and its potentiation of the antiherpes activity of acyclovir. *J. Antimicrob. Chemother.* **1997**, *39*, 821–824. [CrossRef] [PubMed]

40. Kurokawa, M.; Nagasaka, K.; Hirabayashi, T.; Uyama, S.-I.; Sato, H.; Kageyama, T.; Kadota, S.; Ohyama, H.; Hozumi, T.; Namba, T.; et al. Efficacy of traditional herbal medicines in combination with acyclovir against herpes simplex virus type 1 infection in vitro and in vivo. *Antivir. Res.* **1995**, *27*, 19–37. [CrossRef]

41. Mucsi, I.; Gyulai, Z.; Béládi, I. Combined effects of flavonoids and acyclovir against herpesviruses in cell cultures. *Acta Microbiol. Hung.* **1992**, *39*, 137–147.

42. Schang, L.M.; Bantly, A.; Knockaert, M.; Shaheen, F.; Meijer, L.; Malim, M.H.; Gray, N.S.; Schaffer, P.A. Pharmacological Cyclin-Dependent Kinase Inhibitors Inhibit Replication of Wild-Type and Drug-Resistant Strains of Herpes Simplex Virus and Human Immunodeficiency Virus Type 1 by Targeting Cellular, Not Viral, Proteins. *J. Virol.* **2002**, *76*, 7874–7882. [CrossRef] [PubMed]

43. Taylor, J.L.; Tom, P.; O'Brien, W.J. Combined effects of interferon-alpha and acyclovir on herpes simplex virus type 1 DNA polymerase and alkaline DNase. *Antivir. Res.* **1998**, *38*, 95–106. [CrossRef]

44. Barnard, D.; Smee, D.; Huffman, J.; Meyerson, L.; Sidwell, R. Antiherpesvirus Activity and Mode of Action of SP-303, a Novel Plant Flavonoid. *Chemotherapy* **1993**, *39*, 203–211. [CrossRef] [PubMed]

45. Docherty, J.J.; Smith, J.S.; Fu, M.M.; Stoner, T.; Booth, T. Effect of topically applied resveratrol on cutaneous herpes simplex virus infections in hairless mice. *Antivir. Res.* **2004**, *61*, 19–26. [CrossRef] [PubMed]

Brief Report

Inhibition of Epstein-Barr Virus Lytic Reactivation by the Atypical Antipsychotic Drug Clozapine

Abbie G. Anderson, Cullen B. Gaffy, Joshua R. Weseli and Kelly L. Gorres *

Department of Chemistry & Biochemistry, University of Wisconsin-La Crosse, 1725 State St., La Crosse, WI 54601, USA; anderson.abbie@uwlax.edu (A.G.A.); gaffy.cullen@uwlax.edu (C.B.G.); weseli.joshua@uwlax.edu (J.R.W.)
* Correspondence: kgorres@uwlax.edu; Tel.: +1-608-785-5284

Received: 2 April 2019; Accepted: 15 May 2019; Published: 17 May 2019

Abstract: Epstein–Barr virus (EBV), a member of the *Herpesviridae* family, maintains a lifelong latent infection in human B cells. Switching from the latent to the lytic phase of its lifecycle allows the virus to replicate and spread. The viral lytic cycle is induced in infected cultured cells by drugs such as sodium butyrate and azacytidine. Lytic reactivation can be inhibited by natural products and pharmaceuticals. The anticonvulsant drugs valproic acid and valpromide inhibit EBV in Burkitt lymphoma cells. Therefore, other drugs that treat neurological and psychological disorders were investigated for effects on EBV lytic reactivation. Clozapine, an atypical antipsychotic drug used to treat schizophrenia and bipolar disorder, was found to inhibit the reactivation of the EBV lytic cycle. Levels of the viral lytic genes BZLF1, BRLF1, and BMLF1 were decreased by treatment with clozapine in induced Burkitt lymphoma cells. The effects on viral gene expression were dependent on the dose of clozapine, yet cells were viable at an inhibitory concentration of clozapine. One metabolite of clozapine—desmethylclozapine—also inhibited EBV lytic reactivation, while another metabolite—clozapine-N-oxide—had no effect. These drugs may be used to study cellular pathways that control the viral lytic switch in order to develop treatments for diseases caused by EBV.

Keywords: Epstein–Barr virus; herpes viruses; lytic gene expression; Burkitt lymphoma cells; clozapine; antipsychotic drug; antiviral drug

---

1. Introduction

Epstein–Barr virus (EBV) is a member of the *Herpesviridae* family and causes infectious mononucleosis. EBV was the first virus discovered to cause cancer in humans. EBV is associated with Burkitt lymphoma, Hodgkin lymphoma, gastric carcinoma, nasopharyngeal carcinoma, and post-transplant lymphoproliferative disorder. After infection with EBV, the virus maintains a lifelong latent infection within the host. The expression of a few viral genes during the latent phase allows the virus to persist. The viral life cycle alternates between two phases: the latent and the lytic phases. During the lytic phase the virus replicates and spreads among cells and hosts.

The lytic phase of the virus can be triggered in latently infected cultured cells by various inducing agents [1]. Sodium butyrate (NaB), a short-chain fatty acid that inhibits histone deacetylases, promotes the reactivation of the lytic cycle (Figure 1) [2]. Although quite different in chemical structure from butyrate, the DNA methyltransferase inhibitors 5-azacytidine and 5-aza-2′-deoxycytidine (dAzaC), and the protein kinase C agonist 12-O-tetradecanoylphorbol-13-acetate (TPA) also induce the EBV lytic cycle [3]. Molecules with diverse structures inhibit reactivation of the EBV lytic cycle by these inducing agents. Some inhibitors are structurally similar to butyrate. Valproate (valproic acid, VPA) and valpromide (VPM) prevent the virus from reactivating into the lytic cycle in Burkitt lymphoma cells [4,5]. VPA and VPM are used clinically as anticonvulsant and mood-stabilizing drugs.

Figure 1. Structures of drugs tested for effects on Epstein-Barr virus (EBV) lytic reactivation: sodium butyrate (NaB), 5-aza-2'-deoxycytidine (dAzaC), 12-*O*-tetradecanoylphorbol-13-acetate (TPA), clozapine, clozapine-N-oxide, and N-desmethylclozapine.

To determine if there is any commonality between the effects of VPA in neurological and psychological conditions and in blocking EBV reactivation, we investigated other drugs used to treat neurological conditions for effects on the EBV lytic cycle. Clozapine, a member of the dibenzodiazepine class, is used in the treatment of schizophrenia and bipolar disorder (Figure 1). Clozapine (Clozaril™) was the first atypical, or second-generation, antipsychotic drug developed [6]. It is therapeutically effective at treating schizophrenic patients who are resistant to typical antipsychotic drugs [7]. We demonstrate here that clozapine and one of its metabolites inhibit the induction of EBV lytic cycle gene expression.

2. Materials and Methods

2.1. Chemicals

Sodium butyrate (NaB; >98%, Aldrich) and 5-aza-2'-deoxycytidine (dAzaC; Chem-Impex, Wood Dale, IL, USA) were dissolved in water. Clozapine (>99%, ApexBio, Houston, TX, USA), clozapine-N-oxide (>98%, ApexBio), N-desmethylclozapine (98%, Santa Cruz Biotech, Santa Cruz, CA, USA), and TPA (99%, AdipoGen, San Diego, CA, USA) were dissolved in DMSO. Drugs were used at concentrations noted in the figures and legends.

2.2. Cell Culture and Chemical Treatments

The HH514-16 human Burkitt lymphoma [8] and Raji cells were cultured in RPMI 1640 + glutamine supplemented with 8% FBS, penicillin (50 U/mL), streptomycin (50 U/mL), and amphotericin B (1 μg/mL). Cells were grown at 37 °C under 5% CO_2. Cells were subcultured to 3–4×10^5 cells/mL two days prior to the experiment. The experiments started with 1×10^6 cells/mL in RPMI 1640 supplemented with 1% FBS. Cells were harvested 24 h post-treatment. Cell death was measured by trypan blue staining and counting using a hemacytometer. In all experiments that investigated EBV reactivation, >90% of the cells were viable.

2.3. Lytic Reactivation by RT-qPCR

Quantitative reverse transcription polymerase chain reaction (RT-qPCR) was used to measure lytic gene expression. RNA was extracted from cells using the ReliaPrep system (Promega, Madison, WI, USA). Primers used to detect the expression of BZLF1 were AGCAGACATTGGTGTTCCAC (forward) and CATTCCTCCAGCGATTCTG (reverse); for BRLF1 they were CCATACAGGACACAACACCTCA (forward) and ACTCCCGGCTGTAAATTCCT (reverse); and for BMLF1, GGAGGAGGATGAAGA

TCCAA (forward) and TTTCTGGGAATCACAAACGA (reverse). The RT-qPCR utilized the iScript SYBR green RT-qPCR kit (Bio-Rad). Expression levels were normalized to 18S RNA, present at consistent levels among cells.

2.4. Statistical Analysis

Data are reported as the average of the number of biological replicates noted in the figure legends. The values are displayed as the mean ± standard deviation. Values are either the fold increase compared to the untreated control or the percent of maximum lytic reactivation by the inducing agent. To determine the differences among treatments, p-values were calculated using a paired t-test in R 3.3.3 of the quantitation cycle (ΔCq) values from RT-qPCR. Significant differences were considered when the p-value < 0.05.

3. Results

3.1. Clozapine Blocked the Induction of EBV Lytic Genes

We investigated the response of EBV to the atypical antipsychotic drug clozapine. The experiments were performed in the HH514-16 EBV-positive Burkitt lymphoma cell line, derived from Jijoye and P3HR1 cells [8]. The degree of viral reactivation was measured by expression of the viral BZLF1 gene, an immediate early gene that encodes a transcription transactivator [9]. The expression of BZLF1 initiates the reactivation of the Epstein–Barr virus lytic cycle. Treatment of cells with the known inducing agent NaB for 24 h caused an approximately 200-fold increase in BZLF1 expression compared to untreated cells (Figure 2A). Clozapine alone did not induce BZLF1 expression or decrease basal levels of expression. When clozapine (50 µM) was added to cells with NaB, BZLF1 induction was significantly decreased (Figure 2A). Induction of another EBV immediate early gene, BRLF1, was also significantly inhibited by clozapine (Figure 2B). To determine if the reduction in these viral immediate early genes by clozapine affects the expression of a downstream lytic gene, levels of the BMLF1 gene that encodes an mRNA export factor were measured [10,11]. Expression of BMLF1 was induced in cells treated with butyrate, as expected. The addition of clozapine plus butyrate decreased BMLF1 expression to levels comparable to untreated cells (Figure 2C).

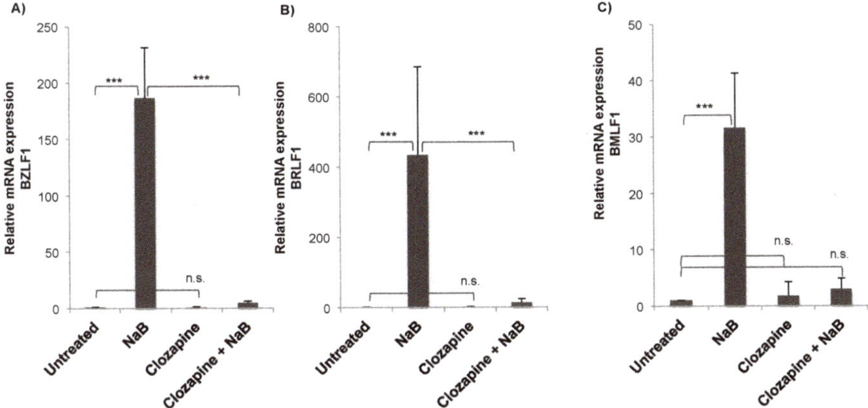

Figure 2. Clozapine inhibited the induction of Epstein–Barr virus (EBV) lytic gene expression. Expression of each gene, (**A**) BZLF1, (**B**) BRLF1, (**C**) BMLF1, was measured by RT-qPCR in untreated HH514-16 cells compared to treatment for 24 h with NaB (3 mM), clozapine (50 µM), or the combination of NaB and clozapine. Values are the average fold induction compared to the untreated control of four or more biological replicates. Error bars show the standard deviation. Differences with a p-value < 0.001 are denoted with ***, p-value < 0.01 with **, and not significantly different (p-value > 0.05) with n.s.

3.2. Dose-Dependent Inhibition by Clozapine

The effects of varying concentrations of clozapine on the reactivation of EBV into the lytic cycle were tested. Clozapine at 2, 10, or 50 µM did not induce BZLF1 viral gene expression in the HH514-16 Burkitt lymphoma cells compared to untreated cells (Figure 3). When added with butyrate, lower concentrations of clozapine (i.e., 2 or 10 µM) inhibited BZLF1 expression by ~40%–50% compared to the level reached in cells treated with only NaB. Clozapine at 50 µM inhibited lytic reactivation by NaB by >95% compared to the reactivation seen with NaB alone.

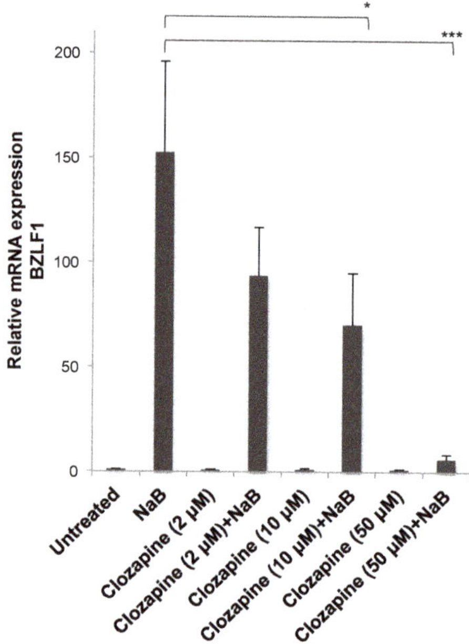

Figure 3. Inhibition of Epstein–Barr virus (EBV) lytic reactivation by clozapine was dose dependent. Clozapine (2, 10, and 50 µM) was tested in the presence and absence of NaB (3 mM) for the effects on BZLF1 expression in HH514-16 cells. Values are the average of seven biological replicates. There was no significant difference between untreated cells and cells treated with only clozapine at any concentration. Differences in BZLF1 expression comparing cells treated with butyrate in the absence and presence of clozapine are marked with * for p-value < 0.05 and *** for a p-value < 0.001.

The concentrations of clozapine that inhibited EBV lytic reactivation, up to 50 µM, did not limit cell growth or the percentage of dead cells when treated for 24 h (Figure 4). With 50 µM clozapine, cells remained >93% ± 4% viable (n = 6) after 48 h and 88% ± 3% viable (n = 3) after 72 h of treatment. Toxicity was observed within 24 h when the clozapine concentration reached 100 µM. Cell toxicity with 100 µM clozapine varied widely among experiments, but the average of 12 replicates resulted in ~40% cell death. When the clozapine concentration reached 200 µM, nearly all of the cells were dead after 24 h in all experiments.

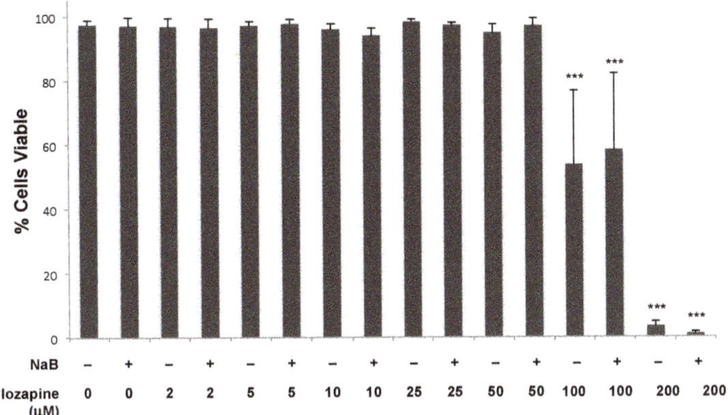

Figure 4. Cells remained viable when treated with 50 µM clozapine for 24 h. Clozapine was tested at concentrations from 2–200 µM in the presence and absence of NaB (3 mM) for the effects on the viability of Burkitt lymphoma cells. Data from four or more biological replicates were averaged, and error bars represent the standard deviation. Conditions are not marked unless significantly different than untreated cells. Differences with a *p*-value < 0.001 are denoted with ***.

3.3. Clozapine Decreased EBV Lytic Induction by dAzaC and TPA

Like butyrate, 5-aza-2′-deoxycytidine (dAzaC) also induces lytic gene expression in HH514-16 cells [3]. Sold under the drug name Decitabine, dAzaC is a DNA methyltransferase inhibitor that is thought to activate EBV by a different mechanism than butyrate [1]. dAzaC (10 µM) was not as potent an activator of BZLF1 expression (~40-fold) as butyrate in HH514-16 cells, but activated the expression of BZLF1 significantly compared to untreated cells (Figure 5). The addition of clozapine (50 µM) at the same time as dAzaC resulted in a 60% decrease in BZLF1 expression compared to dAzaC alone. Clozapine decreased EBV lytic reactivation stimulated by two different lytic inducing agents, but the effectiveness varied. This may have been due to the different mechanisms used by the inducing agents and the shorter length of exposure time required for dAzaC to induce the EBV lytic cycle [12].

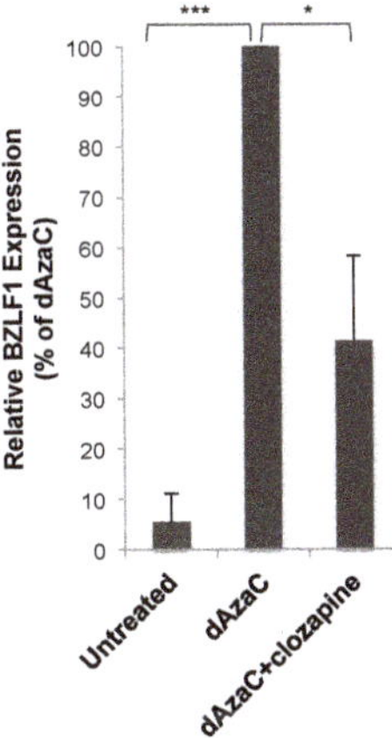

Figure 5. Clozapine decreased EBV lytic BZLF1 expression induced by 5-aza-2′-deoxycytidine (dAzaC). BZLF1 expression was measured in HH514-16 cells after treatment for 24 h with dAzaC (10 µM) alone or combined with clozapine (50 µM) and compared to untreated cells. The average of six biological replicates was plotted as a percentage of BZLF1 expression induced by dAzaC. *p*-value < 0.05 is denoted with *, *p*-value < 0.001 with ***.

To determine the effectiveness of clozapine as an inhibitor in a separate EBV$^+$ cell line, lytic reactivation was tested in Raji cells—a Burkitt lymphoma cell line with a different genetic background than HH514-16 cells. The lytic cycle was induced in the Raji cells by the addition of TPA (20 ng/mL) and detected by the expression of the EBV BRLF1 mRNA (Figure 6). When the cells were treated with clozapine (50 μM) and TPA for 24 h, the induction of BRLF1 expression was blocked. Raji cells treated with clozapine alone showed a 60% decrease, though not statistically significant, in basal levels of BRLF1 expression compared to untreated cells. These results provide evidence that clozapine inhibits EBV lytic reactivation by different classes of inducing agents and in different cell lines.

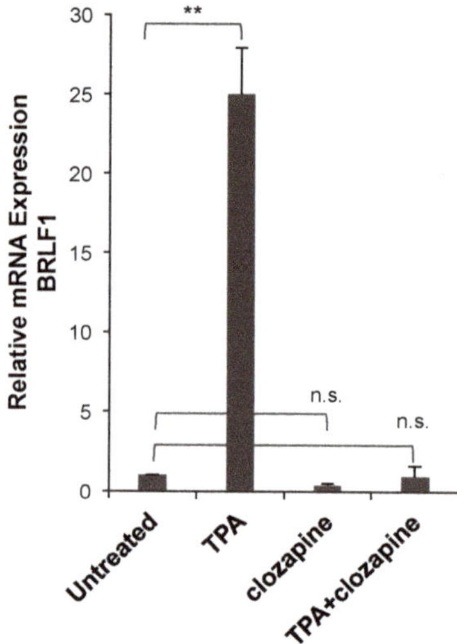

Figure 6. Clozapine inhibited EBV lytic BRLF1 expression in Raji cells. BRLF1 expression was measured in Raji cells after treatment with TPA (20 ng/mL) for 24 h in the absence and presence of clozapine (50 μM). The average BRLF1 expression of three or more biological replicates for each of the treated conditions was compared to untreated cells. Treated conditions are marked not significantly different (n.s.) or different with p-value < 0.01 (**) compared to untreated.

3.4. Metabolites of Clozapine

Two of the major metabolites of clozapine—clozapine-N-oxide (CNO) and N-desmethylclozapine (NDMC; norclozapine) [13]—were tested to determine if the effects of clozapine on EBV lytic reactivation would be altered as the drug was metabolized. Neither clozapine metabolite by itself had any effect on basal levels of BZLF1 expression in HH514-16 Burkitt lymphoma cells (Figure 7). Cells were then treated with butyrate and CNO at 50 μM—the same concentration at which clozapine inhibited BZLF1 expression (Figures 2 and 3). The induction of BZLF1 expression by NaB was the same in the absence or presence of CNO, demonstrating no inhibitory effect by CNO (Figure 7). However, when combined with butyrate, NDMC (50 μM) decreased BZLF1 expression by 90% compared to butyrate alone. Therefore, CNO did not decrease BZLF1 expression, but clozapine and its metabolite NDMC did inhibit EBV lytic gene expression.

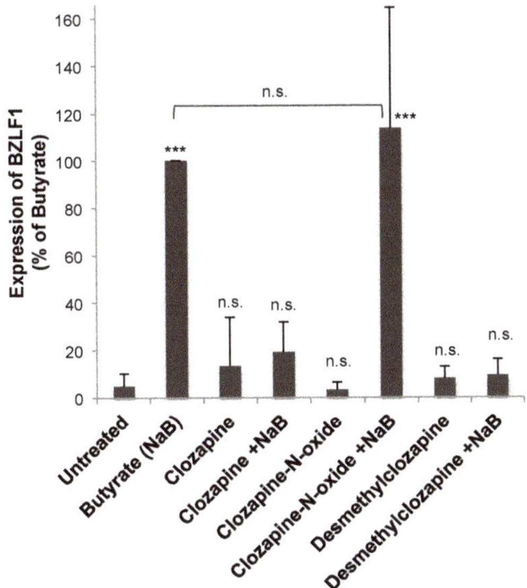

Figure 7. Desmethylclozapine, a metabolite of clozapine, inhibited EBV lytic reactivation, but clozapine-N-oxide did not. HH514-16 cells were treated with clozapine (50 μM), clozapine N-oxide (50 μM), or desmethylclozapine (50 μM) for 24 h in the presence and absence of butyrate (NaB; 3 mM). EBV lytic reactivation was measured by the expression of BZLF1. The averaged data are plotted as a percent of the BZLF1 expression induced by NaB. Data represent the average and standard deviation of five biological replicates. Changes from untreated are marked n.s for not significant, *p*-value < 0.001 is denoted with ***, and *p*-value < 0.05 with *. The NaB and NaB+clozapine-N-oxide were not significantly different.

4. Discussion

4.1. Concentrations of Clozapine in Therapeutic Use

Clozapine, shown here to inhibit expression of EBV lytic genes, is an antipsychotic drug used to treat schizophrenia. The standard dosing for patients is 300–600 mg of clozapine per day [14]. The recommended therapeutic range for clozapine plasma levels range from 350–550 ng/mL for effective treatment. The actual concentrations vary by patient, with factors such as weight and whether the patient smokes influencing this greatly. Studies of patients taking 400 mg/day of clozapine have measured blood concentrations of 40–1911 ng/mL and 84–1088 ng/mL [15]. The maximum plasma level recommended varies between 600 and 2000 ng/mL [16]. The clozapine concentration that inhibited EBV was 50 μM (Figure 3), which is ~8–16-fold higher than plasma concentrations in patients. In experiments conducted in vitro, concentrations of clozapine up to 50 μM had no effect on the viability of HH514–16 Burkitt lymphoma cell line (Figure 5). In another study that used a modified tetrazolium assay to assess the viability of U-937 cells from a patient with histiocytic lymphoma, clozapine had no effect on cell survival after 24 h of treatment with 6250 ng/mL (19 μM) clozapine, and 80% of cells survived when exposed to 12,500 ng/mL (~40 μM) clozapine [17]. No toxicity was observed by clozapine at >50 μM in neutrophils, monocytes, or HL-60 human leukemia cells [18].

4.2. Metabolites of Clozapine

The metabolism of clozapine is catalyzed by the cytochrome P450 enzymes in the liver into two main metabolites: clozapine-n-oxide (CNO) and N-desmethylclozapine (NDMC). NDMC is found in patient plasma at concentrations similar to clozapine, while CNO is much less [13]. CNO is pharmacologically inactive, but has the potential to reverse-metabolize into its parent compound clozapine [19]. No therapeutic benefits of NDMC have been demonstrated for the treatment of schizophrenia [20], but it does have biological effects. While clozapine is an antagonist of the dopamine D_2 receptor, NDMC is a partial agonist in some assays [21]. NDMC is an allosteric agonist at the muscarinic M1 receptor. The muscarinic agonist activity of NDMC can potentiate N-methyl-D-aspartate (NMDA) receptor currents [22]. NDMC has a higher affinity for $5\text{-}HT_{1C}$ and $5\text{-}HT_2$ receptors than clozapine, while CNO is less potent [23]. NDMC is more effective than clozapine as a partial agonist of the $5\text{-}HT_{1C}$ receptor [21]. At higher concentrations, clozapine and NDMC, but not CNO, also antagonize the $GABA_A$ receptor [24]. Overall, clozapine and NDMC are more biologically active than CNO, which correlates with the observed effects on EBV lytic reactivation where clozapine and NDMC inhibited EBV, but CNO did not (Figure 7).

4.3. The Effects of Clozapine on Immune Cells

One of the most potentially critical side effects of clozapine is agranulocytosis—a reduction in granule-containing white blood cells, particularly neutrophils. Due to the potential for agranulocytosis and a high risk of infection, patients taking clozapine require long-term hematology monitoring. A number of mechanisms for the clozapine-induced agranulocytosis have been explored [25,26]. In macrophages, clozapine affects adhesion, phagocytosis, and reactive oxygen species production [27]. Clozapine also alters cytokine production in macrophages [27]. Effects of clozapine on cytokine production have been observed in a number of studies on peripheral blood mononuclear cells or whole blood and in patients with schizophrenia, though reported results have varied, and even contradicted, possibly due to varying cell sources and treatment conditions [28]. Clozapine suppresses interferon-γ production in peripheral blood mononuclear cells and inhibits Th1 cell differentiation [29]. Clozapine inhibits the production of the T-bet transcription factor and enhances mRNA expression of STAT6 and GATA3 [29]. Whole-genome analysis using the T lymphocyte cell line JM-Jurkat treated with clozapine revealed changes in expression of hundreds of mRNAs and miRNAs involved in a number of cellular processes, including cellular metabolism and oxidative stress [30]. The effects of clozapine on B cells have been less well studied.

4.4. Mechanism of Action

The first-generation, or typical, antipsychotic drugs are high-affinity antagonists of the dopamine D_2 receptor. However, the extrapyramidal symptoms (EPSs), such as involuntary movement and muscle control, caused by typical antipsychotics have been attributed to potent dopamine antagonism [31]. The second-generation, atypical, anti-psychotic drugs such as clozapine have less affinity for the D_2 receptors and result in reduced EPS side effects. Atypical drugs also target other dopamine receptors (D_1, D_3, D_4), the $5\text{-}HT_{1A}$, $5\text{-}HT_{2A}$, $5\text{-}HT_{2C}$, $5\text{-}HT_6$, and $5\text{-}HT_7$ receptors for serotonin (5-hydroxytryptamine), as well as muscarinic, adrenergic, and histamine receptors [32]. Clozapine has a high affinity for the serotonin $5\text{-}HT_{2A}$ receptor [33,34]. A theory on the effectiveness of atypical antipsychotics relates a higher ratio of a drug's affinity for the $5\text{-}HT_{2A}$ receptor compared to the dopamine D_2 receptor. The $5\text{-}HT_{1A}$ receptor may also play a role in the antipsychotic effect, as the receptor is stimulated by clozapine and other antipsychotic drugs [35]. Serotonin is not only active in the nervous system, but also affects the immune system. Serotonin increases the mitogen-stimulated proliferation of B-cells, which is dependent on the serotonin $5\text{-}HT_{1A}$ receptor [36]. The $5\text{-}HT_{3A}$ receptor is expressed on B-cells and is differentially expressed in diffuse large B-cell lymphomas (DLBCL) compared to non-neoplastic B cells [37]. The serotonin transporter (SERT) is also expressed on B cells. Culturing Burkitt lymphoma cell lines with serotonin leads to increased serotonin uptake by SERT, decreased DNA synthesis, and apoptosis of Burkitt lymphoma cells, including in EBV-positive cells [38].

The mechanism of action of VPA as an inhibitor of EBV lytic reactivation in lymphoma cells is not known [4,39]. VPA is known to have a number of effects in neurons, however, which of these roles is responsible for the clinical effects of this drug in neurological disorders is uncertain [40]. The proposed mechanisms of VPA and clozapine are not similar, and they do not have a similar chemical structure, so it is possible that the two drugs affect EBV lytic reactivation in different ways. The cellular targets of antipsychotic, anti-epileptic, and mood-stabilizing drugs will continue to be explored in cells infected by EBV.

4.5. Effect of Clozapine and Its Metabolites on Other Viruses

The clozapine metabolite N-desmethylclozapine was found to inhibit replication of dengue virus (DENV) [41]. The inhibition of N-desmethylclozapine was specific to DENV. It did not have an effect on other flaviviruses (i.e., Japanese encephalitis virus, West Nile virus), or on the RNA viruses respiratory syncytial virus and rotavirus. Inhibition occurred at an early step in the viral life cycle prior to viral replication. Two other metabolites of clozapine, 8-OH-deschloro-clozapine and 8-OH-desmethylclozapine, inhibit human immunodeficiency virus (HIV) type 1 [42]. Neither clozapine, the primary compound, nor desmethylclozapine show any antiviral effects, suggesting that inhibition of HIV is due to the metabolism of clozapine. Clozapine (30 µM) decreased the infection of a human glial cell line by the human polyomavirus JC virus by approximately one-half [43]. The human endogenous retroviruses (HERVs) are associated with schizophrenia and other neurological diseases. Clozapine had no significant effect on the transcription of HERVs in three types of brain cell lines, though VPA upregulated the transcription of many HERVs [44]. Valpromide, an anti-epileptic drug that inhibits EBV lytic reactivation [5], had no effect on vesicular stomatitis virus infection [45]. Valpromide and a structurally-related molecule valnoctamide inhibited infection and replication of the human herpesvirus cytomegalovirus in cell culture, and increased the survival rate of mice infected with mouse cytomegalovirus [45].

In conclusion, the repurposing of antipsychotic drugs as antivirals has the potential for therapeutic use. Understanding how clozapine inhibits EBV may lead to the discovery of cellular pathways that regulate the latent–lytic switch of the virus. Future studies will be aimed at determining the effects of other antipsychotic drugs on EBV lytic reactivation and identifying common molecular targets in cells that may provide information about the mechanisms important in neurological disease and for the viral life cycle.

Author Contributions: K.L.G conceived the research and designed the experiments; A.G.A., C.B.G., and J.R.W. performed the experiments; K.L.G., A.G.A., and C.B.G. analyzed the data; K.L.G. did the statistical analysis; K.L.G., A.G.A., and J.R.W. wrote the paper; K.L.G. edited the paper.

Acknowledgments: We thank McKenna Theine, Jenna Hayes, and Kayla Feehan for technical assistance. This work was funded by the UWL College of Science and Health and UWL Faculty Research Grants to K.L.G., the UWL Eagle Apprentice Program for J.R.W., and UWL Undergraduate Research and Creativity grants to A.G.A. and C.B.G.

Conflicts of Interest: The authors declare no conflict of interest.

References

1. Miller, G.; El-Guindy, A.; Countryman, J.; Ye, J.; Gradoville, L. Lytic Cycle Switches of Oncogenic Human Gammaherpesviruses1. *Adv. Cancer Res.* **2007**, *97*, 81–109. [PubMed]
2. Luka, J.; Kallin, B.; Klein, G. Induction of the Epstein-Barr virus (EBV) cycle in latently infected cells by n-butyrate. *Virology* **1979**, *94*, 228–231. [CrossRef]
3. Ben-Ssson, S.A.; Klein, G. Activation of the epstein-barr virus genome by 5-aza-cytidine in latently infected human lymphoid lines. *Int. J. Cancer* **1981**, *28*, 131–135. [CrossRef]
4. Tuck, D.; Gradoville, L.; Daigle, D.; Gorres, K.; Wang'ondu, R.; Miller, G.; Schulz, V.; Ye, J. Valproic Acid Antagonizes the Capacity of Other Histone Deacetylase Inhibitors To Activate the Epstein-Barr Virus Lytic Cycle. *J. Virol.* **2011**, *85*, 5628–5643.
5. Gorres, K.L.; Daigle, D.; Mohanram, S.; McInerney, G.E.; Lyons, D.E.; Miller, G. Valpromide Inhibits Lytic Cycle Reactivation of Epstein-Barr Virus. *MBio* **2016**, *7*, e00113. [CrossRef]
6. Wenthur, C.J.; Lindsley, C.W. Classics in chemical neuroscience: Clozapine. *ACS Chem. Neurosci.* **2013**, *4*, 1018–1025. [CrossRef]

7. Nucifora, F.C.; Mihaljevic, M.; Lee, B.J.; Sawa, A. Clozapine as a Model for Antipsychotic Development. *Neurotherapeutics* **2017**, *14*, 750–761. [CrossRef]

8. Heston, L.; Rabson, M.; Brown, N.; Miller, G. New Epstein–Barr virus variants from cellular subclones of P3J-HR-1 Burkitt lymphoma. *Nature* **1982**, *295*, 160–163. [CrossRef]

9. Countryman, J.; Jenson, H.; Seibl, R.; Wolf, H.; Miller, G. Polymorphic proteins encoded within BZLF1 of defective and standard Epstein-Barr viruses disrupt latency. *J. Virol.* **1987**, *61*, 3672–3679. [PubMed]

10. Kenney, S.; Holley-Guthrie, E.; Mar, E.C.; Smith, M. The Epstein-Barr virus BMLF1 promoter contains an enhancer element that is responsive to the BZLF1 and BRLF1 transactivators. *J. Virol.* **1989**, *63*, 3878–3883.

11. Buisson, M.; Manet, E.; Trescol-Biemont, M.C.; Gruffat, H.; Durand, B.; Sergeant, A. The Epstein-Barr virus (EBV) early protein EB2 is a posttranscriptional activator expressed under the control of EBV transcription factors EB1 and R. *J. Virol.* **1989**, *63*, 5276–5284. [PubMed]

12. Himmelfarb, S.; Bhaduri-McIntosh, S.; Gradoville, L.; Heston, L.; Miller, G.; Ye, J.; Shedd, D.; Countryman, J. Stimulus Duration and Response Time Independently Influence the Kinetics of Lytic Cycle Reactivation of Epstein-Barr Virus. *J. Virol.* **2009**, *83*, 10694–10709.

13. Centorrino, F.; Baldessarini, R.J.; Kando, J.C.; Frankenburg, F.R.; Volpicelli, S.A.; Flood, J.G. Clozapine and metabolites: concentrations in serum and clinical findings during treatment of chronically psychotic patients. *J. Clin. Psychopharmacol.* **1994**, *14*, 119–125. [CrossRef]

14. Subramanian, S.; Völlm, B.A.; Huband, N. Clozapine dose for schizophrenia. *Cochrane Database Syst. Rev.* **2017**, *2017*. [CrossRef] [PubMed]

15. Mauri, M.C.; Volonteri, L.S.; Colasanti, A.; Fiorentini, A.; De Gaspari, I.F.; Bareggi, S.R. Clinical Pharmacokinetics of Atypical Antipsychotics A Critical Review of the Relationship Between Plasma Concentrations and Clinical Response. *Clin. Pharmacokinet.* **2007**, *46*, 359–388. [CrossRef]

16. Stark, A.; Scott, J. A review of the use of clozapine levels to guide treatment and determine cause of death. *Aust. New Zeal. J. Psychiatry* **2012**, *46*, 816–825. [CrossRef]

17. Heiser, P.; Enning, F.; Krieg, J.-C.; Vedder, H. Effects of haloperidol, clozapine and olanzapine on the survival of human neuronal and immune cells. *J. Psychopharmacol.* **2007**, *21*, 851–856. [CrossRef]

18. Gardner, I.; Leeder, J.S.; Chin, T.; Zahid, N.; Uetrecht, J.P. A comparison of the covalent binding of clozapine and olanzapine to human neutrophils in vitro and in vivo. *Mol. Pharmacol.* **1998**, *53*, 999–1008.

19. Chang, W.-H.; Lin, S.-K.; Lane, H.-Y.; Wei, F.-C.; Hu, W.-H.; Lam, Y.F.; Jann, M.W. Reversible metabolism of clozapine and clozapine N-oxide in schizophrenic patients. *Prog. Neuro-Psychopharmacol. Biol. Psychiatry* **1998**, *22*, 723–739. [CrossRef]

20. Ghosh, A.; Chakraborty, K.; Mattoo, S.K. Newer molecules in the treatment of schizophrenia: A clinical update. *Indian J. Pharmacol.* **2011**, *43*, 105–112.

21. Heusler, P.; Bruins Slot, L.; Tourette, A.; Tardif, S.; Cussac, D. The clozapine metabolite N-desmethylclozapine displays variable activity in diverse functional assays at human dopamine D 2 and serotonin 5-HT 1A receptors. *Eur. J. Pharmacol.* **2011**, *669*, 51–58. [CrossRef]

22. Sur, C.; Mallorga, P.J.; Wittmann, M.; Jacobson, M.A.; Pascarella, D.; Williams, J.B.; Brandish, P.E.; Pettibone, D.J.; Scolnick, E.M.; Conn, P.J. N-desmethylclozapine, an allosteric agonist at muscarinic 1 receptor, potentiates N-methyl- D-aspartate receptor activity. *Proc. Natl. Acad. Sci. USA* **2003**, *100*, 13674–13679. [CrossRef]

23. Kuoppamäki, M.; Syvälahti, E.; Hietala, J. Clozapine and N-desmethylclozapine are potent 5-HT1C receptor antagonists. *Eur. J. Pharmacol.* **1993**, *245*, 179–182. [CrossRef]

24. Kuoppamiiki, M.; Liiddens, H.; Syviilahti, E. Effects of clozapine metabolites and chronic clozapine treatment on rat brain. *Eur. J. Pharmacol.* **1996**, *314*, 319–323.

25. Numata, S.; Umehara, H.; Ohmori, T.; Hashimoto, R. Clozapine Pharmacogenetic Studies in Schizophrenia: Efficacy and Agranulocytosis. *Front. Pharmacol.* **2018**, *9*, 1049. [CrossRef]

26. Wiciński, M.; Węclewicz, M.M. Clozapine-induced agranulocytosis/granulocytopenia. *Curr. Opin. Hematol.* **2018**, *25*, 22–28. [CrossRef]

27. Chen, M.-L.; Wu, S.; Tsai, T.-C.; Wang, L.-K.; Tsai, F.-M. Regulation of macrophage immune responses by antipsychotic drugs. *Immunopharmacol. Immunotoxicol.* **2013**, *35*, 573–580. [CrossRef]

28. Røge, R.; Møller, B.K.; Andersen, C.R.; Correll, C.U.; Nielsen, J. Immunomodulatory effects of clozapine and their clinical implications: What have we learned so far? *Schizophr. Res.* **2012**, *140*, 204–213. [CrossRef]

29. Chen, M.-L.; Tsai, T.-C.; Wang, L.-K.; Lin, Y.-Y.; Tsai, Y.-M.; Lee, M.-C.; Tsai, F.-M. Clozapine inhibits Th1 cell differentiation and causes the suppression of IFN-γ production in peripheral blood mononuclear cells. *Immunopharmacol. Immunotoxicol.* **2012**, *34*, 686–694. [CrossRef]

30. Gardiner, E.; Carroll, A.; Tooney, P.A.; Cairns, M.J. Antipsychotic drug-associated gene–miRNA interaction in T-lymphocytes. *Int. J. Neuropsychopharmacol.* **2014**, *17*, 929–943. [CrossRef]

31. Divac, N.; Prostran, M.; Jakovcevski, I.; Cerovac, N. Second-Generation Antipsychotics and Extrapyramidal Adverse Effects. *Biomed Res. Int.* **2014**, *2014*, 1–6. [CrossRef]

32. Aringhieri, S.; Carli, M.; Kolachalam, S.; Verdesca, V.; Cini, E.; Rossi, M.; McCormick, P.J.; Corsini, G.U.; Maggio, R.; Scarselli, M. Molecular targets of atypical antipsychotics: From mechanism of action to clinical differences. *Pharmacol. Ther.* **2018**, *192*, 20–41. [CrossRef] [PubMed]

33. Meltzer, H.Y. What's atypical about atypical antipsychotic drugs? *Curr. Opin. Pharmacol.* **2004**, *4*, 53–57. [CrossRef]

34. Kuroki, T.; Nagao, N.; Nakahara, T. Neuropharmacology of second-generation antipsychotic drugs: a validity of the serotonin–dopamine hypothesis. *Prog. Brain Res.* **2008**, *172*, 199–212.

35. Meltzer, H.; Massey, B. The role of serotonin receptors in the action of atypical antipsychotic drugs. *Curr. Opin. Pharmacol.* **2011**, *11*, 59–67. [CrossRef]

36. Iken, K.; Chheng, S.; Fargin, A.; Goulet, A.-C.; Kouassi, E. Serotonin Upregulates Mitogen-Stimulated B Lymphocyte Proliferation through 5-HT1AReceptors. *Cell. Immunol.* **1995**, *163*, 1–9. [CrossRef] [PubMed]

37. Rinaldi, A.; Chiaravalli, A.M.; Mian, M.; Zucca, E.; Tibiletti, M.G.; Capella, C.; Bertoni, F. Serotonin Receptor 3A Expression in Normal and Neoplastic B Cells. *Pathobiology* **2010**, *77*, 129–135. [CrossRef] [PubMed]

38. Serafeim, A.; Grafton, G.; Chamba, A.; Gregory, C.D.; Blakely, R.D.; Bowery, N.G.; Barnes, N.M.; Gordon, J. 5-Hydroxytryptamine drives apoptosis in biopsylike Burkitt lymphoma cells: Reversal by selective serotonin reuptake inhibitors. *Blood* **2002**, *99*, 2545–2553. [CrossRef]

39. Gorres, K.L.; Daigle, D.; Mohanram, S.; Miller, G. Activation and Repression of Epstein-Barr Virus and Kaposi's Sarcoma-Associated Herpesvirus Lytic Cycles by Short- and Medium-Chain Fatty Acids. *J. Virol.* **2014**, *88*, 8028–8044. [CrossRef]

40. Monti, B.; Polazzi, E.; Contestabile, A. Biochemical, Molecular and Epigenetic Mechanisms of Valproic Acid Neuroprotection. *Curr. Mol. Pharmacol.* **2010**, *2*, 95–109. [CrossRef]

41. Medigeshi, G.R.; Kumar, R.; Dhamija, E.; Agrawal, T.; Kar, M. N-Desmethylclozapine, Fluoxetine, and Salmeterol Inhibit Postentry Stages of the Dengue Virus Life Cycle. *Antimicrob. Agents Chemother.* **2016**, *60*, 6709–6718. [CrossRef]

42. Jones-Brando, L.V.; Buthod, J.L.; Holland, L.E.; Yolken, R.H.; Fuller Torrey, E. Metabolites of the antipsychotic agent clozapine inhibit the replication of human immunodeficiency virus type 1. *Schizophr. Res.* **1997**, *25*, 63–70. [CrossRef]

43. Baum, S.; Ashok, A.; Gee, G.; Dimitrova, S.; Querbes, W.; Jordan, J.; Atwood, W.J. Early events in the life cycle of JC virus as potential therapeutic targets for the treatment of progressive multifocal leukoencephalopathy. *J. Neurovirol.* **2003**, *9*, 32–37. [CrossRef] [PubMed]

44. Diem, O.; Schäffner, M.; Seifarth, W.; Leib-Mösch, C. Influence of Antipsychotic Drugs on Human Endogenous Retrovirus (HERV) Transcription in Brain Cells. *PLoS ONE* **2012**, *7*, e30054. [CrossRef] [PubMed]

45. Ornaghi, S.; Davis, J.N.; Gorres, K.L.; Miller, G.; Paidas, M.J.; van den Pol, A.N. Mood stabilizers inhibit cytomegalovirus infection. *Virology* **2016**, *499*, 121–135. [CrossRef]

Article

Potential Application of TALENs against Murine Cytomegalovirus Latent Infections

Shiu-Jau Chen [1,2] and Yuan-Chuan Chen [3,4,*]

[1]   Department of Neurosurgery, Mackay Memorial Hospital, Taipei 10449, Taiwan; chenshiujau@gmail.com
[2]   Department of Medicine, Mackay Medicine College, New Taipei City 25245, Taiwan
[3]   Comparative Biochemistry program, University of California, Berkeley, CA 94720, USA
[4]   National Applied Research Laboratories, Taipei 10636, Taiwan
*   Correspondence: yuchuan1022@gmail.com

Received: 22 March 2019; Accepted: 30 April 2019; Published: 3 May 2019

Abstract: Cytomegalovirus (CMV) infections are still a global health problem, because the latent viruses persist in humans and cause recurring diseases. Currently, there are no therapies for CMV latent infections and the therapies for active infections are limited by side effects and other problems. It is impossible to eradicate latent viruses in animals. HCMV (human CMV) is specific to human diseases; however, it is difficult to study HCMV due to its host specificity and long life cycle. Fortunately, MCMV (murine CMV) provides an excellent animal model. Here, three specific pairs of transcription activator-like effector nuclease (TALEN) plasmids (MCMV1–2, 3–4, and 5–6) were constructed to target the MCMV M80/80.5 sequence in order to test their efficacy in blocking MCMV lytic replication in NIH3T3 cell culture. The preliminary data showed that TALEN plasmids demonstrate specific targeting and cleavage in the MCMV M80/80.5 sequence and effectively inhibit MCMV growth in cell culture when the plasmid transfection is prior to the viral infection. The most specific pairs of TALEN plasmids (MCMV3–4) were further used to confirm the negative regulation of latent MCMV replication and gene expression in Balb/c mice. The injection of specific TALEN plasmids caused significant inhibition in the copy number level of immediately early gene (*ie-1*) DNA in five organs of mice, when compared with the controls. The result demonstrated that TALENs potentially provide an effective strategy to remove latent MCMV in animals.

Keywords: cytomegalovirus; latent infection; TALEN; Surveyor nuclease mutation detection assay; *ie-1* gene; quantitative real-time PCR

---

1. Introduction

Studies on the functions of viral genes in human cytomegalovirus (HCMV) replication in vivo and understanding of viral pathogenesis are essential for developing novel drugs and strategies to treat the viral infections, but there are no suitable animal models for HCMV infection at present. Because HCMV proliferates in human cells specifically, grows slowly and has a very long lytic replication cycle in humans, it is quite difficult to study HCMV (genome size 235 kb) gene function and pathogenesis [1]. However, the murine cytomegalovirus (MCMV) provides an excellent animal model for studying the biology of cytomegaloviruses (CMV) through its specific infection of mice. An MCMV genome of 230 kb is predicted to encode more than 170 open reading frames (ORFs), 78 of which have an extensive homology to those of HCMV [1–3]. Moreover, the pathogenesis of MCMV infection in mice is very similar to that of HCMV infection in humans in several aspects such as active infections, establishment of latency, and reactivation after latent infections [1,3–5]. A complete understanding of the biology of MCMV and the function of its genes may provide insights into the pathogenesis of HCMV.

As a structural protein, the HCMV UL80 assembly protein is presumed to function in packaging DNA. A protease is encoded by the N-terminal region of UL80 that cleaves the assembly protein precursor at a site near the C terminus. The MCMV homolog (M80) of the assembly protein of HCMV (UL80) conserves the domain structure and cleavage sites present in HCMV UL80. MCMV is different in that the conserved CD1 and CD2 domains are separated by 100 amino acids, whereas in all other sequenced herpes viruses the two domains are separated by 80 to 84 residues [2]. The MCMV homolog (M80.5) of HCMV, UL80.5 ORF, is referred to code for protease. The ORFs of MCMV M80 and M80.5 are required for the assembly of proteins and proteases (capsid synthesis) and further virion production. The transcriptions of adjoining MCMV M80 and M80.5 ORFs have different start sites; however, they have the same stop site. Therefore, the overlapping region of MCMV M80 and M80.5 (M80/80.5) is an appropriate target site for virus inhibition.

Transcription activator-like effectors (TALEs) are proteins secreted by the plant pathogenic bacteria *Xanthomonas* via a type III secretion system where they infect various plant species [6]. These proteins can bind promoter sequences in the host plant and activate the expression of host genes that aid bacterial infections. TALEs are important virulence factors that act as transcriptional activators in the plant cell nucleus, where they directly bind to DNA via a central domain of tandem repeats [6]. Each TALE contains a central repetitive region consisting of varying numbers of repeat units (about 17.5 repeats) of 34 amino acids [6,7]. The DNA-binding domain contains a highly conserved 34-amino acid sequence, with the exception of the 12th and 13th amino acids [6]. Only the 12th and 13th amino acids in TALEs are changeable and variable; the other amino acids are constant and stable. These two locations' RVDs (Repeat Variable Di-residues) are highly variable and show a strong correlation with a specific nucleotide recognition by different frequencies, for example NI recognizes A (55%), NG recognizes T (50%), NN recognizes G (7%) and HD recognizes C (69%) [6]. DNA transcribes RNA according to the complementary base pairing rule (A = U, G ≡ C) and RNA translates protein according to the standard genetic code (RNA codon table). After the breaking of the code of the DNA-binding specificity of TALEs [6,7], we know that two amino acids can also recognize one nucleotide. The restriction endonuclease *Fok* I, naturally found in the bacterium *Flavobacterium okeanokoites*, consists of an N-terminal specific DNA-binding domain and a C-terminal nonspecific DNA cleavage domain. The DNA-binding domain recognizes the non-palindromic sequence 5′-GGATG-3′ when the catalytic domain cleaves double-stranded DNA nonspecifically at a fixed distance of 9 and 13 nucleotides downstream of the recognition site. *Fok* I exists as an inactive monomer and becomes an active dimer upon binding to its target DNA and in the presence of specific divalent metals [7]. The DNA cleavage domain of *Fok* I functions as a homodimer, requiring two constructs with unique DNA-binding domains for sites in the target genome with proper orientation and spacing. Transcription activator-like effector nucleases (TALENs) are artificial restriction enzymes generated by fusing the specific TALE DNA-binding domain to a nonspecific *Fok* I DNA cleavage domain [8,9] (Figure 1).

TALENs were shown to be a valuable tool for precise genome engineering with low toxicity [10]. Additionally, they have been successfully applied in the genetic engineering of human pluripotent cells [11,12], and the generation of knockout animals, such as nematodes (*Caenorhabditis elegans*) [8], rats [13], and zebrafish [14,15]. TALENs can be used to edit genomes by making double-stranded breaks (DSBs), which cells respond to with two repair mechanisms: non-homologous end joining (NHEJ) or homologous recombination (HR). DNA can be introduced into a genome through NHEJ in the presence of exogenous double-stranded DNA fragments. HR can also introduce foreign DNA at the DSB as the transfected double-stranded sequences are used as templates for the repair enzymes. However, TALENs have some potential problems. If TALENs do not specifically target a unique site within the genome of interest, off-target cleavage may occur [11]. Such off-target cleavage may lead to the production of enough DSB to overcome the repair machinery and consequently result in chromosomal rearrangements and/or cell death [8].

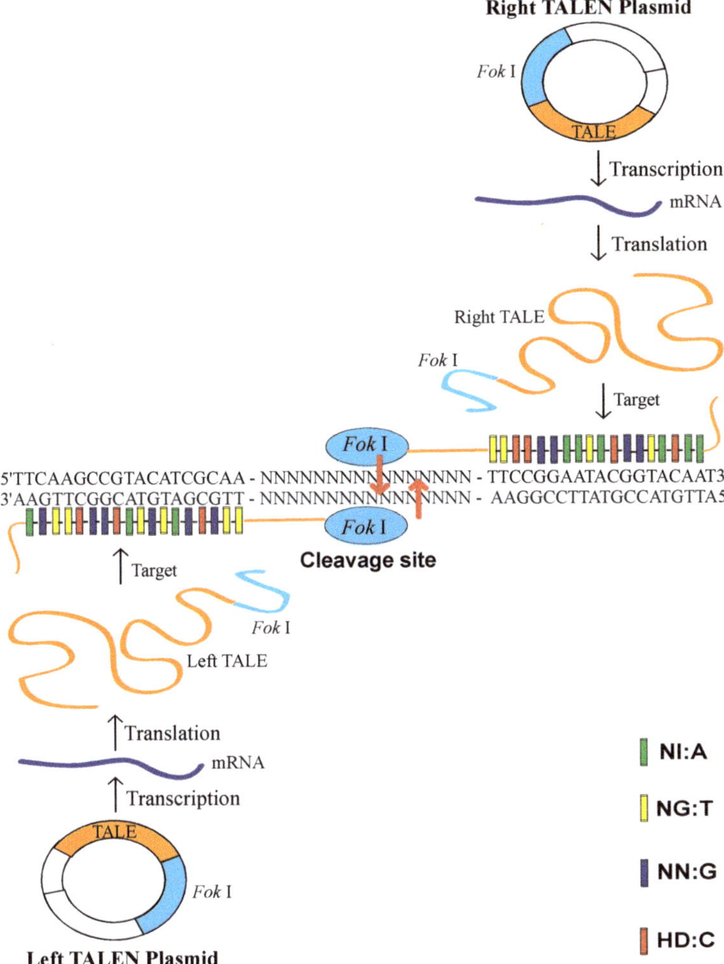

Figure 1. Transcription activator-like effector nuclease (TALEN) targeting and cleavage on the target site. The specific repeat variable di-residue (RVD) used to recognize each base is indicated by shading, as defined in the key (NI: A, NG: T, NN: G, HD: C). The cleavage site is in the adjoining region between the left and right target site. The left transcription activator-like effector (TALE) and the right TALE of TALEN plasmids recognized their target sequence and allowed their associated *Fok* I endonucleases to work as homodimers to cleave the sense strand 9 bp and antisense strand 13 bp downstream of the binding site. The binding of the TALEs to the target sites allows *Fok* I to dimerize and create a double-stranded break (DSB) with sticky ends within the spacer.

Previous reports have suggested that TALENs can be engineered to adapt for an antiviral strategy. For example, TALENs were known to be effective in the inactivation of Hepatitis B virus (HBV) replication in cultured cells and in vivo [16], and in the targeting of the HBV genome [17]. Also, Epstein–Barr virus (EBV)-encoded nuclear antigen-1 (EBNA1) plays a crucial role in EBV episome replication and persistence. TALEN-mediated targeted disruption of EBNA1 was shown to inhibit the growth of EBV-infected cells, hinting at a possible therapeutic application for EBV-associated disorders [18]. TALENs may also provide a new strategy for the treatment of CMV infections.

Latency is a specific phase in viral life cycles, in which viral particles stop producing after infection, but the viral genome has not been completely removed. Proviral latency and episomal latency are two known viral latency models. CMV belongs to the episomal latency model which is essentially quiescent in myeloid progenitor cells, and is reactivated by differentiation, inflammation, immunosuppression or critical diseases [19]. CMV latency has been defined as the absence of infectious viruses, despite the presence of viral DNA. Although the molecular mechanisms by which latency is established and maintained have not been clear, transcriptional control of viral gene expression is very important in controlling viral latency and reactivation. Viral replication is initiated by the expression of *ie* (immediately early) genes. Studies with CMV have suggested that latency is established through the repression of *ie-1* gene expression. *Ie-1* proteins, the first proteins expressed by the virus during productive infection, are transcriptionally regulatory proteins that are required for the induction of early and late gene expression, viral DNA synthesis, and virion production [20–22]. Since MCMV *ie-1* proteins are required for viruses to start replication from latency, the *ie-1* gene is one of the key targets for latency in animal studies. In this study, our goal is to develop an effective strategy to inhibit the growth of MCMV in cell culture and animals specifically, particularly for the removal of latent viruses.

2. Materials and Methods

2.1. Ethics Statement

For all experiments on live mice, we confirm that all methods were carried out in accordance with relevant guidelines and regulations. The protocol for all animal experiments was approved by the Institutional Animal Care and Use Committee (IACUC) of the University of California at Berkeley, USA (Protocol #R240 and #R276). All efforts were made to minimize suffering.

2.2. Viruses and Cell Culture

The MCMV Smith strain and mouse embryonic fibroblast NIH3T3 cells were obtained from the American Type Culture Collection (ATCC, Manassas, VA, USA). The MCMV was grown in NIH3T3 cells (ATCC) NIH3T3 cells were cultured in 500 mL Dulbecco's modified Eagle's medium (DMEM) (ThermFischer Scientific, Waltham, MA, USA) supplemented with 10% Nu-Serum (Coring, Union City, CA, USA), 1% Pen-Strep (100 U/mL of penicillin and 100 µg/mL of streptomycin), 1× MEM essential amino acids (EAA), 1× MEM nonessential amino acids (NEAA) and 12 mL sodium bicarbonate (ThermoFischer Scientific, USA).

2.3. Mice

The three-week-old immunocompetent Balb/c mice were purchased from the Jackson Laboratory, USA and used at four weeks of age.

2.4. Primers and Probes

MCMV M80/80.5 forward primers (5′-CTTGCCTCAGGTGCCCTCTTATTACGGAAT-3′) and reverse primers (5′- ATAAATCACACGTTCACTCCGTTAGTCCGG -3′) were both synthesized by Life technologies, Camarillo, CA, USA.

MCMV *ie-1* forward primers (5′-TCAGCCATCAACTCTGCTACCAAC-3′) and reverse primers (5′-ATCTGAAACAGCCGTATATCATCTTG-3′) were synthesized by Life technologies, USA. TaqMan probes (5′-TTCTCTGTCAGCTAGCCAATGATATCTTCGAGC-3′) were synthesized by Genscript, Nanjing, Jiangsu, China. The probe was labeled at the 5′ end with the reporter dye FAM and at the 3′ end with the quencher dye TAMRA.

2.5. Transcription Activator-Like Effector Nuclease (TALEN) Plasmids

Three specific pairs (left and right) of TALEN plasmids for each targeting site MCMV1-2, 3-4, 5-6, and two nonspecific pairs W1FS-W7R1, KSHV1-2 were constructed by others. They all contain a

CMV promoter, a *Fok* I gene (cleavage domain) and a TALE DNA sequence using pTAL4 Leu (8467 bp, Addgene, Watertown, MA, USA) as the backbone vector. All ten TALEN plasmids are listed in Table 1.

Table 1. List of TALEN plasmids.

TALEN Plasmid	Target DNA Sequence	Nucleotide No.	Target, Gene
MCMV 1	CGGGCCGATCGCCCGCCT	18	MCMV, M80/80.5
MCMV 2	TACAGGGGAGAGAGGAAT	18	MCMV, M80/80.5
MCMV 3	CCAGAACCGATGAGT	15	MCMV, M80/80.5
MCMV 4	GACTCTCAACGAGATCCGC	19	MCMV, M80/80.5
MCMV 5	GGAAGTGGGAGAACCCT	17	MCMV, M80/80.5
MCMV 6	GAGGAAGGGGGGTGAGGCC	19	MCMV, M80/80.5
W1FS	GCTGATTCTTCCCTGTG	17	293T cell, WAS
W7R1	AAGAGTGGATGGAGG	15	293T cell, WAS
KSHV 1	TTACAATGGTGTAGGTG	17	KSHV, RTA
KSHV 2	AGCTCTACGTCCGAAC	16	KSHV, RTA

Abbreviations: MCMV (murine cytomegalovirus), M80/80.5 (the overlapping region of MCMV M80 and M80.5), 293T cell (human embryonic kidney 293 cells), WAS (a specific gene in 293 cells), KSHV (Kaposi's sarcoma-associated herpes virus), RTA (replication and transcription activator gene in KSHV).

2.6. Determination of the Murine Cytomegalovirus (MCMV) Growth Curve in Host Cells

NIH3T3 cells (1.00×10^5 cells/well) were plated in a 12-well format containing 1 mL of growth medium and infected with MCMV (multiplicity of infection, MOI = 0.05) 1 day later (initial titer: 5.00×10^3 pfu/mL). In the preliminary test, the viral titers were 4.20×10^3, 1.40×10^5, 5.10×10^5, 2.70×10^5 pfu/mL at 1, 3, 5 and 7 days post infection, respectively. We found that the viral titers were increasing for 1 to 5 days post infection, but gradually decreasing after 5 days. The results demonstrated that MCMV reached the highest titer at the 5th day post infection.

2.7. Determination of Transfection Efficiency

Lipofectamine™ 2000 transfection reagent was commercially obtained from Life Technologies, USA. The other transfection reagent NKS11, a new lipoid, was synthesized by others. In a safety test, the NKS11 formulation was proved to be nontoxic to Balb/c mice by weight measurement and health observation.

One day before transfection, adherent NIH3T3 cells (1.00×10^5 cells/well) in 1 mL growth medium without antibiotics were plated in a 12-well format so that cells would be 90–95% confluent at the time of transfection. GFP (green fluorescent protein) plasmids (1.6 µg/well, pRK-9-Flag-EGFP, 5520 bp, Addgene, USA) were transfected into the cells to determine the transfection efficiency. The medium was changed after 4–6 h. The growth medium was removed and the cells were washed with phosphate buffered saline (PBS), followed by trypsinizing the cell pellet for 5 min using trypsin (100 µL/well) at 1, 2, and 3 days post transfection, respectively. The cell pellet was resuspended with 1 ml growth medium and cell number was counted using a hemocytometer under the fluorescence microscope.

By the data we obtained, the transfection efficiency increased during the period 1, 2, and 3 days post transfection, and then saturated at the 3rd day. The percentages were about 14.3%, 21.4%, and 21.4% using lipofectamine. The results revealed that the highest transfection efficiency for plasmids in NIH 3T3 cells was about 20–25% and it took about 2–3 days for the plasmids to transfect into cells completely. NKS 11 showed almost the same efficiency for transfection in NIH 3T3 cells.

2.8. Cell Count

One day before transfection, NIH3T3 cells (1.00×10^5 cells/well) were plated in a 12-well format containing 1 mL of growth medium without antibiotics so that adherent cells would be 90–95% confluent at the time of transfection. TALEN plasmids MCMV 1–2, 3–4 or 5–6 were transfected for each well (1.6 µg/well, 0.8 µg for each plasmid), or none were transfected as a negative control using lipofectamine. GFP plasmids (1.6 μg/well) were transfected into the cells as a positive control. The medium was changed after 4–6 h. The growth medium was removed and the cells were washed with PBS. The cell pellet was harvested using trypsin (100 μL/well) at 1, 3, 5 and 7 days post transfection, respectively. The cell pellet was resuspended with 1 mL growth medium. We counted the cell number using a hemocytometer under a light microscope.

2.9. Cell Viability Assay

The viable cells were assayed using the MTT Cell Growth Assay Kit (Sigma-Alderich, Temecula, CA, USA). The trypsinized cells were appropriately diluted to adjust the cell number range of 1000–50,000 cells/well with growth medium. An amount of 0.2 mL of cell dilution was plated into each well in a 96-well format. The cells were incubated at 37 °C overnight. The medium was changed to make the final volume of each well 0.1 mL. We added 0.01 mL AB Solution (MTT) to each well. We incubated the cells at 37 °C for 4 h for cleavage of MTT (3-(4,5-dimethyl-2-thiazolyl)-2,5-diphenyl-2-H-tetrazolium bromide, Thiazolyl Blue Tetrazolium Bromide). We added 0.1 mL isopropanol with 0.04 N HCl to each well. The absorbance was measured on an ELISA plate reader (Bio-rad, Hercules, CA, USA) with a test wavelength of 570 nm and a reference wavelength of 630 nm within 1 h.

2.10. TALEN Plasmid Transfection and MCMV Infection

On the 1st day, NIH3T3 cells (1.00×10^5 cells/well) in 1 mL growth medium without antibiotics were plated in a 12-well format so that adherent cells would be 90–95% confluent at the time of transfection. On the 2nd day, one pair of TALEN plasmids for each well (1.6 μg/well, 0.8 μg for each plasmid) was transfected, or none were transfected as a negative control, into the cells. GFP plasmids (1.6 μg/well) were transfected into the cells as a positive control. The medium was changed after 4–6 h. On the 3rd day, the cells were infected with MCMV (MOI = 0.05). The steps taken on the 2nd and 3rd day would be reversed if MCMV infection was prior to TALEN plasmid transfection. The medium was changed after 1 h. The cells were incubated at 37 °C in a CO_2 incubator for 5–7 days and we changed the medium every three days.

2.11. Virus Titration Assay

NIH3T3 cells grown to 60–70% confluent in 12-well format were prepared for virus titration. At 1, 3, 5, and 7 days post infection, the infected cells together with the medium were harvested and followed by 10 folds of serial dilution. After 1 h of incubation with the dilution at 37 °C in a CO_2 incubator, the prepared cells were overlaid with 2 mL fresh complete medium containing 1% low melting agarose and cultured for 4 to 5 days before the plaques were counted under a light microscope. The viral titer (pfu/mL) was determined by plaque assays. The values of the viral titers were the average of triplicate experiments.

2.12. Harvest of the Total DNA and Amplification of the Target DNA Sequence

The total DNA was harvested from the cell culture including the cell pellet and supernatant using the Blood and Tissue DNeasy kit (Qiagen, Germantown, MD, USA) and quantified by UV260 with a spectrophotometer. We amplified the specific product using MCMVM80/80.5 primers and the total DNA as a template by a polymerase chain reaction (PCR), under the conditions of 300 nM for each primer and 1× Hotstar Taq master mix (Qiagen, USA) in a 50 μL mixture. The thermal cycling conditions were 95 °C for 15 min followed by 35 cycles of 94 °C for 30 s, 60 °C for 30 s, 72 °C for 1 min, and 72 °C for 10 min.

2.13. Surveyor Nuclease Mutation Detection Assay

A Surveyor nuclease mutation detection kit (Surveyor nuclease and G + C control included) was obtained from IDT Integrated Technologies, USA [23].

We amplified wild-type (reference) and mutant (test) total DNA by PCR using MCMV M80/80.5 primers. We mixed equal amounts of reference and test PCR products. We incubated the mixture at 95 °C for 5 min in a beaker filled with 800 mL of water. We then allowed the mixture to denature in order to rehybridize, by heating and cooling it to form heteroduplexes and homoduplexes (finally leaving the water at <30 °C). We treated the annealed mixture with the Surveyor nuclease and incubated at 42 °C for 1 h. The reference PCR product was treated alone as a negative control. DNA fragments were separated by 2% agarose gel electrophoresis [23].

The cleavage efficiency of PCR products was calculated by scanning the signal strength of DNA bands on the UV illuminator. They were indicated by the percentage (%) of extra bands divided by the total bands (major bands + extra bands) in signal strength.

2.14. Latency Establishment, Treatment and Reactivation

In total, 27 Balb/c mice were infected with Smith strain MCMV (1×10^5 pfu/mouse) intraperitoneally (IP), but 3 Balb/c mice were not infected with MCMV. In total, 30 mice were housed to establish their latency for 4–5 months [3,22,24]. We sacrificed 3 mice infected with MCMV to harvest their organs (livers, lungs, spleens, kidneys and salivary glands), in order to test whether MCMV latency was established. The remaining 24 infected Balb/c mice were divided into five experimental groups (3–5 mice/group). They were untreated or treated with TALEN plasmids by tail vein injection 8 times (once/ 5–6 days). The treatment formulation formerly confirmed to be safe for mice was as follows: for total 200 µL injection, 6 µg TALEN plasmids (0.5 µg/µL), 30 µg NKS11 (10 µg/µL), 3.125 mM Sodium Acetate (25 mM, pH5.5) and PBS in each mouse. After treatment, all 24 mice were injected with an immunosuppressive agent cyclophosphamide (Sigma-Alderich, USA) at 150 mg/kg body weight twice (1 dose/5–6 days) to reactivate latency by tail vein injection [25–27]. Five days later, all mice were sacrificed and their organs harvested. We sonicated the organs to harvest the homogenate and total DNA.

2.15. Plaque Assay of MCMV in Mouse Organs

The organ homogenates were prepared by sonication and stored in 10% skim milk at −80 °C. The concentrations of all homogenates were adjusted to 10% (100 mg/mL). The presence of infectious viruses in the livers, lungs, spleens, kidneys, and salivary glands were determined by titrating organ homogenates. Plaque assays were performed by virus titration assay. The values given were calculated as PFUs per mg of tissue.

2.16. Quantitative Real-Time Polymerase Chain Reaction (qPCR) Analysis for DNA Copy Number

The total DNAs were harvested in 200 µL organ homogenates (100 mg/mL) in mice tissue using a Blood and Tissue DNeasy kit. Total DNA was dissolved in 50 µL Buffer AE (Qiagen, USA).

For the generation of a standard curve, the total DNA products were amplified by PCR using MCMV *ie-1* forward and reverse primers and the total MCMV DNA as a template, under the conditions of 500 nM for each primer and 1X Hotstar Taq polymerase master mix (Qiagen, USA) in a 50 µL mixture. The thermal cycling conditions were 95 °C for 15 min followed by 43 cycles of 94 °C for 30 s, 60 °C for 30 s, 72 °C for 1 min, and 72 °C for 10 min. The MCMV *ie-1* DNA PCR product (100 bp, nucleotide no.:181091-181190 in MCMV genome) was isolated to create the DNA dilution standard using the QIAquick Gel Kit Protocol. For absolute quantification of the MCMV DNA copy number in the organs, a standard curve was generated by serial dilutions of MCMV *ie-1* DNA PCR products, such that 1 µL of the standard curve template contained $5 \times 10^1, 5 \times 10^2, 5 \times 10^3, 5 \times 10^4, 5 \times 10^5, 5 \times 10^6$ for *ie-1* DNA copies. The standard curve was obtained by plotting the average threshold cycle (Ct) values against the logarithm of the target template molecules eluted from the MCMV *ie-1* DNA PCR products, followed by a sum of least squares regression analysis. Results were expressed as DNA copies/ mg of tissue. Since DNA yield per mg of tissue differs from tissue to tissue, DNA was extracted from 20 mg of each tissue and all extractions were done in triplicate and the average was used to determine the DNA yield/mg of each tissue type.

All qPCRs were performed with the TaqMan gene expression master mix (2× Hotstar Taq polymerase, Qiagen, USA) using the standard curve assay. Each sample was analyzed in triplicate at a 20 μL volume. For the *ie-1* DNA copy number assay, reaction mixtures contained 150 nM of each MCMV *ie-1* primer and 100 nM of the TaqMan probe. The amplification conditions were 50 °C for 2 min, 95 °C for 10 min, followed by 43 cycles of 95 °C for 15 s and 60 °C for 1 min. Values were calculated as copies per mg of tissue [28].

2.17. Statistical Analysis

The data were analyzed statistically using Microsoft Excel. In all cases, the values were the average of triplicate experiments and indicated as Mean ± SD (standard deviation).

Significant differences between the 2 groups in MCMV DNA copies in each organ of mice were determined using a two-tail Student's t test (type 3). Data were calculated in triplicate and expressed as Mean ± SEM (standard error of the mean). To determine real-time PCR techniques and their relative sensitivity, all real-time PCR data were calculated to DNA copies/mg of tissue. In all cases, a *p* value of ≤0.05 was considered statistically significant.

3. Results

3.1. Construction of TALEN Plasmids

Three specific pairs of TALEN plasmids MCMV1-2, 3-4 and 5-6 were constructed to target the MCMV M80/80.5 overlapping region. However, the TALE DNA sequences of two nonspecific pairs of TALEN plasmids W1FS-W7R1 and KSHV1-2 were both proved to be nonhomologous to MCMV M80/80.5 targeting sites and their MCMV genome match size was less than 10 nucleotides.

Several RVDs were constructed as TALE sequences in TALEN plasmids. Each RVD recognizes one nucleotide. The RVDs of MCMV1-2, 3-4, and 5-6 were designed according to MCMV M80/80.5 ORF. MCMV M80/80.5 forward and reverse primers were designed to cover all the three pairs of targeting sites and their PCR product size was 1048 bp (Figure 2).

```
ATGTCCATCCTGGCGGCCGCCGCGAAACAGAACGCCGTACCCGGCGCTTTAA
CGTCGCCGCAGCAGGCCTTGCCTCAGGTGCCCTCTTATTACGGAATGCCGCC
GGACGGTGTTCAGTATCATCTCCCGCCGCCACCACCACCGCCGTCTCACCAT
CGCGGCGGTGGCGGTACTTTCGATCCCCCTCTCCCACACGGGGGATACGGTC
CTCCCTATCATCACCCGGACGCCTATCGGGGGGGGATACCATCATCCGGATCG
CGACCCGCGCGGCGGAGTGCCCTACGAGGGCTGGTATCGCCCTCGCTACGA
CCCCGCCGGGGACGATCATCCCTCCTACAACAACAGGCGGGGAGACCGCTAT
CGGGCCGATCGCCCGCCTCAGCAGCAGCCCCTCTACAGGGGAGAGAGGAAT
AGGCGGCGGAGCCCTCCCGATTCGGACGACGACGATGACGATGACGACGAG
GATCTCGAAGCCGGCGAGCGTACGGGGGGGCAAGAGAACGCGGCAGAGAGGG
AGCGCCGACTCCGGCAGGAAGAGGAGACGTCGGGGCGCGGCGCCCGACGA
CGATGGCGGTGACCTCTCCCTACCCGGAGAGCGGGGATATCCCAAGCGCACC
GCCGGTGATCATCATCAGTCGGCGCCGCCGGCCTCCAGAACCGATGAGTTC
GGAGAGGTGAGGGCGACTCTCAACGAGATCCGCAAAGACATCTCGCAGATCC
GCGCGGCCGCCAGGGCCGAGGGCAACGGCGCGCGCGAGGACGCGGCCTCG
GTCGGCTCTTCTGACCAGAAATGTGCCGCCCCACCCCCCGGAGCCACCGAGA
TGATGGCCTCGGAACCGCCGGCCGGCGGCACCGTCGTCGCCAGGATGGCCC
TGGATCCGGCCGTGGCCGCCGCGACGGGGCATACCGCCGGGCTGCTTACCG
CCGGGAAATTGGTGAATGCTTCGTGCGAACCGACTCCCATGGAAGTGGGAGA
ACCCTCGGGAGGAGGAACCTCGAGGAAGGGGGGTGAGGCCAGCATGTTGGA
GGTGAACAAACGGATGTTCGTGTCCCTCCTCAATAAAATGGAATGAGTGATTC
AGACACCGGACTAACGGAGTGAACGTGTGATTTAT
```

Figure 2. Targeting sites in the MCMV M80 and M80.5 overlapping region (M80/80.5, nucleotide no. 114434-115507). The targeting sites of TALEN plasmids MCMV 1, 2, 3, 4, 5 and 6 are highlighted in yellow, red, blue, green, purple and gray, respectively. ATG (start codon for M80.5) and TGA (stop codon for M80 and M80.5) are both in red and underlined. MCMV M80/80.5 forward and reverse primers are bolded and underlined.

3.2. The Effect of TALEN Plasmids on Host Cells

We cotransfected TALEN plasmids MCMV1-2, 3-4 or 5-6 for each well, or did not transfect any plasmids as an untransfected (negative) control, using lipofectamine. The cell number was increasing for the first 5 days, although a slight decrease was observed on the 7th day. The growth period for NIH3T3 cells is from 1 to 5 days. All the four growth curves show almost the same trend for 7 days. TALEN plasmids MCMV1-2, 3-4 and 5-6 had no obvious effect on the growth of NIH3T3 cells compared with the untransfected control (Figure 3). Additionally, similar data were available in the cell viability assay (Table 2). These results demonstrated that the growth of cells was not influenced by the TALEN plasmids MCMV1-2, 3-4 and 5-6 for the first 5 days. TALEN plasmids exhibited no significant inhibition or enhancement on host cells.

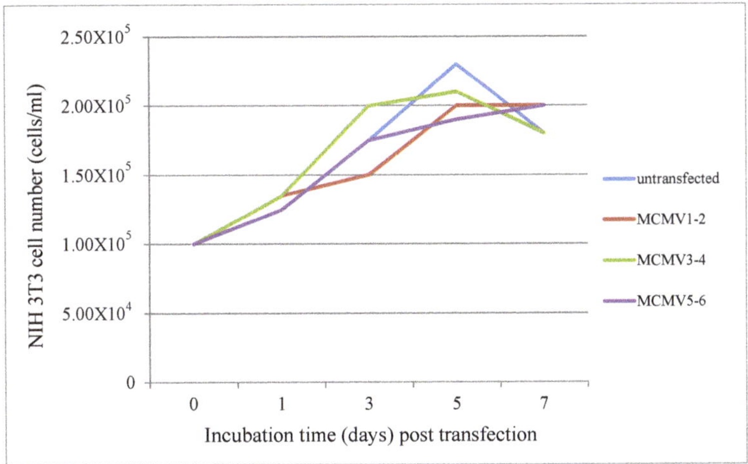

Figure 3. Host cell growth under the effect of TALEN plasmids. NIH 3T3 cells (1.00×10^5 cells/mL) were transfected with TALEN plasmids MCMV1-2, 3-4 or 5-6, respectively. The total cells were harvested and counted 1, 3, 5 and 7 days post transfection, respectively.

Table 2. Cell viability assay (initial NIH 3T3 cell count: 1.00×10^5 cells/mL).

Day Post Transfection	Untransfected	MCMV1-2	MCMV3-4	MCMV5-6
1	$1.20 \times 10^5 \pm 5000$	$1.25 \times 10^5 \pm 5000$	$1.27 \times 10^5 \pm 2890$	$1.17 \times 10^5 \pm 5700$
3	$1.68 \times 10^5 \pm 2890$	$1.42 \times 10^5 \pm 2890$	$1.92 \times 10^5 \pm 7640$	$1.70 \times 10^5 \pm 5000$
5	$2.25 \times 10^5 \pm 5000$	$1.92 \times 10^5 \pm 7640$	$2.03 \times 10^5 \pm 5770$	$1.88 \times 10^5 \pm 2890$
7	$1.72 \times 10^5 \pm 2890$	$1.87 \times 10^5 \pm 5770$	$1.73 \times 10^5 \pm 5770$	$1.85 \times 10^5 \pm 5000$

The cell number is expressed by cells/mL, and all data are indicated as Mean ± SD (standard deviation).

3.3. The Effects of TALEN Plasmids on MCMV Titer

Host NIH3T3 cells were divided into two groups as follows: (1) TALEN plasmid transfection was prior to the MCMV infection by 1 day; (2) MCMV infection was prior to TALEN plasmid transfection by 1 day.

When the specific TALEN plasmid (MCMV1-2, 3-4 and 5-6) transfection was prior to the viral infection, we found that the viral titers decreased by about 65%, 75%, and 25%, 1, 3 and 5 days post infection, respectively, compared with the controls, using lipofectamine as a transfection reagent (Figure 4A–C). Additionally, the viral titers decreased by about 50%, 60%, and 25%, 1, 3 and 5 days post infection, respectively, compared with the controls, using NKS11 as a transfection reagent (Figure 4D–F).

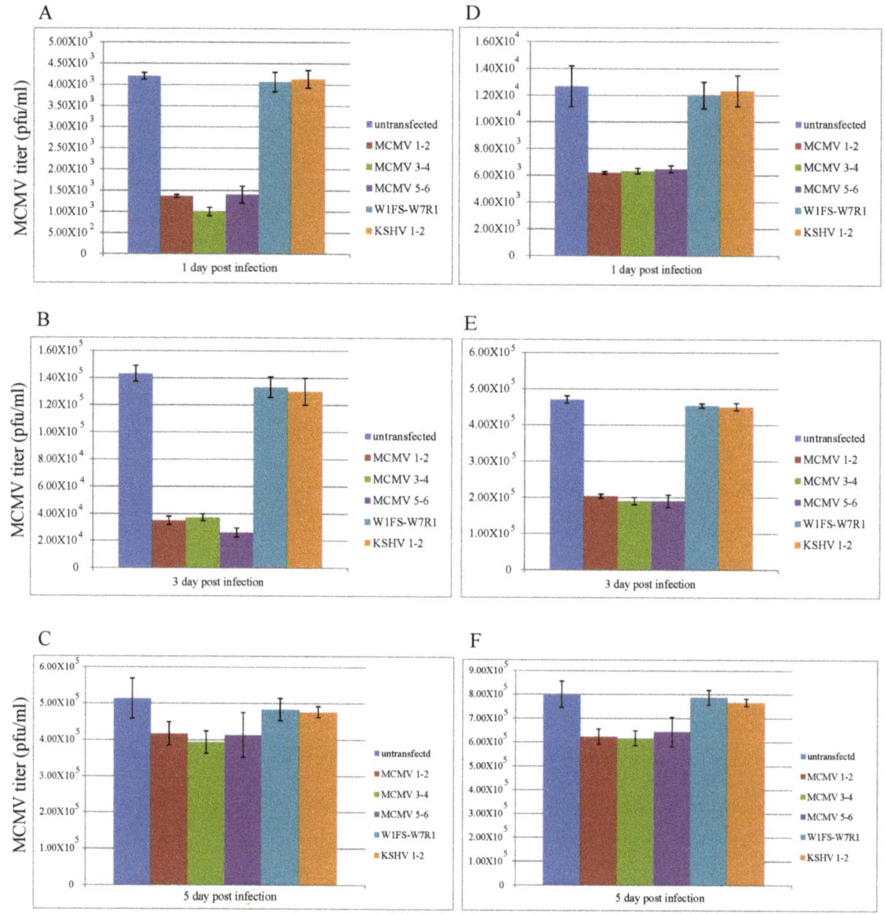

Figure 4. TALEN plasmid transfection was prior to MCMV infection. When using lipofectamine as a transfection reagent, the viral titers were determined 1, 3 and 5 days post infection, respectively, in (**A–C**). When using NKS11 as a transfection reagent, the viral titers were determined 1, 3 and 5 days post infection, respectively, in (**D–F**). All the data are expressed by columns (mean ± standard deviation).

However, if the viral infection was prior to the plasmid transfection, we found that TALEN plasmids resulted in no obvious growth inhibition on MCMV at 1, 3, 5 and 7 days post transfection, compared with the controls using lipofectamine or NKS11 as a transfection reagent (Figure 5).

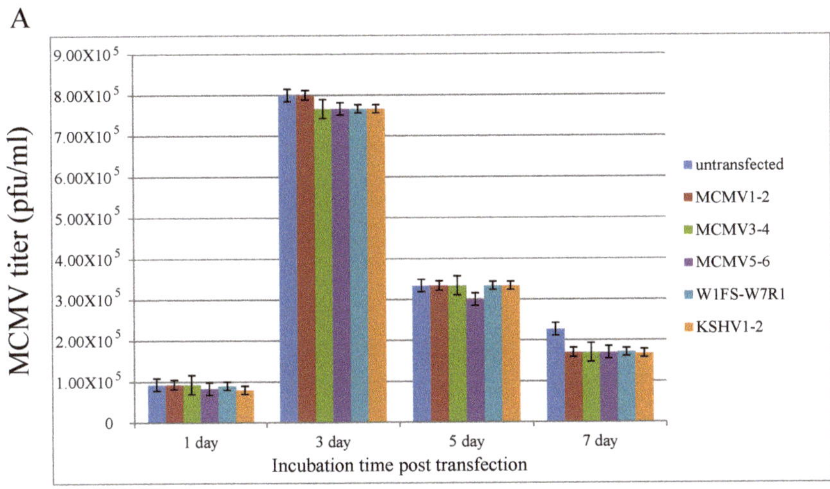

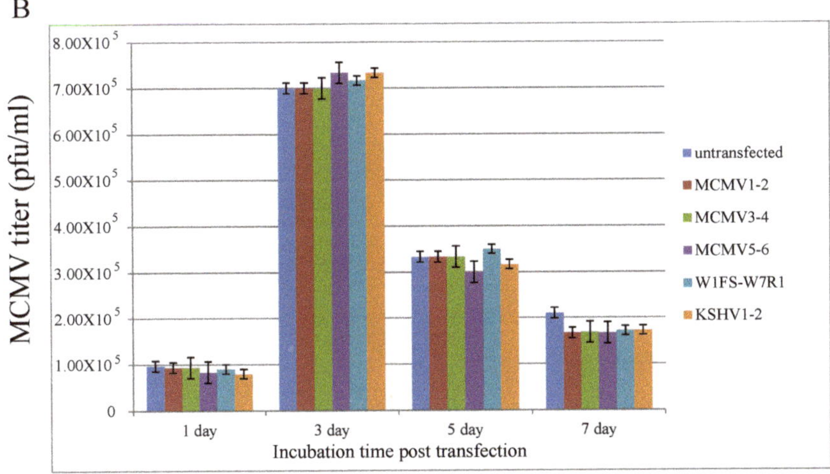

Figure 5. MCMV infection was prior to TALEN plasmid transfection. Viral titers were determined at 1, 3, 5 and 7 days post transfection using lipofectamine (**A**) or NKS11 (**B**), respectively. All the data are expressed by columns (mean ± standard deviation).

3.4. Analysis of the Targeting and Cleavage of TALEN Plasmids

We explored the targeting specificity and cleavage efficiency of TALEN plasmids for MCMV using the Surveyor nuclease mutation detection assay (Figure 6) [23]. NIH3T3 cells were either treated with both lipofectamine and specific TALEN plasmids (MCMV1-2, 3-4 and 5-6), nonspecific controls (W1FS-W7R1 and KSHV1-2) or only lipofectamine as a negative control. At 5 days post infection (or transfection), we harvested their total DNA from cell culture and used the total DNA as templates to amplify their M80/80.5 PCR products. In theory, specific TALEN plasmids MCMV1-2, 3-4 and 5-6 can specifically target MCMV 80/80.5 coding sequences and cleave their PCR product (1048 bp) to produce two extra DNA bands 322 and 726 bp, 608 and 440 bp, 923 and 125 bp, respectively.

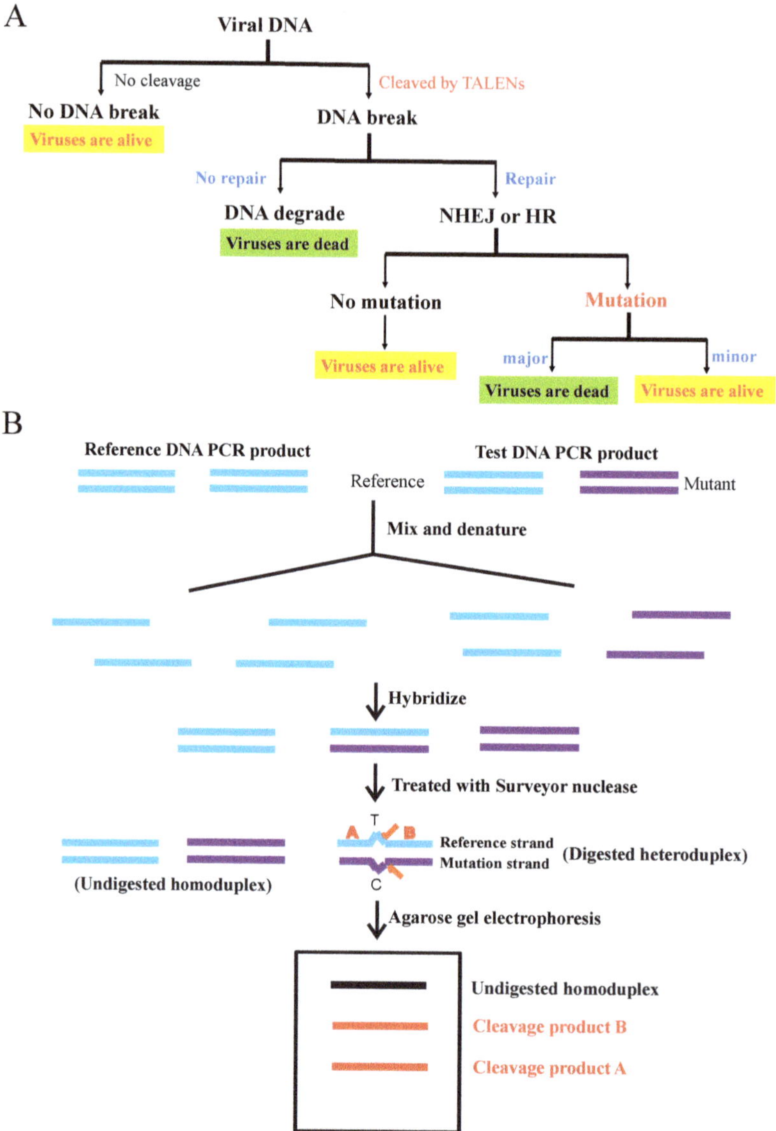

Figure 6. (A) Viral DNA targeting and cleavage: Once TALENs target and cleave the viral DNA, they will create a double stranded break (DSB). Either repaired by homologous recombination (HR) or non-homologous end joining (NHEJ), a mild mutation such as mismatch may appear. If viral DNA is mutated seriously or not repaired, viruses will be dead (highlighted in green). Viruses without cleavage, viruses without mutations, and viruses only with mild mutations are all alive (highlighted in yellow). **(B)** Surveyor nuclease mutation detection assay: Total DNA of all alive viruses are harvested and amplified by the polymerase chain reaction (PCR), followed by the denaturation, hybridization and analysis of the PCR products. Surveyor nuclease is a mismatch-specific endonuclease.

When the plasmid transfection was prior to the viral infection, the PCR products synthesized from the total DNAs targeted by TALEN plasmids were almost all the same (Figure 7A). However, we could just about clearly observe that MCMV3-4 produced two extra DNA bands (608 and 440 bp), in addition to the homoduplex bands (1048 bp). The other two specific pairs (MCMV1-2 and 5-6) did not produce dominant extra bands (only faint bands or smears were seen), which was likely due to weak bands and non-specificity of cleavage (off target). The nonspecific controls (W1FS-W7R1 and KSHV1-2) and the untransfected (negative) control did not produce any extra bands (Figure 7B). This meant that TALEN plasmids MCMV3-4 should be the most specific. The specificity is critical to avoid damaging normal and/or unrelated cells in animals; therefore, we selected MCMV3-4 as the specific treatment in the following animal studies.

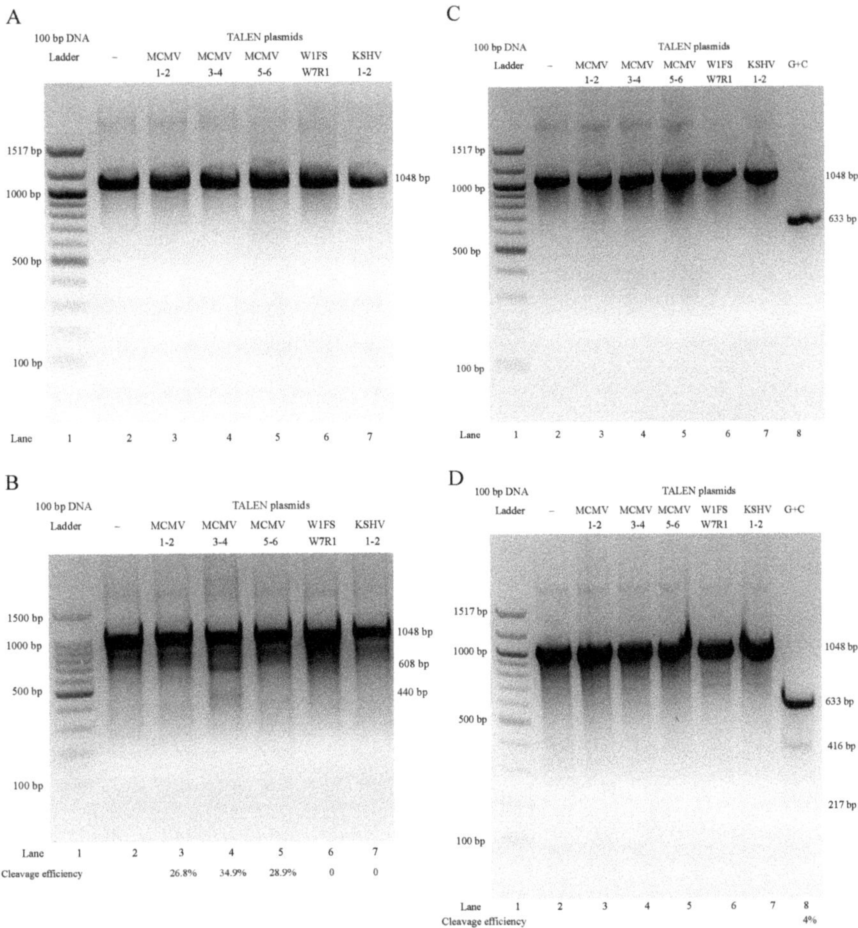

Figure 7. TALEN plasmid transfection was prior to MCMV infection: (**A**) Only homoduplex (or heteroduplex) PCR products (1048 bp) were observed (Lane 2–7). (**B**) Only MCMV3-4 (Lane 4) target and cleave PCR products produced two extra DNA bands (608 and 440 bp) clearly. MCMV1-2 and 5-6 (Lane 3 and 5) could not produce dominant extra bands, compared with the negative control (Lane 2). Neither the nonspecific TALEN plasmids W1FS-W7R1 (Lane 6) nor KSHV1-2 (Lane 7) could produce any extra bands, the same as the negative control (Lane 2). MCMV infection was prior to TALEN plasmid transfection: (**C**) Only homoduplex (or heteroduplex) PCR products (1048 bp) were observed (Lane 2-7). (**D**) No extra bands produced by TALEN plasmids were observed (Lane 3–7). Only positive control G + C (Lane 8) could produce dominant extra bands (416 and 217 bp) in addition to a major band (633 bp), compared with the negative control (Lane 2).

However, if the viral infection was prior to the plasmid transfection, the PCR products synthesized from total DNAs targeted by TALEN plasmids were almost the same (Figure 7C). We could hardly see any extra bands on agarose gel except the positive control G + C (Figure 7D) [23]. This suggested no obvious targeting and cleavage for all PCR products. The results showed that all TALEN plasmids (specific and nonspecific) did not work on the MCMV M80/80.5 coding sequence when the viral infection was prior to the plasmid transfection.

3.5. Establishment of MCMV Latency in Balb/c Mice

Three Balb/c mice were not infected with MCMV and were housed for latency establishment as negative controls. They were neither treated nor reactivated during the experimental process. We could not find any plaques from all five organs homogenates and *ie-1*DNA copy numbers were not detected. The results demonstrated that there were no viruses in the mice's organs originally and no genome in Balb/c mice is homologous to the MCMV *ie-1* gene.

For the plaque assay, it suggested no lytic viruses or only latent viruses were available in mice if the result was negative (no plaques detected). Otherwise, there were lytic viruses in mice if the result of assay was positive. For the following DNA copy number assay, if there were no DNA copies detectable this meant that there were no viruses; if the result was positive, this meant there were latent viruses in the organs.

We could not find any plaques from all five organ homogenates in 3 infected Balb/c mice, but their MCMV *ie-1* DNA copy numbers were all about 10^2 (Table 3). The results demonstrated that there were latent viruses in the organs and MCMV latency had been established in Balb/c mice 4–5 months after MCMV infection.

Table 3. Assay of MCMV-infected mouse organs for latency establishment.

Organ	Liver	Lung	Spleen	Kidney	Salivary Gland
Plaque assay	ND	ND	ND	ND	ND
DNA copies	315 ± 315	535 ± 94	226 ± 174	86 ± 86	67 ± 67

Balb/c mice were infected with viruses, but neither treated nor reactivated. Plaque assay (pfu/mg of tissue): all five organs from three mice were tested. Copy number of *ie-1* (DNA copies/mg of tissue): Mean ± SEM, SEM = SD/ $\sqrt{n}$, n = 3; SEM: standard error of mean; SD: standard deviation; n: sample size; ND: not detected (less than 10 pfu/mL for plaque assay, less than 50 copies/mg tissue for DNA copy number assay).

3.6. TALEN Treatment for Balb/c Mice

In comparison with latent MCMV-infected Balb/c mice without treatment and reactivation (Table 3), the MCMV *ie-1* DNA copies of the untreated but reactivated group (Group 1 in Table 4) increased by about 5–10 folds in the livers and lungs, 3–5 folds in the spleens, 20–25 folds in the kidneys, and about 3—5 folds in the salivary glands, respectively. This suggested that reactivation induced by the immunosuppressive agent (cyclophosphamide) takes effect to increase the viral load in mice.

In Table 4, we could not observe any plaques in the five organs in all five groups of mice. No detectable DNA copies were found in the five organs of all mice in the specific treatment group (Group 2). Despite this, the DNA copies of the untreated group (Group 1) were about 10^3 in the livers, lungs, spleens, kidneys, and 10^2 in the salivary glands. In the other treatment groups (Group 3, 4, 5), the DNA copies ranged from 10 to 10^2 in the livers, lungs, spleens, kidneys, and salivary glands. However, we found that the MCMV *ie-1* DNA copies of the untreated group (Group 1) were more than those of the less specific treatment groups (Group 3, 4) and the nonspecific treatment group (Group 5) by about 10–100 folds in the livers, lungs, spleens, kidneys, and by about 3–10 folds in the salivary glands. Overall, we could conclude that the specific treatment group (Group 2) was the most efficient one to remove viral load in mouse organs.

Table 4. Assay of MCMV-infected mouse organs for TALEN treatment.

Group	1 ($n = 3$)	2 ($n = 4$)	3 ($n = 4$)	4 ($n = 5$)	5 ($n = 5$)
Plaque Assay	ND	ND	ND	ND	ND
Organ	DNA Copies				
Liver	2284 ± 1301	ND ($p = 0.0004$)	32 ± 32 ($p = 0.0005$)	25 ± 25 ($p = 0.0005$)	ND ($p = 0.0004$)
Lung	3777 ± 2865	ND ($p = 0.0001$)	142 ± 82 ($p = 0.0008$)	353 ± 168 ($p = 0.0032$)	311 ± 153 ($p = 0.0041$)
Spleen	1073 ± 587	ND ($p = 0.0041$)	ND ($p = 0.0052$)	ND ($p = 0.0041$)	29 ± 29 ($p = 0.0087$)
Kidney	2033 ± 1413	ND ($p = 0.0046$)	ND ($p = 0.0046$)	134 ± 82 ($p = 0.0309$)	22 ± 22 ($p = 0.0068$)
Salivary gland	379 ± 260	ND ($p = 0.0158$)	97±57 ($p = 0.2470$)	172 ± 127 ($p = 0.2768$)	38 ± 38 ($p = 0.0468$)

p value: comparison of the untreated group (Group 1) versus the treatment groups (Group 2, 3, 4 and 5) respectively; Plaque assay (pfu/mL): five organs from all five groups of mice were tested; Copy number of *ie-1* (DNA copies/mg of tissue): Mean ± SEM, SEM = SD/ $\sqrt{n}$; SEM: standard error of mean; SD: standard deviation; n: sample size; ND: not detected (less than 10 pfu/mL for plaque assay, less than 50 copies/mg tissue for DNA copy number assay); Experimental groups were all reactivated by cyclophosphamide; Group 1: No treatment (only treated with phosphate buffered saline); Group 2: specific treatment (TALEN plasmids MCMV3-4) with transfection reagent NKS11; Group 3: specific treatment (TALEN plasmids MCMV3-4) without transfection reagent NKS11; Group 4: less specific treatment (TALEN plasmids MCMV1-2, 3-4 and 5-6) with transfection reagent NKS11; Group 5: nonspecific treatment (TALEN plasmids KSHV3-4) with transfection reagent NKS11.

4. Discussion

The transfection efficiency of plasmids in NIH3T3 cells was about 20–25% in cell culture. We realized that an elevated level of transfection efficiency could increase the efficacy in inhibiting virus growth in cell culture. Despite this, we did not sort the transfected cells using flow cytometry for the studies in cell culture, because it was not feasible for us to do so in mice. To establish a more similar animal model, we used all the cells including untransfected and transfected cells for our ex vivo studies. To enhance the efficacy, multiple round injections for the formulation transfection during the treatment period are required in animal studies.

Cultured NIH3T3 cells are not specific for latency studies. If the viral infection is prior to the plasmid transfection, NIH3T3 cells are vulnerable to viruses because they do not have the same immune system as animals. In this case, TALENs can hardly protect the host cells from the viral infection, because the viral titer increases rapidly to about 10^5 pfu/mL within 1–3 days. However, if the plasmid transfection is prior to the viral infection, TALEN plasmid copy number might have increased and induced innate immune responses of host cells to secret cytokine or other factors to fight against invading viruses. Therefore, TALENs could inhibit virus growth by about 50–75% when the viral titer was 10^3–10^4 pfu/mL, and about 25% when the viral titer was about 10^5 pfu/mL.

In cell culture, the results for specific target and cleavage efficiency of TALEN plasmids reveal that some of the plasmids work well on MCMV M80/80.5 target sites. Specific TALEN plasmids also demonstrated their effects on the inhibition of virus growth, ranging from 25–75% depending on the viral titer. Although the decreasing level of viral titers was varied under different conditions, they had the same trend for the inhibition of virus growth. We also found that the higher the viral titer, the lower the effect of the TALEN plasmids on virus growth. Our findings indicated that the inhibition effect on MCMV is about 20–25% when the viral titer reaches the highest level (10^5). The reasons might be that the amino acid-nucleotide recognition frequency is not absolute (e.g., NI-A: 55%, NG-T: 50%, NN-G: 7%, HD-C: 69%) and the transfection efficiency is about 20–25% in cell culture.

The specificity, efficiency and biosafety of delivery tools are critical for drug delivery in animal studies and human clinical trials [29]. Lipofectamine, one of the most common transfection reagents used in cell culture, is known to be toxic to animals. Our results have shown that the new transfection reagent NKS11 can work almost as well as lipofectamine for inhibiting virus lytic replication in cell culture, when TALEN plasmid transfection is prior to MCMV infection. Fortunately, NKS11 also proved to be nontoxic to Balb/c mice in the preliminary tests.

The viral *ie-1* DNA copies increased once Balb/c mice were infected with MCMV. An absence of plaques meant that there were no viruses during the lytic cycle. In latency, plaques were not detectable and the *ie-1* DNA copy number was low but detectable. The *ie-1* DNA copies increased once latency was reactivated. In Table 4, the viral load significantly increased after reactivation in all five organs of mice if there was no treatment (Group 1), compared with the treatment groups (Group 2–5). After treatment, the latent MCMV was removed by TALENs, so the viral load was undetected or significantly decreased even though latency was reactivated using an immunosuppressive agent (Group 2–5). Additionally, if the TALENs are only targeting the reactivating viruses and not the latent pools, it is impossible for the specific treatment group (Group 2) to be all ND (not detected) in the *ie-1* DNA copy number assay for all five organs of mice. Thus, we consider that it is possible to remove latent viruses using this strategy.

In animal studies, the less specific and nonspecific treatment groups (Group 3, 4, 5) also worked well in reducing viral load, although they were less efficient than the specific treatment group (Group 2). This might be explained as follows. Firstly, during the multiple-round injection for TALEN plasmid treatment, recognition of foreign DNA in intracellular compartments or in the cytoplasm of host cells sends a signal of pathogenic invasion. In response, the innate immune DNA-sensing pathways start an antimicrobial type I interferon (IFN) (mainly IFNα and β) secretion to act against viruses. Acting in paracrine or autocrine models, IFNs stimulate intracellular and intercellular networks for regulating innate and acquired immunity in mice that are resistant to the viral infections [30–35]. However, there are no data that categorically show that the latent virus has been eradicated and that this is an IFN effect in our studies. Secondly, there are three possible mechanisms for TALEN plasmids to work on DNA, namely cleavage of target DNA, induction of DNA mutation, and inhibition of DNA transcription and translation [17]. Thirdly, treatment is very complex in animals; it is also influenced by other factors such as the individual diversity of mice and the efficiency of tail vein injection.

Currently, it is impossible to eradicate latent CMV viruses in animals, although there are some effective drugs (e.g., Ganciclovir, Valganciclovir, and Foscarnet) for the treatment of active infections. Our data indicated that TALEN plasmids which could specifically target and cleave MCMV M80/80.5 ORF would effectively reduce the viral load in Balb/c mice, so that they resulted in the implicative decrease of *ie-1* DNA copies level. The viral latent infection in humans is a major barrier for effective treatment and also a long-term risk to the host. Although the mechanisms for inhibiting MCMV are still poorly understood, our studies demonstrate that the removal of latent MCMV in animals is possible using TALENs.

Author Contributions: Y.-C.C. designed the protocols of this study. S.-J.C and Y.-C.C. performed experiments and wrote this article.

Funding: This research received no external funding.

Acknowledgments: We are grateful for the grant support from Mackay Memorial Hospital in publishing this article.

Conflicts of Interest: The authors declare no conflict of interest.

References

1. Zhan, X.; Lee, M.; Xiao, J.; Liu, F. Construction and characterization of murine cytomegaloviruses that contain transposon insertions at open reading frames M09 and M83. *J. Virol.* **2000**, *74*, 7411–7421. [CrossRef] [PubMed]
2. Rawlinson, W.D.; Farrell, H.E.; Barrell, B.G. Analysis of the complete DNA sequence of murine cytomegalovirus. *J. Virol.* **1996**, *70*, 8833–8849.
3. Pollock, J.L.; Virgin, H.W. Latency, without persistence, of murine cytomegalovirus in the spleen and kidney. *J. Virol.* **1995**, *69*, 1762–1768. [PubMed]
4. Mercer, J.A.; Wiley, C.A.; Spector, D.H. Pathogenesis of murine cytomegalovirus infections: Identification of infected cells in the spleen during acute and latent infections. *J. Virol.* **1988**, *62*, 987–997. [PubMed]
5. Sinclair, J.; Sissons, P. Latency and reactivation of human cytomegalovirus. *J. Gen. Virol.* **2006**, *87*, 1763–1779. [CrossRef]

6. Boch, J.; Scholze, H.; Schornack, S.; Landgraf, A.; Hahn, S.; Kay, S.; Lahaye, T.; Nickstadt, A.; Bonas, U. Breaking the code of DNA binding specificity of TAL-type III effectors. *Science* **2009**, *326*, 1509–1512. [CrossRef]

7. Li, T.; Huang, S.; Jiang, W.Z.; Wright, D.; Spalding, M.H.; Weeks, D.P.; Yang, B. TAL nucleases (TALNs): Hybrid proteins composed of TAL effectors and FokI DNA-cleavage domain. *Nucleic Acids Res.* **2011**, *39*, 359–372. [CrossRef]

8. Wood, A.J.; Lo, T.W.; Zeitler, B.; Pickle, C.S.; Ralston, E.J.; Lee, A.H.; Amora, R.; Miller, J.C.; Leung, E.; Meng, X.; et al. Targeted genome editing across species using ZFNs and TALEN. *Science* **2011**, *333*, 307. [CrossRef]

9. Christian, M.; Cermak, T.; Doyle, E.L.; Schmidt, C.; Zhang, F.; Hummel, A.; Bogdanove, A.J.; Voytas, D.F. Targeting DNA double-stranded breaks with TAL effector nucleases. *Genetics* **2010**, *186*, 757–761. [CrossRef] [PubMed]

10. Mussolino, C.; Morbitzer, R.; Lütge, F.; Dannemann, N.; Lahaye, T.; Cathomen, T. A novel TALE nuclease scaffold enables high genome editing activity in combination with low toxicity. *Nucleic Acids Res.* **2011**, *39*, 9283–9293. [CrossRef] [PubMed]

11. Hockemeyer, D.; Wang, H.; Kiani, S.; Lai, C.S.; Gao, Q.; Cassady, J.P.; Cost, G.J.; Zhang, L.; Santiago, Y.; Miller, J.C.; et al. Genetic engineering of human pluripotent cells using TALE nucleases. *Nat. Biotechnol.* **2011**, *29*, 731–734. [CrossRef] [PubMed]

12. Luo, Y.; Luo, Y.; Liu, C.; Cerbini, T.; San, H.; Lin, Y.; Chen, G.; Rao, M.S.; Zou, J. Stable enhanced green fluorescent protein expression after differentiation and transplantation of reporter human induced pluripotent stem cells generated by AAVS1 transcription activator-like effector nucleases. *Stem Cells Transl. Med.* **2014**, *3*, 821–835. [CrossRef] [PubMed]

13. Tesson, L.; Tesson, L.; Usal, C.; Ménoret, S.; Leung, E.; Niles, B.J.; Remy, S.; Santiago, Y.; Vincent, A.I.; Meng, X.; et al. Knockout rats generated by embryo microinjection of TALEN. *Nat. Biotechnol.* **2011**, *29*, 695–696. [CrossRef] [PubMed]

14. Huang, P.; Xiao, A.; Zhou, M.; Zhu, Z.; Lin, S.; Zhang, B. Heritable gene targeting in zebrafish using customized TALEN. *Nat. Biotechnol.* **2011**, *29*, 699–700. [CrossRef]

15. Sander, J.D.; Cade, L.; Khayter, C.; Reyon, D.; Peterson, R.T.; Joung, J.K.; Yeh, J.R. Targeted gene disruption in somatic zebrafish cells using engineered TALEN. *Nat. Biotechnol.* **2011**, *29*, 697–698. [CrossRef] [PubMed]

16. Bloom, K.; Ely, A.; Mussolino, C.; Cathomen, T.; Arbuthnot, P. Inactivation of hepatitis B virus replication in cultured cells and in vivo with engineered transcription activator-like effector nucleases. *Mol. Ther.* **2013**, *21*, 1889–1897. [CrossRef] [PubMed]

17. Chen, J.; Zhang, W.; Lin, J.; Wang, F.; Wu, M.; Chen, C.; Zheng, Y.; Peng, X.; Li, J.; Yuan, Z. An efficient antiviral strategy for targeting hepatitis B virus genome using transcription activator-like effector nucleases. *Mol. Ther.* **2014**, *22*, 303–311. [CrossRef] [PubMed]

18. Noh, K.W.; Park, J.; Kang, M.S. Targeted disruption of EBNA1 in EBV-infected cells attenuated cell growth. *BMB Rep.* **2016**, *49*, 226–231. [CrossRef]

19. Dupont, L.; Reeves, M.B. Cytomegalovirus latency and reactivation: Recent insights into an age old problem. *Rev. Med. Virol.* **2016**, *26*, 75–89. [CrossRef] [PubMed]

20. Liu, X.F.; Yan, S.; Abecassis, M.; Hummel, M. Biphasic recruitment of transcriptional repressors to the murine cytomegalovirus major immediate-early promoter during the course of infection in vivo. *J. Virol.* **2010**, *84*, 3631–3643. [CrossRef] [PubMed]

21. Liu, X.F.; Yan, S.; Abecassis, M.; Hummel, M. Establishment of murine cytomegalovirus latency in vivo is associated with changes in histone modifications and recruitment of transcriptional repressors to the major immediate-early promoter. *J. Virol.* **2008**, *82*, 10922–10931. [CrossRef]

22. Reddehase, M.J.; Podlech, J.; Grzimek, N.K. Mouse models of cytomegalovirus latency: Overview. *J. Clin. Virol.* **2002**, *25*, S23–S36. [CrossRef]

23. IDT Integrated Technologies. Surveyor®Mutation Detection Kit for Standard Gel Electrophoresis. User Guide. Revision: 3. Available online: https://sfvideo.blob.core.windows.net/sitefinity/docs/default-source/user-guide-manual/surveyor-kit-for-gel-electrophoresis-user-guide.pdf?sfvrsn=a9123407_6 (accessed on 1 June 2015).

24. Porter, K.R.; Starnes, D.M.; Hamilton, J.D. Reactivation of latent murine cytomegalovirus from kidney. *Kidney Int.* **1985**, *28*, 922–925. [CrossRef]

25. Mayo, D.R.; Amstrong, J.A.; Ho, M. Reactivation of MCMV by cyclophosphamide. *Nature* **1997**, *267*, 721–723. [CrossRef]

26. Schmader, K.E.; Rahija, R.; Porter, K.R.; Daley, G.; Hamilton, J.D. Aging and reactivation of latent murine cytomegalovirus. *J. Infect. Dis.* **1992**, *166*, 1403–1407. [CrossRef] [PubMed]

27. Gönczöl, E.; Danczig, E.; Boldogh, I.; Tóth, T.; Váczi, L. In vivo model for the acute, latent and reactivated phases of cytomegalovirus infections. *Acta Microbiol. Hung.* **1985**, *32*, 39–47. [PubMed]

28. Zhu, J.; Chen, J.; Hai, R.; Tong, T.; Xiao, J.; Zhan, X.; Lu, S.; Liu, F. In vitro and in vivo characterization of a murine cytomegalovirus with a mutation at open reading frame m166. *J. Virol.* **2003**, *77*, 2882–2891. [CrossRef] [PubMed]

29. Chen, Y.C.; Sheng, J.; Trang, P.; Liu, F. Potential application of the CRISPR/Cas9 system against herpesvirus infections. *Viruses* **2018**, *10*, 291. [CrossRef]

30. Kato, K.; Omura, H.; Ishitani, R.; Nureki, O. Cyclic GMP-AMP as an endogenous second messenger in innate Immune Signaling by Cytosolic DNA. *Annu. Rev. Biochem.* **2017**, *86*, 541–566. [CrossRef]

31. Goubau, D.; Deddouche, S.; Reis e Sousa, C. Cytosolic sensing of viruses. *Immunity* **2013**, *38*, 855–869. [CrossRef]

32. Gürtler, C.; Bowie, A.G. Innate immune detection of microbial nucleic acids. *Trends Microbiol.* **2013**, *21*, 413–420. [CrossRef] [PubMed]

33. Paludan, S.R.; Bowie, A.G. Immune sensing of DNA. *Immunity* **2013**, *38*, 870–880. [CrossRef] [PubMed]

34. Abe, T. Innate immune DNA sensing pathways. *Uirusu* **2014**, *64*, 83–94. [CrossRef] [PubMed]

35. Cai, X.; Chiu, Y.H.; Chen, Z.J. The cGAS-cGAMP-STING pathway of cytosolic DNA sensing and signaling. *Mol. Cell* **2014**, *54*, 289–296. [CrossRef] [PubMed]

 viruses

Article

Emergence of Fluoxetine-Resistant Variants during Treatment of Human Pancreatic Cell Cultures Persistently Infected with Coxsackievirus B4

Enagnon Kazali Alidjinou, Antoine Bertin, Famara Sane, Delphine Caloone, Ilka Engelmann and Didier Hober *

Faculté de médecine, Université Lille, CHU Lille, Laboratoire de Virologie EA3610, F-59000 Lille, France;
enagnonkazali.alidjinou@chru-lille.fr (E.K.A.); antoine.bertin@gmail.com (A.B.); famara.sane@chru-lille.fr (F.S.);
delphine.lobert@univ-lille.fr (D.C.); ilka.engelmann@chru-lille.fr (I.E.)
* Correspondence: didier.hober@chru-lille.fr; Tel.: +33-0-3-20-44-66-88; Fax: +33-03-20-44-48-95

Received: 10 April 2019; Accepted: 21 May 2019; Published: 28 May 2019

Abstract: This study reports the antiviral activity of the drug fluoxetine against some enteroviruses (EV). We had previously established a model of persistent coxsackievirus B4 (CVB4) infection in pancreatic cell cultures and demonstrated that fluoxetine could clear the virus from these cultures. We further report the emergence of resistant variants during the treatment with fluoxetine in this model. Four independent persistent CVB4 infections in Panc-1 cells were treated with fluoxetine. The resistance to fluoxetine was investigated in an acute infection model. The 2C region, the putative target of fluoxetine antiviral activity, was sequenced. However, Fluoxetine treatment failed to clear CVB4 in two persistent infections. The resistance to fluoxetine was later confirmed in HEp-2 cells. The decrease in viral titer was significantly lower when cells were inoculated with the virus obtained from persistently infected cultures treated with fluoxetine than those from susceptible mock-treated cultures (0.6 log TCID50/mL versus 4.2 log TCID50/mL, $p < 0.0001$). Some previously described mutations and additional ones within the 2C protein were found in the fluoxetine-resistant isolates. The model of persistent infection is an interesting tool for assessing the emergence of variants resistant to anti-EV molecules. The resistance of EV strains to fluoxetine and its mechanisms require further investigation.

Keywords: enteroviruses; coxsackievirus B4; persistent infection; fluoxetine; resistance; mutations

---

1. Introduction

The genus *Enterovirus* (*Picornaviridae* family) is a large group of small non-enveloped RNA viruses that are involved in several mild or severe acute clinical infections in humans ranging from enteric or respiratory infections, hand-foot-and-mouth disease, or conjunctivitis to acute flaccid paralysis, viral myocarditis, fulminant pancreatitis, or aseptic meningitis [1–3]. Some of these viruses in this group, especially type B coxsackieviruses (CVB) are also known to play a role in the development of chronic diseases, such as chronic myocarditis or type 1 diabetes [4–6]. Enteroviruses (EV) are well known as cytolytic viruses, but they can also establish persistent infections in vitro and in vivo, a mechanism potentially involved in the pathogenesis of chronic diseases [7].

Despite several attempts of library screening and other than a few compounds under investigation, to date no antiviral molecule has been licensed worldwide for the treatment of enteroviral infections that can sometimes be potentially life threatening to humans [8,9].

Fluoxetine is a selective serotonin reuptake inhibitor (SSRI) used for the treatment of depression or other mental disorders. This drug has been reported to display a significant antiviral activity against enteroviruses in vitro, especially *Enterovirus B* and *D* species [10,11]. The putative target of fluoxetine is the nonstructural viral protein 2C, a highly conserved region among enteroviruses. Other well-known enterovirus replication inhibitors such as, guanidine hydrochloride (GuHCl) or TBZE-029 also target 2C protein, even though the mechanism might be different. Some 2C CVB3 resistant mutants have been described with cross-resistance to all these compounds [8,10].

A model of persistent coxsackievirus B4 (CVB4) infection in pancreatic cells was established by our team and represents an interesting tool to study the activity of anti-enteroviral candidate agents, and subsequently the emergence of viral resistance to these molecules. It was previously shown that the treatment with fluoxetine can cure pancreatic cell cultures persistently infected with CVB4 [12].

We further report the emergence of resistant CVB4 variants during the fluoxetine-treatment of human pancreatic cell cultures persistently infected with the virus.

2. Materials and Methods

2.1. Cells and Reagents

HEp-2 cells (BioWhittaker, Walkersville, MD, USA) were grown in minimum essential medium (MEM) supplemented with 10% of fetal calf serum (FCS), 1% of L-glutamine, 1% of nonessential amino acids, and 1% of penicillin and streptomycin. The human ductal cell line Panc-1 (ATCC) was cultured in Dulbecco's modified Eagle's medium (DMEM) supplemented with 10% of FCS, 1% of L-glutamine, and 1% of penicillin and streptomycin.

Fluoxetine chlorhydrate (Lilly France, Fegersheim, France) was dissolved in dimethyl sulfoxide (DMSO) at a final concentration of 5.48 uM and was used in all experiments, as previously reported [12]. Guanidine hydrochloride (GuHCl) was purchased from Sigma-Aldrich (Saint-Quentin-Fallavier, France) and was used at a final concentration of 2 mM.

2.2. Virus and Persistent Infection

The diabetogenic CVB4 E2 strain, provided initially by Ji-Won Yoon (Julia McFarlane Diabetes Research Center, Calgary, Alberta, Canada), was propagated in HEp-2 cells and used to establish CVB4 persistent infections.

The model of persistent CVB4 infection of Panc-1 cells has been previously described [13,14]. Briefly, a 25 cm^2 Nunc cell culture flask (Thermofisher Scientific, Villebon, France) containing an average of 10^6 cells in DMEM was inoculated with CVB4 at a multiplicity of infection (MOI) of 0.01. During the acute lytic infection, the culture medium was regularly changed, and finally a stable equilibrium was found between the viral replication and cell proliferation. The medium was changed twice a week, and cells were scraped and subcultured once a week. The supernatants were collected at different time points (1, 10, 20, 21, 24, 28, and 30 weeks post infection) during the persistent infection.

2.3. Antiviral Activity Testing

The antiviral activity of fluoxetine was evaluated using HEp-2 cells. Cells were seeded in a 96-well cell culture plate at 1.25 × 10^4 cells per well. Cells were inoculated with the virus at a MOI of 0.01, mixed with fluoxetine or DMSO. The plates were incubated at 37 °C, and the cell cultures were observed every day. The supernatants were collected when 100% cytopathic effect (CPE) was observed in DMSO-treated wells.

2.4. Determination of Viral Titer

The viral titer obtained in supernatants was assessed using the end-point dilution assay, particularly the Spearman–Karber statistical method was used to determine the tissue culture 50% infectious dose (TCID50).

2.5. Viral Genome Sequencing

Viral genome was sequenced using the Sanger (population) method. Viral RNA was isolated from 140 µL of viral suspension with the QIAamp Viral RNA Mini Kit (Qiagen, Courtaboeuf, France) following the manufacturer's instructions. The whole 2C region (from nt 4039 to nt 5025) was amplified using the Superscript III One-step reverse transcription-polymerase chain reaction (RT-PCR) system with Platinum Taq DNA polymerase kit (Thermo Fischer Scientific). The amplicons were checked on a 2% agarose gel, purified on Nucleoseq columns (Machery-Nagel, Hœrdt, France), and sequenced using the BigDye Terminator v3.1 Cycle Sequencing Kit (Thermo Fischer Scientific, Les Ulis, France). The sequences were purified with the BigDye XTerminator Purification Kit. Sequenced products were analyzed on a 3500Dx genetic analyzer (Thermo Fischer Scientific). Electrophoregrams were manually edited with Seqscape software v3 (Thermo Fischer Scientific). The sequences of primers are shown in Table 1.

Table 1. Primers used for sequencing.

Primer Name	Forward/Reverse	Primer Sequence (5′–3′)	Nucleotide Position
EXT-1	Forward	CTCAAGCGGAAAGTGTCCCA	3988–4007
INT-1	Reverse	TTTCCCATCAGGGTTCTGGC	4593–4574
INT-2	Forward	GATTGGGCGTTCACTTGCAG	4461–4480
EXT-2	Reverse	ACTGCCTCACTATCCACCGA	5126–5107

2.6. Statistical Analysis

Data were presented as mean ± SD. Graphs and analyses were performed with GraphPad Prism® V6.0 software. Comparisons were performed with the Mann–Whitney U test with the significance set at 0.05.

3. Results

3.1. Persistent CVB4 Infection and Fluoxetine Treatment

Four independent cultures persistently infected with CVB4 (I1, I2, I3, and I4) were established in Panc-1 cells. Culture supernatants were periodically collected and the presence of infectious particles was checked using HEp-2 cells. The viral titers obtained in the supernatants ranged from 6.83 to 7.83 log TCID$_{50}$/mL at one week postinoculation (p.i.), 6.25 to 7.50 log TCID$_{50}$/mL at ten weeks p.i., and 5.50 to 6.50 log TCID$_{50}$/mL at twenty weeks p.i. (see Figure 1).

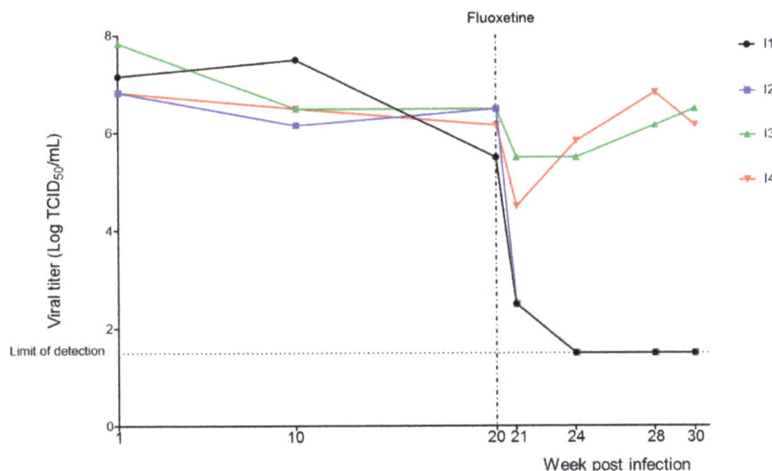

Figure 1. Persistent coxsackievirus B4 (CVB4) infection in Panc-1 cells and treatment with fluoxetine. Four independent persistent CVB4 infections (I1, I2, I3, and I4) were established in Panc-1 cells. At 20 weeks post infection, cells were treated with fluoxetine at 5.48 μM twice a week. The culture supernatants were collected all along the follow-up. Viral titers in supernatants were determined using the end-point dilution assay.

Starting from week 20 p.i., persistently infected cultures were treated twice a week with fluoxetine at 5.48 µM or DMSO, and supernatants were collected periodically to monitor the infection. The levels of infectious viral particles in persistently infected cultures I1 and I2 decreased significantly at one week of treatment (o.t.) i.e., 2.5 log $TCID_{50}$/mL, and the virus was undetectable at four weeks posttreatment. In contrast, for persistently infected cultures I3 and I4, only a slight decrease was observed in viral titers at one week o.t., followed by a rise from the fourth week o.t. The viral titers obtained after 10 weeks of fluoxetine-treatment were still at 6.5 and 6.25 $TCID_{50}$/mL for I3 and I4, respectively (see Figure 1). No significant changes in viral titers were observed in DMSO-treated cultures.

3.2. Investigation of Resistance to Fluoxetine Treatment

To confirm that I3 and I4 were resistant to the treatment, the susceptibility of viral suspensions to fluoxetine was evaluated in a model of acute infection. Various viral suspensions from persistently infected cultures treated with fluoxetine at 5.48 µM or DMSO were collected at week 8 o.t. They were inoculated into HEp-2 cells in the presence of fluoxetine or DMSO. Supernatants were collected on day 3 p.i., and the viral titers were determined. Figure 2 presents the extent of decrease in viral titer in fluoxetine-treated wells, as compared to DMSO-treated ones.

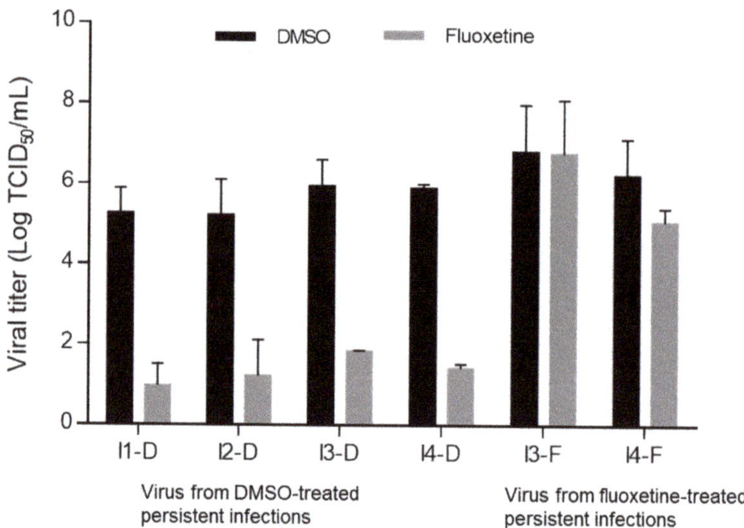

Figure 2. Susceptibility to fluoxetine of virus isolates obtained from treated persistently CVB4-infected cultures. Virus suspensions were collected from persistent CVB4 infections treated with DMSO (I1-D, I2-D, I3-D, and I4-D), or with fluoxetine (I3-F and I4-F) at week 8 of treatment. HEp-2 cells were inoculated with various virus suspensions in the presence of fluoxetine or dimethyl sulfoxide (DMSO). Viral titers were determined 3 days postinoculation. Data are mean ± SD of two independent experiments.

Virus isolates emerging from the four persistently infected cultures treated with DMSO for eight weeks (I1-D, I2-D, I3-D, and I4-D) remained highly susceptible to fluoxetine. Indeed, a mean decrease in viral titer ranging between 4 and 4.5 log TCID50/mL was observed when the isolates were inoculated into HEp-2 cell cultures in the presence of fluoxetine, as compared to the cultures inoculated with virus isolates in the presence of DMSO.

Regarding the virus obtained from persistently infected cultures treated with fluoxetine for 8 weeks, the antiviral activity of fluoxetine was strongly reduced. The mean decrease in viral titer was as low as 0.1 and 1.2 log TCID50/mL for I3-F and I4-F, respectively (see Figure 2).

No viral titer was obtained in persistently infected cultures I1 and I2 treated with fluoxetine (I1-F and I2-F) after 4 weeks of treatment; therefore, they were not tested.

Overall, the mean decrease in viral titer was significantly lower when cells were inoculated with the virus isolated from persistently infected cultures treated with fluoxetine than those from persistently infected cultures treated with DMSO (0.6 log TCID50/mL versus 4.2 log TCID50/mL, $p < 0.0001$).

The same experiments were run using GuHCl instead of fluoxetine. When the viral suspensions were inoculated into HEp-2 cells in presence of GuHCl, a mean decrease of 4.7 log TCID50/mL and 5 log TCID50/mL was observed when cells were inoculated with the virus obtained from persistently infected cultures treated with DMSO and persistently infected cultures I3-F and I4-F treated with fluoxetine, respectively for 8 weeks. There is no statistical difference between these viral titer reductions.

3.3. Mutations in 2C Protein Associated with Resistance to Fluoxetine

Since the 2C viral protein was reported as the target of fluoxetine antiviral activity, we investigated whether the resistance was associated with mutant variants. The sequence of the 2C region was determined in the stock virus, virus obtained from persistently infected cultures before treatment, and virus collected from persistently infected cultures treated with fluoxetine or DMSO treated for 4 and 10 weeks.

All the positions with amino-acid substitution in the different sequences as compared to CVB4 E2 reference strain (NCBI, accession: AF311939.1) are presented in Table 2.

Table 2. Changes observed in 2C sequences.

2C Protein Sequences		Amino-Acid Positions						
		133	188	216	227	229	255	296
CVB4 E2 reference published strain (NCBI, accession: AF311939.1)		A	S	P	I	A	S	R
Laboratory CVB4 E2 stock strain		A	S	P	I	A	S	R
Baseline samples (CVB4 E2 persistent infection, 20 weeks)	I1	A	S	P	I	A	N	R
	I2	A	S	S	I	A	S	R
	I3	A	S	P	I	A	S	G
	I4	A	S	P	I	A	S	G
4 weeks post treatment — Fluoxetine	I1	ND	ND	ND	ND	ND	ND	ND
	I2	ND	ND	ND	ND	ND	ND	ND
	I3	A	S	P	I/V	A	S	G
	I4	A	S	P	I	A	S	G
4 weeks post treatment — DMSO	I1	A	S	P	I	A	N	R
	I2	A	S	S	I	A	S	R
	I3	A	S	P	I	A	S	G
	I4	A	S	P	I	A	S	G
10 weeks posttreatment — Fluoxetine	I1	ND	ND	ND	ND	ND	ND	ND
	I2	ND	ND	ND	ND	ND	ND	ND
	I3	T	S/A	P	V	A/V	S	G
	I4	A/T	S	P	V	A	S	G
10 weeks posttreatment — DMSO	I1	A	S	P	I	A	N	R
	I2	A	S	S	I	A	S	R
	I3	A	S	P	I	A	S	G
	I4	A	S	P	I	A	S	G

ND: Not done, virus undetectable; I3 and I4 are resistant to fluoxetine treatment. AA changes are underlined.

Figure 3 focuses on mutations that appeared in the sequences of virus obtained from the fluoxetine-resistant persistently infected cultures I3-F and I4-F.

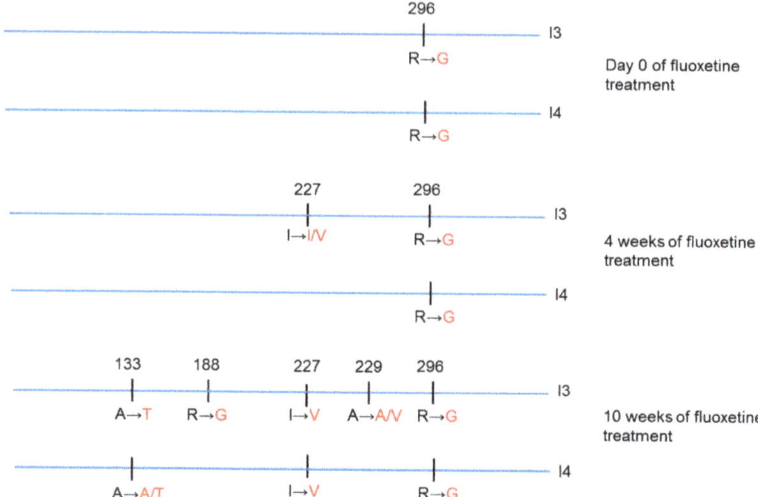

Figure 3. Amino-acid substitutions in fluoxetine-resistant virus. Virus suspensions were collected from persistent CVB4 E2-infected cultures at baseline, and after 4 and 10 weeks of treatment with DMSO or fluoxetine. The sequence of the whole CVB4 E2 2C region (from nt 4039 to nt 5025, 329 aa) was determined using Sanger method. The amino-acid substitutions in the sequences of fluoxetine-resistant viruses (I3 and I4) are shown.

Before fluoxetine treatment, the R296G mutation was observed in I3 and I4, but not in I1, I2, and the initial stock virus. This substitution was present during all follow-ups in the sequences from fluoxetine or DMSO-treated I3 and I4 infections.

During fluoxetine treatment, the I227V mutation was observed in I3-F at week 4 o.t., and in both I3-F and I4-F at week 10 o.t. The A133T emerged in both infections at week 10 o.t. As for mutations R188G and A229V, they were only observed in I3-F infection at week 10 o.t.

Interestingly, these mutations were not observed in the virus obtained from infected cultures treated with DMSO (I1-D, I2-D, I3-D, and I4-D).

4. Discussion

The investigation of existing drugs with well-established safety profiles for new indications is a cheaper and faster strategy to discover new antiviral agents. Indeed, the screening of approved molecule libraries allowed to identify previously unrecognized inhibitors of enterovirus replication, including fluoxetine [11,15].

We have previously shown that fluoxetine can successfully clear persistent CVB4 E2 infection within a month when cultures were treated at 5.48 µM, twice a week [12]. Further experiments described in this study revealed a failure to clear the virus in some persistent CVB4 E2 infections despite a long-term fluoxetine treatment. The lack of susceptibility to fluoxetine in these "resistant" isolates was confirmed in a model of acute infection using HEp-2 cells.

Fluoxetine and other described enterovirus inhibitors were shown to exert antiviral activity by targeting 2C protein that is one of the most conserved and complex nonstructural viral proteins among picornaviruses [11,15]. This protein was reported to be involved in several key events throughout the virus life cycle (different steps of replication, immune evasion …), but its precise role is not fully understood. 2C harbors an N-terminal membrane-binding motif, an adenosine triphosphatase (ATPase) domain, a cysteine-rich motif, and RNA binding sites [16]. The ATPase domain, which belongs to SF3 helicases of the AAA+ ATPase superfamily, contains Walker motifs (motifs A and B) and motif C [17]. ATPase activity has been earlier demonstrated for 2C protein [8]. However, every attempt to determine the helicase activity associated with 2C ATPase has failed until recently when a study provided evidence of this helicase activity in the 2C protein of EV-A71 and CV-A16 [18].

One of the major challenges to overcome during the investigation or the development of an antiviral agent, especially a direct acting agent, is the emergence of resistant mutants. This is particularly true for RNA viruses that usually generate a significant number of mutations during the replication process due to a poor proofreading activity of RNA polymerase [19].

Our model of persistent CVB4 E2 infection in pancreatic cell cultures is attractive to investigate the resistance to antiviral molecules, because it allows multiple and long-term exposition of virus to molecules. This drug pressure promotes the emergence of resistant mutants, which more or less quickly results in failure of virus clearance, depending on the resistance barrier of the drug.

Previous studies have described three residues substitutions in 2C protein (A224V, I227V, A229V), that confer CVB3 resistance to fluoxetine, TBZE-029, GuHCl, and other recently identified 2C targeting inhibitors [8,15]. These substitutions are present in a short stretch of amino acids 224AGSINA229 located immediately at the C terminal of ATPase motif C. The AGSINA motif was found to be conserved in *Enterovirus B* (such as CVB4) and *D* species but is not present in other enteroviruses [10].

In this study, we found two of these substitutions in the fluoxetine-resistant viral suspensions (both I227V and A229V (double population mutant/wild type) in I3-F, and only I227V in I4-F). These changes were not observed before treatment, therefore, have probably emerged under drug-selection pressure. However, given the limit of Sanger sequencing, it cannot be excluded that these mutants preexisted as minor variants. The fluoxetine-resistant mutant used in reported studies (obtained by site-directed mutagenesis) harbored all these three mutations, and the impact of each mutation could not be clearly assessed [10,15]. In addition, the fact that, resistant viruses obtained from persistent infections treated with fluoxetine (I3-F and I4-F) were still susceptible to GuHCl, shows that the effect of these mutations (alone or combined) might depend on the 2C-inhibitor tested. Indeed, de Palma et al. [20]. previously reported a detailed impact of these mutations on the susceptibility to GuHCl. I227V alone did not induce resistance while the combination of I227V + A229V was associated with low-level resistance. Thus, the mutations observed in I3-F and I4-F seems to be insufficient to confer resistance to GuHCl in vitro.

In this report, other mutations not previously described were observed in the 2C ATPase domain of the resistant viruses (A133T and R188G in I3-F, and A133T in I4-F). I3-F, the viral suspension with the most mutations, appears to be the most resistant one to fluoxetine.

Interestingly, the R296G mutation was observed before treatment only in the virus obtained from persistent infections, which would later display resistance to fluoxetine treatment (I3 and I4). Even if viral suspensions from all of the untreated persistently infected cultures (I1, I2, I3, and I4) were susceptible to fluoxetine, the role of R296G mutation in the predisposition of resistance to fluoxetine cannot be excluded. This substitution is located in the zinc finger domain, downstream of the cysteine-rich motif which forms a zinc-binding site [21]. In addition, this residue seems to be conserved in *Enterovirus B* and *D* species (the EV species reported to be susceptible to fluoxetine), and variable in others species. We hypothesized that this mutation might favor other compensatory mutations including those associated with resistance to fluoxetine.

In this study, we focused on the 2C protein; however, the role of changes in other parts of the viral genome cannot be excluded. Site-directed mutagenesis studies are needed to precisely analyze the impact of these new substitutions on the susceptibility of CVB to fluoxetine and other 2C targeting enterovirus inhibitors.

Currently, fluoxetine is not available as a treatment for EV in humans. Nevertheless, the understanding of the inhibition mechanism and resistance profiles can be useful for the design of new compounds [10,22].

5. Conclusions

In conclusion, we took advantage of a model of persistent CVB4 E2 infection to describe the emergence of fluoxetine-resistant variants. In these variants, mutations in the 2C viral protein have been identified and deserve further investigation.

Author Contributions: Conceptualization, E.K.A. and D.H.; methodology, E.K.A., A.B., and D.C.; data analysis, E.K.A., F.S., and I.E.; writing—original draft preparation, E.K.A.; writing—review and editing, I.E., D.H.; supervision, D.H.; funding acquisition, D.H.

Funding: This work was supported by Ministère de l'Education Nationale de la Recherche et de la Technologie, Université Lille 2 (Equipe d'accueil 3610), and Centre Hospitalier Régional et Universitaire de Lille, and by EU FP7 (GA-261441-PEVNET: Persistent virus infection as a cause of pathogenic inflammation in type 1 diabetes—an innovative research program of biobanks and expertise).

Acknowledgments: The authors thank the team of Laboratoire de Virologie EA3610, Université Lille, Faculté de médecine, and CHU de Lille. The authors thank Jennifer Varghese for reading the manuscript.

Conflicts of Interest: The authors declare no conflict of interest. The funder had no role in the design of the study; in the collection, analyses, or interpretation of data; in the writing of the manuscript, or in the decision to publish the results.

References

1. Knowles, N.; Hovi, T.; Hyypiä, T. Picornaviridae. In *Virus Taxonomy: Classification and Nomenclature of Viruses: Ninth Report of the International Committee on Taxonomy of Viruses*; King, A.M.Q., Adams, M.J., Carstens, E.B., Lefkowitz, E.J., Eds.; Elsevier: San Diego, CA, USA, 2012; pp. 855–880.

2. Romero, J.R. Pediatric group B coxsackievirus infections. *Curr. Top. Microbiol. Immunol.* **2008**, *323*, 223–239. [PubMed]

3. Tapparel, C.; Siegrist, F.; Petty, T.J.; Kaiser, L. Picornavirus and enterovirus diversity with associated human diseases. *Infect. Genet. Evol.* **2013**, *14*, 282–293. [CrossRef]

4. Chapman, N.M.; Kim, K.S. Persistent coxsackievirus infection: Enterovirus persistence in chronic myocarditis and dilated cardiomyopathy. *Curr. Top. Microbiol. Immunol.* **2008**, *323*, 275–292. [PubMed]

5. Hober, D.; Alidjinou, E.K. Enteroviral pathogenesis of type 1 diabetes: Queries and answers. *Curr. Opin. Infect. Dis.* **2013**, *26*, 263–269. [CrossRef]

6. Hober, D.; Sauter, P. Pathogenesis of type 1 diabetes mellitus: Interplay between enterovirus and host. *Nat. Rev. Endocrinol.* **2010**, *6*, 279–289. [CrossRef] [PubMed]

7. Alidjinou, E.K.; Sané, F.; Engelmann, I.; Geenen, V.; Hober, D. Enterovirus persistence as a mechanism in the pathogenesis of type 1 diabetes. *Discov. Med.* **2014**, *18*, 273–282.

8. De Palma, A.M.; Vliegen, I.; De Clercq, E.; Neyts, J. Selective inhibitors of picornavirus replication. *Med. Res. Rev.* **2008**, *28*, 823–884. [CrossRef]

9. Thibaut, H.J.; Leyssen, P.; Puerstinger, G.; Muigg, A.; Neyts, J.; De Palma, A.M. Towards the design of combination therapy for the treatment of enterovirus infections. *Antiviral Res.* **2011**, *90*, 213–217. [CrossRef]

10. Ulferts, R.; van der Linden, L.; Thibaut, H.J.; Lanke, K.H.W.; Leyssen, P.; Coutard, B.; De Palma, A.M.; Canard, B.; Neyts, J.; van Kuppeveld, F.J.M. Selective serotonin reuptake inhibitor fluoxetine inhibits replication of human enteroviruses B and D by targeting viral protein 2C. *Antimicrob. Agents Chemother.* **2013**, *57*, 1952–1956. [CrossRef]

11. Zuo, J.; Quinn, K.K.; Kye, S.; Cooper, P.; Damoiseaux, R.; Krogstad, P. Fluoxetine is a potent inhibitor of coxsackievirus replication. *Antimicrob. Agents Chemother.* **2012**, *56*, 4838–4844. [CrossRef]

12. Alidjinou, E.K.; Sané, F.; Bertin, A.; Caloone, D.; Hober, D. Persistent infection of human pancreatic cells with Coxsackievirus B4 is cured by fluoxetine. *Antiviral Res.* **2015**, *116*, 51–54. [CrossRef] [PubMed]

13. Alidjinou, E.K.; Engelmann, I.; Bossu, J.; Villenet, C.; Figeac, M.; Romond, M.-B.; Sané, F.; Hober, D. Persistence of Coxsackievirus B4 in pancreatic ductal-like cells results in cellular and viral changes. *Virulence* **2017**, *8*, 1229–1244. [CrossRef] [PubMed]

14. Sane, F.; Caloone, D.; Gmyr, V.; Engelmann, I.; Belaich, S.; Kerr-Conte, J.; Pattou, F.; Desailloud, R.; Hober, D. Coxsackievirus B4 can infect human pancreas ductal cells and persist in ductal-like cell cultures which results in inhibition of Pdx1 expression and disturbed formation of islet-like cell aggregates. *Cell Mol. Life Sci.* **2013**, *70*, 4169–4180. [CrossRef] [PubMed]

15. Ulferts, R.; de Boer, S.M.; van der Linden, L.; Bauer, L.; Lyoo, H.R.; Maté, M.J.; Lichière, J.; Canard, B.; Lelieveld, D.; Omta, W.; et al. Screening of a Library of FDA-Approved Drugs Identifies Several Enterovirus Replication Inhibitors That Target Viral Protein 2C. *Antimicrob. Agents Chemother.* **2016**, *60*, 2627–2638. [CrossRef] [PubMed]

16. Pfister, T.; Jones, K.W.; Wimmer, E. A cysteine-rich motif in poliovirus protein 2C(ATPase) is involved in RNA replication and binds zinc in vitro. *J. Virol.* **2000**, *74*, 334–343. [CrossRef] [PubMed]

17. Singleton, M.R.; Dillingham, M.S.; Wigley, D.B. Structure and mechanism of helicases and nucleic acid translocases. *Annu. Rev. Biochem.* **2007**, *76*, 23–50. [CrossRef] [PubMed]

18. Xia, H.; Wang, P.; Wang, G.-C.; Yang, J.; Sun, X.; Wu, W.; Qiu, Y.; Shu, T.; Zhao, X.; Yin, L.; et al. Human Enterovirus Nonstructural Protein 2CATPase Functions as Both an RNA Helicase and ATP-Independent RNA Chaperone. *PLoS Pathog.* **2015**, *11*, e1005067. [CrossRef] [PubMed]

19. Smith, E.C. The not-so-infinite malleability of RNA viruses: Viral and cellular determinants of RNA virus mutation rates. *PLoS Pathog.* **2017**, *13*, e1006254. [CrossRef]

20. De Palma, A.M.; Heggermont, W.; Lanke, K.; Coutard, B.; Bergmann, M.; Monforte, A.-M.; Canard, B.; De Clercq, E.; Chimirri, A.; Pürstinger, G.; et al. The thiazolobenzimidazole TBZE-029 inhibits enterovirus replication by targeting a short region immediately downstream from motif C in the nonstructural protein 2C. *J. Virol.* **2008**, *82*, 4720–4730. [CrossRef]

21. Guan, H.; Tian, J.; Qin, B.; Wojdyla, J.A.; Wang, B.; Zhao, Z.; Wang, M.; Cui, S. Crystal structure of 2C helicase from enterovirus 71. *Sci. Adv.* **2017**, *3*, e1602573. [CrossRef]

22. LLerena, A.; Dorado, P.; Berecz, R.; González, A.; Jesús Norberto, M.; de la Rubia, A.; Cáceres, M. Determination of fluoxetine and norfluoxetine in human plasma by high-performance liquid chromatography with ultraviolet detection in psychiatric patients. *J. Chromatogr. B Analyt. Technol. Biomed. Life Sci.* **2003**, *783*, 25–31. [CrossRef]

Article

Antiviral Effect of Epigallocatechin Gallate via Impairing Porcine Circovirus Type 2 Attachment to Host Cell Receptor

Jiarong Li [1], Dongfeng Song [1], Shengnan Wang [1,2], Yadong Dai [2,3], Jiyong Zhou [2,3] and Jinyan Gu [1,*]

[1] Institute of Immunology, College of Veterinary Medicine, Nanjing Agricultural University, Nanjing 210095, China; 2015207025@njau.edu.cn (J.L.); 2017107087@njau.edu.cn (D.S.); wangshengnanwsnj@163.com (S.W.)

[2] MOA Key Laboratory of Animal Virology, Institute of Preventive Veterinary Sciences and Department of Veterinary Medicine, Zhejiang University, Hangzhou 310058, China; 11817028@zju.edu.cn (Y.D.); jyzhou@zju.edu.cn (J.Z.)

[3] Center of Veterinary Medical Sciences, Zhejiang University, Hangzhou 310058, China

* Correspondence: gjy@njau.edu.cn

Received: 14 January 2020; Accepted: 2 February 2020; Published: 4 February 2020

Abstract: The green tea catechin epigallocatechin gallate (EGCG) exhibits antiviral activity against various viruses. Whether EGCG also inhibits the infectivity of circovirus remains unclear. In this study, we demonstrated the antiviral effect of EGCG on porcine circovirus type 2 (PCV2). EGCG targets PCV2 virions directly and blocks the attachment of virions to host cells. The microscale thermophoresis assay showed EGCG could interact with PCV2 capsid protein in vitro with considerable affinity (Kd = 98.03 ± 4.76 µM), thereby interfering with the binding of the capsid to the cell surface receptor heparan sulfate. The molecular docking analysis of capsid–EGCG interaction identified the key amino acids which formed the binding pocket accommodating EGCG. Amino acids ARG51, ASP70, ARG73 and ASP78 of capsid were found to be critical for maintaining the binding, and the arginine residues were also essential for the electrostatic interaction with heparan sulfate. The rescued mutant viruses also confirm the importance of the key amino acids of the capsid to the antiviral effect of EGCG. Our findings suggest that catechins could act as anti-infective agents against circovirus invasion, as well as provide the basic information for the development and synthesis of structure-based anti-circovirus drugs.

Keywords: Porcine circovirus type 2; epigallocatechin gallate; heparan sulfate; antiviral effect; virus attachment

1. Introduction

Circovirus is one of the smallest non-enveloped viruses with a single-stranded circular genomic DNA and classified in the genus *Circovirus* of family *Circoviridae* [1]. It infects various species ranging from plants to animals with different pathogenicities [2]. Porcine circovirus type 2 (PCV2) is a prototype circovirus that causes significant morbidity and mortality in swine [3–5], and is associated with different syndromes, such as the post-weaning multisystemic wasting syndrome [6,7] and reproductive disorder [8]. The genome of PCV2 contains two major open reading frames (ORFs). *ORF1* encodes the replicase protein (Rep) involved in rolling-circle viral DNA replication, while *ORF2* encodes the unique viral structural protein capsid [9], which is involved in diverse and essential biological events during virus infection, such as virion attachment [10].

The primary attachment of the non-enveloped virus on the host cells is based on a low-affinity interaction between viral structural protein and cell surface glycosaminoglycans (GAGs)

receptors [11,12]. GAGs include heparan sulfate (HS), keratan sulfate (KS), and chondroitin sulfate (CS), which are characterized by disaccharide units forming blocks of polysaccharides [13]. Among these GAGs, HS is the most common polysaccharide molecule utilized by viruses for attachment to cell surfaces, such as the hepatitis C virus [14], herpes simplex virus 1 [15], human enterovirus 71 [16] and rabies virus [17]. PCV2 has also been verified to use cell surface HS as attachment receptor, and adopts a distinct non-symmetrical receptor distributed pattern on the virion with a multitude of weak HS binding sites [10,18]. Chemical molecules acting as receptor mimics that interfere with HS-virion interactions might exhibit antiviral activities [19]. However, there are very few reports of PCV2 infection inhibition available for pertaining to such natural small molecule compounds.

The green tea catechins (GTCs) are polyphenolic compounds extracted from the leaves of *Camellia sinensis* [20], and have shown numerous beneficial properties, such as anticancer activity [21], anti-inflammatory ability [22], antioxidative properties [23], antibacterial function [24] and antiviral effects [25]. GTCs are predominantly comprised of four constituents: epicatechin (EC), epigallocatechin, epicatechin gallate, and epigallocatechin gallate (EGCG) [23], among which the EGCG is the major component of GTCs accounting for approximately 59% of the total polyphenols, and it is also the most complicated and principal constituent, possessing the activity against infection of several viruses, such as human immunodeficiency virus (HIV) [26], influenza A virus [27], Zika virus [28], herpes simplex viruses 1 [29], and hepatitis C virus [30]. The mechanisms of EGCG antiviral activity are complicated and diverse, depending on the specific virus infection process or host cell response, one of which is that EGCG interacts with viral structural proteins, consequently blocking the latter from recognizing or binding with cellular receptors. In previous reports, EGCG has been shown to interact with the influenza A virus hemagglutinin, disrupting the binding of envelope glycoprotein to cell surface sialic acid receptor [25]. EGCG inhibited the interaction of HIV envelope glycoprotein gp120 with the cellular CD4 molecule [31]. Regarding the GAGs receptors, EGCG could bind to the herpes simplex virus-1 glycoprotein gB and gD, which should interact with HS and 3-O-sulfated HS of cellular glycans, respectively [29,32].

In previous studies, there have been no reports of catechins inhibiting circovirus invasion, we aimed to investigate the effect of EGCG on PCV2 infection.

2. Materials and Methods

2.1. Cells and Virus

PCV-free Porcine Kidney 15 (PK15) (BH0370, ATCC, Manassas, VA, USA) cells were supplied by China Institute of Veterinary Drugs Control and cultured in Opti-MEM (11095-072, Thermo Fisher, Waltham, MA, USA) with 10% fetal bovine serum (10099-141-FBS, Thermo Fisher). PCV2 strain HZ0201 (GenBank accession no. AY188355.1, $10^{6.5}$TCID$_{50}$/mL) was propagated in PK15 cells [33]. PCV2 virions were enriched using ultra-centrifugation.

2.2. Reagent and Antibodies

Epigallocatechin gallate (EGCG, 99.91% purity, HY-13653, Medchemexpress, Monmouth, NJ, USA), epicatechin (EC, 99.00% purity, HY-N0001, Medchemexpress) and heparin (99.00% purity, H3393, Sigma-Aldrich, St. Louis, MO, USA) were dissolved into ddH2O at a concentration of 50 mM for storage. The monoclonal antibody (mAb) 5E11 against PCV2 capsid was generated in our laboratory [34]. The fluorescein isothiocyanate (FITC) conjugated goat anti-mouse IgG (ab6785, Abcam, Cambridge, MA, USA), horse radish peroxidase (HRP) conjugated goat anti-mouse IgG (074-1802, KPL, Milford, MA, USA) and Mouse IgG1 Isotype Control (564416, BD Horizon™, San Jose, CA, USA) were purchased from Abcam, Kirkegaard & Perry Laboratories and BD Biosciences, respectively.

2.3. Cytotoxicity Assay

The cytotoxicity assay of compounds was performed using enhanced CCK8 kit (C0014, Beyotime, Shanghai, China) according to the standard instructions. PK-15 cells seeded into 96-well plate was treated with different concentrations of EGCG or EC for 12 h. Subsequently, 10 μL CCK8 reagent per well was added into the culture medium and after 1 h of incubation, absorbance value (A450) was measured by a spectrophotometer. The cell viability was calculated as ($A450_{compound}/A450_{mock}$) × 100%.

2.4. Determination of Virus Titer

PK-15 cells seeded in the 96-well plates were added with 10-fold serially diluted virus samples. At 96 hpi (hours post-infection) the virus titers were determined by indirect immunofluorescent assay (IFA) with mAb 5E11 against PCV2 capsid and FITC-conjugated goat anti-mouse IgG. The 50% tissue culture infective dose ($TCID_{50}$) was calculated according to the Reed-Muench method. The multiplicity of infection (MOI) was defined as the number of plaques forming units (PFUs) computed as $PFU = 0.7 \times TCID_{50}$.

2.5. Calculation of EC_{50}

EGCG was serially diluted to 100, 50, 25, 12.5, 6.25 and 3.125 μM, and was added to PK-15 cells which were infected with PCV2 at MOI = 1.0. EGCG was present during whole PCV2 infection. The infected cells without EGCG treatment was set as mock control. $TCID_{50}$ was determined at 72 hpi and the values of inhibition were calculated as $1 - TCID_{50(compound)}/TCID_{50(mock)} \times 100\%$. The values of inhibition and EGCG concentration were used to establish a dose response curve and further calculate the EC_{50} (concentration for 50% of maximal effect).

2.6. Infectivity Assay

PCV2 samples were pretreated with chemical compound at 37 °C for 10 min, then inoculated into near-confluent PK-15 cell monolayers at a MOI = 1.0 at 37 °C. Virus inoculum was removed after 1 h and the infected cells were maintained in medium containing the compound. Virus infectivity was assessed by titer determination at various time points. Meanwhile, at 72 hpi the virions in the infected cells were detected by IFA with mAb 5E11 FITC-conjugated goat anti-mouse IgG; expression level of the PCV2 capsid protein was measured by western immunoblotting using 5E11 antibody and HRP- goat anti-mouse IgG; copy numbers of the viral genome were determined with quantitative real-time PCR (qRT-PCR) with forward primer (TACTGCTGTGAGTACCTTGTTGGA) and reverse primer (TCTGCATTTTCCCGCTCACT) targeted to the 359~483 bp sequence (124 bps) of PCV2 *Rep* gene, and the probe was designed as 'FAM-AGTCTGGTGACCGTTGTTGCAGAGCAGCAC-BHQ'. The pMD18-*Rep* plasmid was serially diluted 10-fold to $10^3 \sim 10^7$ to generate the standard curve and the copy numbers of samples were quantified using absolute quantification method [35].

2.7. Time of Addition and Dilution Experiment

PK-15 cell was pre-exposed to EGCG or Opti-MEM medium at 37 °C for 60 min, then washed and infected with PCV2 at a MOI = 1.0 at 37 °C for 1 h before the inocula were removed, the samples were maintained without EGCG and the infectivity was assessed at 72 hpi. Alternatively, the PCV2 was premixed with EGCG under the same condition before being added to cells, the samples were maintained and determined with same way. For post-treatment samples, cells were first inoculated with PCV2 at a MOI = 1.0 at 37 °C, followed by 100 μM EGCG treatment maintained for 4 h, 12 h and 24 h, respectively, the samples were tested at 72 hpi. For dilution experiments, approximately MOI = 10.0 of enriched virions were pretreated with 100 μM EGCG for 60 min at 37 °C, and then diluted 10-fold before being added to cell monolayers, infected cell samples were maintained without EGCG after the inocula were removed 1 h. Infectivity of samples was assessed at 72 hpi.

2.8. Binding Assay and Flow Cytometry

PCV2 virions enriched with ultra-centrifugation were added to PK-15 cells with MOI = 5.0 in the presence of 100 µM EGCG. After binding for 1 h or 4 h at 4 °C, the mixture was then washed off and the cells were harvested. The amount of the attached virions was determined using qPCR. The ratio of attached cells was measured by flow cytometry, the trypsin digested PK15 cells were resuspended and incubated with the mAb 5E11 for 1.5 h at 37 °C, followed by staining with FITC-conjugated goat anti-mouse IgG for 1 h, mouse IgG1 was used as isotype control, finally the samples were measured by BD Accuri™ C6 flow cytometer (BD bioscience, San Jose, CA, USA) and the results were analyzed by FlowJo X software (Version 10.0.7, FlowJo, San Jose, CA, USA).

2.9. Recombinant Protein Expression and Purification

The engineered *dcapsid* (capsid with N-terminal nuclear localization sequence deleted) gene of PCV2 strain HZ0201 was cloned and inserted into pET28 vector encoded with a hexahistidine tag at the N-terminus. The dcapsid mutants with alanine substituted for specific amino acids were constructed using site-specific mutagenesis PCR technology. The vectors were transformed into *Escherichia coli* BL21 (DE3) and the protein expression was induced using 1 mM Isopropyl β-D-1-Thiogalactopyranoside at 16 °C overnight. The recombinant proteins were purified using Ni-NTA super-flow matrix (30410, Qiagen, Germantown, MD, USA), followed by PD10 column (17-0851-01, GE healthcare, Los Angeles, CA, USA) to remove the excess imidazole. The concentration of recombinant protein was measured using the BCA method and the protein samples were stored at −80 °C.

2.10. Heparin Column Chromatography

The purified recombinant PCV2 dcapsid were bound to a 5-mL HiTrap™ Heparin-Sepharose HP Column (17040701, GE Healthcare, Los Angeles, CA, USA) in loading buffer (10 mM sodium phosphate, 0.3 M NaCl, pH 7.4) at a flow rate of 1 mL per minute. The column was washed with 10 mL loading buffer to remove excess molecules, and thereafter eluted with soluble heparin, EGCG or EC competitor in loading buffer. The remaining bound protein was ultimately eluted with 3 M NaCl in loading buffer because the electrostatic interactions between Heparin-Sepharose and heparin-binding proteins could be disrupted by high salt concentrations. The original and eluted fractions were diluted to equal volumes, and analyzed using Western immunoblotting with the mAb 5E11 and HRP- goat anti-mouse IgG. The intensities of the immunoreactive protein bands were measured using Image J software (Version 1.52 g, ImageJ, Bethesda, MD, USA).

2.11. Microscale Thermophoresis Assay

The purified dcapsid proteins were labelled with the Cy5 fluorophore based on the chemical reaction between the [108]cysteine of capsid and Cy5-maleimide (PA25031, GE Healthcare Life Sciences). The labelled molecules were separated from the unreacted and excess dyes with a PD Mini-Trap G-25 desalting column, and mixed at 10–50 nM concentration with 2 fold serially dilutions of EGCG, EC or heparin in reaction buffer (150 mM NaCl, 50 mM Tris-HCl, 10 mM $MgCl_2$, 0.05% Tween-20, pH 7.4). The binding reactions were fixed into equal volume and incubated at room temperature for 15 min. The mixtures were then enclosed in standard pretreated glass capillaries and analyzed with MST machine (Monolith NT.115, NanoTemper Technologies, Munich, Bavaria, Germany), and the Kd values of each reaction were computed by NanoTemper Analysis software (Version 2.1, NanoTemper Technologies, Munich, Bavaria, Germany).

2.12. Construction of PCV2 Infectious Clone and Virus Rescue

The whole genome of PCV2 strain HZ0201 was cloned and inserted into pMD18-T vector (Takara, Tokyo, Japan). Site-specific mutagenesis was performed using PCR to construct the mutants. The linear genomes of the WT and mutants were extracted, digested by *Eco* RI restriction enzyme (NEB), cyclized by T4 DNA ligase (Takara), and transfected into PK15 cells for virus rescue with the jetPRIME® in vitro transfection reagent (Polyplus-transfection, New York, NY, USA). The transfected cells were maintained for 72 h, then continuously passaged and cultured. The rescued virus samples were harvested and inoculated to PK15 cells and the infectivity was evaluated with immunoblotting in an indirect immunofluorescence assay. The virus titers were calculated according to the Reed-Muench method.

2.13. Statistical Analysis

All results were presented as mean ± standard deviation (SD) based on three independent experiments. Significant differences between experimental and control groups were analyzed using Student's *t*-test. The differences were considered significant or highly significant at *p*-values < 0.05 and < 0.01, respectively.

3. Results

3.1. EGCG Inhibits the Infectivity of PCV2

To assess the effect of EGCG on the infectivity of PCV2, the cytotoxicity on PK15 cells of the compounds was first examined using the CCK8 assay. The results showed that EGCG at a concentration of 100 µM did not generate significant cytotoxicity to PK-15 cells with a duration time of 24 h, 72 h, and 120 h, and EGCG at a concentration of 200 µM could produce cytotoxicity to cells, which limited the maximum concentration of EGCG at 100 µM (Figure 1b). The infectivity of PCV2 was significantly inhibited at the 100 µM EGCG treatment during the whole infective process, the virus titer of PCV2 infected cells with EGCG treatment showed a significant decrease at various time points compared to that without EGCG treatment (Figure 1c). Inhibition of PCV2 infectivity was also manifested by reduced level of viral capsid protein expression measured by immunoblotting (Figure 1d), and the reduction of viral genome copy number detected by qPCR (Figure 1e) at 72 hpi with the 100 µM EGCG treatment. The inhibitory effect of PCV2 infection was still detectable for the 10 µM EGCG treatment (Figure 1d,e). Furthermore, the cells infected with PCV2 at MOI = 1.0 were detected with IFA, which showed that the quantity of the PCV2 infected cells with the 100 µM EGCG treatment was obviously lower than that without EGCG treatment (Figure 1f). The original catechin EC displayed no considerable anti-infective effect on PCV2 under similar treatment conditions (Figure 1c–f). In addition, the EC_{50} of the antiviral effect of EGCG on PCV2 was further determined to be 37.79 ± 1.64 µM, which was consistent with its inhibitory concentration of 100 µM. Accordingly, EGCG could exert a prominent antiviral effect against the PCV2 infection.

3.2. EGCG Exerts the Antiviral Effect via Directly Targeting Virions

The antiviral effect of EGCG might be exerted by targeting PCV2 virions, such as the effect on certain viral nucleic acids and proteins, or targeting host cells [23]. To assess this issue, the time of addition assay was preformed, the cells were pretreated with EGCG and then inoculated with PCV2, or the PCV2 was premixed with EGCG before seeding to cells. Surprisingly, the inhibitory effect could not be detected with immunoblotting in the pre-treated cells group (Figure 2a), which was consistent with results of viral nucleic acid and titer determination (Figure 2b,c). Meanwhile, the prominent antiviral effect could be displayed in the pre-treated virions group with high EGCG concentration (100 µM), but non-sensitive to the low EGCG concentration (10 µM) pretreatment (Figure 2a–c), suggesting the compound might interact with virions. To rule out the possibility that the EGCG might also act on some cellular factors activated only in the presence of virions or some intracellular signaling pathways induced by the virus stimulation, we performed the dilution assay. Briefly, virus samples were pretreated at a high EGCG concentration (100 µM), and thereafter diluted 10-fold before being inoculated on the cells, ensuring the cells were treated with compound at 10-fold lower concentrations (10 µM) than the viral particles; in such a way, the inhibitory phenomenon would disappear if the compound really targeted host cells only in the presence of virions. However, the antiviral effect could still be detected with immunoblotting, qPCR and virus titer determination (Figure 2d–f). Considering that the pretreatment of cells with 10 µM EGCG could not trigger antiviral activity (Figure 2a–c), this inhibition was attributed to the effect of 100 µM EGCG on virions, further supporting that EGCG performed an antiviral effect via directly targeting virions.

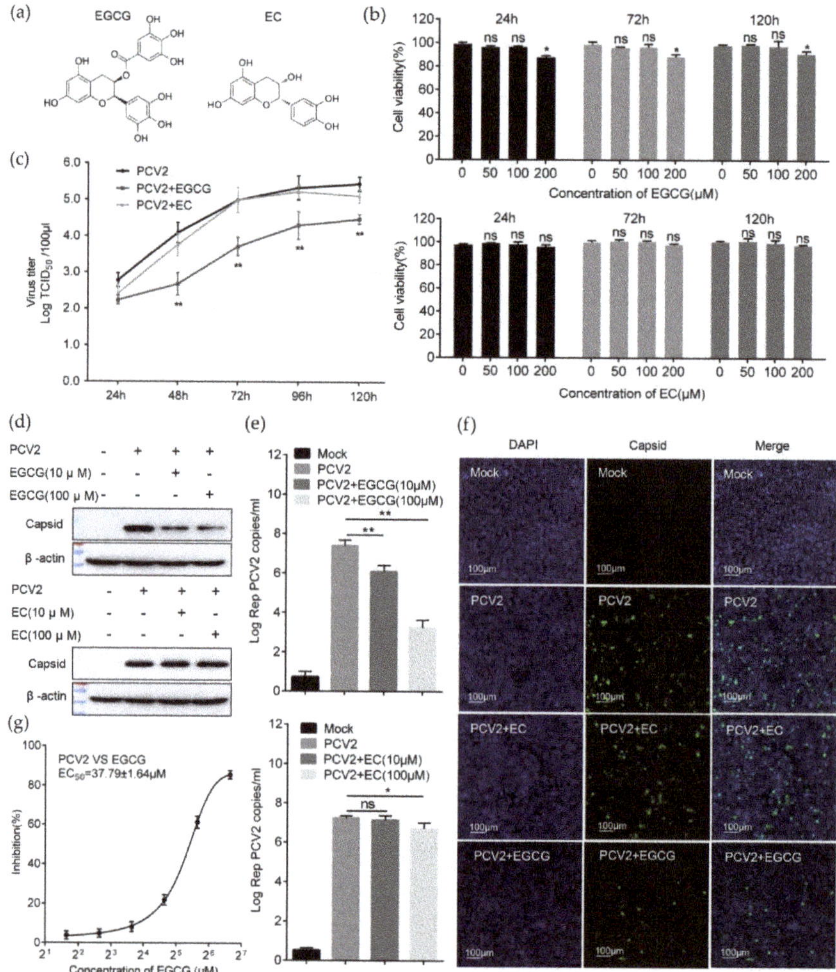

Figure 1. EGCG exhibits an anti-viral effect on PCV2 infection. (**a**) Molecular structure of EGCG and EC. (**b**) Cytotoxicity assay of soluble EGCG and EC. PK-15 cells were added with a concentration of 0, 10, 50, 100 and 200 µM EGCG or EC and cultured for 24 h, 72 h and 120 h, respectively. Cell vitality was detected by CCK8 assay and calculated as ($A450_{compound}$ / $A450_{mock}$) × 100%. (**c**) One-step growth curve of PCV2 with EGCG treatment. PK-15 cells were inoculated with PCV2 at MOI = 1.0 and cultured with 100 µM EGCG or EC treatment. $TCID_{50}$ at various time points were determined by IFA and calculated according to the Reed-Muench method. (**d–f**) EGCG inhibited the infectivity of PCV2. PK-15 cells were inoculated with PCV2 at MOI=1.0 with EGCG or EC addition. (**d**) The expression inhibition of PCV2 capsid protein was detected by immunoblotting at 72 hpi. (**e**) PCV2 genome copies were measured by real-time PCR. (**f**) Indirect IFA detection of PCV2 infected cells. (**g**) Dose response curve and EC_{50} of the EGCG's antiviral effect on PCV2. The mean ± SD of three independent experiments was compared using Student's *t*-test (n.s., not significant; * $p < 0.05$; ** $p < 0.01$).

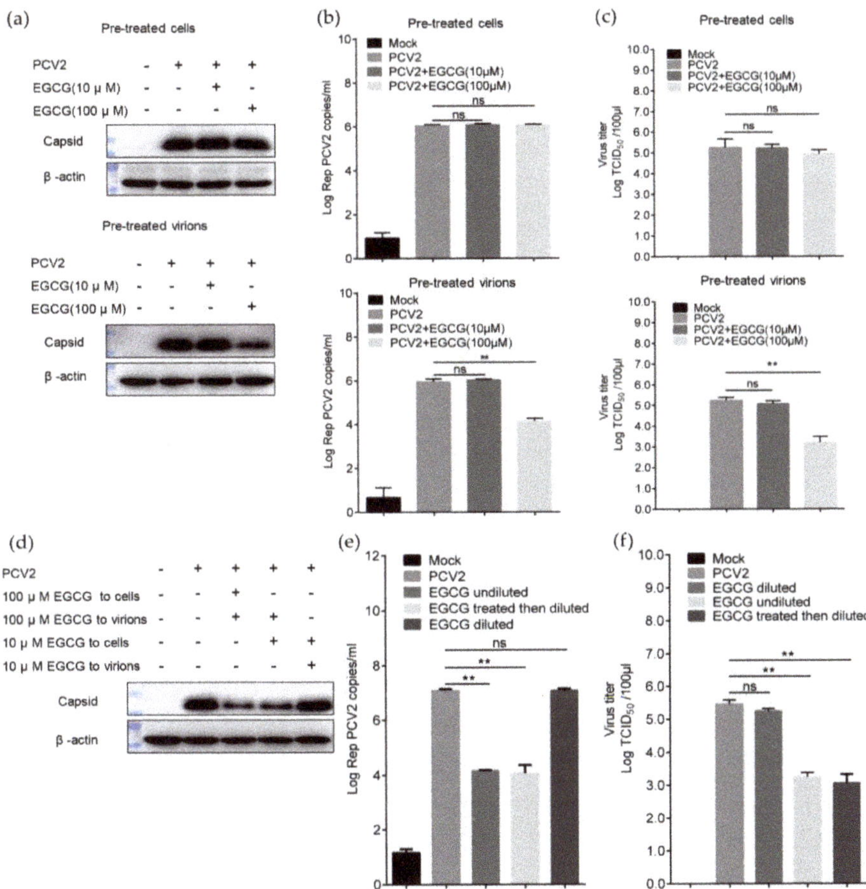

Figure 2. EGCG inhibits the PCV2 infectivity by directly targeting virions. (**a–c**) Cells were pre-exposed to 10 μM or 100 μM EGCG for 60 min at 37 °C, then infected with PCV2 at MOI = 1.0 for 1 h, alternatively, virus was pre-mixed with EGCG under the same condition before being added to the cells, then the cells were cultured without EGCG. At 72 hpi PCV2 infectivity was assessed by viral protein expression (**a**), viral genome copies (**b**), and virus titers (**c**). (**d–f**) Virions were pretreated with 100 μM EGCG and thereafter diluted 10-fold prior to infecting cells, which were cultured without EGCG and tested with immunoblotting (**d**), qPCR (**e**), and virus titers determination (**f**). Three independent experiments were performed and the results were compared using Student's *t*-test (n.s., not significant; ** $p < 0.01$).

3.3. EGCG Impairs the Attachment of PCV2 to Host Cells

To investigate which stage of the EGCG antiviral effect could be active during the PCV2 infective process, EGCG was added post-infection and incubated for various time. The results showed that compared with of the EGCG whole process treatment PCV2 infection, a significant inhibitory effect could not be detected with treatment for 4 h or 12 h after PCV2 infection, only a slight effect was observed when the post treatment duration reached 24 h (Figure 3a), indicating that EGCG might work at the early stage of infection course. Thereby, the binding assay was performed, the virions were added to the cell surface at 4 °C to achieve the binding event for 1 h, and the results indicated that bound PCV2 particles would decrease significantly in cells with the EGCG treatment measured by qRT-PCR (Figure 3b). The cells attached by the virions could be detected by flow cytometry [17], and its proportion also dropped from ~70% to ~55% in cells with the EGCG treatment for 1 h (Figure 3c). This tendency was also detectable when the binding time was extended to 4 h to maximize virions binding, and the proportion of PCV2-attached cells reduced from about 85% to about 67% (Figure 3b,c). Taken together, these findings demonstrated that EGCG inhibited the binding of PCV2 to host cells, but exerted no distinct effect on downstream infection.

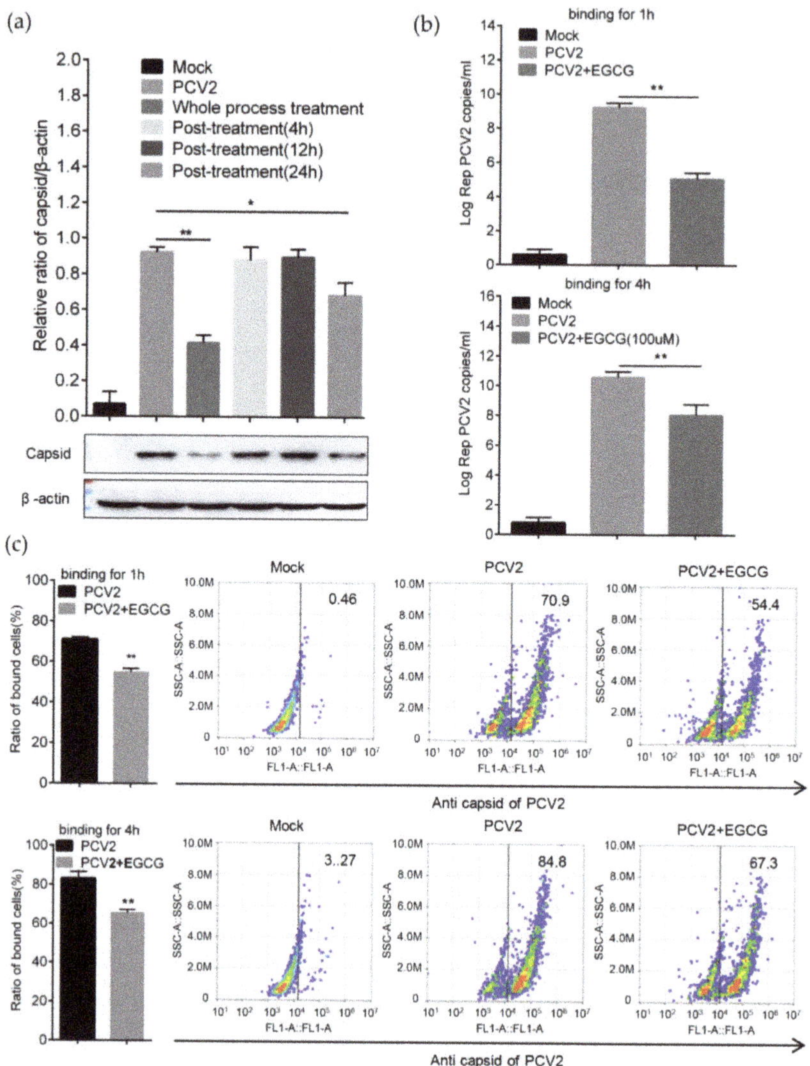

Figure 3. EGCG prevents PCV2 virions binding to host cells. (**a**) Cells were inoculated with PCV2 at MOI = 1.0 with 100 µM EGCG treatment during all the phases, or followed by post-treatment of EGCG for 4 h, 12 h and 24 h, respectively, after which the cells were maintained without EGCG until determined with immunoblotting at 72 hpi. (**b,c**) Virus was added to PK-15 cells at MOI = 5.0 in the presence of 100 µM EGCG. After the binding for 1 h or 4 h at 4 °C. the genome copy numbers of the attached virions were determined using the qPCR method (**b**); PCV2-attached cells were measured by flow cytometry (**c**). Mean ± SD of three independent experiments were shown and Student's *t*-test was conducted for statistical analysis (* $p < 0.05$; ** $p < 0.01$).

3.4. EGCG Competitively Inhibits the Interaction between Capsid and Heparan Sulfate

PCV2 virions attached to host cells via direct interaction between viral capsid protein and cell surface heparan sulfate [10]. Therefore, we hypothesized that EGCG might also bind to capsid, and competitively block the interaction between PCV2 capsid and HS. The microscale thermophoresis assay (MST) was utilized to evaluate the affinity between compounds and protein by measuring the dissociation constant (Kd). The Kd between dcapsid and EGCG was 98.03 ± 4.76 µM, indicating that EGCG could indeed bind to capsid (Figure 4a), and capsid's affinity to EGCG was comparable to the affinity for heparin (an HS analog similar in structure to the sulfated domain of HS), or even a little stronger than the latter (Kd = 120.67 ± 4.73 µM) (Figure 4b). However, the EC, which did not exhibit the inhibitory activity against PCV2 infection, could also bind to capsid with a 10-fold lower affinity than EGCG (Kd = 951.33 ± 6.81 µM) (Figure 4c). These findings implied that the interaction between PCV2 capsid and HS might be hindered by EGCG. This was confirmed by heparin column chromatography assay. Briefly, the dcapsid protein was pre-loaded onto a heparin-Sepharose column, then eluted by the various competitor reagents. The results showed that approximately 40% protein fraction (42.33 ± 3.52%) could be eluted by the EGCG at a concentration of 0.05 mg/mL, similar to the eluted level of heparin (36.00 ± 6.25%), while in case of EC only 10.33 ± 4.50% protein sample was eluted from the heparin column (Figure 4d), which was significantly reduced compared to the level of EGCG or heparin competitor reagents. The eluted fraction was further increased when the concentration of competitor reached 0.5 mg/mL. In that case, the heparin and EGCG could elute 89.33 ± 4.16% and 90.00 ± 5.29% of total protein, respectively (Figure 4e), and the difference between then was still not significant. In summary, EGCG could directly interact with PCV2 capsid, competitively inhibiting the binding of latter to HS.

3.5. Identification of Key Amino Acids in PCV2 Capsid Contributing to the Interaction with EGCG

To further investigate the detailed structural character of the interaction between EGCG and capsid, the flexible docking model of the complex was established with AutoDock software 4.2 (Olson Lab, http://autodock.scripps.edu). The crystal structure of PCV2 dcapsid was obtained as PDB accession ID:3R0R and the whole domain of protein was docked with ligand. Several docking model structures were obtained, and the one with the lowest interface energy was adopted for further analysis, which showed that each EGCG molecule could be attached to a capsid monomer at a distinct pocket, which was completely exposed to the surface of PCV2 particles according to the revealed structure of the PCV2 VLPs [18], and the pocket was formed by some critical amino acids (Figure 5a), among which ARG-51, GLY 54, ASP-70, ARG-73, ASN-75, and ASP-78 were predicted to engage in hydrogen bonds with EGCG (Figure 5b). We established an amino acid sequence alignment to assess the conservation of these amino acid. It was demonstrated that all of these 6 amino acids were highly conserved in PCV2 ranging from PCV2a to PCV2f except for ASN-75 (Figure 5c), suggesting their critical importance for the interaction between capsid and EGCG molecule in the various subtypes of PCV2.

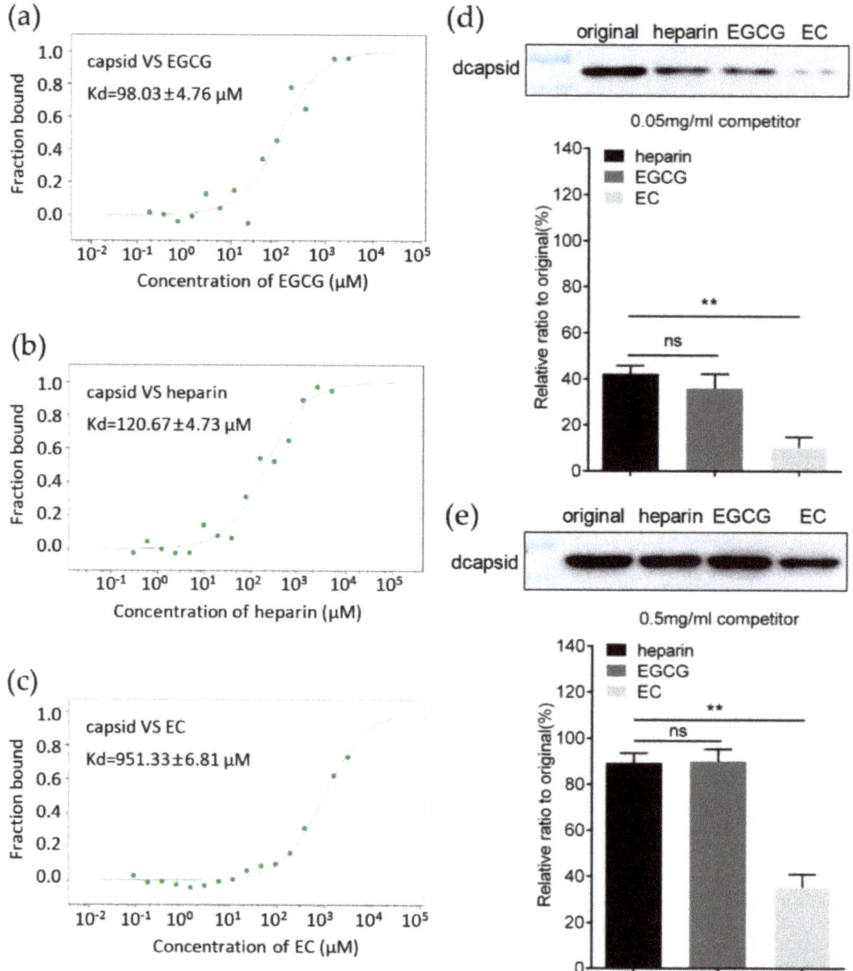

Figure 4. EGCG competitively inhibited capsid binding to heparan sulfate. (**a–c**) EGCG could interact with capsid or with heparin. The dcapsid protein was labeled with the Cy5 fluorophore and incubated with two-fold serial dilutions of EGCG or heprarin (10–50 nM), and then the Kd values of the dcapsid were measured for EGCG (**a**), heparin (**b**), and EC (**c**) were measured by MST. (**d,e**) EGCG eluted dcapsid protein bound to a heparin column with similar efficiency as heparin. Heparin-Sepharose HP column was pre-loaded with dcapsid and eluted with 0.05 (**d**) or 0.5 (**e**) mg/mL soluble competitor reagents (heparin, EGCG and EC). The eluted samples were analyzed using SDS-PAGE and immunoblotting. Three independent experiments were conducted. The results are presented as Mean ± SD and compared using Student's *t*-test (n.s., not significant; * $p < 0.05$; ** $p < 0.01$).

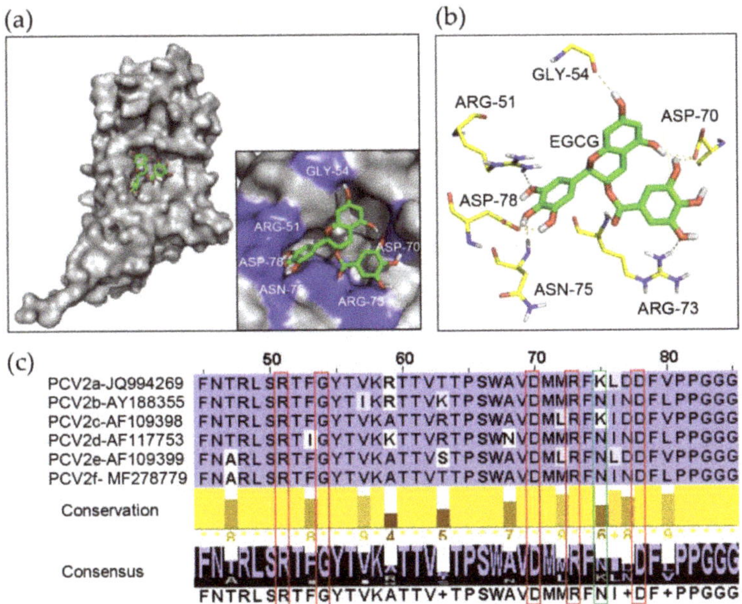

Figure 5. Molecular docking model of interaction between EGCG and PCV2 Capsid. (**a**) Structural diagram of EGCG-capsid complex based on the flexible docking model. The amino acids involved in the formation of the pocket accommodating EGCG molecule (green) were labeled blue in the dcapsid (light grey) in the zoom graph. (**b**) The putative amino acids (yellow) in dcapsid that could establish hydrogen bonds with EGCG (green) based on the docking model. The predicted hydrogen bonds are shown as a yellow dotted line. (**c**) The conservation of the key amino acids in the capsid to maintain the interaction among PCV2 subtypes. All the amino acid sequence data were obtained from GenBank. Among the predicted amino acids which might form hydrogen bonds with EGCG, the ones with high conservation (ARG-51, GLY 54, ASP-70, ARG-73 and ASP-78) are highlighted with a red frame, and the rest (ASN-75) are labeled with a green frame.

To test this hypothesis, and to identify the key amino acids for the interaction, we made a single-point mutation of these amino acids to alanine (A) residue and measured the affinity of each mutant to EGCG, the results of MST showed that the Kd of different mutants and EGCG could fluctuate significantly compared to the wild type (WT) capsid; among all the mutants, the D70A and D78A showed the loweest affinity to EGCG with Kd values of 863.67 ± 7.23 μM and 1061.33 ± 8.08 μM, respectively (Figure 6c,f), almost 10 times lower than the Kd of WT to EGCG (Kd = 98.03 ± 4.76 μM) (Figure 4a). Additionally, mutants aimed at arginine (R) residues also exhibited comparable reduction in affinity to EGCG, the Kd values of R51A and R73A were 591.00 ± 5.57 μM and 971.00 ± 9.54 μM, respectively (Figure 6a,d). On the contrary, the affinity of G54A and N75A mutants did not change significantly (Figure 6b,e). These results indicated that ARG-51, ARG-73, ASP-70 and ASP-78 were critical for the interaction of EGCG with the PCV2 capsid. To further assess the effect of these amino acids on the binding properties of capsid to HS, the heparin column chromatography assay of different mutants was performed and it showed that the eluted level of R51A, D70A, R73A and D78A mutants was remarkably reduced, with 0.5 mg/mL EGCG, especially in the case of D70A and D78A mutants, which is consistent with the results of Kd determination. Meanwhile, the mutation targeted to arginine residues (R51A and R73A) displayed the most significant eluted reduction with 0.5 mg/mL heparin (Figure 6g). Taken together, these results indicated that the ARG-51, ARG-73, ASP-70 and ASP-78 of PCV2 capsid protein were crucial for its interaction with EGCG, and the two arginine residues were also essential for binding with cell surface HS.

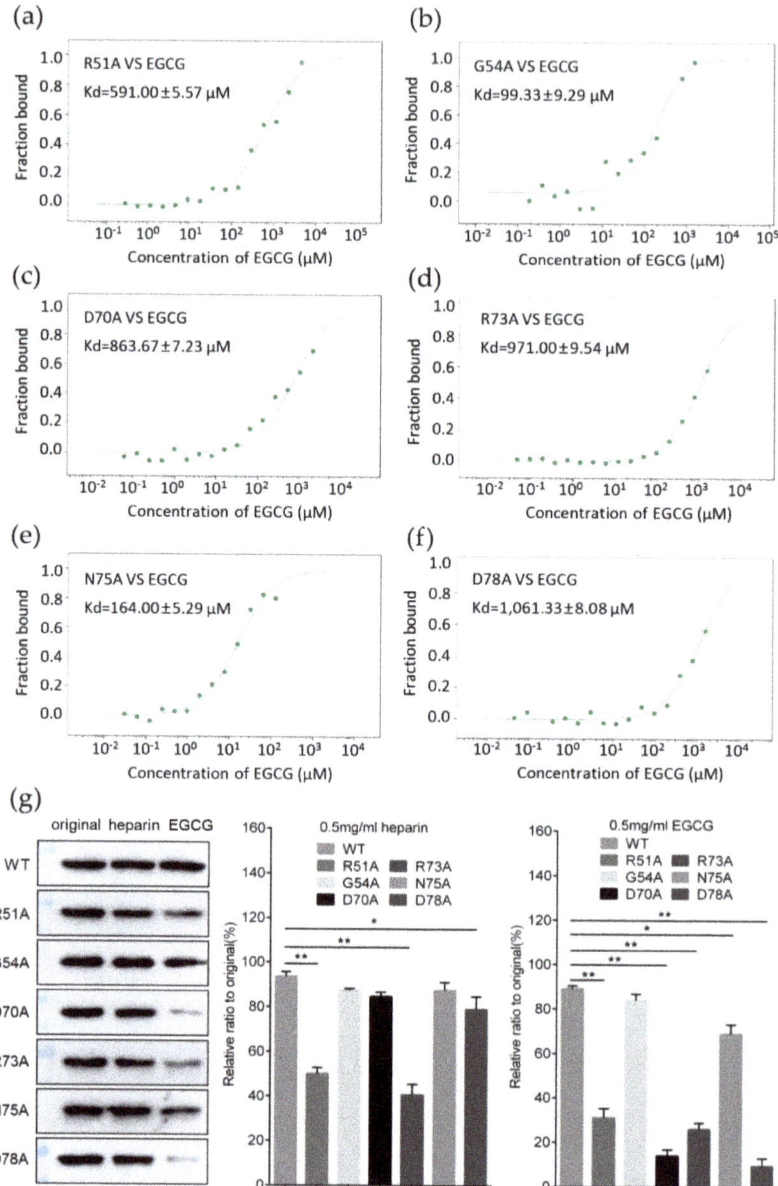

Figure 6. Verification of key amino acids involved in the interaction of EGCG with capsid. (**a–f**) The affinity impairment of capsid to EGCG induced by the key amino acid mutation. The Cy5 fluorophore labeled dcapsid mutants were incubated with two-fold serially dilutions of EGCG at room temperature for 15 min before being tested with MST, and the Kd values between EGCG and capsid mutants R51A (**a**), G54A (**b**), D70A (**c**), R73A (**d**), N75A (**e**) and D78A (**f**) were measured, respectively. (**g**) The eluted level of capsid mutants bound to the heparin column. Heparin column preloaded with dcapsid mutants was eluted with 0.5 mg/mL soluble heparin or EGCG. The eluted level of each mutants was analyzed using immunoblotting. Mean ± SD of three independent experiments were statistically analyzed using Student's *t*-test (* $p < 0.05$; ** $p < 0.01$).

3.6. Replacement of Key Amino Acids in Capsid Weakens the Antiviral Effect of EGCG

To further confirm the role of the key amino acids in capsid towards the antiviral activity of EGCG, we tried to use virus rescue strategy to generate mutant viruses and evaluate the regulation in EGCG's antiviral activity. The circularized PCV2 genome with R51A, G54A, D70A, R73A, N75A, and D78A mutations were transfected into PK-15 cells, respectively. Also, the genome with the A135A mutation was designated as a positive control. Transfected cells were serially passaged and cultured to harvest the rescued virus particles, which were then inoculated into PK-15 cells to test the infectivity. The results of IFA showed that PCV2 (G54A), PCV2 (D70A), PCV2 (N75A) and PCV2 (D78A) could be detectable at 72 hpi, similar to PCV2 (WT) and PCV2 (A135A) (Figure 7a). The results of immunoblotting also confirmed the capsid expression of these mutant viruses at 72 hpi (Figure 7b). One-step growth curves of PCV2 (G54A), PCV2 (D70A), PCV2 (N75A), and PCV2 (D78A) also displayed a similar replication ability to wild-type PCV2 (A135A) (Figure 7c). Surprisingly, PCV2 (G51A) and PCV2 (G73A) could not be rescued and obtained with the same operation (Figure 7a–c), which might be due to the replacement of these key arginine residues related to HS binding ability would cause the impairment of virus binding to cell surface, which eventually lead to the inability of virions to complete the infection process [18].

We next performed the infectivity assay of mutant viruses with 100 µM EGCG treatment during the whole infective process. The titers of mutant viruses were detected at 72 hpi and the results showed that EGCG could inhibit the infectivity of PCV2 (G54A) and PCV2 (N75A) (Figure 7f,h), similar to its antiviral effect on PCV2 (WT) and PCV2 (A135A) (Figure 7d,e). In contrast, the antiviral effect of EGCG could not be detected during the infection of PCV2 (D70A) and PCV2 (D78A) (Figure 7g,i), These findings are consistent with the results of affinity determinations (Figure 6c,f,g), indicating that ASP70 and ASP78 of capsid are essential for binding of PCV virions to EGCG.

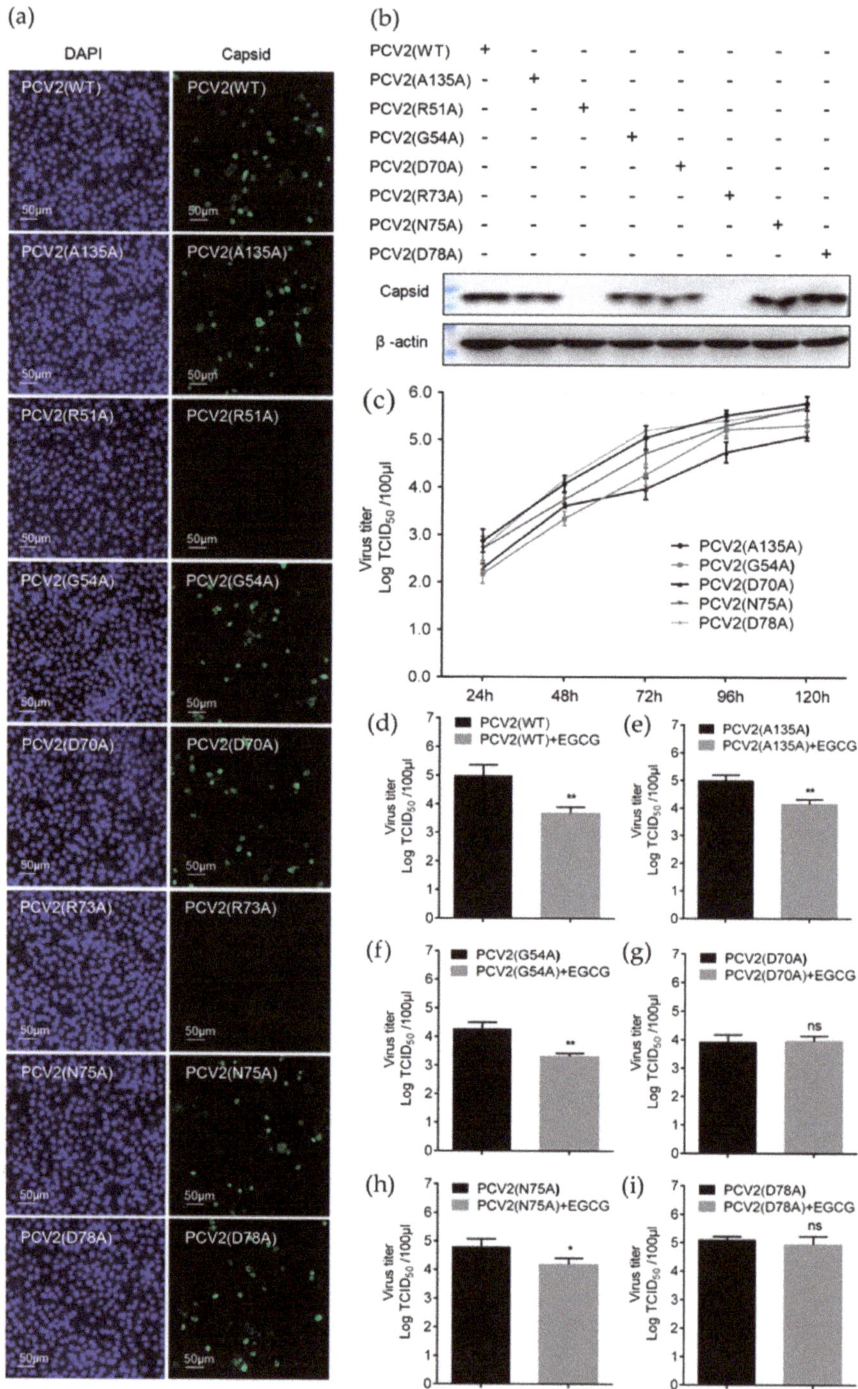

Figure 7. Regulation of EGCG antiviral activity by replacement of key amino acids in capsid. (**a,b**) The infectivity of rescued PCV2 mutants. The rescued PCV2 mutant virus was inoculated into PK-15 cells at

MOI = 1.0; at 72 hpi the cell samples were detected by IFA (**a**), and the capsid expression was measured by immunoblotting (**b**). (**c**) One-step growth curve of PCV2 mutants. PK-15 cells were inoculated with mutant viruses at MOI = 1.0, $TCID_{50}$ at various time points were determined by IFA and calculated according to the Reed-Muench method. (**d–i**) Antiviral effect of EGCG on PCV2 mutants. PK-15 cells were infected by mutant viruses at MOI = 1.0 with 100 μM EGCG treatment during the whole infective process and the titers were detected by IFA at 72 hpi. Mean ± SD of three independent experiments were statistically analyzed using Student's *t*-test (* $p < 0.05$; ** $p < 0.01$).

4. Discussion

As the most prominent component of GTCs, EGCG displayed broad and distinguished anti-viral activity compared to other GTC members [36]. Here, we showed that EGCG exhibited an inhibitory effect on the PCV2 infection at the concentration of 100 μM, without causing significant toxicity to the host cells. EGCG acted directly on PCV2 virions rather than any cellular factors with or without the stimulation of virions, and EGCG inhibited virion attachment to host cells by competitively interacting with capsid with cell surface receptor, but it did not affect other post-binding stages.

Heparan sulfate is a linear, synthetic, acidic polysaccharide belonging to the glycosaminoglycan (GAG) family, which is widely distributed on the surface and in extracellular matrix of various mammalian cells [37]. HS was attached to the protein core and consisted of repeating 1 to 4 linked disaccharide units, one of which is an α-D-glucosamine residue and the other is an anionic acid. HS chains acquired negative charge though epimerization and terminal sulfation processes, thereby facilitating HS in establishing electrostatic interaction with positively charged motifs and residues [38], including signaling pathway proteins, growth factors, plasma proteins, immune modulators and viral structural proteins [39]. In general, these protein ligands possessed distinct HS binding region, which acted as a complementary structure for heparin-protein interactions [40]. Based on the known ligands' structural details, the molecular modeling of protein-glycosaminoglycan revealed that HS binding regions formed a hydrophilic pocket wrapped around and folded over heparin oligosaccharides. The key factors to produce such hydrophilic pocket were the clusters contributed by basic amino acids, especially arginines and lysines [41,42].

The EGCG interacted with PCV2 capsid. Due to the considerable affinity, this interaction could competitively impair the binding of capsid to HS, thereby inhibiting the attachment of virions to the host cell surface. Meanwhile, the EC could not interact with the virus capsid, and hence did not exert an antiviral effect. Structurally, EGCG possessed numerous hydroxyl radicals attached on an extra benzene ring which is not the case for EC (Figure 1a), and the result of molecular docking showed that these additional free radicals were involved in forming hydrogen bonds with ASP70 and ARG-73 of capsid. Indeed, the mutagenesis assay demonstrated that aspartic acids (ASP70 and ASP78) and arginines (ARG51 and ARG73) were essential for EGCG- capsid interaction. It is also worth noticing that the arginine residues of GAG ligand were also important for the formation of a center with a high positive charge density, which could electrostatically interact with the acidic groups of HS; this was supported by the result of the replacement of ARG51 and ARG73 which remarkably attenuated the affinity of capsid to HS. Based on these findings, we conclude that the interaction of EGCG with PCV2 capsid might primarily depend on the formation of hydrogen bonds by specific positively charged amino acids in capsid, such as ARG51 and ARG73, while newly formed links might thus interfere with the original electrostatic interaction between capsid and HS, resulting in the impairment of virus attachment to the cell surface.

In previous studies, EGCG showed similar antiviral activity against various viruses, EGCG inhibited the infection of PRRSV with an EC_{50} of 48.2–63.09 μM in vitro [43], and EGCG exhibited an antiviral effect against JEV with an IC_{50} of 4.9–20 μM in vitro [44]. In our study, the EC_{50} of EGCG against PCV2 in vitro was calculated as 37.79 ± 1.64 μM. These comparable findings suggested that the antiviral activity of EGCG in vitro did not reach the efficacy of traditional antiviral drugs, such as ribavirin. It is worth noting that although this level of potency of EGCG at this stage is not of practical value in the treatment of clinical virus infections, the identification of key amino acids of viral proteins that bind EGCG is valuable for the development of structure-based anti-virus drugs, such as the EGCG-based synthetic EGCG palmitate, whose EC_{50} against PRRSV is decreased to 5.86–12.69 μM, which is nearly 5 times lower than for the original EGCG [45]. On the other hand, more research data focusing on in vivo antiviral effects of EGCG and detailed structural analysis of EGCG-protein interaction should be studied in future research to facilitate a more comprehensive assessment of the antiviral potential of EGCG.

Author Contributions: Conceptualization, J.L., J.G. and J.Z; methodology, J.L.; software, J.L.; validation, D.S., S.W.; formal analysis, J.L. and D.S.; investigation, D.S.; resources, Y.D.; data curation, J.L. and J.G.; writing—original draft preparation, J.L.; writing—review and editing, J.G. and J.Z.; visualization, J.L. and D.S.; supervision, J.G. and J.Z.; project administration, J.Z.; funding acquisition, J.Z. All authors have read and agreed to the published version of the manuscript.

Funding: This research was funded by the National Key Research & Development Program of China (Grant No. 2017YFD0500101 & 2015BAD12B01) and the National Natural Science Foundation of China (Grant No. 31402198).

Acknowledgments: We thank the National Key Laboratory of Natural Medicines for their support of MST instruments and technology, and also the instrumental and technical assistance on heparin column chromatography from technicians of general electric (GE) healthcare life sciences.

Conflicts of Interest: The authors declare no competing interests. The funders had no role in the design of the study; in the collection, analyses, or interpretation of data; in the writing of the manuscript, or in the decision to publish the results.

References

1. Pringle, C.R. Virus taxonomy at the XIth International Congress of Virology, Sydney, Australia, 1999. *Arch. Virol.* **1999**, *144*, 2065–2070. [CrossRef]

2. Li, J.; Xing, G.; Zhang, C.; Yang, H.; Li, G.; Wang, N.; Wang, R.; Sun, H.; Shi, Z.; Lei, J.; et al. Cross-species transmission resulted in the emergence and establishment of circovirus in pig. *Infect. Genet. Evol. J. Mol. Epidemiol. Evol. Genet. Infect. Dis.* **2019**, *75*, 103973. [CrossRef]

3. Finsterbusch, T.; Mankertz, A. Porcine circoviruses—Small but powerful. *Virus Res.* **2009**, *143*, 177–183. [CrossRef]

4. Ouyang, T.; Zhang, X.; Liu, X.; Ren, L. Co-Infection of Swine with Porcine Circovirus Type 2 and Other Swine Viruses. *Viruses* **2019**, *11*, 185. [CrossRef] [PubMed]

5. Afolabi, K.O.; Iweriebor, B.C.; Okoh, A.I.; Obi, L.C. Global Status of Porcine circovirus Type 2 and Its Associated Diseases in Sub-Saharan Africa. *Adv. Virol.* **2017**, *2017*, 6807964. [CrossRef] [PubMed]

6. Hamel, A.L.; Lin, L.L.; Nayar, G.P. Nucleotide sequence of porcine circovirus associated with postweaning multisystemic wasting syndrome in pigs. *J. Virol.* **1998**, *72*, 5262–5267. [CrossRef] [PubMed]

7. Allan, G.M.; Kennedy, S.; McNeilly, F.; Foster, J.C.; Ellis, J.A.; Krakowka, S.J.; Meehan, B.M.; Adair, B.M. Experimental reproduction of severe wasting disease by co-infection of pigs with porcine circovirus and porcine parvovirus. *J. Comp. Pathol.* **1999**, *121*, 1–11. [CrossRef]

8. Sanchez, R.E., Jr.; Nauwynck, H.J.; McNeilly, F.; Allan, G.M.; Pensaert, M.B. Porcine circovirus 2 infection in swine foetuses inoculated at different stages of gestation. *Vet. Microbiol.* **2001**, *83*, 169–176. [CrossRef]

9. Nawagitgul, P.; Morozov, I.; Bolin, S.R.; Harms, P.A.; Sorden, S.D.; Paul, P.S. Open reading frame 2 of porcine circovirus type 2 encodes a major capsid protein. *J. Gen. Virol.* **2000**, *81*, 2281–2287. [CrossRef]

10. Misinzo, G.; Delputte, P.L.; Meerts, P.; Lefebvre, D.J.; Nauwynck, H.J. Porcine circovirus 2 uses heparan sulfate and chondroitin sulfate B glycosaminoglycans as receptors for its attachment to host cells. *J. Virol.* **2006**, *80*, 3487–3494. [CrossRef]

11. Rostand, K.S.; Esko, J.D. Microbial adherence to and invasion through proteoglycans. *Infect. Immun.* **1997**, *65*, 1–8. [CrossRef] [PubMed]

12. Cagno, V.; Tseligka, E.D.; Jones, S.T.; Tapparel, C. Heparan Sulfate Proteoglycans and Viral Attachment: True Receptors or Adaptation Bias? *Viruses* **2019**, *11*, 596. [CrossRef] [PubMed]

13. Prydz, K.; Dalen, K.T. Synthesis and sorting of proteoglycans. *J. Cell Sci.* **2000**, *113*, 193–205. [PubMed]

14. Xu, Y.; Martinez, P.; Seron, K.; Luo, G.; Allain, F.; Dubuisson, J.; Belouzard, S. Characterization of hepatitis C virus interaction with heparan sulfate proteoglycans. *J. Virol.* **2015**, *89*, 3846–3858. [CrossRef]

15. Shieh, M.T.; WuDunn, D.; Montgomery, R.I.; Esko, J.D.; Spear, P.G. Cell surface receptors for herpes simplex virus are heparan sulfate proteoglycans. *J. Cell Biol.* **1992**, *116*, 1273–1281. [CrossRef]

16. Tan, C.W.; Poh, C.L.; Sam, I.C.; Chan, Y.F. Enterovirus 71 uses cell surface heparan sulfate glycosaminoglycan as an attachment receptor. *J. Virol.* **2013**, *87*, 611–620. [CrossRef]

17. Sasaki, M.; Anindita, P.D.; Ito, N.; Sugiyama, M.; Carr, M.; Fukuhara, H.; Ose, T.; Maenaka, K.; Takada, A.; Hall, W.W.; et al. The Role of Heparan Sulfate Proteoglycans as an Attachment Factor for Rabies Virus Entry and Infection. *J. Infect. Dis.* **2018**, *217*, 1740–1749. [CrossRef]

18. Dhindwal, S.; Avila, B.; Feng, S.; Khayat, R. Porcine Circovirus 2 Uses a Multitude of weak binding sites to interact with heparan sulfate, and the interactions do not follow the symmetry of the Capsid. *J. Virol.* **2019**, *93*. [CrossRef]

19. Lin, L.T.; Chen, T.Y.; Lin, S.C.; Chung, C.Y.; Lin, T.C.; Wang, G.H.; Anderson, R.; Lin, C.C.; Richardson, C.D. Broad-spectrum antiviral activity of chebulagic acid and punicalagin against viruses that use glycosaminoglycans for entry. *BMC Microbiol.* **2013**, *13*, 187. [CrossRef]

20. Narotzki, B.; Reznick, A.Z.; Aizenbud, D.; Levy, Y. Green tea: A promising natural product in oral health. *Arch. Oral Biol.* **2012**, *57*, 429–435. [CrossRef]

21. Yang, C.S.; Wang, H. Cancer Preventive Activities of Tea Catechins. *Molecules* **2016**, *21*, 1679. [CrossRef]

22. Gonzalez, Y.; Torres-Mendoza, D.; Jones, G.E.; Fernandez, P.L. Marine Diterpenoids as Potential Anti-Inflammatory Agents. *Mediat. Inflamm.* **2015**, *2015*, 263543. [CrossRef] [PubMed]

23. Chacko, S.M.; Thambi, P.T.; Kuttan, R.; Nishigaki, I. Beneficial effects of green tea: A literature review. *Chin. Med.* **2010**, *5*, 13. [CrossRef] [PubMed]

24. Kono, K.; Tatara, I.; Takeda, S.; Arakawa, K.; Hara, Y. Antibacterial activity of epigallocatechin gallate against methicillin-resistant Staphylococcus aureus. Kansenshogaku zasshi. *J. Jpn. Assoc. Infect. Dis.* **1994**, *68*, 1518–1522. [CrossRef]

25. Song, J.M.; Lee, K.H.; Seong, B.L. Antiviral effect of catechins in green tea on influenza virus. *Antivir. Res.* **2005**, *68*, 66–74. [CrossRef]

26. Yamaguchi, K.; Honda, M.; Ikigai, H.; Hara, Y.; Shimamura, T. Inhibitory effects of (-)-epigallocatechin gallate on the life cycle of human immunodeficiency virus type 1 (HIV-1). *Antivir. Res.* **2002**, *53*, 19–34. [CrossRef]

27. Nakayama, M.; Suzuki, K.; Toda, M.; Okubo, S.; Hara, Y.; Shimamura, T. Inhibition of the infectivity of influenza virus by tea polyphenols. *Antivir. Res.* **1993**, *21*, 289–299. [CrossRef]

28. Carneiro, B.M.; Batista, M.N.; Ana, C.B.; Nogueira, M.L.; Rahal, P. The green tea molecule EGCG inhibits Zika virus entry. *Virology* **2016**, *496*, 215–218. [CrossRef]

29. Isaacs, C.E.; Wen, G.Y.; Xu, W.; Jia, J.H.; Rohan, L.; Corbo, C.; Di Maggio, V.; Jenkins, E.C., Jr.; Hillier, S. Epigallocatechin gallate inactivates clinical isolates of herpes simplex virus. *Antimicrob. Agents Chemother.* **2008**, *52*, 962–970. [CrossRef]

30. Ciesek, S.; von Hahn, T.; Colpitts, C.C.; Schang, L.M.; Friesland, M.; Steinmann, J.; Manns, M.P.; Ott, M.; Wedemeyer, H.; Meuleman, P.; et al. The green tea polyphenol, epigallocatechin-3-gallate, inhibits hepatitis C virus entry. *Hepatology* **2011**, *54*, 1947–1955. [CrossRef]

31. Kawai, K.; Tsuno, N.H.; Kitayama, J.; Okaji, Y.; Yazawa, K.; Asakage, M.; Hori, N.; Watanabe, T.; Takahashi, K.; Nagawa, H. Epigallocatechin gallate, the main component of tea polyphenol, binds to CD4 and interferes with gp120 binding. *J. Allergy Clin. Immunol.* **2003**, *112*, 951–957. [CrossRef]

32. O'Donnell, C.D.; Kovacs, M.; Akhtar, J.; Valyi-Nagy, T.; Shukla, D. Expanding the role of 3-O sulfated heparan sulfate in herpes simplex virus type-1 entry. *Virology* **2010**, *397*, 389–398. [CrossRef]

33. Zhou, J.Y.; Chen, Q.X.; Ye, J.X.; Shen, H.G.; Chen, T.F.; Shang, S.B. Serological investigation and genomic characterization of PCV2 isolates from different geographic regions of Zhejiang province in China. *Vet. Res. Commun.* **2006**, *30*, 205–220. [CrossRef] [PubMed]

34. Zhou, J.Y.; Shang, S.B.; Gong, H.; Chen, Q.X.; Wu, J.X.; Shen, H.G.; Chen, T.F.; Guo, J.Q. In vitro expression, monoclonal antibody and bioactivity for capsid protein of porcine circovirus type II without nuclear localization signal. *J. Biotechnol.* **2005**, *118*, 201–211. [CrossRef] [PubMed]

35. Hjulsager, C.K.; Grau-Roma, L.; Sibila, M.; Enoe, C.; Larsen, L.; Segales, J. Inter-laboratory and inter-assay comparison on two real-time PCR techniques for quantification of PCV2 nucleic acid extracted from field samples. *Vet. Microbiol.* **2009**, *133*, 172–178. [CrossRef] [PubMed]

36. Xu, J.; Xu, Z.; Zheng, W. A Review of the Antiviral Role of Green Tea Catechins. *Molecules* **2017**, *22*, 1337. [CrossRef]

37. Kreuger, J.; Kjellen, L. Heparan sulfate biosynthesis: Regulation and variability. *J. Histochem. Cytochem.* **2012**, *60*, 898–907. [CrossRef]

38. Xu, D.; Esko, J.D. Demystifying heparan sulfate-protein interactions. *Annu. Rev. Biochem.* **2014**, *83*, 129–157. [CrossRef]

39. Billings, P.C.; Pacifici, M. Interactions of signaling proteins, growth factors and other proteins with heparan sulfate: Mechanisms and mysteries. *Connect. Tissue Res.* **2015**, *56*, 272–280. [CrossRef]

40. Cardin, A.D.; Weintraub, H.J. Molecular modeling of protein-glycosaminoglycan interactions. *Arteriosclerosis* **1989**, *9*, 21–32. [CrossRef]

41. Shriver, Z.; Capila, I.; Venkataraman, G.; Sasisekharan, R. Heparin and heparan sulfate: Analyzing structure and microheterogeneity. *Handb. Exp. Pharmacol.* **2012**, 159–176. [CrossRef]

42. Zulueta, M.M.; Lin, S.Y.; Hu, Y.P.; Hung, S.C. Synthetic heparin and heparan sulfate oligosaccharides and their protein interactions. *Curr. Opin. Chem. Biol.* **2013**, *17*, 1023–1029. [CrossRef] [PubMed]

43. Ge, M.; Xiao, Y.; Chen, H.; Luo, F.; Du, G.; Zeng, F. Multiple antiviral approaches of (-)-epigallocatechin-3-gallate (EGCG) against porcine reproductive and respiratory syndrome virus infection in vitro. *Antivir. Res.* **2018**, *158*, 52–62. [CrossRef] [PubMed]

44. Wang, C.Y.; Hour, M.J.; Lai, H.C.; Chen, C.H.; Chang, P.J.; Huang, S.H.; Lin, C.W. Epigallocatechin-3-gallate inhibits the early stages of Japanese encephalitis virus infection. *Virus Res.* **2018**, *253*, 140–146. [CrossRef] [PubMed]

45. Zhao, C.; Liu, S.; Li, C.; Yang, L.; Zu, Y. In vitro evaluation of the antiviral activity of the synthetic epigallocatechin gallate analog-epigallocatechin gallate (EGCG) palmitate against porcine reproductive and respiratory syndrome virus. *Viruses* **2014**, *6*, 938–950. [CrossRef] [PubMed]

Article

Ginsenoside Rg1 Suppresses Type 2 PRRSV Infection via NF-κB Signaling Pathway In Vitro, and Provides Partial Protection against HP-PRRSV in Piglet

Zhi-qing Yu [1,2], He-you Yi [1,2], Jun Ma [1,2], Ying-fang Wei [1,2], Meng-kai Cai [1,2], Qi Li [1,2], Chen-xiao Qin [1,2], Yong-jie Chen [2], Xiao-liang Han [1,2], Ru-ting Zhong [1,2], Yao Chen [3], Guan Liang [1,2], Qiwei Deng [1,2], Kegong Tian [4,5], Heng Wang [1,2,*] and Gui-hong Zhang [1,2,*]

[1] College of Veterinary Medicine, South China Agricultural University, Guangzhou 510462, China; zhiqingyu@stu.scau.edu.cn (Z.-q.Y.); heyouyi@stu.scau.edu.cn (H.-y.Y.); 18814116634@163.com (J.M.); yingfangwe@163.com (Y.-f.W.); caimengkai@126.com (M.-k.C.); Qili@126.com (Q.L.); qinchenxiaol@163.com (C.-x.Q.); XiaoliangHan@scau.edu.cn (X.-l.H.); RutingZhong@scau.edu.cn (R.-t.Z.); Liangguan@scau.edu.cn (G.L.); dengqiwei@scau.edu.cn (Q.D.)
[2] Key Laboratory of Zoonosis Prevention and Control of Guangdong Province, Guangzhou 510462, China; vetcyj@163.com
[3] School of Life Science and Engineering, Foshan University, Foshan 528225, China; chenyao1991scau@foxmail.com
[4] College of Animal Science and Veterinary Medicine, Henan Agricultural University, Zhengzhou 450002, China; 123456@163.com
[5] OIE Reference Laboratory for PRRS in China, China Animal Disease Control Center, Beijing 100125, China
[*] Correspondence: wangheng2009@scau.edu.cn (H.W.); guihongzh@scau.edu.cn (G.-h.Z.)

Received: 15 October 2019; Accepted: 7 November 2019; Published: 9 November 2019

Abstract: Porcine reproductive and respiratory syndrome virus (PRRSV) is a huge threat to the modern pig industry, and current vaccine prevention strategies could not provide full protection against it. Therefore, exploring new anti-PRRSV strategies is urgently needed. Ginsenoside Rg1, derived from ginseng and notoginseng, is shown to exert anti-inflammatory, neuronal apoptosis-suppressing and anti-oxidant effects. Here we demonstrate Rg1-inhibited PRRSV infection both in Marc-145 cells and porcine alveolar macrophages (PAMs) in a dose-dependent manner. Rg1 treatment affected multiple steps of the PRRSV lifecycle, including virus attachment, replication and release at concentrations of 10 or 50 μM. Meanwhile, Rg1 exhibited broad inhibitory activities against Type 2 PRRSV, including highly pathogenic PRRSV (HP-PRRSV) XH-GD and JXA1, NADC-30-like strain HNLY and classical strain VR2332. Mechanistically, Rg1 reduced mRNA levels of the pro-inflammatory cytokines, including IL-1β, IL-8, IL-6 and TNF-α, and decreased NF-κB signaling activation triggered by PRRSV infection. Furthermore, 4-week old piglets intramuscularly treated with Rg1 after being challenged with the HP-PRRSV JXA1 strain display moderate lung injury, decreased viral load in serum and tissues, and an improved survival rate. Collectively, our study provides research basis and supportive clinical data for using Ginsenoside Rg1 in PRRSV therapies in swine.

Keywords: porcine reproductive and respiratory syndrome virus; ginsenoside Rg1; antiviral activity; pro-inflammatory factor; NF-κB signaling pathway

---

1. Introduction

Porcine reproductive and respiratory syndrome (PRRS), characterized by respiratory distress, reproductive failure in pregnant sows and high mortality in piglets [1], is one of the most epidemic porcine infectious diseases that cause huge economic losses in the worldwide pig industry. Porcine reproductive and respiratory syndrome virus (PRRSV) is a positive-sense, single-stranded RNA virus, and it belongs to the Arteriviridae family [2]. PRRSV is divided into two genotypes, including European strains (type 1) and North American strains (type 2) [3]. Type 2 PRRSV is dominant in China for decades, and it is further classified into nine lineages based on the nucleotide sequence of the ORF5 gene [4,5].

Marc-145 cells, purchased from the American Type Culture Collection (ATCC, Manassas, Virginia, USA) and saved in our lab, were cultured in Dulbecco's Minimum Essential Medium (DMEM, Biological Industries, Kibbutz Beit Haemek, Israel), supplemented with 10% fetal bovine serum (FBS, Biological Industries, Kibbutz Beit Haemek, Israel) at 37 °C with 5% CO_2. Porcine alveolar macrophages (PAMs) were collected from the fresh lungs of 4-week-old Large-White piglets, which were free of the PRRSV and anti-PRRSV antibody, and prepared as previously described [6]. PAMs were grown in RPMI 1640 (Gibco, UT, USA) which contains 10% FBS and 100 IU/mL penicillin and 100 µg/mL streptomycin.

In recent decades, the emergence of a novel PRRSV strain and worldwide transmission attracted increasing attention [7–11]. Current major preventive strategies focus on the vaccine application. However, the poorly studied immunosuppression of pigs to PRRSV infection, virus evolution, multiple-recombination event between wild-type strain and Modified Live Virus (MLV), and currently licensed vaccines, fail to offer effective protection against the challenge of a heterogeneous strain, posing a great challenge to vaccine development. Thus, a new strategy for controlling this infectious disease is urgently needed.

Many natural compounds and herbal components have been confirmed to possess antiviral activities in Traditional Chinese Medicine (TCM). Natural herbal extracts contain many bioactive compounds featuring anti-inflammatory, antiviral and immune-regulatory activities, especially flavaspidic acid AB [12], glycyrrhizin [13] and platycodin D [14], which have been demonstrated to suppress PRRSV infection in vitro. The development of novel drugs might become an effective means to fight the global PRRS epidemic. However, the in vivo study of confirming TCM as a potential natural and effective anti-PPRSV agent was poorly described. Therefore, further studies in swine are necessary.

Ginsenosides are biologically-active components of Panax ginseng and Panax notoginseng saponins that were widely used as a traditional herbal tonic in China for a thousand years, and ginsenoside Rg1 is a major bioactive component therein [15] (Figure 1A). A previous study demonstrates that Rg1 treatment enhances the immune responses induced by recombinant Toxoplasma gondii SAG1 antigen [16], and Rg1 could be used as an adjuvant to promote both T helper (Th) 1 and Th2 responses [17,18]. Treatment with Rg1 is found to significantly relieve the cellular inflammatory response in neurons [19] and reduce the expression of TNF-α, IL-1β and IL-6 in vivo [20]. However, the antiviral activity of Rg1 is poorly described. PRRSV infection results in the release of IL-1β, IL-6, IL-8 and TNF-α, and these pro-inflammatory cytokines contribute to the development of excessive systemic inflammatory reactions and pathological injury [6]. Thus, we set out to determine whether Rg1 has anti-PRRSV effects and possesses protective effects against viruses-induced injury.

Here we demonstrate that Rg1 exhibits an antiviral effect against a broad range of type 2 PRRSV in Marc-145 cells and PAMs, and Rg1 treatment reduces the mRNA levels of several pro-inflammatory cytokines triggered by PRRSV infection, and also inhibits the activation of the NF-κB signaling pathway. More importantly, piglets treated with Rg1 display decreased viremia, alleviated lung injury and increased survival rate after challenging with HP-PRRSV JXA1. Together, these data suggested that Rg1 might be a potential natural compound that could be applied in a PRRSV control strategy.

2. Materials and Methods

2.1. Cells and Virus

PRRSV strains, including the classical VR2332 strain (GenBank accession no. U87392.3; lineage 5.1), highly pathogenic XH-GD (GenBank accession no. EU624117; lineage 8.7) [21], and the NADC30-like strain HNLY (isolated in a sow with reproductive problem and saved in our lab; lineage 1), were saved in our lab. Highly pathogenic JXA1 (GenBank accession no. EF112445.1; lineage 8.7) was generously offered by Professor Tian [9].

All of the PRRSV strains were propagated and titrated in Marc-145 cells. Virus titers of each strain were calculated by using a Reed-Muench method.

2.2. Antibodies, Chemicals and Reagents

The mouse monoclonal antibodies against PRRSV N protein were purchased from MEDIAN Diagnostics (Korea). Rabbit monoclonal antibodies directed against Phospho-P65, Phospho-IκBα, and mouse monoclonal antibodies directed against P65 and IκBα were purchased from Cell Signaling Technology (Beverly, MA, USA). Goat anti-rabbit IgG antibody and goat anti-mouse IgG antibody were from LI-COR Biosciences (Lincoln, NE, USA). Glyceraldehyde 3-phosphate dehydrogenase (GAPDH) antibody was purchased from MBL Beijing Biotech (Beijing, China). Goat anti-Mouse IgG (H+L) Cross-Adsorbed Secondary Antibody (Alexa Fluor 594 and 488) were purchased from Thermo Fisher Scientific (Waltham, MA, USA).

Ginsenoside Rg1 (≥98.0%) was purchased from Chengdu Biopurify Phytochemicals (Chengdu, China). LPS (lipopolysaccharide from Escherichia coli 0111:B4) was purchased from Sigma-Aldrich (MA, USA). Rg1 was dissolved in dimethylsulfoxide (DMSO, Sigma-Aldrich) and diluted with DMEM before use. The final concentration of DMSO in the cell culture medium was less than 0.4%.

2.3. Quantitative Real-Time PCR

Total cellular RNA was extracted using a total RNA rapid extraction kit (Fastagen, Shanghai, China) according to the instructions, and 1 μg RNA of each sample was subsequently reverse transcribed to cDNA with a reverse transcription kit (TaKaRa, Dalian, China) according to the manuals. The acquired cDNA was then used as the template in a qPCR assay by using TB Green® Premix Ex Taq™ II (Tli RNaseH Plus) or Premix Ex Taq™ (Probe q-PCR)(Takara Biomedical Technology, Beijing) in CFX96 Real-time polymerase chain reaction system (qPCR) (Bio-Rad, CA, USA). The abundance of individual gene mRNA transcripts in each sample was measured three times, and GAPDH mRNA was used as the endogenous loading control. The sequences of primers and probe are listed in Table 1. Relative mRNA expression of each target gene was calculated by the $2^{-\Delta\Delta CT}$ method.

Table 1. Sequences of the primers and probe used for Real-time polymerase chain reaction (PCR).

Name	Primer Sequence (5′-3′)
PRRSV Nsp9	F: CCTGCAATTGTCCGCTGGTTTG R: GACGACAGGCCACCTCTCTTAG
GAPDH	F: GCAAAGACTGAACCCACTAATT R: TTGCCTCTGTTGTTACTTGGAG
Marc-145 IL-6	F: GAGGCACTGGCAGAAAAC R: TGCAGGAACTGGATCAGGAC
PAM IL-6	F: CCTTCAGTCCAGTCGCCTTCTC R: CATCACCTTTGGCATCTTCTTC
Marc-145 IL-8	F: AGGACAAGAGCCAGGAAG R: CTGCACCTTCACACAGAGC
PAM IL-8	F: CACTGTGAAAATTCAGAAATCATTGT R: CTTCACAAATACCTGCACAACC
Marc-145 TNF-α	F: TCTGTCTGCTGCACTTTGGAGTG R: TTGAGGGTTTGCTACAACATGG

Table 1. *Cont.*

Name	Primer Sequence (5′-3′)
PAM TNFα	F: TGGTGGTGCCGACAGATGG R: GGCTGATGGTGTGAGTGAGG
Marc-145 IL-1β	F: GGAAGACAAATTGCATGG R: CCCAACTGGTACATCAGC
PAM IL-1β	F: ACCTGGACCTTGGTTCTCTG R: CATCTGCCTGATGCTCTTG
Probe JXA1 Nsp9	ACTGCTGCCACGACTTACTGGTCACGCAGT

F: forward primer; R: reverse primer.

2.4. Cell Proliferation and Cytotoxicity Assay

The cytotoxicity of Rg1 was analyzed by using the WST-1 Cell Proliferation and Cytotoxicity Assay Kit (Beyotime). Marc-145 cells (2×10^4 cells per well) and PAMs (1×10^5 cells per well) were seeded in 96-well plates and grown at 37 °C. Cells were incubated with medium supplemented with different concentrations of Rg1 and further incubated for 48 h. The absorbance of each well was read at 450 nm with a reference wavelength of 630 nm using a microplate reader (Thermo Fisher Scientific, MA, USA).

For the cell proliferation assay, Marc-145 cells (5×10^3 cells per well) were seeded in 96-well plates and cultured without FBS at 37 °C for 12 h and then treated with indicated concentrations of Rg1. After 24 h of Rg1 incubation, the relative proliferation was evaluated using the WST-1 kit according to the instructions. The absorbance was read at 450 nm with a reference wavelength of 630 nm using a microplate reader (Thermo Fisher Scientific, MA, USA).

2.5. Antiviral Activity Assay

To analyze the effect of Rg1 on PRRSV infection, Marc-145 cells and PAMs were grown to 70%–80% confluence respectively in six-well plates. PRRSV strains (0.1 MOI) diluted in DMEM or RPMI 1640 were incubated with Marc-145 cells or PAMs at 37 °C for 1 h. The supernatants were removed, and the cells were washed twice with PBS. Then, fresh DMEM containing different concentrations of each compound was added and incubated at 37 °C in 5% CO_2. After treatment, the cells and supernatants were collected at the indicated time points of post-infection. The supernatants were used to titrate the production of progeny virus, and the viral titers were defined and calculated as $TCID_{50}$/mL [22]. The cell plates were washed with PBS and harvested for immunofluorescence assay (IFA), qRT-PCR, and western blotting analysis. The 50% effective concentrations (EC_{50}) value (the concentration of Rg1 required to protect 50% cells from infection) of Rg1 against different type 2 PRRSV strains was determined as previously described [14], and calculated with the GraphPad Prism 7.0 software.

2.6. Immunofluorescence Staining

Marc-145 cells were grown on coverslips and then infected with PRRSV at 37 °C for 1 h. Cells were washed with PBS after infection and cultured with or without Rg1 (200 μM) for 24 h. Following Rg1 treatment, the cells were fixed with paraformaldehyde for 30 min at 4 °C. The fixed cells were permeabilized with 0.1% Triton X-100 in PBS for 5 min, blocked with 3% bovine serum albumin in PBS for 2 h, and endogenous proteins were directly stained with the respective antibodies. For IFA, mouse anti-N protein antibody and goat anti-mouse IgG Alexa Fluor 488 were used as primary and secondary antibodies respectively. Mouse monoclonal antibodies directed against P65 and goat anti-mouse IgG Alexa Fluor 594 were used in confocal assay. The nuclei were stained with 4′,6-diamidino-2-phenylindole (DAPI). Immunofluorescence was captured by Leica DMI 4000B fluorescence microscope (Leica, Wetzlar, Germany). The cover slips were mounted onto glass slides using PBS containing 50% glycerol. Confocal images were obtained using a laser scanning confocal microscope (Olympus, Japan).

2.7. Western Blotting

The total cell samples were washed twice with ice-cold PBS and then lysed in RIPA lysis buffer (Beyotime) with 1% phosphatase inhibitor cocktail (APExBIO, Houston, USA). The protein samples were resolved with sodium dodecyl sulfate–10% polyacrylamide gel electrophoresis (SDS-PAGE) and transferred to polyvinylidene difluoride (PVDF) membranes (Millipore, Billerica, MA, USA). The PVDF membranes were blocked with 5% BSA in Tris-buffered saline containing Tween 20 and then incubated with the primary antibodies. Goat anti-mouse or anti-rabbit IgG (LI-COR Biosciences) were used as the secondary antibodies. An Odyssey Infrared Imaging System (LICOR, CT, USA) was used to analyze the PVDF membranes.

2.8. Rg1 Treatment on PRRSV Life Cycle Assay

Marc-145 cells were grown to 70%–80% confluence in six-well plate at 37 °C in 5% CO_2. Cells were infected with PRRSV. Four steps in PRRSV life cycle including attachment, internalization, replication and release were analyzed at different time-points after post-infection and the pre-treatment of Rg1 on PRRSV infection was also analyzed as previously described [14].

2.9. Animal Experiment

Thirty-two four-week old piglets, which were free of African swine fever virus (ASFV), PRRSV, antibody against PRRSV, pseudorabies virus (PRV) and swine influenza virus (SIV), were randomly divided into four groups with eight piglets in each group. For each group, five piglets were randomly selected and raised together as a sub-group used for collecting samples and recording clinical performance, morbidity and mortality, while the other three piglets were raised in another pigsty and euthanized at seven days post-challenge for pathological detection. The groups PRRSV and PRRSV+Rg1 were challenged intranasally (1 mL) and intramuscularly (1 mL) with the HP-PRRSV JXA1 strain (5×10^5 $TCID_{50}$), and the Mock group was inoculated in the same way with the same volume of DMEM and continued administration for 10 days, and then used as a negative control. At 12 h post infection, the group PRRSV+Rg1 was injected intramuscularly with Rg1 (10 mg/kg/day) and continued administration for 10 days. The PRRSV group was injected with 1 mL DMEM at 12 h post-infection and continued administration for 10 days, and then used as a challenge control group. The group Rg1 was injected with Rg1 (10 mg/kg/day) and continued administration for 10 days, being used as the drug control group. All of the piglets were planned to be monitored for 14 days after the challenge. Clinical signs and rectal temperatures of all of the groups were evaluated and scored daily. Blood samples of each piglet were obtained by venipuncture at 0, 2, 4, 7, 10 and 14 days post-infection (dpi) for the detection of the PRRSV viral load and anti-PRRSV antibody (Idexx PRRSV Elisa Kit, Idexx, US). The tissue sample of lung, thymus and lymph node were collected at 7 and 15 dpi. Tissue (1g) and serum (100 μL) samples were used to extract total RNA. RNA extraction and cDNA synthesis were performed as previous report [23]. Real-time quantification PCR was carried out on a CFX96TM real-time system (Bio-Rad, USA). Standard, serially-diluted PRRSV strain JXA1 (10^0–10^7 $TCID_{50}$/mL) was used to generate a standard curve (slope = -3.717; R^2 = 0.995) [24,25]. All pigs were euthanized for pathological detection at 15 dpi.

2.10. Clinical Performance and Gross Lesions of Lung

The clinical performance of all of the piglets after the HP-PRRSV JXA1 challenge were observed and scored daily, as described previously [26]. Briefly, the scores, ranged from 0 to 5, represent the severity of clinical symptoms, which including behavior, appetite and respiration. The total score for clinical performance was calculated by the sum of the evaluation index daily. The lung injury of each group was evaluated by necropsy at 7 dpi and 15 dpi. Macroscopic gross lesions of each lobe were determined as the percentage of lung with visible pneumonia and scored, while the obtained lung tissue of each group were used to perform histological pathology analysis as described [27].

2.11. Ethics Statement

Animal experiments in this study were approved by Laboratory Animal Committee of South China Agricultural University (No. 2019C001). All piglets were raised in the animal facility of SCAU, and all operations were in accordance with the animal ethics guidelines and approved protocols.

2.12. Statistical Analysis

All data of each assay represents at least two separate experiments and were determined in triplicate. The results collected from triplicate determinations were analyzed as the means ± standard deviations (SD). Data difference of each experiment was analyzed by one-way Analysis of Variance (ANOVA) followed by the Tukey's t-test in GraphPad Prism 7.0 software (San Diego, CA). * $p < 0.05$, ** $p < 0.01$, *** $p < 0.001$ and **** $p < 0.0001$ were considered to be statistically significant at different levels.

3. Results

3.1. Ginsenoside Rg1 Treatment Supressed PRRSV Replication in Marc-145 Cells and PAMs

The cytotoxicity of ginsenoside Rg1 (Rg1) on Marc-145 cells and PAMs was analyzed by WST-1 assay. Rg1 does not impair Marc-145 and PAM cell viability at a concentration as high as 400 μM (Figure 1B,C). In the course of the experiment, we notice that Marc-145 cells cultured with Rg1 exhibit better cellular morphology under the serum deprivation. Therefore, whether Rg1 affects cell proliferation was analyzed in Marc-145 cells. The results indicate that Rg1 does not influence Marc-145 cell proliferation significantly at doses from 5 μM to 400 μM, and cell proliferation increases in 800 μM and 1600 μM (Figure 1D). These results suggest that Rg1 has minimal cytotoxicity on Marc-145 cells and PAMs within the tested doses.

To determine the anti-PRRSV activity of Rg1, Marc-145 cells and PAMs were infected with PRRSV XH-GD (0.1 MOI) for 1 h and then treated with indicated concentrations of Rg1 for 24 h. As shown in Figure 1E, treatment with Rg1 results in a significant dose-dependent reduction in PRRSV Nsp9 mRNA levels both in Marc-145 cells and PAMs. The expression level of N protein, evaluated by western blotting, decreases in proportion to the amount of Rg1 used in the treatment (Figure 1F). This result indicates that Rg1 treatment inhibits PRRSV replication, and 10 μM Rg1 could inhibit Nsp9 mRNA and N protein expression. Therefore, PRRSV inhibition kinetics by Rg1 in Marc-145 cells and PAMs are further analyzed at 10 μM and 50 μM. Nsp9 mRNA expression, representing the PRRSV replication rate in the treated groups, was compared to that in the DMSO-treated control (Figure 1G). The results reveal that the decrease in Nsp9 mRNA is more pronounced from 24 h.p.i. (hours post infection) to 48 h.p.i. in PAMs, and from 36 h.p.i. to 72 h.p.i. in Marc-145 cells. Collectively, these results suggest that Rg1 treatment suppresses PRRSV infection.

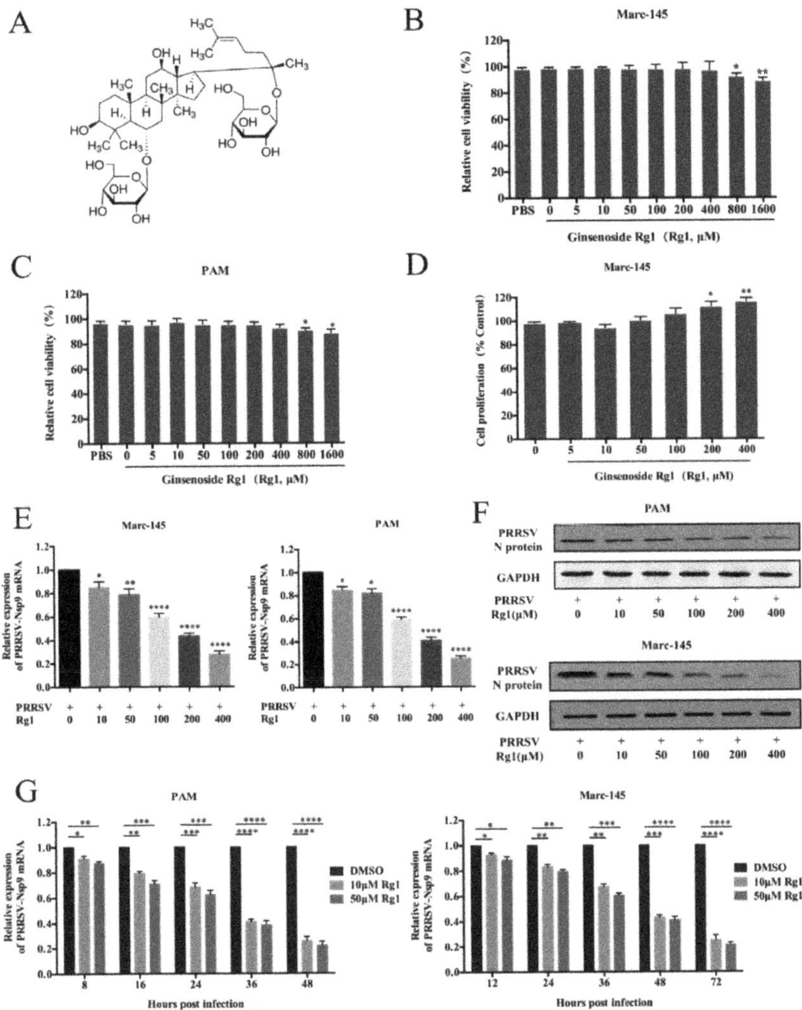

Figure 1. Cytotoxicity and anti-PRRSV activity of Rg1 in Marc-145 cells and PAMs. (**A**) The chemical structures of ginsenoside Rg1 (Rg1). (**B**,**C**) Cytotoxicity of Rg1 in Marc-145 cells (**B**) and PAMs (**C**) were analyzed by using the WST-1 assay. Results are showed as the relative cell viability of PAMs or Marc-145 cells cultured without Rg1 (set as 100%). (**D**) Rg1 affects the proliferation of Marc-145 cells. Cells were seeded and cultured without FBS for 12 h and the medium was replaced with DMEM contains 0, 5, 10, 50, 100, 200 and 400 µM Rg1 respectively. (**E**,**F**) PRRSV XH-GD (0.1 MOI) infected Marc-145 cells or PAMs for 1 h at 37 °C, and then cells were cultured in DMEM or RPMI 1640 supplemented with 2% FBS and indicated concentrations of Rg1. The samples were collected at 48 hpi to analyze PRRSV Nsp9 mRNA levels in different groups by RT-PCR (**E**). N protein expression levels in cells treated with different concentrations of Rg1 were detected by western blot (**F**). (**G**) Marc-145 cells and PAMs infected with PRRSV XH-GD (0.1 MOI) for 1 h at 37 °C and then cultured in fresh medium supplemented with 10 or 50 µM Rg1. The expression levels of PRRSV Nsp9 in Marc-145 cells and PAMs were detected by RT-PCR analysis at the indicated time points. Each data represents results of three independent experiments (means ± SD). Significant differences compared with the control group are denoted by * ($p < 0.05$), ** ($p < 0.01$), *** ($p < 0.001$) and **** ($p < 0.0001$).

3.2. Rg1 Treatment Affect PRRSV Attachment, Replication and Release in Marc-145 Cells

To explore the effects of Rg1 upon the PRRSV life cycle, virus entry was analyzed in cell attachment (from 0–2 h.p.i.) and internalization assays (from 2–5 h.p.i.), virus replication was analyzed between 6–10 h.p.i. and virus assembly or release was determined after 12 h.p.i., based on the ratio of Nsp9 mRNA in infected cells and supernatant. The experimental strategy was as previously described [14]. The pre-treatment of Marc-145 with Rg1 does not reduce viral RNA levels, suggesting that Rg1 does not affect the susceptibility of Marc-145 cell to PRRSV significantly. For virus attachment, the experiment was designed to allow virus binding, but not cellular internalization, and the results show that the Nsp9 mRNA levels in cell lysates are reduced both in the 10 µM and 50 µM treatment groups (Figure 2A). Virus internalization was analyzed from 2–5 h.p.i. in the presence of Rg1, and the Nsp9 expression levels only decrease in the 50 µM group (Figure 2A). The inhibitory effect on PRRSV replication is significant in both the 10 µM and 50 µM groups (Figure 2A). As described previously, PRRSV progeny viruses are released 8 h.p.i. [13,28]. In addition, our results of PRRSV inhibition kinetics by Rg1 in Marc-145 cells indicate that viral mRNA significantly reduces from 12 h.p.i. to 72 h.p.i. (Figure 1G). Thus, we evaluated the virus release rate with Rg1 treatment at 12 h.p.i. Marc-145 cells were infected for 12 h at 37 °C, and the cells were then cultured in DMEM containing 10 µM or 50 µM Rg1 for another 2 h. The Nsp9 mRNA levels in infected cells and culture supernatants were quantified by RT-PCR, and the ratio of cell/supernatant Nsp9 corresponds to the virus progeny release rate. The results indicate that this PRRSV virus released from Marc-145 cells is remarkably suppressed by Rg1 treatment (Figure 2B).

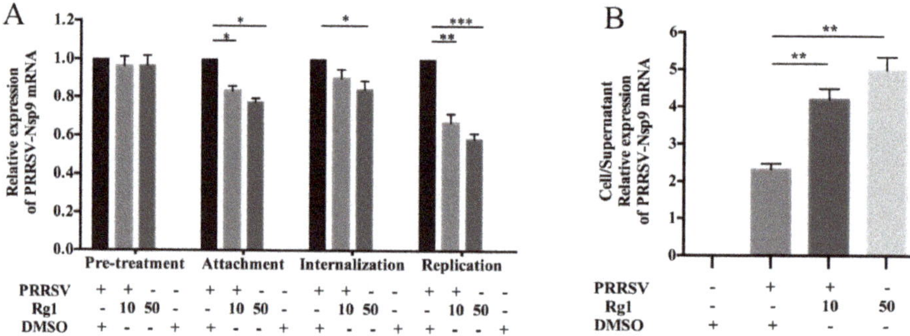

Figure 2. Inhibitory effects on the virus lifecycle of Rg1 in Marc-145 cells. In the Pre-treatment assay, Marc-145 cells were pretreated with DMEM supplemented with 10 or 50 µM Rg1 for 2 h, then cells were washed twice with PBS before being infected with type 2 PRRSV XH-GD (0.1 MOI), and then samples were collected at 48 h.p.i. For the attachment and internalization assay, Marc-145 cells were pre-cultivated at 4 °C for 1 h and then infected with virus (0.1 MOI) at 4 °C for 2 h. During virus attachment upon PRRSV infection, cells were cultured with DMEM or DMEM containing 10 or 50 µM Rg1 to analyze PRRSV Nsp9 mRNA level. Marc-145 cells were infected with XH-GD at 4 °C for 2 h and then cultured with or without Rg1 for 3 h at 37 °C. To avoid interference of other steps of viral lifecycle on replication assay, Marc-145 cells were infected with XH-GD for 6 h and then incubated with DMEM with or without 10 or 50 µM Rg1 at 37 °C, and samples were collected at 4 h.p.i. In all of the trials, GAPDH was used as a housekeeping gene for normalization, and cells treated with 0.4% DMSO was used as a reference control. (**A**) The effect of Rg1 on viral attachment, internalization, replication, and Rg1 pretreatment was analyzed by evaluating Nsp9 mRNA expression levels. (**B**) The effect of Rg1 on PRRSV release was detected by the ratio of Nsp9 RNA copy numbers in the supernatant and the cell lysate detected by qPCR. The analysis above was performed in triplicate. Statistical significance is denoted by * $p < 0.05$, ** $p < 0.01$, and *** $p < 0.001$.

3.3. The Anti-PRRSV Ability of Rg1 Treatment Was Effective in Marc-145 Cells and PAMs Infected with the HP-PRRSV, NADC30-Like and Classical Strains

In view of the high genetic variation among strains of type 2 PRRSV, we tested whether Rg1 possesses antiviral activity against broad lineages of strains. The antiviral effects of Rg1 against PRRSV strains, including JXA1 (HP-PRRSV; lineage 8.7), XH-GD (HP-PRRSV; lineage 8.7), VR2332 (classical strain; lineage 5.1) and HNLY (NADC30-like strain, lineage 1), in Marc-145 cells were examined by IFA and EC_{50}. As shown in Figure 3A, the four lineages of PRRSV infection in Marc-145 cells were significantly inhibited by Rg1 in a dose-dependent manner. Therefore, EC_{50} of Rg1 against JXA1, XH-GD, HNLY and VR2332 infection in Marc-145 cells was calculated respectively. Results indicate that the values vary significantly, ranging from 55.05 to 94.21 µM among the four PRRSV strains by analyzing infection rate from IFA images (Table 2). To evaluate the inhibition effect of Rg1 on virus replication of four PRRSV, the growth curves of these strains were generated in Marc-145 cells. The results indicate that the inhibitory effect of Rg1 on the production of progeny virus is mainly seen at the plateau phase, and the antiviral activity is most obvious at 50 µM in all four PRRSV strains (Figure 3B). Moreover, the growth curves of the four PRRSV strains treated with Rg1 was further determined in PAMs, and the results display a similar decreasing trend (Figure 3C). These data indicate that Rg1 possesses an inhibitory effect on PRRSV infection and that this effect is observed in a broad range of PRRSV lineages.

Table 2. Inhibitory activity of Ginsenoside Rg1 against PRRSV infection in MARC-145 cells.

	PRRSV Strain			
	XH-GD	JXA1	HNLY	VR2332
EC_{50} (µM) [a]	75.05 ± 13.52	71.33 ± 13.43	94.21 ± 8.27	55.05 ± 4.535

[a] the concentration required to protect 50% cells from PRRSV infection by counting cells from IFA images.

3.4. Rg1 Treatment Significantly Reduced the Pro-Inflammatory Cytokine mRNA Levels Induced by PRRSV Infection in Both Marc-145 Cells and PAMs

In view of the pro-inflammatory cytokines triggered by PRRSV, this contributes to its pathogenicity, and the role Rg1 plays in relieving inflammatory responses [19]. RT-PCR was used to determine whether Rg1 treatment alleviates the expression of pro-inflammatory factors, including IL-1β, IL-6, IL-8 and TNF-α, induced by PRRSV infection. The data indicates that Rg1 treatment significantly ($p < 0.05$) reduces the mRNA levels of pro-inflammatory factors increased by PRRSV infection both in Marc-145 cells and PAMs. In Marc-145 cells, mRNA level of IL-1β, IL-6 and TNF-α are significantly reduced by Rg1 from 12 to 24 h upon PRRSV infection, while, IL-8 is decreased by Rg1 at 18 and 24 h (Figure 4A). Further, pro-inflammatory cytokines assay was performed in PAMs and the results showed that IL-6, IL-8 and TNF-α triggered by PRRSV lowered by Rg1 pronouncedly from 12 to 24 h, while the reduction of mRNA level of IL-1β was seen at 18 and 24 h (Figure 4B). Although the inhibitory effect of Rg1 on each pro-inflammatory factor in Marc-145 cells and PAMs was not strictly consistent, the decline trend was both obvious at 18 h and 24 h. The results demonstrate that Rg1 could relieve the inflammatory responses caused by PRRSV infection via decreasing the mRNA expression of pro-inflammatory cytokines

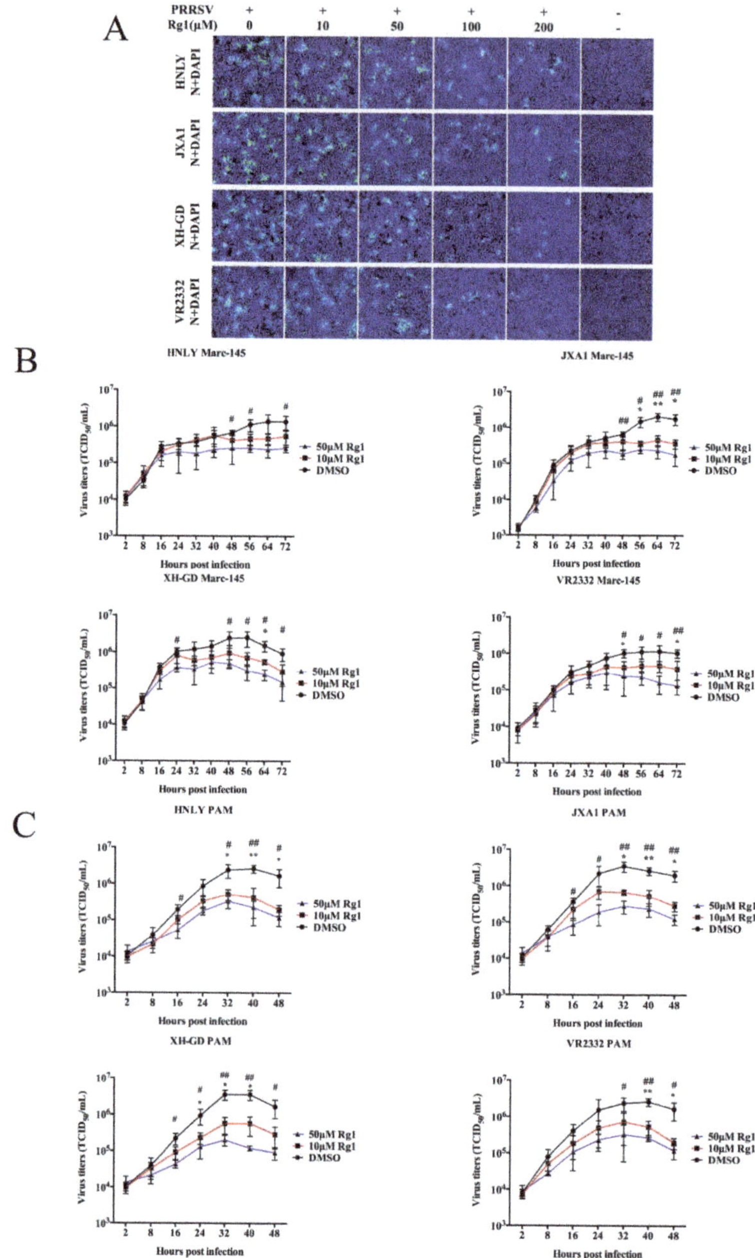

Figure 3. The antiviral activity of Rg1 against different lineages of type 2 PRRSV. (**A**) Antiviral activity of Rg1 against PRRSV strains (HP-PRRSV XH-GD and JXA1, classical VR2332 and NADC30-like strain HNLY) was determined in Marc-145 cells by IFA. Marc-145 cells were seeded in 12-well plates and infected with four type 2 PRRSV strain (0.1 MOI) respectively, and then incubated with DMEM supplemented with indicated concentration of Rg1. N protein was used as indicator of PRRSV infection, and the IFA detection of it was performed at 48 h.p.i. by using mouse anti-N protein antibody and goat anti-mouse IgG Alexa Fluor. Nuclei were counterstained with DAPI (blue). These images above

represent three independent IFA trials with similar results. Magnification, 100 ×. (**B**,**C**) The inhibitory effect of Rg1 on PRRSV replication in Marc-145 cells (**B**), and PAMs (**C**). PRRSV replication was analyzed by virus growth curve. Marc-145 cells and PAMs were seeded in 6-well plates and infected with four PRRSV strain (0.1 MOI) for 1 h at 37 °C respectively and then cultured with DMEM or RPMI 1640 supplemented with 10 or 50 μM Rg1 or DMSO. Cell supernatants (200 μL) of each well were collected at indicated hours of post-infection. Growth assays for each group were performed in triplicate, and the resulting titers were determined as $TCID_{50}/_{mL}$ (the 50% tissue culture infectious dose per mL) and the data are shown as the means ± SD. T-test was applied to perform statistical analysis. Statistical significance between 10 μM Rg1 and DMSO is denoted by $*\, p < 0.05$ and $**\, p < 0.01$, and significance between 50 μM Rg1 and DMSO is denoted by $\#\, p < 0.05$ and $\#\#\, p < 0.01$.

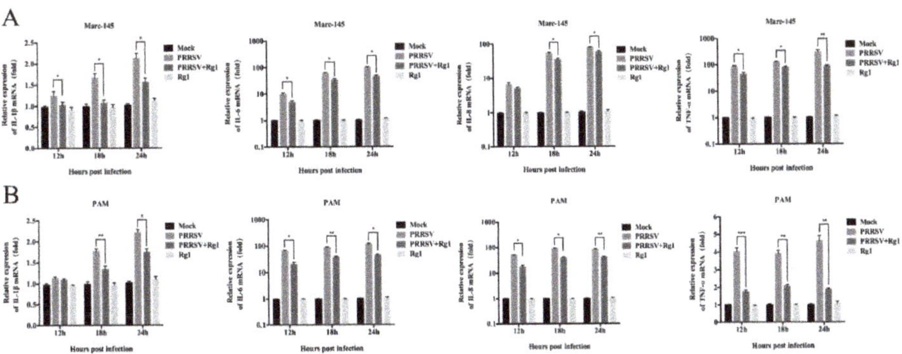

Figure 4. Rg1 suppresses inflammatory cytokines mRNA expression in infected PAMs and Marc-145 cells. (**A**) For PRRSV+Rg1 group, the XH-GD strain (0.1 MOI) infected Marc-145 cells for 1 h and then cultured in DMEM supplemented with Rg1 (200 μM), and cells infected with virus or treated with Rg1 (200 μM) were termed as PRRSV group and Rg1 group, respectively. Cells in the mock group were grown in DMEM containing 0.4% DMSO. Cell samples were collected to extract total RNA at 12, 18, and 24 h.p.i. The relative expression of IL-6, IL-8, IL-1β and TNF-α was analyzed by RT-PCR. GAPDH was used as internal control to normalize values. (**B**) PAMs were cultured with RPMI 1640 and treated as described above. The data of each trial represents three independent experiments and the values are shown as the means ± SD. T-test was applied to perform statistical analysis and the significance was indicated by asterisk in the graphs. Statistical significance is denoted by $*\, p < 0.05$, $**\, p < 0.01$, and $***\, p < 0.001$.

3.5. PRRSV-Infection Triggered NF-κB Activation Was Inhibited by Rg1 Treatment in Marc-145 Cells

In the results above, Rg1 decreases the PRRSV-triggered mRNA level of pro-inflammatory cytokine. It is known that NF-κB is one of the key transcription factors regulating the production of pro-inflammatory factors, such as IL-6 and IL-8 [29]. PRRSV is proven to activate the NF-κB pathway to enhance viral replication [30]. Therefore, the effect of Rg1 treatment on PRRSV mediated NF-κB activation was analyzed by western blotting and immunofluorescence staining, and LPS stimulation, which activates NF-κB, was used as a control to show whether it associated with virus replication or indirectly through cellular response. We find that the increased phosphorylation of p65 induced by PRRSV infection or LPS is weakened by Rg1 treatment, and it also inhibits IκB degradation (Figure 5A). Although the phosphorylation of IκB-α, which is related to IκB degradation, was not significantly enhanced by PRRSV infection compared with mock, it was reduced by Rg1 treatment (Figure 5A). Moreover, the cellular localization of NF-κB was determined by p65 immunofluorescence staining (red) in Marc-145 cells, and the nuclei were stained with DAPI (blue) (Figure 5B). The results are consistent with the phosphorylation level of p65 by western blotting. In the Mock and Rg1 groups, P65 is mainly localized in cytoplasm and the phosphorylation level is lower. PRRSV triggers P65 phosphorylation to facilitate its translocation into nuclear, and this process is significantly reduced after Rg1 treatment. Taken together, these data indicate that Rg1 treatment decreases PRRSV-mediated NF-κB activation and IκB degradation, and a schematic of this event is presented in Figure 5C.

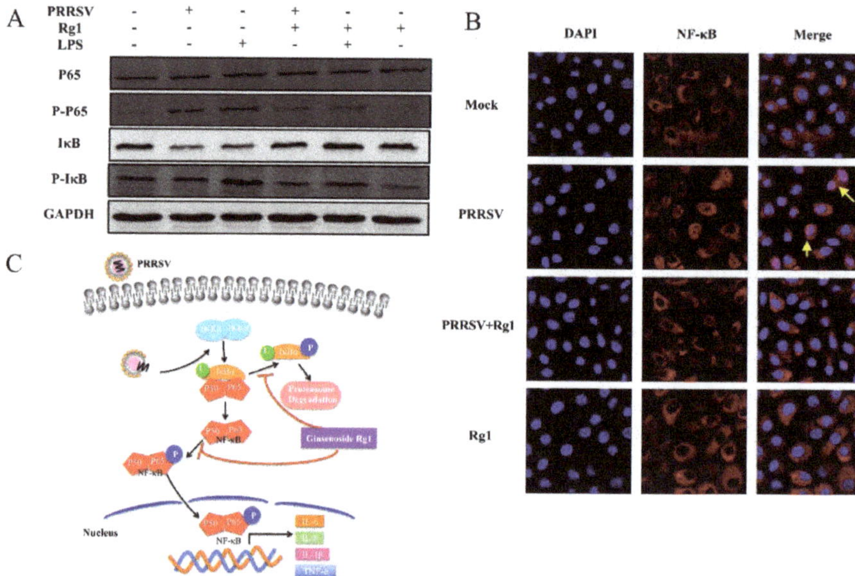

Figure 5. Rg1 inhibits the NF-κB pathway activated by PRRSV infection. (**A**) The expression and phosphorylation level of proteins involved in NF-κB pathway are analyzed in uninfected (Mock) and PRRSV XH-GD (0.1 MOI) infected Marc-145 cells treated with or without Rg1 (200 μM), samples were collected at 24 h.p.i. As a positive control, cells were cultured in DMEM and supplemented with LPS (2.5 μg/mL) at 37 °C for 6 h, and then the medium was changed to medium containing 0 or 200 μM Rg1 for 18 h. The western blotting data of each target protein represents three independent experiments with similar results. (**B**) Marc-145 cells were grown on glass cover slips and cultured in medium at 37 °C for 24 h, and then infected with PRRSV XH-GD (0.1 MOI). After virus infection, cells were incubated in fresh DMEM supplemented with or without 200 μM Rg1 for 24 h. Cells were washed twice with PBS and performed immunostaining by using anti-P65 antibody and red-fluorescent Alexa Fluor 594-conjugated goat anti-mouse IgG antibody. Nuclei were counterstained with DAPI. P65 protein deposited in the nucleus was indicated by yellow arrow. (**C**) Schematic model of Rg1 affect NF-κB signaling pathway upon PRRSV infection.

3.6. Rg1 Treatment Exhibits Antiviral Activity in Piglets

Given the antiviral activity of Rg1 against type 2 PRRSV that we analyzed in vitro, we wondered whether Rg1 could be used in antiviral therapy in piglets. HP-PRRSV JXA1 (EF112445) strain is a prototypical HP-PRRSV strain, characterized by discontinuous deletion of 30 amino acids in nonstructural protein 2, causes typical clinical symptom and pathological changes with high morbidity and mortality in piglets [9]. Here, we have evaluated the antiviral activity of Rg1 against JXA1 strain in vitro and calculated the EC$_{50}$ of it (Table 2). Therefore, HP-PRRSV JXA1 was used to evaluate the anti-PRRSV effect of Rg1 in piglets.

3.6.1. Clinical Signs and Mortality

After being challenged with virulent HP-PRRSV JXA1, all piglets in the PRRSV group (piglets challenged with JXA1) displayed high rectal temperature (Figure 6A) and exhibited typical severe clinical signs, including in-appetence, lethargy, dyspnea, periocular and eyelid edema and hyperspasmia. Piglets treated with Rg1 after JXA1 challenge (PRRSV+Rg1 group) showed moderate anorexia and depression. Animals in Rg1 group (piglets only treated with Rg1) and mock group (piglets only treated with DMEM) behaved normally during the course of the experiment. The scores of the evaluation of clinical signs in the PRRSV+Rg1 group were significantly lower than PRRSV group at 7 d.p.i. ($p < 0.05$) (Table 3), a time-point of peak period of morbidity and mortality. All of the piglets displayed increased rectal temperature higher than 40 °C after challenge with HP-PRRSV JXA1, while piglets treated with Rg1 showed gradual descending since 8 dpi (Figure 6A). The survival rate of piglets in PRRSV group is 0% to 11 dpi, however, 40% animals in the PRRSV+Rg1 group (piglets treated with Rg1 after JXA1 challenge) survived to 14 dpi (Figure 6B). Since piglets showed different loss of appetite during infection course, the body weight of PRRSV and PRRSV+Rg1 displayed obviously drop after challenge, while the body weight of piglets in PRRSV+Rg1 displayed slower drop trend and begin to rise at 10 dpi (Figure 6C).

Table 3. The scores of clinical signs and lung lesions of the different groups at 7 d.p.i.[a].

Groups	Clinical Signs Scores (±S.D.) [b]	Lung Lesions Scores (±S.D.) [c]
PRRSV (Challenge control)	13.270 ± 1.8931^1	76.0 ± 18.171^1
PRRSV+Rg1	8.807 ± 2.5302^1	34.0 ± 11.402^1
Mock	0^2	0^2
Rg1	0^2	0^2

[a] Values followed by letters 1 represents significant difference ($p < 0.05$) between PRRSV and PRRSV+Rg1; letter 2 represents significant difference ($p < 0.05$) between Mock/Rg1 and PRRSV/PRRSV+Rg1. [b] The clinical sign score was calculated by sum of behavior performance, appetite, respiration according to the extent of severity. [c] Analyzing the percentage of the macroscopic lesion features of pneumonia in entire lung.

3.6.2. Pathological Examination

As previously described, piglets infected with JXA1 began to die within 6–8 d.p.i. [9], and animals in the PRRSV group start to die at 6 d.p.i. in our study. Therefore, three living pigs of each group were euthanized for pathological examination at 7 d.p.i. to evaluate whether Rg1 treatment relieved lung injury.

The lungs of pigs in PRRSV+Rg1 group showed fewer pathological lesions and got significantly lower scores of macroscopic injury of lungs than those in the PRRSV group as shown in Table 3 ($p < 0.05$). Further histological examination of lung tissue showed that PRRSV+Rg1 group exhibited moderate interstitial pneumonia, while the PRRSV group were characterized by thickened alveolar walls, interstitial fibro-plastic proliferation and intensive mononuclear cell infiltration, which revealed the severe viral pneumonia (Figure 6D). At 15 dpi, all of the living pigs were euthanized for pathological examination. Due to no piglet survived at 14 dpi in the PRRSV group, the deceased pigs at 10 dpi of it were collected to perform pathological examination. The results displayed the same trend as 7 dpi. In the PRRSV+Rg1 group, the survived piglets at 15 dpi exhibited mild pneumonia compared with that of the mock group and Rg1 control group (Figure 6D).

3.6.3. Viremia and Tissue Viral Load

Blood samples of the piglets were collected at 2, 4, 7, 10, 14 dpi to determine viral load and N protein antibody of PRRSV. And, the viral load in tissues, include lung, lymph node and thymus were determined by qRT-PCR as described previously [31]. Since all of the animals in the PRRSV group died at 11 d.p.i., there was no viremia and tissue viral load analysis on it at 14 or 15 h.p.i. The data revealed that piglets treated with Rg1 showed significantly lower viremia, indicated by antibody against PRRSV and viral load in serum, in the blood than those in PRRSV group at 7 ($p < 0.05$) and 10 dpi ($p < 0.01$)

(Figure 6E,F). Meanwhile, PRRSV+Rg1 group displayed significantly reduced viral load in the lung, lymph node and thymus at 7 dpi ($p < 0.01$) (Figure 6G).

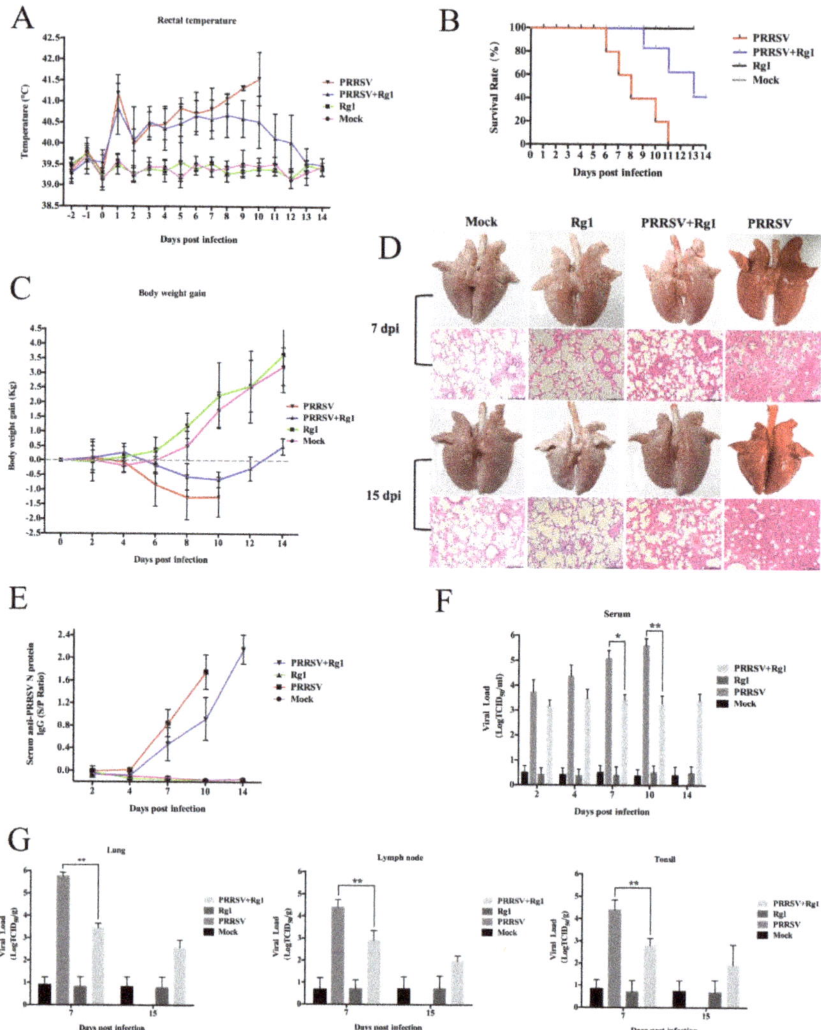

Figure 6. Rg1 exhibits anti-PRRSV activity in 4-week-piglet. (**A**) Daily rectal temperature of the pigs in the PRRSV, PRRSV+Rg1, Rg1 and mock groups. Rectal temperature reach or beyond 40 °C was defined as fever. (**B**) The mortality of each group was recorded daily and calculated as survival rate until 14 dpi. (**C**) The body weight gain of different groups during the experiment. (**D**) Severe lung lesions in the PRRSV group characterized by swelling, congestion, fibrosis, and inflammatory cell aggregates; however, in the PRRSV+Rg1 group, these index were moderate. Due to no piglet survived at 14 dpi in the challenge control group (PRRSV), the lung of the deceased pigs at 10 dpi was used to perform pathological analysis. (**E**) The anti-PRRSV antibody levels in serum at different time-points. The value of S/p ratio $\geq$ 0.4 was considered antibody positive. (**F**) The level of PRRSV mRNA in the serum was measured by real-time PCR. (**G**) The expression level of PRRSV mRNA level in lungs, lymph node and thymus was measured by qRT-PCR. Each tissue sample was measured three times, and the error bars represent the standard deviations of samples.

4. Discussion

PRRSV was identified in Europe in 1991 and emerged in United States in 1992 [32,33]. This virus is classified into European genotype (type 1) and North American genotype (type 2) [3], and the widespread outbreaks of PPRS in China are associated with constant evolution of viruses through high frequency of recombination and immune suppression events in recent decades. The protective immunity to PRRS elicited by current vaccines is effective only against homologous infections and exhibits partial protection from heterologous PRRSV. New antiviral therapeutic strategies became an urgent need. In the present study, we demonstrate that Rg1 suppressed broad lineages of type 2 PRRSV infection, including HP-PRRSV, classical strain and NADC30-like strains, both in Marc-145 cells and PAMs. It suggests that Rg1, as a natural herbal molecular, could be applied in broad-spectrum anti-PRRSV medicament.

PRRSV infection triggers the up-regulated release of IL-1β, IL-6, IL-8 and TNF-α [6,34,35]. These pro-inflammatory factors contribute to the stimulation of protective immune responses, while it also leads to the development of excessive systemic inflammatory reactions and causing inflammatory lesions [36]. Extensive studies have illustrated the potent anti-inflammatory effect of Rg1 treatment in various diseases by regulating inflammatory cytokine expression [37–39]. However, the anti-viral activity of Rg1 associated with modulating inflammatory response was poorly described. In our study, the results reveal that the expression level of several pro-inflammatory factors, including IL-1β, IL-6, IL-8 and TNF-α, is significantly reduced by Rg1 treatment in Marc-145 cell and PAMs upon PRRSV infection. It suggested that Rg1 moderate the PRRSV-induced inflammatory responses by decreasing mRNA expression of pro-inflammatory cytokines in vitro.

PRRSV infects host via membrane receptor mediated endocytosis process including virus attachment and binding, membrane fusion, and followed internalization [40,41]. Once the viral genome, single strand positive-sense RNA, is released into the cytoplasm, it substantially process translation to generate replication and transcription complex. Here, we performed analysis to determine whether Rg1 affect PRRSV lifecycle. The results indicated that pre-treatment of Rg1 could not affect virus host-cell tropism. Marc-145 cells treated with Rg1 during different period of infection process were analyzed respectively, the results showed virus attachment, internalization, and release were impaired (Figure 2). Meanwhile, the inhibitory effect or Rg1 on PRRSV replication stage was more obvious.

NF-κB, key transcription factor that regulates the activation of inflammatory cytokines, can be activated by virus infection, viral gene expression or by LPS stimulation [29], and it could be exploited by influenza viruses or type 1 HIV to sustain a high viral replication [42,43]. In view of the role of Rg1 played in suppressing pro-inflammatory factor expression triggered by PRRSV (Figure 4) and PRRSV-induced NF-κB activation facilitated its replication was demonstrated in previous report [30]. Therefore, we wondered Rg1 inhibited virus replication were associated with cellular process indirectly. In the present study, we identified that Rg1 alleviated PRRSV infection induced IκB degradation, phosphorylation level of p65 and p65 nuclear aggregation, important factors contribute to NF-κB activation, in Marc-145 cells (Figure 5). Moreover, our data indicated that Rg1 treatment suppressed NF-κB activity in LPS-treated Marc-145 cells (Figure 5), which was consistent with a previous study in mouse RAW 264.7 cells and macrophages [44]. This result suggested that the decreased NF-κB activation in Marc-145 cells, which treated with Rg1 upon PRRSV infection, was not due to a reduction in virus infection, it involved in the interaction between Rg1 and cellular process. Overall, the inhibitory effect on both IκB degradation and NF-κB nuclear translocation signaling contributes to the anti-PRRSV replication activity of Rg1. Possibly, it can be speculated that Rg1 mediated cellular process contributes to its anti-viral activity and it might be widely applied in other viral infection.

Recently, multiple Chinese traditional medicines were shown to possess anti-PRRSV ability in vitro [12–14]. However, the antiviral activities of these identified natural herbal extracts, active ingredients, compound or drugs have not been further evaluated in vivo. In present study, piglets challenged with HP-PRRSV JXA1 under Rg1 treatment showed increased survival rate, moderate pneumonia and lower serum and tissue viral loads compared to those in the PRRSV group (challenge control) (Figure 6). It suggests Rg1 might be a natural anti-PRRSV agent that could be considered as an adjuvant therapy in the pig herd production. Furthermore, it has been confirmed that ginseng extract inhibited virus infection including influenza virus and hepatitis C virus, and Rg1, which purified from the roots or stems of Panax notoginseng (PN) and Panax ginseng (PG), could also be used as an immunoadjuvant to improve immune responses [45,46]. In the current status, there are several Modified live Virus (MLV), which induced immune response did confer protection to vaccinated animals against homologous PRRSV challenge, licensed in various countries and extensively applied in pig farms [47,48]. However, existing evidence suggests that these MLV vaccines of type 1 or type 2 both stimulates limited humoral and cellular immunity and fail to against heterogeneous strain [49,50]. Therefore, it would be possible that ginseng extract used as a natural supplement in feeding management for disease prevention. Besides, it might be used in a combination with some direct-acting antivirals to achieve widely effective disease prevention or as immune-adjuvant of vaccination to reach immune enhancement. For the proper application of it, the pharmacokinetic data of ginseng extract in swine model needs further systematic study. Meanwhile, the optimum dosage and administration mode should be defined reasonably to avoid drug residues in meat production. Considering the economic benefit and most of the nutrient component were similar between PN and PG, PN might possess more potential value of application in the scaled raising of pigs.

In summary, our study demonstrated that ginsenoside Rg1 with low cytotoxicity and possess anti-PRRSV activity both in vitro and in piglets. And, it suppressed different lineages of type 2 PRRSV infections. These findings not only provide new insights into the molecular mechanism of Rg1 against PRRSV infection but also suggest a potential immune-modulatory and anti-viral agent in the control of the PRRS.

Author Contributions: Conceptualization, G.-h.Z. and Z.-q.Y.; Methodology, Z.-q.Y. and Y.C.; Software, H.-y.Y.; Validation, Q.L. and R.-t.Z.; Formal Analysis, M.-k.C.; Investigation, Z.-q.Y., H.-y.Y., J.M., C.-x.Q., G.L., Q.D., Y.-j.C., and X.-l.H.; Resources, G.-h.Z., H.W. and K.T.; Data Curation, Y.-f.W.; Writing—Original Draft Preparation, Z.-q.Y.; Writing—Review & Editing, G.-h.Z.; Visualization, Z.-q.Y.; Supervision, H.W.; Project Administration, G.-h.Z. and Z.-q.Y.

Funding: This project was supported by the National Natural Science Foundation of China (Grant No. 31872489), the Modern Agro-industry Technology Research System (Grant number CARS-35) and the Natural Science Foundation of Guangdong province (Grant No. 2018A030313712).

Conflicts of Interest: The authors declare no conflict of interest.

References

1.	Music, N.; Gagnon, C.A. The role of porcine reproductive and respiratory syndrome (PRRS) virus structural and non-structural proteins in virus pathogenesis. *Anim. Health Res. Rev.* **2010**, *11*, 135–163. [CrossRef] [PubMed]

2.	Kappes, M.A.; Faaberg, K.S. PRRSV structure, replication and recombination: Origin of phenotype and genotype diversity. *Virology* **2015**, *479*, 475–486. [CrossRef] [PubMed]

3.	Nelsen, C.J.; Murtaugh, M.P.; Faaberg, K.S. Porcine reproductive and respiratory syndrome virus comparison: Divergent evolution on two continents. *J. Virol.* **1999**, *73*, 270–280. [PubMed]

4.	Shi, M.; Lam, T.T.; Hon, C.C.; Murtaugh, M.P.; Davies, P.R.; Hui, R.K.; Li, J.; Wong, L.T.; Yip, C.W.; Jiang, J.W.; et al. Phylogeny-based evolutionary, demographical, and geographical dissection of North American type 2 porcine reproductive and respiratory syndrome viruses. *J. Virol.* **2010**, *84*, 8700–8711. [CrossRef] [PubMed]

5.	Shi, M.; Lemey, P.; Singh Brar, M.; Suchard, M.A.; Murtaugh, M.P.; Carman, S.; D'Allaire, S.; Delisle, B.; Lambert, M.E.; Gagnon, C.A.; et al. The spread of type 2 Porcine Reproductive and Respiratory Syndrome Virus (PRRSV) in North America: A phylogeographic approach. *Virology* **2013**, *447*, 146–154. [CrossRef] [PubMed]

6.	Ait-Ali, T.; Wilson, A.D.; Westcott, D.G.; Clapperton, M.; Waterfall, M.; Mellencamp, M.A.; Drew, T.W.; Bishop, S.C.; Archibald, A.L. Innate immune responses to replication of porcine reproductive and respiratory syndrome virus in isolated Swine alveolar macrophages. *Viral Immunol.* **2007**, *20*, 105–118. [CrossRef] [PubMed]

7.   Sun, Y.K.; Li, Q.; Yu, Z.Q.; Han, X.L.; Wei, Y.F.; Ji, C.H.; Lu, G.; Ma, C.Q.; Zhang, G.H.; Wang, H. Emergence of novel recombination lineage 3 of porcine reproductive and respiratory syndrome viruses in Southern China. *Transbound. Emerg. Dis.* **2019**, *66*, 578–587. [CrossRef]

8.   Sun, Z.; Wang, J.; Bai, X.; Ji, G.; Yan, H.; Li, Y.; Wang, Y.; Tan, F.; Xiao, Y.; Li, X.; et al. Pathogenicity comparison between highly pathogenic and NADC30-like porcine reproductive and respiratory syndrome virus. *Arch. Virol.* **2016**, *161*, 2257–2261. [CrossRef]

9.   Tian, K.; Yu, X.; Zhao, T.; Feng, Y.; Cao, Z.; Wang, C.; Hu, Y.; Chen, X.; Hu, D.; Tian, X.; et al. Emergence of fatal PRRSV variants: Unparalleled outbreaks of atypical PRRS in China and molecular dissection of the unique hallmark. *PLoS ONE* **2007**, *2*, e526. [CrossRef]

10.   Tong, G.Z.; Zhou, Y.J.; Hao, X.F.; Tian, Z.J.; An, T.Q.; Qiu, H.J. Highly pathogenic porcine reproductive and respiratory syndrome, China. *Emerg. Infect. Dis.* **2007**, *13*, 1434–1436. [CrossRef]

11.   Zhou, L.; Wang, Z.; Ding, Y.; Ge, X.; Guo, X.; Yang, H. NADC30-like Strain of Porcine Reproductive and Respiratory Syndrome Virus, China. *Emerg. Infect. Dis.* **2015**, *21*, 2256–2257. [CrossRef] [PubMed]

12.   Yang, Q.; Gao, L.; Si, J.; Sun, Y.; Liu, J.; Cao, L.; Feng, W.H. Inhibition of porcine reproductive and respiratory syndrome virus replication by flavaspidic acid AB. *Antivir. Res.* **2013**, *97*, 66–73. [CrossRef] [PubMed]

13.   Duan, E.; Wang, D.; Fang, L.; Ma, J.; Luo, J.; Chen, H.; Li, K.; Xiao, S. Suppression of porcine reproductive and respiratory virus proliferation by glycyrrhizin. *Antivir. Res.* **2015**, *120*, 122–125. [CrossRef] [PubMed]

14.   Zhang, M.; Du, T.; Long, F.; Yang, X.; Sun, Y.; Duan, M.; Zhang, G.; Liu, Y.; Zhou, E.M.; Chen, W.; et al. Platycodin D Suppresses Type 2 Porcine Reproductive and Respiratory Syndrome Virus in Primary and Established Cell Lines. *Viruses* **2018**, *10*, 657. [CrossRef] [PubMed]

15.   Im, D.S.; Nah, S.Y. Yin and Yang of ginseng pharmacology: Ginsenosides vs gintonin. *Acta Pharm. Sin.* **2013**, *34*, 1367–1373. [CrossRef] [PubMed]

16.   Qu, D.F.; Yu, H.J.; Liu, Z.; Zhang, D.F.; Zhou, Q.J.; Zhang, H.L.; Du, A.F. Ginsenoside Rg1 enhances immune response induced by recombinant Toxoplasma gondii SAG1 antigen. *Vet. Parasitol.* **2011**, *179*, 28–34. [CrossRef] [PubMed]

17.   Su, F.; Yuan, L.; Zhang, L.; Hu, S. Ginsenosides Rg1 and Re act as adjuvant via TLR4 signaling pathway. *Vaccine* **2012**, *30*, 4106–4112. [CrossRef] [PubMed]

18.   Sun, J.; Song, X.; Hu, S. Ginsenoside Rg1 and aluminum hydroxide synergistically promote immune responses to ovalbumin in BALB/c mice. *Clin. Vaccine Immunol.* **2008**, *15*, 303–307. [CrossRef] [PubMed]

19.   Liu, Q.; Kou, J.P.; Yu, B.Y. Ginsenoside Rg1 protects against hydrogen peroxide-induced cell death in PC12 cells via inhibiting NF-kappaB activation. *Neurochem. Int.* **2011**, *58*, 119–125. [CrossRef]

20.   Chen, Y.Q.; Rong, L.; Qiao, J.O. Antiinflammatory effects of Panax notoginseng saponins ameliorate acute lung injury induced by oleic acid and lipopolysaccharide in rats. *Mol. Med. Rep.* **2014**, *10*, 1400–1408. [CrossRef]

21.   Chen, Y.; He, S.; Sun, L.; Luo, Y.; Sun, Y.; Xie, J.; Zhou, P.; Su, S.; Zhang, G. Genetic variation, pathogenicity, and immunogenicity of highly pathogenic porcine reproductive and respiratory syndrome virus strain XH-GD at different passage levels. *Arch. Virol.* **2016**, *161*, 77–86. [CrossRef] [PubMed]

22.   Ma, S.C.; Du, J.; But, P.P.; Deng, X.L.; Zhang, Y.W.; Ooi, V.E.; Xu, H.X.; Lee, S.H.; Lee, S.F. Antiviral Chinese medicinal herbs against respiratory syncytial virus. *J. Ethnopharmacol.* **2002**, *79*, 205–211. [CrossRef]

23.   Xie, J.X.; Cui, T.T.; Cui, J.; Chen, Y.; Zhang, M.Z.; Zhou, P.; Deng, S.C.; Su, S.; Zhang, G.H. Epidemiological and evolutionary characteristics of the PRRSV in Southern China from 2010 to 2013. *Microb. Pathog.* **2014**, *75*, 7–15. [CrossRef] [PubMed]

24.   Liu, X.; Li, Y.F.; Lu, Q.; Bai, J.; Wang, X.Y.; Jiang, P. A new porcine reproductive and respiratory syndrome virus strain with highly conserved molecular characteristics in its parental and attenuated strains. *Virus Genes* **2014**, *49*, 259–268. [CrossRef] [PubMed]

25.   Zhao, K.; Gao, J.C.; Xiong, J.Y.; Guo, J.C.; Yang, Y.B.; Jiang, C.G.; Tang, Y.D.; Tian, Z.J.; Cai, X.H.; Tong, G.Z.; et al. Two Residues in NSP9 Contribute to the Enhanced Replication and Pathogenicity of Highly Pathogenic Porcine Reproductive and Respiratory Syndrome Virus. *J. Virol.* **2018**, *92*, e02209-17. [CrossRef] [PubMed]

26.   Wang, Y.; Liang, Y.; Han, J.; Burkhart, K.M.; Vaughn, E.M.; Roof, M.B.; Faaberg, K.S. Attenuation of porcine reproductive and respiratory syndrome virus strain NM184 using chimeric construction with vaccine sequence. *Virology* **2008**, *371*, 418–429. [CrossRef]

27. Halbur, P.G.; Paul, P.S.; Meng, X.J.; Lum, M.A.; Andrews, J.J.; Rathje, J.A. Comparative pathogenicity of nine US porcine reproductive and respiratory syndrome virus (PRRSV) isolates in a five-week-old cesarean-derived, colostrum-deprived pig model. *J. Vet. Diagn. Investig.* **1996**, *8*, 11–20. [CrossRef]
28. Nauwynck, H.J.; Duan, X.; Favoreel, H.W.; Van Oostveldt, P.; Pensaert, M.B. Entry of porcine reproductive and respiratory syndrome virus into porcine alveolar macrophages via receptor-mediated endocytosis. *J. Gen. Virol.* **1999**, *80 Pt 2*, 297–305. [CrossRef]
29. Hayden, M.S.; Ghosh, S. Shared principles in NF-kappaB signaling. *Cell* **2008**, *132*, 344–362. [CrossRef]
30. Wang, D.; Cao, L.; Xu, Z.; Fang, L.; Zhong, Y.; Chen, Q.; Luo, R.; Chen, H.; Li, K.; Xiao, S. MiR-125b reduces porcine reproductive and respiratory syndrome virus replication by negatively regulating the NF-kappaB pathway. *PLoS ONE* **2013**, *8*, e55838.
31. Chen, Y.; Yu, Z.Q.; Yi, H.Y.; Wei, Y.F.; Han, X.L.; Li, Q.; Ji, C.H.; Huang, J.M.; Deng, Q.W.; Liu, Y.X.; et al. The phosphorylation of the N protein could affect PRRSV virulence in vivo. *Vet. Microbiol.* **2019**, *231*, 226–231. [CrossRef] [PubMed]
32. Benfield, D.A.; Nelson, E.; Collins, J.E.; Harris, L.; Goyal, S.M.; Robison, D.; Christianson, W.T.; Morrison, R.B.; Gorcyca, D.; Chladek, D. Characterization of swine infertility and respiratory syndrome (SIRS) virus (isolate ATCC VR-2332). *J. Vet. Diagn. Investig.* **1992**, *4*, 127–133. [CrossRef] [PubMed]
33. Wensvoort, G.; Terpstra, C.; Pol, J.M.; ter Laak, E.A.; Bloemraad, M.; de Kluyver, E.P.; Kragten, C.; van Buiten, L.; den Besten, A.; Wagenaar, F.; et al. Mystery swine disease in The Netherlands: The isolation of Lelystad virus. *Vet. Q.* **1991**, *13*, 121–130. [CrossRef] [PubMed]
34. Lee, S.M.; Kleiboeker, S.B. Porcine arterivirus activates the NF-kappaB pathway through IkappaB degradation. *Virology* **2005**, *342*, 47–59. [CrossRef] [PubMed]
35. Lunney, J.K.; Fritz, E.R.; Reecy, J.M.; Kuhar, D.; Prucnal, E.; Molina, R.; Christopher-Hennings, J.; Zimmerman, J.; Rowland, R.R. Interleukin-8, interleukin-1beta, and interferon-gamma levels are linked to PRRS virus clearance. *Viral. Immunol.* **2010**, *23*, 127–134. [CrossRef] [PubMed]
36. Thanawongnuwech, R.; Thacker, B.; Halbur, P.; Thacker, E.L. Increased production of proinflammatory cytokines following infection with porcine reproductive and respiratory syndrome virus and Mycoplasma hyopneumoniae. *Clin. Diagn. Lab Immunol.* **2004**, *11*, 901–908. [CrossRef]
37. Bao, S.; Zou, Y.; Wang, B.; Li, Y.; Zhu, J.; Luo, Y.; Li, J. Ginsenoside Rg1 improves lipopolysaccharide-induced acute lung injury by inhibiting inflammatory responses and modulating infiltration of M2 macrophages. *Int. Immunopharm.* **2015**, *28*, 429–434. [CrossRef]
38. Cao, L.; Zou, Y.; Zhu, J.; Fan, X.; Li, J. Ginsenoside Rg1 attenuates concanavalin A-induced hepatitis in mice through inhibition of cytokine secretion and lymphocyte infiltration. *Mol. Cell. Biochem.* **2013**, *380*, 203–210. [CrossRef]
39. Ning, C.; Gao, X.; Wang, C.; Huo, X.; Liu, Z.; Sun, H.; Yang, X.; Sun, P.; Ma, X.; Meng, Q.; et al. Protective effects of ginsenoside Rg1 against lipopolysaccharide/d-galactosamine-induced acute liver injury in mice through inhibiting toll-like receptor 4 signaling pathway. *Int. Immunopharm.* **2018**, *61*, 266–276. [CrossRef]
40. Vanderheijden, N.; Delputte, P.L.; Favoreel, H.W.; Vandekerckhove, J.; Van Damme, J.; van Woensel, P.A.; Nauwynck, H.J. Involvement of sialoadhesin in entry of porcine reproductive and respiratory syndrome virus into porcine alveolar macrophages. *J. Virol.* **2003**, *77*, 8207–8215. [CrossRef]
41. Van Breedam, W.; Delputte, P.L.; Van Gorp, H.; Misinzo, G.; Vanderheijden, N.; Duan, X.; Nauwynck, H.J. Porcine reproductive and respiratory syndrome virus entry into the porcine macrophage. *J. Gen. Virol.* **2010**, *91 Pt 7*, 1659–1667. [CrossRef]
42. Nimmerjahn, F.; Dudziak, D.; Dirmeier, U.; Hobom, G.; Riedel, A.; Schlee, M.; Staudt, L.M.; Rosenwald, A.; Behrends, U.; Bornkamm, G.W.; et al. Active NF-kappaB signalling is a prerequisite for influenza virus infection. *J. Gen. Virol.* **2004**, *85 Pt 8*, 2347–2356. [CrossRef]
43. Williams, S.A.; Kwon, H.; Chen, L.F.; Greene, W.C. Sustained induction of NF-kappa B is required for efficient expression of latent human immunodeficiency virus type 1. *J. Virol.* **2007**, *81*, 6043–6056. [CrossRef] [PubMed]
44. Wang, Y.; Liu, Y.; Zhang, X.Y.; Xu, L.H.; Ouyang, D.Y.; Liu, K.P.; Pan, H.; He, J.; He, X.H. Ginsenoside Rg1 regulates innate immune responses in macrophages through differentially modulating the NF-kappaB and PI3K/Akt/mTOR pathways. *Int. Immunopharm.* **2014**, *23*, 77–84. [CrossRef] [PubMed]

45. Bi, S.; Chi, X.; Zhang, Y.; Ma, X.; Liang, S.; Wang, Y.; Hu, S.H. Ginsenoside Rg1 enhanced immune responses to infectious bursal disease vaccine in chickens with oxidative stress induced by cyclophosphamide. *Poult. Sci.* **2018**, *97*, 2698–2707. [CrossRef] [PubMed]

46. Yuan, D.; Yuan, Q.; Cui, Q.; Liu, C.; Zhou, Z.; Zhao, H.; Dun, Y.; Wang, T.; Zhang, C. Vaccine adjuvant ginsenoside Rg1 enhances immune responses against hepatitis B surface antigen in mice. *Can. J. Physiol. Pharm.* **2016**, *94*, 676–681. [CrossRef] [PubMed]

47. Lopez, O.J.; Osorio, F.A. Role of neutralizing antibodies in PRRSV protective immunity. *Vet. Immunol. Immunopathol.* **2004**, *102*, 155–163. [CrossRef]

48. Charerntantanakul, W. Porcine reproductive and respiratory syndrome virus vaccines: Immunogenicity, efficacy and safety aspects. *World J. Virol.* **2012**, *1*, 23–30. [CrossRef]

49. Zuckermann, F.A.; Garcia, E.A.; Luque, I.D.; Christopher-Hennings, J.; Doster, A.; Brito, M.; Osorio, F. Assessment of the efficacy of commercial porcine reproductive and respiratory syndrome virus (PRRSV) vaccines based on measurement of serologic response, frequency of gamma-IFN-producing cells and virological parameters of protection upon challenge. *Vet. Microbiol.* **2007**, *123*, 69–85. [CrossRef]

50. Li, X.; Galliher-Beckley, A.; Pappan, L.; Trible, B.; Kerrigan, M.; Beck, A.; Hesse, R.; Blecha, F.; Nietfeld, J.C.; Rowland, R.R.; et al. Comparison of Host Immune Responses to Homologous and Heterologous Type II Porcine Reproductive and Respiratory Syndrome Virus (PRRSV) Challenge in Vaccinated and Unvaccinated Pigs. *Biomed. Res. Int.* **2014**, *2014*, 416727. [CrossRef]

Article

The 40 kDa Linear Polyethylenimine Inhibits Porcine Reproductive and Respiratory Syndrome Virus Infection by Blocking Its Attachment to Permissive Cells

Jie Wang [1,2,†], Jie Li [1,2,†], Nana Wang [1,2], Qi Ji [1,2], Mingshuo Li [1,2], Yuchen Nan [1,2], En-Min Zhou [1,2], Yanjin Zhang [3] and Chunyan Wu [1,2,*]

1   Department of Preventive Veterinary Medicine, College of Veterinary Medicine, Northwest A&F University, Yangling 712100, China; 18829350677@163.com (J.W.); lijie2016c@163.com (J.L.); m17629097868@163.com (N.W.); jiqi10102@163.com (Q.J.); LiMingS1996@163.com (M.L.); nanyuchen2015@nwsuaf.edu.cn (Y.N.); zhouem@nwsuaf.edu.cn (E.-M.Z.)
2   Scientific Observing and Experimental Station of Veterinary Pharmacology and Veterinary Biotechnology, Ministry of Agriculture, Yangling 712100, China
3   Molecular Virology Laboratory, VA-MD College of Veterinary Medicine, University of Maryland, College Park, MD 20742, USA; zhangyj@umd.edu
*   Correspondence: chunyanwu@nwsuaf.edu.cn
†   These authors contributed equally to this work.

Received: 26 July 2019; Accepted: 17 September 2019; Published: 19 September 2019

Abstract: Porcine reproductive and respiratory syndrome (PRRS) is one of the most economically devastating infectious diseases in pigs worldwide. The causative agent is the PRRS virus (PRRSV). In this study, we explored polyethylenimine (PEI), a cationic polymer with different forms (linear or branched), to inhibit the replication of PRRSV. Our results demonstrate that the linear but not the 40 kDa branched PEI, or the 25 kDa linear PEI, were well tolerated in cultured cells and exhibited a broad-spectrum inhibition of heterogeneous *PRRSV-2* isolates in both MARC-145 cells and primary porcine pulmonary alveolar macrophages (PAMs). Further analysis suggests that PEI could prevent the attachment of PRRSV virions to the susceptible cells. Notably, PEI had a minimal effect on PRRSV internalization in MARC-145 cells, whereas PEI promoted the internalization of PRRSV virions in PAMs, which suggests that these two types of cells might have different internalization processes of PRRSV virions. In conclusion, our data demonstrate that PEI could be used as a novel inhibitor against PRRSV.

Keywords: PRRSV; polyethylenimine; PEI; virion internalization; antiviral; endocytosis

---

1. Introduction

Porcine reproductive and respiratory syndrome virus (PRRSV) is an enveloped, single-stranded, positive-sense RNA virus, which belongs to the genus *Porartevirus*, of the family *Arteriviridae* [1,2]. The genome of PRRSV is approximately 15 kb in length and contains over 10 open reading frames (ORFs) [3]. The ORF1a and ORF1b of PRRSV account for two-thirds of its genome and encode all non-structural proteins, which are required for viral replication, while ORFs 2–7 encode structural proteins [3]. Based on sequence similarity, all PRRSV isolates within the genus *Porartevirus* were further classified into two species: *PRRSV-1* and *PRRSV-2* [1,2]. *PRRSV-1* and *PRRSV-2* strains share approximately 60% nucleotide sequence identity and exhibit serotype differences [4,5]. PRRSV infection in pigs is highly restricted to cells of monocyte-macrophage lineages, such as pulmonary alveolar macrophages (PAMs) [6,7]. It is believed that PRRSV infection in susceptible cells is mediated by several cellular receptors or factors [8], such as heparin sulfate (HS) [9], vimentin [10], Cluster of differentiation (CD)151 [11], porcine CD163 (CD163) [12], sialoadhesin (CD169) [13], Dendritic Cell-Specific Intercellular adhesion molecule-3-Grabbing Non-integrin(DC-SIGN, also called CD209) [14] and myosin-9 heavy chain(MYH9) [15]. However, only CD163 and MYH9 are indispensable for PRRSV infection [15,16].

Current strategies for PRRS control are inadequate, although substantial efforts have been dedicated to controlling this disease. Since the first report of PRRS in the United States in 1987, PRRS remains one of the major challenges to the global swine industry, while vaccines against PRRS have been commercially available for over two decades [17–19]. This suggests the urgent need for novel methods for PRRSV control and prevention, as well as a deep understanding of viral–host interactions between PRRSV and its permissive cells.

Cationic polymers, which have been used as DNA transfection reagents via facilitating the entry of DNA to target cells [20,21], are also known to have potent antimicrobial activity [22,23]. Polyethylenimine (PEI) is one such cationic polymer, and its novel anti-viral activity has been investigated in several studies [24–27]. Since PEI can be either a linear or branched molecule with a broad molecular weight range, different PEI molecules have great divergence on modulating viral replication. *N,N*-dodecyl, methyl-PEI (DMPEI), synthesized from a commercial 750 kDa branched PEI, has been demonstrated to significantly reduce the infectivity of wild-type and drug-resistant influenza A virus [24], as well as HSV-1 and HSV-2 [25]. Meanwhile, the 25 kDa linear PEI is highly effective against human papillomaviruses (HPVs) and human cytomegaloviruses (HCMVs) [26]. However, another report suggests that the 70 kDa branched PEI accelerates HIV-1 infection despite its inhibition of HIV-1 attachment to cells [27]. This paradoxical effect of PEI on HIV infection is not fully understood. The effect of PEI on PRRSV proliferation is not known.

The objective of this study was to explore PEI to inhibit the replication of PRRSV. Our results demonstrate that the 40 kDa linear PEI, but not branched PEI, exhibited a broad-spectrum inhibition of heterogeneous *PRRSV-2* isolates in both MARC-145 cells and primary porcine pulmonary alveolar macrophages (PAMs). Furthermore, we investigated mechanisms of the 40 kDa linear PEI in the inhibition of PRRSV. Our results showed that PEI could reduce the attachment of PRRSV to the susceptible cells. PEI had a minimal effect in the internalization of PRRSV in MARC-145 cells but promoted the internalization of PRRSV into PAMs. This suggests that the internalization process of PRRSV virions into these two types of cells may be different.

2. Materials and Methods

2.1. Cells, Viruses, and Chemicals

PAMs were collected from 6-week-old PRRSV-negative pigs, as previously described [28]. PAMs were maintained in RPMI 1640 medium (Biological Industries, Beit HaEmek, Israel) supplemented with 10% FBS (Biological Industries). PRRSV-permissive MARC-145 cells were maintained in Dulbecco's Modified Eagle Medium (DMEM) (Biological Industries) supplemented with 10% FBS as well.

The *PRRSV-2* virus isolates used in this study were VR-2332 (GenBank: EF536003.1), and highly pathogenic PRRSV (HP-PRRSV) isolates SD16 (GenBank: JX087437.1), JXA1 (GenBank: EF112445.1), GD-HD (GenBank: KP793736.1), and NADC30-like isolate HNhx (GenBank: KX766379). All these PRRSV isolates were used to inoculate MARC-145 cells or PAMs at a multiplicity of infection (MOI) as indicated. The median tissue culture infectious dose (TCID$_{50}$) of all the PRRSV isolates (except the NADC30-like isolate HNhx) were titrated in MARC-145 cells as previously described [29]. Propagation and titration of the NADC30-like isolate HNhx were conducted in PAMs.

Three different forms of PEIs were used in this study: a branched PEI and two linear PEIs. The branched PEI (average molecular weight of 75 kDa) was purchased from Sigma-Aldrich (St. Louis, MO, USA) and dissolved in molecular-grade water (Thermo Fisher Scientific, Waltham, MA, USA) at a final concentration of 1 mg/mL. Two linear PEIs (molecular weights of 25 and 40 kDa) were purchased from PolySciences (Warrington, UK) and solubilized in molecular-grade water as well. To make the solution of the 25 kDa linear PEI, the PEI was added to molecular-grade water and then heated to 95 °C until the chemical was completely dissolved. All PEI solutions were sterilized by filtration through a 0.2 μm filter (Millipore, Burlington, MA, USA) and stored at −20 °C before use except the 25 kDa linear PEI solution (stored at room temperature before use to prevent the precipitation of the PEI).

2.2. Cell Viability and Cytotoxicity

Cell viability and cytotoxicity were evaluated using the CellTiter-Glo® Luminescent Cell Viability Assay Kit (Promega, Madison, WI, USA) by following the manufacturer's instructions. Briefly, MARC-145 cells or PAMs were seeded into 24-well plates at a density of 5×10^4 cells/well or 4×10^5 cells/well and cultured for 24 h. Then, fresh medium containing the PEI at the indicated concentration was added to the cells. After incubation for another 24 h, the cells were lysed by using 1× passive cell lysis buffer (Promega). The cell lysate was transferred into a 96-well black polystyrene microplate (Corning Inc, Corning, NY, USA). CellTiter-Glo® reagent was added and mixed with the cell lysate. After incubation for 10 min at room temperature (RT), luminescence signal was determined with VICTORX™ X5 Multilabel Reader (Perkin-Elmer Life and Analytical Sciences, Wellesley, MA, USA).

2.3. PRRSV Inhibition Assay

MARC-145 cells were seeded into 24-well plates at a density of 5×10^4 cells/well and cultured for 24 h. The cells were inoculated with a mixture of PRRSV (0.1 or 1 MOI) in plain DMEM and linear PEI or branched PEI (with indicated concentration). The cells were then incubated at 37 °C for 5 h. The mixture containing PEIs and virions was discarded and the cells were rinsed with phosphate-buffered saline (PBS), pH7.2, followed by culturing for another 24 h before further experiments. Subsequently, cells were rinsed with PBS twice and prepared for immunofluorescence assay (IFA), or harvested for Western blot and qPCR analyses. Cell culture supernatant containing progeny virus were further titrated to evaluate the infectious virions as well.

To evaluate the antiviral activity of PEIs in PAMs, cells were seeded into 24-well plates at a density of 5×10^5 cells/well and incubated for 6 h before further experiments. A fresh RPMI 1640 medium containing PRRSV (0.01 MOI) and PEI was added to the cells. After 1 h inoculation, PAMs were rinsed with PBS to remove the virus and PEI mixture unbosund and cultured for another 24 h before further analysis with qPCR, Western blot, and virus titration.

2.4. Immunofluorescence Assay (IFA)

MARC-145 cells in cell culture plates were fixed with 4% paraformaldehyde (Sigma-Aldrich), permeabilized with PBS containing 0.5% Triton X-100 (Sigma-Aldrich) and blocked with PBS contained 1% BSA (Sigma-Aldrich). Then the cells were stained with monoclonal antibody (mAb) against PRRSV-N protein (Clone No. 6D10, made in-house) as previously described [30]. Specific binding between the antibody and its target was detected using Alexa Fluor®488-labeled goat anti-mouse IgG (Thermo Fisher Scientific, Waltham, MA, USA). Cellular nuclei were counterstained by 4,6-diamidino-2-phenylindole (DAPI) (Thermo Fisher Scientific) at 37 °C for 10 min and observed under a Leica DM1000 fluorescence microscope (Leica Microsystems, Wetzlar, Germany). All images were captured and processed using Leica Application Suite X (Version 1.0. Leica Microsystems). IFA-positive cells from representative images were quantified using ImageJ software (Version 1.52a, National Institute of Health, Bethesda, MD, USA)

2.5. Reverse Transcription and Quantitative PCR (RT-qPCR)

Total RNA was extracted from cells using TRizol Reagent (Thermo Fisher Scientific) and reverse-transcribed using the PrimeScript RT reagent Kit (TaKaRa, Dalian, China) following the manufacturer's instructions. The qPCR was conducted using 2×RealStar Green Fast Mixture 2× PCR (GenStar, Beijing, China). Transcripts of the β-actin were also amplified and used to normalize total

RNA input. The relative quantification of target gene expression was calculated with the $2^{-\Delta\Delta Ct}$ method. Primers used for qPCR and their sequences are listed in Table 1.

Table 1. Primers and corresponding sequence used for qPCR.

Gene	Forward Primer (5′–3′)	Reverse Primer (5′–3′)
PRRSV-M	TGGGGAGTGTACTCAGCCAT	AATGTACTTGCGGCCTAGCA
β-actin	GGCATCCACGAAACTACCTT	TGATCTCCTTCTGCATCCTG

2.6. Western Blot Analysis

Cells were lysed by 1× Laemmli sample buffer (Bio-Rad Laboratories, Hercules, CA, USA) prior to SDS-PAGE. Proteins were separated using 12% SDS-PAGE gels, followed by transferring to a PVDF membrane as previously described [31]. After blocking, membranes were probed with homemade Mab-6D10 against the PRRSV-N protein and Mab against β-actin (Abcam, Cambridge, MA, USA). Specific binding between antibodies and their targets was detected using HRP-conjugated goat anti-mouse IgG (Thermo Fisher Scientific) and revealed with ECL substrate (Beyotime, Jiangsu, China). Chemiluminescence signal acquisition was conducted using a ChemiDoc MP Imaging System (Bio-Rad Laboratories) and analyzed using ImageLab software (Version 5.1, Bio-Rad Laboratories).

2.7. Analysis of Virus Attachment and Internalization

MARC-145 cells or PAMs were pre-incubated with the 40 kDa linear PEI for 1 h at 37 °C or left untreated (control), followed by pre-chilling at 4 °C for 30 min prior to PRRSV inoculation. The PRRSV-JXA1 strain was used to inoculate the cells at 4 °C for 1 h at the indicated MOI. Similarly, pre-chilled MARC-145 cells or PAMs were co-incubated with a mixture of PRRSV-JXA1 virus and the 40 kDa linear PEI at 4 °C for 1 h or PRRSV-JXA1 virus only. Subsequently, the cells with the indicated treatment were washed three times with cold PBS to remove unbound virions before being harvested for RT-qPCR to evaluate viral RNA level from the virions bound.

For the virion internalization assay, PRRSV-JXA1 at the indicated MOI was used to inoculate MARC-145 cells or PAMs at 4 °C for 1 h to allow the sufficient attachment of virions to the cell surface without triggering endocytosis as previously instructed [32]. After removing unbound viral particles by washing cells with cold PBS for three times, fresh medium containing the 40 kDa linear PEI was used to treat MARC-145 or PAMs followed by shifting cells to 37 °C to trigger endocytosis, which allows the internalization of bound PRRSV virions on the cell surface. One hour later, the cells were washed with cold PBS again and further incubated with 50 μg proteinase K (Sigma-Aldrich) for 45 min at 4 °C to remove non-internalized virions located on the cell surface. After the inactivation of proteinase K by a protease inhibitor cocktail (Roche, Basel, Switzerland), the cells were harvested for RT-qPCR analysis of the viral RNA of internalized PRRSV virions.

2.8. Statistical Analysis

All experiments were conducted with at least three independent replicates. The results were analyzed using GraphPad Prism version 5.0 (GraphPad Software, San Diego, CA, USA). Statistical significance was determined by student's *t*-test if two groups were compared. A *P* value less than 0.05 was considered statistically significant.

3. Result

3.1. Evaluation of Cytotoxicity of Polyethylenimine in MARC-145 and PAM Cells

In this study, three different forms of Polyethylenimine (PEI) were studied for their effect on PRRSV replication in both the continuous cell line MARC-145 and the primary PAM cells. They are two linear PEIs (40 and 25 kDa) and one branched PEI (75 kDa). Their chemical structures are illustrated in

Figure 1A–C. The cytotoxicity assay showed that no obvious adverse effect in MARC-145 cells was observed for all three forms of PEI up to 10 µg/mL (Figure 1D). However, in the PAMs, the branched PEI was less toxic than the linear PEI, while the 25 kDa linear PEI was more toxic than the 40 kDa PEI (Figure 1E).

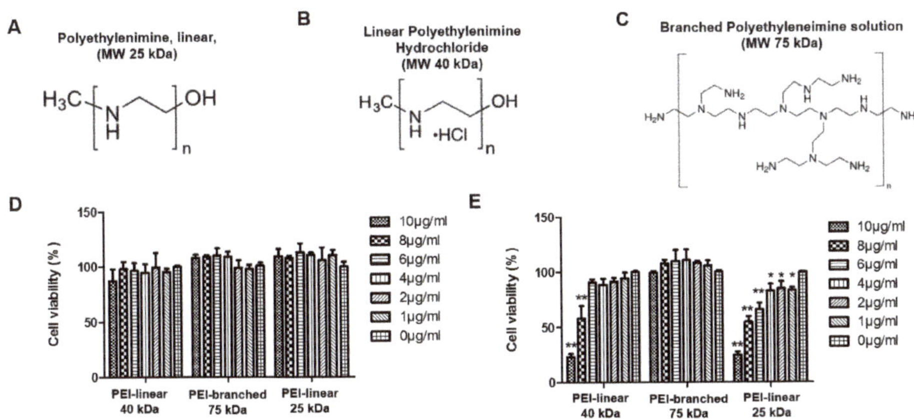

Figure 1. Illustration of the structures of different molecules of polyethylenimine (PEI) and the evaluation of in vitro cytotoxicity. (**A**) Structure of linear PEI with a molecular weight of 25 kDa; (**B**) Structure of linear PEI-hydrochlorides with a molecular weight of 40 kDa; (**C**) Structure of branched PEI with a molecular weight of 75 kDa; (**D**) Cell viability assay of MARC-145 cells incubated in the presence of PEI at the indicated concentrations at 37 °C for 24 h. (**E**) Cell viability assay of pulmonary alveolar macrophage (PAM) cells incubated in the presence of PEI at the indicated concentrations at 37 °C for 24 h. All experiments were repeated at least three times, and the data were presented as the mean ± SD, which is further subjected to Student's *t*-test. Significant differences between indicated groups were marked by "*" ($P < 0.05$) and "**" ($P < 0.01$).

Meanwhile, it is notable that the cytotoxic effect of PEI in these two types of cells was different. In MARC-145 cells treated with branched or the 25 kDa linear PEI, a slight increase in cell viability was observed along with an increasing dose of PEI (Figure 1D). On the contrary, the linear PEI had a cytotoxic effect in PAMs along with the incremental concentration. The cell viability of PAMs was significantly impaired by the 40 kDa linear PEI at a concentration higher than 8 µg/mL and by the 25 kDa PEI at a concentration as low as 1 µg/mL. However, it appears branched PEI was non-toxic to PAMs at the concentrations within the 10 µg/mL range (Figure 1E). The CC_{50} of different PEIs in PAMs were calculated to be 7.389, 8.273 and 23.831 µg/mL for the 25 kDa linear PEI, the 40 kDa linear PEI and the 75 kDa branched PEI, respectively, (Figures S1 and S2). Based on the cytotoxicity assay, the 40 kDa linear PEI and the 75 kDa branched PEI were selected for the following experiments in MARC-145 cells and PAMs.

3.2. The 40 KDa Linear PEI Effectively Inhibits PRRSV Replication in MARC-145 Cells

We next analyzed the effect of the 40 kDa linear PEI (hereby and thereafter, PEI-linear) and the 75 kDa branched PEI (hereby and thereafter: PEI-branch) on PRRSV replication in MARC-145 cells with a range of concentrations from 1 to 8 µg/mL based on the cytotoxicity assay. IFA results showed that the PEI-linear led to a dose-dependent inhibition of PRRSV-SD16 infection, as evidenced by a reduction in numbers of PRRSV-positive cells, whereas only weak inhibition could be observed in MARC-145 cells treated with PEI-branch at a concentration of 8 µg/mL (Figure 2A). By using ImageJ to analyze the IFA-positive cells, the median effective concentration (EC_{50}) of the PEI-linear and PEI-branch was calculated to be 2.435 and 6.883 µg/mL, respectively (Figure S3).

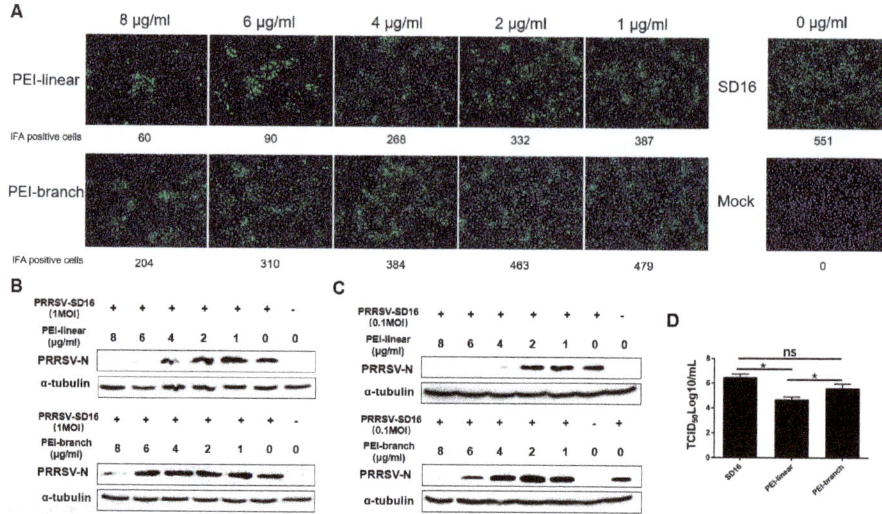

Figure 2. The 40 kDa linear PEI is more effective in inhibiting porcine reproductive and respiratory syndrome virus (PRRSV) replication than the branched PEI. (**A**). Immunofluorescence assay (IFA) of MARC-145 cells that were infected with PRRSV-SD16 in the presence of PEI. The virus at an MOI of 1 was diluted in DMEM and mixed with 40 kDa PEI-linear or PEI-branch at the indicated concentrations for 1 h before being inoculated to the cells. IFA with anti-PRRSV-N Mab-6D10 was done 24 hpi. MARC-145 cells infected with PRRSV-SD16 alone and non-infected cells were included as controls. Quantification of IFA-positive cells was conducted using ImageJ software. (**B**). Western blotting of MARC-145 cells that were infected with PRRSV-SD16 at an MOI of 1 in the presence of PEI. (**C**) Western blot of MARC-145 cells that were infected with PRRSV-SD16 at an MOI of 0.1 in the presence of PEI. (**D**) Progeny PRRSV virions from the cell culture supernatant of MARC-145 cells that were infected with PRRSV-SD16 at an MOI of 0.1 in the presence of PEI at a concentration of 6 µg/mL. Cell culture supernatants were harvested at 24 hpi and titrated for TCID50/mL. All experiments were repeated at least three times, and the data were presented as the mean ± SD along with subjection of Student's *t*-test. Significant differences between indicated groups were marked by "*" (*P* < 0.05). No significant difference is marked as "ns".

To further evaluate the inhibitory effect of both PEIs on PRRSV replication in MARC-145 cells, the level of the nucleocapsid (N) protein in virus-infected cells was determined by Western blot. Similar to the trends of the IFA results, an incremental inhibition of N expression was correlated with an increasing concentration of PEI, and N was below the detection level in cells treated with linear-PEI at the concentration of 6 µg/mL. In contrast, PEI-branch had no significant inhibitory effect on PRRSV replication in MARC-145 cells until the concentration was increased to 8 µg/mL (Figure 2B). Moreover, if PRRSV virus was used to inoculate the cells at 0.1 MOI but treated with a similar PEI dose, the PEI-linear at the concentration of 4 µg/mL could block the PRRSV replication (Figure 2C). Meanwhile, the significant inhibition of PRRSV replication by PEI-branch could be observed at the concentration of 6 µg/mL (Figure 2C).

To further confirm and quantify the effect of PEI on PRRSV, the progeny virions from the cell culture supernatant of MARC-145 cells infected with PRRSV with treatment of two forms of PEI were titrated (Figure 2D). Consistent with the Immunofluorescence assay (IFA) and Western Blotting (WB) results, linear PEI treatment resulted in a 1.5 Log_{10} reduction of infectious virus particles in the medium. Therefore, compared to PEI-linear, the inhibition of PRRSV by PEI-branch was much weaker. In summary, the results indicate that the 40 kDa linear PEI exerted a stronger inhibition on PRRSV replication than the PEI-branch.

3.3. The PEI-linear Demonstrates Broad inhibition of Heterogeneous PRRSV-2 Isolates

As a chemical reagent, the inhibitory effect of PEI-linear on PRRSV is expected to be broad spectrum. We next tested whether this PEI could inhibit other *PRRSV-2* isolates in MARC-145 cells. Our results showed that the replication of other *PRRSV-2* isolates (JXA1, GD-HD, and VR-2332) was significantly inhibited by PEI in a dose-dependent manner (Figure 3A–C). Moreover, the progeny virus from cell culture supernatants were titrated. Based on our data, it appears that PEI-linear demonstrated the strongest inhibition to PRRSV-2 VR-2332 for at least two Log_{10} reductions of infectious viral particles (Figure 3D). Notably, the NADC30-like isolate replicated poorly in MARC-145 with a viral titer below 10^5 $TCID_{50}$/mL (Figure 3D). The inhibitory effect of PEI against different PRRSV strains was not uniform as the titers of certain viral isolates such as JXA1 were higher than the others. Taken together, these data showed that PEI-linear could effectively inhibit several PRRSV-2 isolates tested in MARC-145 cells.

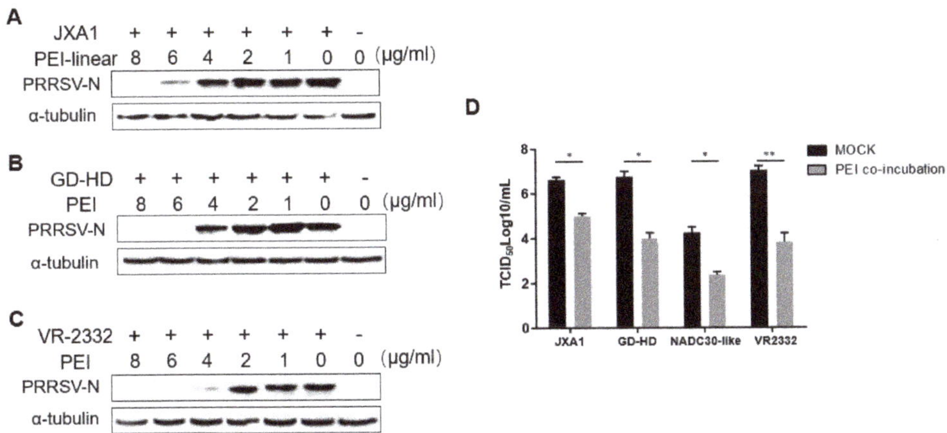

Figure 3. PEI demonstrates broad inhibition of heterogeneous *PRRSV-2* isolates in MARC-145 cells. MARC-145 cells were incubated with a mixture containing 1 MOI of PRRSV JXA1 (**A**), GD-HD (**B**) and VR-2332 (**C**) mixed with the 40 kDa linear PEI at the indicated concentrations for 1 h. Then the cells were washed and cultured for an additional 24 h. Replication of PRRSV was determined in Western blot with Mab-6D10 against N protein. (**D**) PRRSV yields in MARC-145 cells infected with the mixture of different PRRSV virus strains at 0.1 MOI and PEI-linear (6 µg/mL). The mixture was removed and replaced by a fresh medium. The cells were incubated for 24 h before the cell culture supernatant was harvested for titration. All experiments were repeated at least three times, and the data were presented as the mean ± SD, which is further subjected to Student's t-test. Significant differences between indicated groups were marked by "*" ($P < 0.05$) and "**" ($P < 0.01$).

3.4. The PEI-Linear Blocks PRRSV Replication in PAMs

Since PAMs are the primary target cells for PRRSV infection in vivo, we also evaluated the antiviral activity of PEI-linear in PAMs. According to the cell viability assay results (Figure 1), PEI-linear at a concentration of 6 µg/mL was used in PAMs. As our preliminary results demonstrated the attachment and infection efficiency of PRRSV to PAMs is nearly as high as 1000-fold in comparison to MARC-145 cells [33]. Therefore, 10^6 PAMs were inoculated with the PRRSV-2 strains: SD16, JXA1, GD-HD, and NADC30-like, at 0.01 MOI due to the high susceptibility of PAMs to PRRSV. Our results demonstrated that PEI inhibited the replication of the heterogeneous PRRSV isolates in PAMs (Figure 4A). Meanwhile, the RT-qPCR results confirmed the PEI inhibition of PRRSV in PAMs as well (Figure 4B). Furthermore, the titration results showed that PEI-linear inhibited all the PRRSV isolates tested (Figure 4C). Interestingly, it appears that the NADC30-like isolate replicated similarly in

PAMs as the other isolates (Figure 4C). Taken together, these data demonstrated that PEI-linear inhibits PRRSV infection in PAMs as well.

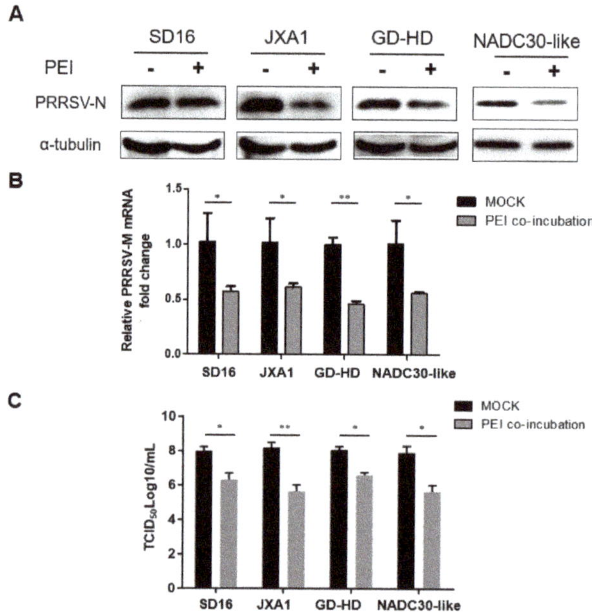

Figure 4. PEI-linear inhibits the replication of PRRSV-2 isolates in PAMs. (**A**). PAMs were inoculated with the virus and PEI mixture containing the indicated PRRSV strains (SD16, JXA1, GD-HD, NADC30-Like, 0.01 MOI) and PEI-linear 6 μg/mL for 1 h. The cells were then washed with PBS buffer and incubated for another 24 h. PRRSV N level was determined with Western blot. (**B**). PRRSV RNA levels in PAMs inoculated with the virus and PEI mixture containing the indicated PRRSV strains (0.01 MOI) and 6 μg/mL linear PEI. (**C**) Progeny PRRSV virions in cell culture supernatant of PAMs infected with different PRRSV strains at an MOI of 0.01 in the presence of linear PEI (6 μg/mL) for 24 h. All experiments were repeated at least three times, and the data were presented as the mean ± SD, which is further subjected to Student's t-test. Significant differences between indicated groups were marked by "*" ($P < 0.05$) and "**" ($P < 0.01$).

3.5. The PEI-Linear Blocks the Attachment of PRRSV Virions but Not Internalization in MARC-145 Cells

It has been reported that PEI mediated the blockage of the attachment of HPV and HCMV to surface of susceptible cells and this probably occurs through competitively binding the same attachment sites on cells between PEI and virions [26]. Thus, we next determined whether PEI prevents the attachment of PRRSV virions to MARC-145 cells. We first tested whether the inhibition of PRRSV infection by PEI was due to the reduced virion attachment. After the inoculation of MARC-145 cells with the virus and PEI mixture, the cells were maintained at 4 °C without triggering endocytosis but allowed virion attachment. Then, the attached virions were quantified by RT-qPCR. The results showed that PEI reduced the PRRSV virion attachment to MARC-145 cells at both 5 and 10 MOI when PEI and the virus were added at the same time (Figure 5A). To further confirm this result, we pre-incubated cells with PEI-linear first, followed by inoculating the cells with PRRSV (5 or 10 MOI) at 4 °C. The RT-qPCR analysis suggests that there was reduced binding of viral particles to MARC-145 cells as well (Figure 5B).

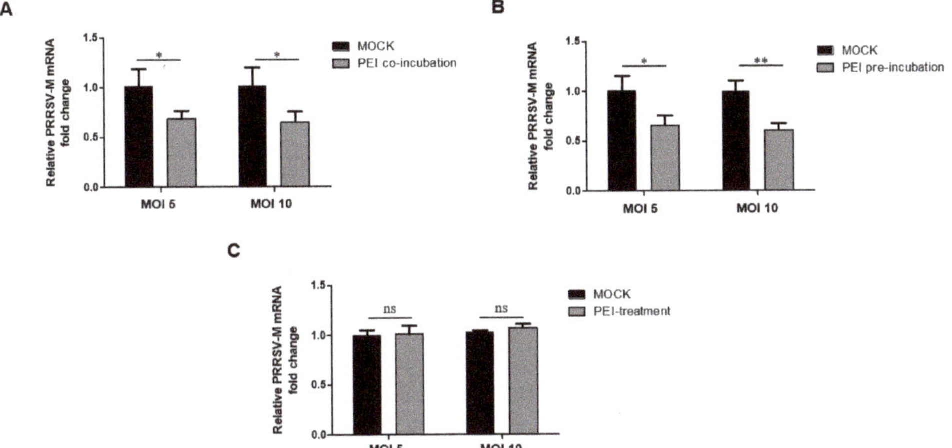

Figure 5. PEI inhibits PRRSV virion attachment to MARC-145 cells but not virion internalization. (**A**). PEI inhibits PRRSV virions attachment to cells when co-incubated with PRRSV to MARC-145 cells. PRRSV RNA levels in MARC-145 cells that were incubated with a pre-chilling virus and PEI mixture containing an MOI of 5 or 10 PRRSV-JXA1 strain mixed with 6 μg/mLPEI-linear for 2 h at 4 °C. Then, the cells were harvested for RT-qPCR analysis. MARC-145 cells inoculated with the virus only without PEI were included as a control. (**B**). Pre-incubation of MARC-145 cells with PEI inhibits virion attachment. PRRSV RNA levels in MARC-145 cells that were pre-incubated with 6 μg PEI for 1 h, followed by chilling at 4 °C for 30 min and inoculation with an MOI of 5 or 10 PRRSV-JXA1 strain at 4 °C for 1 h. (**C**). PEI treatment of MARC-145 cells does not affect virion internalization. MARC-145 cells were inoculated with an MOI of 5 or 10 PRRSV-JXA1 for 1 h at 4 °C, followed by washing with cold PBS, treatment with PEI, temperature shift to 37 °C to trigger virion internalization via endocytosis, and removing non-internalized virions with protease K treatment. Then the cells were harvested to evaluate the PRRSV RNA level to determine internalized virions. All experiments were repeated at least three times, and the data were presented as the mean ± SD, which is further subjected to Student's *t*-test. Significant differences between indicated groups were marked by "*" ($P < 0.05$) and "**" ($P < 0.01$).

To determine whether the PEI affects PRRSV internalization, we inoculated MARC-145 cells with the virus first for attachment at 4 °C, followed by PEI treatment and temperature shift to 37 °C to trigger endocytosis-mediated internalization of the virions. The viral particles remaining on cell surface were removed by treatment with protease K. So only internalized virions were harvested for qPCR. The data suggest that PEI could not block the internalization of attached virions as shown by similar viral RNA levels in both PEI-treated and mock-treated cells (Figure 5C). Taken together, these data suggest that the 40 kDa linear PEI inhibits PRRSV via blocking virion attachment to MARC-145 cells.

3.6. The 40 kDa Linear PEI Blocks the Attachment of PRRSV Virions to PAMs via Acting on the Virus but Not Cells

Since PAMs are the major target of PRRSV in vivo, we investigated whether PEI has a similar inhibition mechanism to PRRSV in PAMs as that of MARC-145 cells. Similar to MARC-145 cells, thw co-administration of PRRSV and PEI to PAMs reduced the binding of PRRSV virions to PAMs (Figure 6A). Moreover, it is notable that the susceptibility of PAMs to PRRSV is much higher than MARC-145 cells, since the inoculation of PRRSV at 0.01 MOI to PAMs was sufficient for RT-qPCR analysis of attached virions but it required at least an MOI of 5 of the virus to inoculate the MARC-145 cells for RT-qPCR analysis of attached virions (below the detection limit in MARC-145 cells if below 1 MOI). Meanwhile, if PAMs were pre-incubated with PEI-linear, followed by inoculation with PRRSV-JXA1 at 4 °C, there was a significant increase in virion binding compared with PAMs without

PEI pretreatment (Figure 6B), which is different from that of MARC-145 cells. Moreover, if PAMs were inoculated with PRRSV before PEI treatment, followed by a temperature shift to trigger endocytosis, the analysis of the internalized PRRSV virions showed that PEI also promoted the internalization of attached virions as well (Figure 6C). Therefore, the data suggest that PEI-linear may act on the virus rather than cells to prevent the binding of the virus to PAMs. Meanwhile, the pre-treatment of PAMs with PEI even promotes PRRSV attachment and internalization to PAMs. Taken together, these data suggest that the 40 kDa PEI inhibits PRRSV attachment and is capable of conferring the broad inhibition of heterogeneous *PRRSV-2* isolates.

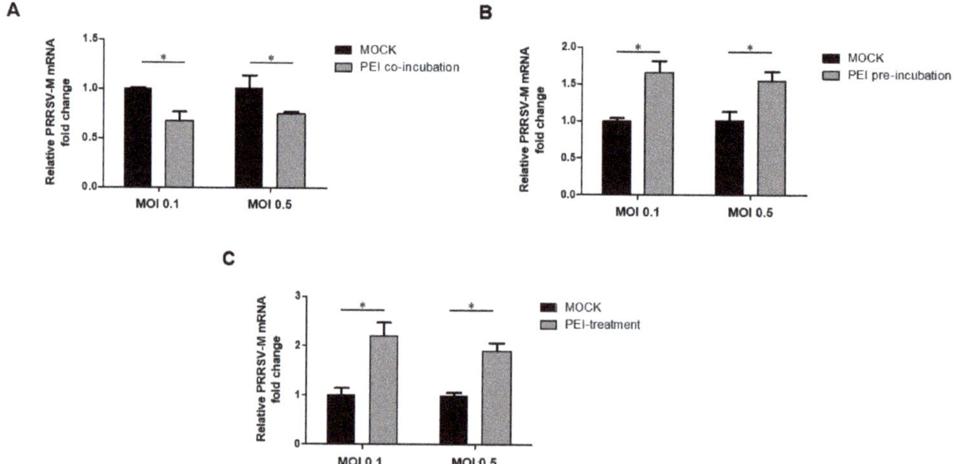

Figure 6. PEI inhibits the attachment of PRRSV virion to PAMs but promotes PRRSV internalization into PAMs. (**A**). PEI prevents virion attachment to cells when co-incubated with PAMs. PRRSV RNA levels in PAMs that were incubated with a pre-chilling virus and PEI mixture containing an MOI of 0.1 or 0.5 PRRSV-JXA1 virus and 6 µg/mL PEI-linear for 2 h at 4 °C to allow attachment. Cells inoculated with the virus only were included as a control. (**B**). Pre-incubation of PAMs with PEI enhances virion attachment to PAMs. PRRSV RNA levels in PAMs that were pre-incubated with 6 µg PEI-linear for 1 h, followed by chilling at 4 °C for 30 mins and inoculation with an MOI of 0.1 or 0.5 of PRRSV-JXA1 virus at 4 °C for 1 h. (**C**). PEI treatment of PAMs enhances virion internalization into PAMs. PRRSV RNA levels in PAMs that were inoculated with an MOI of 0.1 or 0.5 PRRSV-JXA1 for 1 h at 4 °C, followed by washing with cold PBS, treatment with 6 µg PEI and temperature shift to 37 °C to trigger virion internalization via endocytosis. All experiments were repeated at least three times, and the data were presented as the mean ± SD, which is further subjected to Student's *t*-test. Significant differences between indicated groups were marked by "*" ($P < 0.05$) and "**" ($P < 0.01$).

4. Discussion

PEI, also known as polyaziridine, is a polymer with a repeating structure that is composed of the amine group along with a two-carbon aliphatic CH_2CH_2 spacer. It was reported that PEI could be in linear, branched, and dendrimeric forms [34], while only linear and branched molecules of PEI are available from commercial suppliers. PEI has been used for many applications, usually due to its polycationic character. For cell biology research, PEI has been used as a transfection reagent [35]. Also, PEI is explored for its antiviral activity. For HPV, the 25 kDa linear PEI blocks the primary attachment of HPV16 and HCMV to susceptible cells [26]. It appears that PEI preincubation with cells blocks HPV and HCMV binding to their primary receptor HSPG [26]. Moreover, the antiviral activity of PEI has been investigated in vivo as well. Intranasal administration of the 25 kDa linear PEI suppresses influenza virus infection in mice [36]. The 3610 Da branched PEI combined with liposomes was shown to strongly enhance antiviral efficiency against herpes simplex virus type 2 in a mouse model [37].

Currently, the inhibition mechanism of PEIs against viral infections is still not fully understood and appears to depend on blocking virion attachment to susceptible cells. As a result, PEI potentially interferes with electrostatic interactions between viral surface proteins and host receptors [26]. However, it was reported that branched PEI with a molecular mass of 70 kDa accelerated HIV-1 infection despite its inhibition of HIV-1 attachment to cells [27], which may be partially due to the promotion of the exposure of co-receptor and/or viral entry into cells via the influence of cell membrane fluidity [27].

In our study, the 40 kDa linear PEI was shown to be effective against PRRSV infection through the inhibition of PRRSV virion attachment to MARC-145 cells, the most frequently used cell line for PRRSV study in vitro. Meanwhile, PEI treatment of MARC-145 cells does not affect the internalization of attached PRRSV virions on the cell surface. This inhibition mechanism appears to be consistent with a previous observation for HPV and HCMV [26]. Conversely, the inhibition of PRRSV in PAMs by the PEI was weaker than that in MARC-145 cells when the same dose of PEI was used. Notably, based on our data, although PEI prevented the attachment of PRRSV virions to PAMs when the PEI and virus were co-administrated simultaneously, further analysis suggests that the pre-treatment of PAMs with PEI enhanced the attachment of PRRSV virions. Moreover, PEI enhanced the internalization of PRRSV virions, which appears to be similar to the scenario in HIV but different from PRRSV in MARC-145 cells. Therefore, these data suggest that the exact procedures of the attachment and internalization of PRRSV virions in MARC-145 cells and PAMs may be different, which could also be supported by the different susceptibility of both cells to PRRSV as well. Moreover, how such a difference affects PRRSV pathogenesis or antibody-mediated viral neutralization (mainly via blocking virion attachment to susceptible cells) remains unclear. Meanwhile, when acting as a transfection agent, PEI condenses DNA into positively charged particles, which bind to anionic cell surface residues to deliver the PEI/DNA complex into the cells via endocytosis [38,39]. Once inside the endosome, amines of PEI result in an influx of counter-ions and a lowering of the osmotic potential. Therefore, osmotic swelling disrupts the vesicle to release the PEI/DNA complex into the cytoplasm [38,39]. It has been known that PRRSV virions enter PAMs via clathrin-mediated endocytosis [32]. It is possible that PEI promotes endocytosis after PRRSV virion attachment or disrupts the endosome vesicle to release PRRSV RNA into the cytoplasm in PAMs to enhance the virion internalization. However, why such a scenario in PAMs is different from MARC-145 cells remains unclear.

On the one hand, PK-15 cells (immortalized swine kidney cell line is not susceptible for PRRSV) stably expressing porcine CD163 (the essential receptor for PRRSV infection) support the complete PRRSV replication cycle [40]. On the other hand, PK-15 cells stably expressing CD169 (a putative receptor involved in PRRSV internalization) support the internalization of PRRSV virions but not the uncoating of nucleocapsid and fusion with the endocytic vesicle membrane [41]. These data imply that the endocytosis-mediated internalization and uncoating procedure of PRRSV particles in PAMs and immortalized cell lines such as PK-15 and MARC-145 is different. Such a difference may contribute to the different inhibitory effect of the 40 kDa linear PEI on PRRSV virions.

The cytotoxicity of different PEI forms was evaluated in this study as well. A previous report suggested that branched (25 kDa) and linear (750 kDa) PEI can both induce membrane damage and initiate apoptosis in human cell lines [42]. Based on our observation, the cytotoxicity of PEI appears to be cell type and structure dependent. Both linear PEIs demonstrated better tolerance in MARC-145 cells than primary PAMs, while branched PEI had a much better tolerance than linear PEI in PAMs. However, the reason for the differences remains elusive. This implies that a certain modification or optimization of PEI molecular structure may be explored to reduce cytotoxicity while maintaining comparable antiviral activity.

Mechanically, PEI blocks the binding of virus to its receptors to prevent the attachment of virions to susceptible cells as demonstrated for HPV16 and HCMV [26]. This is also consistent with our observation that PEI prevents PRRSV attachment to MARC-145 cells. The difference in PEI inhibition of PRRSV in MARC-145 and PAM cells suggests that the interaction between PRRSV virions and receptors in these two types of cells might be different. Several membrane proteins have been identified as potential receptors for PRRSV infection in permissive cells, such as heparin sulfate (HS) [9], vimentin [10], CD151 [11], porcine CD163 [12], sialoadhesin (CD169) [13], DC-SIGN (CD209) [14] and MYH9 [15]. However, how these receptors cooperate to mediate the PRRSV infection remains elusive.

For HIV, branched PEI enhanced HIV-1 infection partially due to its promotion of the exposure of co-receptor and/or viral entry into cells via the influence of cell membrane fluidity [27]. Since PRRSV might use several receptors in PAMs in vivo, it is possible that PEI may promote the exposure of co-receptor or receptors and influence the cell membrane fluidity of PAMs.

Moreover, it is also notable that the inhibitory effect of the 40 kDa linear PEI on different *PRRSV-2* isolates is variable, especially in MARC-145 cells. This is somehow unusual as the PEI inhibition of PRRSV should be non-specific and such an inhibition on different PRRSV isolates should be similar. It is unclear whether PEI blocks the interaction of PRRSV virions with the receptors mentioned above indiscriminately or whether PEI preferably blocks the interaction of PRRSV virions with certain receptors from the list above. It is possible that there is a different receptor preference among various PRRSV isolates, as evidenced by the observation that the replacement of the fifth scavenger receptor cysteine-rich domain (SRCR5) of porcine CD163 with the SRCR5 domain of the human CD163-like homolog (CD163Li) only confers resistance to PRRSV-1 but not to PRRSV-2 [43]. Therefore, it is possible that, due to an uneven PEI blockage to the interaction of PRRSV virions to different receptors along with a strain-specific receptor preference of various PRRSV isolates, the PEI cannot confer a consistent inhibition among different PRRSV isolates. Moreover, it appears that different steps and variable outcomes of PEI mediated inhibition between MARC-145 cells and PAMs suggest that there is a difference in the attachment and internalization of PRRSV virions on these two cells. The impact of such a difference on our understanding of PRRSV virion attachment and internalization as well as virus neutralization and pathogenesis in vivo require further investigation.

In conclusion, the 40 kDa linear PEI demonstrates an inhibitory effect against different PRRSV-2 isolates both in MARC-145 cells and PAMs with tolerable cytotoxicity. Further study is needed to investigate whether PEI can confer the inhibition of PRRSV infection in swine. Meanwhile, the potential different internalization procedure of PRRSV in PAMs and MARC-145 cells warrants further investigation.

Supplementary Materials: The following are available online at http://www.mdpi.com/1999-4915/11/9/876/s1, Figure S1: Calculation of median cytotoxic concentration (CC50) for two linear polyethylenimines (PEIs) in PAMs, Figure S2: Evaluation of cytotoxicity of branched PEI in PAMs and calculation of CC50, Figure S3: Calculation of median effective concentration (EC50) of PEI in MARC-145 cells.

Author Contributions: Data curation, J.W., J.L., N.W., Q.J., M.L. and C.W.; Funding acquisition, Y.N. and C.W.; Methodology, Y.N.; Project administration, C.W.; Resources, E.-M.Z.; Supervision, E.-M.Z. and C.W.; Writing—original draft, J.W., J.L. and Y.N.; Writing—review and editing, Y.N., Y.Z. and C.W.

Funding: This work was supported by a grant from the National Key Research and Development Program of China awarded to Y.N. and C.W. (Grant No. 2017YFD0501004), a grant from the Key Research and Development Program of Shaanxi Province (Grant No. 2017NY-156) and the Fundamental Research Funds for the Central Universities awarded to C.W.

Acknowledgments: We thank Gaiping Zhang of Henan Agriculture University for providing the NADC30-like isolate.

Conflicts of Interest: The authors declare no conflict of interest.

References

1. Kuhn, J.H.; Lauck, M.; Bailey, A.L.; Shchetinin, A.M.; Vishnevskaya, T.V.; Bao, Y.; Ng, T.F.; LeBreton, M.; Schneider, B.S.; Gillis, A.; et al. Reorganization and expansion of the nidoviral family Arteriviridae. *Arch. Virol.* **2016**, *161*, 755–768. [CrossRef] [PubMed]

2. Adams, M.J.; Lefkowitz, E.J.; King, A.M.; Harrach, B.; Harrison, R.L.; Knowles, N.J.; Kropinski, A.M.; Krupovic, M.; Kuhn, J.H.; Mushegian, A.R.; et al. Ratification vote on taxonomic proposals to the International Committee on Taxonomy of Viruses (2016). *Arch. Virol.* **2016**, *161*, 2921–2949. [CrossRef] [PubMed]

3. Lunney, J.K.; Fang, Y.; Ladinig, A.; Chen, N.; Li, Y.; Rowland, B.; Renukaradhya, G.J. Porcine Reproductive and Respiratory Syndrome Virus (PRRSV): Pathogenesis and Interaction with the Immune System. *Annu. Rev. Anim. Biosci.* **2016**, *4*, 129–154. [CrossRef] [PubMed]

4. Forsberg, R. Divergence time of porcine reproductive and respiratory syndrome virus subtypes. *Mol. Biol. Evol.* **2005**, *22*, 2131–2134. [CrossRef]

5. Van Woensel, P.A.; Liefkens, K.; Demaret, S. Effect on viraemia of an American and a European serotype PRRSV vaccine after challenge with European wild-type strains of the virus. *Vet. Rec.* **1998**, *142*, 510–512. [CrossRef] [PubMed]

6. Morgan, S.B.; Frossard, J.P.; Pallares, F.J.; Gough, J.; Stadejek, T.; Graham, S.P.; Steinbach, F.; Drew, T.W.; Salguero, F.J. Pathology and Virus Distribution in the Lung and Lymphoid Tissues of Pigs Experimentally Inoculated with Three Distinct Type 1 PRRS Virus Isolates of Varying Pathogenicity. *Transbound. Emerg. Dis.* **2016**, *63*, 285–295. [CrossRef] [PubMed]

7. Albina, E.; Carrat, C.; Charley, B. Interferon-alpha response to swine arterivirus (PoAV), the porcine reproductive and respiratory syndrome virus. *J. Interferon Cytokine Res.* **1998**, *18*, 485–490. [CrossRef]

8. Shi, C.; Liu, Y.; Ding, Y.; Zhang, Y.; Zhang, J. PRRSV receptors and their roles in virus infection. *Arch. Microbiol.* **2015**, *197*, 503–512. [CrossRef]

9. Delputte, P.L.; Vanderheijden, N.; Nauwynck, H.J.; Pensaert, M.B. Involvement of the matrix protein in attachment of porcine reproductive and respiratory syndrome virus to a heparinlike receptor on porcine alveolar macrophages. *J. Virol.* **2002**, *76*, 4312–4320. [CrossRef]

10. Kim, J.K.; Fahad, A.M.; Shanmukhappa, K.; Kapil, S. Defining the cellular target(s) of porcine reproductive and respiratory syndrome virus blocking monoclonal antibody 7G10. *J. Virol.* **2006**, *80*, 689–696. [CrossRef]

11. Wu, J.; Peng, X.; Zhou, A.; Qiao, M.; Wu, H.; Xiao, H.; Liu, G.; Zheng, X.; Zhang, S.; Mei, S. MiR-506 inhibits PRRSV replication in MARC-145 cells via CD151. *Mol. Cell. Biochem.* **2014**, *394*, 275–281. [CrossRef]

12. Guo, L.; Niu, J.; Yu, H.; Gu, W.; Li, R.; Luo, X.; Huang, M.; Tian, Z.; Feng, L.; Wang, Y. Modulation of CD163 expression by metalloprotease ADAM17 regulates porcine reproductive and respiratory syndrome virus entry. *J. Virol.* **2014**, *88*, 10448–10458. [CrossRef] [PubMed]

13. Delputte, P.L.; van Breedam, W.; Delrue, I.; Oetke, C.; Crocker, P.R.; Nauwynck, H.J. Porcine arterivirus attachment to the macrophage-specific receptor sialoadhesin is dependent on the sialic acid-binding activity of the N-terminal immunoglobulin domain of sialoadhesin. *J. Virol.* **2007**, *81*, 9546–9550. [CrossRef] [PubMed]

14. Pineyro, P.E.; Subramaniam, S.; Kenney, S.P.; Heffron, C.L.; Gimenez-Lirola, L.G.; Meng, X.J. Modulation of Proinflammatory Cytokines in Monocyte-Derived Dendritic Cells by Porcine Reproductive and Respiratory Syndrome Virus Through Interaction with the Porcine Intercellular-Adhesion-Molecule-3-Grabbing Nonintegrin. *Viral Immunol.* **2016**, *29*, 546–556. [CrossRef] [PubMed]

15. Gao, J.; Xiao, S.; Xiao, Y.; Wang, X.; Zhang, C.; Zhao, Q.; Nan, Y.; Huang, B.; Liu, H.; Liu, N.; et al. MYH9 is an Essential Factor for Porcine Reproductive and Respiratory Syndrome Virus Infection. *Sci. Rep.* **2016**, *6*, 25120. [CrossRef] [PubMed]

16. Burkard, C.; Lillico, S.G.; Reid, E.; Jackson, B.; Mileham, A.J.; Ait-Ali, T.; Whitelaw, C.B.; Archibald, A.L. Precision engineering for PRRSV resistance in pigs: Macrophages from genome edited pigs lacking CD163 SRCR5 domain are fully resistant to both PRRSV genotypes while maintaining biological function. *PLoS Pathog.* **2017**, *13*, e1006206. [CrossRef] [PubMed]

17. Rowland, R.R.; Lunney, J.; Dekkers, J. Control of porcine reproductive and respiratory syndrome (PRRS) through genetic improvements in disease resistance and tolerance. *Front. Genet.* **2012**, *3*, 260. [CrossRef] [PubMed]

18. Tian, K.; Yu, X.; Zhao, T.; Feng, Y.; Cao, Z.; Wang, C.; Hu, Y.; Chen, X.; Hu, D.; Tian, X.; et al. Emergence of fatal PRRSV variants: Unparalleled outbreaks of atypical PRRS in China and molecular dissection of the unique hallmark. *PLoS ONE* **2007**, *2*, e526. [CrossRef] [PubMed]

19. Butler, J.E.; Lager, K.M.; Golde, W.; Faaberg, K.S.; Sinkora, M.; Loving, C.; Zhang, Y.I. Porcine reproductive and respiratory syndrome (PRRS): An immune dysregulatory pandemic. *Immunol. Res.* **2014**, *59*, 81–108. [CrossRef]

20. Zhang, C.; Ji, J.; Shi, X.; Zheng, X.; Wang, X.; Feng, F. Synthesis of Structurally Defined Cationic Polythiophenes for DNA Binding and Gene Delivery. *ACS Appl. Mater. Interfaces* **2018**, *10*, 4519–4529. [CrossRef]

21. Zhao, L.; Li, Y.; Pei, D.; Huang, Q.; Zhang, H.; Yang, Z.; Li, F.; Shi, T. Glycopolymers/PEI complexes as serum-tolerant vectors for enhanced gene delivery to hepatocytes. *Carbohydr. Polym.* **2019**, *205*, 167–175. [CrossRef] [PubMed]

22. Park, D.; Wang, J.; Klibanov, A.M. One-step, painting-like coating procedures to make surfaces highly and permanently bactericidal. *Biotechnol. Prog.* **2006**, *22*, 584–589. [CrossRef] [PubMed]

23. Santos, M.R.E.; Mendonca, P.V.; Almeida, M.C.; Branco, R.; Serra, A.C.; Morais, P.V.; Coelho, J.F.J. Increasing the Antimicrobial Activity of Amphiphilic Cationic Copolymers by the Facile Synthesis of High Molecular Weight Stars by Supplemental Activator and Reducing Agent Atom Transfer Radical Polymerization. *Biomacromolecules* **2019**, *20*, 1146–1156. [CrossRef] [PubMed]

24. Haldar, J.; Chen, J.; Tumpey, T.M.; Gubareva, L.V.; Klibanov, A.M. Hydrophobic polycationic coatings inactivate wild-type and zanamivir- and/or oseltamivir-resistant human and avian influenza viruses. *Biotechnol. Lett.* **2008**, *30*, 475–479. [CrossRef] [PubMed]

25. Larson, A.M.; Oh, H.S.; Knipe, D.M.; Klibanov, A.M. Decreasing herpes simplex viral infectivity in solution by surface-immobilized and suspended N,N-dodecyl,methyl-polyethylenimine. *Pharm. Res.* **2013**, *30*, 25–31. [CrossRef]

26. Spoden, G.A.; Besold, K.; Krauter, S.; Plachter, B.; Hanik, N.; Kilbinger, A.F.; Lambert, C.; Florin, L. Polyethylenimine is a strong inhibitor of human papillomavirus and cytomegalovirus infection. *Antimicrob. Agents Chemother.* **2012**, *56*, 75–82. [CrossRef] [PubMed]

27. Owada, T.; Miyashita, Y.; Motomura, T.; Onishi, M.; Yamashita, S.; Yamamoto, N. Enhancement of human immunodeficiency virus type 1 (HIV-1) infection via increased membrane fluidity by a cationic polymer. *Microbiol. Immunol.* **1998**, *42*, 97–107. [CrossRef] [PubMed]

28. Xiao, S.; Zhang, A.; Zhang, C.; Ni, H.; Gao, J.; Wang, C.; Zhao, Q.; Wang, X.; Ma, C.; Liu, H.; et al. Heme oxygenase-1 acts as an antiviral factor for porcine reproductive and respiratory syndrome virus infection and over-expression inhibits virus replication in vitro. *Antivir. Res.* **2014**, *110C*, 60–69. [CrossRef]

29. Zhang, Y.J.; Stein, D.A.; Fan, S.M.; Wang, K.Y.; Kroeker, A.D.; Meng, X.J.; Iversen, P.L.; Matson, D.O. Suppression of porcine reproductive and respiratory syndrome virus replication by morpholino antisense oligomers. *Vet. Microbiol.* **2006**, *117*, 117–129. [CrossRef]

30. Li, Q.; Wang, X.; Wang, C.; Yu, Y.; Wang, G.; Gao, J.; Liu, H.; Xie, H.; Huang, B.; Li, Z.; et al. Intracellular expression of an anti-idiotypic antibody single-chain variable fragment reduces porcine reproductive and respiratory syndrome virus infection in MARC-145 cells. *Antivir. Ther.* **2016**, *21*, 161–170. [CrossRef]

31. Mu, Y.; Li, L.; Zhang, B.; Huang, B.; Gao, J.; Wang, X.; Wang, C.; Xiao, S.; Zhao, Q.; Sun, Y.; et al. Glycoprotein 5 of porcine reproductive and respiratory syndrome virus strain SD16 inhibits viral replication and causes G2/M cell cycle arrest, but does not induce cellular apoptosis in Marc-145 cells. *Virology* **2015**, *484*, 136–145. [CrossRef] [PubMed]

32. Nauwynck, H.J.; Duan, X.; Favoreel, H.W.; van Oostveldt, P.; Pensaert, M.B. Entry of porcine reproductive and respiratory syndrome virus into porcine alveolar macrophages via receptor-mediated endocytosis. *J. Gen. Virol.* **1999**, *80*, 297–305. [CrossRef] [PubMed]

33. Wang, J.; Li, J.; Wang, N.; Wu, C. Northwest A&F University, Yangling, Shannxi, China. Material and Methods used in the same way as Section 2.7 of this manuscript, 2019, (Material not intended for publication).

34. Yemul, O.; Imae, T. Synthesis and characterization of poly(ethyleneimine) dendrimers. *Colloid Polym. Sci.* **2008**, *286*, 747–752. [CrossRef]

35. Longo, P.A.; Kavran, J.M.; Kim, M.S.; Leahy, D.J. Transient mammalian cell transfection with polyethylenimine (PEI). *Methods Enzym.* **2013**, *529*, 227–240.

36. He, B.; Fu, Y.; Xia, S.; Yu, F.; Wang, Q.; Lu, L.; Jiang, S. Intranasal application of polyethyleneimine suppresses influenza virus infection in mice. *Emerg. Microbes Infect.* **2016**, *5*, e41. [CrossRef] [PubMed]

37. Maitani, Y.; Ishigaki, K.; Nakazawa, Y.; Aragane, D.; Akimoto, T.; Iwamizu, M.; Kai, T.; Hayashi, K. Polyethylenimine combined with liposomes and with decreased numbers of primary amine residues strongly enhanced therapeutic antiviral efficiency against herpes simplex virus type 2 in a mouse model. *J. Control. Release* **2013**, *166*, 139–146. [CrossRef]

38. Rudolph, C.; Lausier, J.; Naundorf, S.; Muller, R.H.; Rosenecker, J. In vivo gene delivery to the lung using polyethylenimine and fractured polyamidoamine dendrimers. *J. Gene Med.* **2000**, *2*, 269–278. [CrossRef]

39. Akinc, A.; Thomas, M.; Klibanov, A.M.; Langer, R. Exploring polyethylenimine-mediated DNA transfection and the proton sponge hypothesis. *J. Gene Med.* **2005**, *7*, 657–663. [CrossRef]

40. Wang, T.; Fang, L.; Zhao, F.; Wang, D.; Xiao, S. Exosomes Mediate Intercellular Transmission of Porcine Reproductive and Respiratory Syndrome Virus. *J. Virol.* **2018**, *92*. [CrossRef]
41. Vanderheijden, N.; Delputte, P.L.; Favoreel, H.W.; Vandekerckhove, J.; van Damme, J.; van Woensel, P.A.; Nauwynck, H.J. Involvement of sialoadhesin in entry of porcine reproductive and respiratory syndrome virus into porcine alveolar macrophages. *J. Virol.* **2003**, *77*, 8207–8215. [CrossRef]
42. Moghimi, S.M.; Symonds, P.; Murray, J.C.; Hunter, A.C.; Debska, G.; Szewczyk, A. A two-stage poly(ethylenimine)-mediated cytotoxicity: Implications for gene transfer/therapy. *Mol. Ther.* **2015**, *11*, 990–995. [CrossRef] [PubMed]
43. Wells, K.D.; Bardot, R.; Whitworth, K.M.; Trible, B.R.; Fang, Y.; Mileham, A.; Kerrigan, M.A.; Samuel, M.S.; Prather, R.S.; Rowland, R.R. Replacement of Porcine CD163 Scavenger Receptor Cysteine-Rich Domain 5 with a CD163-Like Homolog Confers Resistance of Pigs to Genotype 1 but Not Genotype 2 Porcine Reproductive and Respiratory Syndrome Virus. *J. Virol.* **2017**, *91*, e01521-16. [CrossRef] [PubMed]

MDPI
St. Alban-Anlage 66
4052 Basel
Switzerland
Tel. +41 61 683 77 34
Fax +41 61 302 89 18
www.mdpi.com

Viruses Editorial Office
E-mail: viruses@mdpi.com
www.mdpi.com/journal/viruses